ECOSYSTEMS OF THE WORLD 9A

HEATHLANDS AND RELATED SHRUBLANDS

DESCRIPTIVE STUDIES

ECOSYSTEMS OF THE WORLD

Editor in Chief:

D.W. Goodall

CSIRO Division of Land Resources Management, Wembley, W.A. (Australia)

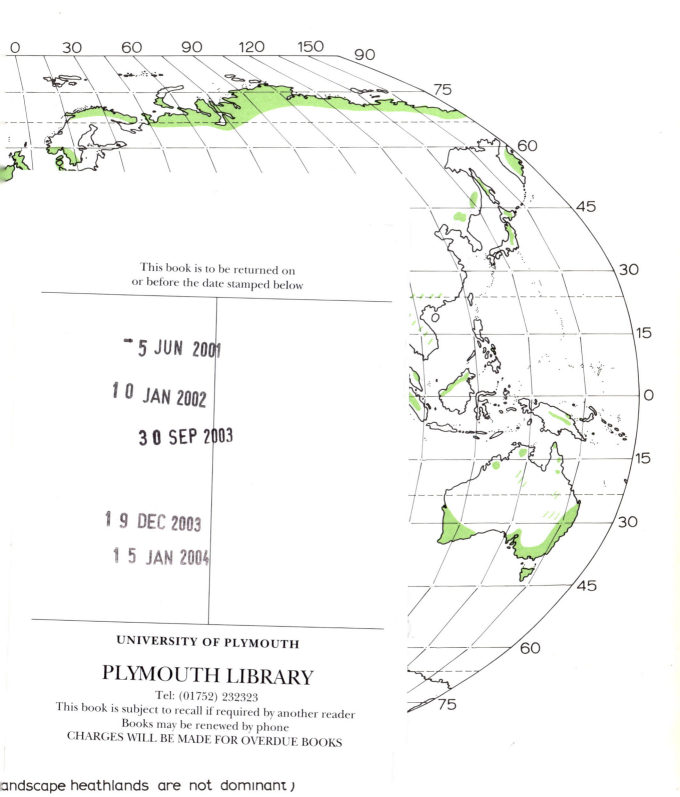

andscape heathlands are not dominant)

ECOSYSTEMS OF THE WORLD 9A

HEATHLANDS AND RELATED SHRUBLANDS

DESCRIPTIVE STUDIES

Edited by

R.L. Specht

Department of Botany
University of Queensland
St. Lucia, Qld. (Australia)

ELSEVIER SCIENTIFIC PUBLISHING COMPANY

Amsterdam — Oxford — New York 1979

ELSEVIER SCIENTIFIC PUBLISHING COMPANY
335 Jan van Galenstraat
P.O. Box 211, 1000 AE Amsterdam, The Netherlands

Distributors for the United States and Canada:

ELSEVIER NORTH-HOLLAND INC.
52 Vanderbilt Avenue
New York, N.Y. 10017, U.S.A.

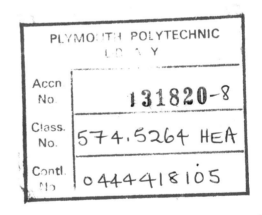

Library of Congress Cataloging in Publication Data

Main entry under title:

Heathlands and related shrublands.

(Ecosystems of the world ; 9A-   )
Includes bibliographical references and index
CONTENTS: A. Descriptive studies.
    1. Heath ecology. 2. Moors and heaths.
I. Specht, Raymond Louis. II. Series.
QH541.5.M6H42   574.5'2632   79-13938
ISBN 0-444-41810-5

Printed in The Netherlands

PREFACE

Heathlands and related ecosystems may be found in all parts of the world — from the tropics to the polar regions, from lowland to subalpine altitudes. It appears that heathlands evolved from ancient ecosystems which had spread throughout the component continents of Gondwanaland (South America, Africa, India, Australasia and Antarctica) and adjacent North America. Later, the continental plates of Africa (including the Mediterranean–Iranian region), the Indian subcontinent, and Australasia drifted northward and impinged on the Eurasian continental plate. From these fragments of the Gondwanan Region, the heathland ecosystems expanded from Africa into Western Europe and from there across the northern part of Eurasia and North America; from India into southeastern Asia and northward to Japan; from Australasia (including New Guinea) westward into the Malay Archipelago.

Heathlands, wherever they occur, are invariably found on oligotrophic soils, especially deficient in phosphorus and nitrogen. A distinctive sclerophyllous flora, remarkably adapted to fire, has evolved to occupy these nutrient-poor habitats. Many ancient families of gymnosperms and angiosperms, including the "heath" families Ericaceae and Epacridaceae, coexist in the complex heathland vegetation. Although many of the heathland plants can be cultivated on more fertile soils, only a few of the characteristic heathland taxa are found growing naturally in other ecosystems. Animals, being more mobile, are usually not confined to the heathland ecosystems; those species dependent on the ecosystem must be able to cope with both nutritional and seasonal water stress, as well as with fire.

Historically, the study of heathland ecosystems began in Europe — the work of Graebner (Volume 5 of the Series on *Die Vegetation der Erde*, 1895) being a classic. The oligotrophic *Calluna–Erica* heathlands of northwestern Europe do not extend onto the extensive calcareous soils around the Mediterranean where a quite different suite of plant communities, termed *maquis* (*macchia*) and *garrigue* has developed; these, together with the *chaparral* and *matorral* of North and South America, will be covered in Volume 11 of this series.

The belief that heathlands were confined to the cool-temperate climatic region typical of northwestern Europe was firmly embedded in the minds of early European ecologists. Imagine the dilemma of Diels in describing the oligotrophic ecosystems in the Mediterranean-type climate of southwestern Australia (Volume 7 of the Series on *Die Vegetation der Erde*, 1906) — were they equivalent to the *Calluna* heathlands or to macchia? The same problem faced ecologists in describing the vegetation on the oligotrophic soils in the Mediterranean climate of South Africa; here, the large number of *Erica* species added to the confusion. The South Africans finally resolved the problem by calling the vegetation by the Afrikaans term *fynbos*.

The fynbos of South Africa is essentially confined to the Mediterranean-type climate of Cape Province. To the north of the Cape, heathlands are found in subalpine regions within the tropics and subtropics. On coastal lowlands, small areas of heathland dominated by *Philippia simii*[1] have been reported on poorly drained soils in tropical Moçambique, *c*. latitude 17°S (*Flora Zambesiaca Supplement*, edited by H. Wild and A. Fernandes, M.O. Collins, Salisbury, 71 pp. in 1967). The *geoxylic suffrutex* plants (with large lignotubers), endemic in seasonally waterlogged sandy soils in the Zambesian Region, may have evolved in the "edaphic grasslands" as a response to the unfavourable edaphic conditions provided by the extremely oligotrophic soils (White, F., 1976. The underground forests of Africa. *Gardens' Bull.*, Singapore, 29: 57–71) — an origin paralleling that

---

[1] See also Tinley, K.L., 1977. *Framework of the Gorongosa Ecosystem, Moçambique*. Thesis, University of Pretoria, Pretoria, 186 pp.

of the heathlands, in general, and of the "bush islands" in the Guianas (Chapter 21, this volume).

Heathlands, allied to the fynbos, are also found in the islands between Africa and South America, and extend onto the latter continent.

In southern Australia, the heathland flora, related to the South African fynbos, extends from lowland to subalpine altitudes. The same flora continues northwards through the subtropics into the tropics where it is found in isolated pockets throughout New Guinea and the Malay Archipelago on both lowland and subalpine sites. Remnants of the Australasian heathlands are found throughout New Zealand, even in the subantarctic conditions of Auckland Island.

Typically, heathland vegetation on oligotrophic soils is depauperate (less than 1–2 m tall), composed of many ancient and/or primitive genera of angiosperms which have sclerophyllous, often very small, leaves. Throughout the world, the presence of genera belonging to the Order Ericales is considered to be a characteristic attribute of heathland; but many other "heathland" families coexist with the Ericales which may be rare or even absent in some stands.

The heathland flora is often found as an understorey with trees or tall shrubs forming an open-forest, woodland, or tall shrubland. It has been debated whether true heathlands (without trees and tall shrubs) are artefacts induced by man. Man-induced heathlands are common, but natural stands of treeless dry-heath and wet-heath are found in lowland and alpine areas throughout the world. In subtropical to tropical climates true heathlands are found in alpine areas and on those lowland, oligotrophic habitats where seasonal waterlogging is extreme; as water stress is reduced the "heathland" flora can grow progressively taller, even becoming a *Heidewald*, a closed-forest of sclerophyllous, "heath" plants, 30 m tall.

Thus, the purpose of this volume is to study the ecology of those plant communities which are found on oligotrophic habitats throughout the world, be they typical heathlands, related shrublands or even, in extreme cases, "heath-forests" (Heidewald).

Almost nothing is known of the distribution, structure and ecology of heathlands and related shrublands which must be found in the subalpine/alpine regions of tropical South America and in Southeast Asia where, respectively, at least 200 and 600 species of *Gaultheria*, *Gaylussacia*, *Rhododendron* and *Vaccinium* have been recorded. Over the last decade, A.S. Weston (pers. comm., 1978) has studied the ericaceous communities epiphytic in the tops of trees of the upper montane rain forest (an oligotrophic habitat, also described for Borneo and New Guinea in Chapter 12, this volume), and the heathy *paramo* and bogs which occur above 3000 m in Costa Rica; these studies will be reported elsewhere.

It must be stressed that, in many parts of the world, little detailed research has been published on characteristic heathland ecosystems. Most of the chapters on animal ecology, some of the information on plant ecology and on the eco-physiology of heathland vegetation, included in this volume, report original, unpublished research still in progress.

Part A of this volume surveys the extent, variation and ecology of heathlands and related shrublands, throughout the various continents. Homologous vegetation developed on the oligotrophic sands of tropical South America is also examined. In Part B, the phenological, morphological and physiological characteristics of the heathland ecosystems, and their component plants and animals, are surveyed in relation to the annual sequence of climate, fire, nutritional and water stress. Finally, the problems of conservation of these fragile ecosystems are studied.

As editor, I would like to thank Dr. D.W. Goodall, the Editor in Chief, for inviting me to contribute this volume to the series on *Ecosystems of the World* and providing expert advice on the problems of editing. The 52 authors of the 49 articles in this volume have all contributed to the project with an enthusiasm, alacrity and excellence which have rendered my task as editor most enjoyable. Miss Lyndell Danaher, Mrs. Marion Greenfield, Mrs. Lynn Jessup and Mrs. Merrilyn Rorrison gave considerable assistance in the preparation of manuscripts, diagrams and the compilation of the indexes.

R.L. SPECHT

Professor of Botany
University of Queensland

LIST OF CONTRIBUTORS TO VOLUME 9

W. ARMSTRONG
Department of Plant Biology
University of Hull
Hull HU6 7RX (Great Britain)

ROLF Y. BERG
Botanical Garden and Museum
University of Oslo
Oslo 5 (Norway)

R.C. BIGALKE
Faculty of Forestry
University of Stellenbosch
Stellenbosch 7600 (Republic of South Africa)

L.C. BLISS
Department of Botany
University of Washington
Seattle, Wash. 98195 (U.S.A.)

C.J. BURROWS
Department of Botany
University of Canterbury
Christchurch (New Zealand)

ELLA O. CAMPBELL
Department of Botany
Massey University
Palmerston North (New Zealand)

P.C. CATLING
CSIRO Division of Wildlife Research
P.O. Box 84
Lyneham, A.C.T. 2602 (Australia)

S.B. CHAPMAN
Institute of Terrestrial Ecology
Furzebrook Research Station
Wareham
Dorset BH20 5AS (Great Britain)

NORMAN L. CHRISTENSEN
Department of Botany
Duke University
Durham, N.C. 27706 (U.S.A.)

H.T. CLIFFORD
Department of Botany
University of Queensland
St. Lucia, Qld. 4067 (Australia)

R.G. COLEMAN
Department of Botany
University of Queensland
St. Lucia, Qld. 4067 (Australia)

D.J. CONNOR
School of Agriculture
La Trobe University
Bundoora, Vic. 3083 (Australia)

ALAN COOPER
School of Environmental Sciences
Ulster Polytechnic
Newtonabbey BT37 0QB (Northern Ireland)

D. DOLEY
Department of Botany
University of Queensland
St. Lucia, Qld. 4067 (Australia)

WENDY E. DRAKE
Department of Botany
University of Queensland
St. Lucia, Qld. 4067 (Australia)

PETER D. DWYER
Department of Zoology
University of Queensland
St. Lucia, Qld. 4067 (Australia)

S.J. EDMONDS
South Australian Museum
North Terrace
Adelaide, S.A. 5000 (Australia)

A.E. ESLER
Botany Division
D.S.I.R.
Mt. Albert Research Centre
120 Mt. Albert Road
Auckland (New Zealand)

A.S. GEORGE
Western Australian Herbarium
Department of Agriculture
George Road
South Perth, W.A. 6151 (Australia)

A.M. GILL
CSIRO Division of Plant Industry
P.O. Box 1600
Canberra City, A.C.T. 2601 (Australia)

C.H. GIMINGHAM
Department of Botany
University of Aberdeen
St. Machar Drive
Aberdeen AB9 2UD (Great Britain)

A.R. GLENN
School of Environmental and Life Sciences
Murdoch University
Murdoch, W.A. 6153 (Australia)

R.H. GROVES
CSIRO Division of Plant Industry
P.O. Box 1600
Canberra City, A.C.T. 2601 (Australia)

A.J.M. HOPKINS
Western Australian Wildlife Research Centre
Mullaloo Drive
Wanneroo, W.A. 6065 (Australia)

GLEN J. INGRAM
Queensland Museum
Gregory Terrace
Brisbane, Qld. 4000 (Australia)

JIRO KIKKAWA
Department of Zoology
University of Queensland
St. Lucia, Qld. 4067 (Australia)

D.J.B. KILLICK
Botanical Research Institute
Private Bag X101
Pretoria 0001 (Republic of South Africa)

H. KLINGE
Abteilung Tropenökologie
Max-Planck-Institut für Limnologie
Plön/Holstein (Germany G.F.R.)

F.J. KRUGER
Forest Research Institute
Department of Forestry
Pretoria West 0002 (Republic of South Africa)

A.V.A. KWOLEK
Department of Plant Sciences
The University of Leeds
Leeds LS2 9JT (Great Britain)

BYRON B. LAMONT
Department of Biology
Western Australian Institute of Technology
Hayman Road
South Bentley, W.A. 6102 (Australia)

SILAS LITTLE
Northeastern Forest Experiment Station
P.O. Box 4
New Lisbon, N.J. 08064 (U.S.A.)

A.R. MAIN
Department of Zoology
University of Western Australia
Nedlands, W.A. 6009 (Australia)

N. MALAJCZUK
CSIRO Division of Land Resources Management
Private Bag P.O.
Wembley, W.A. 6014 (Australia)

N.G. MARCHANT
Western Australian Herbarium
Department of Agriculture
George Road
South Perth, W.A. 6151 (Australia)

D.R. McQUEEN
Department of Botany
Victoria University of Wellington
Private Bag
Wellington (New Zealand)

E. MEDINA
Centro de Ecología
Instituto Venezolano de Investigaciones Científicas
Caracas (Venezuela)

D.M. MOORE
Department of Botany
University of Reading
Whiteknights
Reading RG6 2AS (Great Britain)

A.E. NEWSOME
CSIRO Division of Wildlife Research
P.O. Box 84
Lyneham, A.C.T. 2602 (Australia)

P.G. OZANNE
CSIRO Division of Land Resources Management
Private Bag P.O.
Wembley, W.A. 6014 (Australia)

C.N. PAGE
Royal Botanic Gardens
Edinburgh EH3 5LR (Great Britain)

R.F. PARSONS
Department of Botany
La Trobe University
Bundoora, Vic. 3083 (Australia)

R.W. ROGERS
Department of Botany
University of Queensland
St. Lucia, Qld. 4067 (Australia)

M. M. SPECHT
Department of Botany
University of Queensland
St. Lucia, Qld. 4067 (Australia)

R.L. SPECHT
Department of Botany
University of Queensland
St. Lucia, Qld. 4067 (Australia)

P. WARDLE
Botany Division
D.S.I.R.
Private Bag
Christchurch (New Zealand)

N.R. WEBB
Institute of Terrestrial Ecology
Furzebrook Research Station
Wareham BH20 5AS (Great Britain)

WALTER E. WESTMAN
Department of Geography
University of California
Los Angeles, Calif. 90024 (U.S.A.)

R.H. WHITTAKER
Division of Biological Sciences
Cornell University
Ithaca, N.Y. 14853 (U.S.A.)

J.S. WOMERSLEY
82 Richmond Road
Westbourne Park, S.A. 5041 (Australia)

H.W. WOOLHOUSE
Department of Plant Sciences
The University of Leeds
Leeds LS2 9JT (Great Britain)

D.J. YATES
Department of Botany
University of Queensland
St. Lucia, Qld. 4067 (Australia)

CONTENTS OF VOLUME 9A[1]

---

[1] For short contents of Volume 9B, see p. XIII.

CONTENTS OF VOLUME 9B

Chapter 1

HEATHLANDS AND RELATED SHRUBLANDS OF THE WORLD[1]

R.L. SPECHT

INTRODUCTION

It is just over a hundred years since Haeckel (1869) first defined the discipline of ecology. Since that time, the interrelated complex of the biome (the community of producer, consumer and decomposer organisms) and the environment has been termed the ecosystem (Tansley, 1935). Frequently in the past, only processes operating within a small section of an ecosystem have been studied. Today, more ecologists are attempting to study the ecosystem as an entity, to understand how it operates as a total system and how small perturbations may produce changes in sections of the ecosystem and, in turn, the whole ecosystem.

The present volume deals with a complex of ecosystems dominated by sclerophyllous shrubs. It will be shown below that a closely related series of ecosystems containing the heathland flora may be found from the tropics to the subarctic, from lowland to subalpine regions (see the end-paper maps). The climatic range is probably one of the widest of any of the world's biomes. The ecological problems associated with such a world-wide distribution are vast. In most regions, the heathland ecosystem is clearly subdivided into two distinct entities — dry-heathland on well-drained soil and wet-heathland on seasonally waterlogged soil. Soil nutrient status appears to be the only common environmental factor controlling the distribution of these ecosystem types; they are always found on the most infertile soils.

Probably as an evolutionary response to low nutrient status, the heathland communities are evergreen and sclerophyllous. As tannins, resins and essential oils are usual components of sclerophyllous leaves and stems, the vegetation is very susceptible to fire during periods of water stress. The component species of the heathland have evolved remarkable strategies of fire survival and regeneration.

This volume attempts to explore the origin, biogeography, structure and ecology, the ecophysiology of water-balance and nutrition, the role of fire, and the problems of long-term conservation of these widespread ecosystems.

THE DEFINITION OF HEATHLAND

In 1914 Rübel investigated the original meaning of the vernacular term, heath. It is clear that the Germanic word — Heide, hed, heath — means an **uncultivated stretch of land**; the economic nature of the land is implied in the concept, regardless of the vegetation. It is purely coincidental that in many areas of northwestern Germany and in Great Britain the vegetation is dominated by Ericaceae. In less oceanic parts of German-speaking Europe the uncultivated plains, the Heiden, are of a different character. The vegetation of the local Heiden is no longer ericaceous but dry meadow, prairie or subalpine meadow — termed Heidewiese or Steppenheide by early ecologists.

After investigating the original meaning of a number of vernacular terms used by different nations to describe vegetation, Rübel was forced to conclude that most had the same original meaning of **uncultivated land**. Heath, Heide, hed, lande, steppe, desert, Wüste, prairie, macchia, maquis, garigue (or garrigue), tomillares, phrygana, etc., all mean **waste land**, irrespective of the different plant

---

[1] Manuscript completed May, 1977.

communities found growing on these areas. When the popular names are brought into scientific use, the terms vary widely in their application, from one meaning to a totally different meaning according to different language groups. As well, the interpretation of the vegetation varies according to the experience of individual observers. The vernacular terms become too imprecise for universal use in phytogeography.

At the end of his essay, Rübel suggested that the term heath (Heide, hed, lande) be restricted phytogeographically to the ericoid-leaved bushland found in Britain, northwestern Germany, southwestern France, the Canary Islands, and Cape Province in South Africa. In the last locality, the Afrikaans term for the vegetation is *fynbos*, the **fine-leaved bushland**, containing about 832 species of Ericaceae, which covers the sandstone mountains of the southwestern Cape Province.

On the evidence then available to him, Schimper (1903) found it difficult to consider the fynbos of the Mediterranean climatic region of South Africa and the Heide of cold-temperate Europe as the same formation. Graebner (1895, also 1901 and 1925) had described the heathlands of northern Germany as being dominated by heather, *Calluna vulgaris*, over large tracts of country which possessed both sandy and peaty (moor) soils. He stressed the point that although the moor soils were physically wetter than the sandy heathland soil, both soils tended to be "physiologically dry" during certain periods of the year. Moreover, both soils were very poor in mineral foods, a major factor favouring the growth of *Calluna*, and the few ericaceous and graminoid plants associated with it.

Schimper (1903, p. 657) noted that the evergreen heather imparts to the heathland some likeness to the poorest sclerophyllous formations of warmer districts with moist winter (the Mediterranean-type climate). However, on the evidence then available, he was forced to conclude that the formations were different — the sclerophyllous formations of the world were dependent on the seasonal fluctuations in climate, not on the nature of the soil, whereas the distribution of the heathlands of Europe was controlled edaphically, being confined to sandy and peaty soil.

Information from many areas outside Europe was rather fragmentary when Schimper (1903) wrote his book on plant geography. In particular,

he cited only one major paper from southern Australia — a chapter written by Schomburgk (1876) as part of a book on 'the geography and history of South Australia. Schimper quoted verbatim almost all of that part of Schomburgk's article (pp. 211–219) which dealt with the plant communities found in the vicinity of Adelaide. Unfortunately, Schomburgk's description of the vegetation was somewhat confused, possibly because he did not fully appreciate the ecological patterns present in the South Australian vegetation. His article should have stated clearly that four major plant formations (Specht, 1972) could be seen in the area:

(1) An open-forest of *Eucalyptus* spp. with an understorey of heath species (including many Epacridaceae) — on infertile soils.

(2) A savannah–woodland or savannah–forest dominated by scattered small, shady trees of *Eucalyptus*, *Casuarina*, *Myoporum*, with some shrubby *Acacia*, *Bursaria* and *Pittosporum*. The ground stratum was composed largely of graminoids, herbs, forbs and geophytes. The soils on which the savannah woodland was developed were mostly rich and arable — in marked contrast to the infertile soils which supported the previous formation.

(3) A grassland or savannah formation, with trees essentially absent or very sparse, was found on gilgai soils (termed "Bay of Biscay lands") and on fertile soils in the drier and colder upland areas north of Adelaide.

(4) A mallee scrub, with considerable variation in structure, stretched as a monotonous vegetation over much of the lower rainfall belt of the southern third of South Australia. The soils on which the mallee scrub flourished were mostly solonized sands and loams, often containing a considerable amount of calcium carbonate as particles, concretions or hardpans in the subsoil. The understorey varied from herbaceous to heathy, depending on the fertility of the surface soil.

Instead, Schomburgk tended to confuse the evidence, interweaving the four communities under his three headings: Forest Land Region, Scrub Land Region, and Grass Region. It was only after Schimper's book was published in 1903 that Diels visited Western Australia and clearly defined the plant communities, *Savannen-Wald*, *Sklerophyllen-Wald* and its structural variants,

Sklerophyll-Gebüsch and *Sand-Heiden* (Diels, 1906). The Australian *Sklerophyllen*[1] (heath) flora (sometimes with an overstorey of trees and shrubs) was later shown to be restricted in its distribution to soils very low in plant nutrients (Andrews, 1916; Adamson and Osborn 1924; Wood 1939; Specht and Perry, 1948; Specht et al., 1961). In effect the Sklerophyllen flora of southern Australia is confined to infertile soils in the same way as the distribution of *Calluna* heathland is controlled edaphically in northeastern Europe. To strengthen the analogy, the Australian Sklerophyllen flora contains many members of the heath family Epacridaceae. This family is, in general, more advanced than the family Ericaceae (including *Calluna*), but the two families have many morphological and floral characters in common (Hutchison, 1973).

It should be stressed that the Sklerophyllen flora of Australia is not confined to the southern part of the continent where the climate is like that experienced in the Mediterranean-type regions of the world. The flora flourishes in the uniformly wet climate of the eastern coast, at least from Tasmania in the south to the Tropic of Capricorn in the north. It also extends into the monsoonal region across northern Australia where a wet summer alternates with a dry winter. Examples of the heathland flora are found in New Guinea and northward to Malesia[2] (Van Steenis, 1957); the flora also extends to New Caledonia and New Zealand. Wherever it occurs, the vegetation is restricted in its distribution to poor soils and contains several members of the heath family Epacridaceae, although these species are rarely dominants in the flora.

An overstorey of trees and shrubs is usually present with the heathland flora. However, in certain seasonally droughty sites on deep sand or on groundwater podzols or peats, overstorey species fail to establish; the Sklerophyllen flora without trees is then left as a true heathland — the Sand-Heide of Diels (1906) and the "wet-heath" of Tate (1883). An identical heathland is also seen in subalpine habitats (on peaty, siliceous or possibly kaolinitic soils) on the mountains of Indo-Malaya, Australia and New Zealand. Genera typical of lowland heathlands are found in these subalpine heathlands. As well, members of the family Ericaceae are often codominant — *Agapetes*,

Gaultheria, *Pernettya*, *Rhododendron*, *Vaccinium* and *Wittsteinia*. Schimper (1903, p. 508) recognised that the sclerophyllous nature of the extensive bush woods on the summits of the mountains of the Malay Archipelago was essentially the same as the lowland sclerophyllous woodlands of the Mediterranean regions of the world. Similarly, Warming (1909) recognised that the *Rhododendron* bushland (containing also *Vaccinium* and *Calluna*) of the European Alps and Pyrenees was structurally allied to the dwarf-shrub heathland of northern Germany.

It is thus clear from the above evidence that the Sklerophyllen flora is widely spread from lowland to subalpine regions, from the tropics to the subarctic. As Johnson and Briggs (1975) have so well argued, the Mediterranean myth concerning the distribution of the Sklerophyllen flora is no longer tenable. Similarly, the *Calluna* heathland of Europe must be considered as part of the whole world continuum; it is not unique.

The heathlands of the world are united by: (1) their evergreen sclerophyllous nature; (2) the presence, but not necessarily the dominance, of the heath families — Diapensiaceae, Empetraceae, Epacridaceae, Ericaceae, Grubbiaceae, Prionotaceae, Vacciniaceae — in the stand; and (3) their ecological restriction to soils very low in plant nutrients. These infertile soils may be well-drained (supporting dry-heathland or "sand-heath") or seasonally waterlogged (supporting wet-heathland); they may be found in lowland or subalpine climates.

It has been a source of some confusion that the word "heath" has been applied not only to a vegetation type, but also to some of the plant species that dominate it in Europe — particularly *Erica* spp. In fact, in North America the shift in meaning has gone so far that "heath" much more often means a plant than a type of vegetation, as Whittaker explains in Chapter 17 of this volume. Accordingly for the sake of consistency and to avoid confusion, heath will here be used for a member of those families[3] listed in the previous

---

[1] By "Sklerophyllen", Diels was referring to the sclerophyllous (heath-like) **understorey**, and not to the eucalypt overstorey, as has often mistakenly been assumed (Specht, 1970).

[2] For a definition of this biogeographical region, see Ch. 12.

[3] Celastrales: Empetraceae. Ericales: Diapensiaceae, Epacridaceae, Ericaceae, Prionotaceae, Vacciniaceae. Santalales: Grubbiaceae.

paragraph, "heathland" for the vegetation type, and "heathland species" for the many diverse species and families which together comprise the "heathland" community.

In most areas in which heathlands occur, both fertile and infertile soils may be found. Grasses and herbs flourish on the fertile soils, whereas sclerophyllous plants are characteristic of the infertile soils. The two floras are clearly segregated in lowland Australia into Sklerophyllen and Savannen communities; in subalpine areas, heathlands are found adjacent to alpine grasslands and alpine savannah woodlands (Costin, 1954). In South Africa, the *coastal rhenosterbosveld* with a dense ground stratum of grass, is found on fertile clayey soils near Cape Town, whereas, nearby, *coastal fynbos* (or "*macchia*") grows on leached, inland sand dunes; the main body of fynbos is found on the sandstone mountain ranges (Acocks, 1975). The same transition from heathland (on infertile soils) to meadow (on fertile soils) is also seen in Europe especially in mountainous places (Warming, 1909, p. 211).

Small trees or tall shrubs are a normal component of both the Sklerophyllen and the Savannen communities. The tree stratum may have been destroyed by man using fire and/or grazing, thus leaving the heathland or herbaceous understorey — an historical sequence which is well documented in Europe (Specht, 1969; Gimingham, 1972). In areas less affected by man, true structural heathlands (without trees and tall shrubs) are usually of limited area on deep sandy or seasonally water-logged soil. Most extensive areas of fynbos in South Africa have developed on shallow sandstone soil, seasonally saturated with water during winter and spring; coastal fynbos grows on deep sandy soils, seasonally droughted.

In Australia, treeless savannahs are found in the same climatic zones as the Sklerophyllen flora, but on fertile, deep-cracking clays (allied to chernozem soils) in which drought conditions appear to be too severe for the survival of most dicotyledonous seedlings possessing a single tap-root.

THE "MEDITERRANEAN" VEGETATION

The evidence, presented above, indicates that heathlands and related shrublands are widespread

from the subarctic to the tropical regions of the world, but virtually confined to infertile soils. Sclerophyllous shrubs and subshrubs (nanophanerophytes and chamaephytes) dominate the ground stratum; seasonal geophytes are often present, but grasses and herbs (hemicryptophytes and therophytes) are rare. In contrast, the ground stratum of plant communities on richer soils is largely herbaceous, composed of seasonal grasses, herbs, and forbs (hemicryptophytes, geophytes and therophytes).

This distinction between Sklerophyllen and Savannen ground strata is also applicable to the vegetation of the Mediterranean-type regions of the world (Specht, 1969) — the nature of the ground stratum alternates between heathy and herbaceous, depending on the fertility of the soil. If an overstorey of trees or tall shrubs is present, the understorey may be suppressed as the community matures. Typical "Mediterranean" communities which belong to the two classes are shown in Table 1.1. Transitional communities do occur, but are by no means common. A few heathland species may survive amongst the grasslands on slightly more fertile soils (e.g. *Astroloma humifusa*, *Eutaxia microphylla*, *Lepidosperma* spp. in southern Australia). Poor management of marginal grassland areas, by fire and grazing, may lead to the expansion of heathland species — for instance in the Amatole Mountains of South Africa (Trollope, 1970, 1973).

It should be stressed that if trees and/or tall shrubs are present as an overstorey of either the Sklerophyllen or Savannen ground stratum, the leaves of these taller species are invariably evergreen and sclerophyllous, no matter what the ground stratum. In contrast to the ground stratum, sclerophylly of the upper stratum does not appear to be edaphically controlled; nor is the character confined to the Mediterranean regions of the world as implied by Schimper (1903) and later Warming (1909). In Australia, about 600 species of each of the genera *Eucalyptus* and *Acacia* tend to dominate the landscape from the monsoonal north to the Mediterranean south, from the uniform rainfall of the east coast to the arid environment of the centre (Specht, 1970); all these species possess sclerophyllous leaves. It is only the tropical and temperate rain forests and the grassland communities which possess mesophyllous leaves; the shrub

TABLE 1.1

Typical "Mediterranean" plant communities with (1) a Sklerophyllen (heath) and (2) a Savannen (herbaceous) ground stratum

Locality	Ground stratum		References
	Sklerophyllen	Savannen	
Southern France	*Cistus crispus–* *Erica cinerea* lande	*Quercus ilex* woodland *Quercus coccifera* garigue–maquis (macchia)	Braun-Blanquet et al. (1952) Specht (1969)
California	*Adenostoma* chaparral	*Quercus* woodland *Ceanothus–* *Arctostaphylos* chaparral	Cooper (1922) Sampson (1944) Specht (1969)
South Africa	fynbos (macchia) coastal fynbos	coastal rhenosterbosveld (*Elytropappus*)	Acocks (1975) Werger et al. (1972)
Australia	sclerophyll forest sclerophyll woodland tree-heathland mallee-heathland dry-heathland wet-heathland	savannah–forest savannah–woodland mallee–scrub grassland	Specht (1969, 1973)
Chile	?	matorral (?)	

steppes of the temperate arid zone, dominated by the genera *Atriplex* and *Maireana* (*Kochia*), possess semi-succulent leaves. Sclerophylly of the leaves of the evergreen tree/shrub stratum thus seems to be an evolutionary development in response to seasonal water stress.

It may eventually be shown that the evergreen, sclerophyllous nature of the tree/shrub canopy is also basically induced by limiting mineral nutrition (Loveless, 1961, 1962; Beadle, 1968). For example, the evergreen, sclerophyllous trees which dominate the forests and woodlands of seasonally dry tropical Australia grow on infertile lateritic red earths, lateritic podzols, and allied soils. In contrast, in a similar seasonal tropical climate in Africa, the umbrella-shaped trees which dominate the fertile savannah and open-forest communities are seasonally sclerophyllous (xerophilous), the trees being mostly bare during dry weather (Schimper, 1903).

Hence, in the Mediterranean regions of the world, the tree/shrub stratum possesses evergreen sclerophyllous leaves, probably developed in re-

sponse to water stress. The ground stratum may be either evergreen and sclerophyllous (heathy) or seasonally herbaceous, depending on the fertility of the soil on which the community grows.

STRUCTURE OF HEATHLAND COMMUNITIES

Life forms

True heathlands (without emergent trees or tall shrubs) are characterised by a dense to mid-dense assemblage of evergreen, sclerophyllous shrubs and subshrubs (*nanophanerophytes* and *chamaephytes*).

Seasonal *hemicryptophytes* and *therophytes*, characteristic of savannah vegetation, are usually sparsely represented in heathlands. Instead, plants which may be classified best as *evergreen sclerophyllous hemicryptophytes* are often very important. The members of the monocotyledonous families Cyperaceae (*Lepidosperma* spp., *Tetraria* spp.) Iridaceae (*Patersonia* spp.), Liliaceae (*Dianella* spp.), Restionaceae (all species), Xanthorrhoeaceae (many species, especially *Lomandra* spp.) possess long-lived, sclerophyllous leaves which remain functional for at least two years; all these species regenerate seasonally from perennating buds located on rhizomes and culms at or near the surface of the soil — the hemicryptophytic habit defined by Raunkiaer (1934). Evergreen hemicryptophytes become increasingly important in wet-heathlands, producing a structure which may be best described as a "graminoid-heathland".

Geophytic plants (Amaryllidaceae, Iridaceae, Liliaceae, Orchidaceae, and a few dicotyledonous families such as Droseraceae) form a seasonal component of most heathlands.

Parasitic epiphytes, such as *Cassytha* or *Cuscuta*, are found in some stands.

Leaf characteristics

The leaves of all long-lived species of the heathland are *evergreen* and sclerophyllous (Fig. 1.1). The leaves have typically thick cuticles, sunken stomata, and usually thick-walled cells, often lignified, sometimes silicified (Hamilton, 1927; McLuckie and Petrie, 1927; Paterson, 1961; Watson, 1962, 1964, 1965; Rao, 1971). The cells usually contain tannins, resins and essential oils. Sclerophylly may be an evolutionary character produced in response to the low level of mineral nutrients (especially phosphorus) in the soils on which the heathland plants flourish (Loveless, 1961, 1962; Beadle, 1968). In certain species, the normal development of the sclerophyllous leaf from the mesophytic juvenile leaf on young shoots may be delayed considerably by the application of phosphate fertilizer to the heathland (Specht, 1963).

The evergreen, sclerophyllous leaves of many of the heathland species are small with a surface area less than 25 mm² (the *leptophyll* class of Raunkiaer, 1934). The leptophyllous leaves of a number of species are linear. In some of these species with linear leaves the edges of the leaf are rolled inwards on the undersurface and almost meet along the median line, leaving only a very narrow longitudinal slit; the outer upper surface is hard and waxy and does not allow for any transpiration. This leaf-form is characteristic of most species of *Erica* and *Calluna* and hence is referred to as "ericoid". Heathland species of families other than

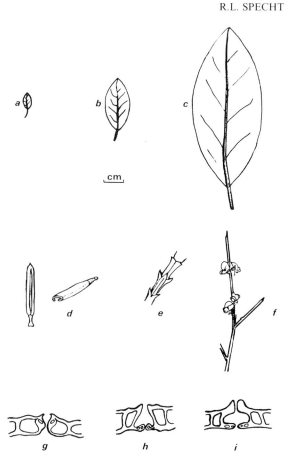

Fig. 1.1. Leaf characteristics of sclerophyllous plants: leptophyll (less than a); nanophyll (between a and b); microphyll (between b and c); ericoid (d); cupressoid (e); aphyllous (f). Stomata of *Acrotriche prostrata* (Epacridaceae) (g), *Hakea hookerana* (Proteaceae) (h) and *H. macrocarpa* (i). (After Hamilton, 1927; Paterson, 1961; Rao, 1971.)

the Ericaceae may possess ericoid leaves, e.g. some heathland members of the Dilleniaceae, Fabaceae and Myrtaceae. Some species may even possess leaves grooved above, not below.

Plants with "cupressoid" leaves are often found in heathlands — gymnospermous plants such as *Callitris*, *Juniperus*, *Microstrobos*, *Widdringtonia*, etc., and some of the Asteraceae, e.g. *Olearia*, possess these minute leptophyllous leaves. Heathlands may contain many species with sclerophyllous leaves larger than leptophyll — often half the heathland flora in the Southern Hemisphere possess leaves which may be classed as *nanophyll* (25–225 mm²). Dominant, long-lived, nanophanerophyte species may have sclerophyllous leaves of even larger size — *microphyll* (2–20 cm²) and *mesophyll* (20–182 cm²). The leaves of emergent trees and tall shrubs (microphanerophytes of Raunkiaer) are usually in the microphyll–mesophyll size-class.

It would appear from aerodynamic studies that the breadth of the leaf is an important ecological attribute in controlling heat loss from the leaf by convection, the amount of heat energy convected being inversely proportional to the square root of the breadth of the leaf. A narrow leaf loses heat faster than a broad leaf and its leaf temperature remains near ambient air temperature even under full sunlight; a broad leaf, in contrast, may develop a leaf temperature some degrees above the ambient air temperature. Fosberg (1967) introduced the terms *narrow-sclerophyll* (needle-like or narrow-linear leaves) and *broad-sclerophyll* (other than narrow-linear in outline). Plant taxonomists classify leaves with a length:breadth ratio of more than 12:1 as narrow-linear[1]; a narrow-linear leaf 10 cm long may be almost 1 cm wide. To avoid including exceptionally broad leaves in Fosberg's category narrow-sclerophyll, the breadth of "narrow" sclerophyllous leaves is, in this volume, limited to 5 mm. The leaves of many heathland species are thus sclerophyllous, small in size (leptophyll to nanophyll) and narrow in breadth (narrow-sclerophyll).

Fire modifications

The high concentrations of tannins, resins and essential oils in the sclerophyllous leaves of many heathland plants increase the flammability of the community during periods of water stress. Fire is thus an integral part of heathland ecosystems.

Many species, razed in a fire, regenerate from underground root stocks (*lignotubers*) or sprout from *epicormic buds* buried deep in fire-resistant stems. Other species regenerate as seedlings from the wealth of seed either stored in the soil, or encased in hard, woody fruits which only release seed when heated in a bush fire. This *bradysporous* habit of seed release is characteristic of a number of genera of the families Casuarinaceae and Proteaceae.

Community structure

It is clear from the above discussion that heathlands (without emergent trees or tall shrubs) are characterised by a dense to mid-dense ????? of nanophanerophytes, chamaephytes and evergreen hemicryptophytes, all of which possess evergreen, sclerophyllous leaves. The leaves of most component species are small (in the size-classes cupressoid, leptophyll, nanophyll), sometimes ericoid. Long-lived nanophanerophytes and emergent microphanerophytes may have larger leaves (microphyll–mesophyll in size). Geophytes often appear seasonally in the ground stratum.

Members of the heath group of families (Ericales) are usually present, but not necessarily dominant, within the heathland stands (Table 1.2). However where this order is absent, for example in lowland, tropical South America, equivalent plant communities have been developed, by parallel or convergent evolution, on infertile sandy soils.

The height and foliage projective cover of the heathlands (both dry-heathlands and wet-heathlands) vary considerably with habitat. For convenience, they may be subdivided on the height of the uppermost stratum as follows:

shrubs >2 m	*scrub*[2]
shrubs 1–2 m	*tall heathland*
shrubs 25–100 cm	*heathland*
shrubs <25 cm	*dwarf heathland*

In the case of wet-heathlands, growing on seasonally waterlogged soils, evergreen hemicryptophytes (especially Restionaceae, Cyperaceae and some Poaceae) may become co-dominant with the heathland shrubs and subshrubs. It seems best to refer to this type of heathland as:

shrubs and graminoids co-dominant	*graminoid-heathland*

In extreme wet-heathlands, usually on peaty soil, the graminoid component is suppressed in favour of *Sphagnum*.

shrubs and *Sphagnum* co-dominant	*bog-heathland*

---

[1] Systematics Association Committee, 1962. *Taxon*, 11: 246.
[2] Scrub, as here defined, may be found on both calcareous and acid soils; the Mediterranean maquis (or macchia) vegetation found on calcareous soils is structurally scrub. Similarly, the calciphile garigue vegetation is structurally equivalent to heathland vegetation restricted to oligotrophic habitats.

TABLE 1.2

Number of families and genera of characteristic heathland species of angiosperms and gymnosperms found in various parts of the world

Region	All orders		Order Ericales[1]		Other heathland families[2]	
	families	genera	families	genera	families	genera
Arctic	15	26	3	9	1	1
North America	36	66	4	21	2	3
South America						
Guyana–Surinam (*muri*)	20	29	–	–	–	–
Venezuela–Rio Negro						
(*caatingas* and *campinas*)	16	24+	–	–	–	–
Southern Oceanic						
(wet-heathland)	26	48	1	2	1	1
Europe (NW)	18	40	2	12	1	1
South Africa	55 + 1[3]	158 + 19[3]	1	5 + 3[3]	1	1 + 1[3]
Australia	68	342	3	33	–	–
New Guinea						
(high-altitude heathlands)	32	64	3	5	–	–
Borneo						
Padang heathland	37	64	2	2	–	–
Kerangas forest	103	380	2	5	–	–
New Caledonia	50	110	1	3	–	–
New Zealand	26	61	2	7	–	–

[1] Order Ericales include the families Clethraceae, Diapensiaceae, Epacridaceae, Ericaceae, Prionotaceae, Vacciniaceae.

[2] Families Cyrillaceae, Empetraceae, Grubbiaceae.

[3] Families and genera recorded by Acocks (1975), additional to those listed by Kruger (Chapter 2).

The density of the heathland will also vary. Specht (1970) has used foliage projective cover (F.P.C.)[1] of the uppermost stratum to subdivide heathlands into **closed** (or dense) with an F.P.C. of 70–100% and **open** (or mid-dense) with an F.P.C. of 30–70%.

True heathlands, as described above, may have been a minor part of the landscape in many parts of the world before man wrought destruction with axe, fire, grazing animal and plough. Apart from the dry-heathlands and wet-heathlands of extreme habitats, trees and/or tall shrubs usually overtopped the heathland flora. Depending on the seasonal water balance of the ecosystem, the trees and tall shrubs may be absent, scattered, sparsely distributed or reasonably dense. A typical heathland or savannah understorey may be found under all these structural formations, depending on the fertility of the soil. The terms "sclerophyll forest", "sclerophyll woodland", "tree-heath", "mallee-heath" have been used to describe such heathy formations in Australia. The various plant formations in which heathland vegetation may be found either as a dominant stratum or as an understorey are shown in Table 1.3.

[1] Foliage projective cover is the percentage of land covered by foliage vertically above it.

TABLE 1.3

Structural formations of heathlands and related communities (after Specht et al., 1974)

Life form of upper stratum	Foliage projective cover of upper stratum (%)			
	100–70 (dense)	70–30 (mid-dense)	30–10 (sparse)	10–0 (very sparse)
Trees[1] > 10 m	–	open-forest with heathy understorey[2]	woodland with heathy understorey[2]	–
Small trees < 10 m	–	low open-forest with heathy understorey[2]	low woodland with heathy understorey[2]	low open-woodland with heathy understorey[3]
Tall shrubs[1] > 2 m	closed-scrub	open-scrub	tall shrubland with heathy understorey	tall open-shrubland with heathy understorey[3]
Shrubs 1–2 m	tall closed-heathland	tall open-heathland	–	–
Shrubs 25–100 cm	low closed-heathland	low open-heathland	–	–
Shrubs < 25 cm	–	–	dwarf heathland[2]	dwarf open-heathland[2]
Shrubs and graminoids co-dominant	closed graminoid-heathland	open graminoid-heathland	–	–
Shrubs and *Sphagnum* co-dominant	closed bog-heathland	open bog-heathland	–	–
Graminoids	closed-herbland[3]	open-herbland[3]	–	–

[1] A tree is defined as a woody plant, usually with a single stem; a shrub is a woody plant, frequently with many stems arising at or near the base.

[2] Alternative names: Sclerophyll forest = open-forest with heathy (sclerophyll) understorey; sclerophyll woodland = woodland with heathy (sclerophyll) understorey; tree-heathland = low open-woodland with heathy understorey; shrub-heathland = tall open-shrubland with heathy understorey; mallee-heathland = tall open-shrubland of mallee eucalypts with heathy understorey; fellfield = dwarf heathland.

[3] Closed- or open-herbland containing restiads, sedges, *Lomandra* spp., etc., but few grasses.

ORIGIN OF HEATHLAND COMMUNITIES

There is now firm evidence from several independent lines of geophysical research, that the continents of the world were closely united in the Permian and probably well into the Triassic period (Fig. 1.2). Over a long period of time, land connections were thus available for intercontinental migration of the then existing flora and fauna across the whole of the super-continent termed Góndwanaland and into south-western Europe and North America.

By mid-Cretaceous, the continental plates of North and South America had begun to drift westward from Europe and Africa (Fig. 1.3). Antarctica, together with India and Australasia, drifted eastward from Africa. The region which was to become Italy, Greece, Turkey and Iran was part of the African plate, still separated from Eurasia. The European section of the Eurasian plate lay slightly to the south of its present position, though the Borneo-Malay Peninsula, at the far southeast of Eurasia, lay, as today, on the Equator (Haile et al., 1977).

Later in the Cretaceous, the Indian plate broke away from the Antarctic plate and began its drift northward across the Indian Ocean (Fig. 1.3). Eventually, in the Miocene, the Indian subcontinent and the northern part of the African plate made contact with the Eurasian plate; the series of mighty mountain ranges from the Swiss Alps to the Indian Himalayas resulted (Fig. 1.4).

Antarctica and Australasia drifted southeastward during the Cretaceous and Paleocene periods. The continental plate of Australasia then moved northward at an estimated rate of 66 ±5 mm per year (Wellman and McDougall, 1974). New Zealand, and islands such as New Caledonia,

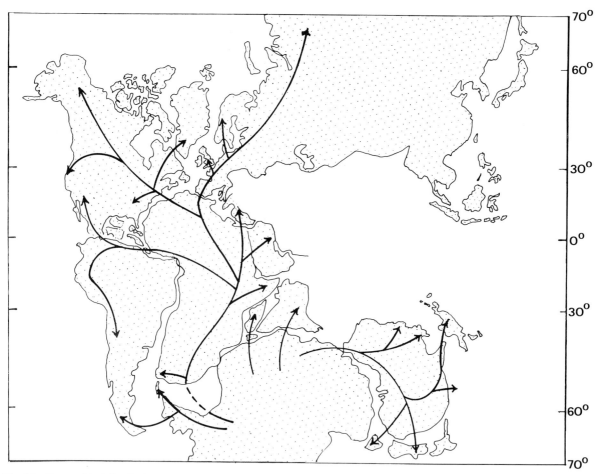

Fig. 1.2. Geographical distribution of continents in the Triassic/Jurassic periods (after Smith and Briden, 1977), showing the probable migration routes of the heathland flora.

separated from the Australian continent early in the Tertiary. New Guinea, however, has remained part of the Australian plate. The Indo-Malayan Archipelago of southeast Asia came into contact with the New Guinea section of Australasia in the late Miocene (Smith and Briden, 1977), during which time elements of the flora and fauna of the two regions could have been exchanged. However, the Australian–New Guinea plate has continued its northward drift, thus breaking the close contact which apparently existed with the Indo-Malayan Archipelago during the Miocene (Fig. 1.4).

Many angiosperm families are represented in the fossil records of the late Cretaceous and Palaeocene (Dettmann, 1973: Axelrod, 1975); before then, fossil evidence is fragmentary. Melville (1969, 1971, 1973, 1975) considered that the archaic vegetative features[1] of many characteristic

members of the heathlands of Australia and South Africa link them with the *Glossopteris* flora widespread across the Permian super-continent of Gondwanaland. To support his argument, Melville (1975) noted that fossil leaves, very closely resembling leaves of the heath family Epacridaceae in both size and venation, have been recorded from the Permian of New South Wales. After considering the fossil record and the vascular anatomy of the flowers of the heathland representatives of the family Proteaceae, Melville (1975) suggested that the Permian pro-angiosperm ancestors of the Proteaceae were "probably shrubs or small trees living in a temperate climate, bearing leaves of the

---

[1] The inflorescence of several species of the genera *Acrotriche* and *Melichrus* in the heath family Epacridaceae are borne on the old wood (cauliflory). This attribute is regarded as primitive.

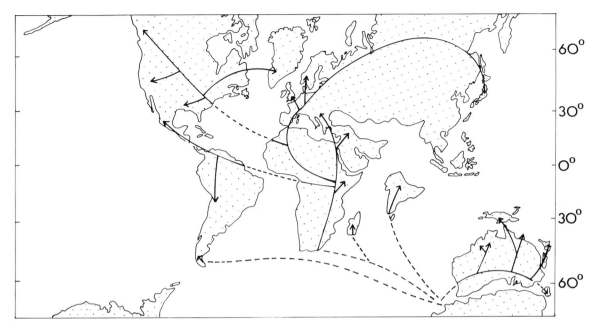

Fig. 1.3. Geographical distribution of continents in the Paleocene period (after Smith and Briden, 1977), showing the probable migration routes of the heathland flora.

Gangamopteris type. Their reproductive structures terminated the branch tips and were probably on short lateral branches. The reproductive unit consisted of four free scales of the *Eretmonia* type, each having one to three dichotomous trusses of microsporangia, followed by four gynoecial units also free from one another. The gynoecial unit consisted of a gangamopteroid blade probably with the venation pattern already somewhat reduced and giving rise to a smaller scale bearing a dichotomous truss of ovules".

Melville (1975) further observed that the evolutionary development of many angiospermous families characteristic of heathland vegetation (such as Proteaceae, Restionaceae, Fabaceae and Asteraceae) must have been initiated early in the Permian for the observed parallel evolution to have taken place subsequently on separate fragments of Gondwanaland.

In their monograph on the evolution and classification of the Proteaceae, Johnson and Briggs (1975) concluded that "separate ancestors of the subfamilies must have evolved in Gondwanaland before the final breaking away of any of its constituent land masses, implying an origin of the family well before any fossils attributable to it are known". The moist-forest genera of

the family Proteaceae had apparently evolved from primitive ancestors which already lived in moist-forest vegetation. The sclerophyllous genera of the family arose later, probably about the early Tertiary, from the warm moist-forest flora.

Axelrod (1975) came to the same conclusion that sclerophyll vegetation of the Madrean–Tethyan region (from California and Mexico eastward to the western Himalayan region) had evolved, by the middle Eocene, from the older oak–laurel–palm forests of tropical to subtropical climates. The evolution of sclerophylly, and many related attributes, was considered to be a response to increasing aridity.

Biogeographical evidence suggests that sclerophylly and other attributes of the heathland flora had evolved long before the early Tertiary. The sclerophyllous flora must have co-existed with the moist-forest flora and developed independently. It appears that, in the Southern Hemisphere, the mesothermic moist climate of much of the Mesozoic promoted closed-forests in which various angiospermous families grew in association with southern conifers. Such ecological conditions usually favour mesophytic plants, but not on deep infertile sandy soils. In Borneo, and several other places in Malesia, a low sclerophyllous vegetation

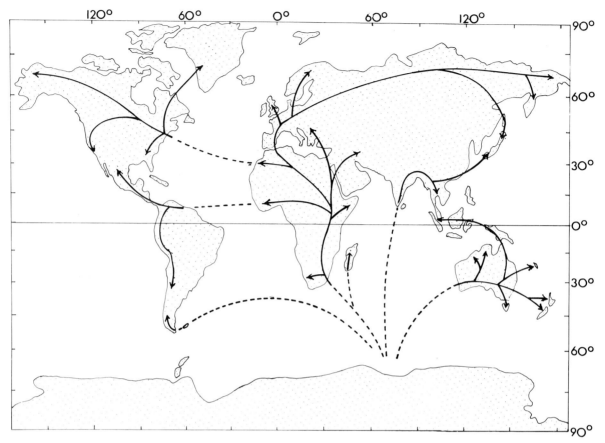

Fig. 1.4. Present-day distribution of the continents, showing the probable migration routes of the heathland flora. Migration of the Australasian heathland flora from New Guinea into the western section of Malesia could have occurred during the late Miocene (Smith and Briden, 1977).

termed *padang* is found on sandy areas amongst a rather luxuriant tropical rain forest (Van Steenis, 1957, 1958). Under primary conditions, these lowland sands are covered in dense forest, physiognomically identical with rain forest but with thinner stems and sclerophyllous leaves. In Borneo, the primary rain forest on sands is dominated by Dipterocarpaceae (mainly *Shorea* spp.) and many small-leafed Myrtaceae (*Baeckea*, etc.), *Casuarina sumatrana*, *Styphelia* (Epacridaceae), *Gonystylus*, *Dacrydrium elatum*, *Agathis*, etc. Winkler (1914) called the community *Heidewald*, a term adopted by Richards in 1936. When the forest type on these sterile sands is destroyed, the regrowth is very slow and it takes a long time to build up something like the original forest. The destroyed localities, termed padang grounds, are covered with a heath-like, easily flammable brushwood with open spaces of sand between the low shrubs. The flora consists of:

Myrtaceae (*Baeckea frutescens*, *Leptospermum*, *Tristania*, *Eugenia*, etc.), *Styphelia*, *Vaccinium*, *Casuarina*, Clusiaceae (*Calophyllum* and *Garcinia*), *Cratoxylum*, *Eurya*, *Timonius*, *Bromheadia* and other terrestrial orchids, *Nepenthes*, *Drosera* and *Xyris*, together with mosses under the bushes. A similar heathland is found on sandy waterlogged soils in the Jardine River area near the tropical northeastern tip of Australia (Lavarack and Stanton, 1977). Here the heathland grades into a dense scrub of heathland species, which becomes taller as the depth of sand above the water table increases. In well-drained sites, the sclerophyllous community grades into a mesophytic, broad-leafed rain forest (see Specht, Chapter 6).

The environmental conditions in which Heidewald and Heide have developed in present-day Malesia appear to be similar to the mesothermic, moist environment of much of the Mesozoic.

Mesophytic closed-forest (containing conifers, and the progenitors of the angiospermous rain-forest flora) probably covered much of the Mesozoic landscape. In contrast, on highly infertile, seasonally waterlogged, sandy soils (developed on alluvial sands and on siliceous rocks) a sclerophyllous vegetation developed.

The sclerophyllous flora of Malesia is not confined to the small areas of lowland infertile sands found amongst the extensive rain forests. It is also seen above 3500 m as a subalpine shrubbery which becomes heath-like as the alpine zone is reached (Van Steenis, 1957). Conifers (such as *Araucaria*, *Dacrydium*, *Libocedrus* and *Podocarpus*) are common; the numbers of Ericaceae and Vacciniaceae (*Rhododendron*, *Vaccinium*), Epacridaceae (*Styphelia*), Myrtaceae (*Leptospermum*, *Xanthomyrtus*), Orchidaceae and Asteraceae (*Olearia*) are high. Compared with the lowland sclerophyllous vegetation, the climate of the subalpine shrubbery is much cooler, but still uniformly moist throughout the year. The soils, however, are again infertile — peaty or sandy (on granitic or sandstone rocks) in marked contrast to the richer soils on which the mesophytic rain forest flourishes at lower altitudes. Under the cool wet climate at this altitude, even basaltic rocks tend to weather into kaolinite — a clay mineral which tends to bind phosphate and molybdate anions in an unavailable form within its crystal structure. The subalpine basaltic soil can be almost as infertile as the peaty and sandy soils and, thus, also support many heath elements.

A number of the terrestrial heathland species found in subalpine zones in Malesia extend as epiphytes into the montane and primary rain forests of lower altitudes (Sleumer, 1966–67; Specht and Womersley, Chapter 12). The epiphytic plants still possess coriaceous leaves and are attached to the trees by lignotubers in an environment which must be described as oligotrophic.

Here again in Malesia, we see the distinctive ecological attributes which must have been characteristic of the embryonic sclerophyll flora of the Mesozoic. Soils of low fertility must have been the key to this development. The nutritional physiology evolved by the sclerophyllous flora would have restricted the distribution of the flora to infertile soils (of limited extent); the flora would have found considerable difficulty in surviving on the widespread, more fertile soils occupied by mesophytic closed-communities (Specht, 1963; Heddle and Specht, 1975).

The southern part of the super-continent of Gondwanaland appears to have been the most likely place for the initial development of the sclerophyllous flora. For example, the sub family Ericoideae, which shows a large concentration of species in South Africa, has only a few representatives in Central and North Africa and in Europe (Baker and Oliver, 1967). As all five subfamilies, as well as most tribes, of the family Proteaceae are represented in Australia, that part of Gondwanaland would appear to be the most likely place of initial diversification in this family (Johnson and Briggs, 1975).

If the physiological tolerances of today's sclerophyllous flora are a guide, a considerable temperature gradient must have existed across this large land mass. As shown by the following few examples, present-day heath families are found in a very wide range of habitats (Croizat, 1952; Sleumer, 1964, 1966, 1967; Hutchison, 1969, 1973; Willis, 1973).

The *family Empetraceae* occurs in the mountains and cold temperate regions of the Northern Hemisphere, the Andes, the Falkland Islands and Tristan da Cunha.

The *family Vacciniaceae* is found in mountainous regions of tropical America, Southeast Asia and Malesia but is rare in tropical Africa and temperate regions; the European *Vaccinium myrtillus* is common in hilly districts.

The *family Ericaceae:* The genus *Rhododendron* has main centres in the mountains of Southeast Asia and Malesia (extending to northeastern Australia), Caucasus, European Alps, and temperate North America. A few species are found in arctic regions. The genus produces light seed, easily carried by wind; it quickly colonises waste land in favourable tropical montane or cool temperate climates, e.g. Malesia, United Kingdom. It appears to have spread from the Himalayas to Japan. The *subfamily Ericoideae* is confined to Africa, the Mediterranean and Europe, and outlying islands. The genus *Cassiope* is Himalayan as well as circumpolar.

The *family Epacridaceae* occurs mainly in extra-tropical Australia but extends to tropical Malesia, New Caledonia, New Zealand.

The *family Prionotaceae* is found in highland southeastern Australia and sub-antarctic South America.

The mean annual temperature of lowland sites, in which the world heathlands are found, ranges from 5 to 28° (Table 1.4). Subalpine heathlands apparently exist in the lower part of this temperature range (viz. 6 to 11°C). However, much lower temperatures (1 to 2°C) may occur at the 2000-m altitude in the European Alps, where Schröter (1932) considered that scattered trees gave way to alpine *Krummholz* and ericaceous scrub (see Coetzee, 1967).

It is difficult to visualise the climatic conditions which existed over Gondwanaland when the sclerophyllous (heath) flora was evolving. It seems probable from biogeographical evidence that the families Empetraceae, Ericaceae (Arbuteae, Gaultherieae and Rhododendroideae), Prionotaceae, Vacciniaceae, etc., were cool temperate heaths which are now confined to the montane regions which developed around the circumference of Gondwanaland as it drifted apart.

The Afro-European subfamily Ericoideae prob-

ably developed in a temperate climate with a mean annual temperature ranging from 6 to 16°C. Members of this subfamily were later confined to subalpine regions of East Africa when tropical Africa became much warmer.

The Australian family Epacridaceae appears to have been much more plastic in its evolution. It is now found in habitats with an annual temperature range of 6 to 28°C.

In the Southern Hemisphere, heathlands are composed of many primitive or ancient dicotyledonous and monocotyledonous families (Hutchison, 1969, 1973) — all of which must have evolved as Gondwanaland was splitting into present-day continents:

DICOTYLEDONS

Casuarinales	Casuarinaceae
Celastrales	Cyrillaceae, Empetraceae, Stackhousiaceae
Dilleniales	Dilleniaceae (*Hibbertia*)
Ericales	Clethraceae, Diapensiaceae, Epacridaceae, Ericaceae, Prionotaceae, Vacciniaceae

TABLE 1.4

Mean annual temperatures of lowland and subalpine localities in which heathlands have been described

Country	Altitude of subalpine zone (m)	Mean annual temperature (°C)		Reference
		subalpine	lowland	
Europe (Western)	±2000	<1	6.5–13.5	Schröter (1932), Gimingham (1972)
U.S.A. (Eastern)	1400–2000	12.5–20.8		
South Africa			8.0–16.5	Acocks (1975), Killick (1963)
Cape	1500–2017	6–8		
Cathedral Peak	1830–2895	6–12		
East Africa	3300–3550	6–8	–	Hedberg (1951), Coetzee (1967)
Malesia	3500–4150	8–11	26–28	Van Steenis (1957)
Australia				
Tropical	1500	?	22–28	Specht (Chapter 6)
Subtropical	–	–	16–22	Specht (Chapter 6)
Temperate	1250–2230	6–8	12–16	Specht (Chapter 6)
New Zealand (South Island)	700+	6–8	10–15	Burrows et al. (Chapter 13)
Tristan da Cunha	–	–	14	Wace and Holdgate (1958)
Falkland Is.	–	–	5.5	Wace (1960)

Temperature data extracted from Meteorological Office (1958), extrapolated linearly from lower altitudes to obtain estimates of subalpine temperatures.

Euphorbiales	Euphorbiaceae (Stenolobieae)
Goodeniales	Brunoniaceae, Goodeniaceae, Stylidiaceae
Hamamelidales	Bruniaceae
Leguminales	Fabaceae (Genisteae, Podalyrieae), Mimosaceae
Myrtales	Myrtaceae
Olacales	Olacaceae
Pittosporales	Byblidaceae, Pittosporaceae (Billardiereae), Roridulaceae, Tremandraceae
Polygalales	Polygalaceae
Proteales	Proteaceae
Rhamnales	Rhamnaceae (Rhamneae)
Rosales	Rosaceae (*Adenostoma, Cliffortia*)
Rutales	Rutaceae
Santalales	Grubbiaceae, Santalaceae
Sarraceniales	Droseraceae, Sarraceniaceae
Thymelaeales	Penaeaceae, Thymelaeaceae (*Passerina, Pimelea*)

MONOCOTYLEDONS

Agavales	Xanthorrhoeaceae
Cyperales	Cyperaceae (*Lepidosperma*)
Haemodorales	Haemodoraceae, Hypoxidaceae
Iridales	Iridaceae
Juncales	Centrolepidaceae, Restionaceae
Liliales	Liliaceae
Orchidales	Orchidaceae (Diurideae)

Only a few of these families have penetrated into Indo-Malaya, Europe, and North and South America.

In general, the mesothermal moist climate characteristic of much of the Mesozoic continued, after a fall in temperature in the late Cretaceous, well into the Tertiary. Hornibrook (1971) summarised the New Zealand studies of Devereux (1967) and others and concluded that, for much of the Tertiary, the climate was about 10°C hotter than the temperatures recorded in Wellington (lat. 41°S) today. It was only during the last stages of the Miocene and into the Pliocene that thermal conditions fell towards those experienced at present.

Tropical conditions existed in southern Australia during the middle and late Eocene, when tropical mangrove plant communities flourished near Perth, W.A. (lat. 32°S today, but lat. 60°S in the Eocene), according to the palaeo-ecological studies of Churchill (1973). The mangrove palm *Nypa* and other tropical Indo-Malayan elements recorded in the Eocene London Clay flora of latitude 50 to 52°N in England (Chandler, 1964) suggest that the tropics then occupied a belt that was some two to three times wider than at present.

It would appear that there were many infertile soils — podzolised sands, sandstone and granitic soils, montane and lowland peats, montane basaltic soils (high in kaolinite) — on which the sclerophyllous flora could evolve through the Tertiary. As well, under the subtropical–tropical climate, extensive areas of lateritic soils (low in plant nutrients) were developed across much of Africa, India, and Australia, and also in South America, especially in Brazil and the Guianas (Prescott and Pendleton, 1952). It appears that the process of lateritisation started late in the Cretaceous, continued through much of the Tertiary, but was at its maximum in the Pliocene at the end of the Tertiary (Stephens, 1971). It is widely inferred that the seasonally oscillating water table associated with the major lateritic surfaces was a consequence not only of low relief but also of the wet conditions of the mid Tertiary. A mesothermal climate apparently favoured the rates of the physical and chemical reactions involved in the development of the lateritic soil profile. Sclerophyllous plant communities, thus, had many infertile and seasonally waterlogged sites on which to expand and evolve.

The Gondwanaland sclerophyllous flora was not to remain under the influence of uniformly wet (but seasonally waterlogged) conditions. Seasonal aridity had appeared in the climate of the Madrean–Tethyan region from California and Mexico eastward to Pakistan by the middle Eocene (Axelrod, 1975). In the Afro-European continents, this region of seasonal aridity apparently developed between the humid, subtropical belt of Africa and humid temperate Europe (lat. 30 to 45°N). Seasonal waterlogging in a uniformly wet climate had already induced a seasonal rhythm in the evergreen, sclerophyllous flora. It was no difficulty for the flora to adapt to seasonal aridity provided the aridity was not too severe. There are many examples throughout the world of this transition from wet-heathland to dry-heathland. Much of the flora — trees, shrubs and evergreen hemicryptophytes — made the transition on infertile soils (Table 1.1). On soils of marginal fertility or on highly calcareous soils (in which plant nutrients such as iron, manganese, copper, cobalt are often in forms unavailable to plants), only a few evergreen sclerophyllous trees and shrubs made the transition to seasonal aridity; seasonal grasses and herbs occupied the ground layer instead (Table 1.1).

The Quaternary has been a period of considerable climatic fluctuations. Seasonal aridity has tended to increase over the world. In the Mediterranean climate of southern Australia and in South Africa, the larger shrubs and small trees of the sclerophyll communities retain a subtropical summer growth rhythm (Specht and Rayson, 1957). In many parts of the world, aridity has become so extreme that the sclerophyllous vegetation has become extinct except in small montane or moist lowland pockets. However, in the Pleistocene considerable areas of coastal sands, blown off exposed continental shelves, accumulated around Australia, near Cape Town in South Africa, and along the eastern coast of U.S.A. These dunes and sand plains presented new infertile soils onto which the sclerophyll flora was able to migrate.

CONCLUSION

The heathland flora apparently evolved early in the Mesozoic in the southern part of Gondwanaland, just as the super-continent was breaking into its component continents. In lowland areas, the sclerophyllous vegetation must have developed on small areas of seasonally waterlogged, infertile sandy and peaty soils within well-drained fertile closed-forest communities. Similar infertile, seasonally waterlogged soils would also have occurred in subalpine localities and there supported a wet-heathland.

The Mesozoic sclerophyllous flora, both dicotyledons and monocotyledons, evolved into a large number of primitive angiosperm families. A few sclerophyllous conifers and ferns were also included. Some of these families spread to the periphery of Gondwanaland where they now occupy subalpine and cool temperate localities. Other families have spread across two or more continents. Most genera have remained confined to infertile soils; a few have produced species able to colonise more fertile soils.

During the Tertiary, the continents in the Southern Hemisphere experienced humid subtropical conditions over much of their latitudinal length. Extensive areas of highly leached soils, such as lateritic podzols, were formed, especially in the Pliocene; these soils would favour the expansion of the wet-heathland flora.

Seasonal aridity appeared in the Madrean–Tethyan region (from California and Mexico to Pakistan) early in the Tertiary. Similar conditions developed over much of the Southern Hemisphere in the late Tertiary and especially during the Quaternary. The seasonal rhythms already present in the evergreen, seasonally waterlogged, wet-heathland vegetation, were easily adapted to seasonal aridity. Provided the aridity was not too severe, dry-heathland vegetation, with many species in common with wet-heathland, developed whenever the soil was infertile. A few sclerophyllous shrubs and trees were able to extend their range onto slightly more fertile soils in these seasonally dry climates. Most of the sclerophyllous ground stratum, however, was not successful in this new environment, and was easily supplanted by seasonal grasses and herbs.

Today, lowland infertile soils in many parts of the world support dry-heathland on well-drained sites; if they are seasonally waterlogged, wet-heathland is formed, Sclerophyllous trees and tall shrubs are often present as an overstorey. In well-watered sites, the sclerophyllous trees may be so dense that the heathland shrubs in the understorey are excluded. An altitudinal gap then occurs between lowland and subalpine regions where heathland re-appears.

Evergreen sclerophylly appears to be an evolutionary adaptation in the Mesozoic to low levels of plant nutrients especially phosphate. It appeared first, apparently, on seasonally waterlogged soils where the wet-heathland vegetation was subjected to physiological drought for a portion of the year. It was a simple transition for the wet-heathland flora to adjust to the seasonal drought conditions which developed in many regions of the world during the Tertiary and Quaternary.

REFERENCES

Acocks, J.P.H., 1975. Veld types of South Africa. *Mem. Bot. Surv. S. Afr.*, No. 40: 128 pp. (2nd ed.).

Adamson, R.S. and Osborn, T.G.B., 1924. The ecology of *Eucalyptus* forests of the Mount Lofty Ranges (Adelaide District), South Australia. *Trans R. Soc. S. Aust.*, 48: 87–144.

Andrews, E.C., 1916. The geological history of the Australian flowering plants. *Am. J. Sci.*, 249: 171–232.

Axelrod, D.I., 1975. Evolution and biogeography of Madrean–Tethyan sclerophyll vegetation. *Ann. Mo. Bot. Gard.*, 62: 280–334.

Baker, H.A. and Oliver, E.G.H., 1967. *Ericas in Southern Africa.* Purnell and Sons, Cape Town, 180 pp.

Beadle, N.C.W., 1968. Some aspects of the ecology and physiology of Australian xeromorphic plants. *Aust. J. Sci.*, 30: 348–355.

Braun-Blanquet, J., Roussine, N. and Nègre, R., 1952. *Les Groupements Végétaux de la France Méditerranéenne.* Cent. Natl. Rech. Sci., Montpellier, 297 pp.

Chandler, M.E.J., 1964. *The Lower Tertiary Floras of Southern England. IV. A Summary and Survey of Findings in the Light of Recent Botanical Observations.* Br. Mus. Nat. Hist., London, 151 pp.

Churchill, D.M., 1973. The ecological significance of tropical mangroves in the Early Tertiary floras of southern Australia. *Spec. Publ. Geol. Soc. Aust.*, 4: 79–86.

Coetzee, J.A., 1967. *Palaeoecology of Africa, 3. Pollen Analytical Studies in East and Southern Africa.* A.A. Balkema, Cape Town, 146 pp.

Cooper, W.S., 1922. The broad-sclerophyll vegetation of California; an ecological study of the chaparral and its related communities. *Publ. Carneg. Inst.*, No. 319; 124 pp.

Costin, A.B., 1954. *A Study of the Ecosystems of the Monaro Region of New South Wales.* Government Printer, Sydney, N.S.W., 860 pp.

Croizat, L., 1952. *Manual of Phytogeography.* Junk, The Hague, 695 pp.

Dettmann, M.E., 1973. Angiospermous pollen from Albian to Turonian sediments of eastern Australia. *Spec. Publ. Geol. Soc. Aust.*, 4: 3–34.

Devereux, I., 1967. Oxygen isotope palaeotemperature measurements on New Zealand Tertiary fossils. *N. Z. J. Sci.*, 10: 988–1011.

Diels, L., 1906. *Die Vegetation der Erde, 7. Die Pflanzenwelt von West-Australian südlich des Wendekreises.* Engelmann, Leipzig, 413 pp.

Fosberg, F.R., 1967. A classification of vegetation for general purposes. In: G.F. Peterken (Editor), *Guide to the Check Sheet for IBP Areas.* IBP Handbook No. 4, Blackwell, Oxford, pp. 73–120.

Gimingham, C.H., 1972. *Ecology of Heathlands.* Chapman and Hall, London, 266 pp.

Graebner, P., 1895. Studien über die norddeutsche Heide. *Englers Bot. Jahrb.*, 20.

Graebner, P., 1901, 1925, *Die Vegetation der Erde, 5. Die Heide Norddeutschlands.* Engelmann, Leipzig, 277 pp. (1901: 1st ed.; 1925: 2nd ed.).

Haeckel, E., 1869. Entwicklungsgang und Aufgaben der Zoologie. *Jenaische Z.*, 5: 353.

Haile, N.S., McElhinny, M.W. and McDougall, I., 1977. Palaeomagnetic data and radiometric ages from the Cretaceous of West Kalimantan (Borneo) and their significance in interpreting regional structure. *J. Geol. Soc. Lond.*, 133: 133–144.

Hamilton, A.G., 1927. The xerophytic structure of the leaf in the Australian Proteaceae. Part 1. *Proc. Linn. Soc. N.S.W.*, 52: 258–274.

Hedberg, O., 1951. Vegetation belts of the East African mountains. *Sven. Bot. Tidskr.*, 45: 140–202.

Heddle, E.M. and Specht, R.L., 1975. Dark Island heath (Ninety-Mile Plain, South Australia). VIII. The effect of fertilizers on composition and growth, 1950–1972. *Aust. J. Bot.*, 151–164.

Hornibrook, N de B., 1971. New Zealand Tertiary climate. *N.Z. Geol. Surv. Rep.*, No. 47: 19 pp.

Hutchison, J., 1969. *Evolution and Phylogeny of Flowering Plants. Dicotyledons: Facts and Theory.* Academic Press, London, New York, 717 pp.

Hutchison, J., 1973. *The Families of Flowering Plants.* Clarendon Press, Oxford, 3rd ed., 968 pp.

Johnson, L.A.S. and Briggs, B.G., 1975. On the Proteaceae — the evolution and classification of a southern family. *Bot. J. Linn. Soc.*, 70: 83–182.

Killick, D.J.B., 1963. An account of the plant ecology of the Cathedral Peak area of the Natal Drakensberg. *Mem. Bot. Surv. S. Afr.*, No. 34: 178 pp.

Lavarack, P.S. and Stanton, J.P., 1977. Vegetation of the Jardine River Catchment and adjacent coastal areas. *Proc. R. Soc. Qld.*, 88: 39–48.

Loveless, A.R., 1961. A nutritional interpretation of sclerophylly based on differences in the chemical composition of sclerophyllous and mesophytic leaves. *Ann. Bot. (Lond.)*, N.S., 25: 168–184.

Loveless, A.R., 1962. Further evidence to support a nutritional interpretation of sclerophylly. *Ann. Bot. (Lond.)*, N.S., 26: 551–561.

McLuckie, J. and Petrie, A.H.K., 1927. An ecological study of the flora of Mount Wilson. Part IV. Habitat factors and plant response. *Proc. Linn. Soc. N.S.W.*, 52: 161–184.

Melville, R., 1969. Leaf venation patterns and the origin of the angiosperms. *Nature*, 244: 121–125.

Melville, R., 1971. Some general principles of leaf evolution. *S. Afr. J. Sci.*, 67: 310–316.

Melville, R., 1973. Relict plants in the Australian flora and their conservation. In: A.B. Costin and R.H. Groves (Editors), *Nature Conservation in the Pacific.* Aust. Natl. Univ. Press, Canberra, A.C.T., pp. 83–90.

Melville, R., 1975. The distribution of Australian relict plants and its bearing on angiosperm evolution. *Bot. J. Linn. Soc.*, 71: 67–88.

Meteorological Office (Air Ministry), 1958. *Tables of Temperature, Relative Humidity and Precipitation for the World.* Parts I–VI. H.M.S.O., London, 699 pp.

Paterson, B.R., 1961. Systematic studies of the anatomy of the genus *Acrotiche* R.Br. *Aust. J. Bot.*, 9: 197–208.

Prescott, J.A. and Pendleton, R.L., 1952. Laterite and lateritic soils. *Comm. Bur. Soil Şci. Tech. Commun.*, No. 47: 51 pp.

Rao, C.V., 1971. Proteaceae. *Counc. Sci. Ind. Res., India, Bot.*

Monogr., No. 6: 208 pp.

Raunkiaer, C., 1934. *The Life Forms of Plants and Statistical Plant Geography.* Oxford University Press, Oxford, 632 pp.

Richards, P.W., 1936. Ecological observations on the rainforest of Mount Dulit, Sarawak. Part I. *J. Ecol.*, 24: 1–37.

Rübel, E.A., 1914. Heath and steppe, macchia and garigue. *J. Ecol.*, 2: 232–237.

Sampson, A.W., 1944. Plant succession on burned chaparral lands in northern California. *Bull. Calif. Agric. Exp. Station*, No. 685: 144 pp.

Schimper, A.F.W., 1903. *Plant-geography upon a Physiological Basis.* Clarendon Press, Oxford, 893 pp.

Schomburgk, R., 1876. Flora of South Australia. In: W. Harcus (Editor), *South Australia: Its History, Resources and Productions.* Sampson Low and Co., London, pp. 205–280.

Schröter, C., 1932. *Kleiner Führer durch die Pflanzenwelt der Alpen.* Albert Raustein, Zürich, 80 pp.

Sleumer, H., 1964. Epacridaceae. *Flora Malesiana*, Ser. 1, 6: 422–444.

Sleumer, H., 1966, 1967. Ericaceae. *Flora Malesiana*, Ser. 1, 6: 469–668; 669–914.

Smith, A.G. and Briden, J.C., 1977. *Mesozoic and Cenozoic Paleocontinental Maps.* Cambridge University Press, Cambridge, 63 pp.

Specht, R.L., 1963. Dark Island heath (Ninety-Mile Plain, South Australia). VII. The effect of fertilizers on composition and growth, 1950–1960. *Aust. J. Bot.*, 11: 67–94.

Specht, R.L., 1969. A comparison of the sclerophyllous vegetation characteristic of Mediterranean type climates in France, California and southern Australia. I. Structure, morphology and succession. II. Dry matter, energy and nutrient accumulation. *Aust. J. Bot.*, 17: 277–292; 293–308.

Specht, R.L., 1970. Vegetation. In: G.W. Leeper (Editor), *The Australian Environment.* C.S.I.R.O. — Melbourne University Press, Melbourne, Vic. 4th ed., pp. 44–67.

Specht, R.L., 1972. *Vegetation of South Australia.* Government Printer, Adelaide, S.A., 2nd ed., 328 pp.

Specht, R.L., 1973. Structure and functional response of ecosystems in the Mediterranean climate of Australia. In: *Ecological Studies, 7 Mediterranean Climate Ecosystems.* Springer-Verlag, Berlin, pp. 113–120.

Specht, R.L. and Perry, R.A., 1948. The plant ecology of part of the Mount Lofty Ranges (1). *Trans. R. Soc. S. Aust.*, 72: 91–132.

Specht, R.L. and Rayson, P., 1957. Dark Island heath (Ninety-Mile Plain, South Australia). I. Definition of the ecosystem. *Aust. J. Bot.*, 5: 52–85.

Specht, R.L., Brownell, P.F. and Hewitt, P.N., 1961. The plant ecology of the Mount Lofty Ranges, South Australia. 2. The distribution of *Eucalyptus elaeophora. Trans. R. Soc. S. Aust.*, 85: 155–176.

Specht, R.L., Roe, E.M. and Boughton, V.H. (Editors), 1974. Conservation of major plant communities in Australia and Papua New Guinea. *Aust. J. Bot. Suppl.*, No. 7: 667 pp.

Stephens, C.G., 1971. Laterite and silcrete in Australia. *Geoderma*, 5: 5–52.

Tansley, A.G., 1935. The use and abuse of vegetational concepts and terms. *Ecology*, 16: 284–307.

Tate, R., 1883. The botany of Kangaroo Island. *Trans. R. Soc. S. Aust.*, 6: 116–171.

Trollope, W.S.W., 1970. *A Consideration of Macchia (Fynbos) Encroachment in South Africa and an Investigation with Methods of Macchia Eradication in the Amatole Mountains.* Thesis, University of Natal, Pietermaritzburg.

Trollope, W.S.W., 1973. Fire as a method of controlling macchia (fynbos) vegetation on the Amatole Mountains of the Eastern Cape. *Proc. Grassland Soc. S. Afr.*, 8: 35–41.

Van Steenis, C.G.G.J., 1957. Outline of vegetation types in Indonesia and some adjacent regions. *Proc. Eighth Pac. Sci. Congr., 1953.*, 4: 61–97.

Van Steenis, C.C.G.J., 1958. Condition and cause in ecological interpretation. *Blumea*, Suppl. IV: 93–95.

Wace, N. M., 1960. The botany of the southern oceanic islands. *Proc. R. Soc. Lond.*, Ser. B, 152: 475–490.

Wace, N.M. and Holdgate, M.W., 1958. The vegetation of Tristan da Cunha. *J. Ecol.*, 46: 593–620.

Warming, E., 1909. *Oecology of Plants.* Clarendon Press, Oxford, 422 pp.

Watson, L., 1962. The taxonomic significance of stomatal distribution and morphology in Epacridaceae. *New Phytol.*, 61: 36–40.

Watson, L. 1964. The taxonomic significance of certain anatomical observations on Ericaceae. The Ericoideae, *Calluna* and *Cassiope. New Phytol.*, 63: 274–280.

Watson, L., 1965. The taxonomic significance of certain morphological variations among Ericaceae. *J. Linn. Soc. Lond. Bot.*, 59: 111–125.

Wellman, P. and McDougall, I., 1974. Cainozoic igneous activity in eastern Australia. *Tectonophysics*, 23: 49–65.

Werger, M.J.A., Kruger, F.J. and Taylor, H.C., 1972. A phytosociological study of the Cape Fynbos and other vegetation at Jonkershoek, Stellenbosch. *Bothalia*, 10: 599–614.

Willis, J.C. (revised by Airy Shaw, H.K.), 1973, *A Dictionary of the Flowering Plants and Ferns.* Cambridge University Press, Cambridge, 8th ed., 1245 pp.

Winkler, H., 1914. Die Pflanzendecke Südost-Borneos. *Bot. Jahrb.*, 50 (Festband): 188–208.

Wood, J. G., 1939. Ecological concepts and nomenclature. *Trans. R. Soc. S. Aust.*, 63: 215–223.

Chapter 2

SOUTH AFRICAN HEATHLANDS[1]

F.J. KRUGER

INTRODUCTION

This chapter is confined to a description and discussion of *fynbos*, the most important type of heath found in South Africa. Fynbos is a comprehensive term for the vegetation of a well-defined and limited landscape of the southern Cape Province. It is derived from the Afrikaans terms for "fine" and "bush" and evokes the physiognomy of the typical community. Taylor (1978) has reviewed the use of this term and its synonyms. It is used here in his sense for the typical South African heathlands in a zone of subarid to wet climate, usually of the Mediterranean type, on infertile soils.

Taylor (1978) has identified the following as characteristic of the physiognomy of fynbos communities: (a) the restioid element, which is invariably present and comprises wiry aphyllous hemicryptophytes of the Restionaceae and some Cyperaceae: (b) the ericoid element, which comprises dwarf and low evergreen ericoid shrubs. A frequent, but not constant, feature is a component of taller broad-sclerophyllous shrubs (the proteiod element).

The only constant and differential floristic element is the Restionaceae which may dominate communities or at least the herb stratum in certain habitats. Other typical taxa are the genera *Protea*, *Leucadendron*, (Proteaceae), *Erica* and *Blaeria* (Ericaceae), *Aspalathus* (Fabaceae), *Tetraria* and *Ficinia* (Cyperaceae), *Merxmuellera*, *Pentaschistis* and *Ehrharta* (Poaceae), *Agathosma* (Rutaceae), *Cliffortia* (Rosaceae), *Brunia* and *Berzelia* (Bruniaceae), *Gnidia*, *Struthiola* and *Passerina* (Thymelaeaceae), *Metalasia*, *Helichrysum*, *Stoebe*, *Corymbium* and many others in the Asteraceae, and numerous genera in the Liliaceae, Orchidaceae, and Iridaceae: Acocks (1953) and Taylor (1978) present more comprehensive lists. These elements all vary considerably in their relative importance in the communities.

The flora is marked by a great richness and high degree of endemism (Table 2.1). Numerous species exist as small populations in isolated habitats. In general, communities are distinguished by a strikingly high alpha diversity and equitability and similarly high gamma and delta diversities are found in the vegetation as a whole.

Fynbos is largely devoid of native trees (see p. 29 for exceptions) but is replaced abruptly by evergreen forest in suitable habitats, especially in the humid climate in the George and Knysna areas. In the south and southwest, forest or scrub forest is restricted to sheltered *kloofs* and patches of rock scree (Werger et al., 1972). Toward the east, where summer rainfall régimes predominate, there is a gradual transition to grassland on open sites and a

TABLE 2.1

Species-area data from a systematic sample of ninety 5×10 m quadrats in mountain fynbos at Zachariashoek (33°49'S, 19°02'E); quadrats were subdivided in quarters, and these were nested; all visible vascular plants were listed (from F.J. Kruger and B.W. van Wilgen, unpublished, 1977)

Quadrat size (m²)	Mean number of species		
	12.5	25	50
Full sample	28.1	36.2	43.8
Ten richest quadrats	47.3	60.0	72.4
Ten poorest quadrats (mainly phreatic sites)	10.6	12.9	16.7

---

[1]Manuscript completed February, 1977.

rapid one to low evergreen forest ("Valley Bushveld" in Acocks' terminology) in the kloofs. In arid Mediterranean and transitional Mediterranean climates fynbos is replaced on heavier soils by low open-heathland ("Coastal and Mountain Rhenosterbosveld") succulent shrubland ("Succulent Mountain Scrub") and succulent low open-shrubland ("Karroid Broken Veld" and "Succulent Karroo"). These transitions on the aridity gradient normally coincide with and are reinforced by edaphic changes, from soils derived from quartzite, to more fertile, neutral fine-textured soils on shales or granite. On coastal sands of the arid west coast there is a complex transition from typical fynbos communities to a spiny and somewhat succulent shrubland known as *Strandveld*.

Acocks, in his classification and map of the vegetation of South Africa (Acocks, 1953), distinguished three categories of fynbos: "Coastal Macchia", "Macchia" (including the subcategory "Arid Fynbos"), and "False Macchia"; Taylor (1978) distinguished two: "Coastal Fynbos" (identical with coastal macchia) and "Mountain Fynbos". Macchia and false macchia (Acocks, 1953) are included in the latter type since Acocks has made the distinction between two floristically and physiognomically indistinguishable types on hypotheses about the climaxes. The three categories — coastal fynbos, mountain fynbos, and also arid fynbos — recognised by Taylor are described in this chapter.

Coastal fynbos is confined to the recent sands of the coastal forelands, southwards from near Redelinghuys (32° 28′S, 18° 32′E) to the Cape Flats, on the northern shores of False Bay, where its eastward extension is interrupted by the mountains. It re-appears west of Hermanus and continues in an almost unbroken zone to Mosselbaai. Outliers and fragments occur wherever the rocky coastline or chains of hills are broken by local accumulations of sand (as at the Vishoek gap in the Cape Peninsula and near Betty's Bay), and these extend eastwards to near Cape St. Francis.

Mountain fynbos is confined to the ranges of the Cape Folded Belt, southward from Nieuwoudtville to the Peninsula and Cape Hangklip, and thence eastward to Port Elizabeth and Grahamstown. A northern outlier occurs on granite hills in the Kamiesberg (30°18′S, 18°05′E), and "islands"

occur on high quartzite ridges in the Little Karoo. This formation occupies a relatively small area of about 3,5 million ha. Acocks (1953) and Taylor (1978) recognize arid fynbos as a subcategory. The distribution of fynbos is shown in Fig. 2.1.

ENVIRONMENT

Physiography

The fynbos landscape is dominated by the ranges of the Cape Folded Belt, bounded on the interior by the basin of the Great Karoo. The mountains are flanked seawards by low peneplained erosion surfaces and flat plains of the coastal forelands (Wellington, 1955). In the east, between Mosselbaai and Humansdorp, the mountains fall southwards to a coastal platform with a high rocky coast.

Three major geological formations are present: the late Precambrian Nama System (Malmesbury Formation), the Ordovician to Devonian Cape Supergroup, and Tertiary and Quaternary deposits.

Precambrian rocks consist of shales and siltstones of the Malmesbury Formation which form the Malmesbury plain or Swartland, the major basin of the western coastal forelands. Early Cambrian intrusions of Cape Granite are common. The early intrusions consist of medium- to coarse-grained biotite granite, and are followed by a finer-grained quartz porphyry (Tankard, 1976). The strata of the Cape Supergroup rest unconformably on these rocks. This system comprises strata of the Table Mountain Group, the Bokkeveld Series, and the Witteberg Series. The character of the Cape Folded Belt is conferred mainly by strata of the Table Mountain Group. These are principally quartzites about 3700 m thick, of which the most prominent stratum is the Ordovician Peninsula Formation, white or grey orthoquartzites about 1500 m thick. This is followed by a narrow band of sandy tillite, the Pakhuis Formation, and siltstones and shales of the Cedarberg Formation. The Nardouw Formation, 900 m of Silurian quartzite much like the rocks of the Peninsula Formation, completes the succession. Rust (1967) describes the system in detail.

The Bokkeveld Series contains five sandstone

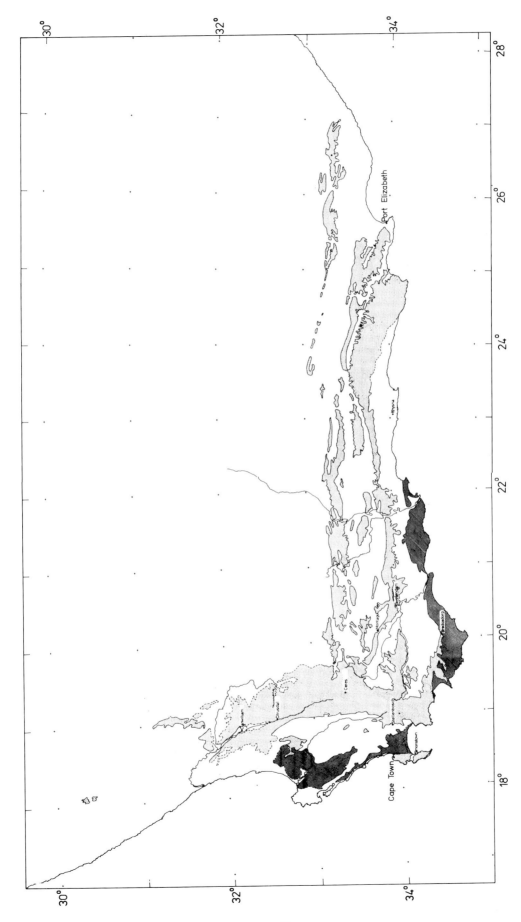

Fig. 2.1. Distribution of fynbos. Heavy shading indicates coastal fynbos and light shading, mountain fynbos (after Acocks, 1953).

horizons but consists mainly of shales. In contrast, the Witteberg Series, while including some shale bands, is composed mainly of hard white quartzites.

Cainozoic formations are located near the coast and comprise marine and aeolian materials deposited during alternating submergences and emergences of the coastline. Miocene and Pliocene coastal limestone formations are prominent in the south near Bredasdorp and near Saldanha Bay. These are overlain in parts by aeolianite. Siesser (1972), Tankard (1976) and Marker (1976) describe the stratigraphy and origin of these formations.

The coastal plain is covered in large part by siliceous dune sands of Recent origin, of which the Cape Flats are typical. Live dunes are present in certain areas but most carry vegetation.

The mountains of the Cape Folded Belt consist of Permo-Triassic anticlines of resistant Table Mountain and Witteberg quartzites. The intervening valleys and basins, in turn, occur on softer Bokkeveld or, occasionally, Malmesbury rocks. Cretaceous Enon conglomerates occur in intermontane valleys and on coastal platforms in the east. These sub-parallel ranges run more or less north–south in the west and east–west in the south. A complex mountain block occurs in the central zone of conflict, between the Hex River and the Hottentots–Holland ranges. Several distinct chains may be recognized (see King, 1963, fig. 55, and Wellington, 1955, figs. 13 and 14). In the west, the most prominent line is that of the Olifants River Mountains, which run approximately southwards, forming successively the Elandskloof, Drakenstein and Hottentots–Holland ranges, the latter breaking through to the coast at Cape Hangklip. Major ranges immediately to the east are the Cedarberg in the north and Koue Bokkeveld, Skurweberg and Groot Winterhoek further south. These ranges are all dominated by strata of the Table Mountain Group. Immediately east, flanking the Karoo, Witteberg quartzite anticlines form the Swartruggens.

Two major chains dominate the ranges of the southern zone. The Langeberg Mountains were formed by a massive fault which runs eastwards almost uninterrupted for about 450 km from Worcester to Humansdorp. These are known in the east as the Outeniekwaberge (Outeniqua) and the Tsitsikama Mountains. To the north, the

Swartberg runs from the Anysberg in the west, more or less parallel with the Langeberg, to form the Baviaanskloof Mountains in the east, these then dividing into the Winterhoek and Elandskloof mountains which decline eastwards and end in the vicinity of Port Elizabeth. The broad intervening valley — the Little Karoo — is broken in the west and centre by anticlines which form the Waboomberg, Warmwaterberg, Rooiberg and others. In the east, the Little Karoo ends in the Kouga Mountains.

Witteberg quartzites form a line of mountains north of the Swartberg and extend eastward to form the Suurberg which terminates near Grahamstown. Various short but prominent ridges break the coastal forelands south of the Langeberg.

Maximum elevations are not great. Some of the highest peaks are the Seweweekspoortberg in the Swartberg (2325 m), Matroosberg in the Hex River Mountains (2249 m), Swartberg (beacon 53, 2130 m) and the Groot Winterhoek peak (2078 m). Nevertheless, local relief is pronounced. Seweweekspoortberg rises 1500 m from the nearby water gap over a distance of 2.9 km. The difference in elevation between the Hex River and the crest of Matroosberg, 5 km to the north, is 1680 m. This youthful landscape, created by faulting which persisted into the Cretaceous, is enhanced by the resistant quartzite layers. Steep slopes, narrow defiles and prominent scarps and crests are characteristic, but the topography is often relieved by extensive subsummit plateaux and basins.

Climatic control of erosion is expressed in broad differences in landform. Two categories may conveniently be distinguished, the first being the zone of warmer, less humid climate and with a more pronounced contrast between winter and summer. Especially in the mountains of the north and west the landscape is dominated by high cliffs and extensive slopes of bouldery debris and fans deposited by mass wastage. Talus and soil creep, and periodic mass wastage, are often active on debris slopes. The second type has a cooler humid climate and is typical of poleward slopes of the southern ranges. Here, slopes are steep but well-vegetated so that cliffs are not prominent and the debris mantle more stable; colluvial fans are rare and small.

In the east the strongly overturned Witteberg layers give the landscape a different character, with few scarps, rounded crests and deep *kloofs*.

The coastal forelands fall naturally into a western and a southern component. They comprise essentially a series of three more or less dissected Tertiary erosion surfaces. Immediately to the west of the mountains lies the Malmesbury surface, a lightly dissected lowland formed on Malmesbury shales. Between this and the coast the shales have been covered by aeolian sands, forming a coastal plain up to 30 km wide in parts, and broken by granite plutons as in the vicinity of Darling and of Saldanha Bay; local ridges and outcrops of Miocene and Pliocene rocks, described above, occur also in the vicinity of Saldanha Bay.

Bokkeveld shales form a series of undulating lowlands south of the Riviersonderend and Langeberg ranges, extending eastward to about Mosselbaai. These are buried by aeolian sands near the coast, to form a narrow plain up to about 25 km wide. In this, outcroppings of coastal limestone form ranges of low hills, especially in the vicinity of Bredasdorp.

A marine terrace at about the 250-m level is much dissected and fragmented in the west but forms an extensive elevated coastal platform east of George, where it is underlain principally by rocks of the Table Mountain Group. This surface, relatively level in parts, hilly elsewhere, has been deeply incised by gorges. This platform is terminated by wide floodplains in the valleys east of Humansdorp and Port Elizabeth.

King (1963) and Wellington (1955) provide extensive descriptions of Cape Folded Belt landscapes.

Climate

The fynbos climate comprises in fact a complex of climates, ranging from typical Mediterranean in the west to humid temperate, uniform rainfall types in the east, and from subarid to perhumid. This range reflects the effects on normal weather systems of the character and arrangement of the major landforms and of the Agulhas and Benguela Currents. Orographic effects of the mountain ranges are reinforced by their alignment more or less parallel with the coasts and by pronounced local relief.

Flowing westward as an extension of the Moçambique Current, the Agulhas Current mod-

erates climates along the south coast, but with decreasing effect westward as its course diverges from the coast west of Port Elizabeth; nevertheless, its effect is felt as far west as Cape Agulhas. Cold upwellings of the Benguela Current maintain uniformly low sea temperatures along the west coast. "The temperature and moisture content of the air moving in contact with the contrasting waters of these two major coastal currents are significantly different and profoundly influence the nature of the climates along the adjacent shores. The mild, humid forest climates of the southeast and south coasts are related to the warm Agulhas Current, and the dry summer scrub woodland climate into which these blend along the south and southwest coasts is related to the colder waters beyond Cape Agulhas, particularly the Benguela Current." (Rumney, 1968.)

"The winter circulation of the South-Western Cape is associated with disturbances in the circumpolar westerly winds, taking the form of a succession of eastward-moving cyclones (depressions) and anticyclones. Originating in areas of cyclogenesis far to the south and west of Southern Africa, these disturbances first bring rain to the south-western and later to the south and south-eastern coasts and may even extend far inland. Fronts are associated with the depressions: warm fronts are diffuse, difficult to recognise and almost impossible to follow; cold fronts are more usual, sharper and more easily recognised. Following the passage of a cold front, winds back from northwest to west and south-west, pressure starts rising, temperatures fall and instability showers and storms may occur. Most winter rain in the South-Western Cape however, occurs in association with north-westerly pre-frontal winds." (Jackson and Tyson, 1971). Modified polar air-masses may reach the continent during deep depressions followed by an anticyclone from the South Atlantic; considerably lower temperatures are then experienced and snow falls on the mountain ranges. *Föhn*-like *bergwinds* often precede winter anticyclones. Dry subsiding air moves off the interior plateau in response to strong coastward pressure gradients. Standing waves arise as the air is drawn across the mountains and strong thermodynamic downwash on the lee of the coastal ranges results in hot turbulent winds. These bergwinds have been described by Tyson (1964); they are characterised by sudden increases in temperature and decreases in humidity, resulting in abnormal fire hazard conditions (see Wicht and De Villiers, 1963).

Summer weather arises primarily as a result of slight southward displacement of the subtropical high pressure belt over the oceans. On the surface this belt appears as a dynamic system of anti-cyclones which travel eastward toward and along the coast, blocking westerly cyclones. As a result, the southwestern part of the Cape Province experiences weather characterised by warm, dry conditions with frequent strong southeast winds. Nonetheless, with a thermal low on the interior plateau, mid-latitude depressions which skirt the southwestern coastal region do periodically pen-etrate inland and synoptic situations arise which bring rains to the south coastal regions on many occasions, as they bring wet weather to the eastern half of the subcontinent (Wellington, 1955; Schulze, 1965; Rumney, 1968).

This brief outline of prevailing weather systems must suffice as a framework for the discussion of climatic elements.

Essential features of fynbos climates as reflected in records of different elements have been sum-marised by Wellington (1955), Schulze (1965) and Rumney (1968). Records for representative sta-tions are presented in Appendix I (p. 53). The description which follows is a brief summary of published information, supplemented where neces-sary; the account is hampered by an almost total lack of data from higher elevations.

Schulze and McGee (1978) present the most recent synopsis of solar radiation over southern Africa. Their data for the fynbos zone indicate rates for incoming radiation during winter of between 110 and $120 \times 10^5 \, \text{J m}^{-2} \text{day}^{-1}$. Summer rates amount to $280 \times 10^5 \, \text{J m}^{-2} \text{day}^{-1}$ in the west, decreasing to $240–250 \times 10^5 \, \text{J m}^{-2} \text{day}^{-1}$ in the east. These rates are naturally much influenced by topography, and representative north slopes may receive more than twice as much radiation annually as south slopes in these latitudes (Kruger, 1974). The average annual duration of bright sunshine amounts to between 60 and 70% of the possible, i.e. about six to eight hours per day.

Mean annual screen temperatures range between about 15° and 18°C. Citrusdal, in the arid fynbos, has a high mean temperature of 18.9°C. Upper mountain stations, such as those on Table

Mountain, have means of about 12° to 13°C, though higher stations would be cooler. Maritime influences are strong, so that inland stations at low elevations are considerably warmer (17° to 18°C) than coastal ones (15° to 16°C). Mean tempera-tures of the warmest month (which is normally February) range from about 20.5° to 24°C, except in the Knysna forest region, where they are low as 18° to 20°C. Absolute maxima are relatively low on the coastal forelands, but frequently exceed 40°C inland. Citrusdal, for example, experiences an annual average of 4.2 days when the maximum temperature exceeds 40°C. The absolute maximum recorded over thirteen years is 45°C. Winter temperatures are mild and the region is enclosed largely by the 10° and 12.4°C July isotherms, with rather higher values at the coast. Frosts are rare in the coastward lowlands and valleys, but freezing temperature may be a common feature of certain mountain situations and interior valleys. At De Keur, in the broad, level subsummit basin of the Koue Bokkeveld, an average of about 25 frost days is recorded annually; the absolute minimum tem-perature in an eleven-year record is −3.9°C. Misgund-oos, situated in a basin of the Langkloof intermontane valley, experiences 47.3 frost days annually, with an absolute minimum of −7.8°C, but Langkloof Experimental Farm, on rising ground in the same area, experiences few frost days.

The annual range of mean temperature is low at the coast (about 4.5°C), but increases to about 11°C inland.

The fynbos zone is enclosed inland mainly by the 300-mm rainfall isohyet, but occurs on coastal sites with mean annual rainfall as little as 250-mm (as near Saldanha Bay). There are pronounced rainfall gradients in the mountain environments: Schulze (1965) suggests changes at a rate of 1800 mm per 1000 m elevation change. The maximum mean precipitation recorded is 3300 mm per annum on the Dwarsberg Plateau (1300 m), Jonkershoek State Forest.

Summer anticyclones usually produce cloud on the mountains. This sometimes induces orographic rain in the higher mountains of the south-west. As a consequence, rainfall gradients in the mountains are complex and not explained entirely by elevation change and rain-shadow effects (Wicht et al., 1969).

The annual march of rainfall changes considerably from west to east. The proportion of rain falling in the winter half-year ranges from almost 90% at Citrusdal to 38% at Grahamstown, with the transition from a winter rainfall régime (more than 60% precipitation in the winter half-year) to constant rainfall occurring at about 20°E, i.e. near Swellendam.

In most instances, precipation is principally in the form of rain. Snowfalls occur at higher elevations (above about 1000 m) on about five occasions annually, but snow persists only on sheltered sites above about 1500 m, where accumulations may lie as long as a fortnight. Upper mountain sites do receive a significant input in the form of mist precipitation, especially during anticyclonic southeaster weather when considerable orographic cloud is induced (normally without rain). This precipitation is demonstrably significant (Marloth, 1904; Nagel, 1956) but it has not been satisfactorily measured (Kerfoot, 1968). Nagel (1962) derived an empirical equation to calculate (possibly underestimate) the amount of mist precipitation, and used it to indicate that, in the extreme case on Table Mountain (1087 m) mist precipitation amounted to about 5700 mm yr^{-1}, as opposed to rainfall of about 1900 mm yr^{-1}. The ratio of mist to rain was 2.4 in winter and 4.7 in summer. A significant proportion of mist precipitation occurred in the absence of rain, but this was not the case on a ridge (840 m) at Jonkershoek where rainless orographic cloud is rare. Kruger (1974) showed from a three-year record of a Grunow mist gauge in the Jakkalsrivier catchment (at 870 m) that mist precipitation in the absence of rainfall occurred on about 20 to 36 days a year, and most of such events occurred in summer; this station experiences frequent rainless southeast cloud. Marloth (1907a) has observed that summer orographic cloud on mountain crests is frequent even on inland ranges such as the Swartberg.

Mist precipitation evidently plays an important role in moderating summer drought at upper elevations and may be a highly significant input in the hydrological cycle.

Precipitation intensities are generally low. Wicht et al. (1969) showed that, over a twenty-year period, an average of about 5% of the rainfall at Jonkershoek occurred as storms with intensities in excess of 18 mm h^{-1} measured over 15-min units.

The highest intensity recorded was 59.4 mm h^{-1}. Extreme daily amounts reach about 100 mm, but about 95% of rainfall occurs in daily falls of less than 36 mm. High intensity rains tend to be associated with convective storms, which are more common along the interior mountain ranges where thunderstorms are frequent in spring and autumn.

Soils

There is no adequate synthesis of fynbos pedology and the account which follows is necessarily incomplete.

Soil types are strongly correlated with parent material but are uniformly structureless, acid or very acid and have a very low base saturation; many are podzols and most are podzolised to some degree. Those on quartzites are loamy sands to sands, while sandy loams and sandy clay loams are found on granites or shales and these, though leached, are normally more fertile. Most profiles drain freely, but some are seasonally or perennially wet or waterlogged due to obstructions to lateral drainage, or by virtue of their location in depressions.

Soils are classified according to the South African binomial soil classification system (Soil Classification Working Group, 1977). Soil *forms*[1] are recognised on the basis of defined diagnostic topsoil and subsoil horizon sequences, and subdivided in *series* on subsidiary characters which normally reflect pedogenesis. Some representative profile descriptions appear of the predominant fynbos soil forms appear in Appendix II (p. 56).

Fynbos soils of the coastal forelands

Soil patterns on the Cainozoic deposits to which coastal fynbos is restricted are relatively simple. Soils on deep sands are normally of the Fernwood form: near the coast sands are calcareous and these

---

[1] In the South African system the soil form is often equivalent in rank to soil order in traditional systems and in the USDA system (the Lamotte form, for example, is equivalent to Podzols or Spodosols) but others include soils which would be classified in two or more different orders in the USDA system (the Hutton form, for example includes Oxisols, Ultisols, Entisols and Inceptisols). This is because of use of different diagnostic criteria at the higher levels of classification in different systems. The South African forms are approximately equivalent to FAO soil units.

belong to the Langebaan series (medium sands), or the Soetvlei series, the latter on flats and depressions where seasonally high water tables permit an abnormal accumulation of organic material in the A horizon. Older dunes are neutral to acid, and, if the latter, qualify as members of the Fernwood series. Under suitable soil-moisture conditions, yellowish sands with a non-calcareous B horizon develop (Ellis, 1973).

Soils of limestone hills and outcrops usually consist simply of a dark grey to black A horizon (less than 0.5 m deep) on hard calcrete, although the profile is often reduced to mere pockets of rather humic accumulations in hollows and crevices. These are soils of the Mispah form. Where soils have developed *in situ* the profile is normally alkaline, sometimes neutral (Kalbank series). A horizons of soils in young transported sands are acid (Ellis, 1973), and the limestone can be considered inactive as a parent material.

Where quartzites of the Cape System underlie the forelands the fynbos soils are acid and resemble inland forms.

Soils of the mountains and foothills

Mountain fynbos occurs exclusively on acid, highly leached soils. The dominant soils are shallow (less than 0.5 m) and skeletal, and profile development is often virtually absent. Most profiles consist of structureless sands or sandy loams with dark grey or grey-brown A horizons, normally containing less than 2% organic carbon (orthic A). The most common forms are Mispah (acid Mispah series), Glenrosa (acid Oribi series), and Cartref (Waterridge series).

Where material accumulates on plateaux and necks or on concave footslopes, deeper (0.5–2.0 m) profiles show a greater degree of differentiation. In the warmer western and interior mountains, with less effective precipitation, B horizons retain a yellow, reddish or brown colour. Soils of the Hutton and Clovelly forms are common on lower hill-slopes. Sands on plateaux are deep (over 1 m) and are of the Fernwood form (Fernwood series), though on wetter sites darker organic-rich A horizons are the rule (Warrington series). On lower pediments or terraces, lateral drainage intensifies leaching and illuviation and typical podzols of the Lamotte form are the rule. Podzolisation is pronounced in the cooler more humid southern

ecosystems. Deep, pale strongly eluviated horizons underlie the A horizon where there is any significant depth of accumulation of material, but there is often no illuviated ferrihumic B, possibly because of low content of iron and bases in the rock or parent material: humic acids are leached out of the profile with the result that streams and rivers are stained a dark reddish-brown. Where profiles are perennially wet through restricted water movement — usually local pediments on the strike slopes — peat-like organic horizons develop, and these are soils of the Champagne form.

On terraces along drainage lines deep undifferentiated sandy alluvial profiles of the Dundee form are the rule.

Soils on granite or shale differ principally in texture. Impeded drainage results in gleying or plinthification: soils of the Avalon form are an example of this type. In humid situations B horizons may be structured, in which case a common toposequence would be Mispah–Glenrosa–Swartland–Sterkspruit, whereas in drier situations the sequence is commonly Mispah–Glenrosa–Hutton–Clovelly.

Foothill soils generally differ from mountain soils in that they often occur on old land surfaces that have been subjected to a greater degree of weathering under palaeo-environmental conditions. Sands are similar to mountain podzols, but lateritic soils are characteristic.

Fynbos soils sometimes occur outside the main vegetation zone on isolated Tertiary silcrete or laterite cappings.

Chemical characteristics

Many coastal fynbos soils differ from inland soil forms in that they are base-saturated, often containing free lime, with relatively high levels of phosphorus and nitrogen, but there is a rapid gradient of increasing acidity inland, as noted above.

Mountain fynbos soils are universally poor, with base saturation levels of around 20 to 40% (ratio of net exchangeable cations to cation exchange capacity), though Neethling (1970) has recorded levels as low as 8% in the Tsitsikama region. In a typical humid mountain catchment, net exchangeable cations in the A horizon ranged from about 2 to 80 μeq g^{-1}, and cation exchange capacity from about 5 to 400 μeq g^{-1}, the variation being principally correlated with organic carbon content (Kruger, 1974). Soils contain negligible amounts of extractable phosphorus (about 1 to μ g^{-1}) and of total nitrogen (0.1 to 0.2%). The C/N ratio is normally in the range from 15 to 25, although Neethling (1970) has recorded higher ratios.

Granite soils, though equally highly leached, often have high concentrations of nutrients. Various samples from granite profiles in the Jonkershoek Valley showed the following range of values in analytic parameters (data from Joubert, 1965):

Phosphorus	3–40 μg g^{-1}
Nitrogen	0.1–0.4%
Organic carbon	2–12%

Fire

In common with other heathlands, fynbos is subject to recurrent fires of variable frequency.

Human agency is a major cause of fires, especially in areas close to principal population centres, but it is an old influence. Khoisan pastoralists were noted to set the fynbos alight, probably for pasture control, in pre-European times and immediately after settlement (Mossop, 1927; West, 1965). This practice was possibly the rule for the 1500 years or so (Schweitzer and Scott, 1973) that pastoralists had been in the region. Hearths have been dated at 40000 years B.P. but this reflects rather the limitations of the radiocar-

bon dating system than the probable antiquity of man's use of fire, and he is likely to have been a cause of fire in the region for the more than 100000 years (possibly more than 700000: Klein, 1974) during which he has been there.

Natural causes of fire (lightning and falling rocks: Wicht, 1945) are common, and fires of this origin continue as the most important in remoter parts (see Table 2.2). Natural fires must be an ancient factor. Hendey (1973) has reported burnt bones in a late Pliocene to early Pleistocene context, suggesting veld fires at that time.

In the west, fynbos fires are most common in summer, but may occur any time between September and April, with occasional burns in winter. The fire season in the south is less distinct, and major winter fires are common during bergwind conditions; the incidence of fires peaks in December to March, and again in June to September (Le Roux, 1969).

The mean frequency of occurrence of fires in fynbos can only be inferred. There are few good historical records, and the present situation is thoroughly confused by vastly increased ignition sources in most areas and long-standing fire exclusion policies in others. However, most fynbos communities do not burn readily until they have reached an average of four years, when sufficient cured fuel has accumulated (Kruger, 1977). On the other hand, communities which have remained unburnt for more than forty to fifty years are scarce, even in carefully protected areas. Since the

TABLE 2.2

Incidence and extent of wildfires in the Cedarberg State Forest (79 000 ha) during the period 1958 to 1974, inclusive (from Andrag, 1977)

Cause	Number of fires reported					Mean area per fire (km^2)	Proportion of total burnt area (%)
	summer	autumn	winter	spring	total		
Lightning	9	4	1	5	19	3.2	17
Rolling rocks	5	2	0	5	12	15.5	52
Escape from prescribed burn	0	0	1	3	4	8.8	10
From outside	0	3	0	1	4	1.7	2
Negligence of public	1	2	0	1	4	6.5	7
Honey hunters	–	1	–	–	1	18.2	5
Unknown	2	3	3	1	9	2.9	7
Total	*17*	*15*	*5*	*16*	*53*		
Mean area per fire (km^2)	10.9	7.5	1.8	3.2	6.8		
Proportion of total burnt area (%)	52	32	0	14			

primary youth period of many obligate seed regenerating shrubs is about six years, it appears that fires would have occurred at intervals of from about six to thirty or forty years. Fire frequency probably varied at random except where cultural influences predominated, in which case intervals would most likely have been shorter on average.

There are no data available on fire intensities. Qualitative evidence, however, indicates that intensities are by and large lower than in Californian chaparral for example. Fuel particles with a diameter greater than about 6 to 10 mm are seldom consumed. High-intensity behavioral features, such as intense fire-whirls, are seldom observed. Ashbeds are absent or scarce after the normal fire, and surface soil is seldom oxidised.

Fires are normally relatively slow, though rates of advance of about 1 km h^{-1} have been recorded during experiments, and a wildfire in the Cedarberg has been observed advancing at about 4 to 5 km h^{-1} in certain phases. Nonetheless, large wildfires are common and burns of 50 to 100 km^2 occur almost annually in remoter mountain areas.

Synopsis

Fynbos cuts across the boundaries of the normal climate types; for example, it occurs in Köppen's (1923) types **Csa**, **Csb** and **Cfb**. The climate types defined in various systems do not consistently correspond with biome types (Bailey, 1958; Rumney, 1968; Schulze and McGee, 1978) and this is confirmed here.

Aschmann (1973) has defined the climate of Mediterranean-type shrublands by strict criteria. Winter precipitation should equal or exceed 65% of the total; effective precipitation, as measured by the Bailey (1958) index (EP), ranges from 3.5 to 8.6, that is, from a minimum rainfall of 275 mm yr^{-1} in cool coastal situations, or 350 mm yr^{-1} at warmer interior stations, to a maximum of about 900 mm. Winters would not be severe, and less than 3% of the year would experience temperatures of 0°C and lower. The climate of the fynbos zone meets the temperature criteria, but extends beyond the limits of precipitation variables, both with respect to amount and to seasonality. The effective precipitation indices for two Table Mountain stations are 15.9 and 19.1. Many other stations in mountains and wetter valleys have indices over 10 and these

stations, on other continents, would have a forest cover.

At the semi-arid end of the gradient (as at Citrusdal, with EP=3.0) fynbos is found on quartzites and sands, but is displaced by other vegetation types on granites and shales in the EP range from 3.0 to about 6.0. In contrast, forest occurs in the Knysna region in a zone not immediately distinguished by climate. Here, EP values range from 7.4 (George) to 10.9 (Deepwalls). In fact, Acocks' (1953) representation of the forest (Veld Type 4) is to some extent hypothetical since much, if not most, of the area is in fact covered by fynbos. Thus, though the forest zone experiences relatively low summer temperatures and is probably more humid, due to favourable maritime influences, than fynbos stations with the same effective precipitation indices, there is little evidence of macroclimatic differences between adjacent forest and fynbos ecosystems.

The distribution of fynbos correlates best with an adequate winter precipitation, and with podzols or podzolisation, which are evident as far east as Grahamstown (Martin, 1965). Granite and shale soils on the coastal forelands, where fynbos is replaced with coastal rhenosterbosveld, are base-saturated or oversaturated (Ellis, 1973). It appears that a rainfall of at least 600 mm (EP=5.5–6.0) is required to allow sufficient leaching to permit fynbos to survive on these substrates. The factors governing the boundary between forest and fynbos are obscure. Maps produced by Neethling (1970) show that forest occurs on soil types as shallow, highly leached and infertile as those on which fynbos occurs, so that the governing factor appears to lie in soil moisture régimes or a complex of soil physical variables. The common factor in the fynbos zone, aside from low to very low levels of soil nutrients, is a seasonally severe soil moisture deficit, even on situations where the profile is periodically waterlogged. This influence is probably the one which by and large precludes the growth of native trees and its effect is reinforced by periodic fires.

VEGETATION

Introduction

The account which follows is intended to supplement the broad description of fynbos provided by Taylor (1978). He has collated most of the published and unpublished information and has contributed much from his own records and observations. This account is also descriptive but aims at a more precise portrayal of community structure and pattern where possible. It has been necessary to draw heavily on unpublished data. Analytic records for selected communities (Appendix III, p. 58), unless otherwise indicated, were gathered recently by the author and colleagues. These records consist of lists of all species found on 0.1 ha (20 × 50 m) samples, with visual estimates of canopy cover and rapid estimates of height for each species. Species were assigned to life- and growth-form classes in the field.

Synusiae

Phanerophytes

Trees. The fynbos tree flora is noted for rarity and endemism. *Widdringtonia cedarbergensis* and *W. schwarzii* occasionally form open-forests or woodlands, but populations are normally very sparse and restricted to special habitats on cliffs, in boulder fields and, occasionally, along watercourses, apparently where fires are less frequent than in the surrounding areas. Both species have narrow ranges, the former occurring over some 250 km^2 in the Cedarberg and the latter as scattered, isolated populations in the Kouga and Baviaanskloof Mountains. In both cases, trees are killed relatively easily by fire and rely for survival on seed. Seed ripens over about thirty months, and all or nearly all is shed over a period of weeks. This is in marked contrast with *Widdringtonia nodiflora*, which has the capacity to sprout after fire and bears its seeds in serotinous cones. Primary youth periods are in the order of twenty to fifty years, and life-spans range to 250 years or more in the case of *W. cedarbergensis*, possibly somewhat less in the case of *W. schwarzii*. These trees appear ill-adapted to the fynbos environment; Hubbard (1937) and Lückhoff (1963, 1971) describe their biogeography, ecology and conservation.

Leucadendron argenteum forms a low woodland with heathland understorey on cool, moist eastern and poleward slopes with granite or shale soils on the Cape Peninsula and Stellenbosch area. The species does not have the capacity to resprout after fire, but the cambium is well protected by a thick

(1.0–1.5 cm) corky bark and this, with a capacity for relatively rapid height growth, enables many individuals in a population to survive fire. Fruit are retained in cone-like heads "... for some years ..." after ripening (Williams, 1972). Seedlings establish readily after a burn so that populations, where burnt, comprise even-aged subpopulations. Longevity is apparently not in excess of a century.

A few other species form small trees under favourable conditions. These include *Maytenus oleoides* and *Protea arborea*, which may attain about 6 m, and both of which have protective bark and the capacity to resprout from epicormic buds. *Olea africana* reaches the stature of a tree where the vegetation is long protected against fire, as does *Widdringtonia nodiflora*. In the south and extreme east, respectively, *Aloe ferox* and *Encephalartos longifolius*, small rosulate trees, are found occasionally in specialised habitats.

Riparian communities are dominated by species which form small trees where fire is excluded for twenty or more years. Such species include *Metrosideros angustifolia*, *Podalyria calyptrata*, *Psoralea pinnata* and *Salix mucronata*. Others, such as *Brabejum stellatifolium*, grow tall but do not lose the shrub habit.

Mid-height and tall shrubs. The distinction between mid-height and tall shrubs is not useful here, since a given species may fall into either class depending on age and there is often no clear segregation of the two classes in the community; they are therefore combined in one synusia for this description.

Shrubs in this category are predominantly broad-sclerophylls of the Proteaceae. The flora of this synusia is relatively small and the upper stratum of scrub or tall-heathland communities is usually dominated by one or two species. Taylor (1978) lists some of the typical species and describes their distribution. Mature broad-sclerophylls normally reach a height of about 3 to 4 m, but some (*Protea mundii*) may reach 6 or 7 m. Crowns may be slender to somewhat sphaeroid, depending on population densities, but are mostly regular with shoots that ascend at an acute angle, reflecting their sympodial, multifurcate branching habit. Species are normally microphyllous, sometimes mesophyllous (*Protea arborea*, *P. lorifolia*), entire (*Protea*, *Leucadendron*, *Maytenus*) or nearly so (*Leucospermum*), and set at an angle to the branch, curving upwards. Leaf surfaces may be glabrous, hairy, coriaceous or glaucous.

Most of these shrubs are killed in fires and rely on seed for regeneration. In *Protea* and many *Leucadendron* species seed are normally retained on the plant, in protective capitula or cones respectively, for several years after ripening. These organs open within days after a fire. In other cases (*Leucospermum*, *Cliffortia*, *Aspalathus*) seed is released promptly on ripening. This seed is presumably hard and able to remain dormant for some years; Van Staden (1966) found pronounced dormancy in fully developed seed of *Leucospermum cordifolium*, but there is no information on seed longevity. Williams (1972) adduced data which indicated heat stimulation (at temperatures up to 100°C) of germination in nut-seeded *Leucadendron* spp., species which release their ripe fruits.

Germination and establishment occur freely after a fire but are rare or absent at other times (Levyns, 1970; Rourke, 1972; Williams, 1972). Seedling mortality is low, and even in dense populations natural thinning rates remain low until plants approach the end of their life-spans. Longevity in most instances is in the order of thirty to fifty years (e.g. Rourke, 1972). Primary youth periods range from about four to eight years with some inter- and intraspecific variation, reflecting both ecological and genetic factors (Jordaan, 1949; Rourke, 1972; F.J. Kruger, pers. obs. — see Table 2.3).

Some species are not readily killed by fire

TABLE 2.3

Primary youth periods of some seed-regenerating fynbos species: percentage of population at given age with ripe seed (F.J. Kruger, pers. obs.).

Species	Age of population (years)							
	1	2	3	4	5	6	7	8
Erica sessiliflora	0	0	0	0	1.0	21.0	–	–
Leucadendron concavum	0	0	0	0	37.5	–	–	–
L. salicifolium	0	0	0	0	1.0	64.0	100.0	
L. xanthoconus	0	0	0	0	0	5.0	45.0	–
Protea lacticolor (moist site)	0	0	0	1.0	6.0	19.0	85.0	–
P. lacticolor (dry site)	0	0	0	0	0	1.0	28.0	–
P. mundii	0	0	0	0	0	1.0	7.0	–
P. stokoei	0	0	0	0	0	0.8	14.2	–
Roridula gorgonias	0	0	0	0	17.0	73.0	–	–

because of protective bark and heat-tolerant foliage, although if the entire crown is scorched the plant dies. As in *Leucadendron argenteum*, populations consist of even-aged subpopulations, reflecting intermittent seed reproduction after burns. Examples are *Leucospermum conocarpodendron*, *Protea laurifolia*, and *P. lorifolia*.

Protea arborea ("waboom"), a tall shrub or small tree with a crooked stem and irregular crown, is an element of a distinct community throughout most of the fynbos zone. Leaves are grey-green, glaucous, elliptic and, with a mean area of about 30 to 40 cm², considerably larger than those of most other fynbos shrubs. Adult plants are normally up to 3 or 4 m tall but under favourable circumstances may reach 5 to 6 m, with relatively clear boles of up to 60 cm diameter below the first branch. The species has the capacity to regenerate from epicormic buds throughout the stem and thicker shoots, or from lignotubers or both. Seed reproduction occurs freely after a burn, and seedlings develop a lignotuber within two or three years. Young shrubs develop a caespitose habit after repeated burns and do not exceed about 0.5–1.0 m for several years, until one shoot gains apical dominance and develops to form a shrub of mature form. This combination of adaptive traits and features of the life cycle is rare among fynbos shrubs.

Phreatic communities are often dominated by tall narrow-sclerophyllous shrubs of the Bruniaceae; species such as *Berzelia lanuginosa* and *Brunia alopecuroides* are typical, and these have life-cycles and adaptive traits much like those of typical broad-sclerophylls. On cool, moist slopes tall and mid-height, multi-stemmed narrow-sclerophyll Bruniaceae, such as *Brunia nodiflora*, *Berzelia abrotanoides*, and *B. intermedia*, are often prominent or dominant. These resprout after a burn. Mature Bruniaceae of this synusia have erect, slender stems with a small, tufted canopy of fine shoots and short, pinoid leaves. Seed is small (about 1 mm) and shed on ripening or a year or two thereafter.

Other narrow-sclerophylls in this synusia include Ericaceae (*E. patersonia*, *E. sessiliflora*, *E. versicolor*, and others), Proteaceae (*Paranomus* spp., with terete-dissected leaves), Fabaceae (*Aspalathus* spp.), Rosaceae (*Cliffortia* spp.) and a number of other taxa. These tend to reach maximum importance on sheltered slopes at higher elevations, where broad-sclerophylls in turn tend to drop out.

Tall semi-parasitic shrubs of the genus *Thesium* are sometimes common. These often are semi-succulent and have acicular leaves or are aphyllous with photosynthetic stems; *Thesium euphorbioides* has broad succulent leaves.

The synusia often contains opportunistic fire-followers which sometimes dominate the regenerating community, only to die off four to eight years later having flowered for one or a few seasons — *Aspalathus* spp. (Adamson, 1935; Dahlgren, 1963), *Othonna quinquedentata* (Van der Merwe, 1966), *Euryops abrotanifolius* (Adamson, 1935).

Population densities in this synusia vary considerably but the common seed-regenerating shrubs are often highly gregarious, with densities up to 4000 to 5000 ha^{-1}. Sprouting shrubs are seldom dense; *Protea arborea* adult plants occur at densities of about 100 to 400 ha^{-1}.

There are almost no deciduous species, spiny stems are rare, but some taxa (*Cliffortia*, *Aspalathus*) often have spiny leaves.

Low shrubs. This heterogeneous synusia includes a range of shrub growth forms and is normally rich in species.

Ericoid and other leptophyll caespitose shrubs are prominent, although those that rely on seed for regeneration are short-lived (ten to twenty years) and begin to disappear gradually as communities age. Ericaceae usually dominate the synusia but are replaced by other taxa in certain habitats. A number of multi-stemmed lignotuberous shrubs occur, some of which are leptophylls (*Erica coccinea*, *Nebelia paleacea*, *Phylica* spp.) though most are broad-sclerophylls (*Leucadendron* spp., *Leucospermum* spp., *Psoralea* spp.)

In some communities shrubs of the Asteraceae come to dominate this synusia. This may be due to a local controlling factor, as in the case of dense *Stoebe plumosa* ("slangbos") in *Protea arborea* communities on steep, relatively mobile talus slopes, where aerial biomasses of 10 to 12 ha^{-1} have been measured. In other cases, a combination of burning and overgrazing apparently causes proliferation of *Elytropappus rhinocerotis* ("rhenosterbos") in fynbos communities (Levyns, 1929; Adamson, 1938a). These highly resinous, cupressoid shrubs provide conditions for intense fires.

A common element of many communities is a group of sub-ligneous suffrutescent shrubs, particularly of the genera *Helichrysum* and *Helipterum*. These are normally at maximum abundance up to eight years after a fire, during which they flower prolifically. Leaves of many of these species are covered in a dense indumentum.

As noted, there is a great variety of leaf morphologies in this synusia. Evergreen species dominate, but a handful of drought-deciduous shrubs (*Asparagus thunbergianus*, *Montinia caryophyllacea*, *Rhus rosmarinifolia*) are sometimes prominent. Shrubs with spiny leaves are frequent, and those with spiny stems (*Asparagus* spp.) occasional. Most leptophyll species produce copious quantities of light, small (1 mm and less in diameter) seed which are shed annually. Martin (1966) has suggested that, in *Erica chamissonis*, the seed is not long-lived and regeneration after a fire depends on dispersal from populations outside the burn. Other Ericaceae, however, seem to regenerate from viable seed in or on the soil, since abundant seedlings are often observed in the first year after fire in situations remote from adult populations. Levyns (1935) has shown *Elytropappus rhinocerotis* seed to have a longevity of some seven years; it seems that many low shrubs have small seeds which survive on or in the soil, and that this reservoir ensures regeneration after fire. Some taxa (Fabaceae, Rutaceae) produce larger, hard seed which apparently have greater longevity and are stimulated by heat (e.g. Blommaert, 1972).

As in other synusia, seedling establishment is rare in mature communities and restricted largely to the period after fire. Some Ericaceae delay germination and establishment (Adamson, 1935; Martin, 1966) but most low-shrub seedlings appear in the first winter and spring after a burn. Youth periods are slightly shorter than in the case of taller shrubs. Michell (1922) found *Erica mauritanica* seedlings flowering for the first time about thirty months after a summer fire (see also Table 2.2).

Chamaephytes

A wide variety of growth forms is included in this synusia. True dwarf shrubs of the Ericaceae, Rhamnaceae (*Phylica*) and others occur in most communities but dominate only in a few extreme habitats. Marloth (1902) has noted the frequency of compact, accumbent, and reptant shrubs (and other alpine growth forms) on high peaks in exposed situations at elevations in excess of about 1800 m, but there is no true alpine or subalpine zone (Adamson, 1938a). Shrubs of this form may be found at elevations of around 500 m (cf. *Erica banksia, E. parvula*). Procumbent shrubs, especially in the Fabaceae (*Aspalathus, Indigofera*) are common on sandy flats after fire.

Proteaceae are represented by acaulescent rosette shrubs (*Protea scolopendrifolia*), rhizomatous or reptant species (*P. acaulos*), and low, circular spreading shrubs (*Protea witzenbergiana*). Succulent dwarf shrubs (Mesembryanthemaceae, *Crassula*) are present after fire in mountain fynbos (Adamson, 1935), or as a permanent component on exposed rocky situations and in arid fynbos communities. As in low shrub synusia, suffrutescent semi-shrubs, especially of the Asteraceae and Selaginaceae, are sometimes frequent, as are semiparasitic dwarf *Thesium* and *Thesidium* species.

Diffuse caespitose or spreading herbs (*Carpacoce*) and creeping herbs (*Centella*) may sometimes be abundant, especially in wet communities.

Leaf morphology is highly variable, but narrow-sclerophylls tend to dominate.

Hemicryptophytes

This synusia is prominent in all but very dense, senescent closed-heathland or closed-shrub communities and is usually very rich in species, especially graminoids. It is the synusia which most clearly distinguishes fynbos from its Mediterranean and New World analogues (Kruger, 1977).

Restionaceae often dominate graminoid hemicryptophytes almost to the exclusion of other taxa, especially on infertile rocky sites at high elevations and on sites that are seasonally waterlogged and subjected to drought. Growth forms vary considerably. Tall (to 2 m) coarse tufted plants, with stem diameters up to 5–7 mm, are frequent on rocky, subarid sites on the one hand, and on boggy areas on the other. Most communities, however, contain a variety of forms, ranging from rhizomatous, scapose forms through to large tussocks, and varying in height from about 25 to 100 cm. In bogs and some high elevation communities species of Restionaceae may form large hummocks up to 70 cm high, with the same diameter (*Restio perplexus, Hypolaena crinalis*).

These are the restioid component of fynbos, but some species of Cyperaceae are included in the category. Aphyllous Cyperaceae are found in bogs and on seepage steps; examples are *Neesenbeckia punctoria* and *Epischoenus* spp.

Other Cyperaceae are invariably present, mainly as tufted hemicryptophytes about 25 to 50 cm tall. Tufted plants with culms to 2 or 3 m high are a striking component of some communities (*Tetraria bromoides, T. thermalis*).

Grasses tend to increase in importance with increasing soil fertility and increasing proportion of summer rainfall. Toward the east they tend to dominate the synusia, and this trend is reinforced by frequent burning (Martin, 1966; Trollope, 1973). Nevertheless, most fynbos communities include at least two or three perennial grass species.

Restionaceae are truly evergreen, with shoots that remain green until those of the next season have grown almost to their full length. Sedges and grasses are evergreen in most communities but the incidence of deciduousness increases along the aridity gradient.

Most, but not all, hemicryptophytes resprout after a burn. Species that are killed (*Chondropetalum hookeranum*) regenerate freely from seed. They confer a bright green colour on communities up to about three years old, but, where they dominate older communities, the conspicuous colours are bronze to reddish- or greyish-brown.

Forbs comprise a wide variety of growth forms, no one of which may be considered typical.

Geophytes

Geophytes are often practically undetectable, many species remaining completely dormant in the interval between fires (Hall, 1959; Levyns, 1966; Martin, 1966) while others are invisible in summer. Any list of geophytes is therefore deceptive but available data indicate a wide range of species and growth forms, with a tendency toward greater frequency and diversity with increasing soil fertility, increasing aridity, and more frequent fires. A sample in a firebreak on granite soil at Jonkershoek, for example, contained about 2500 individuals (9000 ramets) of *Watsonia pyramidata* per hectare, with a total biomass (aerial and subterranean) of 350 kg ha^{-1} (F. J. Kruger, pers. obs.).

Periodicity in fynbos geophytes varies considerably (see Vol. B, Ch. 1). Most are deciduous, but evergreen geophytes (*Aristea*, *Corymbium*) are prominent in certain communities. Larger species appear to have great longevity; *Watsonia pyramidata* clones reach ages in excess of a century, and ramets reach about thirty years.

Therophytes

Arid communities have a rich spring annual flora. In good seasons therophytes can be the major cover component. In other fynbos types annuals are rare and the few species present (usually in the genera *Heliophila*, *Aira*, and *Pentaschistis*, and families Asteraceae and Campanulaceae) are normally fire weeds (Adamson, 1935; Martin, 1966).

Heterotrophic vascular plants

Cassytha ciliolata is common in communities at lower elevations and may sometimes dominate the upper shrub stratum. *Cuscuta* spp. occur occasionally, but fynbos communities on the whole are free of parasitic vines.

Root parasites in genera such as *Harveya*, *Hyobanche*, and *Mystropetalon* are common but never abundant and are usually overlooked.

Life-form spectra

A number of authors (e.g. Adamson, 1931; Martin, 1965; Van der Merwe, 1966) have published life-form spectra for different fynbos communities but since definitions of life-form cate-

gories have seldom been uniformly applied it is difficult to make comparisons. Table 2.4 indicates the life-form spectra for a range of fynbos communities; it has been abstracted from the site descriptions in Appendix III (p. 58), each of which is based on analysis of a sample 20 × 50 m. Table 2.5 illustrates crudely the relationship between structural components and environmental variables. [The interaction table indicates observed strong positive (+) and negative (−) responses of life-form categories to high or low levels of what are seen to be variables having the most important control of community structure.]

Stratification

Fynbos life forms do not associate readily in distinct strata but three layers may be distinguished in mature scrub and tall-heathland communities. The upper shrub stratum varies considerably in cover but under favourable conditions, particularly where moisture is freely available, it is normally dense. Several shrub species occur in this stratum, but there is normally a taller and often dominant component of one or a few species. As noted above, leptophylls tend to dominate on phreatic sites, but elsewhere broad-sclerophyllous shrubs are the rule. This layer is commonly 2 to 3 m tall, becoming taller with age where the dominants regenerate from seed. In certain communities, such as those dominated by *Protea mundii*, the shrubs may reach 4 to 5 m or more in height. If dense, the upper

TABLE 2.4

Life-form spectra for a range of fynbos communities

Life-form category	Number of species							
	mountain open-scrub	mountain tall shrubland with heathland	mountain low open-heathland	mountain open graminoid-heathland	mountain closed-herbland	arid tall shrubland with heathland	arid open graminoid-heathland	coastal tall open-heathland
Trees	0	0	0	0	0	0	0	0
Mid-height and tall shrubs	10	11	1	2	0	5	6	3
Low shrubs	19	38	16	19	5	10	13	17
Dwarf shrubs	7	12	4	13	2	2	17	5
Graminoid hemicryptophytes	13	17	23	22	13	5	14	8
Forbs	3	7	1	5	2	5	2	1
Geophytes	18	34	7	10	9	15	9	4
Therophytes	1	7	0	0	0	18	8	1
Heterotrophic plants	1	0	0	0	0	0	0	0
Total number of species	72	126	52	71	31	60	69	39

TABLE 2.5

Interaction table showing approximate relationships between environmental variables and some lifeforms in fynbos communities

Life form	Environmental variables															
	rainfall		proportion winter rainfall		seasonal soil moisture deficit		water-logging		elevation		radiation		soil fertility		fire frequency	
	high	low	high	low	strong	weak	strong	weak	high	low	high	low	high	low	high	low
Trees	·	·	·	·	·	+	−	·	·	·	·	·	·	−	−	·
Shrubs >1 m	·	+	·	·	·	·	−	·	·	+	+	·	+	·	−	·
Low and dwarf shrubs	+	·	·	·	·	·	−	·	(+)	·	·	+	·	·	·	·
Restioid hemicryptophytes	+	·	+	·	+	·	+	·	+	·	·	·	·	+	+	·
Grasses	·	·	·	+	·	·	−	·	·	·	·	·	+	·	+	·
Forbs	·	·	·	·	·	+	·	·	·	·	·	·	+	·	·	·
Geophytes	·	+	·	·	·	·	·	·	·	·	·	·	+	·	+	·
Annuals	·	+	·	·	·	·	−	·	·	·	+	·	+	−	+	·

shrub stratum suppresses other layers, sometime almost completely when lower plants are buried in heavy litter fall.

The upper shrub stratum is generally absent from communities at higher elevations and is a feature mainly of the vegetation at attitudes below 1000 m (Taylor, 1978). Its incidence and density is also governed by fire régime and the layer is reduced or eliminated by fires at short intervals.

A dense middle layer comprises a great diversity of species and life forms except where suppressed by taller plants. Major components include low shrubs of Ericaceae, Proteaceae, Rhamnaceae and others, and taller hemicryptophytes.

Low hemicryptophytes, chamaephytes, forbs and geophytes compose the lower layer which is also usually rich in species. In arid fynbos, this layer is sparse in the dry season, being composed mainly of seasonal geophytes and annuals.

Taylor (1978) has described these strata in some detail. He recognized only these three layers, but it should be noted that many communities include a ground layer of rosette herbs, mosses and lichens.

VEGETATION ZONATION AND COMMUNITY TYPES

Introduction

The major categories of fynbos — mountain, arid, and coastal — are distinguished principally on habitat. Physiognomic differences exist, but there is a gradual transition along habitat gradients and there is no rigorous study upon which to base distinctions. This classification will be maintained here, although Acocks (1953) and Taylor (1978) recognize that these categories obscure major observable classes within them. There is no attempt in this account at a comprehensive enumeration of the important alliances or associations within each major type, mainly because no overall survey has been completed. Examples will serve to indicate the range of types encountered.

Mountain fynbos

There is a great diversity of communities in mountain fynbos, reflecting the influence of geographic factors and of a complex landscape. Taylor (1969) recognized six mountain fynbos associations on an area of 78 km² (and six other associations). Werger et al. (1972) defined five fynbos and three scrub and forest communities on about 2 km², while Kruger (1974) distinguished four major community categories which included at least eleven subordinate types, on about 1.5 km². (Fig. 2.2). A portion of an area of 0.8 km² near Grahamstown carried a heathland alliance in which Martin (1965) distinguished nine communities. Boucher (1972) found twenty mountain fynbos community types in his study area of

115 km². There are few other detailed community studies available, and the information from extant studies is difficult to collate.

Taylor (1978) approached this complexity by recognizing two broad if indefinite vegetation zones, the proteoid zone of the lower mountain slopes and foothills, and the ericoid–restioid zone of higher elevations. A third category, hygrophilous fynbos, cuts across these zones in habitats with perennial soil moisture. He describes these types extensively. The proteoid zone, characterized by the presence of mid-height and tall broadsclerophyllous shrubs, is restricted largely to elevations less than about 1000 m. There is considerable variation in structure and composition, but communities comprise the three layers described above. In the ericoid–restioid zone the upper stratum is absent and communities normally contain ericoid and other leptophyllous shrubs, ranging in height from about 20 cm to

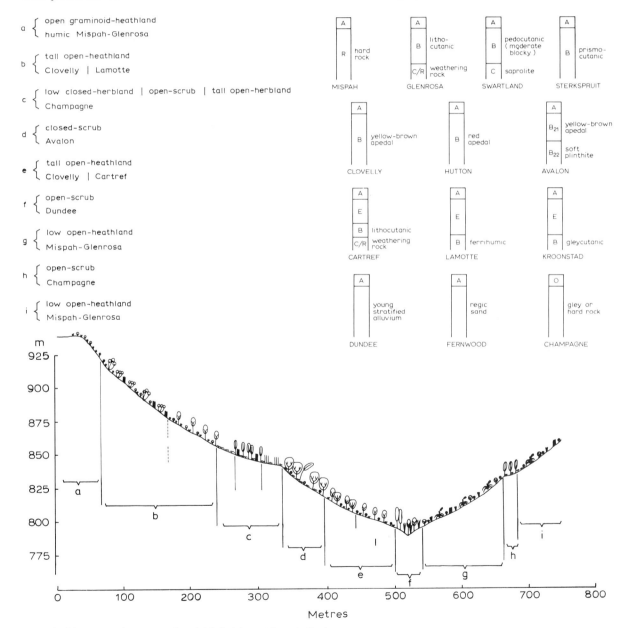

Fig. 2.2. Diagrammatic cross-section, Jakkalsrivier catchment, Grabouw.

1.5 m. Shrubs are sparse in certain habitats and there the vegetation is normally dominated by Restionaceae. Hygrophilous fynbos includes riverine communities and those of boggy sites, as well as the mixed scrub of steep humid poleward slopes.

Here, some dominant structural formations of the major habitat types will be enumerated and described briefly within the framework of broad habitat types.

Drier slopes and foothills

The warmer and well-drained slopes and foothills, where summer soil-moisture deficits are the rule, are typically the zone of tall sclerophyllous shrubs. Rainfall is generally in the range 500 to 1000 mm yr^{-1}, but may reach 1800 mm. Soils are usually rocky phases of the Mispah, Glenrosa and Clovelly forms. Several structural formations occur, of which some examples are discussed below.

Broad-sclerophyllous scrub or open-scrub (Fig. 2.3). Tall broad-sclerophyllous shrubs dominate the upper stratum of most communities of this zone,

except east of about Port Elizabeth, where tall shrubs are limited to local habitats. Cover of this stratum varies, normally increasing with rainfall, and decreasing to the point of disappearance with increasing fire frequency; it is normally mid-dense but dense (though seldom closed) where rainfall is high. Typical species in this layer include *Protea laurifolia*, *P. neriifolia*, *P. repens* and other *Protea* spp., and *Leucadendron laureolum* and *L. eucalyptifolium*. The communities are usually three-layered, when mature (Adamson, 1931, 1938a; Taylor, 1978) with the upper stratum dominated by one or a few species (although other relatively tall shrubs are found under the canopy, as in the case of *Brunia nodiflora* in Appendix III, Site 1). Lower layers are usually floristically and structurally diverse with composition varying considerably along geographic and environmental gradients. Density of these layers is determined partly by age of the stand.

The community analyzed as Site 1 (Appendix III) serves as an example of this type. Others have been described by Adamson (1931), an unnamed climax community near Hermanus, dominated by *Leucadendron xanthoconus* (=*L. salignum*) and by

Fig. 2.3. Broad-sclerophyllous scrub communities on granite foothills in Langrivier, Jonkershoek. *Protea neriifolia* is the dominant shrub on the slopes. Note *Widdringtonia nodiflora* in right foreground and middle distance. Scrub gives way to low heathland at about 900 to 1000 m, below the quartzite cliffs.

Fig. 2.4. *Protea arborea* tall broad-sclerophyllous open-shrubland, Assegaaiboschkloof, near Franschhoek (photo: Department of Forestry).

Taylor (1969), the *Protea lepidocarpodendron* tall fynbos association. Information from Adamson's example (65 species in the census) may be summarized as follows:

(a) **Upper stratum**: six species of microphanerophytes of which five species are broad-sclerophyllous proteaceae, dominated by *Leucadendron xanthoconus*.

(b) **Middle stratum**: comprising mainly leptophyllous ericoid nanophanerophytes (fifteen species, dominated by *Stoebe fusca*) on the one hand, and aphyllous graminoid hemicryptophytes, with greater cover, on the other (seven species, with dominance shared between *Restio cuspidatus*, *R. filiformis*, *Leptocarpus asper*, and *Hypolaena digitata*); he notes the conspicuous presence in this stratum of frequent shrubs of "... . characteristic form with a single elongated stem, which is either unbranched or only produces branches above the

general level of the vegetation . . ."; *Anthospermum aethiopicum* and *Phaenocoma prolifera* are examples.

(c) **Lower stratum**: a discontinuous layer, with about thirty species of a variety of life forms, and with low importance values.

Tall broad-sclerophyllous shrubland or open-shrubland with heathland (Fig. 2.4). *Protea arborea* characterizes a distinct formation (waboomveld; Taylor, 1978) which occurs through most of the mountain fynbos zone (see Appendix III, Site 2, for an analysis). Waboomveld tolerates a wide rainfall range, from about 300 mm to 1800 mm yr^{-1}, and is found at elevations from near sea level to about 1200 m. Slopes with relatively mobile, but not loose, talus, or coarse colluvial fans are typical of the habitat; stones and boulders are often found to accumulate against the upper side of *P. arborea* trunks. Soils are normally of the Clovelly form and are free-draining, but the position of *P. arborea* communities in the landscape suggests that summer soil moisture is augmented through lateral drainage.

Protea arborea is normally the only prominent tall shrub in the community, although other species with similar responses to fire (*Heeria argentea*, *Maytenus oleoides*) are sometimes conspicuous. At Jonkershoek, certain of the waboomveld communities long protected (more than thirty years) against fire have abundant tall well-grown *Kiggelaria africana* and *Olea africana*, bird-distributed species whose seedlings are often found under *P. arborea* crowns. This suggests that succession may be relatively quick in this type under certain conditions. Otherwise, the structure of these communities is much as indicated in the Site 2 description in Appendix III, with lower

strata often notably diverse. Multi-stemmed broad-sclerophyllous low shrubs (*Leucadendron salignum* and others) and cupressoid low shrubs (*Stoebe plumosa* and *Elytropappus* spp.) are usually typical. Poaceae are important, and *Themeda triandra*, *Cymbopogon marginatus* and *Ehrharta* spp. are abundant in most communities. In the northern fynbos ranges (Groot Winterhoek to Cedarberg, and Elandskloof Mountains), *Stoebe plumosa* comes to dominate the lower stratum in older communities on slopes with active talus creep, and comprise up to 10 to 12 tonnes ha^{-1} of fine, highly resinous fuel (F. J. Kruger, pers. obs.). Because of the relatively abundant Poaceae in waboomveld it is often regularly grazed and burnt, and this usually results in a sense low shrub stratum dominated by *Elytropappus rhinocerotis*.

P. arborea was heavily utilized for tan-bark periodically until around the turn of the century

Fig. 2.5. Low open graminoid-heathland: an example of the *Tetraria thermalis–Hypodiscus aristatus* community type on Nuweberg State Forest, near Grabouw. *T. thermalis* at lower right, and scattered through the scene. Restionaceae include *Restio dispar*, *R. egregius* (both taller than the average plants) and numerous others. Dominant low narrow-sclerophyllous shrub is *Sympieza articulata*.

(Lückhoff, 1971). Many communities must have been much disturbed and may still reflect the effects of exploitation.

Examples of *Protea arborea* communities have been described by Taylor (1969) as a *P. arborea* pseudo-savannah association, and by Werger et al. (1972) as the *Protea arborea — Rhus augustifolia* community.

Low ericoid open-heath or open graminoid-heath. (Fig. 2.5). Taylor (1963, 1978) recognized a distinct type of low fynbos (*bergpalmietveld*) which occurs on extremely rocky, infertile Mispah, Glenrosa and Cartref soils subject to drought, from Table Mountain and the Hottentots–Holland mountains eastward to the Bredasdorp area and the mountains near Riviersonderend, but possibly as far east as the Langeberg. Rainfall ranges from about 400 to 1200 mm yr^{-1}. This type, which has a distinct floristic composition (Kruger, 1974 and Glyphis et al., 1978: *Tetraria thermalis–Hypodiscus aristatus* community), is also of relatively uniform structure. Taller broad-sclerophyllous shrubs are prominent in certain phases of the type but are rare or absent in most instances, possibly being excluded by extreme infertility and pronounced summer soil-moisture deficits. Ericoid low shrubs are important, usually ranging from mid-dense to dense in mature communities with one species dominant. The type is characterised by scattered *Tetraria thermalis*, a broad-sclerophyllous rhizomatous sedge conspicuous for its glossy yellow-green foliage, and with culms up to 2 m tall. Other tall rhizomatous graminoids such as *Willldenowia argentea* (up to 1.5–1.7 m tall) and *Restio egregius* (0.9–1.3 m) are conspicuous emergents. Prominent shrubs in the canopy include such species as *Sympieza articulata*, *Erica corifolia*, *Erica hispidula* and *Erica pulchella*. Some unusual shrubs include *Erica fascicularis*, which rises almost unbranched to about 2 m in height, and the endemics *Grubbia tomentosa* and *Retzia capensis*. The latter two species, and *Mimetes cucullatus*, are among the few multi-stemmed shrubs found in this type. The dominant low ericoid shrubs form a layer about 0.5 to 0.7 m tall and are slightly overtopped by erect tufted Restionaceae like *Hypodiscus aristatus* and *Chondropetalum hookeranum*, which together may often be as important as the shrubs.

Low and dwarf narrow-sclerophyll shrubs and tufted hemicryptophytes form a lower stratum about 0.2 to 0.4 m tall.

The data in Appendix III, Site 3, serve as typical of the type described above, but the low ericoid heathlands of drier mountain slopes vary considerably in composition. Under favourable conditions species of *Erica* may form a dense canopy, and such low closed-heaths may often be found on the southern footslopes of the Langeberg; as a rule, however, the diagnostic stratum is mid-dense. Low ericoid heathlands are particularly prominent on the foothills and well-drained slopes of the Riviersonderend and Langeberg ranges (Fig. 2.6), and although many species of Ericaceae are present, one species normally dominates in a given community (see for example, Muir, 1929). A widespread form of low heathland, highly variable in composition, is characterized by a sparse upper stratum of low to mid-height multi-stemmed broad-sclerophyllous shrubs, especially *Leucadendron salignum*. East of Swellendam, however, *Leucospermum cuneiforme* becomes equally prominent in this layer, especially on soils of old land surfaces. This form of heathland is typical of the pediments and foothills of the fynbos zone but is also found at higher elevations, especially on the gently sloping verges of subsummit plateaux and basins. Low shrubs are once more ericoid, with a somewhat variable cover; with injudicious grazing and burning they are replaced with *Elytropappus glandulosus*, especially at lower elevations.

Cool moist mountain slopes

Well-watered, cool slopes on the Langeberg and eastward to about Port Elizabeth are characterized mainly by mixed sclerophyllous scrub communities included in the category of "Hygrophilous Macchia or Fynbos" by Phillips (1931) and Taylor (1978), and "Wet Sclerophyll Bush" by Adamson (1938a). Phillips (1931) suggested that communities in this habitat are seral to forest.

Adamson (1938a) remarks on the prevalence of soft leaves and this is true of such members as *Brunia*, *Berzelia*, some *Erica* species, and low forbs. Sclerophyllous species do, however, remain prominent.

Probably the most striking features of these formations are their stature, their extreme density, indefinite stratification, and great diversity of species and growth forms.

Fig. 2.6. Low ericoid heathland on south slopes of the Langeberg on Grootvadersbosch State Forest near Heidelberg. Dominant shrub is *Erica melanthera*. Staff is 2 m high.

Mixed sclerophyllous scrub. Very dense, vigorous relatively uniform communities clothe the slopes of the Langeberg chain. The upper stratum, 3 to 4 m tall normally, but up to about 7 m if protected for twenty to thirty years, is dominated by a mixture of broad-sclerophylls such as *Leucadendron eucalyptifolium* and *Laurophyllous capensis* and narrow-sclerophyll Bruniaceae such as *Berzelia intermedia*. Canopy cover ranges from about 50 to 70%. Thickets of tall and mid-height shrubs (*Erica*, *Penaea*, *Phylica*) and tall graminoid herbs (such as *Restio foliosus*, *Cannomois virgata*, *Tetraria bromoides* and *Tetraria involucrata*) form a diffuse middle layer about 1.5 to 2.5 m tall — although the shrubs may reach 3 to 4 m in old communities. Below this layer, tufted hemicryptophytes, low shrubs and herbs maintain a cover of some 10 to 25%.

This formation occurs in a belt between about 250 to 300 m and 400 to 600 m on poleward slopes, but extends to about 900 m on drier aspects. It alternates with dry broad-sclerophyllous scrub with changes in aspect and in soil-moisture régime, apparently depending on adequate summer soil moisture from lateral drainage.

Broad-sclerophyllous scrub. Most broad-sclerophyllous scrub communities occur on two types of site: steep cool slopes at and near drainage lines, especially on shales, and level, perennially moist erosion surfaces on the southern coastal platform. Dominant species are *Protea lacticolor*, *P. aurea*, *P. mundii*, or *P. punctata*. These shrubs are relatively long-lived and *P. mundii* reaches a height of 6 to 8 m at an age of forty to fifty years. Lower strata are suppressed in old stands, but otherwise resemble those of other communities in this zone.

Upper mountain slopes and crests

In cool moist montane habitats, from about 600 to 800 m upwards, broad-sclerophyllous scrub and tall heathland communities are displaced by narrow-sclerophyllous shrub and restioid communities, although they may occur at higher elevations on sheltered sites and in dry habitats. These high-altitude communities correspond with Taylor's (1978) ericoid–restioid zone. A simplified sequence within this includes a lower, tall narrow-sclerophyllous heathland zone, a middle, low narrow-sclerophyllous heathland and graminoid-heathland zone and an upper herbland zone; but this montane vegetation is much complicated by variations in aspect and substrate. Soils qualify mainly as humic phases of the Mispah and Glenrosa forms. Some typical formations are described below.

Tall narrow-sclerophyllous heathland. In humid southern ecosystems the zone of broad-sclerophylls is usually succeeded by tall heathland communities. Especially in the Langeberg, these have a dense upper stratum of multi-stemmed or caespitose Ericaceae, Bruniaceae, Penaeaceae and *Cliffortia*, about 1 to 1.5 m tall. Westwards and northwards, this stratum is mid-dense to sparse. Cyperaceae and Restionaceae form dense, poorly stratified layers below this.

Low narrow-sclerophyllous heathland and graminoid-heathland (Fig. 2.7). These are the dominant formations on slopes in the montane zone. Composition varies considerably; Ericaceae and Bruniaceae once more form a low shrub stratum, varying from mid-dense to dense under humid conditions. Rutaceae and Rosaceae (*Cliffortia*) are common associates and sometimes dominate. Taller leptophyllous shrubs (*Cliffortia*, Asteraceae, Bruniaceae) often form a sparse

Fig. 2.7. Low graminoid heathland on south slopes of Jakkalsrivier catchment (*Willdenowia sulcata–Erica brevifolia* community). The dense, dark-toned shrub in the upper left quarter of the plot is *Nebelia paleacea*. At ten years age, the low ericoid shrubs have not yet reached their full height. Emergent graminoid is *Elegia racemosa*.

emergent stratum 1.5 to 2.5 m tall, while broad-sclerophylls such as *Leucadendron gandogeri* sometimes occur scattered through the vegetation. Tufted and rhizomatous hemicryptophytes are usually dense and taller members of the category normally mingle with the shrub stratum.

Site 4 (Appendix III) is a community which is perhaps not an unusual member of this structural formation. However, there are few studies in this zone from which to draw comparative data.

On extreme sites, for example where soils are very shallow, ericoid shrubs and graminoid herbs become very stunted and communities belong to the dwarf heathland structural formations. These are normally an extreme form of nearby low heathland communities and do not constitute a major recognizable zone. Although Marloth (1902) and Taylor (1978) have emphasized the incidence of alpine life forms at uppermost elevations they seldom dominate communities so that no distinct alpine or subalpine zone may be distinguished.

Herbland. With increasing exposure toward mountain crests low shrubs steadily decrease, becoming sparse to absent or limited to favourable micro-habitats, the prevailing communities here are normally dominated by caespitose hemicryptophytes of the Restionaceae but vary considerably in composition and structure. Species such as *Restio fraternus*, *Chondropetalum paniculatum*, *Chondropetalum ebracteatum* and *Elegia neesii* are often prominent as somewhat spreading but stiff tussocks 0.4 to 0.8 m tall, although the herb stratum is seldom dominated by more than one or two species.

Herblands are the rule on the shale band (Cedarberg Formation) at elevations over about 1200 m, especially on the inland mountains. A rather striking form of shale-band herbland is shown in Fig. 2.8; in this case the dominant plant is *Hypolaena crinalis*, a lax tangled hemicryptophyte which forms hummocks about 0.4 to 0.5 m high. Scattered among these is the tufted *Protea caes-*

Fig. 2.8. Herbland dominated by *Hypolaena crinalis*. On the shaleband near Sneeukop, Hottentots–Holland Mountains. See text.

pitosa, a low shrub with sprawling stems, and *Erica racemosa* and *Nebelia sphaerocephala* emerge to about 1.0 to 1.4 m.

Plateaux and depressions

Plateaux and gentle subsummit basins with deep soils of the Fernwood and Champagne forms are normally subjected to alternate waterlogging and drought, and are frequently frosty in winter. Flats and gentle depressions at lower elevations experience a similar soil-moisture régime, and although the climate is different these habitats usually have communities of the same structural formations as on the high-elevation sites. A variety of formations may be found but shrubs are typically rare or absent, except on rising ground or rocky patches. Herblands are the rule but communities are much poorer in species than those of the herblands on upper mountain slopes. On the broad scale there is usually a gradient of vegetation height with increasing soil wetness, from dwarf herbland to tall communities.

Dwarf open-herbland communities on level frosted plateaux of the Cedarberg and Olifants River ranges are typically dominated by tufted wiry hemicryptophytes such as *Restio curviramis* and *Pentaschistis curvifolia* interspersed with scattered caespitose (*Erica senilis*, *Rafnia* spp.) and creeping (*Aspalathus* spp.) dwarf shrubs. Low cupressoid shrubs, especially *Stoebe plumosa*, may become prominent in the absence of fire (see also Taylor, 1978). As internal drainage improves with rockiness or local relief shrubs become more prominent; these dwarf herblands are consequently surrounded by a zone of heathland communities.

Taylor (1969) describes an intermediate formation from seasonally waterlogged flats and depressions at low elevations in the Cape of Good Hope Nature Reserve. This type, called the "Restionaceous Tussock Marsh Association", is a low closed-herbland dominated by tufted *Elegia parviflora* about 30 cm high with taller tufted Restionaceae emerging to about 0.60 to 0.90 m.

High-level plateaux and depressions are often wet; communities of such habitats are described below.

Phreatic sites

Perennially moist or wet soils are typically found on slopes where rock outcrops cause lateral drainage water to rise into the soil profile, and on parts of subsummit plateaux and flats where drainage from adjacent slopes is sufficient to maintain moisture levels through the year, as well as along streams and rivers. These habitats are occupied by a range of communities whose composition and structure varies considerably with the exposure of the site, fertility, and fire history, from forest to herbland. Many communities are of the type called "Hygrophilous Macchia or Fynbos" by Phillips (1931) and Taylor (1978) and "Wet Sclerophyll Bush" by Adamson (1938a).

Broad-sclerophyllous closed-scrub. Riparian sites typically are occupied by communities dominated by tall broad-sclerophyllous shrubs. These have a relatively simple structure with tall sclerophylls forming a continuous canopy at about 5.0 to 6.0 m. The most important species are *Brabejum stellatifolium* (mesophyll) and *Metrosideros angustifolia* (nanophyll), but the composition of the canopy may vary quite considerably and will often include forest species such as *Cunonia capensis* and *Ilex mitis*. There is little undergrowth below a dense canopy and lower strata may be reduced to a handful of ferns (*Blechnum capense*, *B. punctulatum*), graminoids like *Ehrharta erecta* and the twining *Asparagus scandens*. Where the canopy is somewhat broken, and at the fringes of dense thickets, a variety of species is found. Typical shrubs are *Freylinia lanceolata* (nanophyll), *Erica caffra* (ericoid) and *Myrica serrata* (microphyll); these are somewhat lower than the former shrubs, at about 1.5 to 3.0 m tall. In these exposed areas, tall plumose Restionaceae are characteristic (e.g. *Restio subverticillatus*, *Elegia capensis*, *Leptocarpus paniculatus* and *Cannomois virgata* — about 1.8 to 3.0 m tall).

Taylor (1978) and Werger et al. (1972) provide further information on the typical scrub.

The structure and composition of riparian vegetation varies more or less continuously from forest through the typical formation to a tall herbland. It seems that where the substrate is unstable, and where high-intensity fires recur, shrubs lose dominance and are replaced by tall restioids; but all shrubs in these communities resprout vigorously and are not easilt displaced by herbaceous vegetation.

Fig. 2.9. Narrow-sclerophyllous closed-scrub dominated by *Brunia alopecuroides*. The erect graminoid is *Chondropetalum mucronatum*. Jakkalsrivier catchment, Grabouw.

Narrow-sclerophyllous and mixed closed-scrub and open-scrub. Perennially moist or waterlogged habitats along high-level drainage lines or seepage zones carry a variety of scrub formations. A common set of species, such as *Elegia thyrsifera*, *Restio purpurascens* and *Cliffortia graminea*, tends to occur exclusively in these communities and serves for ready identification (Werger et al., 1972; Kruger, 1974). An example of a simple formation is that dominated by *Brunia alopecuroides*, a common community of the small, gently sloping bogs on infertile quartzites from the Hottentots–Holland mountains eastward (Fig. 2.9). This comprises a tall shrub stratum, up to about 4 m tall in old communities, in which *B. alopecuroides* is almost the only species, and below which occur abundant erect shoots of the rhizomatous *Elegia thyrsifera* and *Chondropetalum mucronatum*, about 1.7 to 2.0 m tall. The lower stratum is dominated by tuss-

ocks, mainly *Restio ambiguus*, and hummocks of *Hypolaena crinalis* (from Kruger, 1974).

Composition of scrub formations on bogs varies quite considerably. A common type of the south-western mountains from Table Mountain north-wards to the Elandskloof Mountains is a mixed sclerophyllous open-scrub characterized by the broad-sclerophyllous *Osmitopsis asteriscoides* and *Leucadendron salicifolium* and narrow-sclerophyllous *Berzelia lanuginosa* in the tall shrub stratum. These communities are characteristically richer than the *Brunia alopecuroides* type, and have a tangled mat of herbs and creeping shrubs (Restionaceae, Cyperaceae, *Erica parviflora*, *Cliffortia graminea*) forming a lower layer from which emerge massive, tall (1.5–2.0 m) tussocks of *Neesenbeckia punctoria* and *Merxmuellera cincta* (Fig. 2.10; see also Werger et al., 1972; Taylor, 1978).

Fig. 2.10. Mixed closed-scrub, Zachariashoek, near Paarl. Tall slender shrub in foreground is *Osmitopsis asteriscoides*, the other, with a light-toned, bushy crown, is *Leucadendron salicifolium*. Large tussock to left of figure is *Neesenbeckia punctoria*. Dominant herb in ground layer is *Pteridium aquilinum*, not typical of phreatic sites.

Phreatic scrub formations are often bordered by herblands on bogs, since, it seems, a slight change of moisture régime in any direction can result in exclusion of shrubs. Some of these are discussed below.

Restiad herblands. Herblands occur on phreatic sites apparently under two sets of conditions i.e. where waterlogging is interrupted by brief soil droughts, and under extreme, uninterrupted waterlogging. The community at Site 5 (Appendix III, Fig. 2.11) is an example of the first kind; while poor in species relative to most other fynbos communities it is nevertheless much more diverse than are the herblands of bogs. This type often occurs adjacent to the *Brunia alopecuroides* communities described above. At the other extreme, i.e. increased soil moisture, the *Brunia* and other shrubs are eliminated and a herbland with the residuum of tall rhizomatous graminoids (*Elegia thyrsifera*, *Chondropetalum mucronatum*) emergent from a tussock layer (*Restio ambiguus*, *R. bifidus*, *Tetraria flexuosa*) is the rule.

Fig. 2.12 illustrates the herblands of the Dwarsberg Plateau at Jonkershoek, which serves as example of high-elevation bog herblands. The raingauge in the photograph (at 1234 m elevation) is that with the highest recorded rainfall in South Africa — 3330 mm yr^{-1}. Presumably there is also considerable precipitation from the orographic cloud induced by persistent summer southeasters. The site is exposed to frequent northwest gales in

Fig. 2.11. Low herbland dominated by *Tetraria flexuosa*, Jakkalsrivier, near Grabouw.

Fig. 2.12. Herbland communities on the Dwarsberg Plateau, Jonkershoek.

winter. Two to four snowfalls occur regularly each year, but snow seldom lies more than fourteen days. Some data on the typical phreatic communities appear in Appendix III (Site 6) (in this case, the data were not collected by survey on 20×50 m plots and were assembled by examination of undefined areas in typical stands). In the photograph, the community immediately to the left of the figures at the raingauge is the *Chondropetalum esterhuyseniana* community, tall closed-herbland consisting almost entirely of one species, a coarse tussock 1.3 to 1.5 m tall. It occurs on rock outcrops where water flows throughout the year. The light-toned vegetation to the right of the figures is *Epischoenus adnatus* low closed-herbland, with spreading tussocks of *Epischoenus* scattered among hummocks of *Hypolaena crinalis* both up to 0.6 m high. In the right foreground is the dark *Chondropetalum esterhuyseniana* community once more. *Elegia intermedia* tall herbland appears as the taller, light-toned community in the centre of the depression.

In the left foreground and right middle distance are *Chondropetalum paniculatum* herbland communities (distinguished by scattered, rounded tussocks) and the dark low community in the middle distance at left is dwarf heathland with *Erica fastigiata*, *Restio bifarius* and *Restio echinatus* as typical species.

Eastern mountains and hills

The fynbos of the eastern extremity of the Winterhoek Mountains, the Suurberg and the hills toward Grahamstown differs markedly from that toward the west in that Poaceae of many species dominate the hemicryptophytes. *Themeda triandra* is normally the dominant species in this synusia, but there is usually a significant component of species such as *Tristachya hispida* and *Alloteropsis semialata* whose centres of distribution lie eastward and northward of the fynbos zone. Nevertheless, the vegetation retains the character of fynbos in that Restionaceae persist. Low shrubs, especially narrow-sclerophylls, are prominent where fire is relatively infrequent but even with frequent burns some multi-stemmed sclerophyllous shrubs such as *Leucospermum cuneiforme*, *Leucadendron salignum*, *Phylica* spp. and *Cliffortia* spp., which resprout from rootstocks, remain common though relatively sparse. With prolonged absence of fire narrow-

sclerophyllous shrubs form a dense canopy, normally dominated by one or a few species (Martin, 1965).

The dominant structural formations are low heathlands and graminoid-heathlands. Of the nine communities listed by Martin (1965), in Grahamstown Nature Reserve at least five belong to these categories: that analysed as Site 7 (Appendix III) serves as an example of low closed-heathland dominated by *Erica demissa*. He also describes a transitional type, the *Festuca–Bobartia* heathland, which does not qualify as fynbos due to the absence of Restionaceae and paucity of sclerophyllous shrubs.

Jessop and Jacot Guillarmod (1969) describe a form of low graminoid-heathland called "Macchia-grassveld", that resembles the vegetation which Martin describes, and adjoins *Acacia karroo* low woodland. Von Gadow (1977) describes vegetation patterns in the Suurberg, where open-heathland and graminoid-heathland communities form zones which alternate with grassland and low forest communities on a north–south transect from one crest across the intervening slopes and ravines to the next crest.

The few descriptions available preclude a thorough account of what appear to be complex vegetation patterns in the eastern zone. Not all the fynbos communities are low heaths; in the Suurberg, tall *Leucospermum cuneiforme* shrubland with heathland has been found in certain habitats, and *Protea neriifolia* tall closed-scrub occurs as very restricted communities along drainage lines near ridge crests (F. J. Kruger, pers. obs.). The abundance of grasses relative to Restionaceae is probably in part due to the high proportion of summer rainfall, and partly due to a régime of frequent fire as suggested by Martin (1966) and demonstrated by Trollope (1973); but different communities are likely to differ in their response to fire, as shown by Trollope's experiments. Therefore the peculiarities of community structure in eastern fynbos should not all be ascribed to the fire régime.

Fynbos communities have been reported east of the area described above, in the Amatole Mountains and in the Drakensberg (Story, 1952; Killick, 1963) but although many of these are dominated by broad-sclerophyllous and by narrow-sclerophyllous shrubs the restioid element is absent and the communities do not qualify as fynbos as here defined; these communities are dealt with by Killick in Chapter 4.

Arid fynbos

Taylor (1978) has maintained the category of arid fynbos on the grounds of its "... clear-cut geographical range and specialized habitat ...". Following Adamson (1938a) he recognized low total cover (less than 50%) as the most distinctive physiognomic feature, although other important features include a simple structure, with indefinite stratification, and few Restionaceae and other aphyllous hemicryptophytes which "... are only locally conspicuous". Geophytes and annuals are common, as are succulents. However, he differs from Adamson (1938a) in stating that narrow-sclerophylls are preponderant and proteoids few. Adamson also maintained that this type of fynbos, which he called "Dry Sclerophyll Bush", occurred in the zone with rainfall from 380 to 500 mm yr^{-1}; Taylor disagrees indicating that arid fynbos occurs in areas with around 250 mm rainfall per annum. and somewhat less. Taylor implies that the coastal fynbos of the dry north is not a form of arid fynbos, but Adamson includes it in his "Dry Sclerophyll Bush".

Open-fynbos communities of the kind described by Taylor are found on deep sands and rocky slopes in areas of the interior with rainfall ranging from about 400 mm down to 250 mm and, in places, 200 mm per annum. Soils are derived from Table Mountain or Witteberg quartzites, or, in the Kamiesberg, granite. This is in fact a zone of arid to semi-arid climate. Fynbos communities on coastal sand plains in similar rainfall zones are more luxuriant, possibly because of the moderating maritime influences, and will not here be considered to belong to arid fynbos (see also below).

The arid fynbos environment will differ in many ways from that of normal mountain fynbos. For example, soils ought to be less thoroughly leached, although casual inspection of profiles has not suggested that the soils are base-saturated. Fires certainly occur much less frequently, but this may be partly because most arid fynbos lands are grazed.

Many arid fynbos communities do not strictly qualify as heathland formations, because total plant cover is too low. Nevertheless, they are included because of the obvious floristic affinity with mountain fynbos, and because the transition from one type to the other is very gradual. However, it is not possible to do much more than note the existence of the type, once more because of scarcity of information, and to highlight some features by means of a few examples.

Some of the communities described from the Kamiesberg by Adamson (1938b) belong to arid fynbos, but his descriptions are superficial. Taylor (1978) describes a community on the eastern footslopes of the Cedarberg characterized by an indistinct middle layer of narrow-sclerophyllous shrubs about 0.5 to 1.5 m tall (*Anthospermum*, *Diosma*, *Eriocephalus*, *Passerina* and *Phylica* spp.), a sparse emergent stratum (1.5–2.5 m) with *Protea laurifolia*, *Protea glabra* (driest sites) and *Cannomois dregei* the most conspicuous species, and a "... very open ground layer .... of small succulents, ericoids and restioids"; total cover amounted to between 33 and 45%. A similar community sampled in the same region resembled this in structure, but included two species of stem-succulent *Euphorbia* (F.J. Kruger and H.C. Taylor, pers. obs.).

The communities analysed at Sites 8 and 9 (Appendix III) represent contrasting formations, one on deep sands between Citrusdal and Clanwilliam, and the other on a steep south-facing rock outcrop near Ladismith (see also Fig. 2.13 and 2.14). The first is dominated by broad-sclerophyllous shrubs, the second by narrow-sclerophyllous shrubs and (less so) Restionaceae. These by no means represent the full range of arid fynbos formations, which include tall open-herblands (as on deep sands south of Klawer), and often dominated by *Willdenowia striata*, to low open graminoid-heathlands.

The data in Table 2.3 and in Appendix III (Sites 8 and 9) illustrate some of the general features of arid fynbos communities. There are relatively large numbers of annual species but no more geophyte species than are found in mountain fynbos communities; but the relative importance of annuals and geophytes is greater than in mountain fynbos communities, especially in the semi-arid zone in the Cedarberg area and northward. Here, geophytes and especially annuals may cover the open spaces between perennials in a good spring, but communities have not been sampled under these

Fig. 2.13. Arid fynbos, Clanwilliam District. *Leucadendron loranthifolium* in right foreground, *L. pubescens* in left foreground (male) and right-centre (female). Note annual herb (*Ursinia* sp.) in centre. This is Site 8 in the Appendix.

conditions. However, in one sample a single species of annual (*Heliophila* sp.) comprised 5% of total crown volume (F.J. Kruger and H.C. Taylor, pers. obs.). Succulent species, also relatively numerous, are seldom of great relative importance in the community; the highest score noted was about 5% of total crown volume in the Ladismith sample. the relative importance of broad-sclerophyllous shrubs, narrow-sclerophyllous shrubs and graminoid herbs is variable, and any one may dominate, as in other fynbos types. Deciduous hemicryptophytes are common, and drought-deciduous shrubs occasional. Community diversity is not markedly lower than that of mountain fynbos communities.

Arid fynbos is something of an enigmatic vegetation type in that it is fairly readily recognized in the field but difficult to define. it contains a fairly distinct flora which includes fynbos species such as *Protea laurifolia*, *Phylica lachneoides*, *Cannomois*

dregei, *C. parviflora*, *Hypodiscus striatus*, *Restio fruticosus* and *Willdenowia striata*, but all of these tend to occur from mountain crest to foot in the interior ranges, extending beyond the limits of arid climate (some Restionaceae, like *R. fruticosus*, adopt giant forms in the arid zone). A fairly diverse "Karoid" flora (Mesembryanthemaceae, *Eriocephalus*, *Relhania*) is subordinate in the community, until increasing aridity results in transition to such types as mountain renosterbosveld.

Coastal fynbos

Coastal fynbos is structurally like mountain fynbos and has the same plant families, but occurs on the limestones and sands of the coastal forelands. However, Taylor (1978) noted that species composition differed between the two types and

Fig. 2.14. Open graminoid-heathland in arid fynbos near Ladismith. This is Site 9 in the Appendix.

that two major subdivisions could be recognized within coastal fynbos on biogeographic grounds, i.e. the fynbos on marine sands of the west coast (Cape Flats to Elandsbaai), and that on the limestones eastward from Danger Point to near Mosselbaai. The former is ericoid and open with Restionaceae in lower strata, while the latter has an upper proteoid stratum with ericoids and Restionaceae below; both types have more grass species and annuals than mountain fynbos.

Coastal fynbos on limestones

Acocks (1953) recognized a third type, i.e. a dwarf fynbos of the flats near Elim, but this will be included here as limestone fynbos.

Judging from a classification and description by Van der Merwe (1976) the tall open-heathland of Site 10 (Appendix III) appears to be representative of community structure on the eastern limestones

(see also Fig. 2.15); structurally, this community is like any other tall open-heathland of the mountain fynbos and includes some species typical of the latter vegetation (*Erica plukenetii*, *Hypodiscus striatus*), but also a complement of strictly coastal fynbos species such as *Protea obtusifolia*, the other *Erica* spp., *Chascanum cernuum*, and *Thamnochortus fraternus*.

A tall open-scrubland with heathland on a sand about 0.5 m deep over level calcrete east of Danger Point differed in composition from Site 10 (F.J. Kruger, pers. obs.); the following strata occurred: (a) emergents: *Leucadendron coniferum*, 2.0 to 3.0 m, 1% cover; (b) mid-height shrubs, 0.9 to 1.1 m tall, 40% cover, dominated by *Psoralea fruticans* and *Protea obtusifolia* (broad-sclerophylls) and *Agathosma* sp. (narrow-sclerophyll); (c) lower stratum of graminoid herbs and chamaephytes, 45% cover, 0.2 to 0.5 m tall, dominated by *Restio*

Fig. 2.15. Tall open-heathland on limestones (coastal fynbos). This is Site 10 in the Appendix.

eleocharis and a mixture of shrubs and sub-shrubs; (d) a ground layer of annuals and geophytes, with cover amounting to 10%.

On the whole, the dominant structural formations on limestones appear to be open-heaths of various forms but Taylor (1969) describes a dwarf-heathland, the "Dwarf Mixed Fynbos Sub-Association", at the Cape of Good Hope, suggests that similar formations may be found around Bredasdorp, and reports limestone dwarf-heathland on the Cape Flats (Taylor, 1972). He also describes a *Leucadendron coniferum* broad-sclerophyllous closed-scrub, poor in species, which may occur elsewhere in the limestone forelands.

Fynbos of coastal sands

A great variety of structural formations is encountered on coastal sands, but although there are descriptions from Taylor (1969, 1972), Boucher (1972) and Boucher and Jarman (1977), few analytic data are available. Taylor (1972), in an account of the vegetation of the Cape Flats, describes structural formations which range from tall narrow-sclerophyllous closed-heathland (domi-

nated in places by *Metalasia muricata*, or by *Passerina* spp.), to tall open graminoid-heathland with *Passerina vulgaris* and *Thamnochortus erectus*, and low heaths. Communities on unconsolidated sands characteristically included a number of vigorous rhizomatous herbs such as *Ehrharta villosa*, and soboliferous shrubs like *Myrica cordifolia*. Muir (1929), Taylor (1969) and Boucher and Jarman (1977) also describe what appears to be a typical coastal fynbos formation, a tall open-herbland dominated by stiff erect tufts of *Thamnochortus erectus* or of *Thamnochortus spicigerus*, about 1.2 to 1.5 m tall. These communities often are poor in species, with a very simple structure.

General

Taylor (1978) discusses the biogeography of coastal fynbos extensively and suggests that many of its features unique in the fynbos zone are due to disturbance resulting from fluctuating shorelines, on the one hand, and to the past influence of the then diverted Moçambique Current on the west coast environment. In any event both he and Acocks (1953) stress the tropical element of the flora, most obviously represented by low clumps of broad-sclerophyllous shrubs such as *Euclea racemosa* and *Rhus lucida*, and, on the coast, wind-sheared clumps of stunted, sclerophyllous trees like *Pterocelastrus tricuspidatus* and *Cassine maritima*. Low forest, often dominated by *Sideroxylon inerme* may be found in some sheltered localities. It is often suggested that some fynbos is seral to these scrub and forest communities, and that elsewhere coastal fynbos has replaced a community richer in grasses (e.g. Acocks, 1953).

Because of the often unconsolidated substrate and its location, the coastal fynbos ecosystem is subject to more frequent disturbance than other types. Burning, overgrazing, exploitation for firewood and disturbance with development of resorts are said to have resulted in the extensive live dune systems of certain parts (Walsh, 1968). Phillips (1931) and Taylor (1972) have described the littoral succession briefly, but there is no information on colonization and fixing of interior dunes, although experience shows that native species recolonize easily if the sand can be stabilized.

Taylor (1972) provides a brief account of suc-

cession after fire and his data indicate, first, that
native species revegetate burnt areas readily in the
absence of further disturbance, and that community response to fire is, in broad terms, like that
of mountain fynbos.

Finally, it should be noted that much of the
coastal fynbos in the west occurs in the 200 to 300
mm yr^{-1} rainfall zone; but since it is relatively
luxurious (Boucher and Jarman, 1977, report total
cover ranging from 65 to 80%) it does not belong to
arid fynbos as described above.

COMMUNITY DYNAMICS

This account is confined to a discussion of
succession in mountain fynbos because of the
dearth of information on the other types.

The most obvious successional change in fynbos
is succession after fire; this has been dealt with by
Gill and Groves (Vol. B, Ch. 7) and it remains
merely to add certain points of detail. Martin (1966)
has examined the mode of regeneration and aspects
of the life cycles of the species he dealt with, and Van
der Merwe (1966) classified all species in his study

area by regeneration type. In the latter case, two-
thirds of the 448 vascular plant species examined
regenerated vegetatively, the proportion ranging
from about 60 to 90% in different communities.
These data tend to confirm the anticipated relationship between life form and mode of regeneration, all geophytes and almost all hemicryptophytes being resprouters, while just less than
half of chamaephytes, and just more than half of
phanerophytes rely entirely on seed (except that all
trees resprout). On the other hand, most species of
dominant shrub do not resprout, and this includes
members of Proteaceae, Ericaceae, Rosaceae
(*Cliffortia*) and Fabaceae (*Aspalathus*). There are
many exceptions to the rules. For example,
important hemicryptophytes like *Chondropetalum
hookeranum* are completely reliant on seed for
regeneration (F.J. Kruger, pers. obs.).
Nevertheless, with such a large proportion of the
fynbos flora having the capacity to resprout,
community recovery after a fire is rapid, with
shoots appearing within days (Wicht, 1948;
Martin, 1966). Initial regrowth of the large persistent herbaceous component is rapid and occurs
at a rate of about 1000 to 4000 kg ha^{-1} yr^{-1}, so that

TABLE 2.6

Summary of successional phases in an ideal tall heathland or scrub fynbos community

Successional phase	Years after fire	Character and composition of the community
1. Immediate post-fire	0 to 1	annuals present (in many communities, confined to this phase); all species capable of vegetative regeneration resprout; reproductive response, especially in geophytes, may be complete (some *Cyrtanthus* spp., *Haemanthus canaliculatus*, *Moraea ramosissima*, some Orchidaceae) or incomplete (most other geophytic species); most germination occurs
2. Regenerating	2 to 4 or 5	communities dominated by graminoid herbs; sprouting species attain reproductive maturity; opportunistic shrubs (*Aspalathus* spp., *Othonna quinquedentata*, *Euryops abrotanifolius*) set seed and die; longer-lived shrubs begin to emerge from the canopy
3. Maturing	4 or 5 to about 10	all species reach reproductive maturity; tall shrubs emerge and adopt ascending branch habit
4. Mature	10 to 30	tall shrubs attain full rounded form, maximum flowering activity; seed-regenerating low shrubs (e.g. *Erica* spp.) begin to die; litter accumulates and lower herbaceous strata reduced in importance; no germination
5. Senscent	30+	accelerated mortality among seed-regenerating tall shrubs; canopy opens, some seed reproduction of tall shrubs; some immigration of forest precursors on relatively fertiie, relatively moist sites

canopy cover returns to about four-fifths of pre-fire levels in less than 24 months (Kruger, 1977, and unpublished).

Some features of pyric succession in mountain fynbos have been abstracted from various sources (Levyns, 1929, 1935; Adamson, 1935; Wicht, 1948; Hall, 1959; Martin, 1966; Van der Merwe, 1966) and are summarised in Table 2.6. The phenomenon of senescence, here defined as sudden increased rates of mortality in dominant shrubs at advanced age, tending to local extinction, is reflected in the records of populations of *Protea neriifolia* in Langrivier, Jonkershoek, and commonly observed in old fynbos. It has been the source of embarassment to conservation agencies who initially sought to conserve such species as *Serruria florida* and *Orothamnus zeyheri* by protecting them against fire, only to have the species steadily decline.

This description of pyric succession does not take account of differences in fire régime nor of changes in succession rates with differences in habitat. An integrated account on this scale, and also of other vegetation change induced by grazing, with or without burning, and other extraneous influences, would be difficult with the present lack of information and of a conceptual framework; some information is obtainable from the reports quoted above and those in Levyns (1929, 1935, 1956) and Wicht (1945).

CONCLUSION

Fynbos, as a broad category of heathlands, is remarkable for a number of features. Firstly, it extends along considerable habitat gradients, with a greater than tenfold change in rainfall, for example, without major changes in vegetation composition and structure. In other instances, apparently subtle changes in habitat result in abrupt transitions to forest, to succulent shrublands, or to other, quite different, formations. It seems that the limits of the type are fairly well defined by relatively high winter rainfall (at least about 40 mm per month), with no more than about the same rate in summer, in combination with low soil fertility, i.e. communities are stressed by summer soil-moisture deficits and by nutrient deficiency, the latter being probably due in part to leaching through high winter rainfall.

The communities which occupy this habitat, and presumably have evolved under these conditions, are with some notable exceptions among the richest plant communities in the world, and many are structurally complex, with a great diversity of growth forms. It is interesting to speculate as to what extent these structural features reflect a response to the environment. It would also be useful to determine whether the Australian heathlands, whose habitats resemble those of fynbos in their low soil fertility but are unlike in that they extend to tropical climates, are such close analogues that it is more profitable to speculate in terms of relict vegetations surviving on relictual soils in spite of climatic change, as suggested for the Australian situation. In any event, plant responses to environmental stress and periodic fire, and maintenance of community diversity under these conditions, will remain as profitable subjects for research at least to contribute to conservation of these unique communities.

APPENDIX I: CLIMATIC RECORDS[1]

Station	Season[2]	Daily sunshine (h)	Air temperature (°C) Mean daily		Precipitation (mm)	Pan evaporation (mm)	Frosts (days < 0°C)
			maximum	minimum			
Citrusdal Experimental Farm	summer	10.5	32.0	16.8	15	1120	0
32°34'S, 18°59'E 250 m	autumn	7.8	26.9	13.0	74	620	0
	winter	5.6	19.0	6.4	192	224	0.6
	spring	8.7	25.9	11.6	52	763	0
	year	8.1	25.9	11.9	333	2727	0.6
Tygerhoek Experimental Farm	summer	8.1	28.2	14.8	75	.	0
34°09'S, 19°54'E 168 m	autumn	5.8	23.8	11.7	123	.	0.1

Appendix I: Climatic records[1] *(continued)*

Station	Season[2]	Daily sunshine (h)	Air temperature (°C) Mean daily		Precipitation (mm)	Pan evaporation (mm)	Frosts (days < 0°C)
			maximum	minimum			
	winter	5.6	18.0	5.4	157	.	4.1
	spring	6.7	22.9	9.6	93	.	0.4
	year	6.5	23.2	10.4	448	.	4.6
Misgund-Oos Experimental Farm 33°45'S, 23°29'E 701 m	summer	.	25.5	11.0	92	.	0
	autumn	.	22.0	7.2	138	.	7.5
	winter	.	17.0	1.9	123	.	34.4
	spring	.	21.3	6.5	104	.	5.4
	year	.	21.4	6.6	456	.	47.3
Grahamstown 33°18'S, 26°32'E 539 m	summer	.	26.4	13.6	191	.	0
	autumn	.	23.7	10.6	188	.	0.5
	winter	.	19.7	4.9	102	.	4.6
	spring	.	22.4	9.6	216	.	0.1
	year	.	23.0	9.7	697	.	5.2
La Plaisant Experimental Farm 33°27'S, 19°12'E 290 m	summer	.	29.8	15.2	41	908	.
	autumn	.	24.5	11.9	148	471	.
	winter	.	17.0	6.9	290	235	.
	spring	.	23.3	10.6	114	599	.
	year	.	23.7	11.2	593	2213	.
De Keur Experimental Farm 32°58'S, 19°18'E 290 m	summer	.	26.9	11.5	25	849	.
	autumn	.	22.1	8.3	136	450	1.4
	winter	.	14.3	2.5	300	190	20.1
	spring	.	20.6	6.4	112	523	3.8
	year	.	20.9	7.1	573	2012	25.3
Jakkalsrivier 34°09'S, 19°09'E 655 m	summer	7.3	23.6	12.1	110	528	0
	autumn	5.1	20.0	10.5	194	294	0
	winter	3.7	14.5	6.4	343	153	0.6
	spring	5.9	18.2	8.1	176	358	0.2
	year	5.5	19.1	9.2	824	1333	0.8
Elgin Experimental Farm 34°08'S, 19°02'E 305 m	summer	8.4	25.2	12.7	90	585	0
	autumn	6.4	22.1	9.5	194	304	0.2
	winter	5.2	16.5	5.2	456	182	7.9
	spring	7.1	20.6	8.5	179	411	0.7
	year	6.8	21.1	9.0	919	1482	8.8
Deepwalls 33°57'S, 23°10'E 519 m	summer	.	22.8	13.5	317	.	0
	autumn	.	21.0	12.2	284	.	0
	winter	.	17.4	8.4	239	.	0.1
	spring	.	19.1	10.0	374	.	0
	year	.	20.1	11.0	1214	.	0.1
Jonkershoek (Biesievlei) 33°58'S, 18°57'E 282 m	summer	9.2	27.3	13.7	105	739[3]	0
	autumn	5.9	22.6	10.6	346	344[3]	0
	winter	4.4	16.7	6.4	576	120[3]	0
	spring	6.9	21.4	9.4	246	459[3]	0
	year	6.6	22.0	10.0	1273	1662[3]	0

Appendix I: Climatic records[1] *(continued)*

Station	Season[2]	Daily sunshine (h)	Air temperature (°C) Mean daily		Precipi- tation (mm)	Pan evaporation (mm)	Frosts (days < 0°C)
			maximum	minimum			
Table Mountain House	summer	.	20.1	12.0	186	.	.
33°59'S, 18°24'E 761 m	autumn	.	17.3	10.6	433	.	.
	winter	.	12.6	6.8	771	.	.
	spring	.	15.5	8.3	390	.	.
	year	.	16.4	9.4	1780	.	
Cape St. Blaize	summer	.	24.2	17.5	92	.	0
34°11'S, 22°09'E 60 m	autumn	.	22.1	15.2	111	.	0
	winter	.	19.3	11.3	95	.	0
	spring	.	20.5	13.6	119	.	0
	year	.	21.5	14.4	417	.	0
Port Elizabeth	summer	.	24.9	16.1	106		0
33°59'S, 25°36'E 58 m	autumn	.	23.1	12.9	154	.	0.1
	winter	.	19.8	7.6	146	.	0.1
	spring	.	21.1	11.9	170	.	0
	year	.	22.3	12.2	576	.	0.2
Danger Point	summer	.	21.7	15.3	65	.	.
34°37'S, 19°18'E 28m	autumn	.	19.6	13.6	127	.	.
	winter	.	16.9	10.4	218	.	.
	spring	.	18.4	12.5	134	.	.
	year	.	19.1	12.9	544	.	.

[1] These data are derived from publications of the South African Weather Bureau and Department of Agricultural Technical Service, and from records of the Department of Forestry.

[2] Summer: December to February; autumn: March to May; winter: June to August; spring: September to November.

[3] Adjusted Symons tank measurements.

APPENDIX II: REPRESENTATIVE FYNBOS SOIL PROFILES

Horizon	Depth (m)	Colour[1]	Clay (%)	Texture[2]	S[3] (μeq g^{-1})	CEC[4] (μeq g^{-1})	pH (water)	pH (at 1 eq l^{-1})	Organic (%)	Total N (%)	Phosphorus[3] (μg g^{-1})
A. COASTAL FYNBOS											
1. Mispah form.[4] Muden series[4] (calcareous). Shelly marine terrace with a variant of coast-shelf fynbos. Cape of Good Hope Nature Reserve (from Taylor, 1969).											
A_1	0 -?	10YR5/1 grey	7.4	Lm Sa	–	–	8.0	–	2.20	–	–
C/A_2	? -0.22	10YR6/1 grey	5.8	c Sa	–	–	8.1	–	1.65	–	–
R	0.22+										
2. Mispah form. Muden series. Limestone hill with open sclerophyllous scrub. De Poort, Bredasdorp.											
A	0 -0.20	black	9.0	Lm Sa	255.8	–	6.8	6.1	4.35	0.26	44
R	0.20+										
3. Fernwood form. Langebaan series (calcareous). Inland dune with *Thamnochortus erectus* dune fynbos. Cape of Good Hope Nature Reserve (from Taylor, 1969).											
A_1	0 -?	10YR3/1 very dark grey	3.6	m Sa	free lime	20.0	8.1	–	0.73	–	–
C_1	? -1.50	10YR6/2 light brown-grey	2.0	m Sa	free lime	1.5	–	–	0.15	–	–
B. MOUNTAIN FYNBOS											
1. Glenrosa form. Oribi Series. Sandstone on hillslope, with low-heath. Jakkalsrivier Catchment Area, Grabouw.											
A_1	0 -0.35	very dark grey	6	Sa	29.9	–	4.5	3.5	3.49	0.16	2
C	0.35+	saprolite									
2. Swartland form. Rosehill series. Slope of upper terrace on Bokkeveld shales, old field. Korenterivier, Garcia State Forest (from Neethling, 1970).											
A_1	0 -0.30	10YR3/2 very dark greyish-brown	16.1	m Sa Lm	52.9	73.9	5.6	4.3	1.46	0.14	–
B_1	0.30-0.50	7.5YR3/2 dark brown	36.8	Cl Lm	70.7	98.1	6.0	6.2	0.95	0.10	–
B_2	0.50-1.00	7.5YR3/2 brown	30.3	Sa Cl Lm	52.7	84.7	4.7	4.9	0.69	0.10	–
C/R	1.00 +	saprolite									
3. Sterkspruit form. Stanford series. Lower hillslope, sandy drift material on granite, with short fynbos (Outeniqua Experimental Farm, George) (from Lambrechts, 1964).											
A_{11}	0 -0.20	10YR2/2 very dark brown	14.0	f Sa Lm	10.9	66.8	4.3	4.3	1.86	0.11	–
A_{12}	0.20-0.35	10YR3/3 dark brown	14.2	f Sa Lm	8.2	49.8	4.4	4.4	1.00	0.06	–
A_3	0.35-0.48	10YR3/4 dark yellowish brown	15.5	gr. f Sa Lm	9.7	47.0	4.4	4.4	0.71	0.05	–
B_{21}	0.48-0.70	5YR4/8 yellowish red	45.9	Cl	47.6	121.4	4.1	4.9	0.59	0.05	–

Horizon	Depth (m)	Colour		Texture			pH (H$_2$O)	pH	S		P
B_{31}	0.70–0.94	7.5YR5/6 strong brown	34.3	Sa Cl Lm	50.8	93.2	—	4.4	0.25	0.02	—
C	0.94–1.16	10YR6/4 light yellowish brown	21.9	gr. Lm	44.9	60.9	—	4.8	0.09	0.02	—
R	1.16+										

4. Clovelly form. Oatsdale series. Lower hillslope, colluvial granite, shale and milonite, pine plantation. Jonkershoek State Forest (from Joubert, 1965)

Horizon	Depth (m)	Colour		Texture			pH (H$_2$O)	pH	S		P
A_1	0–0.45	10YR2/2 very dark brown	19.2	f Sa Lm			4.9	4.0	—	0.21	29
A_2	0.45–0.76	10YR4/4 dark yellowish brown	—	f Sa Lm			—	—	—	—	—
B_2	0.76–1.10	7.5YR5/6 strong brown	29.6	Sa Cl			4.9	3.7	—	0.02	—
B_3	1.10–1.80	7.5YR5/6 strong brown	—	Cl			—	—	—	—	—

5. Avalon form. Ruston series. Hillslope with tillite outcrops, with open-scrub. Jakkalsrivier Catchment Area, Grabouw.

Horizon	Depth (m)	Colour		Texture			pH (H$_2$O)	pH	S		P
A_1	0–0.30	dark brown	14	Sa Lm	19.1		5.9	4.6	1.76	0.11	1
B_{21}	0.30–0.70	brownish-yellow	18	Sa Lm	5.4		5.5	4.8	0.71	0.03	1
B_{22}	0.70–1.00	brownish-yellow, mottled	16	Sa Lm	1.8		5.2	4.8	0.36	0.03	1
R	1.00+										

6. Cartref form. Waterridge series. Hillslope on sandstone with low heath. Jakkalsrivier Catchment Area, Grabouw.

Horizon	Depth (m)	Colour		Texture			pH (H$_2$O)	pH	S		P
A_1	0–0.30	very dark grey	6	Sa	31.3		4.5	3.4	2.77	0.10	1
E	0.30–1.40	pale grey	6	Sa	2.1		5.1	4.0	0.09	0.01	1
R	1.40+										

7. Lamotte form. Lamotte series. Midslope of alluvial terrace: well-drained colluvial–alluvial sand (from Lambrechts, 1975).

Horizon	Depth (m)	Colour		Texture			pH (H$_2$O)	pH	S		P
A_1	0–0.20	10YR6/2 light brownish grey	trace	c Sa	1.06			5.35	0.40	—	38
E	0.20–1.00	10YR8/1 white	trace	c Sa	1.2			5.25	0.00	—	6
B_{2hir}	1.00–1.20	5YR2.5/2 dark reddish brown	trace	c Sa	7.5			4.75	1.04	—	47
C	1.20+	10YR7/6 yellow	—	c Sa	—			—	—	—	—

[1] Colour notations: Colour notations are those of the *Munsell Soil Color Charts* (1954), Munsell Color Co. Inc., Baltimore, Md., U.S.A.

[2] Texture abbreviations: c=coarse, Cl=clay, f=fine, gr.=gritty, Lm=loam, m=medium, Sa=sand.

[3] Note on analytical methods: Net extractable cations (S) and cation exchange capacity (CEC) were determined from a leachate with ammonium acetate (1 eq l^{-1}), except in the case of profiles A_1, B_1 and B_6, where a solution at 0.1 eq l^{-1} was used. Soil P was determined by Bray No. 2 extract (strong acid) in the case of profiles A_1, B_1, B_5 and B_6, by the method of Jackson (1962) (H_2SO_4 extract) in the case of profile B_4.

[4] For a definition of soil **forms** and **series**, see p. 25.

APPENDIX III: ANALYTIC RECORDS FOR SELECTED FYNBOS COMMUNITIES

In each community description, the species are arranged by life-form groups. The canopy volume of each species is expressed as a percentage of the total for all species; and for each life-form group the total is indicated in bold face. Abbreviations used for leaf type and size are:

BS: broad-sclerophyll
NS: narrow-sclerophyll

Gra: graminoid
Lep: leptophyll
Lin: Linear
Mac: macrophyll
Mes: mesophyll
Mic: microphyll

Pin: pinnae or leaflets
Nan: nanophyll
Nee: needle-like
Spi: spiny
Suc: succulent

Aph: aphyllous
Cup: cupressoid
Eri: ericoid

SITE 1

Open-scrub structural formation (mountain fynbos)

Brunia nodiflora–Psoralea rotundifolia community, Jonkershoek State Forest

Latitude: 33 59 S **Longitude:** 18° 57 E **Elevation:** 490 m **Nearest climate station:** Biesievlei, Jonkershoek
Geology: granite **Soil:** Clovelly form **Age:** 17 years since fire **Total canopy cover:** 95–100%
Structure: Open-scrub **Reference:** enumeration by F.J. Kruger and R.H. Whittaker, October 1975 (see also Werger et al., 1972)

	Leaf type and size	% of total canopy volume
Evergreen, tall shrubs (>2 m)		
Cupressaceae		
Wildringtonia nodiflora	(Cup)	0.1
Proteaceae		
Protea nerifolia	BS (Mic)	47.9
		48.0
Evergreen, mid-height shrubs (1–2 m)		
Anacardiaceae		
Rhus angustifolia	BS (Nan)	trace
R. tomentosa	BS (Mic)	trace
Bruniaceae		
Berzelia lanuginosa	NS (Lep)	0.7
Brunia nodiflora	NS (Lep)	43.1
Rosaceae		
Cliffortia cuneata	BS (Nan)	trace
C. polygonifolia	BS (Nan)	trace
C. ruscifolia	BS (Lep, Spi)	0.1
Rubiaceae		
Anthospermum aethiopicum	NS (Lep)	trace
		43.9

	Leaf type and size	% of total canopy volume
Evergreen, caespitose dwarf-shrubs (<25 cm)		
Asteraceae		
Helichrysum capitellatum	BS (Lep)	trace
H. teretifolium	NS (Lep, Eri)	trace
Osteospermum tomentosum	BS (Nan)	trace
Ericaceae		
Erica cerinthoides	NS (Lep, Eri)	trace
Euphorbiaceae		
Clutia alaternoides	BS (Nan)	trace
Fabaceae		
Psoralea rotundifolia	BS (Nan)	trace
		trace
Evergreen, creeping dwarf-shrubs (<25 cm)		
Rubiaceae		
Galium sp.	NS (Lep)	**trace**
Evergreen, mid-height caespitose graminoid herbs (25–100 cm)		
Cyperaceae		
Ficinia grandiflora	NS (Gra)	trace
Tetraria bromoides	NS (Gra)	0.5

Taxon		
Cassythaceae		
Cassytha ciliolata (vine)	NS (Aph)	trace
Evergreen, low shrubs (50–100 cm)		
Asteraceae		
Helichrysum crispum	BS (Nan)	trace
Stoebe cinerea	NS (Lep, Eri)	trace
Ebenaceae		
Diospyros glabra	BS (Nan)	trace
Ericaceae		
Erica hispidula	NS (Lep, Eri)	0.5
E. plukenetii	NS (Lep, Eri)	1.2
E. sphaeroidea	NS (Lep, Eri)	trace
Fabaceae		
Podalyria racemulosa	BS (Nan)	trace
Proteaceae		
Leucadendron salignum	BS (Nan)	0.5
L. spissifolium	BS (Nan)	trace
Selaginaceae		
Agathelpis dubia	NS (Lep)	trace
		2.2
Drought-deciduous, low shrubs (50–100 cm)		
Liliaceae		
Asparagus thunbergianus (spinose)	(Lep, Nee)	trace
Montiniaceae		
Montinia caryophyllacea	(Nan)	trace
		trace
Evergreen, low shrubs (25–50 cm)		
Ericaceae		
Blaeria dumosa	NS (Eri, Lep)	trace
Eremia totta	NS (Eri, Lep)	trace
Erica articularis	NS (Eri, Lep)	0.1
E. nudiflora	NS (Eri, Lep)	0.1
Rhamnaceae		
Phylica spicata	BS (Nan)	trace
Rutaceae		
Adenandra marginata	BS (Lep)	trace
Diosma hirsuta	NS (Lep)	0.2
		0.4

Taxon		
T. ...ata	NS (Gra)	0.6
T. fimbriolata	NS (Gra)	0.3
Poaceae		
Ehrharta sp. cf. *E. bulbosa*	NS (Gra)	trace
Merxmuellera stricta	NS (Gra)	0.3
Restionaceae		
Hypodiscus albo-aristatus	NS (Aph)	trace
Restio triticeus	NS (Aph)	2.3
		4.0
Evergreen, low caespitose graminoid herbs (<25 cm)		
Cyperaceae		
Ficinia filiformis	NS (Gra)	trace
Liliaceae		
Caesia eckloniana	NS (Gra)	trace
Poaceae		
Anthoxanthum tongo	NS (Gra)	trace
Merxmuellera rufa	NS (Gra)	trace
Schizaeaceae		
Schizaea pectinata	NS (Aph)	trace
		trace
Deciduous rosette forbs (<25 cm)		
Droseraceae		
Drosera trinervia	(Lep)	trace
Schizaeaceae		
Mohria caffrorum	(Pin, Lep)	trace
		trace
Evergreen forbs (<25 cm)		
Hydrocotylaceae		
Centella sp.	NS (Lep)	**trace**
Evergreen, mid-height geophytes (25–100 cm)		
Iridaceae		
Aristea major	BS (Mes)	0.4
Bobartia indica	NS (Gra)	trace
		0.4

SITE 1 (*continued*)

	Leaf type and size	% of total canopy volume		Leaf type and size	% of total canopy volume
Deciduous, mid-height geophytes (25–100 cm)			Deciduous, low geophytes (<25 cm)		
Apiaceae			*Iridaceae*		
Lichtensteinia lacera	(Mes–Mac)	trace	*Geissorhiza ovata*	(Lep)	trace
Asteraceae			*Liliaceae*		
Berkheya herbacea	(Mes)	trace	*Dipidax punctata*	(Nan)	trace
			Liliaceae spp. (3)	(Nan)	trace
Iridaceae			*Oxalidaceae*		
Gladiolus sp.	(Mic)	trace	*Oxalis commutata*	(Nan)	trace
Iridaceae sp.	(Mic)	trace	*O. eckloniana* var. *sonderi*	(Nan)	trace
Watsonia pyramidata	(Mes)	trace	*O. lanata* var. *rosea*	(Nan)	trace
Liliaceae			Indeterminate sp.		trace
Trachyandra hirsuta	(Mic)	trace			trace
Lobeliaceae			Low annuals (<25 cm)		
Cyphia volubilis (twiner)	(Lep)	trace	*Asteraceae*		
		trace	*Senecio cymbalariaefolius*	(Nan)	**trace**

SITE 2

Tall shrubland with heath formation (mountain fynbos)

Protea arborea–Rhus angustifolia community, Jonkershoek State Forest

Latitude: 34°00'S   **Longitude:** 18°57'E   **Elevation:** 285 m   **Nearest climate station:** Biesievlei, Jonkershoek

Precipitation: 1600 mm yr^{-1} **Soil:** Clovelly form   **Age:** 7 years since fire   **Total canopy cover:** 95%

Geology: quartzite colluvium over granite   **Reference:** enumeration by F.J. Kruger, H.C. Taylor and R.H. Whittaker, October 1975 (see also Werger et al., 1972)

Structure: tall shrubland with heath

	Leaf type and size	% of total canopy volume		Leaf type and size	% of total canopy volume
Evergreen, tall shrubs (> 2 m)			Evergreen, low shrubs (25–50 cm)		
Proteaceae			*Asteraceae*		
Protea arborea	BS (Mes)	**36.8**	*Elytropappus glandulosus*	NS (Lep, Eri)	0.6
			Helichrysum crispum	BS (Nan)	trace
Evergreen, mid-height shrubs (1–2 m)			*Senecio pubigerus*	NS (Lep)	trace
Asteraceae			*S.* sp.	NS (Lep)	trace
Chrysanthemoides monilifera	BS (Mic)	0.1	*Stoebe capitata*	NS (Lep, Eri)	1.6
			S. spiralis	NS (Lep, Eri)	0.4
Anacardiaceae			*Campanulaceae*		
Rhus angustifolia	BS (Mic)	0.1	*Roella ciliata*	NS (Lep)	trace
R. tomentosa	BS (Mic)	0.1			

Celastraceae		
Maytenus oleoides	BS (Mic)	0.2
Ebenaceae		
Diospyros glabra	BS (Nan)	1.2
Fabaceae		
Aspalathus cordata	BS (Nan)	0.1
Rosaceae		
Cliffortia cuneata	BS (Nan)	0.1
C. pterocarpa	NS (Lep)	0.1
Rubiaceae		
Anthospermum aethiopicum	NS (Lep)	02.
Santalaceae		
Thesium strictum (partial root parasite)	NS (Lep)	0.2
		2.4

Evergreen, low shrubs (50–100 cm)

Asteraceae		
Metalasia muricata	NS (Lep, Eri)	trace
Stoebe aethiopica	NS (Lep, Eri)	trace
S. plumosa	(Cup)	3.4
Ursinia pinnata	NS (Lep)	2.4
Ericaceae		
Erica bicolor	NS (Lep, Eri)	trace
E. hispidula	NS (Lep, Eri)	trace
E. lucida	NS (Lep, Eri)	trace
E. plukenetii	NS (Lep, Eri)	trace
E. racemosa	NS (Lep, Eri)	trace
E. sphaeroidea	NS (Lep, Eri)	0.8
Fabaceae		
Podalyria montana	BS (Nan)	trace
Psoralea obliqua	BS (Nan)	10.3
Proteaceae		
Leucadendron salignum	BS (Nan)	7.1
		24.0

Drought-deciduous, low shrubs (50–100 cm)

Liliaceae		
Asparagus thunbergianus (spinose)	(Nee)	2.1
Montiniaceae		
Montinia caryophyllacea	(Nan)	1.0
		3.1

Ericaceae		
? *Blaeria ericoides*	NS (Lep, Eri)	trace
Erica calycina var. *periplocaeflora*	NS (Lep, Eri)	trace
E. imbricata	NS (Lep, Eri)	trace
E. nudiflora	NS (Lep, Eri)	trace
Euphorbiaceae		
Clutia alaternoides	BS (Nan)	0.3
Fabaceae		
Aspalathus laricifolius	NS (Lep)	0.3
Podalyria racemulosa	BS (Nan)	0.5
Polygalaceae		
Muraltia heisteria	NS (Lep, Eri)	trace
Rhamnaceae		
Phylica imberbis	NS (Lep, Eri)	0.3
P. spicata	BS (Nan)	0.4
Rutaceae		
Agathosma ciliata	NS (Lep)	2.0
Diosma hirsuta	NS (Lep)	1.5
Santalaceae		
Thesium carinatum	NS (Lep)	trace
Selaginaceae		
Agathelpis dubia	NS (Lep)	0.5
Sterculiaceae		
Hermannia hyssopifolia	BS (Nan)	trace
		8.4

Drought-deciduous, low shrubs (25–50 cm)

Anacardiaceae		
Rhus rosmarinifolia	(Nee)	**0.5**

Evergreen creeping dwarf-shrubs (<25 cm)

Fabaceae		
Psoralea decumbens	BS (Lep)	trace
P. imbricata	BS (Nan)	trace
Proteaceae		
Protea acaulos	BS (Mes)	1.0
		1.0

SITE 2 *(continued)*

	Leaf type and size	% of total canopy volume
Evergreen, caespitose or scapose dwarf-shrubs (<25 cm)		
Asteraceae		
Helichrysum odoratissimum (suffr.)	BS (Nan)	trace
H. rutilans (suffr.)	BS (Nan)	0.3
H. zeyheri (suffr.)	NS (Lep)	trace
Metalasia cephalotes	NS (Lep, Eri)	trace
Senecio paniculatus	NS (Lep)	trace
Crassulaceae		
Crassula fascicularis	(Lep, Suc)	trace
Mesembryanthemaceae		
Lampranthus leipolditii	(Lep, Suc)	trace
Selaginaceae		
Selago spuria (suffr.)	NS (Lep)	trace
Thymelaeaceae		
Gnidia inconspicua	NS (Lep)	0.3
		0.6
Evergreen, caespitose graminoid herbs (25–100 cm)		
Cyperaceae		
Tetraria cuspidata	NS (Gra)	1.5
T. ustulata	NS (Aph)	1.0
Poaceae		
Merxmuellera stricta	NS (Gra)	2.8
Themeda triandra	NS (Gra)	0.8
Restionaceae		
Restio filiformis	NS (Aph)	trace
		6.1
Evergreen, caespitose graminoid herbs (<25 cm)		
Cyperaceae		
Ficinia filiformis	NS (Gra)	0.3
F. nigrescens	NS (Gra)	trace
Liliaceae		
Caesia eckloniana	NS (Gra)	trace

	Leaf type and size	% of total canopy volume
Evergreen forbs (<25 cm)		
Asteraceae		
Leontonyx spathulatus	BS (Nan)	trace
Hydrocotylaceae		
Centella glabrata (reptant)	BS (Nan)	0.4
		0.4
Evergreen geophytes		
Asteraceae		
Corymbium glabrum	BS (Mic)	0.3
C. villosum	BS (Mic)	trace
Gerbera crocea	BS (Mic)	0.2
Iridaceae		
Aristea capitata	BS (Mes)	trace
A. spiralis	BS (Mes)	trace
		0.5
Seasonally green geophytes		
Apiaceae		
Lichtensteinia lacera	(Mic)	trace
Apiaceae sp.	(Mes)	trace
Asteraceae		
Berkheya herbacea	(Mes, Spi)	trace
Haplocarpha lanata	(Mes)	trace
Droseraceae		
Drosera trinervia	(Lep)	trace
Geraniaceae		
Pelargonium pinnatum	(Nan)	trace
Iridaceae		
Gladiolus carneus	(Mic)	trace
G. sp.	(Mic)	trace
Ixia sp.	(Mic)	trace
Lapeirousia corymbosa	(Mic)	trace
Micranthus alopecuroides	(Mic)	trace
Romulea flava	(Mic)	trace
Watsonia pyramidata	(Mes)	0.7

Poaceae

Cymbopogon marginatus	NS (Gra)	2.1
Eragrostis capensis	NS (Gra)	trace
Pentaschistis curvifolia	NS (Gra)	0.6
Plagiochloa uniolae	NS (Gra)	0.3

Restionaceae

Restio cuspidatus	NS (Aph)	2.9
		6.2

Evergreen, sparingly tufted, more or less rhizomatous graminoid herbs (25–100 cm)

Poaceae

Ehrharta calycina	NS (Gra)	0.3

Restionaceae

Leptocarpus distichus	NS (Gra)	trace
Restio gaudichaudianus	NS (Aph)	6.9
Thamnochortus fruticosus	NS (Gra)	0.3

Deciduous forbs with erect or scrambling seasonal leaves (25–100 cm)

Euphorbiaceae

Euphorbia genistoides	(Lep)	trace

Geraniaceae

Pelargonium myrrhifolium	(Pin–Lep)	trace
P. tabulare	(Mic)	trace
		trace

Seasonal rosette forbs (25–100 cm)

Asteraceae

Castalis nudicaulis	(Mic)	trace

Dipsacaceae

Scabiosa columbaria	(Mic)	trace
		trace

Liliaceae

Albuca canadensis	(Mic)	trace
Baeometra uniflora	(Mic)	trace
Dipidax punctata	(Nan)	trace
Lachenalia glaucina	(Mic)	trace
L. sp.	(Mic)	trace
Liliaceae sp.	(Nan)	trace
Trachyandra hirsuta	(Mic)	trace
T. muricata	(Mic)	0.2
Wurmbea sp.	(Nan)	trace

Lobeliaceae

Cyphia volubilis (twiner)	(Lep)	trace

Orchidaceae

Disa sp.	(Mic)	trace

Oxalidaceae

Oxalis bifida	(Nan)	trace
O. commutata	(Lep)	trace
O. versicolor	(Lep)	trace
Indeterminate spp. (2)	(Nan)	trace
		0.9

Annuals

Asteraceae

Helichrysum indicum	(Lep)	trace
Ursinia anthemoides	(Lin)	trace

Campanulaceae sp.	(Lep)	trace

Gentianaceae

Sebaea exacoides	(Lep)	trace

Iridaceae

Aristea africana	(Gra)	trace

Poaceae

Aira cupaniana	(Gra)	trace
Helictotrichon longum	(Gra)	trace
		trace

SITE 3

Low open-heathland formation (mountain fynbos)

Erica corifolia–Restio egregius community, Jakkalsrivier, Grabouw

Latitude: 34°09'S **Longitude:** 19°09'E **Elevation:** 756 m **Nearest climate station:** Jakkalsrivier
Precipitation: 950 mm **Soil:** Mispah/Glenrosa **Total canopy cover:** 80%
Geology: TM quartzite **Age:** 18 years since fire
Structure: low open-heath **Reference:** enumeration by R.A. Haynes, H.C. Taylor and R.H. Whittaker, October 1975 (see also Kruger, 1974)

	Leaf type and size	% of total canopy volume
Evergreen, mid-height shrubs (1–2 m)		
Proteaceae		
Leucadendron xanthoconus	BS (Nan)	0.3
Evergreen, low shrubs (50–100 cm)		
Asteraceae		
Metalasia muricata	NS (Lep, Eri)	0.1
Phaenocoma prolifera	(Cup)	0.2
Ericaceae		
Erica coccinea	NS (Lep, Eri)	0.1
E. corifolia	NS (Lep, Eri)	0.1
E. hispidula	NS (Lep, Eri)	0.6
E. longifolia	NS (Lep, Eri)	1.8
E. sessiliflora	NS (Lep, Eri)	0.2
Sympieza articulata	NS (Lep, Eri)	54.4
Grubbiaceae		
Grubbia tomentosa	BS (Mic)	0.2
Lobeliaceae		
Lobelia pinifolia	NS (Lep)	0.1
Penaeaceae		
Penaea mucronata	BS (Lep)	2.3
Proteaceae		
Mimetes cucullatus	BS (Mic)	0.1
Protea cynaroides	BS (Mes)	0.1
Selaginaceae		
Agathelpis dubia (suffr.)	NS (Lep)	0.1
		60.4
Poaceae		
Pentaschistis colorata	NS (Gra)	3.9
Pseudopentameris macrantha	NS (Gra)	trace
Restionaceae		
Chondropetalum hookeranum	NS (Aph)	0.2
Hypodiscus aristatus	NS (Aph)	3.5
		26.1
Evergreen, mid-height sparingly tufted, more or less rhizomatous graminoid herbs (25–100 cm)		
Restionaceae		
Elegia racemosa	NS (Aph)	0.1
Leptocarpus membranaceus	NS (Aph)	0.1
Restio egregius	NS (Aph)	8.8
R. triticeus	NS (Aph)	1.4
Thamnochortus gracilis	NS (Aph)	0.1
		10.5
Evergreen, low caespitose graminoid herbs (<25 cm)		
Cyperaceae		
Chrysithrix dodii	NS (Gra)	trace
Ficinia filiformis	NS (Gra)	trace
F. zeyheri	NS (Gra)	trace
Tetraria brevicaulis	NS (Gra)	trace
T. microstachys	NS (Gra)	1.0
Poaceae		
Merxmuellera rufa	NS (Gra)	0.4
Pentaschistis steudelii	NS (Gra)	trace

Evergreen, low shrubs (25–50 cm)

Fabaceae		
Amphithalea sp.	NS (Lep)	trace
Rhamnaceae		
Phylica lasiocarpa	NS (Lep, Eri)	trace
		trace

Evergreen, creeping dwarf shrubs (<25 cm)

Fabaceae		
Indigofera gracilis	BS (Lep)	trace
Rubiaceae		
Anthospermum prostratum	NS (Lep)	trace
		trace

Evergreen, caespitose dwarf shrubs (<25 cm)

Campanulaceae		
Roella psammophila	NS (Lep)	trace
Euphorbiaceae		
Clutia polygonoides	NS (Lep)	trace
		trace

Evergreen, mid-height caespitose graminoid herbs (25–100 cm)

Cyperaceae		
Tetraria cuspidata	NS (Gra)	1.9
T. exilis	NS (Gra)	1.9
T. fasciata	NS (Gra)	trace
T. flexuosa	NS (Gra)	14.6
T. thermalis	BS (Gra)	0.1

Restionaceae		
Restio ? cincinnatus	NS (Aph)	trace
Schizaeaceae		
Schizaea pectinata	NS (Aph)	trace
		1.4

Deciduous rosette forbs (<25 cm)

Droseraceae		
Drosera aliciae	(Nan)	**trace**

Seasonally green, mid-height geophytes (25–100 cm)

Iridaceae		
Gladiolus carneus	(Mic)	trace
G. debilis	(Nan)	trace
		trace

Evergreen, low geophytes (<25 cm)

Asteraceae		
Corymbium glabrum	BS (Mic)	trace
C. scabrum	NS (Gra)	trace
Gerbera tomentosa	BS (Mic)	trace
Iridaceae		
Aristea spiralis	BS (Mic)	trace
Bobartia gladiata	NS (Gra)	0.3
		0.3

SITE 4

Graminoid-heathland formation (mountain fynbos)

Willdenowia sulcata–Erica brevifolia community, Jakkalsrivier, Grabouw

Latitude: 34 09′E   **Longitude:** 19 09′E   **Elevation:** 960 m   **Nearest climate station:** Jakkalsrivier
Precipitation: 950 mm   **Soil:** acid Glenrosa   **Age:** 19 years since fire   **Total canopy cover:** 95%
Geology: TM quartzite   **Reference:** enumeration by F.J. Kruger, January 1977
Structure: graminoid-heath

	Leaf type and size	% of total canopy volume		Leaf size and size	% of total canopy volume
Evergreen, mid-height shrubs (1–2 m)			Santalaceae		
			Thesium ericaefolium	NS (Lep)	trace
Bruniaceae					**trace**
Berzelia abrotanoides	NS (Lep)	2.2			
			Evergreen, mid-height caespitose graminoid herbs (25–100 cm)		
Proteaceae					
Leucadendron xanthoconus	BS (Nan)	0.2	Cyperaceae		
		2.4	*Chrysithrix capensis*	NS (Gra)	trace
			Ficinia monticola	NS (Gra)	trace
Evergreen, low shrubs (50–100 cm)			*Tetraria cuspidata*	NS (Gra)	trace
			T. fasciata	NS (Gra)	0.5
Bruniaceae			*T. flexuosa*	NS (Gra)	22.8
Nebelia paleacea	NS (Lep)	6.4			
			Poaceae		
Ericaceae			*Ehrharta tricostata*	NS (Gra)	trace
Erica brevifolia	NS (Lep, Eri)	trace	*Pentaschistis colorata*	NS (Gra)	0.6
E. coccinea	NS (Lep, Eri)	0.2			
E. lutea	NS (Lep, Eri)	trace	Restionaceae		
			Hypodiscus aristatus	NS (Gra)	trace
Fabaceae			*Staberoha cernua*	NS (Gra)	trace
Cyclopia falcata	BS (Lep)	trace	*Thamnochortus similis*	NS (Gra)	trace
					23.9
Iridaceae					
Klattia partita	BS (Nan)	0.3	**Evergreen, mid-height sparingly tufted, more or less rhizomatous graminoid herbs (25–100 cm)**		
Penaeaceae					
Penaea mucronata	BS (Lep)	0.9	Poaceae		
			Festuca scabra	NS (Gra)	trace
Proteaceae					
Protea speciosa	BS (Mes)	0.4	Restionaceae		
P. cynaroides	BS (Mes)	trace	*Chondropetalum deustum*	NS (Aph)	1.2
		8.2	*Elegia racemosa*	NS (Aph)	9.9
			Leptocarpus esterhuyseniae	NS (Aph)	trace
Evergreen, low shrubs (25–50 cm)			*L. membranaceus*	NS (Aph)	7.5
			Restio dispar	NS (Aph)	2.0
Asteraceae			*Willdenowia sulcata*	NS (Aph)	trace
Euryops abrotanifolius	NS (Pin–Lep)	trace			**20.6**
Helichrysum cymosum (suffr.)	BS (Nan)	trace			

Ericaceae		
Aniserica gracilis	NS (Lep, Eri)	0.7
Blaeria dumosa	NS (Lep, Eri)	trace
Erica hispidula	NS (Lep, Eri)	36.4
E. transparens	NS (Lep, Eri)	0.1
Fabaceae		
Aspalathus marginata	BS (Lep)	trace
Proteaceae		
Leucadendron spissifolium	BS (Nan)	0.3
Selaginaceae		
Agathelpis dubia (suffr.)	NS (Lep)	trace
Rhamnaceae		
Phylica sp. cf. *P. diffusa*	NS (Lep, Eri)	0.7
		38.2
Evergreen, creeping dwarf shrubs (<25 cm)		
Ericaceae		
Erica krugeri	NS (Lep, Eri)	trace
Fabaceae		
Argyrolobium lunaris	BS (Nan)	trace
Indigofera gracilis	BS (Lep)	trace
		trace
Evergreen, caespitose dwarf shrubs (<25 cm)		
Asteraceae		
Helichrysum felinum (suffr.)	BS (Nan)	trace
H. pinifolium (suffr.)	NS (Lep, Eri)	trace
Senecio umbellatus (suffr.)	NS (Lep)	trace
Ursinia dentata (suffr.)	NS (Pin–Lep)	trace
Euphorbiaceae		
Clutia polygonoides	NS (Lep)	trace
Myricaceae		
Myrica kraussiana	BS (Mic)	trace
Polygalaceae		
Muraltia hyssopifolia	NS (Lep)	trace
Rutaceae		
Agathosma bifida	NS (Lep)	trace
Thymelaeaceae		
Gnidia linearifolia	NS (Lep)	trace

Evergreen, low caespitose graminoid herbs (<25 cm)		
Cyperaceae		
Tetraria exilis	NS (Gra)	trace
Poaceae		
Ehrharta setacea	NS (Gra)	0.2
Schizaeaceae		
Schizaea pectinata	NS (Aph)	1.0
		1.2
Evergreen, low sparingly tufted more or less rhizomatous graminoid herbs (<25 cm)		
Liliaceae		
Caesia sp.	NS (Gra)	trace
Poaceae		
Merxmuellera rufa	NS (Gra)	trace
		trace
Deciduous rosette forbs (<25 cm)		
Droseraceae		
Drosera aliciae	(Nan)	**trace**
Evergreen forbs (<25 cm)		
Asteraceae		
Osmitopsis afra	BS (Lep)	0.5
Senecio erubescens	(Mic)	trace
Campanulaceae		
Prismatocarpus nitidus	(Lep)	trace
Lobeliaceae		
Lobelia coronopifolia	BS (Lep)	trace
		0.5
Seasonally green, mid-height geophytes (25–100 cm)		
Iridaceae		
Gladiolus brevitubus	(Mic)	trace
G. sp.	(Mic)	trace
Watsonia pyramidata	(Mes)	4.5
		4.5

SITE 4 (continued)

	Leaf type and size	% of total canopy volume
Evergreen, mid-height geophytes (25–100 cm)		
Haemodoraceae		
Dilatris viscosa	(Mic)	trace
Iridaceae		
Aristea spiralis	(Mic)	trace
		trace
Seasonally green, low geophytes (<25 cm)		
Oxalidaceae		
Oxalis sp.	(Nan)	trace

	Leaf type and size	% of total canopy volume
Evergreen, low geophytes (<25 cm)		
Asteraceae		
Gerbera tomentosa	BS (Mic)	trace
Mairia crenata	BS (Mic)	trace
Hydrocotylaceae		
Hermas capitata	BS (Nan)	trace
H. ciliata	(Mic)	trace
		trace

SITE 5

Low closed-herbland formation (mountain fynbos)

Tetratia flexuosa community, Jakkalsrivier, Grabouw

Latitude: 34°09'S **Longitude:** 19°09'E **Soil:** Champagne
Precipitation: 950 mm **Geology:** TM quartzite
Structure: closed-herbland

Nearest climate station: Jakkalsrivier
Elevation: 840 m
Age: 17 years since fire
Total canopy cover: 95%
Reference: enumeration by R.A. Haynes, H.C. Taylor and R.H. Whittaker, October 1975

	Leaf type and size	% of total canopy volume
Evergreen, low shrubs (50–100 cm)		
Proteaceae		
Protea cynaroides	BS (Mes)	trace
Thymelaeaceae		
Gnidia oppositifolia	BS (Lep)	1.1
		1.1
Evergreen, low shrubs (25–50 cm)		
Asteraceae		
Stoebe incana	NS (Lep, Eri)	0.1

	Leaf type and size	% of total canopy volume
Poaceae		
Merxmuellera rufa	NS (Gra)	trace
Pentaschistis curvifolia	NS (Gra)	trace
		trace
Evergreen, sparingly tufted, more or less rhizomatous graminoid herbs (25–100 cm)		
Restionaceae		
Restio dispar	NS (Aph)	15.2

Ericaceae		
Blaeria dumosa	NS (Lep, Eri)	trace
Erica hispidula	NS (Lep, Eri)	1.5
		1.6

Evergreen, caespitose or scapose dwarf-shrubs (<25 cm)

Asteraceae		
Senecio pubigerus	NS (Lep)	trace
Euphorbiaceae		
Clutia polygonoides	NS (Lep)	0.4
		0.4

Evergreen, caespitose graminoid herbs (25–100 cm)

Cyperaceae		
Chrysithrix capensis	NS (Gra)	3.8
Tetraria bolusii	NS (Gra)	16.7
T. flexuosa	NS (Gra)	30.4
Poaceae		
Ehrharta rehmannii?	NS (Gra)	trace
Merxmuellera stricta	NS (Gra)	trace
Pentaschistis colorata	NS (Gra)	3.3
Restionaceae		
Hypodiscus aristatus	NS (Aph)	9.5
Staberoha cernua	NS (Aph)	7.6
		71.3

Evergreen, caespitose graminoid herbs (<25 cm)

Cyperaceae		
Ficinia sp.	NS (Gra)	trace

Evergreen, sparingly tufted, more or less rhizomatous graminoid herbs (<25 cm)

Restionaceae		
Willdenowia humilis	NS (Aph)	**3.8**

Seasonal rosette forbs (<25 cm)

Asteraceae		
Senecio erubescens	(Mic)	trace
Droseraceae		
Drosera aliceae	(Nan)	trace
		trace

Seasonally green geophytes (25–100 cm)

Iridaceae		
Gladiolus sp.	(Mic)	trace
Watsonia pyramidata	(Mes)	5.7
W. schlechteri	(Mes)	0.1
		5.8

Seasonally green geophytes (<25 cm)

Iridaceae		
Geissorhiza sp.	(Nan)	trace
Gladiolus sp.	(Mic)	trace
Liliaceae sp. (1)	(Mic)	trace
Oxalidaceae		
Oxalis sp. cf. *O. commutata*	(Nan)	trace
O. sp. cf. *O. melanostricta*	(Nan)	0.1
Indeterminate sp. (1)	(Mic)	trace
		0.1

SITE 6

Tall closed-herbland communities, Dwarsberg, Jonkershoek (mountain fynbos)

Latitude: 34 00'S **Longitude:** 19 01'E **Elevation:** 1234 m **Nearest climate station:** Biesievlei, Jonkershoek
Precipitation: 3330 mm **Soil:** Champagne **Age:** 35 years since fire **Total canopy cover:** 95–98%
Geology: TM quartzite **Reference:** enumeration by F.J. Kruger, June 1977
Structure: closed-herbland

	Leaf type and size	% of total canopy volume
1. *Chondropetalum esterhuyseniana* community		
Evergreen, tall caespitose graminoid herbs (>100 cm)		
Restionaceae		
Chondropetalum esterhuyseniana	NS (Aph)	**100.0**
Low, deciduous geophytes (<25 cm)		
Oxalidaceae		
Oxalis sp. cf. *O. commutata*	(Nan)	**trace**
2. *Elegia intermedia* community		
Evergreen, tall caespitose graminoid herbs (>100 cm)		
Restionaceae		
Elegia intermedia	NS (Aph)	**98.2**
Evergreen, mid-height caespitose graminoid herbs (25–100 cm)		
Restionaceae		
Hypolaena crinalis (hummocks)	NS (Aph)	**1.8**
Evergreen, mid-height rosette forbs (25–100 cm)		
Asteraceae		
Senecio crispus	(Mic)	trace

	Leaf type and size	% of total canopy volume
3. *Epischoenus adnatus* community		
Evergreen, mid-height caespitose graminoid herbs (25–100 cm)		
Cyperaceae		
Epischoenus adnatus	NS (Aph)	17.8
Poaceae		
Ehrharta setacea	NS (Gra)	13.4
Restionaceae		
Hypolaena crinalis (hummocks)	NS (Aph)	66.8
Restio echinatus	NS (Aph)	trace
		98.0
Evergreen, mid-height rosette forbs (25–100 cm)		
Asteraceae		
Senecio crispus	(Mic)	**2.00**
Evergreen, low caespitose graminoid herbs (<25 cm)		
Poaceae		
Pentaschistis sp.	NS (Gra)	**trace**

SITE 7

Low closed-heathland community (eastern grassy form of mountain fynbos)

Eastern graminoid mountain fynbos (*Erica demissa* heath), Grahamstown Nature Reserve

Latitude: 33°20'S **Longitude:** 23°20'E **Elevation:** ±700 m **Nearest climate station:** Grahamstown

Precipitation: ±600 mm **Soil:** Podzol (Glenrosa/ Cartref form?) **Age:** 22 years since fire **Total canopy cover:** ±90%

Geology: Witteberg quartzite colluvium

Reference: Martin (1965). List of species in representative communities with subjective classification in abundance classes: o=occasional; f=frequent; c=common; a=abundant; d=dominant; l=locally

Structure: low closed-heath

	Leaf type and size	Abundance
Evergreen, mid-height shrubs (1–2 m)		
Rubiaceae		
Anthospermum aethiopicum	NS (Lep)	f–c
Evergreen, low shrubs (25–100 cm)		
Acanthaceae		
Chaetacanthus setiger	?	f
Asteraceae		
Helichrysum felinum	BS (Nan)	f–c
Metalasia muricata	NS (Lep, Eri)	o–f
Ericaceae		
Erica chamissonis	NS (Lep, Eri)	o–c
E. demissa	NS (Lep, Eri)	a–d
Rosaceae		
Cliffortia linearifolia	NS (Lep)	c–d
C. repens	NS (Lep)	o–f
Thymelaeaceae		
Struthiola macowanii	BS (Lep)	o–f
Evergreen, caespitose dwarf-shrubs (<25 cm)		
Asteraceae		
Helichrysum appendiculatum	?	f
H. subglomeratum	?	c–a
Euphorbiaceae		
Clutia heterophylla	BS (Nan)	f
Rubiaceae		
Anthospermum paniculatum	NS (Lep)	o–f

	Leaf type and size	Abundance
Evergreen, mid-height caespitose graminoid herbs (25–100 cm)		
Cyperaceae		
Ficinia stolonifera	NS (Gra)	o–f
Poaceae		
Alloteropsis semialata	NS (Gra)	f–c
Eragrostis curvula	NS (Gra)	o–f
Pentaschistis angustifolia	NS (Gra)	f
Themeda triandra	NS (Gra)	c–a
Restionaceae		
Restio triticeus	NS (Aph)	c
Evergreen, mid-height sparingly tufted more or less rhizomatous graminoid herbs (25–100 cm)		
Restionaceae		
Restio sejunctus	NS (Aph)	c–f
Evergreen, low caespitose graminoid herbs (<25 cm)		
Cyperaceae		
Kobresia spartea	NS (Gra)	c
Deciduous, rosette forbs		
Dipsacaceae		
Scabiosa columbaria	(Mic)	c–f
Evergreen forbs		
Asteraceae		
Helichrysum nudifolium	BS (Mic)	c–a
Senecio othonnaeflorus	?	c

SITE 7 (continued)

	Leaf type and size	Abundance		Leaf type and size	Abundance
Evergreen geophytes			Dennstaedtiaceae		
Asteraceae			*Pteridium aquilinum*	NS (Pin–Lep)	o–c
Gerbera viridifolia	?	f	Hypoxidaceae		
Iridaceae			*Hypoxis obliqua*		
Bobartia burchellii	NS (Gra)	f	Oxalidaceae		
B. indica	NS (Gra)	l–a	*Oxalis imbricata* var. *violacea*	?	f–a
Deciduous geophytes			*O. punctata*	?	f–a
Apiaceae			*O. smithiana*	?	o–f
Pimpinella schlechteri	?	o–f	Annuals		
Asteraceae			Gentianaceae		
Berkheya carduoides	?	f	*Sebaea* sp.	(Lep–Nan)	f
B. decurrens	?	f			

Other species recorded as rare or occasional:

Monocotyledons
Orchidaceae	*Satyrium membranaceum*
Poaceae	*Andropogon appendiculatus*

Dicotyledons
Anacardiaceae	*Rhus dentata, R. eckloniana, R. fastigiata, R. lucida*
Asclepiadaceae	*Anisotoma cordifolia*
Asteraceae	*Disparago ericoides, Haplocarpha scaposa, Helichrysum petiolare*
Campanulaceae	*Wahlenbergia capillacea*
Celastraceae	*Maytenus heterophylla*
Ebenaceae	*Diospyros dichrophylla*
Fabaceae	*Argyrolobium stipulaceum, Psoralea spicata, Tephrosia grandiflora*
Gentianaceae	*Chironia melampyrifolia, C. tetragona*
Myrsinaceae	*Myrsine africana*
Ochnaceae	*Ochna serrulata*
Polygalaceae	*Polygala ohlendorfiana*
Rosaceae	*Rubus pinnatus*
Rubiaceae	*Anthospermum hedyotideum, Burchellia bubalina*
Scrophulariaceae	*Halleria lucida*
Thymelaeaceae	*Struthiola parviflora*

SITE 8

Arid fynbos community, Langfontein, Clanwilliam District

Latitude: 32°20'S **Longitude:** 18°55'E **Elevation:** 250 m **Nearest climate station:** Citrusdal Exp. Farm
Precipitation: 400 mm **Soil:** deep pale reddish sand (? Clovelly) **Age:** ±20 years **Total canopy cover:** ±40%
Geology: TM quartzite
Structure: arid tall shrubland with heathy understorey
Reference: enumeration by F.J. Kruger and H.C. Taylor, August, 1976

	Leaf type and size	% of total canopy volume
Evergreen, mid-height shrubs (1–2 m)		
Proteaceae		
Leucadendron loranthifolium	BS (Nan)	0.2
L. pubescens	BS (Nan)	62.1
Sapindaceae		
Dodonaea viscosa	BS (Nan)	0.3
		62.6
Drought-deciduous, mid-height shrubs (1–2 m)		
Anacardiaceae		
Rhus dissecta	BS (Nan)	2.5
Fabaceae		
Wiborgia sp.	BS (Lep)	0.2
		2.7
Evergreen, low shrubs (50–100 cm)		
Asteraceae		
Eriocephalus sp. cf. *E. africanus*	NS (Lep)	0.1
Fabaceae		
Aspalathus sp.	NS (Lep)	0.1
Polygalaceae		
Neylandtia spinosa (spinescent)	NS (Lep)	0.1
Rosaceae		
Cliffortia sp. cf. *C. pterocarpa*	NS (Lep)	0.2
Thymelaeaceae		
Passerina sp.	NS (Lep, Eri)	10.1
		10.6
Evergreen, low shrubs (25–50 cm)		
Asteraceae		
Arctotis sp.	(Semi-Suc, Mic)	0.1
Chrysocoma sp.	NS (Lep)	trace
Fabaceae		
Aspalathus sp.	NS (Lep)	trace
Proteaceae		
Protea glabra (young plant)	BS (Mic)	trace
		0.1
Evergreen, caespitose dwarf-shrubs (<25 cm)		
Asteraceae		
? *Elytropappus sp.*	NS (Lep, Eri)	trace
Felicia sp. (suffr.)	NS (Lep)	trace
Helichrysum sp.	BS (Lep)	trace
Mesembryanthemaceae sp.	(Suc, Lep)	trace
		trace
Evergreen, tall caespitose graminoid herbs (>100 cm)		
Restionaceae		
Willdenowia striata	NS (Aph)	18.8
Evergreen, mid-height caespitose graminoid herbs (25–100 cm)		
Cyperaceae		
Ficinia sp.	NS (Gra)	**trace**

SITE 8 *(continued)*

	Leaf type and size	% of total canopy volume
Deciduous, or sparingly evergreen, mid-height caespitose graminoid herbs (25–100 cm)		
Poaceae		
Ehrharta sp.	(Gra)	0.1
Eragrostis sp.	(Gra)	trace
		0.1
Deciduous, or sparingly evergreen, mid-height scapose graminoid herbs (25–100 cm)		
Poaceae		
Ehrharta sp.	(Gra)	**0.2**
Deciduous forbs with erect or scrambling leaves (<25 cm)		
Geraniaceae		
Pelargonium sp.	(Mic)	**trace**
Low, seasonal rosette forbs (<25 cm)		
Asteraceae		
Hypochoeris radicata	(Mic)	**trace**
Seasonally green mid-height geophytes (25–100 cm)		
Asclepiadaceae		
Microloma tenuifolia (twiner)	(Nan)	0.2
Lobeliaceae		
Cyphia sp. (twiner)	(Lep)	0.2
Polygonaceae		
Rumex sp.	(Mic)	trace
Indeterminate sp.		trace
		0.4
Seasonally green low geophytes (<25 cm)		
Iridaceae		
Babiana sp.	(Mic)	trace
Moraea sp.	(Mic)	trace
Liliaceae		
Lachenalia mutabilis	(Nan)	trace
Trachyandra sp.	(Mic)	trace

	Leaf type and size	% of total canopy volume
Oxalidaceae		
Oxalis spp.	(Lep–Nan)	trace
Indeterminate spp. (6)		trace
		trace
Seasonal, mid-height annuals (25–100 cm)		
Asteraceae		
Osteospermum sp.	(Mic)	trace
Senecio arenarius	(Mic)	0.8
Ursinia sp.	(Pin–Lep)	1.0
U. sp.	(Pin–Lep)	trace
Brassicaceae		
Heliophila sp.	(Lep)	trace
Indeterminate sp.		trace
		1.8
Seasonal, low annuals (<25 cm)		
Asteraceae		
Gnaphalium sp.	(Nan)	0.2
Indeterminate sp.		trace
Campanulaceae sp.	(Lep)	trace
Crassulaceae spp. (1)	(Lep)	trace
Cyperaceae		
Scirpus sp.	(Gra)	trace
Mesembryanthemaceae		
Drosanthemum sp.	(Suc, Mic)	0.4
Molluginaceae		
? *Pharnaceum* sp.	(Lep)	trace
Poaceae		
Pentaschistis sp.	(Gra)	trace
Indeterminate sp. (2)	(Gra)	trace
Scrophulariaceae sp.	(Lep)	trace
Selaginaceae		
Dischisma ciliatum	(Lep)	0.2
		0.8

SITE 9

Arid fynbos community near Ladismith

Latitude: 33°21'S **Longitude:** 21°18'E
Precipitation: ±250 mm **Soil:** Mispah
Geology: Witteberg quartzite
Structure: open graminoid heath

Elevation: 930 m
Age: undetermined, fire apparently long absent
Reference: enumeration by H.C. Taylor, F.J. Kruger and R.H. Whittaker, October 1975

No representative climate station
Total canopy cover: ±60%

Taxon	Leaf type and size	% of total canopy volume
Evergreen, tall shrubs (>2 m)		
Anacardiaceae		
Rhus undulata	BS (Nan)	0.3
Sapindaceae		
Dodonaea viscosa	BS (Mic)	0.5
Thymelaeaceae		
Passerina filiformis	NS (Eri)	0.8
		1.6
Santalaceae		
Thesium subnudum (partial root parasite)	NS (Lep)	trace
Sterculiaceae		
Hermannia odorata	BS (Nan)	trace
Zygophyllaceae		
Zygophyllum sessilifolium	(Semi-Suc, Nan)	1.0
		4.2
Evergreen, mid-height shrubs (1–2 m)		
Asteraceae		
Athanasia parviflora	(Semi-Suc, Lep)	1.8
Elytropappus rhinocerotis	(Cup)	23.8
		25.6
Drought-deciduous mid-height shrubs (1–2 m)		
Montiniaceae		
Montinia caryophyllacea	nanophyll	**0.3**
Evergreen, low shrubs (50–100 cm)		
Campanulaceae		
Lightfootia diffusa	NS (Lep)	1.6
Celastraceae		
Putterlickia sp. (spiny)	BS (Nan)	trace
Fabaceae		
Aspalathus sp.	NS (Lep)	trace
Mesembryanthemaceae sp.	(Suc, Lep–Nan)	trace
Polygalaceae		
Polygala microlopha	NS (Lep)	1.6
P. myrtifolia	NS (Lep)	trace
Rhamnaceae		
Phylica sp.	NS (Eri)	trace
Evergreen, low shrubs (25–50 cm)		
Tetragoniaceae		
Tetragonia sp. cf. *T. spicata*	(Suc, Mic)	1.0
Evergreen, creeping dwarf-shrubs (<25 cm)		
Mesembryanthemaceae sp. (2)	(Suc, Lep–Nan)	1.0
Evergreen, caespitose dwarf-shrubs (<25 cm)		
Asteraceae		
Chrysocoma tenuifolia (suffr.)	NS (Lep)	trace
Helichrysum interzonale (suffr.)	NS (Eri)	0.5
H. zeyheri (suffr.)	BS (Nan)	trace
H. spp. (2) (suffr.)	BS (Nan)	trace
Relhania squarrosa	BS (Nan)	trace
Senecio umbellatus (suffr.)	NS (Lep)	trace
S. sp.	NS (Lep)	trace
S. sp. cf. *S. aizoides*	(Suc, Nan)	trace
Stoebe sp.	(Cup)	trace
Crassulaceae		
Crassula muscosa	(Suc, Lep)	trace
C. pubescens	(Suc, Nan)	trace
C. sp. cf. *C. mollis*	(Suc, Nan)	trace
C. sp. (2)	(Suc, Nan–Mic)	trace
Mesembryanthemaceae		
Drosanthemum sp.	(Suc, Mic)	trace
Rubiaceae sp.	NS (Lep)	trace
		0.5

SITE 9 (continued)

	Leaf type and size	% of total canopy volume
Evergreen, tall caespitose graminoid herbs (>100 cm)		
Poaceae		
Merxmuellera sp.	NS (Gra)	19.0
Restionaceae		
Restio fruticosus	NS (Aph)	14.4
		33.4
Evergreen, tall sparingly tufted, somewhat rhizomatous graminoid herbs (>100 cm)		
Restionaceae		
Restio gaudichaudianus	NS (Aph)	**28.5**
Evergreen, mid-height caespitose graminoid herbs (25–100 cm)		
Cyperaceae		
Ficinia nigrescens	NS (Gra)	trace
Poaceae		
Cymbopogon marginatus	NS (Gra)	trace
Ehrharta calycina	NS (Gra)	1.3
Lasiochloa longifolia	NS (Gra)	0.4
Merxmuellera disticha	NS (Gra)	0.1
M. stricta	NS (Gra)	trace
Poaceae sp.	NS (Gra)	trace
		1.8
Deciduous, mid-height caespitose graminoid herbs (25–100 cm)		
Poaceae sp.	(Gra)	**trace**
Evergreen, mid-height sparingly tufted, somewhat rhizomatous graminoid herbs (25–100 cm)		
Poaceae		
Ehrharta ramosa	NS (Aph)	trace
Restionaceae sp.	NS (Aph)	trace
		trace

	Leaf type and size	% of total canopy volume
Deciduous, low, rosette forbs (<25 cm)		
Sinopteridaceae		
Cheilanthes sp. cf. C. multifida	(Pin–Lep)	**trace**
Deciduous, mid-height geophytes (25–100 cm)		
Iridaceae		
Tritonia bakeri	(Nan)	trace
Liliaceae		
Albuca sp. cf. A. temuifolia	(Nan)	trace
Ornithogalum sp. cf. O. thyrsoides	(Nan)	trace
Lobeliaceae		
Cyphia sp. (twiner)	(Lep)	trace
		trace
Deciduous, low geophytes (<25 cm)		
Apiaceae		
Annesorrhiza sp.	(Pin–Nan)	trace
Asteraceae		
Senecio sp.	(Lep)	trace
Oxalidaceae		
Oxalis sp.	(Lep)	trace
Indeterminate spp. (2)		trace
		trace
Low annuals (<25 cm)		
Asteraceae		
Gnaphalium sp.	(Lep)	trace
Helichrysum cylindricum	(Lep)	trace
Senecio sp.	(Nan)	trace
Ursinia sp.	(Pin–Lep)	trace
Asteraceae sp.	?	trace

Evergreen, low caespitose graminoid herbs (25 cm)

	Leaf type and size	% of total canopy volume
Cyperaceae		
Ficinia filiformis	NS (Gra)	trace

Deciduous low forbs (<25 cm)

	Leaf type and size	% of total canopy volume
Poaceae		
Aira cupaniana	(Gra)	trace
Poaceae sp.	(Gra)	trace
Brassicaceae		
Heliophila sp.	(Lep)	trace
Cucurbitaceae		
Zehneria scabra	(Nan)	**trace**

SITE 10

Tall open-heathland formation (coastal fynbos)

Limestone fynbos, Die Poort, Bredasdorp

Latitude: 34 35'S **Longitude:** 20 00'E **Soil:** Mispah
Precipitation: 400–500 mm
Geology: coastal limestone
Structure: tall open-heath

Elevation: 60 m
Age: 25+ years since fire
Reference: enumeration by F.J. Kruger, H.C. Taylor and R.H. Whittaker, October 1975

No representative climate station
Total canopy cover: 80%

	Leaf type and size	% of total canopy volume
Evergreen, mid-height shrubs (1–2 m)		
Proteaceae		
Leucadendron meridianum	BS (Nan)	54.7
Protea obtusifolia	BS (Mic)	12.2
Rubiaceae		
Anthospermum aethiopicum	NS (Lep)	0.1
		67.0
Rutaceae		
Adenandra obtusata	BS (Nan)	2.2
Euchaetis bolusii	BS (Nan)	trace
Stilbaceae		
Stilbe ericoides	NS (Eri)	4.3
Thymelaeaceae		
Gnidia sp.	BS (Nan)	trace
		23.4

	Leaf type and size	% of total canopy volume
Evergreen, low shrubs (25–50 cm)		
Asteraceae		
Helichrysum sp.	NS (Lep)	trace
Fabaceae		
Amphithalea sp.	BS (Nan)	trace
Indigofera brachystachya	BS (Nan)	0.4
Geraniaceae		
Pelargonium betulinum	BS (Nan)	trace
Verbenaceae		
Chascanum cernuum	BS (Mic)	1.0
		1.4

	Leaf type and size	% of total canopy volume
Evergreen, low shrubs (50–100 cm)		
Anacardiaceae		
Rhus lucida	BS (Mic)	trace
Asteraceae		
Eroeda capensis	NS (Lep)	0.9
Metalasia muricata	NS (Eri)	6.7
Ericaceae		
Erica mariae	NS (Eri)	trace
E. oblongiflora	NS (Eri)	0.7
E. plukenetii	NS (Eri)	trace
E. scytophylla	NS (Eri)	4.9
Rhamnaceae		
Phylica purpurea	NS (Eri)	3.7

SITE 10 *(continued)*

	Leaf type and size	% of total canopy volume
Evergreen, caespitose dwarf-shrubs (<25 cm)		
Asteraceae		
Senecio burchellii	NS (Lep)	trace
Fabaceae		
Argyrolobium sp.	BS (Nan)	trace
Myricaceae		
Myrica quercifolia	BS (Mic)	0.3
Rubiaceae		
Carpacoce scabra	NS (Lep)	0.6
		0.9
Evergreen, creeping dwarf-shrubs (<25 cm)		
Mesembryanthemaceae		
Carpobrotus acinaciformis	(Suc, Mic)	**trace**
Evergreen, caespitose mid-height graminoid herbs (25–100 cm)		
Cyperaceae		
Tetraria cuspidata	NS (Aph)	0.4
Poaceae		
Pseudopentameris macrantha	NS (Gra)	trace
Restionaceae		
Hypodiscus striatus	NS (Aph)	trace
Thamnochortus fraternus	NS (Aph)	1.2
		1.6
Evergreen, mid-height sparingly tufted, more or less rhizomatous graminoid herbs (25–100 cm)		
Restionaceae		
Chondropetalum microcarpum	NS (Aph)	trace
Elegia sp.	NS (Aph)	1.2
Restio triticeus	NS (Aph)	3.7
		4.9

	Leaf type and size	% of total canopy volume
Evergreen, low caespitose graminoid herbs (<25 cm)		
Cyperaceae		
Ficinia truncata	NS (Gra)	trace
Evergreen, low forbs (<25 cm)		
Hydrocotylaceae		
Centella difformis	BS (Nan)	trace
Evergreen, mid-height geophytes (25–100 cm)		
Iridaceae		
Bobartia sp.	NS (Gra)	trace
Seasonally green, mid-height geophytes (25–100 cm)		
Iridaceae		
Gladiolus sp.	(Nan)	trace
Seasonally green, low geophytes (<25 cm)		
Euphorbiaceae		
Euphorbia sp. cf. *E. silenifolia*	(Nan)	trace
Indeterminate sp.		trace
		trace
Seasonal, low annuals (<25 cm)		
Crassulaceae		
Crassula decumbens	(Lep)	trace

REFERENCES

Acocks, J.P.H., 1953. Veld types of South Africa. *Mem. Bot. Surv. S. Afr.*, 28: 192 pp.

Adamson, R.S., 1931. The plant communities of Table Mountain II. Life-form dominance and succession. *J. Ecol.*, 19: 304–320.

Adamson, R.S., 1935. The plant communities of Table Mountain III. A six years' study of regeneration after burning. *J. Ecol.*, 23: 44–53.

Adamson, R.S., 1938a. *The Vegetation of South Africa*. British Empire Vegetation Committee, London, 235 pp.

Adamson, R.S., 1938b. Notes on the vegetation of the Kamiesberg. *Mem. Bot. Surv. S. Afr.*, 18.

Andrag, R.H., 1977. *'n Natuurbewaringsplan vir die Sederberg-staatsbos met besondere verwysing na die invloed van vuur op Widdringtonia cedarbergensis en na buitelugontspanning*. Thesis, University of Stellenbosch, Stellenbosch.

Aschmann, H., 1973. Distribution and peculiarity of Mediterranean ecosystems. In: F. di Castri and H.A. Mooney (Editors), *Mediterranean Type Ecosystems. Origin and Structure*. Springer Verlag, Berlin, pp. 11–19.

Bailey, H.P., 1958. A simple moisture index based upon a primary law of evaporation. *Geogr. Ann.*, 40: 196–215.

Blommaert, K.L.J., 1972. Buchu seed germination. *J. S. Afr. Bot.*, 38: 237–239.

Boucher, C., 1972. *The Vegetation of the Cape Hangklip Area*. Thesis, University of Cape Town, Cape Town.

Boucher, C. and Jarman, M.L., 1977 The vegetation of the Langebaan area. *Trans. R. Soc. S. Afr.*, 42: 241–272.

Dahlgren, R., 1963. Studies on *Aspalathus*: Phytogeographical aspects. *Bot. Not.*, 116: 431–472.

Ellis, F., 1973. *Soil Studies in the Duiwenhoks River Catchment Area*. Thesis, University of Stellenbosch, Stellenbosch.

Glyphis, J., Moll, E.J. and Campbell, B.M., 1978. Phytosociological studies on Table Mountain, South Africa: 1. The Back Table. *J. S. Afr. Bot.*, 44: 281–289.

Hall, A.V., 1959. Observations on the distribution and ecology of Orchidaceae in the Muizenberg Mountains, Cape Peninsula. *J. S. Afr. Bot.*, 25: 265–278.

Hendey, Q.B., 1973. Fossil occurrences at Langebaanweg, Cape Province. *Nature, Lond.*, 244: 13–14.

Hubbard, C.S., 1937. Observation on the distribution and rate of growth of Clanwilliam Cedar, *Widdringtonia juniperoides* Endl. *S. Afr. J. Sci.*, 33: 572–586.

Jackson, D.S., 1962. Parameters of site for certain growth components of slash pine (*P. elliottii* Engelm.). *Duke Univ. School For. Bull.*, 16.

Jackson, S.P. and Tyson, P.D., 1971. Aspects of weather and climate over Southern Africa. *Environment. Stud., Occas. Pap. 6, Univ. Witwatersrand*.

Jessop, J.P. and Jacot Guillarmod, A., 1969. The vegetation of the Thomas Baines Nature Reserve. *J. S. Afr. Bot.*, 35: 367–392.

Jordaan, P.G., 1949. Aantekeninge oor die voortplanting en brandperiodes vir *Protea mellifera* Thunb. *J. S. Afr. Bot.*, 15: 121–125.

Jordaan, P.G., 1965. Die invloed van 'n winterbrand op die voort-planting van vier soorte van die Proteaceae. *Tydskr. Natuurwet.*, 5: 27–31.

Joubert, G.P.J., 1965. *Die verband tussen opstandsboniteit van Pinus radiata en grondeienskappe*. Thesis, University of Stellenbosch, Stellenbosch.

Kerfoot, O., 1968. Mist precipitation on vegetation. *For. Abstr.*, 29: 8–20.

Killick, D.J.B., 1963. An account of the plant ecology of the Cathedral Peak area of the Natal Drakensberg. *Mem. Bot. Surv. S. Afr.*, 34: 178 pp.

King, L.C., 1963. *South African Scenery, a Textbook of Geomorphology*. Oliver and Boyd, 3rd ed., Edinburgh and London, 308 pp.

Klein, R.G., 1974. Environment and subsistence of prehistoric man in the Southern Cape Province, South Africa. *World Archaeol.*, 5: 249–284.

Köppen, W., 1923. *Die Klimate der Erde*. Bornträger, Berlin, 369 pp.

Kruger, F.J., 1974. *The Physiography and Plant Communities of Jakkalsrivier Catchment*. Thesis, University of Stellenbosch, Stellenbosch.

Kruger, F.J., 1977. A preliminary account of aerial plant biomass in fynbos communities of the Mediterranean type climate zone of the Cape Province. *Bothalia*, 12: 301–307.

Lambrechts, J.J.N., 1964. *A Chemical and Mineralogical Study of a Soil Profile near George*. Thesis, University of Stellenbosch, Stellenbosch.

Lambrechts, J.J.N., 1975. Podzol B horizons in the South-western and Southern Cape soils. *Proc. Sixth. Congr. Soil Sci. Soc. Afr.*

Le Roux, P.J., 1969. *Brandbestryding in Suidp-Kaapland met spesiale verwysing na chemiese metodes van beheer*. Thesis, University of Stellenbosch, Stellenbosch.

Levyns, M.R., 1929. Veld-burning experiments at Ida's Valley, Stellenbosch. *Trans. R. Soc. S. Afr.*, 17: 61–92.

Levyns, M.R., 1935. Veld burning experiments at Oakdale, Riversdale. *Trans. R. Soc. S. Afr.*, 23: 231–243.

Levyns, M.R., 1956. Notes on the biology and distribution of the Rhenoster bush. *S. Afr. J. Sci.*, 52: 141–143.

Levyns, M.R., 1966. *Haemanthus canaliculatus*, a new fire-lily from the Western Cape province. *J. S. Afr. Bot.*, 32: 73–75.

Levyns, M.R., 1970. A revision of the genus *Paranomus* (Proteaceae). *Contrib. Bol. Herb.*, 2: 1–48.

Lückhoff, H.A., 1963. Die Baviaanskloof- of Wollowmore-seder. *For. S. Afr.*, 3: 1–14.

Lückhoff, H.A., 1971. The Clanwilliam cedar (*Widdringtonia cedarbergensis* Marsh). *J. Bot. Soc. S. Afr.*, 57: 17–23.

Marker, M.E., 1976. Aeolianite: Australian and Southern African deposits compared. *Ann. S. Afr. Mus.*, 71: 115–124.

Marloth, R., 1902. Notes on the occurrence of alpine types in the vegetation of the higher peaks of the south-western districts of Cape Colony. *Trans. S. Afr. Philos. Soc.*, 11: 161–168.

Marloth, R., 1904. results of experiments on Table Mountain for ascertaining the amount of moisture deposited from the south east clouds. *Trans. S. Afr. Philos. Soc.*, 14: 403.

Marloth, R., 1907a. On some aspects in the vegetation of South Africa which are due to prevailing winds. *Rep. S. Afr. Assoc. Adv. Sci.*, 1905 and 1906: 215–218.

Marloth, R., 1907b. results of further experiments on Table Mountain for ascertaining the amount of moisture deposited from the south-east clouds. *Trans. S. Afr. Philos. Soc.*, 16: 97–105.

Martin, A.R.H., 1965. Plant ecology of the Grahamstown Nature reserve. I. Primary communities and plant succession. *J. S. Afr. Bot.*, 31: 1–54.

Martin, A.R.H., 1966. The plant ecology of the Grahamstown Nature Reserve. II. Some effects of burning. *J. S. Afr. Bot.*, 32: 1–39.

Michell, M.R., 1922. Some observations on the effects of a bush fire on the vegetation of Signal Hill. *Trans. R. Soc. S. Afr.*, 10: 213–232.

Mossop, E.E., 1927. *Old Cape Highways*. Maskew Miller, Cape Town, 202 pp.

Muir, J., 1929. The vegetation of the Riversdale area, Cape Province. *Mem. Bot. Surv. S. Afr.*, 13.

Nagel, J.F., 1956. Fog precipitation on Table Mountain. *J. R. Meteorol. Soc.*, 82: 452–460.

Nagel, J.F., 1962. Fog precipitation measurements on Africa's south-west coast. *Notos*, 11: 51–60.

Neethling, J.H., 1970. *Classification of Some Forest Soils of the Southern Cape*. Thesis, University of Stellenbosch, Stellenbosch.

Phillips, J.F.V., 1931. Forest-succession and ecology in the Knysna region. *Mem. Bot. Surv. S. Afr.*, 14.

Rourke, J.P., 1972. Taxonomic studies on *Leucospermum* R. Br. *J. S. Afr. Bot. Suppl.*, 8.

Rumney, G.R., 1968. *Climatology and the World's Climate*. Macmillan, New York, N.Y., 656 pp.

Rust, I.C., 1967. *On the Sedimentation of the Table Mountain Group in the Western Cape Province*. Thesis, Univeristy of Stellenbosch, Stellenbosch.

Schulze, B.R., 1965. *Climate of South Africa. Part 8. General Survey*. S.A. Weather Bur. 28, Government Printer, Pretoria, 330 pp.

Schulze, R.E. and McGee, O.S., 1978. Climatic indices and classification in relation to the biogeography of Southern Africa. In: M.J.A. Werger (editor), *The Biogeography and Ecology of Southern Africa*. W. Junk, The Hague, pp. 19–52.

Schweitzer, F.R. and Scott, K.J., 1973. Early occurrence of domestic sheep in sub-Saharan Africa. *Nature, Land.*, 241: 547.

Siesser, W.G., 1972. Petrology of the Cainozoic Coastal Limestones of Cape Province, S.A. *Trans. Geol. Soc. S. Afr.*, 75: 178–189.

Soil Classification Working Group, 1977. *Soil Classification. A Binomial System*. Dep. Agric. Tech. Serv., Pretoria, 152 pp.

Story, R., 1952. Botanical survey of the Keiskammahoek District. *Mem. Bot. Surv. S. Afr.*, 27: 184 pp.

Tankard, A.J., 1976. Pleistocene history and coastal morphology of the Ysterfontein–Elands bay area, Cape Province. *Ann. S. Afr. Mus.*, 69: 73–119.

Taylor, H.C., 1963. A bird's-eye view of the Cape mountain vegetation. *J. Bot. Soc. S. Afr.*, 49: 17–19.

Taylor, H.C., 1969. *A Vegetation Survey of the Cape of Good Hope Nature Reserve*. Thesis, University of Cape Town, Cape Town.

Taylor, H.C., 1972. Notes on the vegetation of the Cape Flats. *Bothalia*, 10: 637/ 646.

Taylor, H.C., 1978. Phytogeography and ecology of Capensis. In: M.J.A. Werger (Editor), *The Biogeography and Ecology of Southern Africa*. W. Junk, The Hague, pp. 171–229.

Trollope, W.S.W., 1973. Fire as a method of controlling macchia (fynbos) vegetation on the Amatole mountains of the Eastern Cape. *Proc. Grassland. Soc. S. Afr.*, 8: 35–41.

Tyson, P.D., 1964. Berg winds of South Africa. *Weather*, 19: 7–11.

Van der Merwe, C., 1976. *'n Plantegroei opname van die De Hoop-Natuurreservaat*. Report, Cape Provincial Department of Nature and Environmental Conservation (unpublished).

Van der Merwe, P., 1966. Die flora van Swartboschkloof, Stellenbosch en die herstel van die soorte na 'n brand. *Ann. Univ. Stellenbosh*, 41, Ser. A., No. 14: 691–736.

Van Staden, J., 1966. Studies on the seed of Proteaceae. *J. S. Afr. Bot.*, 32: 291–298.

Von Gadow, K., 1977. *Management Plan for the Sundays River Catchment Area (Portion of Suurberg State Forest)*. Report, Department of Forestry (unpublished).

Walsh, B.N., 1968. Some notes on the incidence and control of drift-sands along the Caledon, Bredasdorp and Riversdale coastline of South Africa. *Dep. For. Bull.*, 44.

Wellington, J.H., 1955. *Southern Africa, A Geographical Study. Volume I: Physical Geography*. Cambridge University Press, Cambridge.

Werger, M.J.A., Kruger, F.J. and Taylor, H.C., 1972. A phytosociological study of the Cape fynbos and other vegetation at Jonkershoek, Stellenbosch. *Bothalia*, 10: 599–614.

West, O., 1965. *Fire in Vegetation and Its Use in Pasture Management with Special Reference to Tropical and Subtropical Africa*. Commonwealth Bureau of Pastures and Field Crops, Hurley, Berks, 53 pp.

Wicht, C.L., 1945. *Preservation of the vegetation of the South-western Cape*. Spec. Publ., R. Soc. S. Afr., Cape Town, 56 pp.

Wicht, C.L., 1948. A statistically designed experiment to test the effects of veldburning on a sclerophyll scrub community. I. Preliminary account. *Trans. R. Soc. S. Afr.*, 21: 479–501.

Wicht, C.L. and De Villiers, Y.R., 1963. Weerstoestande en brandgevaar by Hermanus. *J. Geogr.*, 2: 25–26.

Wicht, C.L., Meyburgh, J.C. and Boustead, P.G., 1969. Rainfall at the Jonkershoek Forest Hydrological Research Station. *Ann. Univ. Stellenbosch*, 44, Ser. A, No. 1: 1–66.

Williams, I.J.M., 1972. A revision of the genus *Leucadendron*. *Contrib. Bol. Herb.*, 3: 1–425.

Chapter 3

ASPECTS OF VERTEBRATE LIFE IN FYNBOS, SOUTH AFRICA[1]

R.C. BIGALKE

INTRODUCTION

The Southwest Cape (Fig. 3.1) region of southern Africa is considered to be a subdivision of the Ethiopian zoogeographic region (Davis, 1963), but it is not one of major importance. Certainly it is by no means as distinct faunistically as floristically (see Kruger's account of the floristics in Chapter 2). Perhaps of greatest zoogeographical importance is the occurrence of primitive relict invertebrates, the palaeogenic element of the fauna. Their affinities are with groups similarly limited to the tips of the other southern continents. These "old southern" temperate groups (Keast, 1972) include some Diptera, the Megaloptera and Onychophora, some Coleoptera and others. They are discussed by Stuckenberg (1962), Saiz (1973) and Vitali-di Castri (1973). Like the temperate Cape Province flora, these invertebrates are a legacy of Gondwanaland.

AMPHIBIA

The vertebrate fauna of the Southwest Cape lacks comparable distinctiveness. This is largely because it evolved after the separation of the southern continents (Darlington, 1957), but the region is nonetheless a minor centre of endemism for all vertebrate groups. This is most marked in the case of the Amphibia. Poynton (1960) reports 19 of the 26 species to be endemic, a fact which Winterbottom (1968b) relates to the great age of this class, since among the most modern vertebrates, birds and mammals, as we shall see, the degree of endemism is low. Indeed Poynton (1960) considers the South African amphibian fauna to

have only two major constituents, one of them centred on the Southwest Cape.

Greig (1976) gives a somewhat different number of Southwest Cape amphibian species, partly no doubt due to taxonomic changes and partly to differences in boundaries chosen. He lists nine species, belonging to the genera *Xenopus*, *Heleophryne*, *Breviceps*, *Microbatrachella*, *Cacosternum* and *Arthroleptella*, which are believed to be totally confined to the region. Three of them, *Heleophryne rosei*, *Arthroleptella lightfooti* and *Breviceps montanus* (possibly) appear to be confined to fynbos proper, veld type 69 of Acocks (1975). The remaining five (*Heleophryne*, *Bufo*, *Breviceps* and *Hyperolius* spp.) present 'in the Southwest Cape are not confined to it. The extent to which the distribution of these frogs is directly related to vegetation *per se* is questionable, but it is convenient to describe it in terms of established zones.

REPTILIA

The reptile fauna is quite large but poorly known. *Psammobates geometricus* is a rare Southwest Cape endemic tortoise. It may have been distributed mainly in the lowlands once covered with coastal rhenosterbosveld and coastal fynbos (Acocks, 1975, veld types 46 and 47) and now mainly under wheat and vineyards. The other five species of tortoise also occur in adjacent areas along the coast or in the karoo (Greig, 1976). Fitzsimons (1970) lists 32 species of snakes for the western Cape Province. Two subterranean forms,

---

[1] Manuscript completed March, 1977.

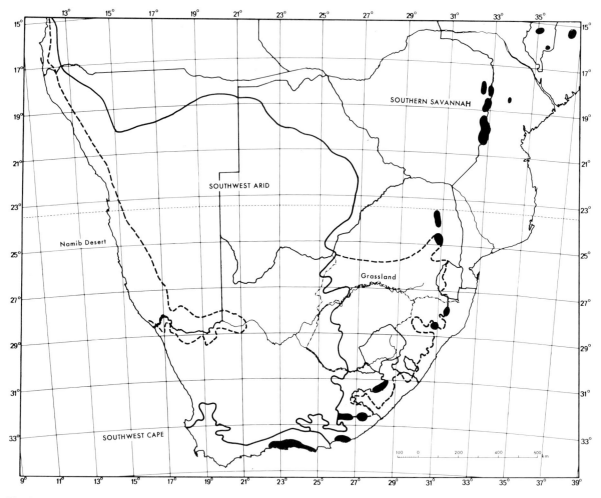

Fig. 3.1. The main biotic zones of southern Africa. Forest patches are indicated in black. (After Meester, 1965.)

Typhlops verticalis and *Leptotyphlops gracilior*, are endemic. Lizard taxonomy is outdated and distributional data are poor. According to Fitzsimons (1943) there are 51 species and subspecies of lizards in the Southwest Cape and 17 of them are endemic. However, as Greig (1976) points out, much of the taxonomy is probably now invalid. Also the variety of lizard habitats available leads one to expect a higher degree of endemism among rupicolous and fossorial species.

AVIFAUNA

Fynbos birds are by far the best-known group. Winterbottom (1968b) has shown that the generalization of Moreau (1966) "It is also noteworthy

that, although the South-west winter-rains corner of the Cape province is so rich in plant endemics .... no species of birds are similarly limited" is only true of the fynbos ("Macchia") *sensu stricto* of Acocks (1975, type 69) in the winter-rainfall region. However to confine attention to the area merely because most rain falls there in winter, he argues, is to obscure the fact that fynbos (*sensu lato*) as a physiognomic vegetation type supports a distinct avifauna with some, albeit not many, endemics.

Winterbottom (1972) treats fynbos ("Macchia") as one of three zoogeographic subdistricts of his "South Temperate District". The other two are "Highveld" and "Karoo". The "Macchia" subdistrict embraces several of Acocks' veld types, the four main ones being "Macchia", "Coastal Macchia", "Coastal rhenosterbosveld" and

"Strandveld" while "Dense Protea", really climax "Macchia", has been treated as an additional subtype in ornithological studies. The "Macchia" subdistrict is more or less equivalent to the Southwest Cape biotic zone. Winterbottom's classification reflects the fact that the avifauna is most strongly related to that of the karoo, and has almost as much in common with that of the highveld grasslands of the Orange Free State, Basutoland and the eastern Cape Province (Winterbottom, 1968a). Although it is probably derived from that of the karoo, this author admits the possibility that the opposite may perhaps be true. In any event, fynbos and karoo faunas have persisted as separate entities throughout the

Pleistocene. There are also a few temperate forest species, and a few that have evolved *in situ*.

The regular fauna (defined as species present in 5% or more of the lists for at least one habitat) of the five main habitats numbers 101 bird species (Winterbottom, 1972). Coastal rhenosterbosveld with 77 species is the richest of them. The dominant species listed by this author are presented in Table 3.1. Small insectivores which forage in thick cover, categorised as the "wren guild" by Cody (1975), are particularly prominent, with five species of warblers (Sylviidae), and one of *Prinia*. Small seed-eaters (Ploceidae, Fringillidae), nectivorous sunbirds (Nectariniidae) and robins and thrushes (Turdidae) are also quite well represented.

TABLE 3.1

Dominant birds of fynbos-type habitats (after Winterbottom, 1972)

Species	"Dense Protea"	Fynbos	Coastal fynbos	Coastal renoster-bosveld	Strand-veld
Game birds, etc.					
Phasianidae					
Francolinus capensis	–	–	×	–	–
Otididae					
Eupodotis afra	–	–	–	×	–
Other non-passerines					
Columbidae					
Streptopelia capicola	–	×	×	×	×
Coliidae					
Colius colius	–	–	–	–	×
Passerines					
Fringillidae					
Emberiza capensis	–	–	–	–	×
Serinus albigularis	–	–	–	–	×
S. *canicollis*	–	×	–	–	–
S. *flaviventris*	–	–	–	–	×
Laniidae					
Lanius collaris	–	–	×	×	×
Malaconotus zeylonus	–	×	×	×	×
Nectariniidae					
Nectarinia chalybea	–	–	×	×	×
N. *famosa*	×	×	–	×	–
N. *violacea*	×	×	–	–	–

TABLE 3.1 *(continued)*

Species	"Dense Protea"	Fynbos	Coastal fynbos	Coastal renoster-bosveld	Strand-veld
Ploceidae					
Passerinae					
Passer melanurus	–	–	×	–	–
Ploceinae					
Euplectes capensis	–	–	–	×	–
Priniidae					
Prinia maculosa	×	×	×	×	×
Promeropidae					
Promerops cafer	×	×	–	–	–
Pycnonotidae					
Pycnonotus capensis	–	–	×	–	–
Sturnidae					
Onychognathus morio	–	×	–	–	–
Spreo bicolor	–	–	×	×	×
Sylviidae					
Apalis thoracica	–	–	×	–	×
Cisticola fulvicapilla	–	×	–	–	–
C. subruficapilla	–	–	×	×	×
Sphenoeacus afer	–	×	×	–	–
Sylvietta rufescens	–	–	–	–	×
Turdidae					
Cossypha caffra	×	×	×	×	×
Erythropygia coryphaeus	–	–	×	×	×
Saxicola torquata	–	–	×	×	–
Zosteropidae					
Zosterops pallidus	×	×	–	–	–

Only the Cape robin (*Cossypha caffra*) and *Prinia maculosa* are dominant in all five habitats. A dove (*Streptopelia*), a shrike (*Malaconotus*), and the Cape bunting (*Emberiza capensis*) are dominant in all but "dense Protea".

There are six endemic birds: the Cape francolin (*Francolinus capensis*), the rufous rock jumper (*Chaetops frenatus*), Victorin's warbler (*Bradypterus victorini*), the orange-breasted sunbird (*Anthobaphes violacea*, syn. *Nectarinia violacea* according to McLachlan and Liversidge, 1970), the Cape sugarbird (*Promerops cafer*) and the white-winged seed-eater (*Serinus leucopterus*). There are also twelve endemic subspecies; but while five of the six endemic species are essentially birds of the true fynbos, this is true of only four of the endemic subspecies (Winterbottom, 1972).

Winterbottom (1966a) has classified regular bird species of the fynbos by feeding habits (Table 3.2). The differences between veld types are not very great. There is an increasing number and percentage of insect-eaters hunting on foot from "Dense Protea" to coastal renosterbosveld. Coastal fynbos and coastal renosterbosveld have a large number of species of mixed diet, indicating the more specialised nature of the fynbos avifauna. "Dense Protea", the climax stage of fynbos and the poorest habitat, has very few insect-eaters foraging on the wing and on foot, as is to be expected in thick bush. The role of Proteaceae, many of them with large, showy flowers, in providing animal food, and their possible dependence on animals for pollination, is of particular interest. Winterbottom (1966a) shows that the six most frequent species in "Dense

TABLE 3.2

Regular fynbos bird species classified by habitat and food (after Winterbottom, 1966a)

Food	"Dense Protea"	Fynbos	Coastal fynbos	Coastal renoster-bosveld
Predators				
large	0	1	1	1
medium	1	2	3	2
small	0	2	3	2
Insect-eaters				
foraging on wing	2	5	6	7
hunting from perch	1	3	1	2
hunting on foot	2	7	11	15
probing flowers	5	4	4	4
probing ground	0	0	1	0
hunting in vegetation	9	9	11	12
hunting on the tracks	0	0	0	0
Seed-eaters	12	6	12	14
Fruit-eaters	3	2	5	2
Birds of mixed diet	3	8	16	14
Total	*33*	*55*	*75*	*75*

Protea" include two endemics, the sugarbird (*Promerops cafer*) and the orange-breasted sunbird (*Nectarinia violacea*), as well as the malachite sunbird (*N. famosa*), the karoo prinia (*Prinia maculosa*), the Cape robin (*Cossypha caffra*) and the Cape white-eye (*Zosterops pallidus*, syn. *Z. virens* according to McLachlan and Liversidge, 1970). The sugarbird, sunbirds and the white-eye feed on nectar and insects, the prinia is insectivorous and the robin omnivorous.

The fynbos avifauna is poor in species compared to several savannah vegetation types, especially in the "other non-passerine" category (Table 3.3). It does not differ much from karoo, however. The paucity of endemics is shown by comparison with *Brachystegia* woodland. Winterbottom (1968b) gives a total of 94 fynbos species to 112 for *Brachystegia* woodland but there are only 6 endemics against 23 in *Brachystegia* woodland. Since the fynbos flora is certainly as old as the flora of *Brachystegia* woodland and probably a good deal older (Levyns, 1961), the paucity of endemics in the former cannot be explained by differences in time available for differentiation.

Let us consider first the total number of bird species. The work of MacArthur (1965) and members of his school (e.g. Cody, 1973) has drawn attention to the close relationship between number of bird species and structural complexity of a habitat. The latter is indicated very simply by vegetation height. Foliage profile, which expresses distribution of vegetation at various heights, is another important index. It is obtained by plotting vegetation density in a horizontal plane against height above the ground. A convenient index is vegetation "half height", that height which divides the area under the foliage profile into two equal sections. Cody (1975) used the logarithms of the principal component of vegetation height and half-height as a single statistic, H, to express structural diversity. Along a gradient from grassland to forest, H increases continuously.

Fynbos vegetation is scrub, lower, structurally less diverse, and thus with lower H values, than steppe or woodland. It must therefore be expected to support fewer bird species than the steppe and woodland communities with which it is compared in Table 3.3. The similarity between the three

TABLE 3.3

Bird communities of sample areas in various African vegetation types (after Winterbottom, 1968b)

Vegetation type	Number of species			
	game birds etc.	other non-passerines	passerines	total
Acacia steppe	6–16	16–29	30–48	63–86
Brachystegia woodland	3	24–30	43–56	71–89
Colophospermum (Mopane) woodland	4–5	14–21	24–26	44–50
Fynbos	3–4	7	27–35	38–45
coastal fynbos	3–8	7–9	28–42	38–59
coastal rhenosterbosveld	7	8	32	47
Karoo	2–6	6–13	23–35	32–51

Note: The first three sets of figures are from Moreau (1966), the last four from Winterbottom (1966a, b).

fynbos veld types, and between them and karoo, in number of bird species is consistent with similarities in vegetation structure.

Three other specific factors may have a bearing on the poverty of the fynbos avifauna. Many birds in the "other non-passerine" group depend on fruit and large insects, which become progressively scarcer as one proceeds away from the equator. Some such as hornbills, barbets, rollers and woodpeckers nest in holes in trees, which are rare in the Southwest Cape (Winterbottom, 1968b). Finally Winterbottom (1968a) has pointed out that fynbos is poor in seeds suitable for and available to birds.

How does the Cape avifauna compare with that of other Mediterranean heathlands in number of species, population density and number of endemics? Cody (1975) studied small plots (1.6–4 ha) situated along gradients from short to tall vegetation in the Cape Province, California and Chile. The number of bird species, S, in a plot is described as "alpha diversity" or "species packing level" and is a measure of basic diversity in a uniform habitat. Values from all three regions were in close agreement for plots at similar points along the H gradient. Bird diversity in Cape fynbos is no lower than in other comparable Mediterranean heathlands. In one respect it is indeed richer, since scrub habitats in the H range 0.4 to 1 were found to

support from 1 to 5 more bird species in the Cape Province than in California or Chile.

Comparison of "beta diversity" revealed an interesting difference. This is a measure of the rate at which species are lost and gained as the observer moves systematically between different habitats, or "the diversity component attributable to habitat change between census sites" (Cody, 1975). Beta-diversity graphs show a high, early peak in Cape habitats ($S=38$; $H=0.65$), whereas in California and Chile the peak comes successively later and lower. The interpretation of this observation is that species replace one another rapidly as one moves from one scrub habitat to another in the Cape province. The implication is that bird species there have become stereotyped to a narrow habitat range. The niche breadth was also found to be narrowest there. On the other hand birds of low beta-diversity faunas, notably Chile, have the largest niche breadths and are behaviourally more plastic. This enables them, for instance, to use exotic *Eucalyptus* and *Pinus* plantations which are almost completely disregarded by Cape birds.

The abundance of birds — total population density — in comparable habitats of equal H was also found to be remarkably similar in the three regions in which Cody (1975) worked. This he interprets as the product of similar overall productivity.

We have yet to deal with possible reasons for the scarcity of endemic birds in the Cape Province. Cody (1975) has examined the extent of "source areas" adjacent to Mediterranean heathlands, i.e. areas of similar habitat from which appropriate colonists could be provided in case of extinctions. He finds a clear correlation between area of source habitat and alpha diversity, that is more species in larger areas of similar habitat.

Furthermore, beta diversity at a given H correlates quite well with the proportion that a given habitat contributes to the total source area. In the case of southern Africa there are large areas of short-scrub habitats, and the Mediterranean zone supports a large number of scrub insectivores (Silviidae and Priniidae). These are serially arrayed over habitats, and thus show high beta diversity. Chile and California have less scrub, and there are many fewer of these habitat-specialist insectivores in ecologically equivalent rôles.

These observations may be interpreted to suggest that there have been sufficient opportunities for faunal exchange between the Cape Province heathlands and extensive adjacent areas of structurally similar habitat to preclude the development of a high degree of endemism. Winterbottom's observation that the fynbos avifauna is most closely related to that of the karoo, and also related to that of the highveld, has already been noted. Put in another way, the uniqueness of the fynbos has not led to the evolution of a comparably unique avifauna mainly because the vegetation is not structurally unique, and because the region has been in communication with large source areas of comparable structure. Winterbottom (1968b) suggests further that the low rate of endemism is likely to be due to the same causes "whatever they are, which have been responsible for the reduction in size of the avifauna as one proceeds from low latitudes to high; with the relatively lesser ecological diversity of the vegetation type as a contributory factor."

It is also significant that four of the six endemic bird species of the fynbos are, or allied to, montane forms (Winterbottom, 1968a). Most of the "Macchia" veld type of Acocks is montane. Two of them, *Promerops cafer* and *Anthobaphes violacea*, may well have evolved with the Proteaceae and Ericaceae. Broekhuysen (1966) in particular has drawn attention to the dependence of the sugarbird

(*P. cafer*) on proteas and of the orange-breasted sunbird (*A. violacea*) on ericas. The implication is that only fynbos *sensu stricto* on the mountains of the Cape Folded Belt had sufficient unique floristic characteristics to stimulate the evolution of a few endemic species associated with the flora.

The contention of Levyns (1964) that Cape Province fynbos is in origin a mountain flora which migrated from Central Africa is supported by the fact that five of its dominant birds (*Cossypha caffra*, *Sphenoeacus afer*, *Onychognathus morio*, *Serinus canicollis* and *Emberiza capensis*) are distributed, outside the fynbos, in mountainous areas further north. Of the remaining species, three (*Prinia maculosa*, *Malaconotus zeylonus* and *Zosterops pallidus*) are south temperate endemics; and the other two (*Streptopelia capicola* and *Cisticola fulvicapilla*) are widespread in woodland in East, as well as South Africa (Winterbottom, 1972).

FYNBOS MAMMALS

Far less is known about mammals than birds. Approximately 90 of the total southern African fauna of about 280 species occur in the Southwest Cape. Of the four faunal elements, savannah, arid, forest and archaic, which Meester (1965) distinguishes in the southern African mammal fauna, savannah and arid elements predominate and there are some archaic groups. As might be expected, forest species are essentially restricted to the relict forest patches which happen to lie within the fynbos region. All orders and families of mammals present in southern Africa are (or were) found in the Southwest Cape except hedgehogs (Erinaceidae), pangolins (Pholidota), giraffe (Giraffidae), squirrels (Sciuridae, except for one exotic species), the rock rat (Petromyidae), the spring hare (Pedetidae) and cane rats (Thryonomyidae).

Insectivores include only one of seven South African elephant shrew (Macroscelididae) species, *Elephantulus edwardi*. It is also found in parts of the karoo and appears to live among rocks. The family is mainly one of savannahs. Four species of shrews (Soricidae) occur. *Myosorex varius* is a member of a primitive, tropical genus which prefers moist regions with dense vegetation (Meester, 1958). It is distributed from the Transvaal highveld through eastern South Africa to the Southwest Cape and up the west coast into the Southwest Arid zone (see Fig. 3.1) to near the Orange River. The distribution is frequently montane and thick streambank vegetation is preferred. Its presence on the arid west coast appears to depend on the dense succulent vegetation which is supported near the sea by mists. *Crocidura flavescens* has a similar distribution except that it is not found up the west coast. Meester (1962) notes that its range falls entirely within a zone with a mean annual rainfall of at least 500 mm, and most of it receives 750 mm. There is no relationship with vegetation type, but dense cover is a prerequisite and the species often inhabits broken country. *C. cyanea* has a wide distribution in southern Africa except in very arid regions such as the Namib Desert and the Kalahari. *Suncus varilla* is also quite widespread (Meester and Lambrechts, 1971). None of the shrews of the Southwest Cape are therefore primarily Mediterranean or fynbos species, and their presence seems to be correlated with high rainfall and densely vegetation microhabitats.

Golden moles (Chrysochloridae) are archaic southern African animals which probably evolved there. It is not surprising to find them well represented in this region. *Chrysochloris asiatica* is a Southwest Cape endemic extending slightly into the southwestern part of the Southwest Arid zone. *Cryptochloris zyli*, known only from the vicinity of Lamberts Bay on the West coast, is probably best considered marginal to the area since there is a closely related Southwest Arid species further north. *Eremitalpa granti* ranges from Lamberts Bay into the Southwest Arid as far north as the Namib and is therefore also marginal. In all three cases sandy soil appears to be the key factor in the habitat, and vegetation type seems irrelevant. *Chrysochloris* is the only one of these three genera with a wider distribution. One species is found in the northern Congo, Uganda and Tanzania and may have migrated from the south during the Pleistocene (Meester, 1965). The genus *Chlorotalpa*, with a wide southern distribution, also has one equatorial species. In the Southwest Cape *C. duthiae* occurs in the coast zone from Knysna to Port Elizabeth. *Amblysomus* is found in well-watered parts of South Africa in forest and open habitats. *A. hottentotus* is widespread in the eastern half of the country, and also extends into the

Southwest Cape. *A. iris* from the southeastern Transvaal, Natal and the eastern Cape Province is also found in the Southwest Cape in the vicinity of George and Knysna. It appears doubtful whether subterranean insectivores and carnivores such as the golden moles are distributed in relation to major vegetation types or particular seasonal rainfall régimes, although the abundance of their prey may be indirectly related to these factors.

Bats of the Southwest Cape comprise one fruit bat and about fourteen species of Microchiroptera (Hayman and Edwards Hill, 1971). *Rousettus aegyptiacus* occupies virtually the whole of sub-Saharan Africa, the eastern Mediterranean and Arabia, and is also successful in the Cape Province. It has some significance as a pest in fruit orchards there, but also feeds on wild fruits. The insectivorous bats include *Nycteris thebaica* (Nycteridae); two species of *Rhinolophus* (Rhinolophidae); *Platymops* (*Sauromys*) *petrophilus* and, marginally, *Tadarida aegyptiaca* (Molossidae). The large family Vespertilionidae is well represented with about nine species. One of two species of *Myotis*, *M. lesueri*, is endemic. *Eptesicus melckorum*, *E. hottentotus* and the endemic *E.* (*Rhinopterus*) *notius* complete the list of proper Southwest Cape forms, but *E. capensis* is marginal in the northwest. A further three species are known from the southern coastal belt as far west as about Knysna, namely *Pipistrellus kuhli*, *Miniopterus schreibersi* and *Kerivoula lanosa*.

Except for the two endemics, these bats are all more or less widely distributed in southern Africa or further afield. That so many species are present in this region may be associated with roosting facilities in caves and rock crevices in the Southwest Cape mountains, but adequate food supplies must of course also be assumed. That these are large may perhaps be inferred from the seasonal abundance of insectivorous birds such as swallows, and in particular the European migrant *Hirundo rustica* (Winterbottom, 1964).

As southern Africa is poor in primate species compared with the tropics, so the primate fauna of the Southwest Cape is limited compared to the rest of the subcontinent. The forest monkey *Cercopithecus mitis* ranges only as far south as the eastern Cape Province and *C. aethiops* is a savannah species which does not appear to have been able to colonise the Southwest Cape. Only the baboon *Papio ursinus* is found in the Southwest Cape. As elsewhere in its wide range, it is here successful and adaptable. Geophytes which are such an important constituent of fynbos appear to be a major food item.

The red rock hare (*Pronolagus rupestris*) is common in rocky situations in the Southwest Cape and in many other parts of southern Africa, and beyond to Kenya (Petter, 1971). *Lepus saxatilis* tends to inhabit hills and *L. capensis* flatter plains country elsewhere in South Africa, and this pattern of habitat separation may also be true of the Southwest Cape where both occur. All three lagomorphs are widespread outside the Mediterranean zone, and are not dependent on fynbos in particular.

The rodent fauna comprises mole rats (Bathyergidae), a porcupine (Hystricidae), dormice (Muscardinidae), Muridae and Cricetidae. The absence of squirrels may be explained by the restriction of ground squirrels to arid areas and the marginal entry of forest forms to the subtropical northeast. There is but a small array of savannah squirrels in the north, widely separated from the Southwest Cape. Here the exotic *Sciurus carolinensis* has not apparently found acceptable habitat far from cultivated farms and exotic plantations (Millar, unpubl.). Cane rats (Thryonomyidae) are tropical, and in South Africa extend only to the eastern Cape province.

Bathyergus is the only genus in the family Bathyergidae with enlarged foreclaws, presumably in adaptation to burrowing in sand. *B. suillus* is a giant form of coast dunes and sandy soils from Knysna through the Southwest Cape to the west coast. Soil quality and underground food supplies rather than vegetation type are likely to be important determinants of distribution. *Georychus capensis* and other members of the family use incisors for digging, and the feet are not specialised. This species is well distributed in the south and south-western coastal areas, with relict populations in Natal and eastern Transvaal (De Graaff, 1971). *Cryptomys hottentotus* is a widespread form which is also found in the Southwest Cape.

The porcupine *Hystrix africae-australis*, known throughout the southern half of Africa, is also successful in the Mediterranean climate of the Cape Province (Moffett, 1973). Dormice (Muscardinidae) are represented by the ubiquitous

and common *Graphiurus murinus* and marginally in the northwest by *G. ocularis*, a rare and poorly known form.

Davis (1962) lists 21 species of Muridae and Cricetidae from the Southwest Cape. *Praomys* (*Myomyscus*) *verreauxi*, Verreaux's mouse, is the only endemic murid. It occupies scrubby hill slopes, and, near Knysna, forest margins (Davis, 1974). The shaggy swamp rat (*Dasymys incomtus*) is considered a relict, with a subspecifically distinct population of a species which typically occupies moist, grassy habitats in woodlands of the Southern Savannah zone (see Fig. 3.1). Other murids are marginal, entering the Southwest Cape but with their main areas of distribution in the Southern Savannah or Southwest Arid zones. They are *Mus minutoides*, *Praomys* (*Mastomys*) *natalensis* (only to Plettenberg Bay), *Acomys subspinosus* and two common species, *Aethomys namaquensis* and *Rhabdomys pumilio*. The commensals *Mus musculus*, *Rattus norvegicus* and *R. rattus* are also present.

One member of the Cricetidae, the gerbil *Tatera afra* is endemic to the Southwest Cape where it occupies sandy areas. Relict populations of the Cape pouched mouse (*Saccostomus campestris*), *Steatomys krebsi*, the white-footed rat (*Mystromys albicaudatus*), and the vlei rats (*Otomys laminatus* and *O. saundersiae*) occur in the region. All other cricetids have a savannah or arid distribution, or both, and occur marginally in the region. They are two species of *Dendromus* (climbing mice) *Otomys irroratus*, *Parotomys brantsi*, and two gerbils, *Desmodillus auricularis* and *Gerbillurus paeba*. *Otomys unisulcatus* is a species of the karoo and Southwest Cape.

Carnivores of all families occupy the Southwest Cape. The black-backed jackal (*Canis mesomelas*) is the most widespread of the Canidae, although it does not appear to be numerous. The Cape fox (*Vulpes chama*) and the bat-eared fox (*Otocyon megalotis*) enter marginally from adjacent arid areas (Moffett, 1973). Mustelids, in decreasing order of abundance, are the polecat (*Ictonyx striatus*), clawless otter (*Aonyx capensis*), and ratel (*Mellivora capensis*). There are two very common viverrids, the Cape grey mongoose (*Herpestes pulverulentus*) and the water mongoose (*Atilax paludinosus*). Both the widespread small-spotted genet (*Genetta genetta*) and the large-spotted genet

(*G. tigrina*) of the Southern Savannah are present. The aardwolf (*Proteles cristatus*; Protelidae) enters this region; both South African hyaenas, *H. brunnea* and the spotted hyaena *Crocuta crocutta*, were present in historic times, but are now extinct. This is also true of the lion, but leopards (*Panthera pardus*) are still common, as is the caracal (*Felis caracal*) and the wild cat (*F. libyca*). The serval (*F. serval*) appears to have extended into the Southwest Cape and may still be present, but scarce, in remote mountains (Moffett, 1973).

The ubiquitous aardvark (*Orycteropus afer*; Tubulidentata) also occupies the region. Elephants (*Loxodonta africana*; Proboscidea) were found when colonists settled the Cape but did not long survive the new régime. *Procavia capensis* is the only one of four species of Procaviidae (Hyracoidea) to extend into the fynbos, where it is probably as successful as in any other of the multitude of environments occupied. Two members of the Perissodactyla, mountain zebra (*Equus zebra*) and black rhinoceros (*Diceros bicornis*), were members of the fauna historically, the former widespread in the mountains and the latter entering marginally from the Southwest Arid zone.

Artiodactyls include the bushpig (*Potamochoerus porcus*; Suidae) of forested habitats in the south as far west as about George, the hippopotamus *H. amphibius* which lived in every river (Du Plessis, 1969), and a surprising variety of antelope (Bovidae). Two of them are (or were) endemic to part of the southern Cape Province between the mountains and the sea, the extinct bluebuck (*Hippotragus leucophaeus*), and *Damaliscus dorcas dorcas*, the bontebok. Bontebok are now preserved in a National Park and on a few farms in the coastal rhenosterbosveld of the southern Cape Province.

Widely distributed species now extinct in the Southwest Cape are eland (*Taurotragus oryx*), buffalo (*Syncerus caffer*), and hartebeest (*Alcelaphus buselaphus*). Common and more or less widespread species at present are a number of small antelope: vaal ribbok (*Pelea capreolus*), three neotragines — steenbok (*Raphicerus campestris*), grysbok (*R. melanotis*) and klipspringer (*Oreotragus oreotragus*), and grey duiker (*Sylvicapra grimmia*). The grysbok is the only one virtually confined to fynbos (Manson, 1974). Forests and relict forest patches bring blue duiker

(*Cephalophus monticola*) as far west as George and bushbuck (*Tragelaphus scriptus*) even further, to Swellendam along the escarpment and to Hermanus on the coast.

The predominance of mainly browsing species, eland and the small antelope, is clearly related to the scrubby nature of fynbos. Grass is not only scarce but of poor quality. Louw (1969) found natural grazing in the western and southwestern Cape Province to be of limited food value. Crude protein and phosphorus levels are too low for optimal production and in some cases deficient, by accepted nutritional standards for domestic animals. Copper, cobalt and manganese are generally deficient, and all deficiencies are more severe in the dry summer months than in winter. Woody plants are most suitable as food for small antelope since, as Jarman (1974) points out, their small size and relatively smaller muzzles enable them to select plant parts with the highest food value. Their low absolute daily intake makes it feasible to spend the time which selective feeding demands. Eland probably moved about over various veld types, as they do today where opportunities for free movement still exist, and were thus able to obtain food of good quality, browse or non-browse. Buffalo were probably confined to grassy river valleys, and hartebeest are likely to have frequented the relatively grassy rhenosterbosveld on the flats.

ENDEMIC MAMMALS

As is the case with birds, endemism is clearly not a feature of the Southwest Cape mammal fauna. To recapitulate, there is an endemic golden mole (*Chrysochloris asiatica*), two bats of somewhat doubtful status [*Myotis lesueuri* and *Eptescus* (*Rhinopterus*) *notius*], the dune mole rat (*Bathyergus suillus*), a murid [*Praomys* (*Myomyscus*) *verreauxi*], and the gerbil (*Tatera afra*, Cricetidae). The fossorial mole rat and the burrowing gerbil are both apparently associated with sandy soil rather than with a particular vegetation type. There are also two large endemic antelope, the extinct bluebuck (*Hippotragus leucophaeus*) and the bontebok (*Damaliscus dorcas dorcas*), both of which ranged only over a portion of the southern part of the region between the mountains and the sea (Du Plessis, 1969) mainly in fairly grassy coastal rhenosterbosveld and coastal fynbos, and not in fynbos *sensu stricto*. The grysbok(*Raphicerus melanotis*) is virtually a Southwest Cape endemic; elsewhere it inhabits dense scrub in mesic areas, similar in structure to fynbos.

AGE OF FYNBOS MAMMALS

It is interesting to analyse the composition of the Southwest Cape mammalian fauna in the light of Di Castri's (1973) observation that the historical heterogeneity of Mediterranean climate ecosystems is greater than in any other terrestrial ecosystem type of the world. Cooke (1972) lists a number of ancient surviving groups that evolved from stocks believed to have been present before Africa became isolated from Eurasia in the Paleocene. They include the following taxa which are represented in the Southwest Cape: elephant shrews, golden moles, primates, bathyergids, aardvark, hyraxes and elephants. Late Oligocene arrivals in Africa present in the region are: cricetids, canids, viverrids, felids, pigs, rhinoceroses and bovids. The Bovidae may be separated into old forms which probably evolved in Africa, and a younger Afro-Asiatic stratum (Wells, 1957). Most antelope of the Southwest Cape: Cephalophinae (grey duiker), Neotragini (grysbok, steenbok, klipspringer), Peleini (vaal ribbok, *Pelea*) and Alcelaphini (bontebok, hartebeest), belong to the old group. The Afro-Asiatic stratum is however also represented by the bushbuck (Tragelaphini) and mountain reedbuck (*Redunca fulvorufula*; Reduncini).

The latest immigrants in the Miocene include hares, murids, mustelids, hyaenas, equids and hippopotamus, all present in the Southwest Cape. Clearly then, all these mammal groups irrespective of their age have occupied the Mediterranean heathlands of the Cape Province, and the mixed age of components of this and similar ecosystems is confirmed.

ECOLOGICAL CLASSIFICATION OF MAMMALS

In Table 3.4 the mammals of southern Africa (data from Meester et al., 1964) and of the Southwest Cape are classified according to adap-

TABLE 3.4

Mammals of Southern Africa and the Southwest Cape classified by adaptive zones

Adaptive zone[1]	Approx. number of species	
	Southern Africa[2]	Southwest Cape
1. Small terrestrial insect-eaters and predators (soricids, erinaceids, macroscelidids)	23	5
2. Fossorial mole-like insectivores (Chrysochlorids)	12	4+2
3. Specialist ant- and termite-feeders (aardvark, pangolin, aardwolf)	3	2
4. Small to medium terrestrial omnivores (smaller mongooses, terrestrial mustelids, *Nandinia, Viverra*)	16	4
5. "Typical" rats and mice	51	21
6. Rabbit-sized herbivores (hyracoids, leporids, *Petromus, Pedetes*)	13	4
7. Medium to large terrestrial herbivores	46	14+3[3]
8. Subterranean herbivores (mole-rats, porcupine)	9	4
9. Arboreal herbivores and omnivores (primates, sciurids, tree hyraxes)	15	1
10. Weasel- to fox-sized terrestrial carnivores (large mongooses, genets, jackals, small cats)	12	6
11. Large carnivores and scavengers	8	6
12. Fruit- and blossom-eating bats	7	1
13. Insectivorous bats	60	14

[1] Adaptive zones after Keast (1972).
[2] Number of southern African species following Meester et al. (1964).
[3] Marginal.

tive zones or ecological categories (after Keast, 1972). It is difficult to be sure whether the number of Southwest Cape species is simply correlated with the area of the region or whether some groups are relatively under-represented.

One might perhaps have expected rather more species of small terrestrial insect-eaters and predators. As we have seen, shrews favour moist environments with dense vegetation, and the species present in this region are more or less widely distributed in other parts of South Africa where these habitat requirements are met. Presumably a basic inability of the group to adapt to arid conditions precluded the evolution of local species for drier parts of the fynbos. It is more difficult to explain why there is only one elephant shrew, since there are other species in arid parts of South Africa. One can do little but speculate that summer drought conditions might severely limit terrestrial insect populations.

In contrast, the fossorial chrysochlorids are particularly well represented. The age of the group may play a role, and speciation may well antedate the fynbos and Mediterranean climatic conditions. Raven (1973) states that summer-dry climates only appeared in the Pleistocene.

Small and medium-sized terrestrial omnivores are limited to two mongooses, *Ictonyx*, and the ratel. Limited food supplies may restrict the variety of these small, partly predatory animals. As will be shown below, population densities of small mammals appear to be low and the possibility of limited insect food supplies has already been mentioned.

Subterranean herbivores are well represented and this is very probably correlated with the abundance of geophytes with underground storage organs. Arboreal mammals are extremely restricted, with only the baboon present in the Southwest Cape. It is a simple matter to relate this to the fact that primates, squirrels and tree hyraxes are all centred on the tropics, as well as to the absence of arboreal habitats. The well-known phenomenon of tropical plants and animals ranging down the eastern side of the continent and progressively decreasing in variety leads to most dropping out in or before the eastern Cape Province while a few extend along the southern coast belt to Knysna and George. Only a very few species extend westward to the Southwest Cape, and arboreal mammals are not amongst them. The presence of only one species of fruit bat in the region is similarly explicable.

POPULATION DENSITIES

Instructive though it is to analyse faunal composition and diversity, the abundance of various mammals is much more important for an understanding of ecosystem function. Unfortunately, virtually nothing is known about population densities. It is a matter of observation that larger mammals and especially antelope are seldom seen even in remote fynbos reserves. Young burned veld attracts some species such as vaal ribbok (*Pelea*) and klipspringer (*Oreotragus*) — and presumably other larger species now extinct — while old, climax fynbos supports very little mammal life. Animal biomass varies with age of vegetation after a fire. As there are now only small antelope, present-day maxima can however at best be very low. What they might have been originally with the full complement of large herbivores is unknown. In view of the low nutritive value of the vegetation for livestock, mentioned above, herbivore biomass cannot however have been very large.

Small mammal communities of rodents and insectivores studied by De Hoogh (1968), Toes (1974), Stuart (in prep.) and Bigalke and Pepler (pers. obs.) in the Southwest Cape consist of four to six prominent species (one of them an insectivore) with two or three more abundant than the rest. Fleming (1975) reports that most tropical habitats, except those at higher altitudes, usually contain ten to sixteen species of rodents and one to six marsupials or insectivores coexisting. In an arid savannah study area in the southern Kalahari, Nel and Rautenbach (1975) recorded thirteen small rodent species, two of them abundant, six common and five rare. Small mammal communities in fynbos habitats are apparently considerably less diverse.

De Hoogh estimated the collective density of four rodent species in his plots in the Cedarberg Mountain range to be 17 ha^{-1} (5% confidence limits 8 to 21 ha^{-1}). From his data a biomass of approximately 527 g ha^{-1} (248 to 651 g ha^{-1} at 5% level) can be calculated. Trapping success was 26 rodents and 6 insectivores, a total of 32 animals in 1944 trap nights or 0.016 animals per trap night.

Toes (1974) gives no data on population density per unit area in the Jonkershoek Valley, Stellenbosch, but from her results trapping success rates between 0.017 and 0.033 can be calculated, suggesting densities ranging from about the same as those of De Hoogh to twice as great. She found sites facing northeast to have higher populations than cool moist sites facing southwest. Young, frequently burned sites had higher populations than old dense fynbos. In 31-year-old climax fynbos on a southwestern aspect only 17 small mammals were caught in 968 trap nights spread over eleven weeks. The most productive site was north-facing short, open 3-year-old veld where the total catch for the same effort was 32 animals. It must be borne in mind that annual fluctuations were not taken into account, although Toes noted a decrease in trap success as the summer progressed.

These populations are small. In a temperate environment in Britain with only two species, wood mice and bank moles, mean densities fluctuated about 20 ha^{-1}, while as many as 361 rodents ha^{-1} were present in a tropical grass/bush habitat in Zaire (Delaney, 1974). Biomass figures for fynbos are so limited that they can hardly be used for comparison, but they appear low. Fleming (1975) quotes figures from the Ivory Coast of about 470 g ha^{-1} and 1000 g ha^{-1} in burned and un-burned savannah respectively and 14 000 to 15 000 g ha^{-1} for the area in Zaire mentioned above.

Even less can be said about the rôle of small mammals in fynbos community dynamics. De Hoogh (1960) found that rodents in the Cedarberg could consume the entire seasonal seed crop of the rare cedar *Widdringtonia cedarbergensis* in a few weeks, and were probably a significant factor influencing regeneration. *Aethomys namaquensis* has been observed climbing in *Protea* bushes (Toes, 1974) and may play a role as a seed predator similar to the role it plays in consuming seed stocks of the exotic proteaceous shrub *Hakea sericea* (S. Neser, pers. comm., 1976).

The apparently low mammal biomasses suggest that these animals do not consume much of the primary production and do not therefore play a significant rôle in energy and nutrient cycling.

SUMMARY AND CONCLUSION

The distribution of rainfall is no doubt the ecological factor of overriding importance in determining the biotic pattern of southern Africa, as Stuckenberg (1969) has clearly shown. We follow this author in recognising a fundamental division of the biota of southern Africa into dry western, humid eastern, and part dry, part humid southern sections. As he points out, this is reflected in the trichotomy of the flora into (1) karoo; (2) tropical or subtropical forest, savannah and grassland; and (3) fynbos — a situation which, according to Levyns (1962), probably extended back over most of the Tertiary. Since the continent does not extend far enough south for temperature effects to become of overriding importance, the thermal pattern is secondary and modifies distribution mainly in the east and south where highlands and mountains are present.

The thermal environment must be expected to be of most direct importance for poikilothermic animals, and it is indeed these groups that are most distinctive in the Southwest Cape. The invertebrate fauna includes ancient southern taxa restricted to the tips of the southern continents as a legacy of Gondwanaland. In spite of the stresses of summer drought, conditions are apparently tolerable in mesic montane microclimates. The importance of the Southwest Cape as a major centre of endemism for amphibia has been attributed to the great age of the group. Presumably the temperature character-istics of the climate must also be significant, but Stuckenberg (1969) points out that the Effective Temperature [1] of this region has no special features. "Admittedly the ET pattern there must be far more complex than the present map scale and data reveal, but on the whole, from the thermal point of view, the Southwest Cape has much in common with the upper Natal Midlands (ET 14°–15°C), and does not deserve to be distinguished as 'temperate', as some authors have done. it follows that much of

---

[1] The Effective Temperature (ET) is defined by Bailey (1960) as:

$$ET = \frac{8T + 14AR}{AR + 8}$$

where T is the mean annual temperature and AR the difference between the mean temperatures of the warmest and coldest months, both in °C.

the biological uniqueness of that region must be due to its winter rainfall." However, he adds that the disposition of the mountain ranges is also important, with mesic seaward aspects and arid interior slopes differing greatly. Summer aridity imposes particular stresses. Nonetheless it is difficult to explain the high degree of endemism of amphibia without accepting that at least some habitats in the Southwest Cape are moist enough and cool enough for a distinctive fauna to have developed. As far as reptiles are concerned there are a number of endemics, but the precise picture is obscured by confused taxonomy and inadequate distributional data. The Southwest Cape is not in any event a significant endocentre, which fits in well with Stuckenberg's (1969) standpoint since snakes more than other reptiles are directly dependent on thermal energy sources, and their distribution patterns, as his maps show, fit significant Effective Temperature isolines very precisely.

In the case of homoeotherms we may expect vegetation as a habitat and source of food — direct or indirect — to influence animal life to a greater extent than climate and weather. The avifauna has been more thoroughly studied than other groups. It is relatively poor in species and is most closely related to that of the karoo, while affinities with the temperate grasslands of the South African highveld are also quite strong. Insectivorous warbler-like forms are particularly prominent. The few endemic birds include two predominantly nectivorous species which may have evolved with the Proteaceae and Ericaceae, and forms which are essentially montane. The limited species diversity is related to distance from the equator but is comparable with that of other Mediterranean regions, and correlates well with height and structural diversity of the vegetation. Population densities are also similar to those in sclerophyll communities of summer-drought areas elsewhere, and presumably reflect similar levels of primary production.

Mammals are a mixture of ancient and younger groups. The fauna has strong affinities with the Southwest Arid and Southern Savannah biotic zones, tropical elements are unimportant and there are few endemic species. Species diversity in mammal communities is low, and the apparently low population densities and biomasses are thought to be related to the limited nutritive value of the vegetation, at least in the dry season.

Fire plays a major role in improving palatability, and burnt areas are attractive to bovids as well as supporting larger small-mammal populations than unburnt stands. Browsing herbivores predominate, and in the past the biggest numbers and variety of large animals were probably confined to river valleys and grassy fynbos on the plains.

REFERENCES

Acocks, J.P.H., 1975. Veld types of South Africa. *Mem. Bot. Surv. S. Afr.*, No. 40: 120 pp. (2nd ed.).

Bailey, H.P., 1960. A method of determining the warmth and temperateness of climate. *Geogr. Ann.*, 42: 1–16.

Broekhuysen, G.J., 1966. The avifauna of the Cape "protea-heath macchia" habitat in South Africa. *Ostrich Suppl.*, 6: 323–334.

Cody, M.L., 1973. Parallel evolution and bird niches In: F. di Castri and H.A. Mooney (Editors), *Mediterranean Type Ecosystems*. Springer-Verlag, Berlin, pp. 307–338.

Cody, M.L., 1975. Towards a theory of continental species diversities: bird distribution over mediterranean habitat gradients. In: M.L. Cody and J.M. Diamond (Editors), *Ecology and Evolution of Communities*. Belknap Press, Harvard University, Cambridge, Mass., pp. 214–257.

Cooke, H.B.S., 1972. The fossil mammal fauna of Africa. In: A. Keast, F.C. Erk and B. Glass (Editors), *Evolution, Mammals and Southern Continents*. State University of New York, Albany, N.Y., pp. 89–139.

Darlington, P.J., 1957. *Zoogeography: The Geographical Distribution of Animals*. John Wiley and Sons, New York, N.Y., 675 pp.

Davis, D.H.S., 1962. Distribution patterns of Southern African muridae, with notes on some of their fossil antecedents. *Ann. Cape Prov. Mus.*, 2: 56–84.

Davis, D.H.S., 1963. Wild rodents as laboratory animals and their contribution to medical research in Southern Africa. *S. Afr. J. Med. Sci.*, 28: 53–69.

Davis, D.H.S., 1974. The distribution of some small Southern African mammals (Mammalia: Insectivora, Rodentia). *Ann. Transvaal Mus.*, 29(9): 135–184.

De Graaff, G., 1971. *Preliminary Identification Manual for African Mammals. 16 Rodentia: Bathyergidae*. Smithsonian Institution, Washington, D.C., 5 pp.

De Hoogh, R.J., 1968. *Report of Practical Work in South Africa*. University of Wageningen, Wageningen, pp. 153.

Delaney, M.J., 1974. *The Ecology of Small Mammals*. Edward Arnold, London, 60 pp.

Di Castri, F., 1973. Animal biogeography and biological niche. In: F. di Castri and H.A. Mooney (Editors), *Mediterranean Type Ecosystems*. Springer-Verlag, Berlin, pp. 279–283.

Du Plessis, S.F., 1969. *The Past and Present Geographical Distribution of the Perissodactyla and Artiodactyla in Southern Africa*. Thesis, University of Pretoria, Pretoria, 82 pp.

Fitzsimons, V.F., 1943. The lizards of South Africa. *Transvaal Mus. Mem.* No. 1: 528 pp. (Pretoria).

Fitzsimons, V.F., 1970. *Snakes of Southern Africa.* Purnell, Cape Town and Johannesburg, 423 pp.

Fleming, T.H., 1975. The role of small mammals in tropical ecosystems. In: F.B. Golly, K. Petruscwicz and L. Repzkawski (Editors), *Small Mammals, Their Productivity and Population Dynamics.* Cambridge University Press, London, 451 pp.

Greig, J.C., 1976. *Herpetology in Southern Cape Fynbos.* Department of Nature and Environmental Conservation, Stellenbosch, 7 pp.

Hayman, R.W. and Edwards Hill, J., 1971. *Preliminary Identification Manual for African Mammals. Chiroptera.* Smithsonian Institution, Washington, D.C., 73 pp.

Jarman, P.J., 1974. The social organisation of antelope in relation to their ecology. *Behavior*, 48 (3 and 4): 215–267.

Keast, A., 1972. Comparison of contemporary mammal faunas of Southern Africa. In: A. Keast, F.C. Erk and B. Glass (Editors), *Evolution, Mammals and Southern Continents.* State University of New York, Albany, N.Y., 543 pp.

Levyns, M.R., 1961. Plants of Southern Africa. In: E. Rosenthal (Editor), *Encyclopaedia of Southern Africa.* Warne, London, 600 pp.

Levyns, M.R., 1962. Past plant migrations in South Africa. *Ann. Cape. Prov. Mus.*, 2: 7–10.

Levyns, M.R., 1964. Migration and origin of the Cape Flora. *Trans. R. Soc. S. Afr.*, 37: 185–213.

Louw, G.N., 1969. The nutritive value of natural grazings in South Africa. *Proc. S. Afr. Soc. Anim. Prod.*, 8: 57–61.

MacArthur, R.H., 1965. Patterns of species diversity. *Biol. Rev.*, 40: 510–533.

McLachlan, G.R. and Liversidge, R., 1970. *Roberts Birds of South Africa.* CNA, Johannesburg, 504 pp.

Manson, J., 1974. *Aspekte van die biologie en gedrag van die Kaapse grysbok*, Raphicerus melanotis *Thunberg.* Thesis, University of Stellenbosch, Stellenbosch, 176 pp.

Meester, J., 1958. Variation in the shrew genus *Myosorex* in Southern Africa. *J. Mammal.*, 39: 325–339.

Meester, J., 1962. The distribution of *Crocidura* Wagler in Southern Africa. *Ann. Cape. Prov. Mus.*, 2: 77–84.

Meester, J., 1965. The origins of the Southern African mammal fauna. *Zool. Afr.*, 1: 87–95.

Meester, J., 1971. *Preliminary Identification Manual for African Mammals. Lipotyphla: Chrysochloridae.* Smithsonian Institution Washington, D.C., 7 pp.

Meester, J. and Von W. Lambrechts, A., 1971. The Southern African species of *Suncus* Ehrenberg (Mammalia: Soricidae). *Ann. Transvaal. Mus.*, 27(1): 1–14.

Meester, J., Davis, D.H.S. and Coetzee, C.G., 1964. *An Interim Classification of Southern African Mammals.* The Zoological Society of Southern Africa and The South African Council for Scientific and Industrial Research, (mimeograph).

Millar, J., unpubl. *Biology of the Grey Squirrel* (Scuiris carolinensis) *in the Western Cape.* Thesis, University of Stellenbosch, Stellenbosch, in prep.

Moffett, R., 1973. *Survey of Wildlife in State Forest Reserves in the Western Cape Forest region 1973* (Mimeograph).

Moreau, R.E., 1966. *The Bird Fauna of Africa and its Islands.* Academic Press, New York and London, 424 pp.

Nel, J.A.J., and Rautenbach, I.L., 1975. Habitat use and community structure of rodents in the Southern Kalahari. *Mammalia*, 39(1): 9–29.

Petter, F., 1971. *Preliminary Identification Manual for African Mammals. Lagomorpha.* Smithsonian Institution Washington, D.C., 7 pp.

Poynton, J.C., 1960. Preliminary note on the zoogeography of the amphibia of Southern Africa. *S. Afr. J. Sci.*, 56: 307–312.

Raven, P.H., 1973. The evolution of Mediterranean floras. In: F. di Castri and H.A. Mooney (Editors), *Mediterranean Type Ecosystems.* Springer-Verlag, Berlin, pp. 213–224.

Saiz, F., 1973. Biogeography of soil beetles in Mediterranean regions. In: F. di Castri and H.A. Mooney (Editors), *Mediterranean Type Ecosystems.* Springer-Verlag, Berlin, pp. 285–294.

Stuart, C.T., 1978. *A Rodent and Insectivore Study in the Jonkershoek Valley, Stellenbosch.* Department of Nature and Environmental Conservation, Stellenbosch, 20 pp. (in prep).

Stuckenberg, B.R., 1962. The distribution of the montane palaeogenic elements in the South African fauna. *Ann. Cape. Prov. Mus.*, 2: 190–205.

Stuckenberg, B.R., 1969. Effective temperature as an ecological factor in Southern Africa. *Zool. Afr.*, 4(2): 145–197.

Toes, E., 1974. *Tellingen van kleine zoogdieren en vogels in de Jonkershoek vallei, Zuid Afrika 1972.* Verslag 171, Landbouwhogeschool, Afd. Natuurbeheer en -behoud, Wageningen, Verslag 171, 65 pp.

Vitali-di Castri, V., 1973. Biogeography of pseudoscorpions in the Mediterranean regions of the world In: F. di Castri and H.A. Mooney (Editors), *Mediterranean Type Ecosystems.* Springer-Verlag, Berlin, pp. 295–305.

Wells, L.H., 1957. Speculations on the paleogeographic distributions of antelopes. *S. Afr. J. Sci.*, 53: 423–424.

Winterbottom, J.M., 1964. The migrations and local movements of some South African birds. In: D.H.S. Davis (Editor), *Ecological Studies in Southern Africa.* Junk, The Hague, 415 pp.

Winterbottom, J.M., 1966. Ecological distribution of birds in the indigenous vegetation of the Southwest Cape. *Ostrich Suppl.*, 37: 76–91.

Winterbottom, J.M., 1966b. The comparative ecology of the birds of some Karoo habitats in the Cape Province. *Ostrich Suppl.*, 37: 109–127.

Winterbottom, J.M., 1968a. Remarks in the possible origin of the avifauna of the South West Cape. *Ostrich Suppl.*, 7: 91–93.

Winterbottom, J.M., 1968b. Remarks on the avifauna of the macchia of the Southern Cape Province. *Rev. Zool. Bot. Afr.*, LXVII(3–4): 221–235.

Winterbottom, J.M., 1972. The ecological distribution of birds in Southern Africa. *Monogr. Percy Fitzpatrick Inst. Afr. Ornithol.*, No. 1: 81 pp. (Cape Town).

Chapter 4

AFRICAN MOUNTAIN HEATHLANDS[1]

D.J.B. KILLICK

INTRODUCTION

In this review the mountain heathlands of Africa are described, excluding those of the southwestern Cape Province in South Africa which have been covered by Kruger in Chapter 2. The mountain heaths which extend from Ethiopia in the north to the southern Cape will be dealt with on a country or regional basis, starting with those in the north and working southwards. The location of most of the mountains mentioned is shown in Fig. 4.1.

In the preparation of this review it has become clear that, on the whole, there is very little detailed ecological information available on African mountain heathlands. Descriptions of the vegetation are invariably superficial; often only the dominant or most important species are mentioned, and little information on the geology, soils and climate of the heathland environment is provided. The reason for this is possibly that, because of their situation at high altitudes, these heathlands are, with some exceptions, uninhabited and undeveloped agriculturally, and therefore there is little reason for detailed studies of the natural vegetation and its environment.

Identifying mountain heathland in the many papers consulted was complicated by the fact that heath was rarely designated as such, but instead was referred to as forest bushland, woodland, scrub, savannah, *macchia*, *fynbos*, and even "altitude grassland". While the literature coverage is fairly comprehensive, it cannot be claimed to be complete.

ETHIOPIA

The heathland vegetation of the Ericaceous belt of Ethiopian mountains is poorly known, particularly the mountains of northern Ethiopia. Scott (1952), an entomologist, has given brief details of the heathland "zone" on Mount Zuquāla, Mount Chillālo, Mount Damōta and the Gughé Highlands in southern Ethiopia. The only heathland plants he mentions are *Erica arborea*, *Helichrysum splendidum*, *Kniphofia thompsonii*, *Alchemilla fischeri* and *Hypericum revolutum*. He encountered no species of *Philippia*.

Hedberg (1971) studied the Ericaceous Belt (3300–3850 m) on Galama Mountain in southern Ethiopia. He writes that ericaceous "scrub", 0.5–1 m high, covers practically all slopes having a reasonably deep and well-drained soil (Fig. 4.2). The most important dominant is *Philippia keniensis* subsp. *abyssinica*, and among its associates are *Erica arborea*, *Blaeria spicata*, *Satureja simensis*, *Bartsia decurva*, *Hebenstreitia dentata*, *Helichrysum splendidum* and *Alchemilla haumanii*. Between the shrubs the vegetation is open and patchy with much bare soil, and consists of *Cerastium octandrum*, *Sagina* spp., *Alchemilla commutata*, *Swertia* sp., *Haplocarpha rueppellii*, *Koeleria cristata* and *Aira caryophyllea*. Hedberg states that there is frequent evidence of burning with resultant soil erosion.

CAMEROONS MOUNTAIN

On the Cameroons Mountain (Mont Cameroun) in West Africa between the upper limit of forest (1675 to 2650 m) and alpine desert (3384 to 4070 m)

---

[1] Manuscript completed October, 1976.

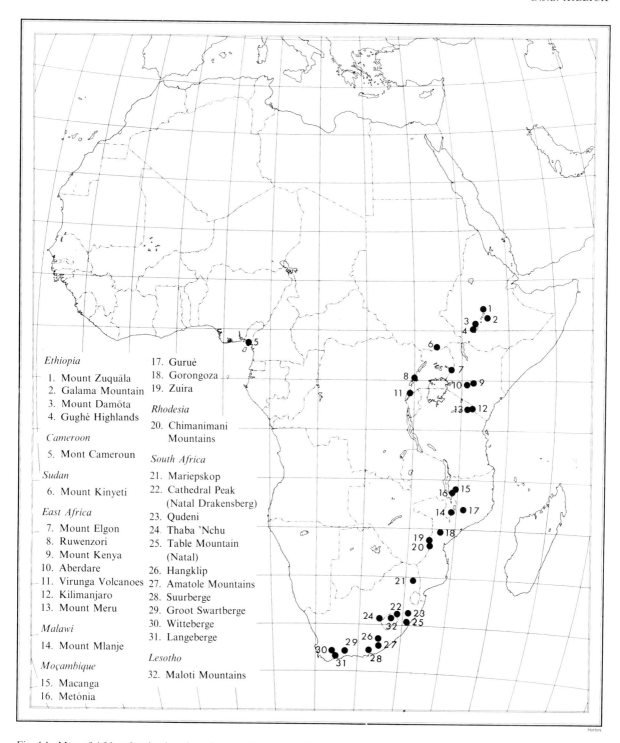

Ethiopia

1. Mount Zuquāla
2. Galama Mountain
3. Mount Damōta
4. Gughé Highlands

Cameroon

5. Mont Cameroun

Sudan

6. Mount Kinyeti

East Africa

7. Mount Elgon
8. Ruwenzori
9. Mount Kenya
10. Aberdare
11. Virunga Volcanoes
12. Kilimanjaro
13. Mount Meru

Malawi

14. Mount Mlanje

Moçambique

15. Macanga
16. Metónia

17. Gurue
18. Gorongoza
19. Zuira

Rhodesia

20. Chimanimani Mountains

South Africa

21. Mariepskop
22. Cathedral Peak (Natal Drakensberg)
23. Qudeni
24. Thaba 'Nchu
25. Table Mountain (Natal)
26. Hangklip
27. Amatole Mountains
28. Suurberge
29. Groot Swartberge
30. Witteberge
31. Langeberge

Lesotho

32. Maloti Mountains

Fig. 4.1. Map of Africa showing location of most of the mountains mentioned in the review.

Fig. 4.2. Part of the Ericaceous Belt of Galama Mountain in southern Ethiopia, looking towards Boraluco. The foreground lies between 3700 and 3750 m. (Photo: O. Hedberg.)

there is "montane grassland" containing low shrubs. This shrubland is clearly heathland and has been described by Richards (1963).

The Cameroons Mountain with its peak Fako at 4070 m is the highest mountain in West Africa. The mountain is still active volcanically, and the lava at altitudes above 1000 m produces brownish to blackish, usually shallow soils.

Information on the climate of the mountain is fragmentary, with no data for altitudes above the station at Buea (985 m). Richards assumes that at 1900 to 2500 m the rainfall is not greater (it may be lower) than at Buea, where the mean annual rainfall is 2897 mm with most rain falling in August. He suggests that the zone of greatest mist and probably humidity is about 1220–2000 m. Working on a lapse rate of about 0.6°C per 100 m, he estimates the mean annual temperature to be about 12°C at 2256 m and about 1°C on the summit. Frost and hail are common at high altitudes, but there appears to be some doubt

whether snow ever lies even near the summit. Richards describes the temperature régime as of the diurnal climatic type.

The shrubs occurring in the grassland on the Cameroons Mountain are locally frequent. The most important shrub is the leptophyllous *Adenocarpus mannii* (Fabaceae), which sometimes forms a low open-heathland between 38 and 75 cm high covering considerable areas. Occasionally *A. mannii* is 1 to 1.5 m high, but then the heathland is even more open. *Hypericum revolutum*, *Philippia mannii* and many other shrubs are also present, but not usually gregariously. The low-growing *Blaeria mannii* is common above 2438 m. The grass understorey is formed chiefly by *Andropogon* spp. and *Loudetia camerunensis*, but above 3048 m the large tussock grasses become less abundant and the fine-leaved grasses, such as species of *Deschampsia* and *Koeleria*, increase in importance.

According to Richards, the factor chiefly responsible for the present grassland areas is not climate or soil, but fire. He speculates that, in the absence of fire, areas too exposed or too cold for trees would probably support heathland, and he relates the present grassland areas to the Ericaceous Belt of East African mountains. He points out that *Adenocarpus mannii* is common to both regions.

SUDAN

Jackson (1956) has described the Ericaceous Zone of the Imatong Mountains lying in southern Sudan on the border with Uganda and rising to 3187 m in Mount Kinyeti. The Ericaceous Zone occupies a small area on the summit of Mount Kinyeti above the line of *Podocarpus* forest between 2900 and 3000 m. The heathlands rarely exceed 2 m in height and comprise as dominants the shrubs *Erica arborea* and *Myrica salicifolia*; other shrubs are *Anthospermum usambarense*, *Senecio* sp., *Alchemilla argyrophylla*, *Gnidia glauca*, *Hypericum revolutum*, *Crassula pentandra* var. *phyturus*, *Erlangea imatongensis*, *Blaeria breviflora* and *Smithia volkensii*. Among the herbs are *Satureja biflora*, *S. simensis*, *Asparagus asiaticus* var. *scaberulus*, *Carduus chamaecephalus* and *Silene burchellii*. The dominant grasses are *Exotheca abyssinica*, *Tripogon major*, *Festuca abyssinica* var. *abyssinica*, *Andropogon thomasii* and *Agrostis kilimandscharica*.

Variations with habitat are also described. Jackson states that the Ericaceous Zone was sometimes burnt in the past, and therefore the vegetation is possibly a fire climax rather than true climax. He mentions that the Ericaceous Zone resembles the lower part of the Ericaceous Zone of East African mountains.

EAST AFRICA

The heathlands of East African mountains are confined to the Ericaceous Belt lying between a lower limit between 2600 and 3400 m and an upper limit between 3500 and 4100 m (Hedberg, 1951). The position of the Ericaceous Belt varies from mountain to mountain and also varies with aspect. The vegetation of the Ericaceous Belt has in the past been referred to as the tree heath belt

(Gyldenstolpe, 1024), heathland type of vegetation (Burtt Davy, 1935), heather forest zone or "*Ericetum*" (Synge, 1937), mountain grassland or heathlands (Edwards, 1940), heath formation (Salt, 1954), heath woodland and heath forest (Ross, 1956), etc.

The mountains of East Africa, with the exception of Ruwenzori, are all of volcanic origin, consequently their most important bed-rocks are lavas, basalts, nepheline-syenite, agglomerates and tuffs, etc. Ruwenzori consists of Archaean rocks such as granite, gneisses, diorites, amphibolites, quartzites, etc. (Hedberg, 1964).

Information about the soils of the Ericaceous belt is limited. Salt (1954) gives details of the soils of the Shira Plateau between 3600 and 4200 m on Mount Kilimanjaro. He states that the surface of the plateau is exposed in parts as rock, chiefly trachydolerite and rhomb-porphyry. Boulders litter the Plateau and small areas are occupied by lapillae and volcanic sand, but by far the greater part of the Plateau is covered by a well-developed soil, generally dark in colour, containing coarse sand particles and peaty in character.

Four rainfall stations on Kilimanjaro are situated within the altitudinal limits of the Ericaceous belt as given by Hedberg (1951) for that mountain, viz. 2800 to 4000 m. The figures in Table 4.1 show that there are two rainy seasons, the "long rains" from March to May and the "short rains" from mid-October to mid-December. Hedberg (1964) states that the ratio between the highest and lowest annual rainfall recorded for any particular station in the Ericaceous Belt lies between 3.6 and 4.9. One year may give a total of 340 mm and another year 1685 mm at the same station. Obviously there are differences in rainfall from one mountain to the other and aspect also plays a part. Some of the precipitation is derived from snow (although the Ericaceous Belt lies below the snow-line) and hail.

Very few temperature data are available for the Ericaceous Belt. It is known that, as in the Alpine Belt, seasonal variations of temperature are almost negligible, whereas diurnal variations are marked. Frost occurs on most nights of the year, while daytime air temperatures can be mild. Salt (1954) cites data obtained between August and October by Klute (1920) above the Machame Escarpment on Kilimanjaro at 4160 m just above the Ericaceous Belt. On only one day was the minimum air

TABLE 4.1

Mean monthly rainfall (mm) at four stations on Kilimanjaro and one on Mount Kenya (from Hedberg, 1964)

	Kilimanjaro				Mount Kenya
Altitude (m):	2850	3800	3050	4000	3650
Number of years:	6–8	10–12	6–8	6–8	2–3
January	127	43	137	97	43
February	137	36	97	76	38
March	259	112	211	145	74
April	338	127	218	198	274
May	155	76	132	109	147
June	99	28	36	28	102
July	76	25	25	18	84
August	58	15	25	15	89
September	97	10	20	13	94
October	132	33	69	48	127
November	102	41	86	76	173
December	173	61	97	76	135
Year	1753	607	1153	899	1380

temperature above 0°C; the mean temperature was −1.8°C and the absolute minimum −4.0°C (unscreened reading, −7.3°C). The maximum air temperature recorded was 10.0°C. Klute recorded a temperature of 39.5°C at noon on a rock surface when air temperature was 8.3°C, a difference of 31.2°C. At night this difference becomes almost nil. Heathland plants must clearly be adapted to cope with the considerable diurnal fluctuations of temperature.

Apart from the main monsoon winds of East Africa, there are local winds characteristic of each high mountain. These "up and down" winds, as Coe (1967) calls them, result in the formation of a daily cycle of cloud formation. In the mornings there is a layer of cloud at the base of the mountains, and from noon onwards the upper parts are capped in clouds. This cloud formation produces a marked difference in insolation between the eastern and western sides of the same mountain and hence affects the snowline; it causes a rapid decrease in temperature and an increase in humidity and provides moisture not measured as rainfall (Hedberg, 1964).

There is very little information about cryonival phenomena in the Ericaceous Belt. Hedberg (1964) records needle-ice at 3800 m on Kilimanjaro.

Fire is an extremely important environmental factor in East Africa. The general effect of fire is the reduction of heathland to grassland and tall heathland to short heathland.

According to Hedberg (1951) the Ericaceous Belt is usually dominated by arborescent or shrubby species of *Philippia*, the commonest being *P. excelsa*. In the lower part of the Belt the genus *Erica* plays an important rôle (*Erica arborea* and other species) with some broad-leaved trees such as *Rapanea* sp. and *Hypericum* spp. also present. The physiognomy of the Belt varies from mountain to mountain: sometimes it is represented by dense *Erica — Philippia* tall heathland ("forest") and sometimes by an open-heathland ("scrub") of the same species. The differences in physiognomy may be attributed to climatic and edaphic differences between mountains and to the effect of fire.

Hedberg describes in broad terms the Ericaceous Belt of seven high mountains in East Africa. For more detailed descriptions, the references cited by Hedberg should be consulted.

Mount Elgon

This mountain is situated on the border of Uganda and Kenya. The Ericaceous Belt extends between about 3000 and 3550 m and consists largely of grassland with scattered trees or shrubs of *Philippia excelsa*, *Erica arborea* and *Stoebe kilimandscharica* forming an open-heathland. Between the patches of grassland is tall heathland consisting of the species already mentioned plus *Hagenia abyssinica*, *Senecio amblyophyllus* and *S. elgonensis*. Hedberg attributes the presence of considerable grasslands in this Belt to fire and grazing. The lower part of this Belt is inhabited by the Elgonis, who own large herds of cattle.

Ruwenzori

Ruwenzori lies on the border between Zaire and Uganda. The Ericaceous Belt forms a wide girdle around the mountain from about 3000 to 3900 m in the Nyamgasani to the south, from 2700 to 3800 m in the west and from 2900 to 3800 m in the east. In the upper part of the Belt the important species are *Philippia johnstonii*, *P. longifolia*, *Senecio ericirosenii* and in the lower part, *Erica arborea*, *E. bequaertii* and *Philippia johnstonii* intermingled with *Podocarpus spp.*, *Hypericum* spp., *Rapanea* spp. and *Hagenia abyssinica*. A most conspicuous feature of the Ericaceous Belt is the ground cover, which consists of thick, swelling carpets of mosses and liverworts, the most important species being *Breutelia stuhlmannii*, *Sphagnum* spp. and *Plagiochloa ericicola*. The branches of the heaths are covered by moss cushions 30 cm deep. All patches of level ground seem to be occupied by *Carex runssorroensis* bogs.

Mount Kenya

On Mount Kenya the Ericaceous Belt lies between 3400 and 3550 m. In its lower part there is a zone of closed *Philippia* tall heathland ("forest") with trees up to 6 m tall. Higher up there is open *Philippia–Erica* heathland with an understorey of wet tussock grassland. Along narrow, sheltered valleys there are often narrow strips of ericaceous (chiefly *Philippia keniensis*) heathland ("scrub") up to at least 4000 m, but these are not usually continuous with the Ericaceous Belt proper and should therefore be treated as part of the Alpine Belt. Hedberg ascribes the shrubby nature of the upper part of the Belt to recurrent fires.

Aberdare

The Aberdare Range is in Kenya between Lake Victoria and Mount Kenya. Hedberg distinguishes two zones in the Ericaceous Belt, the moorland zone and the ericaceous-shrub zone. The moorland zone consists of tussock grassland with scattered patches of bamboo, and trees or shrubs of *Hagenia abyssinica*, *Hypericum revolutum*, *Stoebe kilimandscharica* and *Cliffortia nitidula* var. *aequatoralis*. Unexpectedly this zone is separated from the ericaceous-shrub zone above by zones of bamboo and *Hagenia–Hypericum*, montane elements. Hedberg mentions no constituents for the ericaceous-shrub zone.

Virunga volcanoes

These volcanoes occur on the border of Zaire and Rwanda. The altitudinal limits of the Erica-ceous Belt in this complex vary from mountain to mountain, the absolute limits being between 2600 and 3800 m. The pronounced relief of the young volcanoes with radiating sharp ridges separated by deep and narrow valleys causes diversity of habitat. The ridges are usually occupied by *Erica arborea* and *Philippia johnstonii* often festooned with species of *Usnea*, while the valleys support chiefly *Senecio erici-rosenii* and *Hypericum revolutum*.

Kilimanjaro

Kilimanjaro is situated in Tanzania near the border with Kenya. The Ericaceous Belt, like that of Aberdare, consists of moorland and ericaceous-shrub zones. The moorland zone forms a girdle around most of the mountain between about 2800 and 3250 to 3400 m. It consists of tussock grass-land with scattered trees or shrubs of *Philippia excelsa*, *P. trimera* subsp. *jaegeri* and *Erica ar-borea*. Conspicuous in this zone is *Helichrysum meyeri-johannis*. The dominant shrubs of the ericaceous-shrub zone (Fig. 4.3) between 3250 and 3900 to 4100 m are as in the moorland zone, but intermingled with *Protea kilimandscharica*, *Adenocarpus mannii*, *Anthospermum usambarense* and, in marshes and along streams, *Senecio cottonii* and *S. kilimanjari*. In the lower part of the zone the shrubs are usually 1 to 2 m high and stand fairly close together, while in the upper part they are only 0.5 to 1 m tall and cover only a fraction of the ground, leaving much bare soil between them. Fires are apparently commonplace.

Mount Meru

This mountain lies just southwest of Kilimanjaro in Tanzania. The Ericaceous Belt seems to lie between 3000 and 3600 m to the south and between 3050 and 3450 m in the west. In the saddle between Meru and Longorno, the upper limit reaches 3700 m. In the lower part more or less closed *Philippia* tall heathland ("forest") is developed with *Senecio meruensis* in wet places, and in the higher parts the *"Ericetum"* occurs in patches separated by stretches of loose, dry sand with sparse vegetation. There is evidence of fire.

ANGOLA

The only vegetation type in Angola, which can possibly be called heathland, is found on the

Fig. 4.3. Part of the Ericaceous Belt above 3800 m on Mount Kilimanjaro in Tanzania. The peaks of Mawenzi appear in the background. The heaths are chiefly *Philippia excelsa*, *P. trimera* subsp. *jaegeri* and *Erica arborea*. (Photo: Å. Holm.)

central plateau between 1450 and 2000 m. This heathland occurs in "islands" on the Huambo and Huila Plateaux.

The environmental data which follow are obtained from Barbosa (1970) and Diniz (1973). The soils supporting heathland are ferralitic or psammoferralitic with poor drainage. Mean annual rainfall varies from 100 mm on south-southwest slopes to above 1400 mm on the mountain summits; December is the wettest month. Mean annual temperature and humidity are from 19 to 20°C and from 60 to 70% respectively.

Barbosa (1970) describes the vegetation as "altitude grasslands" or uses the local native name *anharas do alto*. The constituent shrubs are *Parinari capensis*, *Protea* spp., *Faurea* spp., *Philippia benguelensis* and *Myrsine africana*, while the understorey comprises species of *Loudetia*, *Ctenium*, *Newtonia*, *Fimbristylis* and *Xyris*. Diniz (1973) describes the vegetation as *savanna com arbustos* and adds *Syzygium guineense* to Barbosa's list. The community is variable in composition, never very dense, and inclined to occur in patches.

MALAWI

In Malawi Chapman and White (1970) describe montane shrubland and montane shrubby grassland for the uppermost, more exposed, rocky slopes of Mount Mlanje above 1980 m. These communities obviously constitute heathland and in fact are referred to by the authors as belonging to the Ericaceous Zone.

Environmental data for the uppermost slopes above the plateaux are scanty. Two stations on Mount Mlanje, the Tuchila plateau and Lujeri Tea Estate stations, both at 1980 m, record mean annual rainfall figures of 3109 mm and 2055 mm respectively. It is conceivable that above 1980 m the rainfall increases to a maximum and then decreases at the highest altitudes — a normal rainfall pattern for high mountains. Frosts occur regularly each year between June and early September. Chapman (1962) describes the soils above the plateaux as black and peaty.

The following low shrubs, often associated with the grasses *Merxmuellera davyi* and *Eragrostis volkensii*, and the tufted sedge *Coleochloa setifera*, comprise the Mount Mlanje heathland: *Erica whyteana*, *E. johnstoniana*, *Blaeria kiwuensis*, *Phylica tropica*, *Thesium whyteanum*, *Crassula sarcocaulis*, *Diplolophium buchananii*, *Helichrysum densiflorum*, *Muraltia flanaganii*, *Plectranthus crassus*, *P. sanguineus*, *Xerophyta splendens* and *Aloe mawii*. No indication is given of the height, density or relative abundance of the constituent shrubs.

MOÇAMBIQUE

Pedro and Barbosa (1955) describe scrub (*matagais*) and bushland (*matos*) communities for Moçambique, which can probably be referred to heathland. The scrub communities are evergreen and found on plateaux (Macanga and Metónia), and in montane (Mossurize, Manica, Gorongoza, etc) and subalpine (Mossurize, Manica and Gurué) areas. The authors describe the dominants, species of *Philippia*, as sclerophyllous and microleptophyllous.

Under bushland they recognise several different types occurring in the montane zone of Metónia and the subalpine zone of the Chimanimani, Zuira, Gorongoza and Penhalonga Mountains.

Montane zone of Metónia

The heathland in this zone occurs on slopes underlain by a complex formed by granitic gneiss formations and primitive systems and lies between 1000 and 1300 m. The community is low, dense and secondary, and consists of *Uapaca* spp., *Faurea* sp., *Protea* sp. and *Philippia* sp. with a sparse grass cover.

Subalpine zone of Chimanimani and other mountains

Several types of heathland are found in this zone, generally above 1700 m and where mean annual rainfall usually exceeds 2000 mm.

(1) **Secondary heathland.** This type consists of *Widdringtonia nodiflora*, *Philippia benguelensis* and *P. simii*.

(2) ***Widdringtonia nodiflora* heathland**

(3) **Sclerophyll–succulent heathland.** Among the constituents are *Podocarpus* sp., *Leucospermum* sp., *Olea woodiana*, *Coffea salvatrix*, *Widdringtonia*

nodiflora, *Philippia benguelensis*, *Erica johnstoniana* and *Aloe* sp.

(4) ***Erica–Philippia* heathland.** This is a low, streambank community consisting of species of *Erica*, *Philippia*, *Passerina*, *Helichrysum* and Restionaceae. For further details of this and the preceding community see Macedo (1970).

RHODESIA

Phipps and Goodier (1962) describe several types of heathland ("scrub") on the Chimanimani Mountains (1400 m) in the eastern part of Rhodesia. The heathlands are found above 1200 m.

The soils of the heathland region of the Chimanimani Mountains are white and sandy, and derived from quartzite. There are no climatic data, but a few general features are known: rainfall is high (varying between 741 and 2997 mm in the adjacent highlands), falling mainly in summer between November and April; the winter months from May to July are generally dry and frequent night frosts occur, becoming more severe with increasing altitude; humidity is high, with mists common above 1500 m; temperatures are cool, and fire is an important ecological factor.

Phipps and Goodier classify the heathlands of the Chimanimani Mountains into ericaceous and proteaceous types and they represent part of the montane flora.

Ericaceous heathlands

(a) **Streamside heathland.** This community occurs along small streams between 1200 m and the altitudinal limit of streams, and comprises a single species, *Philippia pallidiflora*, a low, rounded shrub 0.6–1.0 m tall.

(b) **Rocky quartzite outcrop heathland.** On quartzite slabs this heathland, lying between 1410 and 2400 m, consists of columnar, well-spaced tall shrubs of *Philippia hexandra* growing up to 4 m tall, but among quartzite crags, where it is more common, *P. hexandra* (and possibly other species of *Philippia*) forms a dense interstitial growth up to 3 m tall. Between 1800 and 2400 m this community becomes widespread and is strongly mixed with other heathland plants such as *Phylica ericoides*,

Passerina montana, *Schistostephium oxylobum* and *Myrsine africana*. Both this and the preceding community are very fire-sensitive, though readily regenerating, and occur on poor, sandy, acid, quartzite soils with a moderate to high water table, but free drainage.

(c) *Erica* heathland. *Erica eylesii*, *E. pleiotricha*, *E. gazensis* and *E. johnstoniana*, up to 0.5 m tall, occasionally form communities in areas partly protected from fire where the soil is too shallow to permit competition of larger plants.

Proteaceous heathlands

(a) *Protea gazensis* heathland. This heathland covers much of the rolling schist grassland between 1350 and 1950 m and consists of *Protea gazensis* shrubs, 1 to 2 m tall, varying in density. Associated plants are *Protea welwitschii* and below 1500 m depauperate *Syzygium cordatum*, *Parinari curatellifolia* subsp. *mobola* and *Faurea saligna*.

(b) *Protea welwitschii–Leucospermum saxosum* heathland. On well-drained rocky quartzite slopes above about 1500 m, away from the streams, steep crags and flattish grassland, a heathland community consisting of fairly close-spaced shrubs 0.6 to 1.2 m tall, is developed in which the dominant species are either *Protea welwitschii* or *Leucospermum saxosum*. Associated shrubs are *Protea gazensis*, *P. crinita*, *Aeschynomene grandistipulata*, *A. leptobotrya* and *Diplophium swynnertonii*; herbs and suffrutices include *Aloe hazeliana*, *Plectranthus chimanimaniensis*, *Lobelia cobaltica*, *Otiophora inyangana*, *Eragrostis longipaniculata* and *Loudetia simplex*. This community appears to be variably fire-sensitive.

(c) Mixed heathland. In certain craggy areas between 1360 and 1800 m there is an apparently fire-resistant heathland dominated by *Philippia* spp., *Olea africana*, *Syzygium* sp. and *Plectranthus* sp., with or without *Triumfetta pilosa* var. *effusa* and *Anthospermum vallicola*.

(d) *Thesium* heathland. This is a minor community occurring in patches of up to 0.4 ha on quartzite only and consisting of the shrub *Thesium* sp. (Wild collection No. 2883).

SOUTH AFRICA

There are three main types of mountain heathland in South Africa, namely montane heathland, subalpine heathland and alpine heathland. They will be discussed in turn.

Montane heathland

The term montane is used here in its broad sense to include submontane. Montane heathland is found on mountains below the upper limit of forest. The mountains may be ranges such as the Drakensberg and Soutpansberg or isolated mountains such as Thaba 'Nchu, Hangklip and Table Mountain (Natal).

There is apparently a type of heathland on the Soutpansberg Mountains in the northern Transvaal. H.C. Taylor (pers. comm., 1976) describes it as "pseudo-fynbos" with dominants such as *Hemizygia obermeryerae*, *Anthospermum ammanioides*, *Psoralea pinnata* and *Helichrysum kraussii* with an upper layer cover of from 60 to 90% chiefly of "ericoids".

Van der Schijff and Schoonraad (1971) describe heathland (fynbos) occurring on the summit plateau of Mariepskop (1946 m) on the Transvaal Drakensberg. The community (Fig. 4.4.) is said to occur in shallow basins surrounded by huge boulders. The species listed as dominant or conspicuous are *Passerina montana*, *Erica leucopelta* vars. *ephebioides* and *luxurians*, *E. woodii*, *Vaccinium exul*, *Cliffortia serpyllifolia*, *C. nitidula* subsp. *pilosa*, *Phylica paniculata*, *Psoralea pinnata* and *Muraltia flanaganii*, often accompanied by *Protea* spp. Associates are numerous.

At Cathedral Peak in the Natal Drakensberg, heathland occurs along the top edge of the Cave Sandstone cliffs at 1800 m. Described as "Cave Sandstone Scrub" by Killick (1963) this heathland, which is fairly open, consists of *Protea roupelliae*, *Myrica pilulifera*, *Cliffortia linearifolia*, *Passerina montana*, *Erica drakensbergensis* and *E. straussiana* and a number of herbs including *Restio fruticosus* and *R. sieberi*.

Roberts (1966b) describes a *Passerina–Cliffortia* association occurring above 1980 m on Thaba 'Nchu, in the eastern Orange Free State. The most important constituents are *Passerina montana*, *Erica maesta*, *Cliffortia nitidula* subsp. *pilosa*,

Fig. 4.4. Montane heathland on Mariepskop, Transvaal Drakensberg. (Photo: P. Vorster.)

Stoebe vulgaris and *Anthospermum tricostatum*. A photograph illustrating the community shows it to be dense heathland about 1.5 m tall.

Edwards (1967) illustrates *Cliffortia nitidula* subsp. *pilosa* fynbos in the Qudeni area of Natal at 1740 m. This heathland, also comprising *Protea subvestita* and *Erica caffrorum*, is seral to forest. On moist cliff faces of Table Mountain in Natal at 762 m *Passerina filiformis* heathland is present (Killick, 1958). Associated with the dominant are *Philippia evansii*, *Anthospermum* sp., *Stoebe vulgaris*, *Psoralea pinnata* and *Ficus* spp.

Roberts (1966a) describes for Hangklip Mountain near Queenstown in the eastern Cape Province what he calls "relics of an earlier heath vegetation". These plants are *Chrysocoma tenuifolia*, an invader from the karoo, *Cliffortia paucistaminea*, *Anthospermum tricostatum*, *Passerina montana* and *Erica schlechteri*. The shrubs are associated with *Bromus leptocladus* and other temperate grasses.

Story (1952) mentions *Cliffortia linearifolia* "macchia" in the Amatole Mountains of the eastern Cape Province (Fig. 4.5). This heathland, not more than 1.2 m high, occurs below 1220 m. If grassland is protected, this heathland takes over very slowly, but if the grassland is heavily grazed it rapidly takes over completely suppressing the grassland. On Gaikaskop, part of the Amatole Range, in areas protected from fire, Roberts (1963) distinguishes a heathland composed of *Felicia filifolia*, *Arrowsmithia styphelioides*, *Cliffortia eriocephalina*, *Erica leucopelta*, *E. caffrorum*, *Passerina montana*, *Phylica* sp., *Stoebe vulgaris* and *Restio* sp.

Acocks (1953) recognises a veld type which he calls "False Macchia" (Fig. 4.6). This veld type stretches for 580 km along the mountains of the southern and eastern Cape Province, e.g. Witteberge, Langeberge, Swartberge, Kouga Mountains and Suurberge, from Montague and Touws River in the west to near Grahamstown in the east, and covers an area of 17 866 ha. Acocks

Fig. 4.5. *Cliffortia linearifolia* heathland showing up white at Hogsback in the Amatole Mountains in the eastern Cape Province. Grasses constitute the understorey.

states that today this veld type (which is clearly heathland) is indistinguishable from the true fynbos of the southwestern Cape province, but that there are indications that in its natural condition it would have been transitional from Dohne *sourveld* to true fynbos. Recently J.P.H. Acocks (pers. comm., 1976) has postulated what he calls a "rotating climax" for the southern Cape Mountains, involving a sequence of short forest, fire, grassland, fynbos, short forest, fire, grassland, fynbos and so on. He states that under present conditions of too frequent burning and selective overgrazing of the grasses, the forest and grassland have almost disappeared and the fynbos stage is tending to become permanent. Unfortunately Acocks does not provide a species list, but Dyer (1937) gives a fairly comprehensive list for the Albany and Bathurst districts in the easternmost part of "False Macchia", to which readers are referred. Dyer states that the heathland is domi-

nated by "sclerophyllous shrubs belonging to the genera *Erica*, *Berzelia*, *Cliffortia*, *Passerina*, *Aspalathus* and *Metalasia*, etc."

Subalpine heathland

This heathland is confined almost entirely to the Subalpine belt of the Drakensberg in the Cape Province and Natal between 1830 and 2895 m. It also occurs in the eastern Cape Province on the Amatole mountains between 1220 and 1938 m. The following account of subalpine heathland in the Drakensberg is derived chiefly from Killick's (1963) researches at Cathedral Peak. More detailed information may be obtained from this publication.

Drakensberg

The soils of the Little Berg, situated below the main Drakensberg Escarpment, are derived from

Fig. 4.6. Montane heathland (false macchia) on the Suurberge in the eastern Cape Province. Genera present: *Elytropappus, Phylica, Erica, Cliffortia Passerina, Leucospermum, Metalasia, Coleonema, Restio, Themeda*, etc. (Photo: J.P.H. Acocks.)

basaltic lavas; the "black clay soils" are deep, with ill-defined horizons. Soil reaction (pH values in KCl) in the A horizon varies from 3.9 under grassland to 5.9 under forest, with slightly lower pH values at depth. Organic matter content is high, 12% in the A horizon not being uncommon, but decreases markedly (to less than 5%) in the subsoil. The high organic matter content contributes to a high value for cation exchange capacity and total absorbed bases. In the subalpine zone, the clay mineral smectite (=montmorillonite) tends to predominate on steep poorly-weathered sites, while kaolinite and gibbsite are the most common clay minerals on the deeply weathered, flatter land surfaces (J.E. Granger and R.W. Fitzpatrick, pers. comm., 1977). The soils generally have a high phosphate-fixing capacity and a high level of exchangeable aluminium, which is toxic to the roots of some plants (C.N. Macvicar, pers. comm., 1977).

Temperature records taken at 1860 m are illustrated in the Deasy (1941) chart in Fig. 4.7. *A*. The curves for mean diurnal fluctuation, *B* and *C*, show that air temperature is "cool" to "mild", and only in the absolute values represented by curves *A* and *D* does the temperature reach "hot" and "cold", The highest temperature recorded is 31.2°C in November 1951 and the lowest −3.6°C in June 1953. Frosts are almost a daily occurrence in winter. The curves for grass minimum temperature (*E* for mean daily minimum, and *F* for absolute minimum) are consistently lower than the corresponding values in the Stevenson Screen.

Wind is an important ecological factor. The *bergwinds*, which blow from the west during late winter and spring, are prevalent at a time when both humidity and soil moisture content are low. Winds are also important because they blow during the dry season when fire hazard is at a maximum.

From Table 4.2 it will be seen that about 85% of

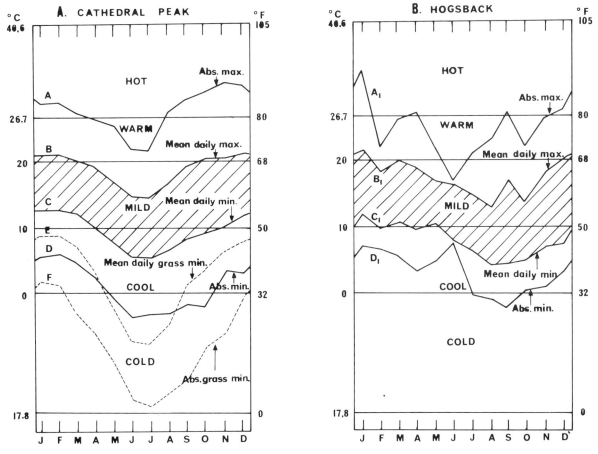

Fig. 4.7. Temperature charts for Cathedral Peak (air temperature, ten years; grass temperature, six years; (A) and Hogsback, Amatole Mountains (B).

the rain falls during the summer months between October and March. Mean annual rainfall varies from 1418 to 2017 mm depending upon altitude.

Snow, which can be expected at any time between April and September, is restricted to the upper part of the Subalpine Belt and rarely reaches the lower parts.

Fire is an extremely important environmental factor in the Drakensberg. Fires may be caused either naturally, by lightning or colliding boulders, or by man. It is probable that natural fires have been a factor of the environment ever since the climate included a dry season.

Subalpine heathland (="fynbos" of Killick, 1963) is the climax community of the Subalpine Belt. Because of recurrent grass fires, it is limited in extent, occurring in situations providing some protection from fire. The most extensive and best

developed stands of heathland are to be found on steep valley and escarpment slopes at the head of the main rivers, for example, the Tseketseke, Indumeni, Mlambonja and Tugela Rivers (Fig. 4.8).

The community consists of shrubs between 0.9 and 3 m tall, the majority of which are evergreen, though some may be deciduous. Most of the constituents have small leaves which are ericoid, elliptic or linear, variously coriaceous and glossy or grey-lanate. The density of the community varies considerably — from shrubs scattered in grassland to an almost impenetrable tangled mass of vegetation.

Heathland in the Drakensberg can be either pure with one species dominant or mixed with several dominants (Fig. 4.9). The most important dominants are probably *Passerina filiformis*, *Philippia*

TABLE 4.2

Mean monthly rainfall in the Cathedral Peak area, South Africa (FD = Forestry Department)

	FD Meteorological Station, Little Berg	FD, IIBr[1], Little Berg	FD, IIAW[1], Little Berg	Organ Pipes Pass, Summit Drakensberg
Altitude (m):	1860	1981	2287	2927
Latitude:	28°59'E	29°0'S	29°0'S	29°1'S
Longitude:	29°14'E	29°13'E	29°13'E	29°11'E
Period (yr):	9	10	6	3
January	197 mm	248 mm	284 mm	342 mm
February	289	299	392	295
March	212	238	284	207
April	57	74	76	49
May	35	38	52	46
June	13	15	24	27
July	13	13	13	1
August	39	39	45	16
September	67	68	93	61
October	108	122	164	109
November	157	183	265	226
December	231	252	325	230
Total	1418	1589	2017	1609

[1] IIBr and IIAW are respectively in the middle and at the top of the same catchment.

evansii and the conifer *Widdringtonia nodiflora.* Pure heathland stands can be formed by the following species:

Passerina filiformis
Philippia evansii
Widdringtonia nodiflora
Passerina montana
Erica ebracteata
Macowania conferta
Buchenroedera lotononoides
Anthospermum aethiopicum

Rhus discolor
Buddleja loricata
Protea dracomontana
P. subvestita
Syncolostemon macranthus
Calpurnia intrusa
Melianthus villosus

Shrubs which do not aggregate to any great extent include:

Senecio haygarthii
Asparagus scandens
A. microraphis
Stoebe vulgaris
Euphorbia epicyparissias
Myrsine africana
Artemisia afra
Psoralea caffra
Polygala myrtifolia
Rhus dentata

Cliffortia spathulata
Diospyros austro-africana
    var. *rubriflora*
Erica straussiana
Anisodontea julii
    subsp. *pannosa*
Berkheya draco
Helichrysum tenax
Lasiosiphon anthylloides
Lotononis trisegmentata

Very characteristic of subalpine heathland is the cycad *Encephalartos ghellinckii*. It is particularly abundant in the Tseketseke Valley.

Subordinate to the shrubs in dense heathland is a layer of grasses, herbs and ferns. The three most constant species are probably *Polystichum* sp., *Cymbopogon validus* and *Berkheya macrocephala.* The remaining species are:

Pellaea quadripinnata
Gleichenia umbraculifera
Festuca costata
Pentaschistis pilosogluma
Agapanthus campanulatus
Scilla natalensis
Eriospermum cooperi
Cyrtanthus erubescens
Anemone fanninii
Ranunculus baurii (usually in wet places)

Gunnera perpensa
Alchemilla natalensis
Geranium pulchrum
Indigofera cuneifolia
I. longebarbata
Sebaea macrophylla
Diclis reptans
Helichrysum setosum
H. cooperi
H. umbraculigerum

Climbers include *Riocreuxia torulosa* var. *tomentosa, Dioscorea sylvatica* and *Clematis brachiata.*

Amatole Mountains

These mountains lie to the north and northwest of Keiskammahoek in the eastern Cape Province and were included in an ecological study of the Keiskammahoek District by Story (1952).

The soils conform very closely to the two geological formations present, the Beaufort Series and Dolerite. The sedimentary soils may be divided

Fig. 4.8. Subalpine heathland showing as dark patches on steep escarpment slopes below the Eastern Buttress of the Natal Drakensberg. *Podocarpus latifolius* forest (Montane Belt) may be seen below the Cave Sandstone cliffs at bottom left.

Fig. 4.9. Mixed subalpine heathland at 2070 m in the Tseketseke Valley, Cathedral Peak area, Natal Drakensberg. Constituents present are *Macowania conferta*, *Encephalartos ghellinckii*, *Widdringtonia nodiflora*, *Diospyros austro-africana* var. *rubriflora*, *Rhus dentata*, *Syncolostemon macranthus* and *Polemannia montana* with a herb layer of *Cymbopogon validus*, *Anemone fanninii*, *Berkheya macrocephala* and *Polystichum* sp.

into shallow grey loams, grey loams on clay and yellowish brown sandy loams on sandstone. The dolerite soils consist of immature black clays, deep red clays and black, well-developed clays.

Story established a temperature station on the slopes of Hogsback at 1615 m. The data (for one year only) are presented in the Deasy chart in Fig. 4.7B. The curves for mean diurnal fluctuation, B_1 and C_1, show that temperature is "cool" to "mild". As at Cathedral Peak, only the absolute values represented by curves A_1 and B_1 reach "hot" or "cold". Frosts are commonplace during winter, sometimes freezing the soil. Mean rainfall varies from 540 mm at 1420 m to 670 mm at 1620 m.

As in the Drakensberg, fire is an important ecological factor in the Amatole Mountains. Story (1952) considered that the mountain slopes above the forests were originally under heathland

(=Story's "macchia"), and that at some time in South Africa's prehistory man started firing the slopes. The result was that heathland became sparse and the fire-resistant grasslands obtained a footing and spread, until the heathland was confined to sheltered places protected from fires. With moderate grazing and frequent fires, these grasslands remained stable. However, as the European populations and their live stock increased, the grasslands were cropped short and fires became weaker and less frequent. Heathland began migrating out of its strongholds, the patches linked, and within a comparatively short time it had re-established itself over much of its former area. Story states that it is doubtful whether natural fires could have occurred sufficiently often to keep the vegetation at the grassland stage.

Subalpine heathland is the climax community of the Subalpine Belt in the Amatole Mountains. According to Story (1952), heathland is dominated by *Cliffortia paucistaminea* growing up to 2 m high and *Erica brownleeae* up to 4 m high (Fig. 4.10). This heathland grows best on south-facing mountain slopes and grows in such an interlacing mass

Fig. 4.10. Subalpine heathland (*Erica brownleeae* and *Cliffortia paucistaminea*) with *Helichrysum argyrophyllum* in the foreground on the Wolf Plateau in the Amatole Mountains.

that it is practically impenetrable by man or live stock. Other constituents of the heathland are *Protea lacticolor*, *Bobartia gracilis*, *Rubus* spp., *Pteridium aquilinum*, *Stoebe* spp., *Passerina* sp., *Helichrysum argyrophyllum* and *Metalasia muricata*.

Alpine heathland

Alpine heathland occurs in the Alpine Belt lying between 2860 and 3484 m. This belt occupies a narrow strip along the top edge of the Drakensberg and extends downwards into Natal, the Cape Province and Lesotho. It is also found on outlying peaks such as Cathedral Peak (3004 m) and the Inner and Outer Horns (3018 and 3009 m respectively) and in Lesotho on the Maluti and other mountains of sufficient height.

The summit plateau of the Drakensberg presents a cheerless, bleak and rather barren-looking picture to the observer. In parts the plateau forms part of the Cretaceous post-Gondwana land-surface, while the higher parts above 3075 m belong to the Gondwana cycle of Jurassic age (King, 1972). Carroll and Bascomb (1967) classify the soils chiefly as lithosols on lava, while Van der Merwe (1941) describes them as "Mountain Black Clays". Soils are shallow (< 45 cm) and sometimes absent. During summer the soils of the summit become very wet and waterlogged, while during winter they are subjected to freezing every night and thawing every day. Cryonival phenomena are common.

The climate of the Alpine Belt, as might be expected, is severe. The temperature régime (Fig. 4.11) is characterised by cool to mild summers and cold winters with absolute temperatures reaching degrees of hot and frigid. The highest temperature recorded in eleven years at Letšeng-la-Draai (sometimes spelt Letšeng-la-Terai) at 3050 m

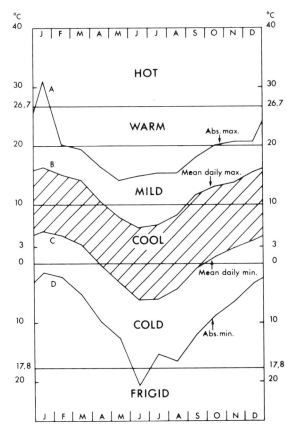

Fig. 4.11. Deasy (1941) temperature chart for Letšeng-la-Draai in northeastern Lesotho.

in northeastern Lesotho is 31°C in January 1972 and the lowest −20.4°C in June 1967. Mean annual temperature is 5.7°C and the estimated mean number of days with frost per year is 183. The snow in winter can lie for periods up to two months, especially on the southern slopes.

Mean annual rainfall varies from 633.9 mm at Station P_1 (3292 m) in the Oxbow Region to 1609 mm at Organ Pipes Pass (2927 m; Table 4.2). Most of the rain falls between October and March during spring and summer.

Relative humidities are high during the wet season when there is an abundance of cloud, but low during the dry season and in early spring. Humidity data are available for Letšeng-la-Draai for the period 1966–1976. The mean relative humidity for January is 54%, the highest monthly mean being 72% in January 1974. The mean humidity for August is 37%, the lowest monthly mean being 18% in August 1969; in September the corresponding figures are 34%, and 21% in

September 1969. Schelpe (1946) recorded a relative humidity of 4% in spring during windy weather on the summit of the Drakensberg. High winds are most common in late winter and spring, and can often attain considerable velocities.

The remarks on fire on p. 109 apply to the Alpine Belt as well.

The climax community on the summit of the Drakensberg is *Erica–Helichrysum* Alpine Heathland, consisting of dwarf and low shrubs 15 to 60 cm high. But for fire, heathland would probably occupy greater areas of the summit; parts of the summit support no heathland at all.

The dominants which cover the largest areas of the summit belong to the genera *Erica* and *Helichrysum*, hence the name "*Erica–Helichrysum* Alpine Heathland". Altogether there are five distinct heathland communities.

(a) *Erica dominans* **heathland**. This community (Fig 4.12) is the most extensive of the heathland communities. The dominant, *Erica dominans*, is a dwarf or low shrub 5 to 45 cm high with minute, leathery, closely adpressed leaves. The plant has an olive-green appearance and attains full flower in October. Occurring on level portions of the summit, it forms fairly dense communities invariably interspersed with alpine grasses.

Other constituents of this community are *Helichrysum trilineatum*, *Erica frigida*, *E. glaphyra*, *E. flanaganii*, *Chrysocoma tenuifolia*, *Thesium imbricatum*, *Cliffortia browniana*, *Gnidia polystachya* var. *congesta*, *Lotononis galpinii*, *Clutia nana*, *Euryops acraeus*, *E. decumbens* and *Anthospermum hispidulum*.

(b) *Erica–Helichrysum* **heathland**. This is a common community usually found above 3200 m. The dominants are *Erica dominans* and *Helichrysum trilineatum*. Casual constituents are as in the preceding community.

(c) *Erica glaphyra* **heathland**. *Erica glaphyra* forms a pure type of heathland on broken promontories at the edge of the summit plateau, a habitat which provides a certain amount of shelter and is fairly moist.

(d) *Helichrysum–Passerina* **heathland**. This community appears to be limited in extent: the author

Fig. 4.12. *Erica dominans* heathland on the summit of the Drakensberg at 3050 m south of Mont aux Sources in Lesotho.

has only seen it at the edge of the escarpment near Castle Buttress. The dominants are *Helichrysum trilineatum* and *Passerina montana*. The habitat is broken, hence the presence of *Merxmuellera stereophylla*, a common summit lithophyte.

(e) Boulder-field heathland

Situated on the summit are fairly large areas supporting boulders varying in their density of aggregation. The habitat is stable and not to be confused with scree. The heathland growing in this habitat is from 0.6 to 1.2 m tall and sometimes quite dense. The tallness of this heathland as compared with the other types is probably due to the protection from fire afforded by the boulders. The dominants are three composites *Athanasia thodei*, *Helichrysum trilineatum* and *Eumorphia sericea*, with *Merxmuellera drakensbergensis* and other alpine grasses filling the intervening gaps. Two ferns often grow in the shade of the boulders, namely *Dryopteris inaequalis* (a depauperate, high-altitude form) and *Woodsia montevidensis* var. *burgessiana*.

LESOTHO

Heathland in Lesotho is not abundant: much of it has been burnt or cut out and used as fuel. However, a pseudo-heathland, dominated by *Chrysocoma tenuifolia* an invader from the Karoo, is common whenever there has been overgrazing or other mismanagement. In 1938 Staples and Hudson estimated that *C. tenuifolia* occupied more than 13% of the mountain area of Lesotho. Two types of heath may be distinguished in the mountain area, namely subalpine heath and alpine heath.

Subalpine heathland

Jacot Guillarmod (1971) refers to isolated patches of *Leucosidea sericea*, *Buddleja loricata*, *Passerina montana* and tall-growing, woody *Erica* species. During a recent visit (December, 1977) to Lesotho the present author observed several types of heath. In the Bushmen's Pass–God Help Me Pass area at about 2530 m the heath, up to 1.5 m tall, is composed of *Philippia* sp. (Killick 4226),

Erica algida, Passerina montana, Buddleja loricata, Pentzia cooperi, Euryops tysonii, Lotononis trisegmentata var. *robusta, Artemisia afra* and *Selago flanaganii*. In the Sehonghong River Valley between Mokhotlong and Sani Pass *Pentzia cooperi* is dominant in parts, while in the Moteng Pass in northeastern Lesotho, the heath consists of *Athanasia thodei, Pentzia cooperi, Buddleja loricata, Clutia natalensis, Helichrysum trilineatum, Rhus* spp., *Artemisia afra, Diospyros* spp. and a number of other shrubs. Most of the subalpine heath types in Lesotho also contain *Chrysocoma tenuifolia* and to a lesser extent *Felicia filifolia*.

Alpine heathland

As for the Drakensberg, South Africa (see p. 112).

REFERENCES

Acocks, J.P.H., 1953. Veld types of South Africa. *Mem. Bot. Surv. S. Afr.*, 28: 192 pp.

Barbosa, L.A. Grandvaux, 1970. *Carta Fitogeográfica de Angola*. Inst. de Invesigação Científica de Angola, Luanda, 323 pp.

Burtt Davy, J., 1935. A sketch of the forest vegetation and flora of tropical Africa. *Emp. For. J.*, 14(2): 191–201.

Carroll, D.M. and Bascomb, C.L., 1967. *Notes on the Soils of Lesotho*. Tech. Bull. No. 1. Land Resources Division, Directorate of Overseas Surveys, Tolworth, 75 pp.

Chapman, J.D., 1962. *The vegetation of the Mlanje Mountains, Nyasaland*. Government Printer, Zomba, 78 pp.

Chapman, J.D. and White, F., 1970. *The Evergreen Forests of Malawi*. Commonwealth Forestry Institute, Oxford, 190 pp.

Coe, M.J., 1967. *The Ecology of the Alpine Zone of Mount Kenya*. W. Junk, The Hague, 136 pp.

Deasy, G.F., 1941. A new type of temperature graph for the geographer. *Mon. Weather Rev.*, 69: 229–232.

Diniz, A. Castantheira, 1973. *Carasteristicas Mesológicas de Angola*. Missão de Inquéritos Agrícolas de Angola, Nova Lisboa, 482 pp.

Dyer, R.A., 1937. The vegetation of the divisions of Albany and Bathurst. *Mem. Bot. Surv. S. Afr.*, 17: 138 pp.

Edwards, D., 1967. A plant ecology survey of the Tugela River Basin, Natal. *Mem. Bot. Surv. S. Afr.*, 36: 285 pp.

Edwards, D.C., 1940. A vegetation map of Kenya with particular reference to grassland types. *J. Ecol.*, 28: 377–385.

Gyldenstolpe, N., 1924. Zoological results of the Swedish expedition to Central Africa 1921. Vertebrate. I. Birds. *K. Sven. Vetensk. Akad. Handl. Tredje Ser.*, 1: 3.

Hedberg, O., 1951. Vegetation belts of the east African mountains. *Sven. Bot. Tidskr.*, 45: 140–202.

Hedberg, O., 1964. Features of afro-alpine plant ecology. *Acta Phytogeogr. Suecia*, 49: 144 pp.

Hedberg, O., 1971. The high mountain flora of the Galama Mountain in Arussi Province, Ethiopia. *Webbia*, 26: 101–128.

Jackson, J.K., 1956. Vegetation of the Imatong Mountains, Sudan. *J. Ecol.*, 44: 341–374.

Jacot Guillarmod, A., 1971. *Flora of Lesotho*. J. Cramer, Lehre, 474 pp.

Killick, D.J.B., 1958. An account of the plant ecology of the Table Mountain area of Pietermaritzburg, Natal. *Mem. Bot. Surv. S. Afr.*, 32: 133 pp.

Killick, D.J.B., 1963. An account of the plant ecology of the Cathedral Peak area of the Natal Drakensberg. *Mem. Bot. Surv. S. Afr.*, 34: 178 pp.

King, L.C., 1972. *The Natal Monocline: explaining the Origin and Scenery of Natal, South Africa*. Geology Department, University of of Natal, Durban, 113 pp.

Klute, F., 1920. *Ergebnisse der Forschungen am Kilimandscharo, 1912*. Berlin.

Macedo, J. de Aguiar, 1970. *Carta de Vegetação da Serra da Gorongosa*. Communicação No. 50, 74 pp. Instituto de Investigacão Agrónomica de Moçambique, Lourenço Marques.

Pedro, J. Gomes and Barbosa, L.A. Grandvaux, 1955. A vegetação. In: *Esboco do Reconhecimento Ecológico-agrícola de Moçambique*, Vol. 2. Memórias e Trabalhos, No. 23. Centro de Investigação Cientifica Algodoeira, Lourenço Marques, pp. 67–224.

Phipps, J.B. and Goodier, R., 1962. A preliminary account of the plant ecology of the Chimanimani Mountains. *J. Ecol.*, 50: 291–319.

Richards, P.W., 1963. Ecological notes on West African vegetation III. The upland forests of Cameroons Mountain. *J. Ecol.*, 51: 529–554.

Roberts, B.R., 1963. A contribution to the ecology of Cathcart and environs with special reference to slope exposure and soil pH. *J. S. Afr. Bot.*, 29: 153–162.

Roberts, B.R., 1966a. Observations on the temperate affinities of the vegetation of Hangklip Mountain near Queenstown, Cape Province. *J. S. Afr. Bot.*, 32: 243–260.

Roberts, B.R., 1966b, *The Ecology of Thaba 'Nchu. A Statistical Study of Vegetation/Habitat Relationships*. Thesis, University of Natal, Pietermaritzburg, 373 pp. (unpublished).

Ross, R., 1956. Some aspects of the vegetation of Ruwenzori. *Webbia*, 11: 451–457.

Salt, G., 1954. A contribution to the ecology of upper Kilimanjaro. *J. Ecol.*, 42: 375–423.

Schelpe, E.A.C.L.E., 1946. *The Plant Ecology of the Cathedral Peak Area*. Thesis, University of South Africa, Pretoria, 185 pp. (unpublished).

Scott, H., 1952. Journey to the Gughé Highlands (southern Ethiopia), 1948–9; biogeographical research at high altitudes. *Proc. Linn. Soc. Lond.*, 163: 85–189.

Staples, R.R. and Hudson, W.K., 1938. *An Ecological Survey of the Mountain Area of Basutoland*. Crown Agents for the Colonies, London, 68 pp.

Story, R., 1952. A botanical survey of the Keiskammahoek district. *Mem. Bot. Surv. S. Afr.*, 27: 184 pp.

Synge, P.M., 1937. *Mountains of the Moon. An Expedition to the Equatorial Mountains of Africa.* Drummond, London, 221 pp.

Van der Merwe, C.R., 1941. Soil groups and subgroups of South Africa. *Dep. Agric. Sci. Bull.*, 231: 316 pp.

Van der Schijff, H.P. and Schoonraad, E., 1971. The flora of the Mariepskop Complex. *Bothalia*, 10: 461–500.

Chapter 5

MACARONESIAN HEATHLANDS[1]

C.N. PAGE

INTRODUCTION

The North Atlantic island groups which together form Macaronesia (the Azores, Madeira, the Canary Islands and Cape Verde Islands) are all of volcanic origin, mountainous height and rugged topography. Although remote from one another, for reasons of geological age and make-up, ocean-icity of climate and palaeofloristic history these archipelagos show strong biotic links with one another. Heathlands form a significant feature of the vegetation of the three more northerly island groups, although they are absent from the southern (and much drier) Cape Verde Islands.

Differing distances from continental masses cause some differences in the altitudinal distribution and floristic composition of heathland vegetation between the archipelagos. In the Azores (lat. 39°N), the Macaronesian heathland vegetation occupies chiefly the lower altitudes. Higher up it is replaced by juniper scrub. Moving southward, in Madeira (lat. 33°N) the lower zones (to about 750 m) are occupied by a coastal scrub, whilst heathland vegetation is well-developed throughout the mountains, ascending to the highest peaks. In the Canary Islands (lat. 28°N) the overall climate is drier. The lower zones (to c. 760 m) are dominated by well-developed xerophyte scrub, and the heathland vegetation begins at the cloud base on the north and northeasterly aspects of the islands. It is replaced above the upper limits of impinging cloud (about 1200 m) by dry *Pinus canariensis* forest (Page, 1974).

Of the three northern Macaronesian archipelagos, the heathland vegetation which has been most thoroughly studied is that of the Canary Islands. This account thus centres principally upon the Canarian heathland vegetation, and draws comparison between this and that of Madeira and the Azores.

PECULIARITIES OF MACARONESIAN HEATHLANDS

Macaronesian heathland vegetation is both floristically poor and vegetationally luxuriant. It exists in an environment of frequent cloud and light precipitation which is so characteristic a feature of the Atlantic, oceanic climate of the Macaronesian mountains. Dominated by *Erica arborea* (*Erica azorica* in the Azores) and *Myrica faya*, the heathland vegetation varies with aspect and exposure, with little change in floristic composition, from a uniform knee- or waist-high sward, to a dense forest reaching 10 m or more. In sheltered pockets heathland vegetation, often already arborescent, merges insensibly into broad-leaved evergreen forests dominated by species of Lauraceae. Because of the floristic poverty of these insular heathland and forest communities and the resulting wide ecological amplitude of their principal species, it is by no means uncommon for members of the two communities to intermingle closely, so that it is difficult to draw hard and fast lines between arborescent heathland and laurel-forest vegetation.

The heathland vegetation is also unusually rich (for heathland) in cryptogams, especially epiphytic lichens, mosses, and terrestrial and epiphytic ferns (Page, 1976). Characteristic but small numbers of species of these, often in large numbers of individuals, form a significant component of the

---

[1] Manuscript completed October, 1976.

117

heathland vegetation, again underlining the close ecological relationship in this mild, moist climate, between heathland vegetation and evergreen forest.

CLIMATIC CONDITIONS OF MACARONESIAN HEATHLANDS

The far more oceanic and hence more equable, moister climate of the northern Macaronesian archipelagos distinguishes their general climate from that of the adjacent African continent at equivalent latitude. The Canary Islands and Madeira particularly are exposed to prevailing northeast trade winds and the cooling influence of the Canaries current. In all the archipelagos, cloud cover is particularly frequent, and the constancy of this cloud, with the frequent light precipitation which it brings wherever it impinges on the topography, is undoubtedly the greatest single factor permitting an evergreen heathland and forest vegetation to flourish. Within the cloud-influenced zone of each island, exposure and porosity of the underlying rock largely determine the balance of local distribution between evergreen heathland and forest. In the Canary Islands the cloud belt normally occupies an altitude between about 800 and 1500 m, occasionally rising to 2000 m.

Associated with the presence of the cloud belt are several environmental factors: air humidity in these islands reaches its maximum value in the Canaries within the cloud belt, dropping both above and below this altitude. The cloud belt is also responsible for great reduction of insolation and hence evaporation at this altitude, and, in addition to the normal temperature drop with altitude, the trade-wind effect is responsible for further temperature differences of over 10°C between north (windward) and south (leeward) coasts of the larger Canary Islands (Bramwell and Bramwell, 1974). Mean temperatures recorded at La Laguna, just below the heathland zone on the north coast of Tenerife give a mean annual temperature range from 13 to 21°C (Ceballos and Ortuño, 1951).

The diurnal range of temperature may however be considerable [reportedly as much as 25°C at Las Cañadas (c. 2200 m) on Tenerife: Bramwell and Bramwell, 1974]. Although undoubtedly considerably less than this in the cloud zone, experience indicates that on most of the islands in summer it is

sufficient to induce a heavy dew-fall, especially when the cloud cover breaks up at night. This dew-fall doubtless helps greatly towards the water balance of the heathland vegetation, where the many fine surface rootlets (especially of *Erica*) seem adapted to take maximum advantage of this moisture.

Rainfall figures for the heathland are few, and probably reflect only a proportion of the incoming precipitation to the community. Nevertheless the highest rainfall figures for Gran Canaria occur between 1000 and 2000 m where they range from 690 to 1160 mm annually (Sunding, 1972). However, the greatest source of regular moisture in the cloud zone is undoubtedly cloud drip from the heathland trees and bushes themselves. Sunding (1972, p. 106) notes that the montane region of Gran Canaria is a region of rapid change: "At one moment one walks in dense wet fog and the next moment finds oneself for a while under clear sky and shining sun". Personal experience amply confirms this rapid change in all the western islands of the archipelago. As moisture-laden cloud drifts through the leaves and twigs of the heathland vegetation, all plant surfaces freely run and drip with abundant water whilst no rain is falling. Sunding (1972, p. 110) quotes that, in experiments in the laurel forest of Tenerife, a collecting vessel placed in the open gave an annual precipitation of 956 mm, compared with a similar vessel placed below trees which gave 3038 mm in the same locality over the same period (i.e. a 218% increase). It is doubtful whether the broad-leaved laurel forests gain as much direct benefit from this cloud drip as do the more exposed heathland vegetation and the lower parts of the pine forests at equivalent altitudes. In both the heathland and pine vegetation the plants even more directly intercept the passing cloud with finely dissected foliage which provides a large surface to condensation (Fig. 5.1). In such vegetation, *Erica* shrubs and pine trees frequently glisten and drip with water even with only modest density of passing cloud. Beneath the shrubs and trees the soil benefits from resulting heavy moisture drip whilst bare ground between trees remains dry and dusty.

This montane Canarian heathland climate compares favourably with that of Madeira and the much more oceanic Azores.

In Madeira, mean monthly temperatures at the coast vary from 17 to 23.4°C, but at altitudes of

Fig. 5.1. The Canary Island heathland vegetation, as well as the lower part of the adjacent native pine forests and evergreen laurel forests, receive much of their annual water balance from "cloud drip" (especially through the otherwise dry summer months). The passing cloud and mists condense directly upon, and drip freely from, all plant surfaces, especially those which are finely dissected and offer a large surface area. On this occasion, following about 10 min passing cloud preceded by bright sunshine and no rain, leaves of *Pinus canariensis* had already collected abundant free water. Gran Canaria, *c.* 1200 m, late spring. (Photo: C.N. Page.)

1400 m and 1700 m, within and just above the heathland zone, Menezes (1914) reports average June temperatures of 13.4°C and 12.2°C respectively. This too is a zone of frequent cloud cover and light precipitation. Sjogren (1972) calculates that the cloud-zone vegetation requires a precipitation of at least 1700 mm per year and permanently high relative humidity values of at least 85%, and that these conditions are met between altitudes of 400 and 1300 m in the north of Madeira where heathland is well-developed.

In the Azores the mountain climate is slightly cooler than that of the Canarian and Madeiran mountains but is similarly characterised by a moderate rainfall spread evenly throughout the year, high relative humidity and small mean annual temperature range (from 6 to 17.5°C at low altitude). Frosts may occur above 1600 m in any month of the year but are rare in summer, and occur regularly above 1860 m in the three coldest months (January to March) (Tutin, 1953). As in the Canary Islands, precipitation in the heathland is light and frequent, and in three stations quoted by Tutin (San Miguel, Fayal and Flores) the number of rain days varies from 147 to 225 per year, with cloud on 360 or 361 days per year.

ROCK AND SOIL TYPES

Rock types in Macaronesia are volcanic and predominantly basaltic. On Gran Canaria, Sunding (1972) records that they are chiefly basaltic types like phonolites, rhyolites, trachytes and olivine basalts. In most other parts of the Canary Islands, generally similar rock types mainly of basalts with layers of volcanic tuff and dolerite dykes predominate. In Madeira soils are again chiefly formed from basaltic rock types (Sjogren, 1972) and in all the Azorean islands except Santa Maria (the easternmost) the soils are also formed chiefly from rocks of alkali-basalt or olivine, and calcifuge species predominate (Marler and Boatman, 1952). Thus almost throughout these island groups, the rocks weather to a similar type of soil; especially in the Canary Islands and Madeira, variation in the physical water-retentive properties of even the same rock type is the prime edaphic influence affecting the balance between evergreen heath and laurel-forest vegetation (Sjogren, 1972; Sunding, 1972).

In the Canary Islands, Hausen (1951) reports that most of the rocks are permeable to water, with very permeable beds of scoriae at the base of nearly every lava sheet. These form the main horizons of ground-water movement, especially when underlain by impermeable tuff layers of fine-grained pyroclastics. The numerous dykes form effective curtains against ground-water movement. Large areas of the Canary Islands are covered with various kinds of loose ejecta, consisting chiefly of lapilli — fragments of compacted lava and pumice. These kinds of material are particularly porous and absorb water quickly, but also help prevent evaporation of moving ground water beneath them.

Broad-leaved hygrophilous laurel forests normally dominate in the Canary Islands wherever the water-retentive capacity of the rock is high, especially where this coincides with the shelter of deep valleys and deep humus-rich soils are formed. Such vegetation occurs extensively over the relatively impermeable basaltic core of the central region of Gomera and in the deep valleys (the "barrancos") of northern La Palma and northeast Tenerife, whilst heathland vegetation regularly occupies the better-drained ridges between. By contrast, on Hierro, where climatic conditions are similar but the rock type is of extremely porous basaltic lavas with scoriae, tuffs and ash beds, laurel forest is replaced by arborescent heathland vegetation virtually throughout, with only a few laurel-forest remanents such as *Ilex canariensis*. On Gran Canaria, Sunding (1972) classifies the heathland zone soils as "Braunlehm", with little stratification, and in the heathland this soil is relatively stable even on steep slopes, due to the many *Erica* rootlets in the upper soil layers.

THE HEATHLAND VEGETATION

In the Canary Islands the heathland has been described and its distribution mapped by Ceballos and Ortuño (1951). It is dominated by a large number of individuals of *Erica arborea* and *Myrica faya* with some *Ilex canariensis*, and this community clothes all the more exposed and better-drained aspects of the topography at the altitude of the cloud belt. At its lower limits it merges rather suddenly into the xerophytic vegetation of the lower zone of the islands. At its upper limits, *Erica arborea* ascends higher than its associated species, and *Erica* bushes gradually give way to increasing dominance of *Pinus canariensis* (from about 1200 m). Within the cloud zone the heathland community of the ridges passes gradually into the forest community of laurels in the sheltered ravines (dominated by the broad-leaved evergreens *Ilex canariensis*, *Laurus azorica*, and *Persea indica*). Here the *Erica* is readily over-topped and shaded

out by the vigour of the growth of the laurels.

Under differing degrees of exposure, the heathland vegetation varies from a metre or so in height to arborescent dimensions (frequently over 10 m). *Erica* forest with *Erica arborea* trees up to 30 m high was described by Knoche (1923) whilst maximum heights of the other dominants are recorded by Sunding (1972) as 20 m for *Myrica faya* and 10 m for *Ilex canariensis*. Besides these dominants, all other members of the heathland community play a minor role. Other species which, according to Sunding (1972, p. 116), serve to characterise the association on Gran Canaria are: *Hypericum canariense* var. *floribundum*, *H. glandulosum*, *Bystropogon canariensis*, *Micromeria varia* var. *citriodora* and the fern *Asplenium onopteris*. Mosses, of which *Hypnum cupressiforme* is the most characteristic, are frequent in the floor vegetation. Other shrubs associated with the lower altitudes of the heathland community [Tenerife (Ceballos and Ortuño, 1951)] include *Cytisus canariensis* var. *ramosisimus*, *C. linifolius* var. *angustifolius*, *Prunus lusitanica* and *Viburnum rugosum*. At higher altitudes on Tenerife there are herbs or shrubs of *Cistus monspeliensis*, *C. vaginatus*, *Hypericum grandiflorum*, *Andenocarpus foliosus*, *Micromeria thymoides* and *Inula viscosa*, and many of these species occur in the heathland vegetation throughout all the western Canary Islands. *Erica scoparia* subsp. *platycodon* is restricted to the crests of the ridges of Tenerife, Gomera and Hierro. On all the islands *Pteridium aquilinum* may dominate locally in the heathland vegetation, amongst which fronds can reach over 4 m (Page, 1976). As in the laurel forests, dry outcropping cliffsides amongst the heathland vegetation allow the establishment of rupestral species (xerophytic or hygrophilous according to exposure). Many endemic Crassulaceae are frequently associated with such habitats, and ferns of open situations may be common: *Asplenium onopteris*, *Adiantum reniforme*, *Notholaena marantae*, *Ceterach aureum*, *Polypodium macaronesicum* and *Davallia canariensis*. Numerous epiphytic lichens and mosses are common, large epiphytic *Polypodium* and *Davallia* ferns occur on larger trunks of *Erica arborea*, and the filmy fern *Hymenophyllum tunbrigense* is present epiphytically in areas of greatest cloud condensation. The club moss *Selaginella denticulata* is frequent throughout the heathland and laurel-forest regions, both terrestrially and rupestrally (Page, 1977).

The life forms of the Canarian plants have been analysed by Lems (1960). The heathland dominants *Erica arborea*, *Myrica faya* and *Ilex canariensis* are all evergreen macrophanerophytes. In leaf size, *Erica* is classified as leptophyll, *Myrica* microphyll, and *Ilex* microphyll/mesophyll. The disseminule of *Erica* is a sclerochore (small, nondescript), whilst those of both *Myrica* and *Ilex* are sarcochores (fleshy).

In Madeira, cloud-zone heathland vegetation exists on the north side of the island between 750 and 1630 m. It shows strong similarities to that of the Canary Islands, but with perhaps an even less clearly defined line between evergreen heathland and evergreen forest. This is the "laurel and heath" zone of Lowe (1868), and the Clethro-Laurion alliance of Sjogren (1972). Bunbury (1857) describes what must have been nearly the original vegetation. This consisted of evergreen woods of *Ocotea foetens*, *Persea indica* and *Laurus azorica*, with some *Clethra arborea*. These trees had an undergrowth of *Erica arborea* and *Erica scoparia*, with *Vaccinium maderense*, *Hypericum grandiflorum*, and "a profusion of ferns". Other plants native to this vegetation include *Heberdenia excelsa*, *Bystropogon maderensis* and *Semele androgyna*, whilst the ferns *Davallia canarensis* and *Polypodium macaronesicum* are epiphytic on the larger trees (Lowe, 1868; Menezes, 1914). In the upper part of this zone the laurel trees become less frequent, and here *Erica arborea* and *Myrica faya* dominate with *Erica scoparia* frequent, and *Laurus azorica*, *Clethra arborea* and *Pteridium aquilinum* present. *Vaccinium maderense* may form extensive low thickets. The cloud-zone vegetation is described by Sjogren (1972, p. 65) as having "a dense shrub-tree layer". Frequent dominants in the ground layer are *Sibthorpia peregrina*, *Senecio maderensis*, *Bystropogon*, *Phyllis nobla*, the fern *Blechnum spicant* and the club moss *Selaginella denticulata*. This vegetation also includes local associations of hygrophilous species on cliffs or crevices in cliffs. The community as a whole is particularly rich in cryptogams. Other conspicuous ferns contributing to the vegetation include *Asplenium onopteris*, *Diplazium caudatum* and *Dryopteris aemula*, the last of which may be

plentiful amongst the *Vaccinium*. *Woodwardia radicans* occurs on stream banks. *Elaphoglossum paleaceum* occurs on volcanic deposits but more frequently on tree trunks, and the filmy fern *Trichomones speciosum* is present, in damp shaded niches. Epiphytic moss communities are frequently present and sometimes there is an epiphyllous moss cover. *Erica cinerea* var. *maderensis* dominates the highest peaks (Lowe, 1868; Menezes, 1914; Sjogren, 1972).

In the Azores, because conditions are unsuitable for the development of a xerophyte zone near the coast comparable to that of the Canary Islands, heathland vegetation extends nearly to the sea and upwards to between 600 and 750 m, the "Faya Zone" (Guppy, 1914, 1917; Tutin, 1953). Here the cloud belt between 600 and 1500 m corresponds more closely to the juniper zone above the heathland. Nevertheless, as in the Canary Islands, the heathland vegetation occurs in a zone of high humidity and frequent light precipitation and the mountain vegetation receives additional water from frequent fog and clouds (Wilmans and Rasbach, 1973). As in the heathland vegetation of the Canary Islands and Madeira, that of the Azores is species-poor in dominant angiosperm vegetation although exceptionally rich in cryptogams. In this vegetation *Erica arborea* is replaced by *Erica azorica*. The vegetation on the island of Pico has been described by Marler and Boatman (1952) and Tutin (1953). The vegetation up to 600 m (the "lower woods") is of *Erica azorica* co-dominated by *Myrica faya* and *Laurus azorica*. Amongst this vegetation Guppy (1914) noted also *Ilex perado*, *Rhamnus latifolius* and *Picconia excelsa*. He records too that *Taxus baccata* is probably native in the higher levels of this zone, although already by 1914 it had been largely or completely felled for timber. The most characteristic shrubs of the lower woods are *Myrsine africana*, *Vaccinium cylindraceum*, *Hypericum foliosum* and *Viburnum linus*. Climbers include *Hedera canariensis* and *Smilax*, and *Rubus fruticosus* is present. The fern *Osmunda regalis* is conspicuous at all levels over 300 m in moist areas.

In the upper reaches of this woodland, *Erica azorica* becomes a more prominent element in the vegetation and eventually, at an altitude which corresponds approximately to the lower frost limit of around 750 m (Tutin, 1953), *Laurus* and *Myrica* disappear, and *Erica azorica* becomes the dominant plant growing to a height of from 4.5 to 6.0 m.

In the "upper woods" above 600 m, the vegetation is described by Tutin (1953) as *Ericetum azoricae* in which *Erica azorica* dominates and *Juniperus brevifolia* replaces *Myrica* as subdominant. Marler and Boatman (1952) have described this vegetation as heath-scrub, rather openly dispersed, with a good ground flora usually dominated by *Calluna vulgaris*, whilst *Vaccinium cylindraceum*, *Agrostis*, *Luzula*, and the ferns *Blechnum spicant* and *Dryopteris aemula* play a significant part. It is clear, however, that throughout this zone the heathland plants have been much reduced in height through interference by man. According to Guppy (1914) *Erica* commonly reached 7.6 m and must once have attained 10 m, whilst *Myrica* and *Laurus* reached 12 m and occasionally 15 m in undisturbed vegetation on Pico. Tutin notes that, from the abundance of epiphytes and epiphyllae, mist and rain are probably even more frequent at this level than they are at sea level.

Indications of the prevailing humidity, especially in the "upper woods", are the relative richness of this community in cryptogams, especially ferns. Tutin notes the abundance of epiphytes, and that the filmy fern *Trichomanes speciosum* is locally abundant in damp, dark places. Other pteridophytes occurring extensively in this vegetation are *Culcita macrocarpa*, *Osmunda regalis*, *Lycopodiella cernua*, *Athyrium filix-femina* and *Blechnum spicant* as terrestrial species, with *Woodwardia radicans* on the sides of gullies. Epiphytic and epiphyllous bryophytes are common, and epiphytic ferns include *Hymenophyllum tunbridgense*, *Elaphoglossum paleaceum* and *Grammitis jungermannioides* (Guppy, 1914; Rasbach et al., 1974). *Pteridium aquilinum* in this community may reach 2 to 3 m (Cunha and Sobrinho, 1940; Tutin, 1953).

This *Ericetum* community reaches 1200 to 1500 m, above which it gives way to an area which was once probably almost pure scrub of *Juniperus brevifolia*, but in which widespread burning and grazing have led to the formation of a *Calluna* heath and grassland. The common species are *Calluna vulgaris*, *Daboecia azorica* and *Thymus caespititius* (Marler and Boatman, 1952; Tutin, 1953).

VEGETATION SUCCESSION

Vegetational succession in the cloud zone ensues after felling of trees, after landslides (often caused on the steep slopes by heavy rain) and on new volcanic outpourings (notably in the Canary Islands).

Colonisation of new bare lava proceeds extremely slowly. Recolonisation of other areas is more rapid and has been studied on Madeira by Sjogren (1972) on surfaces opened by landslides. Succession starts with mosses, which reach a high degree of cover after only three years. There follows a strong development of larger ferns, notably *Dryopteris aemula*, *Diplazium caudatum*, *Pteris serrulata*, *Polystichum setiferum* and *Woodwardia radicans*. The fern fronds form a densely-shading canopy beneath which the original drought-tolerant moss species are replaced by more hygrophilous ones. Natural regeneration of angiosperm vegetation then begins, generally with *Erica scoparia*, *E. arborea* and *Vaccinium maderense*, closely followed by *Clethra arborea* and *Laurus azorica*. This develops to a climax community with little further floristic change. On strongly exposed steep slopes above 100 m, the mosses *Philonotis rigida* and *Saccogyna viticulosa* pioneer, producing a high degree of cover after a few years, when colonisation by vascular plants frequently proceeds by invasion with the club moss *Selaginella denticulata*.

CONCLUSIONS

Clearly, in each of the northern Macaronesian archipelagos, evergreen heathland and evergreen forest vegetation are closely interrelated in their floristic composition and ecology. The distinction between the two — though poor at best — is clearest in the Canary Islands, where there are distinct areas dominated either largely by *Erica arborea* or by tall hygrophilous laurel forest. The two communities merge gradually over broad fronts into one another. The distinction becomes less clear as one moves via Madeira to the Azores, where the laurel forest is more stunted and, as in the Canary Islands, the original heathland vegetation was probably originally largely arborescent. As a result, heathland and laurel vegetation become increasingly intermixed in floristic content, and the heathland vegetation — if the term can still be used — is one which is peculiarly rich in cryptogamic epiphytes, and in which terrestrial ferns play a significant role. In all these areas, and especially in the Canary Islands, widely different vegetation types can occur in close juxtaposition, particularly where cliff faces outcrop in the cloud-zone vegetation. These, as well as stream banks, depending on their degree of exposure or shelter, may include locally well-adapted xerophytic communities or small hygrophilous ones.

In all the Macaronesian islands, the original montane vegetation of heathland and forest has now been very greatly disturbed by man and by grazing animals. The larger heathland plants, as well as the forest ones, have been removed as sources of timber for construction or burning, to the extent of virtual local extinction in the case of *Taxus* in the Azores. Few areas can be found in which this vegetation has not been modified artificially in some way, if not totally destroyed. In the Canary Islands, Ceballos and Ortuño (1951) mapped the natural and actual areas of the vegetation types on Tenerife and Gomera, showing that both the heathland and laurel-forest communities have greatly diminished in extent. On Gran Canaria, where destruction has proceeded at a virtually unprecedented rate, Sunding (1972) suggests that only about 5% of the original cloud-forest vegetation still exists, whilst Bramwell and Bramwell (1974) put the figure at only 1%. Even on Tenerife the figure is reported as less than 10%, and diminishing rapidly. Many of the Macaronesian plants are insular endemics, and the natural vegetation which they form is of palaeobotanical as well as ecological and floristic interest. For the history of many of the Macaronesian flowering plants, as well as that of the Macaronesian pines (Page, 1974) and ferns (Page, 1973 and in press) is one which suggests them to be fragments of a more widespread flora and vegetation of the Tertiary period now virtually extinct in Europe but which have survived as insular relicts on the Macaronesian archipelagos. Conservation of some of this vegetation should be given a high priority.

REFERENCES

Bramwell, D. and Bramwell, Z., 1974. *Wild Flowers of the Canary Islands*. Stanley Thornes, London, 261 pp.

Bunbury, J.F., 1857. Remarks on the botany of Tenerife. *J. Proc. Linn. Soc. Bot.*, 1: 1–35.

Ceballos, L. and Ortuño, F., 1951. *Estudio sobre la Vegetacion y la Flora Forestal de las Canarias Occidentales*. Ministerio de Agricultura, Madrid, 465 pp.

Cunha, G. and Sobrinho, L., 1940. Quelques remarques sur la distribution de la végétation dans l'archipel des Açores. *Bol. Soc. Broteriana*, 14 (Ser. 2A): 1–16.

Guppy, H.B., 1914. Notes on the native plants of the Azores as illustrated on the slopes of the Mountain of Pico. *Kew Bull.*, 9: 305–321.

Guppy, H.B., 1917. *Plants, Seeds and Currents in the West Indies and Azores*. Williams and Norgate, London, 531 pp.

Hansen, A., 1969. Checklist of the vascular plants of the archipelago of Madeira. *Bol. Mus. Municip. Funchal*, 24: 5–62.

Hausen, H., 1951. On the groundwater conditions in the Canary Islands. *Acta Geogr.*, 12(2): 1–12.

Knoche, H., 1923. *Vagandi Mos. Reiseskizzen eines Botanikers. 1. Die Kanarische Inseln*. Strasbourg, 304 pp.

Lems, K., 1960. Floristic botany of the Canary Islands. *Sarracenia*, 5: 1–94.

Lowe, R.T., 1868. *A Manual Floral of Madeira and the adjacent islands of Porto Santo and the Desertas*, 1. Voorst, London, 618 pp.

Marler, P. and Boatman, D.J., 1952. An analysis of the vegetation of the northern slopes of Pico — the Azores. *J. Ecol.*, 40: 143–155.

Menezes, C.A., 1914. *Flora do Archipelago da Madeira*. Junta Agricola da Madeira, Funchal, 282 pp.

Page, C.N., 1973. Ferns, polyploids, and their bearing on the evolution of the Canarian flora. *Monogr. Biol. Canar.*, 4: 83–88.

Page, C.N., 1974. Morphology and affinities of *Pinus canariensis*. *Notes R. Bot. Gard. Edinb.*, 33: 317–323.

Page, C.N., 1976. The taxonomy and phytogeography of bracken — a review. *Bot. J. Linn. Soc.*, 73: 1–34.

Page, C.N. 1977. An ecological survey of the ferns of the Canary Islands. *Br. Fern. Gaz.*, 11, in press.

Page, C.N., in press. Cytology and evolution in the fern flora of the Canary Islands.

Rasbach, H., Rasbach, K. and Reichstein, T., 1974. *Grammitis jungermanniodes* in the Azores. *Br. Fern Gaz.*, 11: 49–52.

Sjogren, P., 1972. Vascular plant communities of Madeira. *Bol. Mus. Municip. Funchal*, 26: 45–125.

Sunding, P., 1972, The vegetation of Gran Canaria. *Skr. Nor. Vidensk.-Akad. Oslo, I Mat.-Naturv. Kl.*, N.S., 29: 1–186.

Tutin, T.G., 1953. The vegetation of the Azores. *J. Ecol.*, 41: 53–61.

Wilmans, O. and Rasbach, H., 1973. Observations on the pteridophytes of São Miguel, Açores. *Br. Fern Gaz.*, 10: 315–329.

Chapter 6

THE SCLEROPHYLLOUS (HEATH) VEGETATION OF AUSTRALIA: THE EASTERN AND CENTRAL STATES[1]

R.L. SPECHT

INTRODUCTION

Plant communities, either containing a large percentage of or dominated by sclerophyllous (heathland) shrubs and sub-shrubs, are widely distributed across the whole of the Australian continent, from the tropics to the temperate regions. The heathland communities are best developed in the higher rainfall regions south of the Tropic of Capricorn from the lowland to subalpine altitudes.

A few relict stands do persist in the dry inland of Australia. Within the tropical zone, a number of extensive heathland communities survive on fine siliceous sands and on skeletal quartzites or sandstones; limited areas are found on subalpine mountain tops.

In 1903 Schimper concluded that the southern Australian heathland flora was remarkably similar in physiognomy, morphology and physiology to the vegetation of other Mediterranean-type climates of the world. Hence, it could be equated with the *maquis* and *garique* around the Mediterranean Sea, the *chaparral* of California and the *matorral* of Chile, as well as the *fynbos* of South Africa. This concept was investigated by Specht (1969, 1973) and is shown in Chapter 1 of this volume to be untenable. The wide geographical distribution of heathland elements throughout the whole continent of Australia negates the idea that it is a Mediterranean flora. Australian heathlands are clearly related to the fynbos of South Africa and to heathlands in many other (non-Mediterranean) parts of the world; it is purely chance that some of them exist today within the region of Mediterranean-type climate.

ATTRIBUTES OF HEATHLAND COMMUNITIES

Community structure

The sclerophyllous (heathland) element of the Australian flora forms a major part of a number of plant formations. In "extreme" habitats (seasonally waterlogged, seasonally droughty, or sub-alpine sites — all very low in plant nutrients), the flora forms low, dense to mid-dense communities which are termed heathlands (Figs. 6.1, 6.2). Considerable variation in structure can be observed. The heathland communities are seen to be, by definition, dense to mid-dense in cover and less than 2 m in height. Habitat terms have been used to subdivide Australian heathlands into "wet-heathlands" (on seasonally waterlogged sites), "dry-" or "sand-heathlands" (on sandy, well-drained sites), and subalpine heaths. In these natural, but restricted, heathland sites, it appears that some environmental or biotic factor tends to operate against the establishment of taller species of tree or shrub.

In some habitats, the heathland communities may grow into a scrub formation with a stature much higher than 2 m (Fig. 6.3).

In most areas of Australia where sclerophyllous shrubs abound, trees or tall shrubs of *Eucalyptus*, *Acacia*, *Casuarina*, *Banksia*, plus a few other genera, have overtopped the heathland flora (Figs. 6.4, 6.5). The density of trees or tall shrubs which have become established varies in a continuum from zero (heathlands), to scattered (low open-woodland, tall open-shrubland), sparse (woodland, low woodland, tall shrubland), or mid-dense

---

[1] Manuscript completed July, 1976.

Fig. 6.1. Open-heathland formation (*Xanthorrhoea australis–Banksia ornata–Casuarina pusilla* alliance) on deep sand near Keith, South Australia. (Photo K.P. Phillips.)

(open-forest or low open-forest, with a heathland understorey[1]). The assemblage of heathland species, characteristic of the understorey of these communities, tends to become sparser as the canopy of the upper strata becomes denser; heathland species appear to flourish best in full sunlight.

Life forms

The heathland communities (with or without an overstorey) are dominated by low shrubs (nanophanerophytes up to 2 m tall) which possess small (leptophyll to nanophyll), evergreen, sclerophyllous leaves and extensive root systems, often arising from lignotubers. Usually several species of nanophanerophyte are co-dominant in the under-

storey. In the open-heathland vegetation growing on deep sand near Keith, South Australia (Specht and Rayson, 1957a), 76 species were recorded on the sand plains: 33 of these species were evergreen nanophanerophytes (of which *Banksia ornata*, *Casuarina pusilla*, *Leptospermum myrsinoides*, and *Xanthorrhoea australis* had the greatest biomass per hectare), 10 were evergreen chamaephytes, 13 evergreen hemicryptophytes and 14 seasonal geophytes; 2 epiphytic species of the parasitic twining plant *Cassytha* were common in the heathland. Perennial grasses (seasonal hemicryptophytes), and therophytes are uncommon or rare in the heathland, except where the communities have been grazed by cattle and fired regularly.

---

[1] The community was termed *"Sklerophyllen-Wald"* by Diels (1906), "dry sclerophyll forest" by later Australian ecologists.

Fig. 6.2. Alpine closed-heathland formation (*Oxylobium ellipticum–Podocarpus lawrencei* alliance) on Sentinel Peak, Kosciusko National Park, New South Wales. *Oxylobium alpestre* and *Grevillea victoriae* are prominent in the heathland community shown in the photograph. (Photo C. Totterdell.)

Leaf characteristics

The leaves of most species common in the sclerophyllous heathland (with or without an overstorey) typically have thick cuticles, sunken stomata, and usually thick-walled cells, often lignified (sometimes silicified) and containing tannins, resins or essential oils. Of 102 species examined in the Dark Island dry-heathland near Keith, South Australia (Specht and Rayson, 1957a), 56 possessed leaves with leaf size less than 25 mm^2 (leptophyll); the leaves of 20 species were in the nanophyll class (25–225 mm^2), while the leaf size of 18 species fell into a category between leptophyll and nanophyll. Only six species (including two long-lived dominants, *Banksia ornata* and

Xanthorrhoea australis) possessed leaves larger than nanophyll (>225 mm^2); three of the other four species were evergreen hemicryptophytes with graminoid leaves; the sixth species was a rare succulent plant. Long-lived tree and shrub species of *Eucalyptus*, *Banksia*, and *Acacia*, which overtop most heathland communities, usually possess even larger leaves (or phyllodes) in the microphyll–mesophyll category (2–20, 20–182 cm^2).

The leaves of several leptophyllous species in the Dark Island heathland may be classed as typically *ericoid* with the upper surface of the leaf hard and waxy, the lower surface deeply grooved; some species even possess leaves grooved above, not below. In general, the ericoid leaf, though present

Fig. 6.3. Closed-scrub formation (*Leptospermum pubescens – Melaleuca squarrosa* alliance) on waterlogged, lowland peat at Eight-Mile Creek, Lower South East of South Australia. (Photo C.M. Eardley.)

in most Australian heathland communities, is by no means as definitive of Australian heathlands as it is in Europe or South Africa. The various inclusions, such as tannins, resins and essential oils, in the leaves of many Australian heathland species increase the flammability of the community during periods of water stress. Fire is thus an integral part of the Australian heathland ecosystems. As shown below, many species, razed in a fire, regenerate from underground root stocks or from seed. Some species regenerate by sprouting new shoots from epicormic buds which are buried deeply in fire-resistant stems.

Root systems

A typical dicotyledonous root system with a tap root (3 to 4 m deep) and an extensive lateral root system, usually no more than 30 cm deep but extending almost 6 m from the central tap root, has been observed in the long-lived heathland plant, *Banksia ornata*, growing on deep sand near Keith, South Australia (Specht and Rayson, 1957b). This plant possesses no underground vegetative buds and is thus particularly vulnerable to fire.

Many other dicotyledonous heathland species form a woody swelling (called a lignotuber) near ground level. The lignotuber is a buried, woody storage organ from which vegetative buds can arise; numerous aerial stems may grow from it and, even though these may be destroyed by fire, the protected lignotuber is capable of producing more buds to enable the plant to regenerate after fire from the woody rootstock. The original tap root may decay with time and secondary vertical roots may arise from the extensive lateral root system.

Fig. 6.4. Dry sclerophyll woodland (woodland of *Eucalyptus baxteri*, with a heathland understorey of *Banksia marginata* and *Leptospermum myrsinoides*) on deep sand between Naracoorte and Bordertown, South Australia. (Photo K.P. Phillips.)

Aerial "sucker" shoots, as well as secondary vertical roots, may also be produced from the lateral root system (Specht and Rayson, 1957b).

An extensive fibrous root system typical of monocotyledons arises from a root stock, rhizome, tuber, or bulb. In some species of *Xanthorrhoea*, contractile roots pull the root stock (a caudex) further into the soil often to a depth of 30 to 40 cm. The buried vegetative apices of all these species are well protected from fire (see Gill and Groves in Chapter 7 of Volume B).

Peculiar rootlet formations have been observed in a number of heathland families. Proteoid roots are developed in the family Proteaceae and in *Viminaria* (Fabaceae); restiad roots in the family Restionaceae; cyperiod or dauciform roots in the family Cyperaceae; nitrogen-fixing, nodulated rootlets in legumes, cyads and *Casuarina* spp.; while mycorrhizas are certainly associated with the roots of the families Myrtaceae, Podocarpaceae and other families (Purnel, 1960; Bergersen and Costin, 1964; Jeffrey, 1967; Bevege, 1968; Lamont, 1972a, b and c, 1973; 1974; see also Lamont and Malajczuk in Chapters 17 and 18 of Volume B).

Haustorial connections have been observed between the roots of semi-parasitic plants (*Exocarpos*, *Leptomeria*, *Euphrasia*, etc) and of other species of the heathland communites.

Bradysporous fruits

Many heathland species release seed as soon as fruits are mature. In contrast, the seeds of a number of genera of the families Proteaceae (for example, *Banksia* and *Hakea*) and Casuarinaceae (*Casuarina*) are retained within densely-packed

Fig. 6.5 Dry sclerophyll forest (open-forest of *Eucalyptus obliqua* and *E. baxteri*, with a heathland understorey of *Leptospermum juniperinum*, *Hakea rostrata*, *Xanthorrhoea semiplana*, *Epacris impressa*) on podzolic soil developed over quartzite on Mount Lofty Summit, South Australia. (Photo H.B.S. Womersley.)

cones or follicles for many years, usually until released by the heat of a bushfire. Because of this bradysporous habit regeneration of these species is, by and large, only possible following fire. The frequency of bush fires is important — a fire before the species sets viable seed can lead to the extinction of the species; a fire, too long delayed, may lead to viable seed being destroyed in the cone-like fruits by wood-boring insects and fungi.

FLORISTIC COMPOSITION

Introduction

The subfamily Ericoideae of the dicotyledonous family Ericaceae derives its vernacular name of heath (German: *Heide*; Swedish: *hed*) from the common name for the wastelands (with their associated plant community — the heathlands) of northern Europe (Rübel, 1914). Historical and palynological evidence suggests that, in many areas, European heathlands are degraded communities resulting from the destruction of woodlands and forests by man. Three species (*Calluna vulgaris*, *Erica cinerea*, and *E. tetralix*) dominate European heathlands, though twelve other species of the Ericoideae may also be present (Gimingham, 1972). The subfamily is strongly represented in the heathlands (fynbos) of South Africa by about 832 species (Baker and Oliver, 1967). It is not found in Australia where, instead, about 333 species of the Epacridaceae, a closely-allied heath family, plus 8 species of other subfamilies of the Ericaceae or Vacciniaceae, are common in the heathlands.

In both South Africa and Australia a number of angiosperm families co-exist with the heath families and, being often markedly different in stature and leaf morphology, modify the physiognomy of the heathlands (with or without an overstorey). In particular, the families Fabaceae (Tribe   Podalyrieae),   Proteaceae   (Tribe

Proteoideae), Restionaceae, and Rhamnaceae (Tribe Rhamneae) co-exist with the heath families in each country.

Gondwanaland flora

There is now a considerable body of evidence that the continents of South America, Africa and Australasia were formerly joined with Antarctica and India to form a single landmass almost until the beginning of the Tertiary Era 65 million years ago. Many elements of the primitive angiosperm flora were apparently widespread across the original Mesozoic continent, later being fragmented when the component landmasses drifted apart. The Australian continent, after breaking away from Antarctica, drifted northwards over the last 70 million years at an average rate of 66 ± 5 mm per year (Wellman and McDougall, 1974).

Tertiary flora

During much of this northward drift during the Tertiary Era, the climate of Australia was warmer than today. Oxygen isotope analyses on Tertiary fossils indicate that the climate of Wellington, New Zealand, ($41°$S) was about $10°$C warmer that at present (Devereux, 1967). At least twice (for quite a long period of time) during the Tertiary, the climate over most of Australia could be classed as subtropical, with a very wet season alternating with a drier (but still wet) season. This led to the widespread development of infertile lateritic soils on the stable peneplain surface of the continent (Jessup, 1961) — a soil formation which, as will be shown below, was ideal for the expansion of the Mesozoic heathland flora.

The fossil record of southern Australia indicates that all the main elements of the present-day flora were represented from the early Tertiary (Cookson, 1945, 1946, 1947, 1950, 1952, 1953, 1954a, b, 1957, 1959; Cookson and Duigan, 1950, 1951; Duigan, 1951; Cookson and Pike, 1953a, b, 1954a, b, 1955; Balme and Churchill, 1959). Most fossil beds contain a mixture of three broad floristic groups which have been classed as *autochthonous* (essentially the heathland flora), *Antarctic* (essentially the *Nothofagus* flora, which in the early Tertiary contained a mixture of both cool temperate and subtropical species — belonging respectively to the

N. fusca–N. menziesii and *N. brassii* types), and *Indo-Malayan* (essentially the tropical and subtropical rain-forest flora).

While the fossil record in the southern parts of Australia shows generally high numbers of *Nothofagus* species over a large part of the Tertiary, this genus seems to be abundant for only a short interval (with a significant maximum in the early Miocene) in southern (Mount Tamborine) to central (Proserpine) Queensland (Hekel, 1972). As most of the fossil *Nothofagus* pollen found in Queensland belongs to the subtropical *N. brassii* group, it is probable that the flora invaded the area when highlands were formed during an intense period of volcanic activity 30–23 million years ago (Wellman and McDougall, 1974). Today, the subtropical McPherson Range on the border of New South Wales and Queensland is the only place in Australia where the three floras still co-exists (Herbert, 1950), and even here the subtropical (tropical highland) species of *Nothofagus* are absent.

No significant fossil deposits of the Tertiary flora have been found in the northern part of Australia. In the absence of such information, one can only speculate on the nature of the Tertiary flora in the light of biogeographical and pedological evidence. Fossil lateritic soils, formed during the Tertiary, are widespread across the north wherever the original peneplain surface has persisted. In places (such as the Kimberleys, Arnhem Land, and the eastern part of Cape York Peninsula), some uplift occurred later and there much of the original lateritic soil has been eroded to expose extensive areas of sandstone or granitic rock very low in plant nutrients.

Four, possibly five, main floristic elements apparently occupied this Tertiary subtropical landscape of northern Australia:

(1) **A subtropical closed-forest (rain forest)**, remnants of which are today seen right across the north of Australia (Specht, 1958a). This vegetation would have been especially well developed along the coastal belt where higher rainfall would have favoured a reasonably continuous water supply throughout the year, either directly by rainfall or by subsoil storage or seepage (Specht et al., 1977). As many of the genera characteristic of this rainforest community have also been recorded in India (Hooker, 1860), it appears that the vegetation

across the north of Australia was formerly continuous with that of the Indian subcontinent when both were united as part of Gondwanaland (Specht, 1958b).

(2) A *Eucalyptus* **open-forest** developed in drier inland areas where the subtropical climate was more seasonal (but not as extreme as monsoonal) than that nearer the northern coastline. Under this climate, lateritic soil developed on the peneplain surface and supported a eucalypt open-forest probably dominated by *Eucalyptus tetrodonta* which is now widespread across the north of Australia (Specht et al., 1977). If present-day flora can be extrapolated back in time the understorey of the eucalypt open-forest on these laterite soils would have been dominated by seasonal, subtropical grasses and herbs. Most of the genera, and even many species, of this understorey are also found today in India and even extend to Africa (Hooker, 1860); they were probably part of the original Gondwanaland flora.

(3) **The Australian heathland flora** was by no means absent from northern Australia, but tended to be confined to the drier localities on sandstone or granitic outcrops and to areas where relatively deep, infertile sands had accumulated. An overstorey of *Eucalyptus* and *Acacia* species plus Indo-Malayan elements was probably present. One may speculate on the reasons why the heathland elements, nowadays common on the lateritic soils of southern Australia, were probably, and are now, absent from the same soils in the north. There is some evidence that root-rot fungi such as *Phytophthora* spp. appear to be much more virulent on the heavy-textured, lateritic earths than on sandy soils lacking finer soil fractions — in spite of the general low nutrient status of both soils. This hypothesis, however, needs to be carefully investigated.

(4) **The subtropical *Nothofagus* flora** was represented in Pliocene deposits in the Papuan area of New Guinea by the subtropical *brassii* subsection of the genus (Cookson and Pike, 1955). About sixteen species of this subsection survive today in New Guinea, five in New Caledonia. As fossil pollen of the *brassii* type has been found in the early Tertiary deposits across southern Australia and even earlier in the Cretaceous of New Zealand, there is a strong probability that members of this subsection of the genus *Nothofagus* were

also found in northeastern Australia during the Tertiary — on the upper slopes of volcanic mountains, such as Mount Bellenden Ker and Mount Bartle Frere near Innisfail. The *brassii* type apparently became prominent on the volcanic hills near Mackay, and the McPherson Ranges on the Queensland–New South Wales border (where a temperate species of *Nothofagus* survives today) for a short tenure during the early Miocene (Hekel, 1972).

(5) **The montane ericaceous flora** is located today on the high peaks of New Zealand, southeastern Australia, northern Queensland and New Guinea, through Malesia, and in the Himalayas bordering India and China. This floral element was probably widespread across the northeastern coastline of the original Gondwanaland (and seems to have extended into the eastern United States). It would have been confined to the summits of high mountains which must have developed along the advancing front of the continental plates of India and Australasia (including New Guinea) as they drifted northward and eastward through the subtropical climate of the Tertiary. The remnant of the ericaceous (*Rhododendron* and *Agapetes*) flora in northern Queensland is now intermixed with subalpine elements of both the Australian heathland and the rain forest on the summits of Mount Bellenden Ker and Mount Bartle Frere.

Quaternary flora

Dramatic changes occurred in the climate of Australia during the Quaternary. Both temperature and rainfall apparently oscillated considerably during the glacial and interglacial stages of the Pleistocene until today the general climate is considerably drier than it was during much of the Tertiary. Almost two-thirds of Australia was reduced to an arid or semi-arid climate. The floristic elements which flourished in the Tertiary subtropical climate now suffered considerable depletion..

Epeirogenic movement, which commenced in the mid-Tertiary, continued into the Quaternary to produce the mountain ranges and escarpments seen today. In all these upland areas, the Tertiary lateritic peneplain was dissected, often eroded completely, to expose the underlying rock. Extensive areas of sandstone, quartzitic, and granitic rock, all low in plant nutrients, were exposed. In southeastern Australia some of these mountains reached elevations which could be classed as subalpine.

In many areas around the coast, considerable quantities of siliceous sand accumulated along the shoreline when strong winds blew sand inland off the continental shelf, exposed when the sea level fell during a glacial period. In some parts of Australia the sand was blown a long distance (up to 200 km) inland.

The climate of the southeastern part of Australia has become much cooler (mean annual temperature less than 15°C) since the Tertiary, probably selecting against the survival of the former tropical heathland flora.

The original subtropical heathland flora of northern Australia has certainly drifted further into tropical regions by approximately 66 km over the last million years (Wellman and McDougall, 1974). However, during that time the climate within the tropical part of Australia has apparently become progressively more monsoonal, with a hot, wet summer season alternating with a much drier, but still warm, winter season. The mean annual temperature now lies between 22 and over 28°C.

It appears that the original subtropical heathland flora of the Tertiary has survived best in the eastern coastal belt (from Bundaberg in Queensland to southern New South Wales), where the rainfall is well distributed throughout the year and the mean annual temperature lies between 15 and 22°C (Fig. 6.6). Inland of this eastern coastal belt, much of present-day Australia lies within the arid to semi-arid zone — the heathland flora only persists in relict pockets. However, in the far southwest of Western Australia (with mean annual temperature 15°C and higher), the winter rainfall is sufficiently high to enable considerable underground storage for use by deep-rooted plants during the dry summer. Here again the heathland elements of the Tertiary have continued to flourish, but the physiognomic dominants of the flora have maintained a summer-growth rhythm characteristic of their original subtropical ancestry (Specht and Rayson, 1957a).

During the long period of time since the beginning of the Tertiary considerable speciation has occurred in the heathland vegetation, until today well over 3700 typical heathland species have been recorded in the Australian flora (Table 6.1).

As shown in Fig. 6.6, about 50% of these species are found in southwest Western Australia and 20% in the Sydney area of New South Wales. The numbers of surviving heathland species falls considerably within the tropics (3 to 6%). In the cooler temperate climate of southeastern Australia, the numbers are only 9 to 14%; most of the heathland plants which have survived in this temperate climate possess a lower temperature threshold for shoot growth.

The figure of 3700 species, cited above as typical Australian heathland taxa, underestimates the actual total surviving today. These species belong to vascular families and tribes most of whose members are restricted to heathlands and related communities (Table 6.1). A few genera of some families listed (e.g. Proteaceae) belong to rain-forest vegetation and have been omitted from the Table. Genera of a number of other families, not included in Table 6.1, contain some species which are

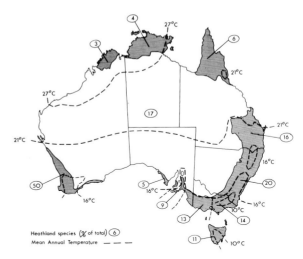

Fig. 6.6. Regional distribution (expressed as a percentage of the total) of 3700 species which may be regarded as typical members of the Australian heathland flora. (As up-to-date statistics are available for only the central coastal region of New South Wales, no information is presented for other areas of this state.) The map is divided into the following geographical regions: Western Australia, northern; Northern Territory, north; Queensland, northeastern; Queensland, southeastern to eastern; New South Wales, central; Victoria, eastern; Tasmania; Victoria, western to South Australia, southeastern; South Australia, Mount Lofty Range and Kangaroo Island; South Australia, Eyre Peninsula; Western Australia, southwestern; Inland Australia.
The figures encircled on the map show the percentage of total Australian heathland species found in each region.

TABLE 6.1

Number of species of characteristic heathland taxa recorded in the various Australian geographical regions illustrated in Fig. 6.6 [floristic information collated from books and articles by Bailey (1899–1902); Black (1943–57); Curtis (1956–67); Willis (1962, 1972); Eichler (1965); Beard (1970); Chippendale (1971); Beadle et al. (1972); Adams et al. (1973); and unpublished information from the Queensland Herbarium (1976)].

Taxa	W.A. N	N.T. N	Qld. NE	Qld. SE-E	N.S.W. C	Vic. E	Tas.	W. Vic., SE S.A.	S.A. Mt. Lofty Kangaroo Is.	S.A. Eyre Pen.	W.A. SW	Inland
CONIFERS												
Cupressaceae												
Actinostrobus and *Callitris* (sand pine species only)	–	–	–	–	–	–	–	1	–	–	3	1
Podocarpaceae												
Podocarpus — dwarf spp.			–	1	1	1	1	–	–	–	1	–
DICOTYLEDONS												
Casuarinaceae												
Casuarina — dwarf spp.	–	–	–	1	3	3	2	4	3	2	8	–
Dilleniaceae (2 genera)	2	13	13	23	22	19	11	14	7	4	60	16
Droseraceae and Byblidaceae												
Drosera spp.	4	3	8	9	6	9	9	8	7	4	38	6
Byblis spp.	1	1	1	1	–	–	–	–	–	–	1	–
Epacridaceae (23 genera)	–	1	7	46	59	48	75	38	25	11	170	16
Ericaceae and Vacciniaceae												
Ericaceae (4 genera)	–	–	1	–	1	1	4	–	–	–	–	–
Agapetes sp.	–	–	1(+1)	–	–	–	–	–	–	–	–	–
Euphorbiaceae												
Tribe Stenolobieae (9 genera)	2	3	4	19	18	13	8	8	8	3	35	27
Fabaceae												
Tribe Podalyrieae (17 genera)	12	10	18	73	95	66	27	57	35	24	252	108
Tribe Genisteae (5 genera)	3	4	6	19	17	18	12	18	7	6	28	15
Myrtaceae												
Tribe Leptospermoideae[1] (32 genera)	20	33	46	86	79	38	22	37	32	21	328	201
Tribe Myrtoideae (2 genera)	1	2	3	1	1	–	–	–	–	–	–	2
Pittosporaceae												
Tribe Billardiereae (3 genera)	–	–	–	3	3	5	3	6	5	3	20	3
Polygalaceae												
Comesperma spp.	1	2	2	7	5	6	5	7	3	4	14	9

Proteaceae												
Tribe Persoonioideae (2 genera)	1	1	1	10	25	10	3	2	1	–	23	10
Tribe Proteoideae (10 genera)	–	–	–	11	15	3	4	4	5	1	113	1
Tribe Grevilleoideae (10 genera)	26	21	29	38	62	32	20	32	21	13	288	86
Rhamnaceae												
Tribe Rhamneae (7 genera)	1	–	4	26	27	36	21	26	22	16	41	20
Rubiaceae												
Opercularia spp.	–	–	1	4	4	4	2	5	4	1	10	2
Rutaceae												
Tribe Boronieae (8 genera)	3	6	14	55	55	27	22	33	22	10	61	41
Scrophulariaceae												
Euphrasia spp.	–	–	–	3	2	4	8	2	2	1	1	–
Stylidiaceae (4 genera)	14	19	13	11	6	8	7	10	6	3	93	20
Tremandraceae (3 genera)	–	–	–	1	6	6	3	3	3	–	17	2
Violaceae												
Hybanthus spp.	2	1	2	3	3	3	–	1	3	2	4	9
MONOCOTYLEDONS												
Cyperaceae												
Lepidosperma spp.	–	–	1	7	14	12	8	12	9	6	28	–
Iridaceae												
Patersonia spp.	–	1	–	3	4	5	2	2	2	–	14	–
Orchidaceae												
Tribe Diurideae (25 genera)	–	2	21	86	161	141	106	125	92	27	139	10
Restionaceae (17 genera)	–	3	6	15	16	9	13	10	7	4	58	3
Xanthorrhoeaceae (8 genera)	1	1	6	17	22	8	4	14	12	6	33	10
Total	94	127	208	579	732	535	402	479	343	172	1881	618

[1] Heathland spp. of *Angophora, Eucalyptus, Metrosideros, Syncarpia* and *Tristania* have been omitted.

characteristic elements of the heathlands, while other species of the same genus may be found in other plant communities (on soils of higher fertility). Characteristic heathland genera belong to about 68 families, of which 16 may be said to be largely heathland families. Altogether, some 342 genera are listed in Table 6.2 as typical of Australian heathlands.

It is clear that southwest Western Australia and the eastern coast of Australia are the centres of survival of the Tertiary heathland flora. An amazing range of heathland families, genera, and species flourish in these areas. Yet, in spite of the immense range of taxa which may be found within a region of say 10 000 km², the number of species found on individual heathland sites is much more restricted. As will be shown below, the numbers of taxa recorded in these heathlands range from 33 to 131. The actual occupation of heathland habitats apparently will permit only 22 to 36 species (on the average) to develop on an area of 8 m² in dry-heathland; fewer species (11 to 25) are found on the same area of wet-heathland (Table 6.3). Seasonally waterlogged soils support heathland of less diversity.

Species–area relationships shown in Table 6.3 for dry- and wet-heathlands of extra-tropical, eastern Australia were recorded for heathlands in their most productive phase of development, only a few years after a bush fire. Specht et al. (1958), in their study of the Dark Island heathland in South Australia, recorded 36 species in the stand after a fire, 20 species in a stand 25 years old, and only 10 species in a 50 year old stand. Similarly, on North Stradbroke Island, Queensland, Specht et al. (1977) noted 72 species on experimental plots three years after a fire, but found only 51 species eight years later. The species–area diversity in heathlands clearly decreases as the regenerating vegetation matures and the physiognomic dominants exert a strong competitive effect on weaker species (Fig. 6.7).

This competitive advantage of the physiognomic dominants has been illustrated on North Stradbroke Island, Queensland, where heathland with no emergents, growing on the exposed first high dune, contained a mean number of 27.0 species per 8 m². Heathland, with emergent *Banksia aemula*, on Mount Hardgrave contained 25.6 species per 8 m² (Table 6.3), while heathland

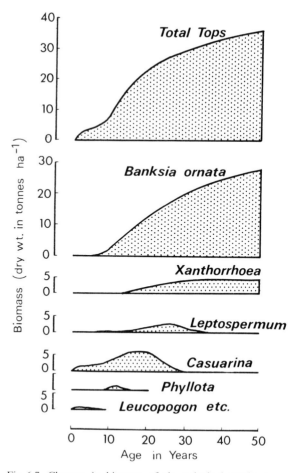

Fig. 6.7. Changes in biomass of the principal species in a heathland as the vegetation regenerates following a bushfire (Dark Island heathland, South Australia; Specht et al. 1958).

of the same age, but with an increasing canopy cover of eucalypt trees, was reduced progressively to 21.5 and 19.0 species per 8 m² as the canopy developed.

ECOLOGICAL DISTRIBUTION

Climate

Heathland vegetation is widespread in Australia from the tropical, monsoonal areas of the northern part of the continent to the Mediterranean-type and temperate climates of southern Australia. The vegetation extends from lowland to subalpine–alpine habitats. Remnants of the orig-

TABLE 6.2

List of genera which may be considered as typical members of Australian heathlands (families or tribes in which most species belong to heathlands and related communities are preceded by an asterisk*)

MOSSES

Sphagnaceae	*Sphagnum*

FERNS AND FERN ALLIES

Dennstaedtiaceae	*Pteridium*
Schizaeaceae	*Schizaea*
Selaginellaceae	*Selaginella*

CYCADS

Zamiaceae	*Macrozamia*

CONIFERS

Cupressaceae	*Actinostrobus, Callitris,* (+*Diselma* in subalpine heaths)
Podocarpaceae	*Podocarpus,* (+*Microcachrys* and *Microstrobos* in subalpine heaths)

DICOTYLEDONS

Apiaceae — Hydrocotylaceae	*Platysace, Trachymene, Xanthosia,* (+*Aciphylla, Actinotus, Dichosciadium, Diplaspis, Oreomyrrhis* in subalpine heaths)
Asteraceae	*Craspedia, Helichrysum, Ixodia, Olearia,* (+*Abrotanella, Celmisia, Ewartia, Pterygopappus* in subalpine heaths)
Baueraceae	*Bauera*
Brunoniaceae	*Brunonia*
Byblidaceae	*Byblis*
Campanulaceae	*Wahlenbergia*
Cassythaceae	*Cassytha*
Casuarinaceae	*Casuarina*
Cephalotaceae	*Cephalotus*
Dicrastylidaceae	*Chloanthes, Cyanostegia, Denisonia, Dicrastylis, Hemiphora, Lachnostachys, Mallophora, Physopsis, Pityrodia*
Dilleniaceae	*Hibbertia, Pachynema*
Donatiaceae	*Donatia* in subalpine heaths
Droseraceae	*Drosera*
Ehretiaceae	*Halgania*
*Epacridaceae — Prionotaceae	*Acrotriche, Andersonia, Archeria, Astroloma, Brachyloma, Choristemon, Coleanthera, Conostephium, Cosmelia, Cyathodes, Dracophyllum, Epacris, Leucopogon, Lissanthe, Lysinema, Melichrus, Monotoca, Needhamiella, Oligarrhena, Pentachondra, Prionotes, Richea, Rupicola, Sphenotoma, Sprengelia, Styphelia, Trochocarpa, Woollsia*
*Ericaceae — Vacciniaceae	*Agapetes, Gaultheria, Pernettya, Rhododendron, Wittsteinia*

*Euphorbiaceae (Tribe Stenolobieae)	*Amperea, Bertya, Beyeria, Micrantheum, Monotaxis, Poranthera, Pseudanthus, Ricinocarpos, Stachystemon*
*Fabaceae (Tribe Podalyrieae)	*Aotus, Brachysema, Burtonia, Chorizema, Daviesia, Dillwynia, Eutaxia, Gastrolobium, Gompholobium, Isotropis, Jacksonia, Mirbelia, Oxylobium, Phyllota, Pultenaea, Sphaerolobium, Viminaria*
*Fabaceae (Tribe Genisteae)	*Bossiaea, Goodia, Hovea, Platylobium, Templetonia*
Gentianaceae	*Gentianella* in subalpine heaths
Goodeniaceae	*Dampiera, Goodenia, Lechenaultia, Scaevola, Velleia*
Haloragaceae	*Glischrocaryon, Gonocarpus, Haloragis*
Lamiaceae	*Hemiandra, Hemigenia, Microcorys, Prostanthera, Westringia*
Lentibulariaceae	*Utricularia*
Loganiaceae	*Logania*
Loranthaceae	*Nuytsia*
Mimosaceae	*Acacia*
*Myrtaceae (Tribe Leptospermoideae)	*Actinodium, Agonis, Angophora, Astartea, Baeckea, Balaustion, Beaufortia, Callistemon, Calothamnus, Calytrix, Calythropsis, Chamaelaucium, Conothamnus, Darwinia, Eremaea, Eucalyptus, Homalocalyx, Homoranthus, Hypocalymma, Kunzea, Lamarchea, Leptospermum, Lhotzkya, Melaleuca, Metrosideros, Micromyrtus, Phymatocarpus, Pileanthus, Regelia, Scholtzia, Sinoga, Syncarpia, Tristania, Thryptomene, Verticordia, Wehlia, Xanthostemon*
*Myrtaceae (Tribe Myrtoideae)	*Austromyrtus, Fenzlia*
Nepenthaceae	*Nepenthes*
Olacaceae	*Olax*
Oleaceae	*Notelaea*
*Pittosporaceae (Tribe Billardiereae)	*Billardiera, Cheiranthera, Marianthus, Pronaya, Sollya*
*Polygalaceae	*Comesperma*
*Proteaceae (Tribe Persoonioideae)	*Bellendena, Persoonia*
*Proteaceae (Tribe Proteoideae)	*Adenanthos, Agastachys, Conospermum, Cenarrhenes, Franklandia, Isopogon, Petrophile, Stirlingia, Symphyonema, Synaphea*
*Proteaceae (Tribe Grevilleoideae)	*Banksia, Dryandra, Grevillea, Hakea, Lambertia, Lomatia,*

TABLE 6.2 *(continued)*

	Orites, Strangea, Telopea, Xylomelum	Ecdeiocoleaceae	*Ecdeiocolea*
Rafflesiaceae	*Pilostyles*	Haemodoraceae	*Anigozanthos, Blancoa, Conostylis, Haemodorum,Macropidia, Phlebocarya, Tribonanthes*
*Rhamnaceae (Tribe Rhamneae)	*Cryptandra, Discaria, Pomaderris, Siegfriedia, Stenanthemum, Spyridium, Trymalium*	Hypoxidaceae	*Hypoxis*
Rubiaceae	*Opercularia*	Iridaceae	*Diplarrena, Libertia, Orthrosanthus,Patersonia, Sisyrinchium*
*Rutaceae (Tribe Boronieae)	*Asterolasia, Boronia, Correa, Eriostemon, Microcybe, Phebalium, Philotheca, Zieria*	Liliaceae	*Agrostocrinum, Anguillaria, Arnocrinum, Arthropodium, Blandfordia, Bulbinopsis, Burchardia, Caesia, Chamaescilla, Corynotheca, Dianella, Dichopogon, Hensmania, Hodgsoniola, Johnsonia, Laxmannia, Sowerbaea, Stawellia, Stypandra, Thysanotus, Tricoryne, Wurmbea (+Astelia, Herpolirion, Milligania in subalpine heaths)*
Santalaceae	*Anthobolus, Choretrum, Exocarpos, Leptomeria, Santalum, Spirogardnera*		
Sapindaceae	*Dodonaea*		
Scrophulariaceae (Tribe Euphrasieae)	*Euphrasia (+Veronica* in subalpine heaths)		
Solanaceae	*Anthocercis, Anthotroche*		
Spigeliaceae	*Mitrasacme*	*Orchidaceae (Tribe Diurideae)	*Acianthus, Adenochilus, Burnettia, Caladenia, Caleana, Calochilus, Chiloglottis, Corybas, Cryptostylis, Diuris, Drakaea, Elythranthera, Epiblema, Eriochilus, Genoplesium, Glossodia, Leptoceras, Lyperanthus, Microtis, Orthoceras, Prasophyllum, Pterostylis, Rimacola, Spiculaea, Thelymitra*
Stackhousiaceae	*Stackhousia*		
Sterculiaceae	*Commersonia, Guichenotia, Lasiopetalum, Rulingia, Thomasia*		
Stylidiaceae	*Forstera, Levenhookia, Phyllachne, Stylidium*		
Thymelaeaceae	*Pimelea (+Drapetes* in subalpine heaths)		
*Tremandraceae	*Platytheca, Tremandra, Tetratheca*	*Restionaceae	*Anarthria, Calorophus, Chaetanthus, Coleocarya, Dielsia, Empodisma, Harperia, Hopkinsia, Hypolaena, Lepidobolus, Leptocarpus, Lepyrodia, Loxocarya, Lyginia, Meeboldina, Onychosepalum, Restio*
Violaceae	*Clelandia, Hybanthus*		
Winteraceae	*Tasmannia* (syn. *Drimys*) in subalpine heaths		
		*Xanthorrhoeaceae	*Acanthocarpus, Baxteria, Borya, Calectasia, Chamaexeros, Dasypogon, Kingia, Lomandra, Xanthorrhoea*
MONOCOTYLEDONS			
Centrolepidaceae	*Aphelia, Brizula, Centrolepis, Hydatella, Trithuria*		
Cyperaceae	*Lepidosperma*, some *Schoenus* spp. (+*Oreobolus* in subalpine heaths)	Xyridaceae	*Xyris*

inal Tertiary flora even survive within the arid to semi-arid centre of Australia.

The major Köppen (1923) climatic types in which heathlands are found in Australia are described in Table 6.4, and their distribution over the continent is shown in Fig. 6.8. The sites listed in Table 6.6 against each of the climatic types refer to Appendices I and II (pp. 146–206), where analyses of the vegetation occurring at these sites are tabulated.

The previous section outlines the climatic stress to which subtropical Tertiary heathlands were subjected during the Quaternary. Today, the best development of heathlands in eastern Australia is seen where the warm climate with uniform rainfall

has some summer months with a mean air temperature over 22°C. In this area, shoot growth of the major species tends to occur whenever the mean temperature rises above 16 to 18°C. Specht and Brouwer (1975) have shown that within the **Cfa** climatic zone, shoot growth is potentially possible for 8 to 12 months of the year. It must be stressed, however, that non-dominant members of the heathland community usually show growth rhythms with temperature thresholds lower than the dominants (Specht and Rayson, 1957a; Maconochie, 1975).

In highland regions within the **Cfa** climatic zone, and south thereof, the **Cfb** climatic zone still receives rainfall throughout the year. However, as

TABLE 6.3

Species/area data (mean number of angiosperm[1] species per quadrat) of several heathland communities in eastern Australia

Locality	Quadrat size (m²)						Reference
	$\frac{1}{4}$	$\frac{1}{2}$	1	2	4	8	
1. Dry-heathlands							
Dark Island Soak, S.A.	8.3	11.5	14.1	17.9	22.4	26.8	Specht and Rayson (1957a)
Wilson's Promontory, Vic.							
Site 1	11.0	15.0	18.5	21.3	24.5	27.2	R.L. Specht (pers. obs., 1964)
Site 2	13.5	17.8	22.0	25.5	29.5	33.1	R.L. Specht (pers. obs., 1964)
Site 3	16.8	20.9	24.8	28.6	32.5	36.1	R.L. Specht (pers. obs., 1964)
Royal National Park, N.S.W.	5.0	7.4	9.7	15.2	18.6	22.6	D.H. Anderson (pers. comm., 1976)
Stradbroke Island, Qld.							
Mt. Hardgrave	6.3	10.7	14.0	17.8	21.9	25.6	R.L. Specht (pers. obs., 1975)
First high dune	10.3	15.0	18.0	21.0	24.0	27.0	R.L. Specht (pers. obs., 1976)
2. Wet-heathlands							
Wilson's Promontory, Vic.	9.1	12.0	14.0	17.2	21.1	24.5	R.L. Specht (pers. obs., 1964)
Royal National Park, N.S.W.	7.1	11.4	13.3	14.6	15.1	16.9	D.H. Anderson (pers. comm., 1976)
Mourawaring Point, N.S.W.	4.3	6.3	8.5	9.7	10.9	11.2	D.H. Anderson (pers. comm., 1976)
Brown Lake, Stradbroke Island, Qld.	6.5	7.4	9.6	11.8	14.3	15.8	Connor and Clifford (1972)

[1] Very few cryptograms have been recorded in Australian heathlands, and there were no gymnosperms in these stands.

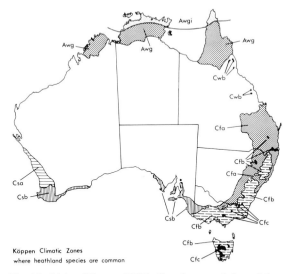

Köppen Climatic Zones
where heathland species are common

Fig. 6.8. Major Köppen (1923) climatic types (adapted from maps by Dick, 1961, and Gentilli, 1972) in which heathland vegetation is found in Australia. The climatic types are defined in Table 6.4.

the climate is cooler, the potential period for shoot growth of subtropical Tertiary survivors is only 2 to 4 summer months (Specht and Brouwer, 1975). In Victoria and Tasmania only a few of the high-temperature group of heathland plants survive; the species which have taken their place as physiognomic dominants appear to possess a lower temperature threshold (13–14°C) for shoot growth (Jones, 1968a).

In the Mediterranean-type climate (**Csa/Csb**) of southwest Western Australia and in South Australia, monthly temperatures tend to favour the survival of those Tertiary heathland species with a high temperature threshold (16–18°C). However, summer drought has operated against these species; they have been able to survive only in those areas where subsoil storage of winter rainfall is available for summer growth (Specht, 1957; Martin and Specht, 1962). The presence of a subsoil water table within the lateritic soils and coastal sand plains has enabled the southwest Western Australian flora to persist and flourish.

A similar pattern of water conservation and subsequent utilisation certainly operates in the monsoonal climate (**Aw**) of eastern Cape York Peninsula, where heathlands grow on deep sandy soils through which water within reach of roots continually seeps, even during the driest months of the year. It is much more difficult to visualise how the evergreen heathland species are able to survive through the dry monsoonal season in the cracks

TABLE 6.4

Major climatic types (Köppen, 1923) in which heathland vegetation is found in Australia

A hot climate with no month below 18°C
  Aw climate with a dry winter
    Awg climate with a temperature maximum before the   Kimberley region, W.A.
          summer solstice   Site 1 (Oenpelli, N.T.)
                                                                 Site 3 (Lizard Island, Cape Flattery, Qld.)

    Awgi climate with a small annual temperature range   Site 2 (Jardine River, Qld.)
         (below 5°C)

B dry climate
  BS semi-arid climate
    BSk mean annual temperature cool (below 18°C)   Site 13 (Billiatt, S.A.)
                                                          Site 15 (Hincks–Murlong, S.A.)
                                                            Goldfields region, W.A.

C warm climate with at least one month below 18°C
  Cf climate with uniform rain
    Cfa climate with hot summer (hottest month above   Site 4 (Shoalwater Bay, Qld.)
         22°C)   Site 5a (Stradbroke Island, Qld.)
                                                        Site 5b (Toolara, Qld.)
                                                        Site 5c (Beerwah, Qld.)
                                                              Site 7 (Bonny Hills, N.S.W.)
                                                           Site 8 (Sydney, N.S.W.)

    Cfb climate with long mild summer (hottest month Site 5d (McPherson Ranges, Qld.–N.S.W.)
         below 22°C and at best four months above 10°C) Site 6 (Mount Kaputar, N.S.W.)
                                                        Site 9 (Wilsons Promontory, Vic.)
                                                        Site 10 (Frankston, Vic.)

    Cfc climate with short mild summer (with less than Site 17 (Mount Kosciusko, N.S.W.)
         four months above 10°C)   Site 18 (Lake Mountain, Vic.)
                                                        Site 19a (Cradle Mountain, Tas.)
                                                       Site 19b (Mount Wellington, Tas.)

  Cs climate with a dry summer
    Csa climate with hot summer (hottest month above Perth–Geraldton region, W.A.
         22°C)

    Csb climate with long mild summer (hottest month Site 11 (Lower South-East, S.A.)
         below 22°C and at least four months above 10°C) Site 12 (Keith, S.A.)
                                                        Site 14 (Mount Compass, S.A.)
                                                       Perth–Albany–Esperance region, W.A.

  Cw climate with a dry winter
    Cwb climate with long mild summer (hottest month Site 16 (Mount Bellenden Ker, Qld.)
         below 22°C and at least four months above 10°C)

and sand flats of the Kimberley and Arnhem Land quartzitic hills. No information is available on the growth rhythms of heathland species growing within the tropical region of Australia.

As has been noted in other parts of the world, Australian heathlands are found both in lowland climates and on high mountain peaks in subalpine climates above the tree line. In intermediate altitudes the dense tree communities appear to have excluded most of the heathland species from the understorey. On high peaks such as Mount Bartle Frere (1611 m) and Mount Bellenden Ker (1491 m) of tropical Queensland, the subalpine climate may be classed as a **Cwb** type. In subtropical areas on the border of Queensland and New South Wales, highland areas only rise to a maximum altitude of 1356 m with a **Cfb** climate. In temperate southeastern Australia and Tasmania, the subalpine **Cfc** climate of the highlands (1200 to 2230 m in Victoria and New South Wales; 1200 to 1617 m in Tasmania) again supports heathland communities; cushion-plants tend to dominate the Tasmanian subalpine heathlands.

Soils

Heathland vegetation in Australia appears to cross all climatic barriers except that of aridity; it is found from temperate to tropical Australia, from lowland to alpine. Wherever the vegetation occurs, the major determining factor controlling its distribution appears to be the nutrient status of the soil.

Since Diels (1906) distinguished *Sklerophyllen-Wald* from *Savannen-Wald* in his study of the vegetation of southwest Western Australia, plant ecologists have become progressively aware of the importance of the edaphic factor in the distribution of vegetation in the humid zones of Australia. As the transition between the two vegetation types described by Diels is often abrupt in the Mount Lofty Ranges of South Australia, Adamson and Osborn (1924) concluded that, although climate was the master factor in controlling the distribution of the forest types in the area, edaphic factors played an important part along the line of junction. In 1939 Wood considered this edaphic problem in depth, and produced three-dimensional diagrams (based on soil phosphorus, nitrogen, and pH, plus rainfall) which showed the clear separation of the Sklerophyllen flora from the Savannen flora. Specht and Perry (1948) and Specht et al. (1961) assembled even more soil analytical data to illustrate the poverty of the soils on which the Sklerophyllen flora flourished (see Table 6.5).

The same edaphic control was stressed by Beadle (1953, 1954, 1962, 1968) for the sclerophyll flora of the Sydney district. The mean values of the phosphorus content of soils under Sklerophyllen scrubs and forests in this area ranged from 23 to 53 p.p.m. (mean 37) and from 53 to 120 p.p.m. (mean 83) respectively — much lower than values from 74 to 266 p.p.m. (mean 139) and from 230 to 720 p.p.m. (mean 430) observed for soils of the dense tree-fern forests and rain forests. Coaldrake and Haydock (1958) collated the phosphorus analyses for 326 surface soils sampled on a 10-chain (200 m) grid from Sklerophyllen communi-

TABLE 6.5

Chemical analyses of surface soils typical of heathland and savannah ecosystems of the Mount Lofty Ranges, South Australia (Specht et al., 1961)

Chemical property	Heathland ecosystems	Savannah ecosystems
pH	5.70 ± 0.07 [1]	6.47 ± 0.11
Calcium carbonate (%)	nil	nil
Total nitrogen (%)	0.074 ± 0.010	0.127 ± 0.012
Total phosphorus (%)	0.007 ± 0.001	0.026 ± 0.004
Total potassium (%)	0.15 ± 0.07	0.29 ± 0.04
Total soluble salts (%)	0.016 ± 0.003	0.022 ± 0.004
Chloride as NaCl (%)	0.006 ± 0.002	0.008 ± 0.002
Exchangeable cations m/eq. 100 g^{-1})	4.89	10.24

[1] Standard error of mean.

nities in southeastern Queensland; the mean values were low, ranging from 30 to 53 p.p.m., with virtually no significant variation between Sklerophyllen vegetation types.

Clearly, the Sklerophyllen (heathland) flora is confined to soils low in plant nutrients. The parent rock on which the soils have developed is often the key to this low nutrient status — sandstone, quartzitic, and granitic rocks, inherently low in plant nutrients, are found in many places in the humid regions of Australia. The nutrient status of soils derived from these rocks has been further reduced by leaching. In humid areas, leaching may be so intense that soils developed on argillaceous rocks, as well as those on siliceous materials, are strongly podzolised and support a Sklerophyllen flora (Specht and Perry, 1948; Specht et al., 1961). As well, large dunes composed of infertile siliceous sand have accumulated along many parts of the Australian coastline, and in some areas have been blown many kilometres inland. These Pleistocene to Recent siliceous dunes have been further leached (podzolised or solonised) to form soil profiles exceedingly low in plant nutrients — much lower than the figures quoted above.

The Australian continent contains many examples of these infertile skeletal, podzolic, or solonetzic siliceous soils, developed either on nutrient-poor basement rock or on Quaternary sand. As well, there are many areas where extensive remnants of lateritic soils, formed during the subtropical climate of the Tertiary (see above), still persist on the landscape (Stace et al., 1968). These fossil soils and their derivatives are very low in nutrient status. If their surface texture is earthy (lateritic red and yellow earths), a grass-herb ground stratum tends to flourish. If their surface texture is sandy (lateritic podzols), the soils usually support a sclerophyllous (heathland) flora.

Residual lateritic podzols are best developed in the south of Australia (southwest Western Australia; Kangaroo Island and the Mount Lofty Ranges in South Australia; Dundas Tableland and Brisbane Ranges in Victoria); north of an arc running from southeastern New South Wales, through Oodnadatta, South Australia to Geraldton in Western Australia only small areas of lateritic podzols are found intermixed with lateritic red earths on extensive tablelands and undulating plains (Stace et al., 1968).

In many subalpine localities in Australia, mineral soils have been formed largely from infertile, siliceous material; alongside, in seasonally waterlogged sites, peaty soil (high moor) has accumulated. As both these soils formed above the tree line contain limited quantities of available plant nutrients, a wide range of heathland species abound.

In Australia all infertile, sandy or peaty soils low in phosphorus support a large assemblage of heathland species. Trees and tall shrubs are generally present, except in small areas of true heathlands where the soils (humus or ground-water podzols, acid to neutral peats) are seasonally waterlogged; these areas have been termed "wet-heathlands" and "subalpine heathlands". Trees and tall shrubs may also be absent or dwarfed in infertile sites, wind-planed by strong sea breezes containing a high percentage of sea salt (Parsons and Gill, 1968). Seasonal drought (in the surface metre of soil) may also prevent the establishment of taller, possibly more mesophytic, tree species in heathland growing on deep, very infertile, sands (Specht and Rayson, 1957a; Groves and Specht, 1965; Jones, 1968a and b; Clifford and Specht, 1979); these areas have been termed "dry-heathlands". There is some suggestion that the extreme infertility of these sands may reduce the chances of inoculation of roots of these tree species with mycorrhizal organisms (Burrell, 1969). The survival of many heathland species appears to depend on a delicate balance between low soil fertility reducing the invasion potential of root-rotting fungi, like *Phytophthora* spp., and yet still enabling more beneficial rhizosphere organisms to survive.

The chemical analyses of all heathland soils show that practically all nutrient elements are in low supply — much lower than the levels recorded for soils which support Savannen communities in the same climatic zone (Table 6.5). The intensive nutritional work of Specht (1963) and Heddle and Specht (1975) on Dark Island heathland, South Australia, has established that, of all nutrient elements, phosphorus is the one most limiting. The addition of this element to the heathland community produces slow, but eventually dramatic, changes to the vegetation, converting the heathland vegetation to a Savannen type.

Soil nutrient status, especially the level of phosphorus, appears to be the major controlling factor in the distribution of heathland vegetation in Australia. Infertile lateritic and other siliceous soils were so widespread across Australia during the subtropical Tertiary that many xeric or seasonally waterlogged niches must have existed in which the heathland flora was not overshadowed by dense forest trees and shrubs. The flora must have occupied the whole of Australia, only being absent from nutrient-rich alluvial and basaltic soils and on calcareous soils developed in areas of southern Australia invaded by Miocene seas.

Fire

Fire is an integral part of the Australian heathland ecosystems. The dense structure of the mature community and the presence of flammable volatile oils and resins in the sclerophyllous leaves of the component species make the heathlands a potential fire hazard during periods of drought. If trees are present in the stand, both ground and crown fires may occur on extremely dry and windy days. Lightning strikes (and, in South Africa, sparks from rocks dislodged during an earthquake) have caused fires in mature heathlands, but man appears to have increased the frequency of fire considerably.

The conflagration may appear disastrous to human eyes, but the vegetation is well-adapted to the ravages of fire (see Gill and Groves in Chapter 7 of Volume B). The aerial stems of shrubby species may be destroyed by fire, but many plants sprout again from lignotubers, rhizomes, corms and bulbs protected in the soil (Beadle, 1940; Specht and Rayson, 1957b). Mallee eucalypts show the same phenomenon, regenerating from large lignotubers or "mallee roots". Other species will release seed from woody fruits only after being heated in a fire; large numbers of seeds are thus available for germination. Seed of other heathland species may have been stored by ants in galleries within their nests (Berg, 1975) and germinate following a fire. In fact, the ash-bed effect of a fire seems to be vital in the germination of many species — the release of plant nutrients, the reduction of seed scavengers, root pathogens, and plant competition all contribute to seed germination and establishment.

Most trees and tall shrubs, which together with a heathland understorey form the Sklerophyllen-Wald (Diels, 1906), possess fire-resistant bark and thus survive fire even though the canopy may be consumed. Numerous epicormic buds soon emerge from the stems, clothing them in green foliage. Gradually, the foliage of the crown assumes dominance again, and the epicormic shoots are shed.

The catastrophic effect of fire is thus soon repaired. A wealth of regenerating plants and seedlings appears. As the stand matures many individuals and even species will disappear (Fig. 6.7). For example, 38 species recorded in Dark Island heathland following a fire were reduced to 20 after twenty-five years, and to 10 fifty years after the fire (Specht et al., 1958). The age of the heathland could have a significant effect on the composition of the stand following the fire.

The biggest danger of fire in heathlands is that the small amount of nutrient constituting the "working capital" of the plant community (Beadle and Burges, 1949) may be depleted by leaching immediately following the fire (before the root systems have been able to reoccupy the whole surface soil).

Animals

Native animals such as the kangaroo and wallaby (*Macropus* spp.), the wombat (*Vombatus ursinus*), emu (*Dromaius novaehollandiae*) and black cockatoo (*Calyptorhynchus funereus*), and a few introduced rabbits (*Oryctolagus cuniculus*), exert a minor grazing pressure on the heathlands (Costin, 1954; Specht and Rayson, 1957a; see also Edmonds and Specht, Vol. B, Ch. 3). The dominant predators are now the introduced fox (*Vulpes vulpes*) and dingo (*Canis familiaris dingo*). Honey-eaters (Meliphagidae) and wrens (Maluridae) are common birds, feeding on nectar and insects and probably aiding cross fertilization of the vegetation (Clifford and Drake, Vol. B, Ch. 5).

Populations of insects inhabiting the heathland shrubs may be classed as foliage-feeders, nectar-feeders, seed-gatherers, stem-suckers and wood-borers; other insects and spiders will be predators and parasites on the above first-order consumers. The populations of each of these groups fluctuate seasonally in response to the seasonal growth and flowering rhythms of the heathland species (see Edmonds and Specht, Vol. B, Ch. 3). Ants play an important role in seed removal and dispersal (Berg,

1975; see also Vol. B, Ch. 6). Collembola, mites worms, nematodes and protozoa, together with decomposing fungi and bacteria act in several ways in breaking leaf litter down (Edmonds and Specht in Vol. B, Ch. 3). In some areas, especially in northern Australia, termites may assist in decomposition.

The dynamics of vertebrate animal populations, the patterns of diversity, distribution and seasonal behaviour of species can only be understood if heathlands are recognised as part of a series of associated habitats. Areas of heathlands are rarely extensive; they give way to woodlands and open-forests, etc., sometimes with a heathy understorey, sometimes with a grassy understorey. As food resources for vertebrates are limited in heathlands, most vertebrates are opportunists, living in other communities and exploiting the heathlands only occasionally during a particular seasonal phase of the vegetation (see Dwyer et al. in Chapter 9; Newsome and Catling in Chapter 10 and Edmonds and Specht in Vol. B, Ch. 3). Only a very few vertebrates may be considered specialists and are able to reside permanently in the heathland.

CONCLUSION

The ancient heathland vegetation of Australia, once widespread across the continent, now exists in disjunct pockets in the higher rainfall areas from the tropical north to the temperate south (Figs. 6.6 and 6.8). Relic fragments of the flora still exist in pockets throughout the arid inland.

Nutritional and climatic stress for well over 100 million years have resulted in considerable speciation within the diverse heathland families and genera. Many heathland species have an extremely limited range; some are widely-spread. Many of the latter species show great plasticity in growth form, often the same taxon ranging in habit from a dwarf shrub to a tree 10 to 30 m in height. As both structure and species-composition change markedly in a continuum over short distances, it is very difficult to delimit and to describe heathland communities. This is especially so in the cooler temperate, mountainous areas of Tasmania. Appendix I (Tables 6.8 to 6.13) summarises the structural formations and species designated by Australian ecologists (see Specht et al., 1974) as

dominant or co-dominant in the upper stratum of the component alliances defined within these formations in the major Köppen climatic zones illustrated in Fig. 6.8.

Detailed analyses of representative areas of these disjunct heathlands are included in Appendix II, Sites 1 to 15 (lowland dry-/and wet-heathlands of northern, eastern and southern Australia) and Sites 16 to 19 (subalpine heathlands of southeastern Australia). George et al. have described lowland heathlands of southwest Western Australia in Chapter 7. These studies show the structural and floristic composition of each heathland (or related shrubland) and their topographic, climatic and edaphic relationships.

The Australian heathland vegetation is distinctive floristically and structurally. It is invariably found on oligotrophic soils very low in phosphorus, nitrogen, and often other plant nutrients. Commonly, an overstorey of tall shrubs or trees is present to form an open-forest, woodland, or tall-shrubland community with heathland vegetation as an understorey. In habitats (at both lowland and subalpine altitudes) where seasonal waterlogging is prolonged, trees and tall shrubs fail to survive and a true heathland, termed **wet-heathland** develops; restiads and sedges often flourish as co-dominants in these wet-heath communities, and in a few cooler environments the bog moss (*Sphagnum*) develops. At the other end of the soil hydrological spectrum, on extremely well-drained sites (such as deep sands or skeletal sandstone or granitic ridges), tall shrubs and trees are also rare or absent — a **dry-heathland** (often termed sand-heathland) survives.

Fig. 6.9 presents a schematic diagram which summarises the ecological relationships (soil nutrition and hydrology) of Australian plant formations in which the heathland vegetation may be considered as an integral part of the plant community. No attempt is made in Fig. 6.9 to show the relationships of the majority of the Australian vegetation to the "heathland communities"; a slightly higher soil-fertility level will enable savannah (grassy) communities to replace the heathland (either as an understorey of the overstorey of trees or shrubs, or to persist alone as grassland). In southern Australia, oligotrophic soils, no matter what the texture of their surface soil, support heathland vegetation (with or without trees); in northern Australia the oligotrophic lateritic earths

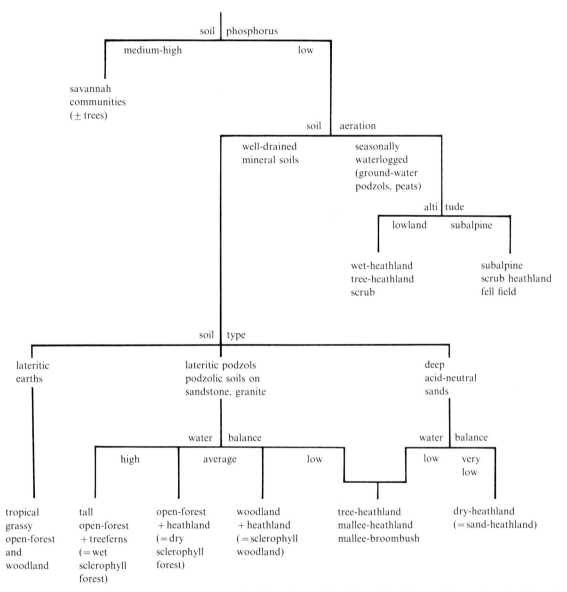

Fig. 6.9. Schematic diagram showing the major ecological relationships (soil fertility and hydrology) of Australian plant formations which possess a characteristic heathy stratum either alone (heathland proper) or as an understorey to trees or shrubs.

tend to support a wealth of tropical grasses, the heathland vegetation being found only where sandy soils have accumulated.

The sclerophylly of the heathland communities appears to have evolved originally in a humid (rain-forest) environment in response to the low mineral nutrient status (particularly for phosphate) of the soil (see also Specht in Chapter 1 and Specht and Womersley in Chapter 12). The attributes which are associated with sclerophylly were later

(in geological time) important for the survival of heathland flora in both seasonally droughted and seasonally waterlogged habitats (see Specht in Volume B, Ch. 10 and 13 and Connor and Doley in Vol. B, Ch. 14). In fact, often the same heathland species may be found in these two extreme habitats and in the intervening mesic sites.

Sclerophylly, and its associated attributes, is by no means attractive to grazing animals (vertebrate and invertebrate alike); it is largely the young, more mesophyllous, shoots and flowers which provide a reasonable food source. On the other hand, sclerophylly has certainly made the flora more susceptible to destruction by fire. After a fire, it is found that pockets of vegetation and many animals survive (Main in Vol. B, Ch. 8). The vegetation has remarkable powers of regeneration (Gill and Groves in Vol. B, Ch. 7) and animals find the young, regenerating shoots much more palatable than the sclerophyllous leaves on older stands.

APPENDIX I: STRUCTURE AND DOMINANT SPECIES IN AUSTRALIAN HEATHLANDS

TABLE 6.8

Structure of heathland communities, and species dominant or co-dominant in the upper stratum of one or more of the heathland alliances found in tropical Australia (Köppen climatic zones **Awg** and **Awgi**) (for further details on structure see Table 1.3; for species composition see Appendix II, Sites 1, 2 and 3, and Specht et al., 1974)

1. Closed-forest[1] ('heath-forest')

   Leptospermum fabricia
   Melaleuca aff. *symphyocarpa*

2. Low closed-forest[1] ('heath-forest')

   Leptospermum fabricia
   Melaleuca aff. *symphyocarpa*

3. Low open-forest[1] ('heath-forest')

   Leptospermum fabricia
   Melaleuca aff. *symphyocarpa*

4. Low woodland[2] with heathy understorey

   Eucalyptus dichromophloia
   E. herbertiana
   E. miniata
   E. phoenicia
   E. tetrodonta

5. Low open-woodland with heathy understorey[3]

   Eucalyptus setosa

6. Closed-scrub[1]

   Acacia crassicarpa
   Leucopogon spp.
   Sinoga lysicephala
   Thryptomene oligandra

7. Open-scrub[1,2]

   Acacia spp.
   Calytrix arborescens
   C. exstipularis
   Melaleuca magnifica
   M. nervosa
   Triodia spp.
   Verticordia cunninghamii

8. Closed-heathland

   Wet-heathland[1]
   Grevillea pteridifolia

9. Open-heathland

   Wet-heathland[1]
   Sinoga lysicephala

   Dry-heathland[2]
   Acacia humifusa
   Calytrix exstipularis
   Melaleuca spp.
   Micraira subulifolia
   Thryptomene oligandra
   Triodia spp.

10. Closed-herbland

    Wetland[1]
    Leptocarpus spathaceus

[1] On ground-water podzols.
[2] On skeletal sandstone soils/leached sands.
[3] On red massive sandy earths in Köppen climatic zone **BShw** (Burra Range between Charters Towers and Hughenden, Queensland).

TABLE 6.9

Structure of heathland communities, and species dominant or co-dominant in the upper stratum of the heathland alliances, found in subtropical eastern Australia (Köppen climatic zone **Cfa**) (for further details on structure see Table 1.3; for species composition see Appendix II, Sites 4, 5A to C, and 7, and Specht et al., 1974)

1. Open-forest with heathy understorey

   On skeletal sandstone
   Angophora costata
   Callitris columellaris
   Eucalyptus cloeziana
   E. crebra
   E. dealbata
   E. maculata
   E. peltata
   E. polycarpa
   E. rossii
   E. tenuipes
   E. watsoniana

   On deep podzolised sand
   Angophora costata
   Eucalyptus acmenoides
   E. intermedia
   E. planchoniana
   E. pilularis
   E. seeana
   E. signata
   E. umbra
   Syncarpia hillii

2. Low open-forest with heathy understorey

   On deep podzolised sand
   Angophora costata
   Eucalyptus acmenoides
   E. bancroftii
   E. intermedia
   E. moluccana
   E. seeana
   E. signata
   Tristania suaveolens

3. Woodland with heathy understorey

   On skeletal sandstone
   Angophora costata
   Eucalyptus citriodora
   E. cloeziana
   E. decorticans
   E. intermedia
   E. peltata
   E. polycarpa
   E. tenuipes
   E. watsoniana
   Tristania suaveolens

   On inland leached sand
   Angophora floribunda
   Callitris columellaris
   Eucalyptus conica
   E. pilligaensis

4. Low woodland with heathy understorey

   On ground-water podzol
   Angophora woodsiana
   Eucalyptus acmenoides
   E. intermedia
   E. robusta
   Tristania suaveolens

5. Low open-woodland with heathy understorey

   On ground-water podzol
   Angophora woodsiana
   Banksia aemula
   Eucalyptus acmenoides
   Melaleuca quinquenervia

6. Closed-scrub

   Wetland
   Leptospermum flavescens

7. Open-scrub ('mallee') with heathy understorey

   On very deep podzolised sand
   Eucalyptus intermedia
   E. planchoniana
   E. signata
   Tristania conferta

8. Closed-heathland
   Wet-heathland
   Baeckea linifolia
   Banksia oblongifolia
   Empodisma minus
   Epacris microphylla
   E. obtusifolia
   Leptospermum flavescens
   L. lanigerum
   L. semibaccatum
   Sprengelia incarnata

9. Open-heathland

   Wet-heathland
   Banksia oblongifolia
   Hypolaena fastigiata
   Leptospermum flavescens
   L. lanigerum
   L. liversidgei
   L. semibaccatum
   Xanthorrhoea spp.

10. Closed-herbland

   Wetland
   Baumea spp.
   Empodisma minus
   Restio pallens

TABLE 6.10

Structure of heathland communities, and species dominant or co-dominant in the upper stratum of one or more of the heathland alliances, found in the warm temperate region of New South Wales, Victoria and Tasmania (Köppen climatic zone **Cfb**) (for further details on structure see Table 1.3; for species composition see Appendix II, Sites 5D, 6, 8 to 10, and Specht et al., 1974)

1. Open-forest with heathy understorey

    On skeletal sandstone
    Angophora costata
    A. floribunda
    Banksia serrata
    Casuarina spp.
    Eucalyptus eximia
    E. gummifera
    E. piperita
    E. punctata
    E. racemosa
    E. sieberi
    E. umbra, etc.

    On granitic, lateritic and sandy soils
    Angophora floribunda
    Eucalyptus amygdalina
    E. baxteri
    E. globulus
    E. gummifera
    E. macrorhyncha
    E. morrisbyi
    E. muellerana
    E. nitida
    E. obliqua
    E. ovata
    E. pulchella
    E. radiata
    E. risdonii
    E. sieberi
    E. st-johnii
    E. tenuiramis
    E. viminalis subsp.

2. Low open-forest with heathy understorey

    On skeletal sandstone
    Angophora bakeri
    A. costata
    Banksia serrata
    Eucalyptus capitellata
    E. eximia
    E. globoidea
    E. gummifera
    E. haemastoma
    E. punctata
    E. racemosa
    E. sieberi

    On granitic, lateritic and sandy soils
    Banksia serrata
    Casuarina stricta
    Eucalyptus baxteri
    E. macrorhyncha
    E. muellerana

    E. nitida
    E. obliqua
    E. ovata
    E. perriniana
    E. radiata
    E. risdonii
    E. rodwayi
    E. tenuiramis
    E. viminalis subsp.

3. Woodland with heathy understorey

    On sandy soil
    Angophora floribunda
    Eucalyptus globulus
    E. gummifera
    E. obliqua
    E. ovata
    E. tenuiramis
    E. viminalis subsp.

4. Low woodland with heathy understorey

    On skeletal sandstone
    Eucalyptus gummifera
    E. haemastoma
    E. piperita
    E. sieberi

    On granitic, lateritic and sandy soils
    Banksia marginata
    B. serrata
    Casuarina stricta
    Eucalyptus aggregata
    E. cinerea
    E. nitida
    E. obliqua
    E. perriniana
    E. radiata
    E. risdonii
    E. st-johnii
    E. tenuiramis
    E. viminalis subsp.

5. Low open-woodland with heathy understorey

    On sandy soils
    Banksia marginata
    Casuarina monilifera
    Eucalyptus cinerea
    E. nitida
    E. st-johnii
    E. viminalis subsp.

TABLE 6.10 *(continued)*

6. Closed-scrub

    Wetland
    Banksia spp.
    Kunzea ambigua
    Leptospermum lanigerum
    Melaleuca ericifolia
    M. squarrosa

7. Open-scrub ('mallee') with heathy understorey

    Banksia spp.
    Eucalyptus alpina
    E. approximans
    E. baxteri
    E. kitsoniana
    E. luehmanniana
    E. multicaulis
    E. nitida
    E. obtusiflora
    E. radiata
    E. stricta

8. Tall shrubland with heathy understorey

    Banksia marginata
    Casuarina monilifera

9. Closed-heathland

    Wetland
    Acacia myrtifolia

    Banksia marginata
    Banksia spp.
    Bauera rubioides
    Leptospermum juniperinum
    L. myrsinoides
    Melaleuca gibbosa
    M. squarrosa
    Sprengelia incarnata

10. Open-heathland

    Dry- and wet-heathlands
    Acacia mucronata
    Acacia spp.
    Agastachys odorata
    Baeckea leptocaulis
    Banksia marginata
    Banksia spp.
    Casuarina monilifera
    C. pusilla
    Casuarina spp.
    Epacris impressa
    E. lanuginosa
    E. petrophila
    Leptospermum myrsinoides
    L. nitidum
    L. sericeum
    Leucopogon spp.
    Melaleuca squamea
    Richea acerosa
    Sprengelia incarnata

TABLE 6.11

Structure of heathland communities, and species dominant or co-dominant in the upper stratum of one or more of the heathland alliances, found in alpine/subalpine regions of New South Wales, Victoria and Tasmania (Köppen climatic zone **Cfc**) (for further details on structure see Table 1.3; for species composition see Appendix II, Sites 16 to 19, and Specht et al., 1974)

1. Low open-forest with heathy understorey

    N.S.W. and Victoria
    Eucalyptus pauciflora

    Tasmania
    Athrotaxis cupressoides
    A. selaginoides
    Eucalyptus archeri
    E. coccifera
    E. gunnii
    E. nitida
    E. pauciflora
    E. rodwayi
    E. subcrenulata
    E. urnigera

2. Low woodland with heathy understorey

    N.S.W. and Victoria
    Eucalyptus pauciflora

    E. perriniana
    E. stellulata

    Tasmania
    Athrotaxis cupressoides
    A. selaginoides
    Eucalyptus archeri
    E. coccifera
    E. gunnii
    E. nitida
    E. pauciflora
    E. rodwayi
    E. subcrenulata
    E. urnigera

3. Low open-woodland with heathy understorey

    N.S.W. and Victoria
    Eucalyptus pauciflora

TABLE 6.11 *(continued)*

Tasmania
Athrotaxis cupressoides
A. selaginoides
Eucalyptus gunnii
E. nitida
E. rodwayi

4. Closed-scrub

N.S.W. and Victoria
Leptospermum flavescens
L. phylicoides
Podocarpus lawrencei

Tasmania
Callistemon viridiflorus
Diselma archeri
Leptospermum lanigerum
Lomatia polymorpha
Microstrobos niphophilus
Nothofagus gunnii
Olearia pinifolia
Orites acicularis
O. diversifolia
Richea scoparia
Tasmannia lanceolata
Telopea truncata
Trochocarpa gunnii

5. Open-scrub

N.S.W. and Victoria
Acacia glaucescens
Bauera rubioides
Callistemon pallidus
Casuarina nana
Eriostemon trachyphyllus
Eucalyptus pauciflora
Leptospermum flavescens
L. phylicoides

Tasmania
Banksia marginata
Callistemon viridiflorus
Eucalyptus nitida
E. vernicosa
Helichrysum hookeri
H. ledifolium
Leptospermum humifusum
L. lanigerum
L. sericeum
Orites acicularis
O. revoluta
Phyllocladus aspleniifolius
Richea pandanifolia
R. scoparia

6. Tall shrubland with heathy understorey

N.S.W. and Victoria
Acacia glaucescens

Casuarina rigida
Eriostemon trachyphyllus
Eucalyptus pauciflora
Leptospermum brevipes
Leptospermum sp. nov.

Tasmania
Richea pandanifolia

7. Closed-heathland

N.S.W. and Victoria
Casuarina rigida
Leptospermum spp.
Oxylobium ellipticum
Phebalium phylicifolium
Podocarpus lawrencei
Prostanthera cuneata

Tasmania
Abrotanella forsterioides
Bauera rubioides
Bellendena montana
Cyathodes dealbata
C. petiolaris
C. straminea
Diselma archeri
Donatia novae-zelandiae
Epacris petrophila
E. serpyllifolia
E. stuartii
Grevillea australis
Helichrysum backhousei
H. hookeri
H. ledifolium
Leptospermum humifusum
L. nitidum
Lissanthe montana
Microcachrys tetragona
Microstrobos niphophilus
Monotoca empetrifolia
Nothofagus gunnii
Olearia algida
O. ledifolia
O. stellulata
Orites acicularis
O. revoluta
Pentachondra pumila
Persoonia gunnii
Pimelea pygmaea
Podocarpus lawrencei
Pterygopappus lawrencei
Richea procera
R. scoparia
R. sprengelioides
Tasmannia lanceolata
Trochocarpa thymifolia
Westringia rubiifolia

TABLE 6.11 *(continued)*

8. Open-heathland

   N.S.W. and Victoria
Casuarina nana
C. rigida
Epacris breviflora
E. serpyllifolia
Kunzea muelleri
Leptospermum spp.
Oxylobium ellipticum
Phebalium phylicifolium
Podocarpus lawrencei
Prostanthera cuneata

   Tasmania
Baeckea gunniana
Bauera rubioides
Bellendena montana
Boronia citriodora
Coprosma nitida
Cyathodes petiolaris
Epacris gunnii
E. petrophila
E. serpyllifolia
Eucalyptus vernicosa
Exocarpos humifusus
Helichrysum backhousei
H. hookeri
H. ledifolium
Leptospermum humifusum
Melaleuca squamea
Olearia algida
O. ledifolia
Orites acicularis
O. revoluta
Podocarpus lawrencei
Richea acerosa
R. scoparia
R. sprengelioides
Sprengelia incarnata
Westringia rubiifolia

9. Closed bog-heathland

   N.S.W. and Victoria
Astelia alpina
Baeckea spp.
Blindia robusta
Epacris breviflora
E. paludosa
E. serpyllifolia
Richea continentis
Sphagnum cristatum

10. Open bog-heathland

   N.S.W. and Victoria
Astelia alpina
Baeckea spp.

Blindia robusta
Epacris breviflora
E. paludosa
E. serpyllifolia
Richea continentis
Sphagnum cristatum

11. Dwarf heathland (fellfield)

   N.S.W.
Colobanthus benthamianus
Coprosma pumila
Epacris petrophila
Pygmea densifolia

   Tasmania
Bellendena montana
Coprosoma nitida
Cyathodes petiolaris
Epacris serpyllifolia
Exocarpos humifusus
Hakea lissosperma
H. microcarpa
Helichrysum backhousei
H. hookeri
Leptospermum humifusum
Olearia algida
O. ledifolia
Orites acicularis
O. revoluta
Persoonia gunnii
Podocarpus lawrencei
Richea scoparia
Tasmannia lanceolata

12. Dwarf open-heathland (fellfield)

   N.S.W.
Colobanthus benthamianus
Coprosma pumila
Epacris petrophila
Pygmea densifolia

   Tasmania
Hakea lissosperma
Olearia ledifolia
Orites acicularis

13. Closed bog-herbland

   N.S.W. and Victoria
Astelia alpina
Carex gaudichaudiana
Empodisma minus
Sphagnum cristatum

   Tasmania
Astelia alpina
Empodisma minus
Sphagnum spp.

TABLE 6.12

Structure of heathland communities, and species dominant or co-dominant in the upper stratum of one or more of the heathland alliances, found in warm temperate South Australia and western Victoria (Köppen climatic zones **Csb**) (for further details on structure see Table 1.3; for species composition see Appendix II, Sites 11 to 15, and Specht et al., 1974)

1. Open-forest with heathy understorey

   On podzolised soils
   Eucalyptus baxteri
   E. cladocalyx
   E. fasciculosa
   E. goniocalyx
   E. macrorhyncha
   E. obliqua

2. Low open-forest with heathy understorey

   On podzolised sandy soils
   Eucalyptus baxteri
   E. cladocalyx
   E. cosmophylla
   E. diversifolia
   E. fasciculosa
   E. goniocalyx
   E. macrorhyncha
   E. nitida
   E. ovata
   E. viminalis subsp.

3. Woodland with heathy understorey

   On podzolised sandy soils
   Eucalyptus fasciculosa
   E. leucoxylon

4. Low woodland with heathy understorey

   On podzolised sandy soils
   Eucalyptus baxteri
   E. fasciculosa
   E. ovata

5. Low open-woodland with heathy understorey

   On podzolised/solonised sandy soils

   Eucalyptus fasciculosa

6. Closed-scrub

   Wetland
   Leptospermum pubescens
   Melaleuca squarrosa

7. Open-scrub ('mallee') with heathy understorey

   On podzolised/solonised sandy soils
   Eucalyptus baxteri
   E. cneorifolia
   E. conglobata
   E. cosmophylla
   E. diversifolia
   E. dumosa subsp. *pileata*
   E. flocktoniae
   E. foecunda
   E. goniocalyx
   E. incrassata
   E. remota
   E. rugosa
   Melaleuca uncinata

8. Tall open-shrubland with heathy understorey

   On deep sands
   Eucalyptus foecunda
   E. incrassata

9. Closed-heathland

   Wet-heathland
   Leptospermum juniperinum
   L. pubescens

10. Open-heathland

   Wet-heathland
   Banksia ornata
   Casuarina paludosa
   Hakea rostrata
   Xanthorrhoea australis

   Saline wet-heathland
   Hakea rugosa
   Melaleuca gibbosa

   Dry-heathland
   Banksia ornata
   Casuarina muellerana
   C. pusilla
   Leptospermum laevigatum var. *minus*
   L. myrsinoides
   Xanthorrhoea australis

TABLE 6.13

Structure of heathland communities, and species dominant or co-dominant in the upper stratum of one or more of the heathland alliances, found in southwestern Australia (Köppen climatic zones **Csa** and **Csb**) (for further details see Chapter 7 in this volume; Specht et al., 1974; and Table 1.3 for definitions of structural formations)

1. Open-forest with heathy understorey

   Agonis flexuosa
   Eucalyptus calophylla
   E. cornuta
   E. diversicolor
   E. dundasii
   E. gomphocephala
   E. marginata

2. Low open-forest with heathy understorey

   Agonis flexuosa
   Banksia attenuata
   B. menziesii
   B. prionotes
   Casuarina fraserana
   Eucalyptus annulata
   E. calophylla
   E. cornuta
   E. falcata
   E. marginata
   E. platypus
   E. spathulata
   E. todtiana

3. Woodland with heathy understorey

   Agonis flexuosa
   Eucalyptus calophylla
   E. cornuta
   E. gomphocephala
   E. marginata
   E. occidentalis

4. Open-woodland with heathy understorey

   Eucalyptus calophylla
   E. marginata

5. Low woodland with heathy understorey

   Agonis flexuosa
   Banksia attenuata
   B. menziesii
   B. prionotes
   Casuarina fraserana
   C. huegeliana
   C. obesa
   Eucalyptus annulata
   E. calophylla
   E. cornuta
   E. erycocorys
   E. le-souefii
   E. marginata
   E. platypus
   E. spathulata

E. todtiana
E. torquata

6. Low open-woodland with heathy understorey

   Agonis flexuosa
   Banksia attenuata
   B. menziesii
   B. prionotes
   Casuarina fraserana
   Eucalyptus calophylla
   E. marginata

7. Closed-scrub

   Acacia cochlearis
   A. cyclops
   A. rostellifera
   Agonis flexuosa
   A. juniperina
   A. linearifolia
   Callitris preissii
   Eucalyptus foecunda
   Melaleuca globifera
   M. huegelii
   M. lanceolata
   Pultenaea reticulata

8. Open-scrub

   Acacia spp.
   Agonis flexuosa
   Banksia attenuata
   B. menziesii
   B. speciosa
   Casuarina fraserana
   Casuarina spp.
   Dryandra spp.
   Eucalyptus rudis
   E. todtiana
   E. uncinata
   Hakea spp.
   Lambertia inermis
   Melaleuca globifera
   M. thyoides
   Melaleuca spp.

9. Tall shrubland

   Acacia spp.
   Actinostrobus arenarius
   Agonis flexuosa
   Banksia ashbyi
   B. attenuata
   B. hookerana
   B. menziesii
   B. prionotes

TABLE 6.13 *(continued)*

B. sceptum	*G. eriostachya*
Casuarina fraserana	*G. leucopteris*
Casuarina spp.	*Xylomelum angustifolium*
Eucalyptus foecunda	
E. todtiana	11. Closed-heathland
E. uncinata	
Grevillea didymobotrya	*Acacia* spp.
G. eriostachya	*Agonis parviceps*
G. leucopteris	*Astartea fascicularis*
Melaleuca spp.	*Casuarina* spp.
Xylomelum angustifolium	*Dryandra* spp.
	Hakea spp.
10. Tall open-shrubland	*Melaleuca preissiana* etc.
Actinostrobus arenarius	12. Open-heathland
Banksia ashbyi	
B. attenuata	*Acacia decipiens*
B. hookerana	*Acacia* spp.
B. menziesii	*Casuarina* spp.
B. prionotes	*Dryandra* spp.
B. sceptum	*Eucalyptus eudesmoides*
Casuarina fraserana	*Hakea* spp.
Eucalyptus calophylla	*Melaleuca* spp.
E. marginata	*Xanthorrhoea reflexa* etc.
E. tetragona	
E. todtiana	13. Low shrubland
Grevillea didymobotrya	
	Viminaria juncea

APPENDIX II: DETAILED FLORISTIC AND ECOLOGICAL RELATIONSHIPS OF REPRESENTATIVE AUSTRALIAN HEATHLANDS

The following pages describe in terms of climate, soil and vegetation a series of 23 sites representative of the range of heathland and similar vegetation in eastern Australia (Western Australia is covered in Chapter 7). The degree of detail varies — in some cases only a species list is available, in others the proportional contribution of each species to the total canopy volume is listed.

The climatic data are tabulated by three-month periods — the mean solar radiation per day; the mean daily maximum and minimum temperatures; the mean total precipitation and pan evaporation over the three-month period; the mean number of days during which frost occurred or snow fell; and the probability of a cyclone during the period.

The following abbreviations are used throughout for leaf type and size, and life-form:

BS:	broad-sclerophyll	Gro:	Grooved above	Ch:	Chamaephyte
NS:	narrow-sclerophyll	Lep:	Leptophyll	Ep:	Parasitic epiphyte
		Mac:	Macrophyll	G:	Geophyte
Aph:	Aphyllous	Mes:	Mesophyll	H:	Hemicryptophyte
Cla:	Cladodes	Mic:	Microphyll	L:	Liana
Cup:	Cupressoid	Nan:	Nanophyll	M:	Microphanerophyte
Eri:	Ericoid	Pin:	Pinnae or leaflets	MM:	Mesophanerophyte
Gra:	Graminoid	Rev:	Revolute	N:	Nanophanerophyte
				Th:	Therophyte

In some sites, the following symbols are used for frequency:

c:	common	o:	occasional	vr:	very rare
f:	frequent	r:	rare	x:	present, frequency unknown
lc:	locally common	vc:	very common	–:	not recorded

The soils are described in these Regional Surveys by use of a nomenclatural code outlined in the *Factual Key for the Recognition of Australian Soils* by Northcote (1971) and in the *Atlas of Australian Soils* (Isbell et al., 1967, 1968; Northcote, 1960, 1962, 1966, 1968).

SITE 1

Oenpelli, Arnhem Land, Northern Territory

Latitude: 12°20'S   **Longitude:** 133°03'E
Diagrammatic cross-section (after Specht, 1958a; Story et al., 1969, 1973):

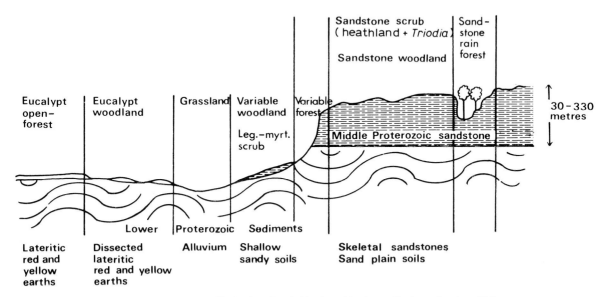

Fig. 6.10. Diagrammatic cross-section, Oenpelli, Arnhem Land, Northern Territory. Horizontal scale *c.* 20 km.

Climate: Köppen **Awg/Aw**

Climatic data	Summer (Dec–Feb)	Autumn (March–May)	Winter (June–Aug)	Spring (Sept–Nov)	Year (Jan–Dec)
Solar radiation[1]	1904 (455)	1854 (443)	1979 (473)	2238 (535)	1996 (477)
Max. air temp. (°C)	33.3	33.1	32.3	36.6	33.8
Min. air temp. (°C)	24.1	22.7	18.3	21.7	21.7
Precipitation (mm)	850	343	4	148	1345
Pan evaporation (mm)	430	445	586	753	2214
Frost (days <0°C)	0	0	0	0	0
Snow (days)	0	0	0	0	0
Cyclones	0.40	0.26	0	0	0.66

[1] In $J\ cm^{-2}\ day^{-1}$, with equivalents in $cal\ cm^{-2}\ day^{-1}$ in parentheses.

Vegetation: Heathland and allied sandstone scrub
Dominant soil: Dissected sandstone plateau; chiefly shallow sands (Uc 1.4, Uc 4.1) and large areas of bare rock
References: Specht, 1958a; Adams et al., 1973; N.B. Byrnes, pers. comm., 1976

Species composition

MICRO- to NANO-PHANEROPHYTES

Arecaceae	*Livistona loriphylla*
Pandanaceae	*Pandanus basedowii*
Blepharocaryaceae	*Blepharocarya depauperata*
Combretaceae	*Terminalia carpentariae*
Euphorbiaceae	*Petalostigma pubescens*

Meliaceae	*Owenia vernicosa* (in crevices)
Moraceae	*Ficus leucotricha, F. opposita, F. scobina*
Myrtaceae	*Calytrix achaeta, C. exstipulata* (syn. *C. microphylla*), *Eucalyptus dichromophloia, E. herbertiana, E. jacobsiana, E. phoenicea, E. tetrodonta, Eugenia eucalyp-*

SITE 1 *(continued)*

Species composition *(continued)*

MICRO- to NANO-PHANEROPHYTES *(continued)*

	toides (in creeks), *Melaleuca* sp., *Tristania psidioides*
Proteaceae	*Grevillea decurrens, G. heliosperma, G. mimosoides, G. pteridiifolia, G. refracta, Hakea arborescens, Persoonia falcata*
Rubiaceae	*Gardenia fucata, G. keartlandii*
Santalaceae	*Exocarpos latifolius*
Sapotaceae	*Planchonella arnhemica, Pouteria sericea*
Simaroubaceae	*Brucea javanica*
Tiliaceae	*Grewia xanthopetala*

NANOPHANEROPHYTES

Araliaceae	*Mackinlaya confusa*
Caesalpiniaceae	*Labichea nitida*
Dicrastylidaceae	*Denisonia* sp., *Pityrodia jamesii, Pityrodia* sp. (aff. *P. obliqua*)
Dilleniaceae	*Hibbertia* sp. (aff. *H. brownii*), *H. dealbata, H. oblongata*
Euphorbiaceae	*Beyeria bickertonensis, Petalostigma quadriloculare, Phyllanthus grandisepalus*
Fabaceae	*Bossiaea bossiaeoides, Brachysema bossiaeoides, Jacksonia thesioides, J. vernicosa, Jacksonia* sp., *Templetonia hookeri*
Hydrocotylaceae	*Platysace arnhemica*
Malvaceae	*Hibiscus arnhemicus, H. symonii, H. zonatus*
Mimosaceae	*Acacia conspersa, A. humifusa, A. lycopodiifolia, A. mountfordiae, A. simsii, A. sublanata*
Myrtaceae	*Baeckea* sp., *Calytrix brachychaeta, C. laricina, Fenzlia obtusa, Melaleuca magnifica, Verticordia cunninghamii, V. decussata*
Olacaceae	*Olax pendula*
Proteaceae	*Grevillea angulata, G. goodii, Grevillea pungens*
Rubiaceae	*Gardenia pyriformis*
Rutaceae	*Boronia affinis*
Santalaceae	*Anthobolus filifolius*
Solanaceae	*Solanum asymmetriphyllum*
Sterculiaceae	*Dicarpidium monoicum*
Ulmaceae	*Trema aspera*

NANOPHANEROPHYTE/LIANA

Apocynaceae	*Alyxia spicata*
Asclepiadaceae	*Sarcostemma australe*
Oleaceae	*Jasminum simplicifolium*
Opiliaceae	*Opilia amentacea*
Passifloraceae	*Adenia heterophylla*
Polygonaceae	*Muehlenbeckia rhyticarya* (rare in crevices)

NANOPHANEROPHYTES — CHAMAEPHYTES

Dilleniaceae	*Hibbertia lepidota, H. tomentosa*
Epacridaceae	*Leucopogon acuminatus*
Euphorbiaceae	*Calycopeplus casuarinoides*
Fabaceae	*Daviesia reclinata, Jacksonia dilatata*
Goodeniaceae	*Dampiera cinerea*
Melastomataceae	*Osbeckia australiana* (in damp areas)
Olacaceae	*Olax aphylla*
Polygalaceae	*Comesperma secundum*
Proteaceae	*Grevillea dryandri*
Rutaceae	*Boronia artemisiifolia, B. grandisepala, B. lanceolata, Zieria* sp.
Sapindaceae	*Distichostemon filamentosus, D. hispidulus*
Sterculiaceae	*Helicteres dentata, Waltheria indica*

CHAMAEPHYTES

Acanthaceae	*Nelsonia brunelloides*
Amaranthaceae	*Ptilotus distans*
Asclepiadaceae	*Hoya australis*
Asteraceae	*Pterocaulon sphacelatum, P. sphaeranthoides*
Convolvulaceae	*Bonamia brevifolia*
Dilleniaceae	*Pachynema junceum*
Euphorbiaceae	*Phyllanthus adamii*
Fabaceae	*Burtonia subulata*
Goodeniaceae	*Goodenia cirrifica, Scaevola angulata*
Haloragaceae	*Gonocarpus acanthocarpus*
Molluginaceae	*Macarthuria apetala, Macarthuria* sp.
Myrtaceae	*Calytrix* sp., *Homalocalyx ericaeus*
Polygalaceae	*Comesperma aphyllum*
Rubiaceae	*Knoxia stricta*

GEOPHYTES

Orchidaceae	*Arthrochilus byrnesii* (rare)
Taccaceae	*Tacca leontopetaloides* (in crevices)
Zingiberaceae	*Curcuma australasica* (in crevices)

HUMMOCK GRASSES

Poaceae	*Triodia microstachya, T. plectrachnoides, T. procera*

HEMICRYPTOPHYTES

Commelinaceae	*Aneilema siliculosum*
Iridaceae	*Patersonia macrantha*
Liliaceae	*Corynotheca lateriflora*
Poaceae	*Aristida browniana, Ectrosia leporina, Eriachne obtusa, Micraira subulifolia, Micraira* sp., *Sorghum plumosum*
Xanthorrhoeaceae	*Lomandra* sp.

SITE 1 *(continued)*

Species composition *(continued)*

CHAMAEPHYTES –- THEROPHYTES

Acanthaceae	*Hypoestes floribunda*
Boraginaceae	*Trichodesma zeylanica*
Caryophyllaceae	*Polycarpaea longiflora*
Cleomaceae	*Cleome viscosa*
Droseraceae	*Drosera burmannii* (in damp areas), *D. indica* (in damp areas), *D. petiolaris* (in damp areas)
Euphorbiaceae	*Euphorbia schizolepis*
Lamiaceae	*Coleus scutellarioides* (in crevices)
Lythraceae	*Rotala roxburghiana* (in damp areas)
Molluginaceae	*Macarthuria* sp. (aff. *M. neocambrica*)
Portulacaceae	*Calandrinia uniflora*
Rubiaceae	*Borreria breviflora*
Scrophulariaceae	*Stemodia coerulea*
Stackhousiaceae	*Stackhousia viminea*

THEROPHYTES

Aizoaceae	*Trianthema pilosa*
Amaranthaceae	*Gomphrena canescens*
Asteraceae	*Sphaeranthus africanus, Vernonia cinerea*

Byblidaceae	*Byblis liniflora*
Campanulaceae	*Wahlenbergia* sp.
Caryophyllaceae	*Polycarpaea corymbosa*
Centrolepidaceae	*Centrolepis exserta*
Cleomaceae	*Cleome tetrandra*
Cucurbitaceae	*Trichosanthes cucumerina* (liana)
Cyperaceae	*Bulbostylis barbata*
Euphorbiaceae	*Monotaxis macrophylla*
Goodeniaceae	*Goodenia armstrongiana*
Lentibulariaceae	*Utricularia fulva* (in damp areas)
Poaceae	*Eriachne ciliata, Pseudopogonatherum contortum* (occas.), *Sorghum* sp., *Thaumastochloa major*
Portulacaceae	*Portulaca bicolor*
Solanaceae	*Solanum clarkiae*
Spigeliaceae	*Mitrasacme indica, M. pygmaea*
Stylidiaceae	*Stylidium* spp. (in damp areas)
Zygophyllaceae	*Tribulus pentandrus*

PARASITIC EPIPHYTES

Cassythaceae	*Cassytha filiformis, C. glabella*

Vegetation: Leguminous–myrtaceous scrub alliance
Dominant soil: Shallow sandy soils at the base of the sandstone escarpment
References: Specht, 1958a; Adams et al., 1973; N.B. Byrnes, pers comm., 1976

Species composition

MICRO- to NANO-PHANEROPHYTES

Mimosaceae	*Acacia alleniana, A. aulacocarpa, A. difficilis, A. dimidiata, A. hemignosta, A. platycarpa, A. plectocarpa, A. torulosa*
Myrtaceae	*Calytrix arborescens, C. exstipulata* (syn. *C. microphylla*), *Eucalyptus tetrodonta, Melaleuca nervosa, Verticordia cunninghamii, Xanthostemon paradoxus*
Proteaceae	*Grevillea pteridiifolia*

HUMMOCK GRASSES

Poaceae	*Triodia microstachya*

HEMICRYPTOPHYTES

Cyperaceae	*Fimbristylis arthrostyloides, Scleria sphacelata*
Platyzomataceae	*Platyzoma microphyllum*
Poaceae	*Ectrosia leporina, E. schultzii, Eriachne obtusa, E. triseta*

THEROPHYTES

Poaceae	*Perotis rara*

SITE 2

Jardine River, Cape York Peninsula, Queensland

Latitude: 11°10′S   **Longitude:** 142°40′E
Diagrammatic cross-section (Lavarack and Stanton, 1977):

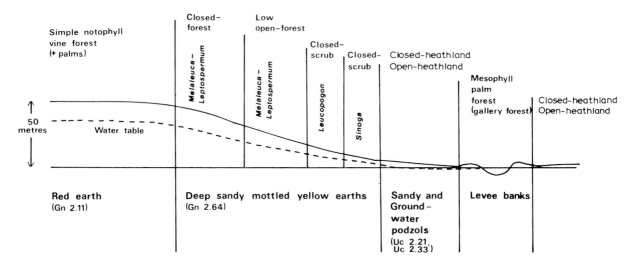

Fig. 6.11. Diagrammatic cross-section, Jardine River, Cape York Peninsula, Queensland. Horizontal scale *c.* 10 km.

Climate: Köppen Awgi

Climatic data [1]	Summer (Dec–Feb)	Autumn (March–May)	Winter (June–Aug)	Spring (Sept–Nov)	Year (Jan–Dec)
Solar radiation [2]	1933 (462)	1695 (405)	1569 (375)	2092 (500)	1820 (435)
Max. air temp. (°C)	31.9	30.4	29.1	32.3	30.9
Min. air temp. (°C)	22.6	21.4	18.0	19.8	20.4
Precipitation (mm)	1041	501	38	93	1673
Pan evaporation (mm)	381	255	283	474	1393
Frost (days <0°C)	0	0	0	0	0
Snow (days)	0	0	0	0	0
Cyclones	0.29	0.17	0	0	0.46

[1] Nearest climate station: McDonnell (11°38′S, 142°28′E).
[2] In J cm^{-2} day^{-1}, with equivalents in cal cm^{-2} day^{-1} in parentheses.

Vegetation: Heathland and allied shrubland
Alliances: (1) closed-forest: *Leptospermum fabricia – Melaleuca* sp. (aff. *M. symphyocarpa*) alliance; (2) low open-forest: *Melaleuca* sp. (aff. *M. symphyocarpa*) – *Leptospermum fabricia* alliance; (3) closed-scrub: *Leucopogon* spp. alliance; (4) closed-scrub: *Sinoga lysicephala* alliance; (5) closed-heathland: *Grevillea pteridiifolia* alliance; (6) open-heathland: *Nepenthes mirabilis – Sinoga lysicephala* alliance
References: Pedley and Isbell, 1971; Lavarack and Stanton, 1977

SITE 2 *(continued)*

Species composition

	Leaf size	Alliance					
		1	2	3	4	5	6
Canopy height (m):		8–15	6–8	5–7	3	0.5–1.5	0.5–1.0
WOODY TREES AND SHRUBS							
Arecaceae							
Calamus sp.	Mes (leaflets)	M	–	–	–	–	–
Blepharocaryaceae							
Blepharocarya involucrigera	Mes (leaflets)	MM	–	–	–	–	–
Casuarinaceae							
Casuarina littoralis	Aph	–	M	–	–	N	–
Cupressaceae							
Callitris columellaris	Cup	–	M	M	–	–	–
Dilleniaceae							
Hibbertia banksii	Mic	–	N	–	–	N	–
Epacridaceae							
Leucopogon (2 spp.)	Nan	–	M	M	–	–	–
Euphorbiaceae							
Choriceras tricorne	Mic	N	N	–	–	N	–
Neoroepera banksii	Nan	–	–	–	–	N	–
Fabaceae							
Gompholobium nitidum	Nan	–	–	–	–	N	–
Jacksonia thesioides	Aph	–	–	–	–	N	–
Lauraceae							
Endiandra glauca	Mes	M	–	–	–	–	–
Mimosaceae							
Acacia calyculata	Mic	–	–	–	–	N	–
A. polystachya	Mes	MM	–	–	–	–	–
A. rothii	Mes	–	M	–	–	N	–
Myrtaceae							
Eucalyptus leptophleba	Mic–Mes	–	M	–	–	–	–
E. polycarpa	Mic–Mes	–	–	M	–	–	–
E. tessellaris	Mic	–	–	M	–	–	–
E. tetrodonta	Mic–Mes	–	–	–	–	N–M	–
Eucalyptus sp.	Mic–Mes	–	M	–	–	N–M	–
Eugenia fibrosa (= *Syzygium fibrosum*)	Mes	MM	–	–	–	–	–
Fenzlia obtusa	Nan	–	–	–	–	N	–
Leptospermum fabricia	Nan/Mic	–	M	M	–	N	–
Melaleuca arcana	Nan–Mic	–	–	–	–	–	N
M. quinquenervia	Mic–Mes	–	–	–	N–M	–	N
Melaleuca sp. (aff. *M. symphyocarpa*)	Mic	MM	M	M	–	N	–
Sinoga lysicephala	Nan	–	–	–	N–M	Ń	N
Thryptomene oligandra	Lep	–	–	–	–	–	N
Tristania longivalvis	Mes	MM	–	–	–	–	–
Pandanaceae							
Pandanus sp.	Mes–Mac	M	–	–	–	–	–
Proteaceae							
Banksia dentata	Mes	–	–	–	N–M	–	N
Grevillea glauca	Mes	–	M	–	–	N–M	–

SITE 2 *(continued)*

Species composition *(continued)*

	Leaf size	Alliance					
		1	2	3	4	5	6
Canopy height (m):		8–15	6–8	5–7	3	0.5–1.5	0.5–1.0
WOODY TREES AND SHRUBS *(continued)*							
G. pteridiifolia	Mic (Pin)	–	–	–	–	N	N
Rubiaceae							
Morinda reticulata	Mic–Mes	–	–	–	–	N	–
Randia sessilis	Mes	M	–	M	–	–	–
Rutaceae							
Boronia bowmanii	Nan (leaflets)	–	–	–	–	N	–
Xanthorrhoeaceae							
Xanthorrhoea johnsonii	Gra	–	N	–	–	N	N
GROUND COVER							
Byblidaceae							
Byblis liniflora		–	–	–	–	–	Th
Cyperaceae							
Gahnia sieberana		–	–	–	H	–	–
Schoenus sparteus		–	–	–	–	H	H
Eriocaulaceae							
Eriocaulon spp.		–	–	–	–	–	Th
Haemodoraceae							
Haemodorum corymbosum		–	–	–	–	–	Th
Lentibulariaceae							
Utricularia chrysantha		–	–	–	–	–	Th
Liliaceae							
Dianella sp.		–	–	H	–	–	–
Nepenthaceae							
Nepenthes mirabilis		–	–	–	Ch	–	Ch
Orchidaceae							
Bromheadia venusta		–	–	–	G	–	G
Restionaceae							
Leptocarpus ramosus		–	–	–	–	–	H
Xanthorrhoeaceae							
Lomandra banksii		–	H	–	–	H	–
PARASITIC EPIPHYTES							
Cassythaceae							
Cassytha sp.		–	–	–	–	Ep	–

SITE 3

Cape Flattery and Lizard Island, Cape York Peninsula, Queensland

Latitude: 14°57'S   **Longitude:** 145°21'E
Diagrammatic cross-section (Byrnes et al., 1977):

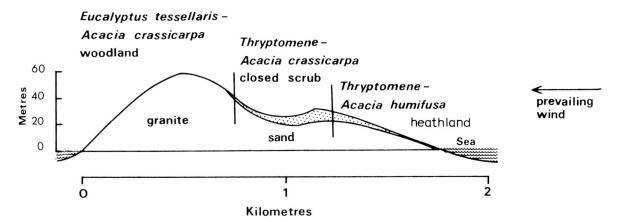

Fig. 6.12. Diagrammatic cross-section, Lizard Island, Cape York Peninsula, Queensland.

Climate: Köppen **Awg**

Climatic data[1]	Summer (Dec–Feb)	Autumn (March–May)	Winter (June–Aug)	Spring (Sept–Nov)	Year (Jan–Dec)
Solar radiation[2]	1996 (477)	1640 (392)	1527 (365)	2105 (503)	1816 (434)
Max. air temp. (°C)	31.5	29.1	26.3	29.4	29.1
Min. air temp. (°C)	24.1	22.6	19.4	22.5	22.2
Precipitation (mm)	867	677	102	78	1724
Pan evaporation (mm)	506	329	300	456	1591
Frost (days <0°C)	0	0	0	0	0
Snow	0	0	0	0	0
Cyclones	0.62	0.31	0	0	0.93

[1] Nearest climate station: Cooktown (15°28'S, 145°17'E).
[2] In J cm^{-2} day^{-1}, with equivalents in cal cm^{-2} day^{-1} in parentheses.

Dominant soils: Deep, bleached, podzolised sand (Uc 2.21, Uc 2.23)
Vegetation: Heathland and allied shrubland
References: Byrnes et al., 1977; T.J. McDonald and G.N.Batianoff, pers. comm., 1976

Species composition

	Lizard Island		Cape Flattery	
Canopy height (m):	closed-scrub 6–8	heathland 0.5–2	closed-scrub 3–6	heathland 0.5–2
WOODY TREES AND SHRUBS				
Apocynaceae *Alyxia spicata*	M	N	M	–
Burseraceae *Canarium australianum*	M	–	–	–

SITE 3 *(continued)*

Species composition *(continued)*

Canopy height (m):	Lizard Island		Cape Flattery	
	closed-scrub 6–8	heathland 0.5–2	closed-scrub 3–6	heathland 0.5–2
WOODY TREES AND SHRUBS *(continued)*				
Caesalpiniaceae				
Labichea buettneriana	–	–	–	N
Celastraceae				
Elaeodendron melanocarpum	M	–	–	–
Chrysobalanaceae				
Parinari nonda	–	N	–	–
Cochlospermaceae				
Cochlospermum gillivraei	M	–	–	–
Combretaceae				
Terminalia muelleri	M	–	–	N
Dicrastylidaceae				
Chloanthes parviflora	–	–	–	Ch/N
Dilleniaceae				
Hibbertia banksii	–	–	–	Ch
Ebenaceae				
Diospyros ferrea var. *reticulata*	M	–	–	–
Epacridaceae				
Leucopogon ruscifolius	–	N	–	N
Euphorbiaceae				
Breynia oblongifolia	–	–	–	N
Neoroepera banksii	–	N	N	–
Fabaceae				
Jacksonia thesioides	–	N	–	N
Malvaceae				
Hibiscus meraukensis	–	N	–	–
H. tiliaceus	–	–	–	N
Mimosaceae				
Acacia crassicarpa	M	–	M	–
A. humifusa	–	N	–	N
A. platycarpa	–	–	M	–
Myrtaceae				
Eugenia suborbicularis (=*Syzygium suborbiculare*)	M	–	–	–
Fenzlia obtusa	–	–	–	N
Leptospermum fabricia	–	–	M	N
Leptospermum sp.	–	–	M	–
Melaleuca arcana	–	–	–	N
M. symphyocarpa	–	–	M	–
M. viridiflora	–	N	–	–
Sinoga lysicephala	–	–	–	N
Thryptomene oligandra	M	N	–	–
Tristania suaveolens	–	–	–	N
Poaceae				
Triodia stenostachya	–	–	–	Ch
Proteaceae				
Persoonia falcata	M	N	–	–

SITE 3 *(continued)*

Species composition *(continued)*

Canopy height (m):	Lizard Island		Cape Flattery	
	closed-scrub 6–8	heathland 0.5–2	closed-scrub 3–6	heathland 0.5–2
WOODY TREES AND SHRUBS *(continued)*				
Rubiaceae				
Diplospora australis	–	–	–	N
Rutaceae				
Boronia alulata	–	–	–	Ch
Eriostemon australasicus var. *banksii*	–	–	–	N
Sapindaceae				
Alectryon tomentosus	M	–	–	–
Cupaniopsis anacardioides	M	–	–	–
Dodonaea viscosa	–	–	–	N
Sapotaceae				
Mimusops elengi	M	–	–	–
Pouteria sericea	M	–	–	–
Ulmaceae				
Trema aspera	M	–	–	–
HERBACEOUS PLANTS				
Acanthaceae				
Justicia procumbens	Ch	–	–	–
Cyperaceae				
Arthrostylis aphylla	–	–	–	H
Fimbristylis recta	–	–	–	H
Gahnia sieberana	–	–	–	H (swamp)
Schoenus sparteus	–	H	–	–
Trachystylis stradbrokensis	–	–	–	H (swamp)
Tricostularia undulata	–	–	–	H (swamp)
Droseraceae				
Drosera indica	–	–	–	Th (swamp)
Fabaceae				
Lamprolobium fruticosum	–	Ch/L	–	–
Lentibulariaceae				
Utricularia chrysantha	–	–	–	Th (swamp)
Liliaceae				
Dianella sp.	–	–	–	H
Menispermaceae				
Pachygone ovata	–	Ch/L	–	–
Poaceae				
Eragrostis interrupta	–	–	–	H
Perotis rara	Th	–	–	–
Themeda australis	–	H	–	–
Portulacaceae				
Calandrinia sp. nov.	–	Th	–	–
Restionaceae				
Leptocarpus sp.	–	–	–	H (swamp)
Restio tetraphyllus	–	–	–	H (swamp)
Xanthorrhoeaceae				
Lomandra banksii	–	H	–	H

SITE 4

Shoalwater Bay, Queensland

Latitude: 22°30'S   **Longitude:** 150°40'E
          Climate: Köppen **Cfa**

Climatic data[1]	Summer (Dec–Feb)	Autumn (March–May)	Winter (June–Aug)	Spring (Sept–Nov)	Year (Jan–Dec)
Solar radiation[2]	2264 (541)	1745 (417)	1494 (357)	2238 (535)	1937 (463)
Max. air temp. (°C)	32.0	28.7	23.9	29.6	28.6
Min. air temp. (°C)	22.1	17.9	11.5	17.4	17.2
Precipitation (mm)	480	190	137	142	949
Pan evaporation (mm)	556	367	294	495	1712
Frost (days ≤0°C)	0	0	0	0	0
Snow (days)	0	0	0	0	0
Cyclones	0.58	0.40	0	0	0.98

[1] Nearest climate station: Rockhampton (23°24'S, 150°30'E).
[2] In $J\ cm^{-2}\ day^{-1}$, with equivalents in $cal\ cm^{-2}\ day^{-1}$ in parentheses.

Dominant soil: Podzolised sand (Uc 2.21)
Vegetation: Heathland and allied shrubland
Reference: Gunn et al., 1972

Species composition

MICRO- to NANO-PHANEROPHYTES

Casuarinaceae	*Casuarina littoralis*
Dilleniaceae	*Hibbertia linearis*
Epacridaceae	*Sprengelia sprengelioides, Styphelia triflora*
Euphorbiaceae	*Petalostigma pubescens, Ricinocarpos pinifolius*
Fabaceae	*Aotus lanigera, Daviesia ulicifolia, D. umbellata, Gompholobium pinnatum, Hovea longifolia, Jacksonia scoparia, Phyllota phylicoides*
Mimosaceae	*Acacia cunninghamii, A. flavescens*
Myrtaceae	*Baeckea imbricata, B. linearis, B. stenophylla, Callistemon viminalis, Eucalyptus acmenoides, E. tessellaris, Fenzlia obtusa, Leptospermum flavescens, Leptospermum sp., Tristania conferta*
Proteaceae	*Banksia integrifolia, B. robur, Grevillea banksii, G. pteridiifolia, Persoonia linearis*
Rutaceae	*Phebalium woombye, Zieria laxiflora*
Santalaceae	*Exocarpos cupressiformis, E. latifolius*
Sapindaceae	*Dodonaea viscosa*
Thymelaeaceae	*Pimelea linifolia*
Xanthorrhoeaceae	*Xanthorrhoea* sp.
Zamiaceae	*Macrozamia miquelii*

CHAMAEPHYTES

Brunoniaceae	*Brunonia australis*
Campanulaceae	*Wahlenbergia* sp.

Dennstaedtiaceae	*Pteridium esculentum*
Droseraceae	*Drosera burmannii*
Goodeniaceae	*Dampiera ferruginea, D. stricta, Velleia spathulata*
Haloragaceae	*Haloragis heterophylla*
Hydrocotylaceae	*Platysace linearifolia*
Schizaeaceae	*Schizaea dichotoma*
Spigeliaceae	*Mitrasacme indica*

GEOPHYTES

Hypoxidaceae	*Hypoxis hygrometrica*
Liliaceae	*Thysanotus tuberosus, Tricoryne elatior*

HEMICRYPTOPHYTES

Cyperaceae	*Baumea rubiginosa, B. teretifolia, Caustis recurvata, Lepidosperma exaltatum, Schoenus brevifolius, S. calostachyus*
Iridaceae	*Patersonia glabrata*
Liliaceae	*Dianella laevis, D. revoluta, Sowerbaea juncea*
Poaceae	*Imperata cylindrica, Themeda australis*
Restionaceae	*Coleocarya gracilis, Empodisma minus, Restio dimorphus*
Xanthorrhoeaceae	*Lomandra filiformis, L. multiflora*

PARASITIC EPIPHYTES

Cassythaceae	*Cassytha filiformis*

SITE 5A

Southeastern Queensland — Mount Hardgrave dry-heathland, North Stradbroke Island

Latitude: 27°30′S   **Longitude:** 153°25′E
Diagrammatic cross-section (Connor and Clifford, 1972; Specht, 1975; Clifford and Specht, 1979):

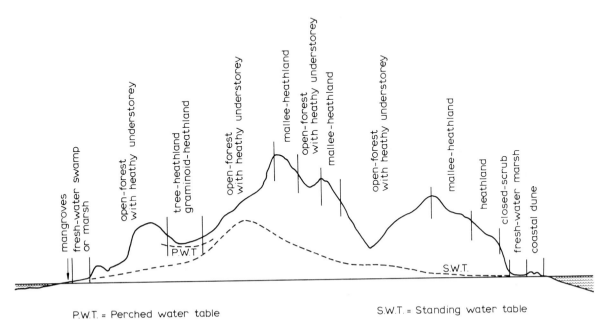

P.W.T. = Perched water table   S.W.T. = Standing water table

Fig. 6.13. Diagrammatic cross-section, North Stradbroke Island, southeastern Queensland. Horizontal scale *c.* 10 km, vertical scale 220 m.

Climate: Köppen **Cfa**

Climatic data [1]	Summer (Dec–Feb)	Autumn (March–May)	Winter (June–Aug)	Spring (Sept–Nov)	Year (Jan–Dec)
Solar radiation [2]	2238 (535)	1561 (373)	1297 (310)	2163 (517)	1816 (434)
Max. air temp. (°C)	27.7	24.7	19.9	24.3	24.2
Min. air temp. (°C)	20.5	17.6	12.1	16.7	16.7
Precipitation (mm)	555	526	276	245	1602
Pan evaporation (mm)	439	257	188	373	1257
Frost (days ≤0°C)	0	0	0	0	0
Snow (days)	0	0	0	0	0
Cyclones	0.51	0.35	0	0	0.86

[1] Nearest climate station: Dunwich, North Stradbroke Island.
[2] In J cm^{-2} day^{-1}, with equivalents in cal cm^{-2} day^{-1} in parentheses.

Dominant soil: Deep, podzolised sands (Uc 2.2)
Vegetation: Dry heathland
Reference: Specht (harvest Dec. 1975); see also Blake, 1968
Age: 15(?) years since fire   **Biomass** (dry wt.): 13.2 tonnes ha^{-1}
Structure: Tall open-heathland

   Scattered, small trees of *Banksia aemula* or mallee eucalypts (*Eucalyptus planchoniana* and *E. intermedia*) emerge from the heathland vegetation in less exposed area adjacent to the sampling point, thus forming a tree-heathland or a mallee-heath structural formation.

SITE 5 *(continued)*

Species composition

Species	Leaf type and size	% of total above-ground biomass	Species	Leaf type and size	% of total above-ground biomass
Evergreen mid-height shrubs (1–2 m)			Rutaceae		
Epacridaceae			*Boronia rosmarinifolia*	NS (Lep, Eri)	0.6
Woollsia pungens	BS (Lep)	10.6	*B. safrolifera*	NS (Pin–Lep)	1.6
Myrtaceae					———
Baeckea linearis	NS (Lep)	11.4			**10.5**
Leptospermum flavescens	BS (Lep)	8.7			
Proteaceae			Evergreen, caespitose graminoid herbs (<25 cm)		
Banksia aemula	BS (Mic)	30.6	Cyperaceae		
		———	*Caustis recurvata*	NS (Aph)	4.5
		61.3	Iridaceae		
			Patersonia sericea	BS (Gra)	0.5
Evergreen, low shrubs (25–100 cm)			Liliaceae		
			Laxmannia gracilis	NS (Gra)	trace
Dilleniaceae			Poaceae		
Hibbertia linearis	NS (Lep)	0.3	*Entolasia stricta*	NS (Gra)	trace
Epacridaceae			Restionaceae		
Brachyloma daphnoides	BS (Lep)	0.2	*Coleocarya gracilis*	NS (Aph)	22.8
Leucopogon leptospermoides	BS (Lep)	1.2	*Hypolaena fastigiata*	NS (Aph)	0.3
L. virgatus	BS (Lep)	0.2	Xanthorrhoeaceae		
Fabaceae			*Lomandra filiformis*	NS (Gra)	trace
Bossiaea heterophylla	(Nan–Aph Mic)	4.6	*L. longifolia*	BS (Gra)	trace
Dillwynia peduncularis	NS (Lep, Gro)	trace			———
Gompholobium pinnatum	NS (Pin–Lep, Eri)	0.1			**28.1**
Phyllota phylicoides	NS (Lep, Eri)	0.6	Heterotrophic, vascular plants		
Mimosaceae			Cassythaceae		
Acacia suaveolens	BS (Mic)	0.1	*Cassytha glabella*	NS (Aph)	
Proteaceae			*C. pubescens*	NS (Aph)	0.1
Petrophile shirleyae	NS (Pin–Lep to Nan, Gro)	0.5			———
Strangea linearis	BS (Nan)	0.5			**0.1**

Other species recorded:

Ferns
  Schizaeaceae   *Schizaea dichotoma*

Monocotyledons
  Cyperaceae   *Lepidosperma laterale, Schoenus* sp.
  Xanthorrhoeaceae *Xanthorrhoea johnsonii*

Dicotyledons
  Epacridaceae   *Epacris pulchella, Monotoca scoparia*
  Euphorbiaceae   *Pseudanthus orientalis, Ricinocarpos pinifolius*

Fabaceae   *Jacksonia stackhousii*
Hydrocotylaceae   *Platysace ericoides, Xanthosia pilosa*
Mimosaceae   *Acacia ulicifolia*
Myrtaceae   *Austromyrtus dulcis, Eucalyptus intermedia* (mallee), *E. planchoniana* (mallee), *Homoranthus virgatus*
Proteaceae   *Conospermum taxifolium, Persoonia linearis*
Rutaceae   *Zieria laxiflora*
Thymelaeaceae   *Pimelea linifolia*
Tremandraceae   *Tetratheca thymifolia*

SITE 5B

Southeastern Queensland — Toolara wet-heathland

Latitude: 26°03'S   **Longitude:** 152°55'E
Diagrammatic cross-section (Coaldrake, 1961):

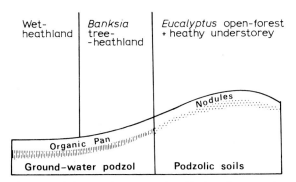

Fig. 6.14. Diagrammatic cross-section, Toolara Area, southeastern Queensland. Horizontal scale *c.* 300 m., vertical scale less than 10 m.

Climate: Köppen **Cfa**   **Climatic data:** see Site 5A above for representative data
Dominant soil: Ground-water podzols (Uc 2.33)
Vegetation: Wet-heathland
Age: 7 years since fire   **Structure:** Low open graminoid-heathland
Biomass (dry wt.): 5.6 t ha^{-1} **Reference:** Groves and Specht, 1965

Species composition

	Leaf type and size	% of total above-ground biomass		Leaf type and size	% of total above-ground biomass
Evergreen, low shrubs (25–100 cm)			*Xanthorrhoea macronema*	BS (Gra)	10.8
					10.8
Epacridaceae					
Epacris pulchella	BS (Lep)	0.2			
			Evergreen, dwarf-shrubs (<25 cm)		
Myrtaceae					
Leptospermum semibaccatum	BS (Lep)	8.3	Proteaceae		
Melaleuca nodosa	NS (Lep)	4.1	*Strangea linearis*	BS (Nan)	1.1
M. thymifolia	NS (Lep)	0.3			**1.1**
Proteaceae					
Banksia oblongifolia	BS (Nan)	36.4			
Petrophile shirleyae	NS (Pin–Lep to Nan, Gro)	1.5	Evergreen, caespitose graminoid herbs (<25 cm)		
			Cyperaceae		
Rutaceae			*Caustis recurvata*	NS (Gra)	
Eriostemon myoporoides ssp.	BS (Nan)		*Cyathochaeta diandra*	NS (Gra)	
queenslandicus		0.5	Restionaceae		**30.0**
		———	*Lepyrodia scariosa*	NS (Aph)	
		51.3	*Leptocarpus tenax*	NS (Aph)	
Rosulate, evergreen geophytes (25–100 cm)					
Xanthorrhoeaceae			Other species [1]		**6.8**

[1] Other species recorded:

Monocotyledons
    Cyperaceae   *Lepidosperma longitudinale, Schoenus brevifolius*

Poaceae   *Aristida warburgii, Eremochloa bimaculata, Eriachne glabrata, Imperata cylindrica* var. *major, Themeda australis*

SITE 5 *(continued)*

Xanthorrhoeaceae	*Lomandra elongata*	Hydrocotylaceae	*Platysace linearifolia*
Xyridaceae	*Xyris operculata*	Lamiaceae	*Westringia tenuicaulis*
		Myrtaceae	*Baeckea stenophylla, Leptospermum liversidgei, Leptospermum whitei*
Dicotyledons		Polygalaceae	*Comesperma denudata*
Dilleniaceae	*Hibbertia vestita*	Proteaceae	*Conospermum taxifolium, Grevillea leiophylla, Persoonia tenuifolia*
Epacridaceae	*Leucopogon pedicellatus*		
Fabaceae	*Daviesia umbellulata, Gompholobium pinnatum*	Stylidiaceae	*Stylidium graminifolium, Stylidium uliginosum*

SITE 5C

Southeastern Queensland — Beerwah wet-heathland

Latitude: 26°51'S   **Longitude:** 152°58'E
Climate: Köppen **Cfa**   **Climatic data:** see Site 5A above for representative data
Dominant soil: Ground-water podzols (Uc 2.33)
Vegetation: Wet-heathland
Reference: Connor and Wilson, 1968
Age: 2–3 years since fire   **Biomass** (dry wt.): 5.73 tonnes ha^{-1}
Structure: Low open graminoid-heathland

Species composition

Species	Leaf type and size	% of total above-ground biomass	Species	Leaf type and size	% of total above-ground biomass
Evergreen, low shrubs (25–100 cm)			*Hibbertia vestita*	NS (Lep)	2.1
Myrtaceae					**2.1**
Angophora woodsiana (dwarf form)	BS (Mic/Mes)	18.3			
			Evergreen, caespitose graminoid herbs (<25 cm)		
Proteaceae			Cyperaceae		
Banksia oblongifolia	BS (Nan)	20.1	*Cyathochaete diandra*	BS (Gra)	3.1
Persoonia tenuifolia	NS (Lep, Gro)	0.3	*Ptilanthelium deustum*	NS (Gra)	23.3
Strangea linearis	BS (Nan)	0.3			
			Liliaceae		
Rutaceae			*Sowerbaea juncea*	NS (Gra)	0.3
Eriostemon myoporoides spp. *queenslandicus*	BS (Nan)	5.4	*Tricoryne elatior*	NS (Gra)	0.3
			Poaceae		
		44.4	*Aristida warburgii*	Gra	3.5
			Themeda australis	Gra	4.4
Rosulate, evergreen geophytes (25–100 cm)			Restionaceae		
Xanthorrhoeaceae			*Lepyrodia scariosa*	NS (Aph)	4.9
Xanthorrhoea macronema	BS (Gra)	7.7	Xanthorrhoeaceae		
		7.7	*Lomandra longifolia*	BS (Gra)	2.3
					42.1
Evergreen, caespitose dwarf-shrubs (<25 cm)					
Dilleniaceae			Other species[1]		**3.7**

[1]Other species recorded:

Algae
    Zygnemataceae   *Zygogonium ericetorum*

Monocotyledons
    Poaceae   *Entolasia stricta, Eragrostis brownii*
    Xanthorrhoeaceae   *Lomandra multiflora*

SITE 5 *(continued)*

Dicotyledons		Goodeniaceae	*Dampiera stricta, Velleia paradoxa*
Epacridaceae	*Epacris pulchella*	Proteaceae	*Grevillea leiophylla, Petrophile shirleyae*
Fabaceae	*Pultenaea myrtoides*		

SITE 5D

Southeastern Queensland — Dave's Creek, Lamington National Park, McPherson Ranges

Latitude: 28°20'S   **Longitude:** 153°00'E   **Altitude:** 854 m
Climate: Köppen **Cfb**

Climatic data [1]	Summer (Dec–Feb)	Autumn (March–May)	Winter (June–Aug)	Spring (Sept–Nov)	Year (Jan–Dec)
Solar radiation [2]	2293 (548)	1590 (380)	1310 (313)	2184 (522)	1845 (441)
Max. air temp. (°C)	25.3	21.3	16.8	22.2	21.4
Min. air temp. (°C)	17.0	13.9	8.8	12.9	13.2
Precipitation (mm)	552	425	222	251	1450
Pan evaporation (mm)	380	219	195	344	1138
Frost (days ≤0°C)	0	0	0.2	0	0.2
Snow (days)	0	0	0	0	0
Cyclones	0.40	0.23	0	0	0.63

[1] Nearest climate station: Tamborine Mount (27°57'S, 153°11'E).
[2] In J cm^{-2} day^{-1}, with equivalents in cal cm^{-2} day^{-1} in parentheses.

Dominant soil: Yellow-brown podzolic soil (Dy 4) on rhyolite
Vegetation: Heathland and allied shrubland
Reference: Jones, 1964

Species composition

Species	Leaf size	Frequency in: heathland	graminoid-heathland	mallee-heathland
MICROPHANEROPHYTES				
Casuarinaceae				
Casuarina torulosa	Lep	–	–	r
Cyatheaceae				
Cyathea australis	Nan	–	–	r
Dilleniaceae				
Hibbertia hexandra	Mic	–	–	c
Mimosaceae				
Acacia longissima	Mic	–	–	r
A. obtusifolia	Mic	c	–	c
Myrtaceae				
Eucalyptus approximans ssp. *codonocarpa*	Mic	c	–	c
E. notabilis	Mes	c	–	–
Leptospermum lanigerum	Lep	c	r	–
L. petersonii	Nan	–	–	c
Tristania laurina	Mic	–	–	c
Proteaceae				
Banksia collina	Mic	c	c	c

SITE 5D *(continued)*

Species composition *(continued)*

Species	Leaf size	Frequency in:		
		heathland	graminoid-heathland	mallee-heathland
MICROPHANEROPHYTES *(continued)*				
B. integrifolia	Mes	c	r	c
Persoonia lanceolata	Mic	c	c	c
NANOPHANEROPHYTES				
Asteraceae				
Olearia heterocarpa	Mic	–	–	c
Casuarinaceae				
Casuarina rigida	Lep	c	c	c
Cupressaceae				
Callitris monticola	Lep	c	–	r
Epacridaceae				
Epacris obtusifolia	Lep	c	c	c
Leucopogon lanceolatus	Mic	c	–	c
L. melaleucoides	Nan	–	–	c
Woollsia pungens	Lep	c	c	c
Fabaceae				
Gompholobium latifolium	Mic	–	–	c
Oxylobium robustum	Mic	–	–	c
Pultenaea retusa	Nan	–	–	c
Lamiaceae				
Prostanthera phylicifolia	Nan	c	c	c
Mimosaceae				
Acacia myrtifolia	Mic	c	c	–
A. ulicifolia	Lep	c	r	r
Myrtaceae				
Baeckea linifolia	Lep	r	c	–
Callistemon montanus	Mic	c	c	c
C. pallidus	Mic	–	c	–
Leptospermum flavescens	Nan	c	–	–
L. microcarpum	Lep	r	–	–
Polygalaceae				
Comesperma retusum	Nan	–	–	c
Proteaceae				
Hakea dactyloides	Mic	c	c	r
Lomatia silaifolia	Nan	r	–	c
Petrophile canescens	Lep	r	–	r
Rhamnaceae				
Pomaderris ferruginea	Mic	r	–	r
Xanthorrhoeaceae				
Xanthorrhoea sp. aff. *X. johnsonii*	Mic/Mes	c	r	c
CHAMAEPHYTES				
Asteraceae				
Hypochoeris radicata	Mic	r	–	–

SITE 5D *(continued)*

Species composition *(continued)*

Species	Leaf size	Frequency in:		
		heathland	graminoid-heathland	mallee-heathland
CHAMAEPHYTES *(continued)*				
Blechnaceae				
Blechnum procerum	Mic	c	–	–
Dennstaedtiaceae				
Pteridium esculentum	Lep	–	–	c
Dilleniaceae				
Hibbertia dentata	Mic	–	–	c
Fabaceae				
Gompholobium pinnatum	Nan	c	–	–
Jacksonia stackhousii	Lep	c	c	–
Pultenaea pycnocephala	Nan	c	–	–
Gleicheniaceae				
Gleichenia microphylla	Lep	–	c	–
G. rupestris	Lep	c	c	r
Sticherus lobatus	Nan	c	c	c
Goodeniaceae				
Goodenia rotundifolia	Nan	c	c	–
Haloragaceae				
Gonocarpus tetragynus	Lep	c	c	c
G. teucrioides	Nan	c	–	–
Hydrocotylaceae				
Xanthosia diffusa	Nan	c	–	r
X. pilosa	Nan	c	–	r
Lindsaeaceae				
Lindsaea linearis	Lep	c	c	c
L. microphylla	Lep	–	c	–
Lycopodiaceae				
Lycopodium deuterodensum	Lep	r	–	–
L. laterale	Lep	–	c	–
Orchidaceae				
Dendrobium kingianum	Mic	c	–	–
Polygalaceae				
Comesperma defoliatum	Lep	c	c	–
Schizaeaceae				
Schizaea bifida	Lep	r	–	–
Violaceae				
Viola hederacea	Nan	–	–	r
HEMICRYPTOPHYTES				
Cyperaceae				
Baumea nuda	Lep	–	c	–
Baumea sp.	Lep	–	c	–
Gahnia insignis	Nan	c	r	c

SITE 5D *(continued)*

Species composition *(continued)*

Species	Leaf size	Frequency in:		
		heathland	graminoid-heathland	mallee-heathland
HEMICRYPTOPHYTES *(continued)*				
G. sieberana	Nan	–	c	–
G. aff. sieberana	Nan	–	–	r
Lepidosperma canescens	Lep	c	c	c
L. laterale	Nan–Mic	c	c	c
Lepidosperma sp.	Lep	–	c	–
Schoenus apogon	Lep	–	c	–
S. maschalinus	Lep	–	c	–
S. melanostachyus	Lep	c	c	c
Tetraria capillaris	Lep	c	c	c
Iridaceae				
Patersonia fragilis	Mic	–	c	–
P. glabrata	Nan	c	c	r
Liliaceae				
Dianella caerulea	Mic	–	–	c
Poaceae				
Danthonia linkii	Lep	r	–	–
Entolasia stricta	Nan	–	–	r
Hierochloe rariflora	Nan	–	–	c
Tetrarrhena sp.	Nan	c	c	c
Restionaceae				
Empodisma minus	Lep	c	c	c
GEOPHYTES				
Droseraceae				
Drosera auriculata	Lep	c	–	–
D. spathulata	Nan	c	c	r
Orchidaceae				
Calochilus campestris	Nan	c	r	–
Chiloglottis sp.	Mic	r	–	–
Cryptostylis erecta	Mes	–	–	c
Cryptostylis sp.	Mes	–	c	–
Gastrodia sesamoides	Nan	–	c	–
Prasophyllum sp.	Lep	c	c	–
Thelymitra pauciflora	Mic	c	c	–
PARASITIC EPIPHYTES				
Cassythaceae				
Cassytha glabella	Lep	c	c	c

SITE 6

Mount Kaputar, New South Wales

Latitude: 30°16′S   **Longitude:** 150°10′E
Altitude: Mount Kaputar 1524 m
          Climate: Köppen **Cfb** (Mount Kaputar); **Cfa** (Narrabri)

Climatic data [1]	Summer (Dec–Feb)	Autumn (March–May)	Winter (June–Aug)	Spring (Sept–Nov)	Year (Jan–Dec)
Solar radiation [2]	2741 (655)	1736 (415)	1268 (303)	2385 (570)	2033 (486)
Max. air temp. (°C)	33.9	26.4	17.9	27.4	26.4
Min. air temp. (°C)	18.7	11.9	4.7	11.6	11.7
Precipitation (mm)	197	131	148	137	613
Pan evaporation (mm)	624	295	178	395	1492
Frost (days ≤0°C)	0	0.6	10.1	1.0	11.7
Snow (days)	0	0	0 [3]	0	0
Cyclones	0	0	0	0	0

[1] Climate station: Narrabri (30°19′S, 149°47′E, altitude 212 m).
[2] In J cm^{-2} day^{-1}, with equivalents in cal cm^{-2} day^{-1} in parentheses.
[3] Occasional snows on Mount Kaputar.

Dominant soil: Plateau remnants at high elevation; rock outcrops; alpine humus soil (Um 7.11) in moister sites, with skeletal (Um 5) and earthy soils (Gn 2) in the drier sites.
Vegetation: Dry sclerophyll forest
Reference: G.J. Harden (unpubl., October 1976) (list for Mount Kaputar only)

Species composition

GYMNOSPERMS

Zamiaceae — *Macrozamia heteromera, M. pauli-guilielmi, M. stenomera*

MONOCOTYLEDONS

Cyperaceae — *Gahnia aspera, Lepidosperma laterale, Schoenus apogon*
Hypoxidaceae — *Hypoxis hygrometrica*
Iridaceae — *Libertia paniculata, Patersonia sericea*
Liliaceae — *Anguillaria dioica, Arthropodium minus, Bulbinopsis bulbosa, Dianella laevis, D. revoluta, Dichopogon fimbriatus, Laxmannia gracilis, Stypandra glauca, Tricoryne elatior*
Orchidaceae — *Caladenia angustata, C. caerulea, C. carnea, C. reticulata, Calochilus robertsonii, Chiloglottis reflexa, Dendrobium speciosum, Dipodium punctatum, Diuris abbreviata, D. aequalis, D. maculata, D. sulphurea, Eriochilus cucullatus, Glossodia major, Lyperanthus suaveolens, Prasophyllum brevilabre, P. odoratum, Pterostylis curta, P. hamata, P. longicurva, P. mutica, P. revoluta, P. rufa, P. truncata, Thelymitra aristata, T. ixioides*
Xanthorrhoeaceae — *Lomandra filiformis, L. longifolia, L. multiflora, Xanthorrhoea resinosa*

DICOTYLEDONS

Asteraceae — *Cassinia aculeata, C. leptocephala, C. theodori, Craspedia uniflora, Gnaphalium japonicum, Helichrysum apiculatum, H. bracteatum, H. diosmifolium, H. obcordatum, H. scorpioides, H. semipapposum, Leptorhynchus squamatus, Olearia elliptica, O. ramosissima, O. ramulosa, O. rosmarinifolia, O. viscidula, Senecio lautus*
Campanulaceae — *Wahlenbergia gracilenta, W. stricta, Wahlenbergia spp.*
Cassythaceae — *Cassytha pubescens*
Dilleniaceae — *Hibbertia acicularis, H. linearis, H. obtusifolia, H. pedunculata, H. serpyllifolia, H. stricta, Hibbertia sp. nov.*
Epacridaceae — *Brachyloma daphnoides, Leucopogon attenuatus, L. microphyllus, L. muticus, L. neo-anglicus, Lissanthe strigosa, Melichrus erubescens, M. urceolatus, Monotoca scoparia*
Euphorbiaceae — *Bertya mitchellii, Beyeria viscosa, Poranthera microphylla*
Fabaceae — *Daviesia genistifolia, D. latifolia, D. ulicifolia, Dillwynia retorta var. phylicoides, Gompholobium huegelii, Hardenbergia violacea, Hovea heterophylla, H. lanceolata, H. linearis, Mirbelia oxylobioidea, M. pungens, Pultenaea boormanii, P. flexilis, P. hartmanii, P. microphylla, P. procumbens, P. retusa, P. scabra, P. setulosa*
Goodeniaceae — *Dampiera purpurea, Goodenia geniculata, G. hederacea, G. ovata, G. rotundifolia, Scaevola sp., Velleia paradoxa*

SITE 6 *(continued)*

Species composition *(continued)*

DICOTYLEDONS *(continued)*

Haloragaceae	*Haloragis serra*
Hypericaceae	*Hypericum gramineum*
Lamiaceae	*Plectranthus parvifolius, Prostanthera cruciflora, P. granitica, P. lasianthos, P. nivea, P. ovalifolia, Westringia eremicola*
Loganiaceae	*Logania albiflora*
Mimosaceae	*Acacia cheelii, A. gunnii, A. lanigera, A. triptera, A. viscidula,* etc.
Myoporaceae	*Myoporum montanum*
Myrtaceae	*Calytrix tetragona, Homoranthus flavescens, Kunzea* aff. *corifolia, K. opposita, Leptospermum attenuatum, L. sericatum, L. flavescens, L. phylicoides, Micromyrtus ciliata*
Olacaceae	*Olax stricta*
Oleaceae	*Notelaea microcarpa*
Oxalidaceae	*Oxalis corniculata*
Pittosporaceae	*Billardiera scandens*
Polygalaceae	*Comesperma sylvestre*
Proteaceae	*Hakea eriantha, H. microcarpa, Lomatia arborescens, Persoonia sericea, Persoonia* sp. aff. *rigida*
Rhamnaceae	*Cryptandra amara* var. *longiflora, C. longistaminea, C. propinqua, Discaria pubescens, Pomaderris* sp.
Rutaceae	*Asterolasia asteriscophora, Boronia* sp., *Correa glabra, C. reflexa* var. *reflexa, Phebalium viridiflorum, Zieria* sp. aff. *aspalathoides, Z. cytisoides*
Santalaceae	*Exocarpos cupressiformis, E. strictus, Santalum acuminatum, S. lanceolatum*
Sapindaceae	*Dodonaea angustissima, D. boroniifolia, D. tenuifolia, D. truncatiales, D. viscosa*
Scrophulariaceae	*Euphrasia collina* spp. *nandewarensis, E. speciosa, Parahebe* sp. aff. *derwentiana, Veronica arvensis, V. calycina, V. gracilis*
Stackhousiaceae	*Stackhousia monogyna, S. viminea*
Stylidiaceae	*Stylidium graminifolium*
Thymelaeaceae	*Pimelea collina, P. curviflora, P. glauca, P. linifolia, P. pauciflora*
Tremandraceae	*Tetratheca ericifolia*
Violaceae	*Hybanthus monopetalus*

SITE 7

Bonny Hills, New South Wales

Latitude: 31° 30′S   **Longitude:** 152° 50′E
Climate: Köppen **Cfa**

Climatic data [1]	Summer (Dec–Feb)	Autumn (March–May)	Winter (June–Aug)	Spring (Sept–Nov)	Year (Jan–Dec)
Solar radiation [2]	2351 (562)	1477 (353)	1021 (244)	2155 (515)	1753 (419)
Max. air temp. (°C)	25.5	22.8	18.3	21.8	22.1
Min. air temp. (°C)	17.6	13.6	7.5	12.4	12.8
Precipitation (mm)	390	497	307	265	1459
Pan evaporation (mm)	316	211	192	263	982
Frost (days ≤0°C)	0	0	0.4	0	0.4
Snow (days)	0	0	0	0	0
Cyclones	0.20	0.13	0	0	0.33

[1] Nearest climate station: Pt. Macquarie (31°35′S, 152°54′E).
[2] In J cm^{-2}day^{-1}, with equivalents in cal cm^{-2} day^{-1} in parentheses.

Dominant soil: Deep podzolic sand-dunes (Uc 2.21, Uc 2.22) with swamplands of Uc 2.2 sands
Vegetation: Sand-heathland
References: Clark, 1975; see also Osborn and Robertson (1939) for details of heathlands at Myall Lakes (32°30′S, 152°25′E)

SITE 7 *(continued)*

Species composition

MICRO- to NANO-PHANEROPHYTES (total cover 31.9%)

Fabaceae	*Dillwynia retorta* (2.0%), *Phyllota phylicoides* (10.1%)
Myrtaceae	*Leptospermum attenuatum* (1.3%), *L. flavescens* (0.2%), *Melaleuca nodosa* (3.2%), *M. sieberi* (0.1%)
Proteaceae	*Banksia ericifolia* (2.1%), *B. serrata* (12.9%)

NANOPHANEROPHYTES (total cover 52.0%)

Baueraceae	*Bauera capitata* (6.1%)
Dilleniaceae	*Hibbertia fasciculata* (1.0%), *H. virgata* (0.1%)
Epacridaceae	*Epacris microphyllus* (0.5%), *E. pulchella* (0.5%), *Leucopogon deformis* (0.1%), *L. ericoides* (0.7%), *L. esquamatus* (0.1%), *L. lanceolatus* (0.1%), *L. leptospermoides* (0.8%), *L. virgatus* (0.7%), *Monotoca scoparia* (1.0%), *Sprengelia incarnata* (0.7%)
Euphorbiaceae	*Pseudanthus orientalis* (1.3%), *Ricinocarpos pinifolius* (0.1%)
Fabaceae	*Aotus ericoides* (trace), *Bossiaea ensata* (trace), *B. heterophylla* (0.4%), *Dillwynia floribunda* (trace)
Goodeniaceae	*Dampiera stricta* (1.0%)
Mimosaceae	*Acacia brunioides* (0.2%), *A. suaveolens* (trace), *A. ulicifolia* (0.1%)
Myrtaceae	*Baeckea imbricata* (0.4%), *B. linearis* (1.2%), *Callistemon pachyphyllus* (0.1%), *Kunzea capitata* (1.3%), *Leptospermum liversidgei* (0.4%), *L. semibaccatum* (0.1%)
Olacaceae	*Olax stricta* (0.3%)
Polygalaceae	*Comesperma defoliatum* (trace), *C. ericinum* (trace)
Proteaceae	*Banksia aspleniifolia* (0.3%), *Conospermum taxifolium* (0.5%), *Hakea* sp. (trace), *Isopogon anemonifolius* (trace), *Persoonia lanceolata* (0.2%), *Petrophile fucifolia* (trace)
Rutaceae	*Boronia falcifolia* (1.7%), *B. pinnata* (3.5%), *Philotheca salsolifolia* (trace), *Zieria* sp. (trace)

Santalaceae	*Exocarpos strictus* (0.2%)
Thymelaeaceae	*Pimelea linifolia* (0.7%)
Xanthorrhoeaceae	*Xanthorrhoea resinosa* ssp. *fulva* (25.6%)

CHAMAEPHYTES (total cover 3.6%)

Blechnaceae	*Blechnum indicum* (trace)
Goodeniaceae	*Goodenia stelligera* (0.7%)
Haloragaceae	*Gonocarpus micranthus* (0.2%)
Hydrocotylaceae	*Platysace ericoides* (0.6%), *Xanthosia pilosa* (0.2%)
Schizaeaceae	*Schizaea bifida* (0.1%)
Selaginellaceae	*Selaginella uliginosa* (0.4%)
Spigeliaceae	*Mitrasacme polymorpha* (0.7%)
Stackhousiaceae	*Stackhousia scoparia* (trace)
Stylidiaceae	*Stylidium graminifolium* (0.6%), *Stylidium debile* (0.1%)

GEOPHYTES (total cover 1.0%)

Droseraceae	*Drosera auriculata* (0.5%), *D. spathulata* (0.1%)
Liliaceae	*Blandfordia grandiflora* (0.4%), *Burchardia umbellata* (trace)

HEMICRYPTOPHYTES (total cover 14.3%)

Cyperaceae	*Caustis flexuosa* (4.9%), *Gahnia sieberana* (trace), *Schoenus imberbis* (0.1%)
Iridaceae	*Patersonia fragilis* (0.5%), *P. glabrata* (0.1%)
Liliaceae	*Laxmannia gracilis* (0.5%), *Sowerbaea juncea* (0.1%)
Poaceae	*Entolasia stricta* (trace), *Panicum simile* (trace)
Restionaceae	*Leptocarpus tenax* (7.6%), *Restio tetraphyllus* (0.1%)
Xanthorrhoeaceae	*Lomandra glauca* (0.3%), *L. longifolia* (0.1%)
Xyridaceae	*Xyris juncea* (trace)

PARASITIC EPIPHYTES (total cover: trace)

Cassythaceae	*Cassytha glabella* (trace)

SITE 8

Sydney, New South Wales

Latitude: 33°51′S   **Longitude:** 151°13′E
               Climate: Köppen **Cfb/Cfa**

Climatic data	Summer (Dec–Feb)	Autumn (March–May)	Winter (June–Aug)	Spring (Sept–Nov)	Year (Jan–Dec)
Solar radiation [1]	2259 (540)	1410 (337)	1079 (258)	2000 (478)	1686 (403)
Max. air temp. (°C)	25.6	22.2	17.2	21.9	21.7
Min. air temp. (°C)	18.1	14.3	8.6	13.2	13.5
Precipitation (mm)	270	384	278	206	1138
Pan evaporation (mm)	449	255	206	372	1282
Frost (days ≤0°C)	0	0	0.2	0	0.2
Snow (days)	0	0	0	0	0
Cyclones	0	0	0	0	0

[1] In J cm^{-2} day^{-1}, with equivalents in cal cm^{-2} day^{-1} in parentheses.

Dominant soils: Dissected sandstone plateau with acid yellow earths (Gn 2.24, Gn 2.34, Gn 2.74) on areas of gentle relief; leached siliceous sands (Uc 1.2, Uc 2.12, Uc 2.2) occur on areas of strong relief; acidic gley soils (Dg 3.81, Dg 4.41) are found on undulating swampy areas of the plateau
Vegetation: (1) dry scrub; (2) dry-heathland; (3) wet-heathland
References: Pidgeon, 1937, 1938, 1940, 1941; Siddiqi et al., 1973

Species composition

	Vegetation type 1	2	3		Vegetation type 1	2	3
MICRO-NANOPHANEROPHYTES				Styphelia longifolia	o	–	–
				S. triflora	c	–	–
Baueraceae				S. tubiflora	o	–	–
Bauera rubioides	–	–	c	Woollsia pungens	lc	–	–
Casuarinaceae							
Casuarina distyla	–	x	x	Euphorbiaceae			
C. nana	–	–	x	Amperea xiphoclada	o	–	–
C. rigida	lc	–	–	Micrantheum ericoides	o	–	x
				Phyllanthus hirtellus	o	–	–
Dicrastylidaceae				Pseudanthus orientalis	–	x	–
Chloanthes stoechadis	r	–	–	Ricinocarpos pinifolius	lc	x	–
Dilleniaceae				Fabaceae			
Hibbertia aspera	–	–	x	Aotus ericoides	–	x	x
H. fasciculata	o	x	–	Bossiaea ensata	c	–	x
H. obtusifolia	–	–	x	B. heterophylla	c	–	–
H. stricta	o	–	–	B. obcordata	o	–	
				B. scolopendria	c	–	–
Epacridaceae				Dillwynia floribunda	r	–	c
Brachyloma daphnoides	c	–	–	D. retorta	c	–	x
Epacris longiflora	lc	–	–	Mirbelia rubiifolia	o	–	x
E. microphylla	c	–	x	Oxylobium ilicifolium	r	–	–
E. obtusifolia	–	–	c	Phyllota phylicoides	o	–	–
E. pulchella	c	–	x	Platylobium formosum	r	–	–
Leucopogon ericoides	lc	–	–	Pultenaea daphnoides	r	–	–
L. esquamatus	r	–	x	P. elliptica	c	–	–
L. microphyllus	c	–	–	P. polifolia	r	–	–
Lissanthe strigosa	r	–	–	P. stipularis	r	–	–
Monotoca scoparia	lc	–	–	Sphaerolobium vimineum	–	–	c
Sprengelia incarnata	–	–	c				

SITE 8 *(continued)*

Species composition *(continued)*

	Vegetation type				Vegetation type		
	1	2	3		1	2	3
MICRO-NANOPHANEROPHYTES *(continued)*				*B. serratifolia* (syn. *B. aemula*)	o	x	x
				B. spinulosa	r	–	x
Viminaria juncea	–	–	c	*Conospermum ellipticum*	o	–	–
				C. ericifolium	o	–	–
Lamiaceae				*C. longifolium*	o	–	–
Hemigenia purpurea	o	–	–	*C. taxifolium*	o	–	–
Westringia rigida	–	–	x	*Grevillea buxifolia*	o	–	–
				G. caleyi	lc	–	–
Mimosaceae				*G. sericea*	o	–	–
Acacia botrycephala	lc	–	–	*G. speciosa*	c	–	–
A. linifolia	r	–	–	*Hakea dactyloides*	c	–	–
A. myrtifolia	lc	–	x	*H. gibbosa*	o	x	x
A. suaveolens	o	–	x	*H. propinqua*	o	–	–
A. ulicifolia	o	–	x	*H. sericea*	c	x	x
				H. teretifolia	o	x	c
Myrtaceae				*Isopogon anemonifolius*	c	–	o
Angophora cordifolia	lc	–	c	*I. anethifolius*	c	–	o
Baeckea crenulata	r	–	c	*Lambertia formosa*	c	–	o
B. densifolia	–	–	c	*Lomatia silaifolia*	o	–	–
B. imbricata	–	–	x	*Persoonia ferruginea*	r	–	–
Callistemon citrinus	–	–	x	*P. lanceolata*	c	–	o
C. lanceolatus	–	–	c	*P. levis*	o	–	c
C. linearis	–	–	c	*Petrophile fucifolia*	–	–	x
Calytrix tetragona	lc	–	–	*P. pulchella*	c	–	–
Darwinia fascicularis	c	–	–				
Eucalyptus camfieldii (mallee)	c^1	–	–	Rutaceae			
E. gummifera (tree)	c^2	–	–	*Boronia ledifolia*	o	–	–
E. haemastoma (tree)	c^2	–	–	*B. pinnata*	o	–	–
E. multicaulis (mallee)	c^1	–	–	*Correa reflexa*	r	–	–
E. obtusiflora (mallee)	c^1	–	–	*Eriostemon australasicus*	r	x	–
E. virgata (mallee)	c^1	–	–	*E. crowei*	r	–	–
Kunzea capitata	c	–	c	*E. hispidulus*	o	–	–
K. corifolia	lc	–	–	*Philotheca australis*	o	–	–
Leptospermum arachnoides	c	–	–	*Zieria pilosa*	o	–	–
L. flavescens	–	x	x				
L. lanigerum	–	–	c	Santalaceae			
L. scoparium	c	–	x	*Leptomeria acida*	–	–	c
L. stellatum	o	–	x				
Melaleuca squarrosa	–	–	c	Sterculiaceae			
Micromyrtus ciliata	lc	–	–	*Lasiopetalum ferrugineum*	lc	–	x
				Rulingia hermanniifolia	–	–	x
Olacaceae							
Olax stricta	–	–	c	Thymelaeaceae			
				Pimelea linifolia	c	–	x
Polygalaceae							
Comesperma ericinum	o	–	–	Tremandraceae			
C. retusum	o	–	–	*Tetratheca ericifolia*	o	–	–
				T. juncea	o	–	–
Proteaceae							
Banksia aspleniifolia (syn.	–	x	x	Xanthorrhoeaceae			
B. oblongifolia)				*Xanthorrhoea hastilis*	lc	–	c
B. ericifolia	c	x	x				
B. latifolia var. *minor* (syn.	r	–	c	CHAMAEPHYTES			
B. oblongifolia)				Dilleniaceae			
B. serrata	o	–	–	*Hibbertia linearis*	lc	–	–

SITE 8 *(continued)*

Species composition *(continued)*

	Vegetation type				Vegetation type		
	1	2	3		1	2	3
CHAMAEPHYTES *(continued)*				**HEMICRYPTOPHYTES**			
Epacridaceae				Cyperaceae			
Dracophyllum secundum	–	–	c	*Baumea teretifolia*	–	–	x
				Caustis flexuosa	lc	–	–
Euphorbiaceae				*C. pentandra*	lc	–	–
Poranthera ericifolia	o	–	–	*C. recurvata*	lc	–	–
				Cyathochaete diandra	–	–	x
Fabaceae				*Lepidosperma filiforme*	o	–	–
Hovea heterophylla	o	–	–	*L. flexuosum*	o	–	c
H. linearis	o	–	–	*L. laterale*	c	x	x
				L. limicola	–	–	x
Goodeniaceae				*Ptilanthelium deustum*	c	–	c
Dampiera purpurea	o	–	–	*Schoenus brevifolius*	–	–	c
D. stricta	lc	–	x	*S. ericetorum*	o	–	x
Goodenia bellidifolia	o	–	x	*S. paludosus*	o	–	x
G. hederacea	o	–	–	*S. turbinatus*	o	–	–
G. stelligera	–	–	x	*S. villosus*	o	–	–
Scaevola ramosissima	o	–	x				
				Iridaceae			
Haloragaceae				*Patersonia glabrata*	o	–	–
Gonocarpus micranthus	–	–	c	*P. sericea*	o	–	–
G. teucrioides	o	x	x				
				Liliaceae			
Hydrocotylaceae				*Dianella caerulea*	c	–	–
Actinotus helianthi	o	–	x	*Sowerbaea juncea*	–	–	c
A. minor	lc	–	c				
Platysace linearifolia	–	x	x	Poaceae			
Trachymene linearis	r	–	x	*Anisopogon avenaceus*	–	x	x
Xanthosia pilosa	o	–	–	*Aristida* sp.	–	–	x
				Entolasia stricta	–	–	x
Lobeliaceae				*Eragrostis* sp.	–	–	x
Lobelia anceps	–	–	c	*Themeda australis*	o	–	x
L. dentata	o	–	–				
L. gracilis	o	–	–	Restionaceae			
				Empodisma minus	–	–	x
Proteaceae				*Hypolaena fastigiata*	lc	x	c
Symphyonema paludosum	–	–	c	*Leptocarpus tenax*	–	–	c
				Lepyrodia scariosa	c	–	c
Rubiaceae				*Restio complanatus*	–	–	c
Opercularia hispida	o	–	–	*R. dimorphus*	o	–	–
O. aspera	o	–	x	*R. fastigiatus*	lc	–	–
Selaginellaceae				Xanthorrhoeaceae			
Selaginella uliginosa	–	–	x	*Lomandra filiformis*	o	–	–
				L. glauca	c	x	x
Spigeliaceae				*L. longifolia*	c	–	x
Mitrasacme polymorpha	lc	–	x	*L. multiflora*	o	–	–
				L. micrantha	–	–	x
Stackhousiaceae				*L. obliqua*	c	–	–
Stackhousia viminea	–	–	c				
				GEOPHYTES			
Stylidiaceae							
Stylidium graminifolium	o	–	–	Droseraceae			
				Drosera peltata	–	–	x
Violaceae							
Hybanthus monopetalus	o	–	–				

SITE 8 *(continued)*

Species composition *(continued)*

	Vegetation type				Vegetation type		
	1	2	3		1	2	3
GEOPHYTES *(continued)*				*Burchardia umbellata*	lc	–	x
				Caesia parviflora	o	–	–
D. pygmaea	–	–	c	*Stypandra caespitosa*	lc	–	–
D. spathulata	–	–	x	*S. umbellata*	lc	–	–
				Thysanotus juncifolius	o	–	–
Haemodoraceae							
Haemodorum planifolium	–	–	x				
H. teretifolium	o	–	o	Orchidaceae			
				Eriochilus cucullatus	–	–	c
Liliaceae				*Thelymitra ixioides*	o	–	–
Blandfordia nobilis	–	–	c				

[1] In mallee-scrub; [2] in tree-scrub.

SITE 9

Wilson's Promontory, Victoria

Latitude: 39°08′S   **Longitude:** 145°25′E
Diagrammatic cross-section (after Parsons, 1966):

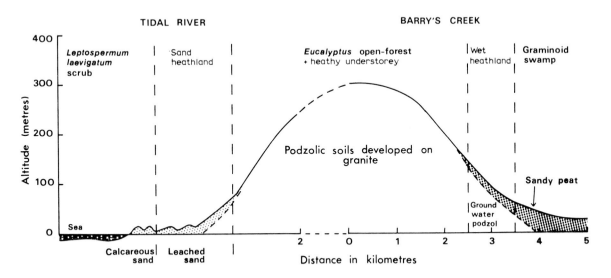

Fig. 6.15. Diagrammatic cross-section, Wilson's Promontory, Victoria.

SITE 9 *(continued)*

Climate: Köppen **Cfb**

Climatic data	Summer (Dec–Feb)	Autumn (March–May)	Winter (June–Aug)	Spring (Sept–Nov)	Year (Jan–Dec)
Solar radiation [1]	2109 (504)	1109 (265)	736 (176)	1720 (411)	1418 (339)
Max. air temp. (°C)	19.3	16.9	12.6	15.5	16.1
Min. air temp. (°C)	13.8	12.6	8.9	10.2	11.4
Precipitation (mm)	162	245	329	244	980
Pan evaporation (mm)	335	177	116	225	853
Frost (days $\leq 0°$C)					
Snow (days)	0	0	0	0	0
Cyclones	0	0	0	0	0

[1] In J cm^{-2} day^{-1}, with equivalents in cal cm^{-2} day^{-1} in parentheses.

Vegetation: Tidal River dry-heathland
Age: 10 years since fire   **Biomass** (dry wt.): 10.9 tonnes ha^{-1}
Structure: Low closed-heathland   **Reference:** Groves and Specht, 1965
Dominant soil: Deep podzolised sand with a pan-like layer below the bleach (Uc 2.3)

Species	Leaf type and size	% of total above-ground biomass	Species	Leaf type and size	% of total above-ground biomass
Evergreen, low shrubs (25–100 cm)			Fabaceae		
			Dillwynia glaberrima	(Lep, Gro)	1.0
Casuarinaceae			Proteaceae		
Casuarina pusilla	NS (Aph)	54.9	*Isopogon ceratophyllus*	BS (Pin–Lep)	2.8
Epacridaceae					———
Epacris impressa	BS (Lep)	1.3			**5.6**
Myrtaceae					
Calytrix tetragona	NS (Lep)	2.4			
Leptospermum myrsinoides	BS (Lep)	20.2	**Evergreen, caespitose graminoid herbs (<25 cm)**		
Proteaceae			Restionaceae		
Banksia marginata	BS (Nan)	1.5	*Hypolaena fastigiata*	NS (Aph)	9.4
Hakea sericea	NS (Nan)	0.9	Cyperaceae		
		———	*Lepidosperma concavum*	BS (Gra)	0.7
		81.2	Iridaceae		
			Patersonia fragilis	BS (Gra)	0.5
Evergreen, dwarf-shrubs (<25 cm)					———
					10.6
Epacridaceae					
Leucopogon ericoides	BS (Lep)	0.6			
L. virgatus	BS (Lep)	1.2	Other species [1]		**2.6**

[1] Other species recorded:

Ferns
Dennstaedtiaceae   *Pteridium esculentum*
Selaginellaceae   *Selaginella uliginosa*

Monocotyledons
Cyperaceae   *Lepidosperma canescens, Schoenus apogon, S. breviculmis*
Liliaceae   *Burchardia umbellata, Chamaescilla corymbosa, Laxmannia sessiliflora,*

Thysanotus patersonii, T. tuberosus
Orchidaceae   *Caladenia caerulea, C. carnea, C. clavigera, C. deformis, Diuris longifolia, Glossodia major, Lyperanthus nigricans, Prasophyllum elatum, Pterostylis vittata, Thelymitra antennifera*
Poaceae   *Danthonia setacea, Stipa semibarbata, Tetrarrhena distichophylla*
Xanthorrhoeaceae   *Lomandra filiformis, Xanthorrhoea australis*

SITE 9 *(continued)*

Dicotyledons

Asteraceae	*Helichrysum baxteri, H. obtusifolium, H. scorpioides, Olearia ciliata*
Dilleniaceae	*Hibbertia acicularis, H. fasciculata, H. procumbens, H. sericea*
Droseraceae	*Drosera auriculata, D. planchonii*
Epacridaceae	*Acrotriche serrulata, Astroloma humifusum, Monotoca scoparia*
Euphorbiaceae	*Amperea xiphoclada*
Fabaceae	*Aotus ericoides, Bossiaea cinerea, B. prostrata, Daviesa ulicifolia, Dillwynia sericea, Gompholobium ecostatum, G. huegelii, Kennedia prostrata, Platylobium obtusangulum, Pultenaea daphnoides*
Gentianaceae	*Centaurium spicatum*
Haloragaceae	*Gonocarpus tetragynus*
Hydrocotylaceae	*Platysace heterophylla, Xanthosia dissecta, X. pusilla, X. tridentata*

Lauraceae (Cassythaceae)	*Cassytha glabella, C. pubescens*
Mimosaceae	*Acacia myrtifolia, A. suaveolens, A. verticillata*
Myrtaceae	*Baeckea ramosissima, Leptospermum juniperinum, L. laevigatum* (an invader)
Oxalidaceae	*Oxalis corniculata*
Pittosporaceae	*Billardiera scandens, Marianthus procumbens*
Polygalaceae	*Comesperma calymega, C. volubile*
Proteaceae	*Hakea ulicina, Persoonia juniperina*
Rhamnaceae	*Spyridium parvifolium*
Rubiaceae	*Opercularia ovata, O. varia*
Rutaceae	*Correa reflexa*
Stackhousiaceae	*Stackhousia monogyna*
Stylidiaceae	*Stylidium graminifolium*
Thymelaeaceae	*Pimelea humilis*
Tremandraceae	*Tetratheca pilosa*
Violaceae	*Viola hederacea*

Vegetation: Darby River tree-heathland (on deep sand)
Age: 10–12 years since fire   **Biomass** (dry wt.): 11.5 tonnes ha^{-1}
Structure: Tree-heathland   **Reference:** Specht (harvest Jan. 1962)
(Scattered, small trees of *Banksia serrata* were not included in the harvest data presented below)
Dominant soil: Deep podzolised sand with a pan-like layer below the bleach (Uc 2.3)

Species	Leaf type and size	% of total above-ground biomass	Species	Leaf type and size	% of total above-ground biomass
Evergreen, low shrubs (25–100 cm)			Dilleniaceae		
Casuarinaceae			*Hibbertia fasciculata*	NS (Lep)	0.2
Casuarina pusilla	NS (Aph)	57.7	*H. virgata*	NS (Lep)	0.1
Epacridaceae			Proteaceae		
Epacris impressa	BS (Lep)	2.6	*Isopogon ceratophyllus*	BS (PinLep)	0.2
Myrtaceae			Rutaceae		
Leptospermum myrsinoides	BS (Lep)	13.3	*Correa reflexa*	BS (Nan)	0.2
		73.6	Tremandraceae		
Rosulate, evergreen geophytes (25–100 cm)			*Tetratheca pilosa*	NS (Lep, Eri)	0.1
Xanthorrhoeaceae					**2.5**
Xanthorrhoea australis	BS (Gra)	11.0	**Evergreen, caespitose graminoid herbs (<25 cm)**		
		11.0	Cyperaceae		
Evergreen, dwarf-shrubs (<25 cm)			*Lepidosperma concavum*	BS (Gra)	1.5
Epacridaceae			Restionaceae		
Leucopogon ericoides	BS (Lep)	0.6	*Hypolaena fastigiata*	NS (Aph)	11.3
L. virgatus	BS (Lep)	0.7			**12.8**
Fabaceae					
Dillwynia glaberrima	NS (Lep, Gro)	0.2			
Platylobium obtusangulum	BS (Nan)	0.2	Other species [1]		**0.1**

[1] Other species recorded:

Monocotyledons	
Poaceae	*Tetrarrhena distichophylla*

Xanthorrhoeaceae	*Lomandra filiformis*
Dicotyledons	
Dilleniaceae	*Hibbertia acicularis, H. procumbens*

SITE 9 *(continued)*

Euphorbiaceae	*Amperea xiphoclada*		Mimosaceae	*Acacia suaveolens*
Fabaceae	*Gompholobium huegelii, Pultenaea daphnoides*		Proteaceae	*Banksia marginata, B. serrata* (seedlings), *B. spinulosa, Hakea sericea, H. ulicina, Persoonia juniperina*
Haloragaceae	*Gonocarpus tetragynus*			
Hydrocotylaceae	*Platysace heterophylla, Xanthosia dissecta, X. pusilla*		Rhamnaceae	*Spyridium parvifolium*
			Thymelaeaceae	*Pimelea humilis*
Lauraceae (Cassythaceae)	*Cassytha glabella*			

Vegetation: Barry Creek's wet-heathland
Age: 6 years since fire   **Biomass** (dry wt.): 8.1 tonnes ha^{-1}
Structure: Low closed-heathland   **Reference:** Groves and Specht, 1965
Dominant soil: Ground-water podzol (Uc 2.33) on granitic soil

Species	Leaf type and size	% of total above-ground biomass	Species	Leaf type and size	% of total above-ground biomass
Evergreen, low shrubs (25–100 cm)			Evergreen dwarf-shrubs (<25 cm)		
Casuarinaceae			Proteaceae		
Casuarina paludosa *C. pusilla* }	NS (Aph)	46.2	*Isopogon ceratophyllus*	BS (Pin–Lep)	2.9
					2.9
Myrtaceae			Evergreen, caespitose graminoid herbs (<25 cm)		
Leptospermum juniperinum	BS (Lep)	10.3			
L. myrsinoides	BS (Lep)	9.3	Cyperaceae		
			Lepidosperma filiforme	NS (Gra) }	19.0
Proteaceae			*L. neesii*	NS (Gra) }	
Banksia marginata	BS (Nan)	2.0			**19.0**
		67.8	Other species[1]		**10.3**

[1] Other species recorded:

Ferns
  Lindsaeaceae   *Lindsaea linearis*
  Selaginellaceae   *Selaginella uliginosa*

Monocotyledons
  Cyperaceae   *Lepidosperma concavum, Schoenus breviculmis*
  Hypoxidaceae   *Hypoxis glabella*
  Iridaceae   *Diplarrhena moraea, Patersonia fragilis, Patersonia glabrata*
  Liliaceae   *Burchardia umbellata, Chamaescilla corymbosa, Dianella revoluta, Laxmannia sessiliflora, Stypandra caespitosa, Thysanotus patersonii*
  Orchidaceae   *Caladenia carnea, C. clavigera, C. deformis, C. menziesii, Diuris longifolia, Glossodia major, Lyperanthus nigricans, Orthoceras strictum, Prasophyllum elatum, Pterostylis barbata, Thelymitra antennifera, T. fuscolutea*
  Poaceae   *Danthonia setacea, Deyeuxia quadriseta, Stipa semibarbata, Tetrarrhena distichophylla*

  Restionaceae   *Hypolaena fastigiata*
  Xanthorrhoeaceae   *Lomandra filiformis, Xanthorrhoea australis, X. minor*

Dicotyledons
  Asteraceae   *Helichrysum obtusifolium, H. scorpioides, Olearia ciliata*
  Campanulaceae   *Wahlenbergia gracilenta, W. gymnoclada*
  Dilleniaceae   *Hibbertia acicularis, H. fasciculata, H. procumbens, H. sericea*
  Droseraceae   *Drosera auriculata, D. planchonii, D. pygmaea*
  Epacridaceae   *Acrotriche serrulata, Astroloma humifusum, Epacris impressa, Leucopogon virgatus, Monotoca scoparia, Sprengelia incarnata*
  Euphorbiaceae   *Amperea xiphoclada*
  Fabaceae   *Bossiaea cinerea, Dillwynia glaberrima, D. sericea, Gompholobium ecostatum, G. huegelii, Hovea heterophylla, Kennedia prostrata, Platylobium obtusangulum, Pultenaea daphnoides, P. hibbertioides, Sphaerolobium vimineum*
  Haloragaceae   *Gonocarpus tetragynus*

SITE 9 *(continued)*

Hydrocotylaceae	*Platysace heterophylla, Xanthosia dissecta, X. pusilla, X. tridentata*	Proteaceae	*Hakea sericea, H. teretifolia, H. ulicina*
		Rhamnaceae	*Spyridium parvifolium*
Lauraceae (Cassythaceae)	*Cassytha glabella, C. pubescens*	Rubiaceae	*Opercularia ovata, O. varia*
		Rutaceae	*Boronia parviflora, Correa reflexa*
Lentibulariaceae	*Polypompholyx tenella*	Scrophulariaceae	*Euphrasia collina*
Lobeliaceae	*Lobelia gibbosa*	Spigeliaceae	*Mitrasacme pilosa*
Mimosaceae	*Acacia myrtifolia, A. suaveolens*	Stackhousiaceae	*Stackhousia monogyna, S. viminea*
Myrtaceae	*Baeckea ramosissima, Eucalyptus baxteri (<2 m) E. radiata (<2 m)*	Stylidiaceae	*Stylidium graminifolium*
		Thymelaeaceae	*Pimelea humilis*
Oxalidaceae	*Oxalis corniculata*	Tremandraceae	*Tetratheca pilosa*
Pittosporaceae	*Billardiera scandens, Marianthus procumbens*	Violaceae	*Viola hederacea*
Polygalaceae	*Comesperma calymega, C. ericinum, C. volubile*		

SITE 10

Frankston, Victoria

Latitude: 38°08'S   **Longitude:** 145°07'E
       Climate: Köppen **Cfb**

Climatic data	Summer (Dec–Feb)	Autumn (March–May)	Winter (June–Aug)	Spring (Sept–Nov)	Year (Jan–Dec)
Solar radiation[1]	2259 (540)	1201 (287)	787 (188)	1757 (420)	1502 (359)
Max. air temp. (°C)	24.3	19.7	13.3	18.2	18.8
Min. air temp. (°C)	12.7	10.5	6.5	9.2	9.7
Precipitation (mm)	135	166	181	170	652
Pan evaporation (mm)	512	284	132	304	1232
Frost (days<0°C)	0	0.1	0.9	0.2	1.2
Snow (days)	0	0	0	0	0
Cyclones	0	0	0	0	0

[1] In $J\,cm^{-2}\,day^{-1}$, with equivalents in $cal\,cm^{-2}\,day^{-1}$ in parentheses.

Dominant soil: Deep podzolised sand with a pan-like layer below the bleach (Uc 2.3)
Vegetation: Dry-heathland
Age: 4–6 years since fire   **Biomass** (dry wt.): 6.7–10.8 tonnes ha^{-1}
Structure: Low closed-heathland   **Reference:** Jones, 1968a, b (also Patton, 1933; Winkworth, 1955)

Species composition

Species	Leaf type and size	% of total above-ground biomass	Species	Leaf type and size	% of total above-ground biomass
Evergreen, low shrubs (50–100 cm)			Evergreen, low shrubs (25–50 cm)		
Casuarinaceae			Epacridaceae		
Casuarina pusilla	NS (Aph)	4.4	*Epacris impressa*	BS (Lep)	0.2
			Leucopogon virgatus	BS (Lep)	4.4
Myrtaceae			*Monotoca scoparia*	BS (Lep)	4.2
Leptospermum myrsinoides	BS (Lep)	55.4	Fabaceae		
			Aotus ericoides	NS (Lep–Eri)	1.5
Proteaceae			*Dillwynia glaberrima*	NS (Lep)	3.4
Banksia marginata	BS (Nan)	1.2			
		61.0			**13.7**

SITE 10 *(continued)*

Species composition *(continued)*

Species	Leaf type and size	% of total above-ground biomass	Species	Leaf type and size	% of total above-ground biomass
Evergreen, caespitose dwarf-shrubs (<25 cm)			Restionaceae		
Dilleniaceae			*Hypolaena fastigiata*	NS (Aph)	9.7
Hibbertia exutiacies	NS (Lep)	0.9			
					24.2
		0.9	Other species[1]		0.2
Evergreen, caespitose graminoid herbs (<25 cm)					
Cyperaceae					
Lepidosperma concavum	BS (Gra)	14.5			

[1] Other species recorded:

Monocotyledons
   Liliaceae   *Burchardia umbellata*
   Orchidaceae   *Diuris* sp., *Lyperanthus nigricans*, *Prasophyllum despectans*
   Poaceae   *Danthonia setacea*
   Xanthorrhoeaceae *Xanthorrhoea minor*

Dicotyledons
   Dilleniaceae   *Hibbertia fasciculata, H. sericea*
   Droseraceae   *Drosera auriculata*

Euphorbiaceae   *Amperea xiphoclada*
Fabaceae   *Dillwynia sericea*
Hydrocotylaceae   *Platysace heterophylla, Xanthosia pusilla*
Lauraceae (Cassythaceae)   *Cassytha glabella*
Mimosaceae   *Acacia oxycedrus, A. suaveolens*
Myrtaceae   *Leptospermum laevigatum* (invader)
Polygalaceae   *Comesperma volubile*
Proteaceae   *Isopogon ceratophyllus*
Rutaceae   *Correa reflexa*
Scrophulariaceae   *Euphrasia collina*
Stylidiaceae   *Stylidium graminifolium*

SITE 11

Lower South-East, South Australia

Latitude: 37°20'S   **Longitude:** 140°40'E
Diagrammatic cross-section (after Crocker, 1944):

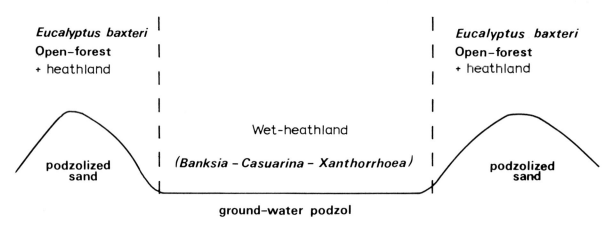

Fig. 6.16. Diagrammatic cross-section, Lower South-East, South Australia. Horizontal scale *c.* 600 m, vertical scale less than 10 m.

SITE 11 *(continued)*

Climate: Köppen **Csb/Cfb**

Climatic data[1]	Summer (Dec–Feb)	Autumn (March–May)	Winter (June–Aug)	Spring (Sept–Nov)	Year (Jan–Dec)
Solar radiation[2]	2251 (538)	1205 (288)	803 (192)	1766 (422)	1506 (360)
Max. air temp. ($^\circ$C)	23.3	19.4	13.9	18.2	18.7
Min. air temp. ($^\circ$C)	11.9	9.7	6.1	8.4	9.1
Precipitation (mm)	92	158	265	167	682
Pan evaporation (mm)	425	205	126	254	1010
Frost (days $\leqslant 0^\circ$C)	0	0	3.4	0.2	3.6
Snow (days)	0	0	0	0	0
Cyclones	0	0	0	0	0

[1] Climate station: Mount Gambier (37°50′S, 140°50′E).
[2] In J cm^{-2} day^{-1}, with equivalents in cal cm^{-2} day^{-1} in parentheses.

Dominant soil: Dunes of leached sand (Uc 2.2) and sand plains of ground-water podzols (Uc 2.33)

Vegetation: Wet-heathland
References: Crocker, 1944; Welbourn and Lange, 1968; see also Gibbons and Downes (1964) for similar ecosystems in southwestern Victoria

Species composition

NANOPHANEROPHYTES

Casuarinaceae	*Casuarina paludosa, C. pusilla*
Dilleniaceae	*Hibbertia fasciculata, H. stricta*
Epacridaceae	*Epacris impressa*
Fabaceae	*Daviesia brevifolia, Dillwynia hispida, Sphaerolobium vimineum*
Mimosaceae	*Acacia verticillata*
Myrtaceae	*Calytrix tetragona, Darwinia micropetala, Leptospermum juniperinum, L. myrsinoides, Melaleuca gibbosa*
Proteaceae	*Banksia marginata, B. ornata, Hakea nodosa, H. rostrata, H. rugosa, Isopogon ceratophyllus*
Thymelaeaceae	*Pimelea flava, P. octophylla*
Xanthorrhoeaceae	*Xanthorrhoea australis*

CHAMAEPHYTES

Asteraceae	*Helichrysum obtusifolium, H. scorpioides*

Epacridaceae	*Leucopogon virgatus*
Fabaceae	*Pultenaea laxiflora, P. tenuifolia.*
Stackhousiaceae	*Stackhousia monogyna*
Stylidiaceae	*Stylidium graminifolium*

HEMICRYPTOPHYTES

Cyperaceae	*Chorizandra enodis, Lepidosperma carphoides, L. congestum, L. laterale, Schoenus apogon, S. brachyphyllus*
Poaceae	*Stipa* sp.
Restionaceae	*Leptocarpus brownii*

GEOPHYTES

Droseraceae	*Drosera planchonii*
Orchidaceae	*Caladenia patersonii*

SITE 12

Keith, South Australia

Latitude: 36°06′S   **Longitude:** 140°31′E
Diagrammatic cross-section (after Coaldrake, 1951, Rayson, 1957):

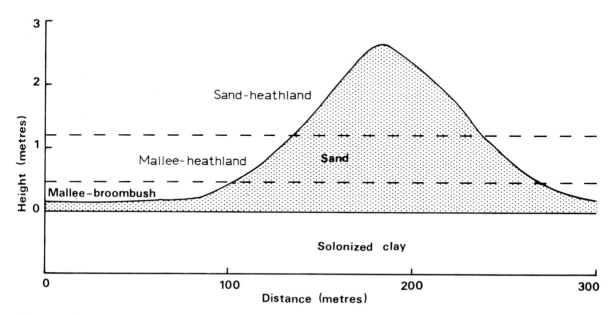

Fig. 6.17. Diagrammatic cross-section, Keith, South Australia.

Climate: Köppen **Csb**

Climatic data	Summer (Dec–Feb)	Autumn (March–May)	Winter (June–Aug)	Spring (Sept–Nov)	Year (Jan–Dec)
Solar radiation[1]	2469 (590)	1339 (320)	929 (222)	1916 (458)	1661 (397)
Max. air temp. (°C)	28.7	22.2	15.2	21.5	21.9
Min. air temp. (°C)	11.3	8.5	5.0	7.6	8.1
Precipitation (mm)	66	107	159	125	457
Pan evaporation (mm)	622	269	145	358	1395
Frost (days ≤0°C)	0	0.2	1.9	0.3	2.4
Snow (days)	0	0	0	0	0
Cyclones	0	0	0	0	0

[1] In J cm^{-2} day^{-1}, with equivalents in cal cm^{-2} day^{-1} in parentheses.

SITE 12 *(continued)*

Dominant soils: Dunes and sandhills with leached sand (Uc 2.2) over solonised clay; sand plains with shallow sandy surface soil over neutral or alkaline yellow mottled clay sub-soils (Dy 5.42, Dy 5.43).

Vegetation: Dark Island dry-heathland
Age: 9 years since fire   **Biomass** (dry wt.): 7.1 tonnes ha^{-1}
Structure: Low open-heathland   **Reference:** Specht et al . 1958

Species	Leaf type and size	% of total above-ground biomass	Species	Leaf type and size	% of total above-ground biomass
Evergreen, low shrubs (50–100 cm)			Dilleniaceae		
			Hibbertia sericea	NS/BS	
Myrtaceae			var. *scabrifolia*	(Lep–Eri)	1.8
Leptospermum myrsinoides	BS (Lep)	1.1	*H. stricta* var. *glabriuscula*	NS (Lep–Eri)	7.0
Proteaceae					
Banksia marginata	BS (Nan)	0.8	Epacridaceae		
B. ornata	BS (Mic)	18.0	*Leucopogon costatus*	BS (Lep)	2.5
		———	*L. woodsii*	BS (Lep)	1.7
		19.9	Fabaceae		
			Gompholobium ecostatum	NS (Pin–Lep)	0.1
Rosulate, evergreen geophytes (50–100 cm)			Myrtaceae		
			Baeckea ericaea	NS (Lep)	1.0
Xanthorrhoeaceae			*Calytrix alpestris*	NS (Lep)	0.8
Xanthorrhoea australis	BS (Gra)	4.7	Rhamnaceae		
		———	*Cryptandra tomentosa*	NS (Lep, Eri)	0.5
		4.7	Rutaceae		
			Boronia caerulescens	NS (Lep)	1.0
Evergreen, low shrubs (25–50 cm)					———
					16.4
Casuarinaceae					
Casuarina pusilla	NS (Aph)	35.4	**Evergreen, caespitose graminoid herbs (<25 cm)**		
Epacridaceae			Cyperaceae		
Astroloma conostephioides	BS (Lep)	0.1	*Lepidosperma carphoides*	NS (Gra)	1.4
			L. laterale	BS (Gra)	1.9
Fabaceae			*Schoenus breviculmis*	NS (Gra)	0.4
Daviesia brevifolia	NS (Aph–thorns)	0.6	Poaceae		
Dillwynia hispida	NS (Lep, Gro)	0.2	*Amphipogon caricinus*	NS (Gra)	0.2
Phyllota pleurandroides }	NS (Lep–Eri)	9.0	Restionaceae		
P. remota }			*Hypolaena fastigiata*	NS (Aph)	4.0
Proteaceae			*Lepidobolus drapetocoleus*	NS (Aph)	0.5
Adenanthos terminalis	NS (Pin–Lep)	2.4	Xanthorrhoeaceae		
Rhamnaceae			*Lomandra juncea*	NS (Gra)	0.1
Spyridium subochreatum var. *laxiusculum*	NS (Lep–Eri)	2.4			———
		———			**8.5**
		50.1	Other species[1]		**0.4**
Evergreen, caespitose dwarf-shrubs (<25 cm)					
Asteraceae					
Helichrysum obtusifolium	NS (Lep, Eri)	trace			

[1] Other species recorded:

Monocotyledons
Liliaceae	*Laxmannia sessiliflora, Thysanotus dichotomus, T. patersonii, Tricoryne elatior*
Orchidaceae	*Caladenia deformis, Diuris maculata,*
	Lyperanthus nigricans, Prasophyllum patens, Thelymitra antennifera
Poaceae	*Danthonia setacea, Neurachne alopecuroidea, Stipa macalpinei, S. semibarbata, Triodia irritans*

SITE 12 *(continued)*

Dicotyledons

Aizoaceae	*Carpobrotus modestus*
Asteraceae	*Helichrysum blandowskianum, Olearia ciliata*
Campanulaceae	*Wahlenbergia gracilenta*
Dilleniaceae	*Hibbertia virgata*
Droseraceae	*Drosera whittakeri*
Epacridaceae	*Acrotriche affinis, Astroloma humifusum, Brachyloma ericoides, Leucopogon clelandii, Styphelia exarrhena*
Euphorbiaceae	*Poranthera microphylla*
Goodeniaceae	*Dampiera marifolia, Goodenia geniculata*
Haloragaceae	*Gonocarpus tetragynus*
Lauraceae (Cassythaceae)	*Cassytha glabella, C. pubescens*
Myrtaceae	*Calytrix tetragona, Kunzea pomifera*
Polygalaceae	*Comesperma calymega*
Proteaceae	*Conospermum patens, Hakea muellerana, H. rostrata, H. ulicina* var. *latifolia, Isopogon ceratophyllus, Persoonia juniperina*
Rubiaceae	*Opercularia scabrida*
Rutaceae	*Correa reflexa*
Santalaceae	*Santalum acuminatum*
Scrophulariaceae	*Euphrasia collina*
Stylidiaceae	*Stylidium graminifolium*
Thymelaeaceae	*Pimelea octophylla, P. phylicoides*

Mosses	*Bryum argenteum* (rare), *Tortella calycina* (common)
Lichens (ground)	*Cladonia aggregata* (common), *C. fimbriata* (common), *C. verticillata* (common)
Lichens (on dead stems)	*Buellia* sp., *Caloplaca* sp., *Candellariella* sp., *Parmelia pruinata, Physcia adscendens, P. aipolia, P. caesia, Ramalina ecklonii, R. pusilla, Usnea scabrida, Xanthoria ectanea*
Algae (in soil depressions between bushes)	*Cylindrocapsa geminella*

Fungi (appearing in springtime)

Ascomycetes	*Geoglossum nigritum, Poronia punctata, Sarcosphaera* sp. (aff. *S. ammophila*)
Basidiomycetes	*Boletus* sp., *Coltricia cinnamomea, C. oblectans, Coriolus sanguineus, Cortinarius fibrillosus, C. subcinnamoneus, Flammulla paludosa, Geastrum campestre, Lycoperdon gunnii, L. spadiceum, L. pusillum, Naucoria arenacolens, N. verona-brunneus, Psathyra sonderiana, Psilocybe subammophila, Tremella mesenterica, Tulostoma albicans*

Vegetation: Dark Island mallee–broombush
Age: 12 years since fire   **Biomass** (dry wt.): 9.2 tonnes ha^{-1}
Structure: Open-scrub with heathy understorey   **Reference:** Specht (1966)

Species	Leaf type and size	% of total above-ground biomass	Species	Leaf type and size	% of total above-ground biomass
Evergreen, tall shrubs (3–4 m)			Evergreen, caespitose dwarf-shrubs (<25 cm)		
Myrtaceae			Dilleniaceae		
Eucalyptus incrassata	BS (Mic)	37.0	*Hibbertia sericea* var. *scabrifolia*	NS/BS (Lep–Eri)	0.3
E. foecunda	BS (Nan–Mic)	28.0	*H. stricta* var. *glabriuscula*	NS (Lep–Eri)	0.2
		65.0			
Evergreen, mid-height shrubs (1–2 m)			Epacridaceae		
Myrtaceae			*Acrotriche affinis*	BS (Lep)	0.2
Baeckea behrii	NS (Lep)	2.0	*Brachyloma ericoides*	BS (Lep)	0.5
Melaleuca uncinata	NS (Lep–Nan)	4.9	*Leucopogon clelandii*	BS (Lep)	0.2
Proteaceae			Fabaceae		
Hakea muellerana	NS (Nan)	5.2	*Dillwynia hispida*	NS (Lep, Gro)	0.3
		12.1	*Pultenaea tenuifolia*	NS (Lep, Gro)	13.6

SITE 12 *(continued)*

Species	Leaf type and size	% of total above-ground biomass	Species	Leaf type and size	% of total above-ground biomass
Mimosaceae			Evergreen, caespitose graminoid herbs (<25 cm)		
Acacia farinosa	BS (Nan)	0.5	Cyperaceae		
A. spinescens	NS (Aph)	3.1	*Lepidosperma carphoides*	NS (Gra)	0.3
Myrtaceae			*L. laterale*	NS (Gra)	0.4
Baeckea crassifolia	BS/NS (Lep)	1.6			
Calytrix tetragona	NS (Lep)	1.5			**0.7**
		22.0	Other species[1]		**0.2**

[1] Other species recorded:

Monocotyledons

Cyperaceae	*Schoenus breviculmis*
Liliaceae	*Dianella revoluta, Tricoryne elatior*
Poaceae	*Stipa* sp.
Restionaceae	*Hypolaena fastigiata, Lepidobolus drapetocoleus*
Xanthorrhoeaceae	*Lomandra juncea*

Dicotyledons

Dilleniaceae	*Hibbertia virgata*
Goodeniaceae	*Dampiera rosmarinifolia, Goodenia geniculata*
Haloragaceae	*Glischrocaryon behrii*
Myrtaceae	*Leptospermum myrsinoides*
Pittosporaceae	*Billardiera sericophora*
Proteaceae	*Grevillea ilicifolia*
Rhamnaceae	*Cryptandra tomentosa, Spyridium eriocephalum, S. subochreatum* var. *laxiusculum*
Rutaceae	*Boronia caerulescens, Eriostemon pungens*

SITE 13

Billiatt, South Australia

Latitude: 34°55′S   **Longitude:** 140°35′E
    Climate: Köppen **Csb/BSk**

Climatic data[1]	Summer (Dec–Feb)	Autumn (March–May)	Winter (June–Aug)	Spring (Sept–Nov)	Year (Jan–Dec)
Solar radiation[2]	2544 (608)	1410 (337)	962 (230)	2000 (478)	1728 (413)
Max. air temp. (°C)	30.0	22.9	15.4	22.7	22.7
Min. air temp (°C)	12.5	8.7	4.4	7.7	8.3
Precipitation (mm)	64	66	97	90	317
Pan evaporation (mm)	661	287	159	384	1491
Frost (days ≤0°C)	0	0	4.8	0.2	5.0
Snow (days)	0	0	0	0	0
Cyclones	0	0	0	0	0

[1] Nearest climate station: Lameroo (35°20′S, 140°30′E).
[2] In J cm^{-2} day^{-1}, with equivalents in cal cm^{-2} day^{-1} in parentheses.

Dominant soil: Dunes of brown sand (Uc 5.11, Uc 5.12) with weak horizon development

Vegetation: Mallee–broombush
Structure: Open-scrub with heathy understorey
Reference: Specht and Cleland, 1963; Specht, 1972; see also Rowan and Downes (1963) for similar ecosystems in northwestern Victoria

Species composition

MESOPHANEROPHYTES

Mimosaceae	*Acacia rigens*
Myrtaceae	*Eucalyptus conglobata* ssp. *anceps,*

E. calycogona, E. dumosa, E. foecunda, E. incrassata, E. socialis, Melaleuca lanceolata

Santalaceae   *Santalum acuminatum, S. murrayanum*

SITE 13 *(continued)*

Species composition *(continued)*

NANOPHANEROPHYTES

Asteraceae	*Olearia lepidophylla*
Casuarinaceae	*Casuarina muellerana, C. pusilla*
Cupressaceae	*Callitris canescens, C. verrucosa*
Dilleniaceae	*Hibbertia stricta, H. virgata*
Ehretiaceae	*Halgania cyanea*
Epacridaceae	*Brachyloma ericoides, Leucopogon cordifolius*
Euphorbiaceae	*Bertya mitchellii*
Fabaceae	*Aotus ericoides, Dillwynia uncinata*
Lamiaceae	*Prostanthera aspalathoides, P. microphylla*
Mimosaceae	*Acacia brachybotrya, A. calamifolia, A. sclerophylla, A. spinescens*
Myrtaceae	*Baeckea behrii, Calytrix tetragona, Leptospermum laevigatum* var. *minus, Melaleuca acuminata, M. uncinata*
Proteaceae	*Grevillea ilicifolia, G. pterosperma*
Rhamnaceae	*Cryptandra amara, C. leucophracta, Spyridium subochreatum*
Rutaceae	*Boronia caerulescens, Phebalium bullatum*
Sapindaceae	*Dodonaea bursariifolia*
Solanaceae	*Anthocercis myosotidea*
Sterculiaceae	*Lasiopetalum behrii*
Thymelaeaceae	*Pimelea stricta*

CLIMBING NANOPHANEROPHYTES

Pittosporaceae	*Billardiera sericophora*

HUMMOCK GRASSES

Poaceae	*Triodia irritans*

CHAMAEPHYTES

Asteraceae	*Brachyscome ciliaris, Helichrysum catadromum, H. leucopsidium, Olearia ciliata, Senecio lautus*
Epacridaceae	*Astroloma humifusum*
Fabaceae	*Eutaxia microphylla*
Goodeniaceae	*Dampiera rosmarinifolia, Goodenia varia*
Haloragaceae	*Glischrocaryon behrii*
Myoporaceae	*Eremophila crassifolia*
Myrtaceae	*Baeckea crassifolia*
Stackhousiaceae	*Stackhousia monogyna*

SUCCULENT CHAMAEPHYTES

Aizoaceae	*Carpobrotus modestus*

HEMICRYPTOPHYTES

Cyperaceae	*Lepidosperma laterale, L. viscidum, Schoenus subaphyllus*
Liliaceae	*Dianella revoluta*
Xanthorrhoeaceae	*Lomandra juncea, L. leucocephala*

GEOPHYTES

Liliaceae	*Bulbinopsis semibarbata, Thysanotus patersonii*

THEROPHYTES

Brassicaceae	*Stenopetalum lineare*
Campanulaceae	*Wahlenbergia* sp.
Crassulaceae	*Crassula colorata, Crassula pedicellosa*
Euphorbiaceae	*Poranthera microphylla*

PARASITIC EPIPHYTES

Cassythaceae	*Cassytha glabella*

SITE 14

Mount Compass, South Australia

Latitude: 35 21'S   **Longitude:** 138 37'E

Climate: Köppen **Csb**

Climatic data[1]	Summer (Dec–Feb)	Autumn (March–May)	Winter (June–Aug)	Spring (Sept–Nov)	Year (Jan–Dec)
Solar radiation[2]	2385 (570)	1331 (318)	900 (215)	1891 (452)	1628 (389)
Max. air temp. (C)	25.5	20.0	13.2	18.7	19.4
Min. air temp. (C)	11.1	8.0	4.7	6.9	7.7
Precipitation (mm)	94	202	350	205	851
Pan evaporation (mm)	536	252	117	320	1225
Frost (days ≤0 C)	0	0.2	3.2	0.2	3.6
Snow (days)	0	0	0	0	0
Cyclones	0	0	0	0	0

[1] Nearest climate station: Myponga (35 24'S, 138 28'E).
[2] In J cm^{-2} day^{-1}, with equivalents in cal cm^{-2} day^{-1} in parentheses.

SITE 14 *(continued)*

Dominant soil: Leached sands (Uc 2.2, Uc 2.3) and acid swamp soils (O)
Vegetation: Mallee-heathland (*Eucalyptus baxteri–E. cosmophylla* alliance)
Reference: Adamson and Osborn, 1924; see also Baldwin and Crocker (1941), Specht and Perry (1948), Specht et al. (1961), and Martin (1961) for studies on adjacent dry sclerophyll forests and heathlands

Species composition

NANOPHANEROPHYTES

Fabaceae	*Viminaria juncea*
Myrtaceae	*Eucalyptus baxteri, E. cosmophylla, E. fasciculosa*

NANOPHANEROPHYTES

Asteraceae	*Ixodia achillaeoides*
Casuarinaceae	*Casuarina muellerana, C. striata*
Dilleniaceae	*Hibbertia exutiacies, H. sericea*
Epacridaceae	*Astroloma conostephioides, Epacris impressa*
Euphorbiaceae	*Amperea xiphoclada, Phyllanthus hirtellus, Poranthera ericoides*
Fabaceae	*Daviesia brevifolia, Dillwynia hispida, D. sericea, Platylobium obtusangulum, Pultenaea graveolens, P. largiflorens, P. villifera, Sphaerolobium vimineum*
Goodeniaceae	*Goodenia ovata*
Loganiaceae	*Logania linifolia*
Mimosaceae	*Acacia myrtifolia, A. verticillata*
Myrtaceae	*Calytrix tetragona, Leptospermum juniperinum, L. myrsinoides, Melaleuca decussata*
Proteaceae	*Adenanthos terminalis, Banksia marginata, B. ornata, Conospermum patens, Grevillea lavandulacea, Hakea rostrata, H. ulicina, Isopogon ceratophyllus, Persoonia juniperina*
Rhamnaceae	*Cryptandra tomentosa, Spyridium coactilifolium, S. spathulatum, S. vexilliferum*
Rutaceae	*Boronia caerulescens, Correa reflexa, Geijera linearifolia, Zieria veronicea*
Santalaceae	*Choretrum glomeratum*
Thymelaeaceae	*Pimelea flava, P. phylicoides, P. octophylla, P. stricta*
Tremandraceae	*Tetratheca pilosa*
Violaceae	*Hybanthus floribundus*
Xanthorrhoeaceae	*Xanthorrhoea semiplana*

CLIMBING NANOPHANEROPHYTES

Pittosporaceae	*Billardiera sericophora, Marianthus bignoniaceus*
Polygalaceae	*Comesperma volubile*

CHAMAEPHYTES

Asteraceae	*Helichrysum apiculatum, H. baxteri, H. blandowskianum, H. scorpioides, H. semipapposum, Leptorhynchos squamatus, Senecio hispidulus*
Epacridaceae	*Acrotriche serrulata, Astroloma humifusum, Leucopogon concurvus, L. virgatus*
Fabaceae	*Gompholobium ecostatum, Kennedia prostrata, Pultenaea pedunculata*
Hydrocotylaceae	*Xanthosia dissecta, X. pusilla*
Myrtaceae	*Baeckea ramosissima*
Scrophulariaceae	*Euphrasia collina*
Stackhousiaceae	*Stackhousia monogyna*
Stylidiaceae	*Stylidium graminifolium*
Thymelaeaceae	*Pimelea glauca*
Violaceae	*Viola hederacea*

HEMICRYPTOPHYTES

Cyperaceae	*Baumea acuta, B. juncea, B. rubiginosa, Gahnia lanigera, Lepidosperma semiteres, Schoenus apogon*
Juncaceae	*Juncus pallidus*
Liliaceae	*Dianella revoluta*
Poaceae	*Amphipogon caricinus, Danthonia racemosa, Neurachne alopecuroidea*
Restionaceae	*Empodisma minus, Hypolaena fastigiata, Lepidobolus drapetocoleus, Leptocarpus brownii, L. tenax*
Xanthorrhoeaceae	*Lomandra micrantha*

GEOPHYTES

Droseraceae	*Drosera planchonii, D. whittakeri*
Liliaceae	*Anguillaria dioica, Arthropodium strictum*
Orchidaceae	*Caladenia carnea, Microtis unifolia, Thelymitra azurea*

THEROPHYTES

Apiaceae	*Daucus glochidiatus*
Campanulaceae	*Wahlenbergia* sp.
Euphorbiaceae	*Poranthera microphylla*

PARASITIC EPIPHYTES

Cassythaceae	*Cassytha glabella, C. melantha*

SITE 15

Hincks-Murlong, Eyre Peninsula, South Australia

Latitude: 33 55'S   **Longitude:** 136 08'E
          Climate: Köppen **Csb/BSk**

Climatic data[1]	Summer (Dec–Feb)	Autumn (March–May)	Winter (June–Aug)	Spring (Sept–Nov)	Year (Jan–Dec)
Solar radiation[2]	2414 (577)	1410 (337)	983 (235)	1966 (470)	1695 (405)
Max. air temp. (°C)	27.2	22.2	15.7	21.7	21.7
Min. air temp. (°C)	14.9	12.2	7.3	10.0	11.1
Precipitation (mm)	54	89	174	105	422
Pan evaporation (mm)	582	315	170	389	1456
Frost (days <0 °C)	0	0.7	9.1	1.2	11.0
Snow (days)	0	0	0	0	0
Cyclones	0	0	0	0	0

[1] Nearest climate station: Ungarra (34 11'S, 136 03'E).
[2] In J cm^{-2} day^{-1}, with equivalents in cal cm^{-2} day^{-1} in parentheses.

Dominant soil: Dune formations of brown sand (Uc 5.11)
Vegetation: Mallee–broombush with some heathland
References: Crocker, 1946; Smith, 1963; Specht and Cleland, 1963; Specht, 1972

Species composition

MICROPHANEROPHYTES

Mimosaceae	*Acacia ligulata, A. rigens*
Myrtaceae	*Eucalyptus calycogona, E. dumosa* ssp. *pileata, E. flocktoniae, E. foecunda, E. incrassata, E. rugosa, Melaleuca lanceolata*
Proteaceae	*Grevillea aspera*
Santalaceae	*Exocarpos sparteus, Santalum acuminatum*

NANOPHANEROPHYTES

Asteraceae	*Ixodia achillaeoides, Olearia lepidophylla*
Casuarinaceae	*Casuarina muellerana*
Cupressaceae	*Callitris canescens, C. verrucosa*
Dilleniaceae	*Hibbertia stricta* var. *glabriuscula, H. virgata*
Ehretiaceae	*Halgania cyanea*
Epacridaceae	*Astroloma conostephioides, Leucopogon cordifolius*
Fabaceae	*Bossiaea walkeri, Daviesia genistifolia, D. pectinata, Dillwynia uncinata, Phyllota remota, Templetonia retusa*
Gyrostemonaceae	*Gyrostemon australasicus*
Lamiaceae	*Prostanthera microphylla, Westringia rigida*
Loganiaceae	*Logania ovata*
Mimosaceae	*Acacia farinosa, A. rivalis, A. spinescens*
Myrtaceae	*Baeckea behrii, Calytrix tetragona, Leptospermum laevigatum* var. *minus, Melaleuca acuminata, M. decussata, M. uncinata, Thryptomene miqueliana, Verticordia wilhelmii*

Proteaceae	*Adenanthos terminalis, Conospermum patens, Grevillea huegelii, G. ilicifolia, Hakea cycloptera, H. muellerana*
Rhamnaceae	*Cryptandra leucophracta, Pomaderris obcordata, Spyridium bifidum, S. subochreatum, S. vexilliferum*
Rutaceae	*Boronia caerulescens, B. inornata, Correa reflexa, Microcybe pauciflora, Phebalium bullatum*
Santalaceae	*Choretrum glomeratum, Exocarpos aphyllus*
Sapindaceae	*Dodonaea bursariifolia, D. hexandra*
Sterculiaceae	*Lasiopetalum behrii*
Thymelaeaceae	*Pimelea octophylla*

CHAMAEPHYTES

Asteraceae	*Brachyscome ciliaris, Helichrysum baxteri, H. catadromum, Olearia ciliata, Vittadinia triloba*
Epacridaceae	*Acrotriche cordata, A. patula, Leucopogon woodsii*
Fabaceae	*Eutaxia microphylla, Pultenaea tenuifolia*
Goodeniaceae	*Goodenia varia*
Haloragaceae	*Glischrocaryon behrii*
Myrtaceae	*Baeckea crassifolia*
Rhamnaceae	*Cryptandra tomentosa*

GEOPHYTES

Liliaceae	*Thysanotus baueri*

SITE 15 *(continued)*

Species composition *(continued)*

HEMICRYPTOPHYTES		Restionaceae	*Loxocarya fasciculata*
		Xanthorrhoeaceae	*Lomandra glauca, L. juncea*
Cyperaceae	*Gahnia deusta, G. lanigera, Lepidosperma carphoides, L. laterale, Schoenus brevi-culmis, S. racemosus*		
		PARASITIC EPIPHYTES	
Liliaceae	*Dianella revoluta*		
Poaceae	*Neurachne alopecuroidea*	Cassythaceae	*Cassytha melantha, C. pubescens*

SITE 16

Mount Bellenden Ker, North Queensland

Latitude: 17°15'S   **Longitude:** 145°52'E   **Altitude:** 1591 m
Climate: Köppen **Cwb**
Dominant soil: Shallow stony loams, chiefly Um 6.2, Um 6.4, and Um 1.43; many rock outcrops on summit
Vegetation: Closed-scrub to open-heathland
References: Domin, 1911; Gibbs, 1917; and unpublished collections made by L.S. Smith (June, 1969) and J.G. Tracey (Aug. 1959)

Species composition

MICROPHANEROPHYTES

		Cyperaceae	*Gahnia sieberana*
		Ericaceae	*Rhododendron lochae*
Apocynaceae	*Alyxia ilicifolia, A. orophila, A. ruscifolia*	Marattiaceae	*Marattia salicina*
Araliaceae	*Mackinlaya macrosciadia, Pentapanax bellendenkeriensis, P. willmottii*	Philydraceae	*Helmholtzia acorifolia*
		Rubiaceae	*Psychotria nematopoda*
Arecaceae	*Calyptrocalyx australasicus, Orania appendiculata*		
		CHAMAEPHYTES	
Balanopaceae	*Balanops australiana*		
Celastraceae	*Hypsophila halleyana*	Blechnaceae	*Blechnum* sp. (aff. *B. procerum*)
Elaeocarpaceae	*Elaeocarpus ferruginiflorus*	Orchidaceae	*Liparis reflexa*
Epacridaceae	*Dracophyllum sayeri, Trochocarpa laurina*		
		LIANAS	
Escalloniaceae	*Argophyllum cryptophlebum, Quintinia fawkneri, Q. quatrefagesii*		
		Dilleniaceae	*Hibbertia scandens*
Euphorbiaceae	*Rockinghamia angustifolia*	Monimiaceae	*Palmeria hypotephra, P. scandens*
Lauraceae	*Cinnamomum propinquum, Cryptocarya* sp. (aff. *C. corrugata*)	Smilacaceae	*Smilax glycyphylla*
		Vacciniaceae	*Agapetes meiniana*
Monimiaceae	*Wilkiea macrooraia*		
Myrsinaceae	*Rapanea* sp. (aff. *R. achradifolia*)	EPIPHYTES	
Myrtaceae	*Austromyrtus metrosideros, Eugenia apodophylla, E. erythrodoxa, E. johnsonii, Leptospermum wooroonooran*		
		Grammitidaceae	*Grammitis billardieri, Scleroglossum wooroonooran*
		Hymenophyllaceae	*Hymenophyllum baileyanum*
Pittosporaceae	*Pittosporum rubiginosum*	Orchidaceae	*Bulbophyllum lilianae, Cadetia taylori*
Proteaceae	*Lomatia fraxinifolia, Musgravea steno-stachya, Orites fragrans*	Polypodiaceae	*Crypsinus simplicissimus*
Rutaceae	*Acronychia chooreechillum*		
Sapotaceae	*Planchonella singuliflora*	PARASITIC EPIPHYTES	

NANOPHANEROPHYTES

		Loranthaceae	*Amyema whitei*
		Viscaceae	*Korthalsella japonica*
Arecaceae	*Linospadix palmeranus*		
Celastraceae	*Hypsophila oblonga* (syn. *Drimys oblonga*)	HEMICRYPTOPHYTES	
Cyatheaceae	*Cyathea rebeccae, C. robertsoniana*	Cyperaceae	*Exocarya scleroides*

SITE 17

Mount Kosciusko, New South Wales

Latitude: 36°26′S   **Longitude:** 148°16′E
Altitude: Mount Kosciusko Summit — 2230 m
     Climate: Köppen **Cfc**

Climatic data[1]	Summer (Dec–Feb)	Autumn (March–May)	Winter (June–Aug)	Spring (Sept–Nov)	Year (Jan–Dec)
Solar radiation[2]	2372 (567)	1326 (317)	833 (199)	1895 (453)	1607 (384)
Max. air temp. (°C)	18.1	11.7	4.3	11.3	11.4
Min. air temp. (°C)	5.6	1.5	−3.4	1.1	1.2
Precipitation (mm)	258	284	343	353	1238
Pan evaporation (mm)	306	176	108	210	800
Frost (days ≤0°C)	7.0	33.0	78.5	39.0	157.5
Snow (days)	1.4	10.8	42.0	17.9	72.1
Cyclones	0	0	0	0	0

[1] Climate station: Kosciusko Chalet.
[2] In J cm^{-2} day^{-1}, with equivalents in cal cm^{-2} day^{-1} in parentheses.

Dominant soils: Dissected plateaux at high elevation; boulder strewn ridges and high plains of organic loamy soils (Um 7.11); small areas of brown podzols (Um 4.2) and sandy soils (Uc) with small swampy valley plains of acid peaty soils (O)

Vegetation: Alpine–subalpine heathland
Structure: Low closed-heathland   **Alliance:** *Oxylobium ellipticum–Podocarpus lawrencei*
References: Costin, 1954, 1957

Species composition

NANOPHANEROPHYTES

Physiognomic dominants

Epacridaceae	*Leucopogon suaveolens*
Fabaceae	*Bossiaea foliosa, Hovea longifolia, Oxylobium ellipticum*
Lamiaceae	*Prostanthera cuneata*
Mimosaceae	*Acacia alpina*
Myrtaceae	*Baeckea gunniana, Callistemon sieberi, Leptospermum phylicoides*
Podocarpaceae	*Podocarpus lawrencei*
Proteaceae	*Orites lancifolia*
Rutaceae	*Phebalium ovatifolium*
Winteraceae	*Tasmannia (Drimys) vickerana*

Associated species

Araliaceae	*Tieghemopanax sambucifolius*
Asteraceae	*Cassinia aculeata, Helichrysum backhousii, H. rosmarinifolium, Olearia flavescens, O. floribunda, O. subrepanda*
Epacridaceae	*Epacris breviflora, E. paludosa, E. robusta Leucopogon attenuatus*
Lamiaceae	*Prostanthera phylicifolia, Westringia rubiifolia* var.
Myrtaceae	*Leptospermum lanigerum*
Proteaceae	*Grevillea victoriae, Hakea lissosperma*
Rutaceae	*Boronia algida, Phebalium ozothamnoides*
Thymelaeaceae	*Pimelea ligustrina*

CHAMAEPHYTES

Physiognomic dominants

Epacridaceae	*Leucopogon montanus*

Associated species

Asteraceae	*Helichrysum hookeri*
Epacridaceae	*Pentachondra pumila, Richea continentis*
Fabaceae	*Oxylobium alpestre*
Goodeniaceae	*Scaevola hookeri*
Myrtaceae	*Kunzea muelleri*
Onagraceae	*Epilobium billardierianum*
Proteaceae	*Grevillea australis*
Rosaceae	*Acaena anserinifolia*
Rubiaceae	*Galium gaudichaudii, Nertera depressa*
Santalaceae	*Exocarpos nanus*
Stylidiaceae	*Stylidium graminifolium*
Thymelaeaceae	*Pimelea alpina, P. axiflora* var. *alpina*
Violaceae	*Hymenanthera dentata*

HEMICRYPTOPHYTES

Associated species

Linaceae	*Linum marginale*
Poaceae	*Danthonia frigida, Poa australis* sp. agg.
Rosaceae	*Alchemilla novae-hollandiae, Geum urbanum*

SITE 17 *(continued)*

Species composition *(continued)*

PTERIDOPHYTES

Associated species
Blechnaceae   *Blechnum penna-marina*

THEROPHYTES

Associated species
Brassicaceae   *Barbarea stricta*

Vegetation: Alpine–subalpine (raised-bog) heathland
Structure: Low open-heathland   **Alliance:** *Epacris paludosa–Sphagnum cristatum*
Reference: Costin, 1954, 1957

Species composition

NANOPHANEROPHYTES

Physiognomic dominants
Epacridaceae   *Epacris paludosa*
Myrtaceae   *Callistemon sieberi*

Associated species
Asteraceae   *Cassinia aculeata, Helichrysum backhousii, Olearia floribunda*
Myrtaceae   *Baeckea gunniana*
Proteaceae   *Hakea microcarpa*
Thymelaeaceae   *Pimelea ligustrina* var. *glabra*

CHAMAEPHYTES

Physiognomic dominants
Epacridaceae   *Epacris serpyllifolia, Richea continentis*

Associated species
Asteraceae   *Parantennaria uniceps, Cotula filicula, Helichrysum hookeri*
Campanulaceae   *Wahlenbergia* sp.
Epacridaceae   *Epacris microphylla, Leucopogon collinus*
Lobeliaceae   *Pratia pedunculata, P. surrepens*
Lycopodiaceae   *Lycopodium fastigiatum, L. selago*
Onagraceae   *Epilobium billardierianum*
Polygalaceae   *Comesperma retusum*
Rubiaceae   *Nertera depressa*
Scrophulariaceae   *Euphrasia glacialis*
Stackhousiaceae   *Stackhousia pulvinaris*
Stylidiaceae   *Stylidium graminifolium*

HEMICRYPTOPHYTES

Physiognomic dominants
Liliaceae   *Astelia alpina* var. *novaehollandiae*

Associated species
Apiaceae   *Oreomyrrhis pulvinifica*
Asteraceae   *Celmisia asteliifolia*
Cyperaceae   *Oreobolus distichus, O. pumilio*
Hydrocotylaceae   *Dichosciadium ranunculaceum*
Plantaginaceae   *Plantago muelleri* (two forms)
Poaceae   *Hemarthria uncinata, Poa australis* sp. agg.

GEOPHYTES

Associated species
Hydrocotylaceae   *Diplaspis hydrocotylea*
Hypoxidaceae   *Hypoxis hygrometrica*
Liliaceae   *Arthropodium milleflorum, Thysanotus tuberosus*
Orchidaceae   *Eriochilus cucullatus, Prasophyllum alpinum, P. suttonii, Prasophyllum* sp., *Thelymitra venosa*

HELOPHYTES

Physiognomic dominants
Cyperaceae   *Carex gaudichaudiana, Carpha nivicola*
Restionaceae   *Restio australis*
Sphagnaceae   *Sphagnum cristatum*

Associated species
Apiaceae   *Aciphylla simplicifolia, Oreomyrrhis eriopoda*
Asteraceae   *Brachycome decipiens, B. radicans, B. scapigera, Craspedia glauca*
Cyperaceae   *Carex blakei, C. cephalotes, C. curta, C. hebes, Schoenus apogon, Scirpus merrillii*
Droseraceae   *Drosera arcturi, D. peltata*
Geraniaceae   *Geranium sessiliflorum*
Haloragaceae   *Gonocarpus micranthus*
Juncaceae   *Juncus antarcticus, J. falcatus*
Lamiaceae   *Prunella vulgaris*
Lentibulariaceae   *Utricularia dichotoma*
Poaceae   *Agrostis hiemalis, Deyeuxia quadriseta*
Ranunculaceae   *Ranunculus pimpinellifolius*
Restionaceae   *Empodisma minus*

THEROPHYTES

Associated species
Gentianaceae   *Gentianella diemensis*
Hypericaceae   *Hypericum japonicum*
Poaceae   *Agrostis aequata, A. muellerana*

SITE 18

Lake Mountain, Victoria

Latitude: 37°25′S   **Longitude:** 145°52′E   **Altitude:** 1400 m

Diagrammatic cross-section (D.H. Ashton and R.L. Specht unpublished survey 1963):

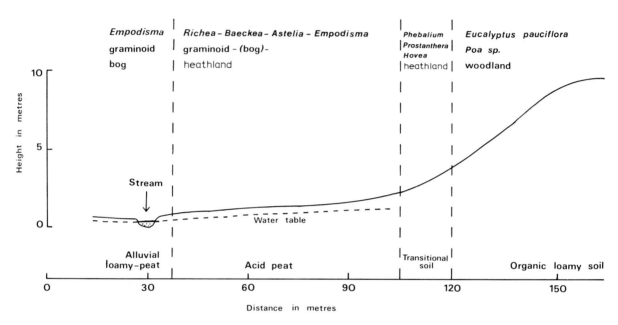

Fig. 6.18. Diagrammatic cross-section, Lake Mountain, Victoria.

Climate: Köppen **Cfc**

Climatic data[1]	Summer (Dec–Feb)	Autumn (March–May)	Winter (June–Aug)	Spring (Sept–Nov)	Year (Jan–Dec)
Solar radiation[2]	2531 (605)	1393 (333)	858 (205)	2000 (478)	1695 (405)
Max. air. temp. (°C)	18.6	11.9	4.4	11.0	11.5
Min. air temp. (°C)	9.9	5.6	−0.3	4.1	4.9
Precipitation (mm)	284	426	712	507	1929
Pan evaporation (mm)	363	177	102	216	858
Frost (days⩽0°C)	1.1	6.8	47.4	15.8	71.1
Snow (days)[3]	0	2.0	20.0	8.0	30.0
Cyclones	0	0	0	0	0

[1] Nearest climate station: Mount Buffalo (36°47′S, 146°46′E).

[2] In J cm^{-2} day^{-1}, with equivalents in cal cm^{-2} day^{-1} in parentheses.

[3] Based on only three years of observations.

SITE 18 *(continued)*

Dominant soils: Organic loamy soils (Um 7.11) grading to acid peaty soils (O).
Vegetation: Subalpine heathland
Age: 24 years since fire   **Biomass** (dry wt.): 2.6 tonnes ha^{-1}
Structure: Low open-heathland   **Reference:** D.H. Ashton and R.L. Specht (harvest Feb. 1963)

Species	Leaf type and size	% of total above-ground biomass	Species	Leaf type and size	% of total above-ground biomass
Evergreen, low shrubs (25–100 cm)			Rubiaceae		
			Asperula gunnii	(Lep)	0.6
Asteraceae					**1.5**
Helichrysum hookeri	BS (Lep)	22.5			
Olearia floribunda	NS (Lep)	0.1	Evergreen, caespitose graminoid herbs (<25 cm)		
Fabaceae			Cyperaceae		
Hovea longifolia var. *montana*	BS (Nan)	4.2	*Carex breviculmis*	NS (Gra)	0.4
Lamiaceae			Liliaceae		
Prostanthera cuneata	BS (Lep)	22.6	*Herpolirion novae-zealandiae*	NS (Gra)	1.7
Rutaceae			Poaceae		
Phebalium phylicifolium	NS (Lep, Eri)	17.5	*Poa australis* var. *alpina*	NS (Gra)	29.1
		66.9	Stylidiaceae		
			Stylidium graminifolium	BS (Gra)	0.2
					31.4
Evergreen forbs (<25 cm)					
Apiaceae			Ground liverworts		
Oreomyrrhis eriopoda	(Pin Lep)	0.1	*Lophocolea heterophylloides*	–	0.2
Asteraceae					**0.2**
Gnaphalium japonicum	(Lep/Nan)	0.7			
Senecio quadridentatus	(Nan)	0.1			

Vegetation: Subalpine bog-heathland
Age: 24 years since fire   **Biomass** (dry wt,): 2.0 tonnes ha^{-1}
Structure: Low open graminoid-heathland   **Reference:** D.H. Ashton and R.L. Specht (harvest Feb. 1963)

Species	Leaf type and size	% of total above-ground biomass	Species	Leaf type and size	% of total above-ground biomass
Evergreen, low shrubs (25–100 cm)			*Asperula gunnii*	BS (Lep)	0.1
Epacridaceae					**0.1**
Epacris paludosa	BS (Lep)	6.1			
Richea continentis	BS (Mic)	24.5	Evergreen, caespitose graminoid herbs (<25 cm)		
Myrtaceae			Asteraceae		
Baeckea gunniana	NS (Lep, Gro)	9.2	*Celmisia asteliifolia*	NS (Gra)	2.7
		39.8	Cyperaceae		
			Oreobolus pumilio	NS (Gra)	1.8
Evergreen forbs (<25 cm)			Liliaceae		
Rubiaceae			*Astelia alpina*	BS (Gra)	12.0

SITE 18 *(continued)*

Species	Leaf type and size	% of total above-ground biomass	Species	Leaf type and size	% of total above-ground biomass
Poaceae			Evergreen, ground mosses		
Poa australis var. *alpina*	NS(Gra)	0.9	Sphagnaceae		
Restionaceae			*Sphagnum cristatum*	(Lep)	0.6
Empodisma minus	NS (Aph)	41.6			
					0.6
		59.0	Miscellaneous species		**0.5**

SITE 19A

Tasmanian subalpine to alpine heathlands — Cradle Mountain, Tasmania

Latitude: 41 38′S   **Longitude:** 145 57′E
Altitude: 1545 m (max. altitude — Mount Ossa 1617 m)
Diagrammatic cross-section (after Jackson 1965):

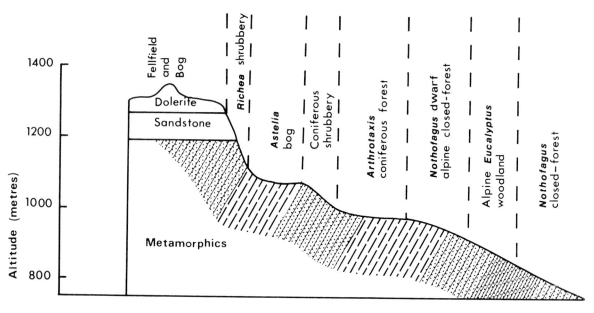

Fig. 6.19. Diagrammatic cross-section, Cradle Mountain, Tasmania. Horizontal scale 4–6 km.

SITE 19 *(continued)*

Climate: Köppen **Cfc**

Climatic data[1]	Summer (Dec–Feb)	Autumn (March–May)	Winter (June–Aug)	Spring (Sept–Nov)	Year (Jan–Dec)
Solar radiation[2]	2142 (512)	1096 (262)	661 (158)	1674 (400)	1393 (333)
Max. air temp. (°C)	17.8	12.0	6.6	12.5	12.2
Min. air temp. (°C)	4.4	2.0	−0.7	0.9	1.7
Precipitation (mm)	464	617	899	750	2730
Pan evaporation (mm)	–	–	–	–	–
Frost (days ⩽0°C)	6.1	29.7	69.9	39.4	145.1
Snow (days)	3.1	8.0	23.9	18.7	53.7
Cyclones	0	0	0	0	0

[1] Climate station: Waldheim Chalet.
[2] In J cm^{-2} day^{-1}, with equivalents in cal cm^{-2} day^{-1} in parentheses.

Dominant soils: Rugged plateaux and mountains at high elevation; alpine humus soils (Um 7.12) in association with acid peaty soils (O), leached sands (Uc 2.3) and shallow sand soils (Uc 4.11)

SITE 19B

Mount Wellington, Tasmania

Latitude: 42°54′S   **Longitude:** 147°17′E   **Altitude:** 1269 m
Diagrammatic cross-section (after Martin, 1940):

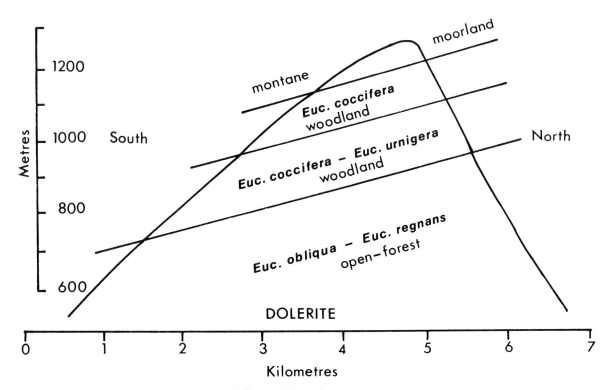

Fig. 6.20. Diagrammatic cross-section, Mount Wellington, Tasmania.

SITE 19B *(continued)*

Climate: Köppen **Cfc**

Climatic data[1]	Summer (Dec–Feb)	Autumn (March–May)	Winter (June–Aug)	Spring (Sept–Nov)	Year (Jan–Dec)
Solar radiation[2]	2125 (508)	1059 (253)	669 (160)	1678 (401)	1385 (331)
Max. air temp. (°C)	15.5	11.8	7.1	11.3	11.4
Min. air temp. (°C)	6.7	4.9	1.8	3.3	4.2
Precipitation (mm)	336	356	313	401	1406
Pan evaporation (mm)	272	154	107	198	731
Frost (days ⩽0°C)	0.5	3.2	19.9	10.5	34.1
Snow (days)	5.6	11.6	23.3	20.9	61.4
Cyclones	0	0	0	0	0

[1] Climate station: The Springs
[2] In $J\ cm^{-2}\ day^{-1}$, with equivalents in $cal\ cm^{-2}\ day^{-1}$ in parentheses.

Dominant soils: Alpine humus soils (Um 7.12) in association with acid peaty soils (O); many rock outcrops, bare rock walls and rock scree

Vegetation: (1) fell-field* (Sutton, 1929); (2) subalpine scrub (Sutton, 1929); (3) austral-montane heathland (Martin, 1940); (4) *Eucalyptus coccifera* scrub (Martin, 1940)

Fell-field (synonymous with *Feldmark* or *Fjaeldmark*) in an open, subglacial community of dwarf-flowering plants, mosses and lichens, usually dominated physiognomically by chamaephytes (Costin, 1954).

Species	Leaf size	Cradle Mountain		Mount Wellington	
		1	2	3	4
NANOPHANEROPHYTES					
Asteraceae					
Helichrysum antennarium	Nan	–	–	o	o
H. backhousii	Nan	o	o	o	r
H. gunnii	Nan (Gro)	–	–	r	–
H. hookeri	Lep	–	–	f	–
H. ledifolium	Lep (Rev)	c	o	vc	f
H. obcordatum	Lep	–	o	–	–
H. thyrsoideum	Nan (Rev)	–	r	–	–
Olearia alpina	Nan	–	r	o	o
O. floribunda	Lep	–	–	o	r
O. ledifolia	Lep–Nan (Rev)	f	–	o	o
O. lepidophylla	Lep	–	–	r	vr
O. obcordata	Nan	–	o	o	–
O. pinifolia	Lep–Nan (Rev)	o	–	o	f
O. stellulata	Mic	–	–	–	f
Baueraceae					
Bauera rubioides	Lep	–	f	f	f
Casuarinaceae					
Casuarina monilifera	(Aph)	–	f	–	–
Cupressaceae					
Diselma archeri	Cup	o	–	–	–

SITE 19B *(continued)*

Species	Leaf size	Cradle Mountain		Mount Wellington	
		1	2	3	4
Epacridaceae					
Archeria hirtella	Lep	r	–	–	–
A. serpyllifolia	Lep	o	o	–	–
Cyathodes glauca	Nan	–	–	–	o
C. juniperina	Lep	o	f	–	o
C. straminea	Lep–Nan	f	–	–	–
Epacris exserta	Lep	r	–	–	–
E. gunnii	Lep	–	f	–	–
E. impressa	Lep	–	f	–	–
E. lanuginosa	Lep	–	f	–	–
E. myrtifolia	Lep	r	o	–	–
E. serpyllifolia	Lep	–	f	f	o
Leucopogon collinus	Lep	o	–	–	–
L. ericoides	Lep	o	–	–	–
L. milliganii	Lep	o	–	–	–
L. montanus	Lep	o	–	o	–
Monotoca empetrifolia	Lep	–	–	o	o
M. scoparia	Lep	–	r	–	–
Richea acerosa	Lep	o	–	vc	c
R. gunnii	Nan–Mic	–	–	c	c
R. pandanifolia	Mac	r	o	–	–
R. procera	Lep–Nan	–	–	o	o
R. scoparia	Mic	c	o	c	c
R. sprengelioides	Lep	f	o	o	o
Sprengelia incarnata	Lep	f	f	r	o
Trochocarpa thymifolia	Lep	–	–	vc	vc
Ericaceae					
Gaultheria hispida	Mic	–	–	–	r
Fabaceae					
Bossiaea cordigera	Lep	–	o	–	–
Oxylobium ellipticum	Nan	f	lc	–	o
Pultenaea juniperina	Lep	–	r	–	–
P. subumbellata	Lep	–	f	–	–
Fagaceae					
Nothofagus gunnii	Nan–Mic	o	–	–	–
Mimosaceae					
Acacia mucronata	Nan–Mic	r	–	–	–
Myrtaceae					
Baeckea gunniana	Lep	–	–	vc	f
B. leptocaulis	Lep	f	lc	–	–
Eucalyptus coccifera	Mic	–	f	–	d
E. johnstonii	Mic	–	f	–	–
E. vernicosa	Nan–Mic	–	r	–	–
Leptospermum humifusum	Lep	o	o	vc	o
L. lanigerum	Lep–Nan	–	o	–	lc
L. scoparium	Nan	o	f	–	–
L. sericeum	Lep–Nan	f	f	–	–
Melaleuca squamea	Lep	f	lc	–	–
Pittosporaceae					
Pittosporum bicolor	Nan–Mic	–	–	–	r

SITE 19B *(continued)*

Species	Leaf size	Cradle Mountain		Mount Wellington	
		1	2	3	4
Podocarpaceae					
Microstrobos niphophilus	Cup	o	–	–	–
Phyllocladus aspleniifolius	Cla (Nan–Mic)	o	–	→	–
Podocarpus lawrencei	Lep	f	–	lc	–
Polygalaceae					
Comesperma retusum	Nan	–	o	–	–
Proteaceae					
Bellendena montana	Nan–Mic	o	c	vc	c
Cenarrhenes nitida	Mic	o	o	–	–
Grevillea australis	Lep (rev)	–	–	–	vr
Lomatia polymorpha	Nan–Mic	–	–	vr	vr
Orites acicularis	Lep–Nan	f	o	vc	c
O. diversifolia	Mic	–	–	–	r
O. revoluta	Nan	f	o	c	c
Persoonia gunnii	Nan–Mic	o	o	–	–
Telopea truncata	Mic–Mes	o	o	–	o
Rubiaceae					
Coprosma nitida	Lep–Nan	f	f	f	c
Rutaceae					
Boronia citriodora	Lep (leaflets)	f	lc	–	–
B. rhomboidea	Lep	o	lc	–	–
Correa lawrenciana	Nan–Mic	–	–	–	o
C. reflexa	Nan–Mic	–	–	–	o
Phebalium oldfieldii	Lep	–	r	–	–
P. squameum	Lep	–	r	–	–
Scrophulariaceae					
Veronica formosa	Lep–Nan	–	–	–	r
Tetracarpaeaceae					
Tetracarpaea tasmanica	Nan	o	o	o	r
Thymelaeaceae					
Pimelea linifolia	Lep	–	o	–	–
P. sericea	Lep	–	–	c	o
Tremandraceae					
Tetratheca pilosa	Lep (Rev)	–	o	–	–
Winteraceae					
Tasmannia (Drimys) lanceolata	Mic Mes	o	o	f	c
CHAMAEPHYTES (cushion plants)					
Asteraceae					
Abrotanella forsterioides	Lep	–	–	o	–
A. scapigera	Lep	f	–	–	–
Ewartia meridithae	Lep	lc	–	–	–
E. planchonii	Lep	r	–	r	–
Pterygopappus lawrencii	Lep	c	–	–	–
Donatiaceae					
Donatia novae-zelandiae	Lep	lc	–	–	–
Epacridaceae					
Dracophyllum minimum	Lep	lc	–	–	–

SITE 19B *(continued)*

Species	Leaf size	Cradle Mountain		Mount Wellington	
		1	2	3	4

CHAMAEPHYTES

Apiaceae
Aciphylla procumbens | | | o | – | – | – |
Oreomyrrhis argentea | | | o | – | r | – |

Asteraceae
Celmisia longifolia | | | f | o | vc | c |
Cotula alpina | | | o | – | – | – |
C. australis | | | o | – | – | – |
C. filicula | | | – | – | r | – |
C. reptans | | | o | – | – | – |
Craspedia alpina | | | – | – | vc | – |
C. glauca | | | f | – | – | – |
Erechtites gunnii | | | – | – | o | r |
Erigeron pappocroma | | | f | – | o | r |
Gnaphalium umbricola | | | o | – | r | r |
Helichrysum apiculatum | | | – | – | c | o |
H. bracteatum var. *viscosum* | | | r | – | – | – |
H. leucopsideum | | | r | – | – | – |
H. milliganii | | | c | – | – | – |
H. pumilum | | | c | – | – | – |
H. scorpioides | | | – | – | c | o |
Lagenifera stipitata | | | r | – | – | – |
Senecio lautus | | | – | – | f | o |
S. pectinatus | | | c | – | – | – |

Boraginaceae
Myosotis suaveolens | | | – | – | r | – |

Brassicaceae
Cardamine dictyosperma | | | o | – | – | – |
C. tenuifolia | | | – | – | f | f |
Cheesemannia radicata | | | r | – | – | – |

Campanulaceae
Wahlenbergia saxicola | | | r | – | o | – |

Caryophyllaceae
Colobanthus apetalus | | | o | – | – | – |

Dilleniaceae
Hibbertia procumbens | | | r | c | – | – |

Droseraceae
Drosera arcturi | | | f | – | f | – |

Epacridaceae
Cyathodes dealbata | | | f | – | o | – |
C. petiolaris | | | f | – | vc | c |
Pentachondra involucrata | | | – | – | o | r |
P. pumila | | | f | – | vc | – |

Ericaceae
Gaultheria depressa | | | o | – | – | – |
Pernettya tasmanica | | | r | – | – | – |

Euphorbiaceae
Poranthera microphylla | | | – | – | o | – |

SITE 19B *(continued)*

Species	Leaf size	Cradle Mountain		Mount Wellington	
		1	2	3	4
Gentianaceae					
Gentianella diemensis		f	o	f	o
Geraniaceae					
Geranium dissectum		o	–	f	f
G. sessiliflorum		–	–	o	o
Pelargonium australe		–	r	–	–
Goodeniaceae					
Scaevola hookeri		–	o	r	–
Haloragaceae					
Gonocarpus micranthus		–	–	r	r
G. teucrioides		–	–	–	r
Hydrocotylaceae					
Actinotus bellidioides		–	f	–	–
A. moorei		r	–	–	–
A. suffocata		r	f	–	–
Dichosciadium ranunculaceum		f	–	–	–
Diplaspis cordifolia		f	–	–	–
Hydrocotyle javanica		o	–	–	–
Hypericaceae					
Hypericum gramineum		o	–	–	–
Onagraceae					
Epilobium billardierianum		r	–	–	–
E. gunnianum		–	–	o	o
E. tasmanicum		–	–	o	–
Oxalidaceae					
Oxalis lactea		o	–	r	o
Plantaginaceae					
Plantago gunnii		r	–	–	–
P. tasmanica		o	r	o	–
P. triantha		o	r	–	–
Podocarpaceae					
Microcachrys tetragona		lc·	–	–	–
Portulacaceae					
Montia australasica		o	–	–	–
Ranunculaceae					
Anemone crassifolia		o	–	–	–
Caltha phylloptera		lc	–	–	–
Ranunculus gunnianus		o	–	–	–
R. lappaceus		r	–	–	–
Rosaceae					
Acaena anserinifolia		–	–	o	o
Rubus gunnianus		o	o	o	o
Rubiaceae					
Coprosma moorei		–	–	r	–
C. pumila		o	–	–	–
Rutaceae					
Boronia nana		–	r	–	–

SITE 19B *(continued)*

Species	Leaf size	Cradle Mountain		Mount Wellington	
		1	2	3	4
Santalaceae					
Exocarpos humifusus		o	r	c	o
Scrophulariaceae					
Euphrasia collina		f	f	vc	c
E. hookeri		o	f	–	–
Ourisia integrifolia		r	–	–	–
Veronica nivea		o	–	r	r
Spigeliaceae					
Mitrasacme montana		o	f	o	–
Stackhousiaceae					
Stackhousia monogyna		–		–	r
Stylidiaceae					
Stylidium graminifolium		f	c	o	–
Thymelaeaceae					
Drapetes tasmanica		r	r	–	–
Violaceae					
Viola hederacea		–	–	–	c
PTERIDOPHYTA					
Aspidiaceae					
Polystichum proliferum		–	–	o	o
Aspleniaceae					
Pleurosorus rutifolius		–		vr	r
Blechnaceae					
Blechnum penna-marina		r	–	o	o
Doodia caudata		–	–	–	r
Gleicheniaceae					
Gleichenia circinnata		f	f	vc	c
Hemionitidaceae					
Anogramma leptophylla		–		r	–
Hymenophyllaceae					
Hymenophyllum flabellatum		–	–	vr	–
Lycopodiaceae					
Lycopodium fastigiatum		–	–	f	f
L. deuterodensum		–	–	o	–
L. laterale		r	–	f	r
L. myrtifolium		–	–	r	–
L. scariosum		r	–	r	–
L. selago		o	–	–	r
Selaginellaceae					
Selaginella uliginosa		–		f	o
HEMICRYPTOPHYTES					
Centrolepidaceae					
Centrolepis monogyna		f	–	–	–
Trithuria filamentosa		r	–	–	–

SITE 19B *(continued)*

Species	Leaf size	Cradle Mountain		Mount Wellington	
		1	2	3	4
Cyperaceae					
Carex gaudichaudiana		vr	–	–	–
Carpha alpina		o	–	vc	o
Gymnoschoenus adjustus		–	f	–	–
Oreobolus pumilio		c	–	f	–
Scirpus antarcticus		o	–	–	–
S. cernuus		o	–	–	–
S. crassiusculus		o	–	f	–
S. inundatus		–	–	o	–
Tetraria capillaris		–	o	–	–
Uncinia compacta		–	–	f	r
U. tenella		–	–	r	–
Iridaceae					
Diplarrena moraea		o	–	–	–
Sisyrinchium pulchellum		–	r	–	–
Patersonia fragilis		–	o	–	–
Juncaceae					
Luzula campestris		f	–	o	–
L. oldfieldii		–	–	c	o
Liliaceae					
Astelia alpina		c	–	vc	lc
Herpolirion novae-zelandiae		r	–	–	–
Milligania densiflora		lc	–	–	–
M. lindoniana		f	–	–	–
M. longifolia		f	–	–	–
Poaceae					
Agrostis parviflora		–	–	r	vr
Danthonia sp.		–	–	r	r
Hierochloe fraseri		–	–	c	o
H. redolens		r	–	o	o
Microlaena stipoides var. *subalpina*		–	–	vr	–
Poa australis sp.		o	f	–	–
Restionaceae					
Empodisma minus		f	c	vc	c
Hypolaena fastigiata		o	o	–	–
Restio australis		o	f	lc	–
R. complanatus		o	f	–	–
Xyridaceae					
Xyris operculata		–	f	–	–
GEOPHYTES					
Hypoxidaceae					
Campynema lineare		f	o	–	–
Liliaceae					
Blandfordia marginata		r	–	–	–
Orchidaceae					
Acianthus caudatus		–	–	r	–
Caladenia angustata		–	–	–	r
Prasophyllum fuscum		–	–	o	–
P. patens		–	–	r	–
Pterostylis cucullata		–	–	vr	vr

REFERENCES

Adams, L.C., Byrnes, N. and Lazarides, M., 1973. Floristics of the Alligator Rivers Area. Part XI (76 pages). In: *Alligator Rivers Region Environmental Fact-finding Study: Physical Features and Vegetation, II.* CSIRO, Canberra, A.C.T.

Adamson, R.S. and Osborn, T.G.B., 1924. The ecology of *Eucalyptus* forests of the Mount Lofty Ranges (Adelaide District), South Australia. *Trans. R. Soc. S. Aust.,* 48: 87–144.

Bailey, F.M., 1899–1902. *The Queensland Flora* Parts I–VI. H.J. Diddams and Co., Brisbane, Qld., 2015 pp.

Baker, H.A. and Oliver, E.G.H., 1967. *Ericas in Southern Africa.* Purnell and Sons, Cape Town, 180 pp.

Baldwin, J.G. and Crocker, R.L., 1941. The soils and vegetation of portion of Kangaroo Island, South Australia. *Trans. R. Soc. S. Aust.,* 65: 263–275.

Balme, B.E. and Churchill, D.M., 1959. Tertiary sediments at Coolgardie, Western Australia. *J. R. Soc. West. Aust.,* 42: 37–43.

Beadle, N.C.W., 1940. Soil temperatures during forest fires and their effect on the survival of vegetation. *J. Ecol.,* 28: 180–192.

Beadle, N.C.W., 1953. The edaphic factor in plant ecology with a special note on soil phosphates. *Ecology,* 34: 426–428.

Beadle, N.C.W., 1954. Soil phosphate and the delimitation of plant communities in eastern Australia. *Ecology,* 35: 370–375.

Beadle, N.C.W., 1962. Soil phosphate and the delimitation of plant communities in eastern Australia. II. *Ecology,* 43: 281–288.

Beadle, N.C.W., 1968. Some aspects of the ecology and physiology of Australian xeromorphic plants. *Aust. J. Sci.,* 30: 348–355.

Beadle, N.C.W. and Burges, A., 1949. Working capital in a plant community. *Aust. J. Sci.,* 11: 207–208.

Beadle, N.C.W., Evans, O.D. and Carolin, R.C., 1972. *Flora of the Sydney Region.* Reed, Sydney, N.S.W., 724 pp.

Beard, J.S. (Editor), 1970. *West Australian Plants.* Society for Growing Australian Plants Publ., Sydney, 2nd ed., 142 pp.

Berg, R.Y., 1975. Myrmecochorous plants in Australia and their dispersal by ants. *Aust. J. Bot.,* 23: 475–508.

Bergersen, F.J. and Costin, A.B., 1964. Root nodules on *Podocarpus lawrencei* and their ecological significance. *Aust. J. Biol. Sci.,* 17: 44–48.

Bevege, D.I., 1968. A rapid technique for clearing tannins and staining intact roots for detection of mycorrhizas caused by *Endogene* spp., and some records of infection in Australasian plants. *Trans. Br. Mycol. Soc.,* 51: 808–810.

Black, J.M., 1943–57. *Flora of South Australia,* I–IV. Government Printer, Adelaide, S.A., 2nd ed., 1008 pp.

Blake, S.T., 1968. The plants and plant communities of Fraser, Moreton and Stradbroke Islands. *Qld. Nat.,* 19: 23–30.

Burrell, J.P., 1969. *The Invasion of Coastal Heathlands of Victoria by* Leptospermum laevigatum. Thesis, University of Melbourne, Melbourne, Vic., 199 pp.

Byrnes, N., Everist, S.L., Reynolds, S.T., Specht, A. and Specht, R.L., 1977. Vegetation of Lizard Island, North Queensland. *Proc. R. Soc. Qld.,* 88: 1–15.

Chippendale, G.M., 1971. Check list of Northern Territory plants. *Proc. Linn. Soc. N.S.W.,* 96: 207–267.

Clark, S.S., 1975. The effect of sand mining on coastal heath vegetation in New South Wales. *Proc. Ecol. Soc. Aust.,* 9: 1–16.

Clifford, H.T. and Specht, R.L., 1979. *The Vegetation of North Stradbroke Island (With Notes on the Fauna of Mangrove and Marine Meadow Ecosystems by Marion M. Specht).* Qld. Univ. Press, St. Lucia, Qld., 141 pp.

Coaldrake, J.E., 1951. The climate, geology, soils, and plant ecology of portion of the County of Buckingham (Ninety-Mile Plain), South Australia. *CSIRO Aust. Bull.,* No. 266: 81 pp.

Coaldrake, J.E., 1961. The ecosystem of the coastal lowlands ('Wallum') of southern Queensland. *CSIRO Aust. Bull.,* No. 283: 148 pp.

Coaldrake, J.E. and Haydock, K.P., 1958. Soil phosphate and vegetal pattern in some natural communities of south eastern Queensland, Australia. *Ecology,* 39: 1–5.

Connor, D.J. and Clifford, H.T., 1972. The vegetation near Brown Lake, North Stradbroke Island. *Proc. R. Soc. Qld.,* 83: 69–82.

Connor, D.J. and Wilson, G.L., 1968. Response of a coastal Queensland heath community to fertilizer application. *Aust. J. Bot.,* 16: 117–123.

Cookson, I.C., 1945. Pollen content of Tertiary deposits. *Aust. J. Sci.,* 7: 149–150.

Cookson, I.C., 1946. Pollens of *Nothofagus* Blume from Tertiary deposits in Australia. *Proc. Linn. Soc. N.S.W.,* 71: 49–63.

Cookson, I.C., 1947. On fossil leaves (Oleaceae) and a new type of fossil pollen grain from Australian brown coal deposits. *Proc. Linn. Soc. N.S.W.,* 72: 183–197.

Cookson, I.C., 1950. Fossil pollen grains of proteaceous type from Tertiary deposits in Australia. *Aust. J. Sci. Res., Ser. B,* 3: 166–177.

Cookson, I.C., 1952. Identification of Tertiary pollen grains with those of New Guinea and New Caledonian beeches. *Nature, Lond.,* 170: 127.

Cookson, I.C., 1953. The identification of the sporomorph *Phyllocladidites* with *Dacrydium* and its distribution in southern Tertiary deposits. *Aust. J. Bot.,* 1: 64–70.

Cookson, I.C., 1954a. The occurrence of an older Tertiary microflora in Western Australia. *Aust. J. Sci.,* 17: 37–38.

Cookson, I.C., 1954b. The Cainozoic occurrence of *Acacia* in Australia. *Aust. J. Bot.,* 2: 52–59.

Cookson, I.C., 1957. On some Australian Tertiary spores and pollen grains that extend the geological and geographical distribution of living genera. *Proc. R. Soc. Vic.,* 69: 41–53.

Cookson, I.C., 1959. Fossil pollen grains of *Nothofagus* from Australia. *Proc. R. Soc. Vic.,* 71: 25–30.

Cookson, I.C. and Duigan, S.L., 1950. Fossil Banksieae from Yallourn, Victoria, with notes on the morphology and anatomy of living species. *Aust. J. Sci. Res., Ser. B,* 3: 133–165.

Cookson, I.C. and Duigan, S.L., 1951. Tertiary Araucariaceae from southeastern Australia, with notes on living species. *Aust. J. Sci. Res., Ser. B,* 4: 415–449.

Cookson, I.C. and Pike, K.M., 1953a. The Tertiary occurrence and distribution of *Podocarpus* (section *Dacrycarpus*) in Australia and Tasmania. *Aust. J. Bot.,* 1: 71–82.

Cookson, I.C. and Pike, K.M., 1953b. A contribution to the Tertiary occurrence of the genus *Dacrydium* in the Australian region. *Aust. J. Bot.*, 1: 474–484.

Cookson, I.C. and Pike, K.M., 1954a. The fossil occurrence of *Phyllocladus* and two other podocarpaceous types in Australia. *Aust. J. Bot.*, 2: 60–68.

Cookson, I.C. and Pike, K.M., 1954b. Some dicotyledonous pollen types from Cainozoic deposits in the Australian region. *Aust. J. Bot.*, 2: 197–219.

Cookson, I.C. and Pike, K.M., 1955. The pollen morphology of *Nothofagus* Bl. subsection Bipartitae Steen. *Aust. J. Bot.*, 3: 197–206.

Costin, A.B., 1954. *A Study of the Ecosystems of the Monaro Region of New South Wales.* Government Printer, Sydney, N.S.W., 860 pp.

Costin, A.B., 1957. The high mountain vegetation of Australia. *Aust. J. Bot.*, 5: 173–189.

Crocker, R.L., 1944. Soil and vegetation relationships in the Lower South-East of South Australia. A study in ecology. *Trans. R. Soc. S. Aust.*, 68: 144–172.

Crocker, R.L., 1946. An introduction to the soils and vegetation of Eyre Peninsula, South Australia. *Trans. R. Soc. S. Aust.*, 70: 83–107.

Curtis, W.M., 1956–67. *The Student's Flora of Tasmania*, 1–3. Government Printer, Hobart, Tasmania, 661 pp.

Devereux, I., 1967. Oxygen isotope palaeotemperature measurements on New Zealand Tertiary fossils. *N.Z. J. Sci.*, 10: 988–1011.

Dick, R.S., 1961. *Climatic Types According to Köppen Scheme (Slightly Modified).* Department Geography, University of Queensland, St. Lucia, Qld., 1 map.

Diels, L., 1906. *Die Vegetation der Erde, 7: Die Pflanzenwelt von West-Australien südlich des Wendekreises.* Engelmann, Leipzig, 413 pp.

Domin, K., 1911. Queensland's plant associations. *Proc. R. Soc. Qld.*, 23: 57–74.

Duigan, S.L., 1951. A catalogue of the Australian Tertiary flora. *Proc. R. Soc. Vic.* 63: 41–56.

Eichler, H., 1965. *Supplement to J.M. Black's Flora of South Australia.* Government Printer, Adelaide, S.A., 385 pp.

Gentilli, J., 1972. *Australian Climatic Patterns.* Thomas Nelson, Melbourne, Vic., 285 pp.

Gibbons, F.R. and Downes, R.G., 1964. A study of the land in southwestern Victoria. *Tech. Comm. Soil Conserv. Auth. Vic.*, No. 3: 289 pp.

Gibbs, L.S., 1917. A contribution to the phyto-geography of Bellenden Ker. I. Introduction. *J. Bot., Lond.*, 55: 297–301.

Gimingham, C.H., 1972. *Ecology of Heathlands.* Chapman and Hall, London, 266 pp.

Groves, R.H. and Specht, R.L., 1965. Growth of heath vegetation. I. Annual growth curves of two heath ecosystems in Australia. *Aust. J. Bot.*, 13: 261–280.

Gunn, R.H., Galloway, R.W., Walker, J., Nix, H.A., McAlpine, J.R. and Richardson, D.P., 1972. Shoalwater Bay area, Queensland. *CSIRO Aust., Div. Land Res. Tech. Mem.*, 72/10: 134 pp.

Heddle, E.M. and Specht, R.L., 1975. Dark Island heath (Ninety-Mile Plain, South Australia). VIII. The effect of fertilizers on composition and growth, 1950–1972. *Aust. J. Bot.*, 23: 151–164.

Hekel, H., 1972. Pollen and spore assemblages from Queensland Tertiary sediments. *Geol. Surv. Qld. Publ.*, No. 355: 48 pp.

Herbert, D.A., 1950. Present day distribution and the geological past. *Vic. Nat.*, 66: 227–232.

Hooker, J.D., 1860. *On the Flora of Australia, Being Part of an Introductory Essay to* Flora Tasmaniae, *The Botany of the Antarctic Voyage of H.M. Discovery Ships* Erebus *and* Terror, *in the Years 1839–1843*, Part III, 1, pp. xxvii–cxxviii.

Isbell, R.F., Thompson, C.H., Hubble, G.D., Beckmann, G.G. and Paton, T.R., 1967. *Atlas of Australian Soils. Explanatory Data for Sheet 4, Brisbane–Charleville–Rockhampton–Clermont Area.* CSIRO Australia — Melbourne University Press, Melbourne, Vic., 164 pp., 1 map.

Isbell, R.F., Webb, A.A. and Murtha, G.G., 1968. *Atlas of Australian Soils. Explanatory Data for Sheet 7, North Queensland.* CSIRO Australia — Melbourne University Press, Melbourne, Vic., 99 pp., 1 map.

Jackson, W.D., 1965. Vegetation. In: J.L. Davies, (Editor), *Atlas of Tasmania.* Lands and Surveys Department, Hobart, Tasmania, pp. 30–35.

Jeffrey, D.W., 1967. Phosphate nutrition of Australian heath plants. I. The importance of proteoid roots in *Banksia* (Proteaceae). *Aust. J. Bot.*, 15: 403–412.

Jessup, R.W., 1961. A Tertiary–Quaternary pedological chronology for the south-eastern portion of the Australian arid zone. *J. Soil Sci.*, 12: 199–213.

Jones, R., 1964. The mountain mallee heath of the McPherson Ranges. *Univ. Qld. Pap., Dep. Bot.*, 4(12): 156–220.

Jones, R., 1968a. The leaf area of an Australian heathland with reference to seasonal changes and the contribution of individual species. *Aust. J. Bot.*, 16: 579–588.

Jones, R., 1968b. Estimating productivity and apparent photosynthesis from differences in consecutive measurements of total living parts of an Australian heathland. *Aust. J. Bot.*, 16: 589–602.

Köppen, W., 1923. *Die Klimate der Erde.* Bornträger, Berlin, 369 pp.

Lamont, B., 1972a. The effect of soil nutrients on the production of proteoid roots by *Hakea* species. *Aust. J. Bot.*, 20: 27–40.

Lamont, B., 1972b. "Proteoid" roots in the legume *Viminaria juncea. Search*, 3: 90–91.

Lamont, B., 1972c. The morphology and anatomy of proteoid roots in the genus *Hakea. Aust. J. Bot.*, 20: 155–174.

Lamont, B., 1973. Factors affecting the distribution of proteoid roots within the root systems of two *Hakea* species. *Aust. J. Bot.*, 21: 165–187.

Lamont, B., 1974. The biology of dauciform roots in the sedge *Cyathochaete avenacea. New Phytol.*, 73: 985–996.

Lavarack, P.S. and Stanton, J.P., 1977. Vegetation of the Jardine River Catchment and adjacent coastal areas. *Proc. R. Soc. Qld.*, 88: 39–48.

Maconochie, J.R., 1975. Shoot and foliage production of five shrub species of *Acacia* and *Hakea* in a dry sclerophyll forest. *Trans. R. Soc. S. Aust.*, 99: 177–181.

Martin, D., 1940. The vegetation of Mt. Wellington, Tasmania. The plant communities and a census of the plants. *Pap. Proc. R. Soc. Tasmania*, 1939: 97–124.

Martin, H.A., 1961. Sclerophyll communities in the Inglewood District, Mount Lofty Ranges, South Australia. Their distribution in relation to micro-environment. *Trans. R. Soc. S. Aust.*, 85: 91–120.

Martin, H.A. and Specht, R.L., 1962. Are mesic communities less drought resistant? A study of moisture relationships in dry sclerophyll forest at Inglewood, South Australia. *Aust. J. Bot.*, 10: 106–118.

Northcote, K.H., 1960. *Atlas of Australian Soils. Explanatory Data for Sheet 1, Port Augusta–Adelaide–Hamilton Area.* CSIRO Australia — Melbourne University Press, Melbourne, Vic., 50 pp., 1 map.

Northcote, K.H., 1962. *Atlas of Australian Soils. Explanatory Data for Sheet 2, Melbourne–Tasmania Area.* CSIRO Australia — Melbourne University Press, Melbourne, 60 pp., 1 map.

Northcote, K.H., 1966. *Atlas of Australian Soils. Explanatory Data for Sheet 3, Canberra–Bourke–Armidale Area.* CSIRO Australia — Melbourne University Press, Melbourne, 107 pp., 1 map.

Northcote, K.H., 1968. *Atlas of Australian Soils. Explanatory Data for Sheet 8, Northern Part of Northern Territory.* CSIRO Australia — Melbourne University Press, Melbourne, Vic., 48 pp., 1 map.

Northcote, K.H., 1971. *A Factual Key for the Recognition of Australian Soils.* CSIRO Australia — Rellim Tech. Publ., Glenside, S.A., 3rd ed., 123 pp.

Osborn, T.G.B. and Robertson, R.N., 1939. A reconnaissance survey of the vegetation of the Myall Lakes. *Proc. Linn. Soc. N.S.W.*, 64: 279–296.

Parsons, R.F., 1966. The soils and vegetation at Tidal River, Wilson's Promontory. *Proc. R. Soc. Vic.*, N.S., 79: 319–354.

Parsons, R.F. and Gill, A.M., 1968. The effects of salt spray on coastal vegetation at Wilson's Promontory, Victoria, Australia. *Proc. R. Soc. Vic.*, 81: 1–10.

Patton, R.T., 1933. Ecological studies in Victoria. The Cheltenham flora. *Proc. R. Soc. Vic.*, 45: 205–218.

Pedley, L. and Isbell, R.F., 1971. Plant communities of Cape York Peninsula. *Proc. R. Soc. Qld.*, 82: 51–74.

Pidgeon, I.M., 1937. The ecology of the central coastal area of New South Wales. Part I. The environment and general features of the vegetation. *Proc. Linn. Soc. N.S.W.*, 62: 315–340.

Pidgeon, I.M., 1938. The ecology of the central coastal area of New South Wales. II. Plant succession on the Hawkesbury sandstone. *Proc. Linn. Soc. N.S.W.*, 63: 1–26.

Pidgeon, I.M., 1940. The ecology of the central coastal area of New South Wales. III. Types of primary succession. *Proc. Linn. Soc. N.S.W.*, 65: 221–249.

Pidgeon, I.M., 1941. The ecology of the central coastal area of New South Wales. IV. Forest types on soils from Hawkesbury sandstone and Wianamatta shale. *Proc. Linn. Soc. N.S.W.*, 66: 113–137.

Purnell, H.M., 1960. Studies of the family Proteaceae. I. Anatomy and morphology of the roots of some Victorian species. *Aust. J. Bot.*, 8: 38–50.

Rayson, P., 1957. Dark Island heath (Ninety-Mile Plain, South Australia). II. The effects of microtopography on climate, soils, and vegetation. *Aust. J. Bot.*, 5: 86–102.

Rowan, J.N. and Downes, R.G. 1963. A study of the land in northwestern Victoria. *Tech. Comm. Soil Conserv. Auth.*

Vic., No. 2: 116 pp.

Rübel, E.A., 1914. Heath and steppe, macchia and garigue. *J. Ecol.*, 2: 232–237.

Schimper, A.F.W., 1903. *Plant-Geography upon a Physiological Basis.* Clarendon Press, Oxford, 839 pp.

Siddiqi, M.Y., Carolin, R.C. and Anderson, D.J., 1973. Studies in the ecology of coastal heath in New South Wales. I. Vegetation structure. *Proc. Linn. Soc. N.S.W.*, 97: 211–224.

Smith, D.F. 1963. The plant ecology of Lower Eyre Peninsula, South Australia. *Trans. R. Soc. S. Aust.*, 87: 93–118.

Specht, R.L., 1957. Dark Island heath (Ninety-Mile Plain, South Australia). V. The water relationships in heath vegetation and pastures on the Makin sand. *Aust. J. Bot.*, 5: 151–172.

Specht, R.L., 1958a. The climate, geology, soils and plant ecology of the northern portion of Arnhem Land. In: R.L. Specht and C.P. Mountford (Editors), *Records of the American–Australian Scientific Expedition to Arnhem Land, 3. Botany and Plant Ecology.* Melbourne University Press, Melbourne, Vic., pp. 333–414.

Specht, R.L., 1958b. The geographical relationships of the flora of Arnhem Land. In R.L. Specht and C.P. Mountford (Editors), *Records of the American–Australian Scientific Expedition to Arnhem Land, 3. Botany and Plant Ecology.* Melbourne University Press, Melbourne, Vic., pp. 415–478.

Specht, R.L., 1963. Dark Island heath (Ninety-Mile Plain, South Australia). VII. The effect of fertilizers on composition and growth, 1950–1960. *Aust. J. Bot.*, 11: 67–94.

Specht, R.L., 1966. The growth and distribution of mallee–broombush (*Eucalyptus incrassata–Melaleuca uncinata* association) and heath vegetation near Dark Island Soak, Ninety-Mile Plain, South Australia. *Aust. J. Bot.*, 14: 361–371.

Specht, R.L., 1969. A comparison of the sclerophyllous vegetation characteristic of Mediterranean type climates in France, California and southern Australia. I. Structure, morphology and succession. II. Dry matter, energy and nutrient accumulation. *Aust. J. Bot.*, 17: 277–292; 293–308.

Specht, R.L., 1970. Vegetation. In: G.W. Leeper (Editor), *The Australian Environment.* CSIRO Australia — Melbourne University Press, Melbourne, Vic., 4th ed., pp. 44–67.

Specht, R.L., 1972. *Vegetation of South Australia.* Government Printer, Adelaide, S.A., 2nd ed., 328 pp.

Specht, R.L., 1973. Structure and functional response of ecosystems in the Mediterranean climate of Australia. In: *Ecological Studies, 7. Mediterranean Climate Ecosystems.* Springer-Verlag, Berlin, pp. 113–120.

Specht, R.L., 1975. Stradbroke Island: A place for teaching biology. *Proc. R. Soc. Qld.*, 86: 81–83.

Specht, R.L. and Brouwer, Y.M., 1975. Seasonal growth of *Eucalyptus* spp. in the Brisbane area of Queensland (with notes on shoot growth and litter fall in other areas of Australia). *Aust. J. Bot.*, 23: 459–474.

Specht, R.L. and Cleland, J.B., 1963. Flora conservation in South Australia. Part II. The preservation of species recorded in South Australia. *Trans. R. Soc. Aust.*, 87: 63–92.

Specht, R.L. and Perry, R.A., 1948. The plant ecology of part of the Mount Lofty Ranges (1). *Trans. R. Soc. S. Aust.*, 72:

91–132.

Specht, R.L. and Rayson, P., 1957a. Dark Island heath (Ninety-Mile Plain, South Australia). I. Definition of the ecosystem. *Aust. J. Bot.*, 5: 52–85.

Specht, R.L. and Rayson, P., 1957b. Dark Island heath (Ninety Mile Plain, South Australia). III. The root systems. *Aust. J. Bot.*, 5: 103–114.

Specht, R.L., Rayson, P. and Jackman, M.E., 1958. Dark Island heath (Ninety-Mile Plain, South Australia). VI. Pyric succession: Changes in composition, coverage, dry weight, and mineral nutrient status. *Aust. J. Bot.*, 6: 59–88.

Specht, R.L., Brownell, P.F. and Hewitt, P.N., 1961. The plant ecology of the Mount Lofty Ranges, South Australia. 2. The distribution of *Eucalyptus elaeophora. Trans. R. Soc. S. Aust.*, 85: 155–176.

Specht, R.L., Roe, E.M. and Boughton, V.H. (Editors), 1974. Conservation of major plant communities in Australia and Papua New Guinea. *Aust. J. Bot., Suppl.*, No. 7: 667 pp.

Specht, R.L., Connor, D.J. and Clifford, H.T., 1977. The heath-savannah problem: The effect of fertilizer on sand-heath vegetation of North Stradbroke Island, Queensland. *Aust. J. Ecol.*, 2: 179–186.

Specht, R.L., Salt, R. B. and Reynolds, S.T., 1977. Vegetation in the vicinity of Weipa, North Queensland. Proc. R. Soc. Qld., 88: 17–38.

Stace, H.C.T., Hubble, G.D., Brewer, R., Northcote, K.H., Sleeman, J.R., Mulcahy, M.J. and Hallsworth, E.G., 1968. *A Handbook of Australian Soils.* Rellim Tech. Publ., Glenside, S.A., 435 pp.

Story, R., 1973. Vegetation of the Alligator Rivers Area. Part IX (53 pp.). In:*Alligator Rivers Region Environmental Fact-Finding Study: Physical Features and Vegetation, II.* CSIRO, Canberra, A.C.T.

Story, R., Williams, M.A.J., Hooper, A.D.L., O'Ferrall, R.E. and McAlpine, J.R., 1969. Lands of the Adelaide–Alligator area, Northern Territory. CSIRO Aust., Land Res. Ser., No. 25: 154 pp.

Sutton, C.S., 1929. A sketch of the vegetation of the Cradle Mountain, Tasmania and a census of its plants. *Pap. Proc. R. Soc. Tas.*, 1928: 132–159.

Welbourn, R.M.E. and Lange, R.T., 1968. An analysis of vegetation on stranded coastal dunes between Robe and Naracoorte, South Australia. *Trans. R. Soc. S. Aust.*, 92: 19–24.

Wellman, P. and McDougall, I., 1974. Cainozoic igneous activity in eastern Australia. *Tectonophysics*, 23: 49–65.

Willis, J.H., 1962. *A Handbook to Plants in Victoria, I. Ferns, Conifers and Monocotyledons.* Melbourne University Press, Melbourne, Vic., 448 pp.

Willis, J.H., 1972. *A Handbook to Plants in Victoria, II. Dicotyledons,* Melbourne University Press, Melbourne, Vic., 832 pp.

Winkworth, R.E., 1955. The use of point quadrats for the analysis of heathland. *Aust. J. Bot.*, 3: 68–81.

Wood, J.G., 1939. Ecological concepts and nomenclature. *Trans. R. Soc. S. Aust.*, 63: 215–223.

Chapter 7

THE HEATHLANDS OF WESTERN AUSTRALIA[1]

A.S. GEORGE, A.J.M. HOPKINS and N.G. MARCHANT

INTRODUCTION

Within Western Australia, heathlands and re-lated shrublands are largely restricted in distri-bution to the southwestern part of the state. These formations include those variously termed "*sand-heath*" (Gardner, 1944), "*quonkan*" (Nelson, 1974), "*gongan*" (Hallam, 1975) and "*kwongan*" (Beard, 1976a) and those with an equivalent understorey of heathland shrubs and with emergent *Acacia*, *Banksia*, *Casuarina* and mallee *Eucalyptus*, termed tall shrubland with heathy understorey (Specht, Chapter 1 of this volume). Formations comprising wind or salt-pruned tall shrub and tree species (e.g. coastal *Agonis flexuosa* open-scrub), as well as swamp and halophytic formations which are floristically unrelated to other heathlands, are not discussed here.

The South West Botanical Province has been variously delimited within the southwest of Western Australia as a triangular-crescentic area of approximately 220,000 km^2 (Fig. 7.1; see Diels, 1906; Gardner, 1944; Burbidge, 1960). The area has long been noted for its floristic richness and a high degree of endemism (Hooker, 1860; Gardner, 1944; Burbidge, 1960); of a total of 3637 species of vascular plants listed for the Province by Beard (1965), some 68% are endemic (Marchant, 1973).

The Province has diverse vegetation types which include tall open-forests, open-forests and wood-lands, often with a shrub understory, as well as heathlands and sedgelands (see, for instance, Smith, 1972, 1973; Beard, 1973, 1976b; Beard and Burns, 1976). Open-forests and woodlands cover almost half the Province, but are considered relatively species-poor, although areas of sand within the forest may be richer in species (Havel

and Batini, 1973). The areas of heathland vege-tation are considered to contribute most to both the species richness of the province and the high degree of endemism. Approximately 50% of the heathland species recorded in Australia (Specht, Chapter 6, this volume) are found in the South West Botanical Province (Beard, 1965); by far the greatest proportion of these species are found in true heathlands. It is these areas of heathlands in the Province which Gardner (1944) regarded as "the real home of the autochthonous flora".

The dominant geological feature of the South West Province is the low-relief Archaean Yilgarn Block which forms part of the Western Australian Shield. Mesozoic rocks outcrop in the Perth Basin north of Perth, while the west and south coasts of the South West Province are flanked by narrow coastal plains of younger sediments.

Most of the older rocks are obscured by a mantle of laterite and associated soils which form exten-sive "sand plains". These sand and gravel plains are probably unique in their great expanse of deeply weathered and leached soil materials of low fertility (Mulcahy, 1973). Sand-plain soils have been shown to be colluvial deposits of local origin largely derived from the ferruginous duricrust of pre-existing laterite. They are predominantly sandy yellow earths with some sands over ironstone gravels. It is on these soils that the heathlands of southwestern Australia are most extensively de-veloped. Limited areas of rock frequently outcrop on the shallowly dipping valley sides, and the broad, shallow valleys usually carry sodic soils with calcareous subsoils (Mulcahy, 1973).

---

[1] Manuscript completed February, 1978

The coastal areas generally consist of highly calcareous, shelly sands which are succeeded inland by slightly podzolised yellow sands, mostly of quartz. These yellow coastal sands are almost always underlain by aeolianite limestone rock which outcrops in many areas between Dongara and Bunbury.

The climate within the Province is predominantly Mediterranean; the small south-coastal region from Pemberton to near Albany with an annual rainfall exceeding 1200 mm is regarded as sub-Mediterranean (UNESCO–FAO, 1963).

Fig. 7.1 shows the boundaries of the South West Botanical Province as described by various authors, together with rainfall isohyets for the region. More detailed climatic data for selected stations within the Province are given in Appendix I (Table 7.5).

VARIATION IN COMPOSITION

Of the three environmental variables which appear to have exerted a major influence on the development and distribution of heathlands in the South West Province, viz soil, climate and landform, the first is considered by far the most important (Gardner, 1944).

The two major regions of these sand plains lie (a) in the north, between the Moore River and Shark Bay, including the Mount Lesueur/Eneabba area, and (b) in the south, between the Stirling Range and Israelite Bay. Small discontinuous areas of sand plain are also scattered over the central part of the southwest on the Western Australian Shield; some of these are discussed by Bettenay and Hingston (1961) and Muir (1976, 1977). Lateritic gravel may occur at the surface or at varying depths under the sand, providing a great variety of habitats in close proximity. Slight variations in soils and landform produce subtle differences in vegetation structure and floristic composition. Some heathlands also occur on calcareous and granitic soils, most commonly in coastal regions.

Climate is apparently important in the development of these heathlands in that they are chiefly confined to the region which receives a regular winter rainfall exceeding 200 mm per annum. There is also a general change in species composition with the reduction in rainfall as one moves further inland. The effect of the pronounced

Mediterranean climate on patterns of growth and flowering in some Western Australian heathlands is discussed by Specht et al. (Vol. B, Ch. 2).

The heathlands of the South West Province are extensive and complex. In Appendix I, summary descriptions of 31 local types are given, which may be regarded as representative of the range of variation encountered. The few ecological studies undertaken on heathlands in Western Australia e.g. Lamont, 1976; Hnatiuk and Hopkins, in prep.) do not permit elucidation of trends in distribution of species in relation to the environmental variables. However, some generalised conclusions have been drawn here regarding variation in structure and composition from the detailed descriptions of vegetation given in the Appendix.

The heathlands of the two major regions are floristically similar at familial level, less so at generic level, and, apart from coastal heathlands, very different at specific level. Species common to the two regions often show infraspecific variation, e.g. *Eucalyptus tetragona*, *Verticordia grandiflora* and *Hakea baxteri*. In view of the general ubiquity of the genus in the South West Province, it is interesting that species of *Eucalyptus* are both fewer and less common in the northern heathlands than in those of the southern. The genera of Proteaceae, on the other hand, are generally well represented in both regions, exceptions being *Franklandia* with one species only in southern heathlands, and *Strangea* with one only in northern heathlands. Each genus also has one species in the forested areas of the South West.

Heathlands of the intervening area, the central part of the South West Botanical province, from east of Perth to beyond Southern Cross, contain some endemic species and show a floristic gradient between the two main regions. For example, species typical of the southern heathlands become fewer with distance from the south coast, and are mostly absent above the latitude of Perth.

The heathlands of coastal dunes contain a number of species which extend throughout the South West Province coastal region, e.g. *Olearia axillaris*, *Acacia rostellifera* and *Spyridium globulosum*. There are, however, differences in abundant species which justify the separation of coastal heathlands into two groups. On the west coast, examples are *Melaleuca acerosa*, *Logania vaginalis*, *Acacia truncata*, *A. lasiocarpa*, *Lechenaultia linarioides* and *Scaevola crassifolia*. Representative south coastal species include *Pimelea ferruginea*, *Melaleuca nesophila*, *Eucalyptus angulosa* and *Scaevola nitida*. A similar distinction can be made between the coastal heathlands on shallow soil over rock in the two regions.

Heathlands are rare in Western Australia beyond the interzone between the South West and Eremean Botanical Provinces, but several areas warrant mention. Between Carnarvon and North West Cape are large areas with a significant content of heathland species, but the widespread inclusion of the typically Eremean *Triodia* (spinifex grass) precludes their classification as true heathlands.

In the Great Victoria Desert, and to a lesser extent the Gibson Desert, a low open-heathland occurs on deep red sand on the lower slopes of dunes and sometimes across swales. It is dominated by *Thryptomene maisonneuvii* (Myrtaceae), which sometimes occurs as a pure stand. More often it is mixed with spinifex (species of either *Triodia* or *Plectrachne*, Poaceae) and becomes heath-spinifex hummock grassland.

Other outliers occur around Twilight Cove, Point Dover, Toolinna and Point Culver, along the western edge of the Great Australia Bight (Nelson, 1974; Beard, 1975). Here tall shrubland with a heathy understorey is found on consolidated dunes of siliceous sand within 15 km of the coast. Most of the species occur also in heathlands west of Israelite Bay (e.g. *Eucalyptus angulosa*, *Beaufortia micrantha*, *Hakea nitida*, *Lysinema ciliatum*), but some are endemic to the area, (e.g. *Grevillea sparsiflora*, *Adenanthos forrestii* and *Styphelia hainesii*).

STRUCTURE

Heathland communities, by definition, are dominated by low shrubs (nanophanerophytes up to 2 m tall) (Specht, Chapter 1). Usually several species of nanophanerophytes are co-dominant in the shrub storey. An analysis of the composition of three low closed-heathlands in relation to Raunkiaer's life-form classes (Table 7.1) emphasises this. More than half the species present in each of the sample areas are nanophanerophytes. The majority of these are in the families Proteaceae and Myrtaceae. Annuals are rare in these habitats, an exception being *Podotheca* (Asteraceae) though they are

TABLE 7.1

Raunkiaer life-form spectrum of three low closed-heathland sites in the Western Australian sand plains

Location:	Eneabba Nature Reserve	Eneabba Area (Victoria Location 10413)	Stirling Range
Sample number[1]:	8	10	15
Nanophanerophytes	67%	59%	61%
Chamaephytes	6%	9%	6%
Hemicryptophytes	20%	22%	18%
Geophytes	5%	7%	12%
Therophytes	0%	1%	1%
Heterotrophs	1%	2%	2%
Lichens and mosses	1%	0%	0%
	100%	*100%*	*100%*

[1] Sample numbers refer to species-richness samples, details of which are given in Table 7.2.

more common in other areas of the South West Province, especially in drier regions.

An analysis of the proportions of species falling into Raunkiaer's leaf-size classes for a heathland area near Eneabba is given by Lamont (1976). Over 50% of the species are leptophyllous, the remainder being nanophyllous and microphyllous. This result is similar to that already discussed for a Dark Island heathland by Specht (Chapter 6).

Few studies have been made in Western Australia of the root profiles of heathland species. Lamont (1976) gives the weights of root material recovered from samples of two different soils in the Eneabba area. The total weights of root materials in the two soils are similar. However, the roots in the deep sand tend to be concentrated in the upper parts of the profile (49% in the top 15 cm) whereas they are more evenly distributed through a 'sand over clay' profile where a clay layer is developed at 55 cm depth (37% of roots in the top 15 cm).

The roots of *Eremaea beaufortioides* (Myrtaceae) have been observed in mining excavations at Eneabba at 10 m depth and are reported to have regenerated foliage from cut roots at 6 m depth (E.A. Griffin, pers. comm., 1977).

SPECIES RICHNESS AND BIOGEOGRAPHY (N.G. Marchant and A.J.M. Hopkins)

The species richness of the two main sand-plain regions of the South West Botanical Province was noted first by Diels (1906) and later by Gardner

(1944) who regarded the two regions as the "cusps" of the Province, in which the "Australian and Antarctic elements" of the flora are richly developed. The evolutionary significance of these two areas is discussed by Marchant (1973).

These observations on richness have not been substantiated in any quantitative way; indeed few data are available to permit any kind of comparison of the species richness of Western Australian heathlands.

Results from the sampling of an island flora (Middle Island, 34°06′S, 123°11′E) will be reported by Hopkins and Trudgen (in prep.), A plot of 0.05 ha in the coastal sand-dune heathland contains 22 species, while one of 0.1 ha in Acacia rostellifera open-scrub contains 12 species. Forest types sampled on this island are considered species-poor relative to their mainland equivalents.

Species-area data for some eastern Australian heathlands given by Specht (Chapter 6) show the mean number of species in quadrats of 8 m^2 ranging from 11.2 to 36.1.

The open-forest and woodland areas comprising almost half the South West Botanical Province are considered relatively species-poor. Average richness for samples of 0.23 ha in Eucalyptus accedens–E. wandoo woodland is 34 species, for E. wandoo woodland 44 species, and for E. marginata–E. calophylla open-forest 60 species (W.A. Loneragan, unpublished data, 1977). The heathy understorey of the forest region contributes most to this richness, and is particularly rich in members of the family Epacridaceae (especially Astroloma and Leucopogon).

Similarly, the understorey component of temperate woodlands and open-forests (excluding rain-forests) in New South Wales and the Australian Capital Territory provided the most important component of species richness values reported by Whittaker (1977). For the 28 plots of 0.1 ha sampled, species numbers range from 29 to 105, with a mean and standard deviation of 47.7 ± 15.5. These Australian forests and woodlands are found to be relatively richer than North American temperate forests and woodlands.

The remainder of this section describes attempts to quantify the species richness of the Western Australian heathlands, to identify patterns of richness, and to relate these patterns to those of geography and soils.

The distribution of species within selected genera in relation to a geographic grid is also examined to gain further insight into biogeographic trends, as indicated by measures of species richness.

Species richness is here regarded as synonymous with "alpha diversity" of Whittaker (1972) and is used to avoid the ambiguity of the term "species diversity" which is often associated with indices based on information theory (Hurlburt, 1971). Species richness of any community can be measured in terms of numbers of species in any appropriately sized sample from within that community.

Survey of species richness

Twenty-five sample sites were chosen throughout the South West Botanical Province — located where possible on National Parks and Nature Reserves. Each site was chosen as representative of the local vegetation in appearance and composition. Where soils varied, plots were selected so as to include only one soil type and further plots were sampled on neighbouring soil types.

The position of sample sites are shown in Fig. 7.1; further details of the sites are presented in Table 7.2.

For preliminary field samples, nested quadrats from 1 m^2 to 1000 m^2 were used. Species–area curves for these samples are shown in Fig. 7.2. While the concept of minimal area may have little validity (Hopkins, 1957), these curves lend support to the use of the information content of species-area curves for making comparisons between communities (Mueller-Dombois and Ellenberg, 1974). Further, since all the curves show a similar trend, the species numbers in single large samples from equivalent communities can be compared. Subsequently, a simplified nested quadrat was used with a 50-m^2 sample included within a 500-m^2 sample area. While the 500-m^2 quadrat exceeds any minimal area requirement for this vegetation type [250 m^2 for 95% of the total number of species (Mueller-Dombois and Ellenberg, 1974, p. 51)] it was selected as a convenient sample size and one which would provide useful comparative data, especially in relation to taxonomic composition of the heathlands sampled.

For all sites sampled, a record was made of each additional species encountered as quadrat size increased. The results of the species richness

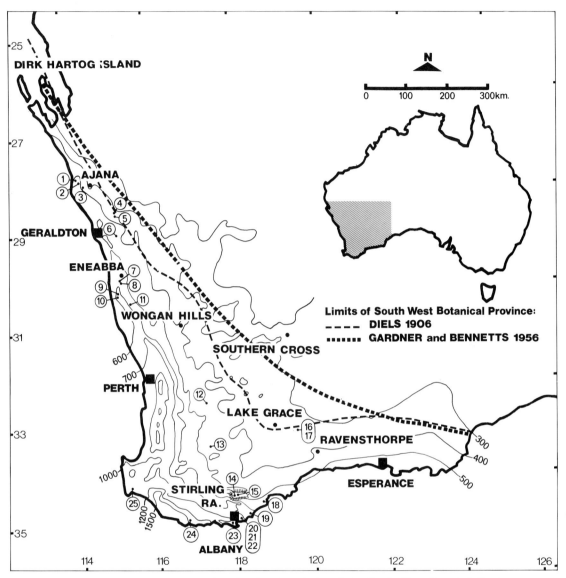

Fig. 7.1. Location map, southwestern Australia, showing boundaries of the South West Botanical Province, annual isohyets and species-richness sampling sites.

sampling are presented in Table 7.3. A number of observations can be made on the basis of the results of this limited sampling.

The sampled heathlands, with a mean of 59.5 species per 500-m² plot, are rich in vascular plant species. The relative nature of this richness cannot be ascertained by direct comparison with those richness values from the literature cited previously, since the areas sampled vary considerably. However, some standardisation of those richness values may be achieved.

If it is assumed that the important heathland component of each of those vegetation types has a species–area relationship similar to the Western Australian heathlands for which species–area curves are given in Fig. 7.2, then a member of that family of curves can be drawn to satisfy each of the species–area values from the literature. The 500-m² solution for each curve can be taken as the approximate richness value for comparison with the data in Table 7.3.

Estimates of richness for plots of 500 m² derived

TABLE 7.2

Location and site details for heathland samples

Sample No.	Location[1]	Vegetation	Soil type	Estimated age since last fire (years)
1	Kalbarri National Park (A27004)	tall closed-heathland	deep, yellow/white sand	7
2	Kalbarri National Park (A27004)	tall closed-heathland	clay over yellow sand	6
3	Kalbarri National Park (A27004)	tall closed-heathland	deep, yellow sand	7
4	East Yuna Nature Reserve (29231)	tall open-heathland	clayey sand over sand	
5	Williamson Road, east of Geraldton, Victoria Location 6067	tall open-heathland	deep, yellow sand	
6	Casuarina Road, east of Geraldton, Victoria Location 10163	tall open-heathland	deep, white sand	10
7	Eneabba Nature Reserve (31030)	tall closed-heathland	deep, white sand	8
8	Eneabba Nature Reserve (31030)	low open-heathland	sand over laterite	
9	Mount Lesueur (15018)	low closed-heathland	clayey sand over laterite	
10	Eneabba area, (Victoria Location 10413)	closed-heathland	grey yellow sand over laterite	
11	Badgingarra National Park (31809)	tall closed-heathland	deep sand over laterite	6–8
12	Tuttanning Nature Reserve (25555)	open-heathland	deep, grey sand	
13[2]	Lime Lake (2091)	open-heathland	sandy loam over clay	
14	Stirling Range National Park (A14792)	mallee-heathland	white sand over laterite	20
15	Stirling Range National Park (A14792)	low open-heathland	clay sand over laterite	6
16[2]	Corackerup Nature Reserve (A26793)	open-heathland	laterite	
17[2]	Corackerup Nature Reserve (A26793)	open-heathland	deep sand over gneiss	
18	Cape Riche (A14943)	tall open-heathland	lateritic sand	
19	Cheynes Beach (29883)	tall closed-heathland	clay sand over laterite	
20	Two Peoples Bay Nature Reserve (A27956)	low closed-heathland	sand over gneiss	15
21	Two Peoples Bay Nature Reserve (A27956)	low closed-heathland	sand over limestone/gneiss	10
22	Two Peoples Bay Nature Reserve (A27956)	low closed-heathland	sand over limestone	6 (+10)
23	Torndirrup National Park (A24258)	closed-heathland	deep, grey sand	10
24	Nornalup National Park (A31362)	low open-heathland	deep loamy sand	6
25	Scott River Common, Augusta district (Sussex Location 4481)	low open-heathland	deep, grey sand	5

[1] The numbers given are location or reserve numbers of the Western Australian Department of Lands and Surveys.
[2] Samples 13, 16, 17 are from unpublished data of K. Newbey.

in this way are: *Acacia rostellifera* open-scrub 9 species; eastern Australian heathlands 30 to 56 species; *Eucalyptus accedens–E. wandoo* woodland 28 species; *E. wandoo* woodland 35 species; *E. marginata–E. calophylla* open-forest 43 species; and eastern Australian open-forests and woodlands (with heathy understorey) 26 to 95 species with a mean of 40 species.

The 25 Western Australian heathlands sampled for this study appear to be generally richer in species than other Australian heathlands and shrublands, and most are richer than open-forests and woodlands of the South West. The range of richness values for these heathlands lies within that for eastern Australian open-forests and woodlands, but the mean value for the heathlands

is substantially higher than that for the open-forests and woodlands.

This species richness of the Western Australian heathlands is not uniform. The sampled heathlands exhibit a considerable range of richness values for plots of 500 m^2, ranging from 37 species (No. 4, east of Geraldton, on the eastern margin of the South West Botanical Province) to 92 species (Nos. 8, 10, both in the Eneabba/Mount Lesueur area). This variation appears to be related to environmental and geographical factors.

Samples from the Eneabba/Mount Lesueur area (Nos. 7–11) and the Stirling Range (Nos. 14, 15) are rich in species, and a further rich node is suggested to the east of Albany (Nos. 18, 19). Eneabba lies in the centre of the northern region of

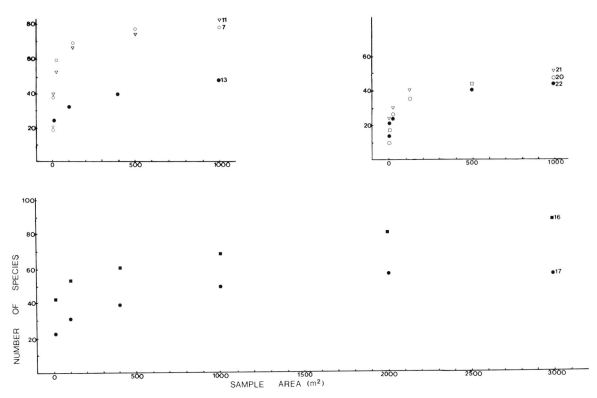

Fig. 7.2 . Species–area curves for selected heathlands based on nested quadrat samples. Site numbers correspond with those in Tables 7.2 and 7.3.

heathlands described earlier, while the Stirling Range is at the western end of the extensive region of southern heathlands. Sampling is inadequate to delimit precisely the nodes of richness.

The three sites of greatest species richness (Nos, 8, 10, and 15) are on sandy soils associated with laterite, while the six species-poor sites (Nos. 4, 5, 17, 20, 21, 22) are on sandy soils. Low species richness is also recorded for two inland heathland sites (Nos. 12, 13) both of which are heathlands of limited extent in central agricultural areas.

Plant-family representation in all samples ranges from 16 families in No. 5 to 26 in samples Nos. 17 and 24. An analysis of family representation is given in Table 7.4. Fourteen commonly encountered families are listed; these account for 50 to 95% of the species recorded. Of the families not listed, perennial species of Poaceae and Asteraceae are frequent. In sample No. 13, of the total of 47 species recorded for 1000 m², five are grasses and six are composites. Two of these grasses and three of these composites are introduced alien species. Other families frequently

recorded in samples are as follows: Apiaceae, Casuarinaceae, Dicrastylidaceae, Euphorbiaceae, Sterculiaceae, Thymelaeaceae and Tremandraceae.

The three families Myrtaceae, Fabaceae and Proteaceae are particularly well represented in heathland and related shrublands of the South West. Fifty percent of the species recorded in the two samples, Nos. 8 and 14 (Eneabba and Stirling Range), belong to these three families.

The Proteaceae is by far the most commonly encountered family in the heathlands sampled. In the three samples with the highest total species count (Nos. 8, 10 and 15) Proteaceae account for 16 to 23%. In the samples with the highest percentage of species of this family (Nos. 8, 9, 11, 14 and 19) the range is 22 to 32%. All these latter samples are on sand or clay over laterite.

Species of Myrtaceae in samples 3, 5, 7 and 17 account for 20 to 31% of the total species in each case. All of these sample sites are on deep yellow or white sand.

These apparent soil-related differences in the development of the two major families, Myrtaceae

TABLE 7.3

Species-richness values for heath samples

Sample No.	Total species per 50 m²	Total species per 500 m²
1	47	70
2	33	47
3	34	48
4	17	37
5	22	42
6	40	66
7	65	77
8	71	92
9	60	67
10	70	92
11	59	71
12	33	44
13	30[1]	41[1]
14	50	71
15	54	85
16	49[1]	62[1]
17	28[1]	41[1]
18	46	72
19	55	77
20	31	44
21	27	40
22	36	44
23	42	56
24	34	51
25	35	50

[1] Extrapolated from species–area curves.

and Proteaceae, may reflect important differences in rooting systems, with species of the former family developing deep roots in sandy soils, while species of the latter family may more frequently develop lignotubers and proteoid roots in lateritic soils.

South coast samples (Nos. 18–25) more frequently show higher proportions of Fabaceae than other samples. These are on varying soils associated with laterite, limestone and gneissic rocks.

The other important Australian heathland family, Epacridaceae, is also recorded as being more frequently represented in south coast samples (Nos. 19–23, 25) as well as one sample from the northern heathland region (No. 7).

Distribution of selected heathland genera

A study of the distribution of species within two typical heathland genera was undertaken to examine further the biogeographic trends revealed by the nested quadrat study.

Distribution data for each species were extracted from records of the Western Australian Herbarium and plotted on a map which was gridded with 96 × 96 km squares, the optimal size suggested by Phipps and Cullen (1976). The number of species

TABLE 7.4

Analysis of taxonomic composition of heathland samples

Sample numbers

	1	2	3	4	5	6	7	8	9	10	11	12	13	14	15	16	17	18	19	20	21	22	23	24	25
Cyperaceae	6	4	6	5	7	6	4	4	10	7	1	14	4	8	8	9	11	8	6	5	5	5	2	4	4
Dilleniaceae	3	2	—	—	—	2	5	5	4	3	7	2	—	3	4	3	2	1	2	2	3	2	2	2	2
Droseraceae	—	—	6	—	—	5	4	—	4	4	4	—	4	3	7	3	—	3	4	—	—	—	—	—	6
Epacridaceae	4	2	6	—	7	6	9	4	3	8	8	5	2	7	8	6	4	8	10	9	10	9	9	4	10
Fabaceae	7	6	4	8	2	9	8	8	9	8	8	5	4	7	9	7	5	17	5	7	10	18	11	6	10
Goodeniaceae	6	2	2	3	2	5	1	—	—	4	1	—	2	1	2	3	3	3	2	2	5	5	7	10	2
Haemodoraceae	6	4	6	—	7	—	—	4	4	2	6	7	—	3	2	3	3	—	3	5	5	3	—	—	4	—
Liliaceae	6	2	4	16	5	8	5	8	9	7	4	2	6	8	5	8	6	3	8	11	3	5	2	4	4
Mimosaceae	1	9	2	5	2	3	—	2	3	—	3	2	4	6	1	3	5	6	3	2	3	2	—	2	—
Myrtaceae	16	17	21	16	31	14	22	18	9	14	14	18	10	13	13	18	20	10	6	16	5	7	4	4	8
Proteaceae	14	17	8	3	7	21	19	23	22	21	26	16	4	30	16	19	9	19	32	18	18	16	5	8	8
Restionaceae	3	4	2	5	5	5	1	5	7	3	3	5	2	4	4	3	2	4	4	5	5	2	4	2	—
Rutaceae	3	—	—	—	3	—	—	3	1	—	2	—	1	—	—	1	1	1	—	—	14	—	2	4
Stylidiaceae	6	4	2	—	—	3	3	3	—	3	2	5	8	1	1	2	4	3	2	—	—	2	4	4	4
Other families	19	27	31	39	25	10	19	16	13	15	13	17	50	5	20	13	25	14	12	18	30	13	50	44	38

Figures are the percentage of species of each family of the total number of species in the 500-m² sample.

occurring in each square was scored and isoflors were plotted following visual inspection. These isoflors, or species-richness isopleths, are lines joining areas containing comparable numbers of species (Cain, 1944). Isoflor maps permit rapid interpretation of patterns of species richness in a geographical area.

Data from the Western Australian Herbarium are limited because collections are neither complete nor thorough. Many areas of heathlands, particularly that around Eneabba, have been poorly collected. The two genera selected were chosen on the basis of their frequent occurrence as low shrubs in heathlands as well as there being relatively numerous collections in the herbarium. Despite the current taxonomic uncertainty of species within each genus, distribution maps of each species showed discrete boundaries indicating reasonably complete distribution data.

The genus *Verticordia* (Myrtaceae) has 49 species in the South West. They are mostly low, woody shrubs (with ericoid leaves) which are frequent in heathlands and related shrublands. Over 1000 herbarium records were examined for this genus. The resulting isoflors are shown in Fig. 7.3.

Data for the genus *Conospermum* (Proteaceae) were used to construct a similar map (Fig. 7.4). This genus of 19 species is also a common component of heathland vegetation. Six hundred herbarium records were utilised.

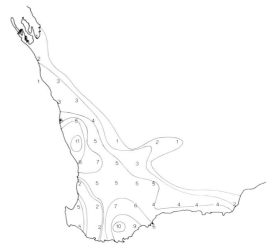

Fig.7.4. Isoflor map for the genus *Conospermum* (Proteaceae). The isoflor interval is 2 species.

A similar isoflor diagram was constructed by Speck (1958) for 426 species of Proteaceae using a 45 × 45 km square grid. Not all of these are heathland species, but his map is reproduced here for comparative purposes (Fig. 7.5). From his map, Speck noted two main centres of richness: the Eneabba/Mount Lesueur area, and the Stirling Range/Fitzgerald River area, together with evidence of a strong coastal influence. The distinct gap between the centres reflects the relative species paucity in open-forest and woodland areas.

The isoflors of the genus *Conospermum* conform with the trends observed for the whole family Proteaceae.

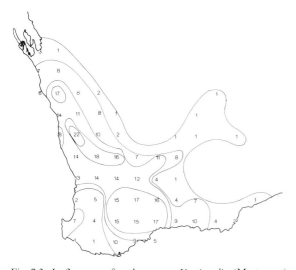

Fig. 7.3. Isoflor map for the genus *Verticordia* (Myrtaceae). The isoflor interval is 5 species.

Fig. 7.5. Isoflor map for the family Proteaceae (redrafted from Speck, 1958). The isoflor interval is 10 species.

The genus *Verticordia* shows a northern centre of richness in the Eneabba/Mount Lesueur area. The southern concentration of species is to the north and east of the Stirling Range. This concentration is not as marked as the northern one. The distribution of species is also strongly affected by the pattern of open-forest and woodland distribution.

The isoflor diagrams for the Proteaceae and *Verticordia* shows nodes of species richness in approximately the same areas which conform with the extensive regions of sand plain with heathland and related shrublands. The isoflors also show the general NW–SE trend of the limit of the South West Botanical Province conforming with the limit marked on Fig. 7.1. The eastwards projection of isoflors at the level of Perth and Southern Cross is almost certainly due to the location of a major highway which has provided a more frequent source of herbarium material.

CONCLUSIONS

It has been considered in the past that the heathlands of southwestern Australia are areas of great floristic richness. We have attempted here to quantify this richness. On the basis of our limited data, these heathlands appear to be typically richer than sclerophyllous vegetation types elsewhere in temperate Australia. The considerable variation in this species richness may be related to the geography and to environmental factors.

Areas of species richness have been located near Eneabba, in the Stirling Ranges, and to the east of Albany. The patterns of species richness are supported by isoflor maps for species of the two heathland genera, *Verticordia* and *Conospermum*.

The richest sites are located on soils composed of sand associated with lateritic material; the poorest sites are on sandy soils, or in inland agricultural areas. Of the two families which are best represented in the heathlands, the Proteaceae appears best developed on lateritic soils, while the Myrtaceae appears to be best represented on sandy soils. This trend may be related to differences in root systems between the two families.

The Mediterranean climate of southwestern Australia would seem to be an important factor in the development of the rich heathland flora. The heathlands of the east coast of Australia are generally not as rich as the western ones, although the former region has substantial areas of equivalent nutrient-poor soils on which sclerophyllous species are richly developed (Beadle, 1966; Johnson and Briggs, 1975). The east coast has never experienced a Mediterranean climate, a feature previously noted by Raven (1971) as being associated with high species numbers and high degrees of endemism.

It is generally accepted that the sclerophyllous flora of Australia evolved from tropical ancestors (Specht and Rayson, 1957; Raven, 1971). Evidence in respect of the family Proteaceae, which has over 600 species in southwestern Australia, supports this view (Johnson and Briggs, 1975). Whatever its origin, the degree of species richness suggests that the southwestern flora has been evolving under a Mediterranean-type climate for some time.

The reasons for the existence of centres of richness in the Eneabba and Stirling Range/Fitzgerald River areas are not known, though some have been postulated (Marchant, 1973). Either these centres represent refugial areas which underwent a great deal of diversification following climatic change, or they represent areas near refugia which provided a variety of ecological niches for a colonising, evolutionarily dynamic flora.

The soils of the so-called sand-plain regions are probably the most varied in the South West. Certainly the habitat variation in the Mount Lesueur/Eneabba and the Stirling Range/Fitzgerald River areas is considerable, These regions must be regarded as having played a key role in the evolution of the numerous shrubby species which are so prominent in the South West flora.

APPENDIX I: DESCRIPTION OF HEATHLANDS AND RELATED SHRUBLANDS (A.S. George)

Within and between the two major regions of heathlands in the southwest of Western Australia, there is considerable variation in both structure and floristics of the heathland vegetation. Descriptions of representative examples of some heathlands are provided in the following section. The examples have been selected to show some of the variation which may be observed.

Examples 1 to 6 and 9 to 11 are considered representative of the northern heathlands, 21 to 23 and 26 to 31 of the southern heathlands, 7 and 8 of the western coastal heathlands and 24 and 25 of the southern coastal heathlands. The remaining examples are from the intervening region. Included are some examples of heathlands with widespread distribution (e.g. 10 and 30) and some heathlands which are restricted (e.g. 6 and 13).

The examples are arranged in a geographical sequence from north to south and from west to east, beginning at Dirk Hartog Island and ending at Israelite Bay. Latitude and longitude are cited for the location of a representative example of most types.

The plant species mentioned are considered representative species for each type. There is no significance in the order of listing. Structural terminology used is consistent with that of Specht (Chapter 1).

Photographs illustrating the structural characteristics of some of the heathland types are provided in Figs. 7.6 to 7.17.

Descriptions of soils are derived from Bettenay et al. (1967), Northcote et al. (1967) and Northcote (1971).

Climatic data for ten selected stations in the South West region are provided in Table 7.5.

It must be emphasised that the examples of heathlands described below represent a selection

TABLE 7.5

Climatic data for ten locations in the southwest of Western Australia

	Location[1]									
	1	2	3	4	5	6	7	8	9	10
Mean daily maximum air temperature (°C)										
Summer (Dec–Feb)			34.5	32.9	29.6	25.0	34	31	28.3	27.9
Autumn (Mar–May)			27.3	25.1	24.4	21.2	26	24	23.1	23.0
Winter (June–Aug)			19.2	16.2	18.1	15.8	17	16	16.7	16.2
Spring (Sept–Nov)			25.5	23.9	22.0	18.9	26	23	22.0	21.0
Year			26.6	24.5	23.5	20.2	26	23	22.5	22.0
Mean daily minimum air temperature (°C)										
Summer (Dec–Feb)			16.6	16.2	18.0	13.2	16	14	13.6	13.6
Autumn (Mar–May)			12.6	12.3	14.3	11.3	11	11	11.3	11.1
Winter (June–Aug)			7.6	6.5	9.7	7.4	5	6	6.9	6.4
Spring (Sept–Nov)			9.7	9.7	12.1	8.9	9	8	8.8	8.7
Year			11.6	11.2	13.5	10.2	11	10	10.2	10.0
Precipitation (mm)										
Summer (Dec–Feb)	16	23	27	31	34	73	44	45	64	69
Autumn (Mar–May)	79	92	118	87	191	194	77	94	110	109
Winter (June–Aug)	184	185	352	181	500	337	110	148	136	180
Spring (Sept–Nov)	21	37	93	49	158	211	48	71	109	127
Year	300	337	590	348	883	815	279	358	419	485
Relative humidity (%, at 9 am)										
Summer (Dec–Feb)			41	59	51	62	44	49	53	59
Autumn (Mar–May)			57	72	65	77	62	66	66	67
Winter (June–Aug)			77	88	77	83	80	83	76	79
Spring (Sept–Nov)			53	67	60	74	52	61	58	63
Year			57	72	63	74	59	65	63	66

[1]*Locations*	*Latitude*	*Longitude*	*Locations*	*Latitude*	*Longitude*
1. Tamala homestead	26°42′S	113°43′E	6. Albany Airport	34°57′S	117°48′E
2. Ajana	27°57′S	114°38′E	7. Southern Cross	31°14′S	119°20′E
3. Eneabba	29°48′S	115°21′E	8. Lake Grace	33°06′S	118°28′E
4. Wongan Hills	30°50′S	118°43′E	9. Ravensthorpe	33°35′S	120°03′E
5. Perth	31°50′S	115°51′E	10. Esperance Downs	33°37′S	121°48′E

from a great number of types into which the southwestern heathlands can be classified. Due to space constraints, the data are tabulated. For the same reason, no attempt has been made to include examples of the sclerophyllous heath-like understoreys which are extensive in many open-forests and woodlands of the South West, e.g. jarrah (*Eucalyptus marginata*) open-forest, and wandoo (*E. wandoo*) woodland.

Descriptions and maps of South West heathlands have appeared in a number of previous works (e.g. Beard, 1969, 1973, 1975, 1976b; Smith 1972, 1973; Kitchener et al., 1975; McKenzie et al., 1973; Nelson, 1974; Beard and Burns, 1976; Muir, 1976, 1977. A general discussion of the floristics of heathland areas, accompanied by numerous colour plates of many of the species present, is included in Erickson et al. (1973).

1.DIRK HARTOG ISLAND (Lat. 25°45'S, Long. 113°00'E; Fig. 7.6)

Landform: Gently undulating sand dunes.
Soil: White calcareous sand.
Structure: Tall closed- to open-heathland and low open-heathland, often mixed with hummock grassland.
Composition: Phanerophytes: *Acacia ligulata* (Mimosaceae); *Thryptomene baeckeacea* (Myrtaceae); *Exocarpos aphyllus* (Santalaceae); *Alyogyne cuneiformis* (Malvaceae). Chamaephyte: *Lechenaultia linarioides* (Goodeniaceae). Hummock grasses: *Plectrachne* sp. inedit. *Triodia plurinervata* (Poaceae). Hemicryptophyte: *Lepidobolus preissianus* (Restionaceae). Therophytes: *Brachycome iberidifolia* (Asteraceae); *Swainsona phacoides* (Fabaceae).

Comment: This example occurs over most of the island as well as on Carrarang and Peron Peninsula on the adjacent mainland where it is poorer floristically. It is the northernmost formation in Western Australia which can be called heathland, though many of its component species extend to North West Cape. Rainfall in the region is variable and somewhat unreliable. For further discussion see Beard (1976a) and Burbidge and George (1978).

2. ZUYTDORP (Lat. 27°00'S, Long. 114°00'E)

Landform: Undulating low hills.
Soil: White calcareous sand over limestone, some rocky outcrops.
Structure: Closed-scrub in swales; open-scrub with tall and low open-heathland on rises.
Composition: In swales — phanerophytes: *Acacia ligulata* (Mimosaceae). On rises — phanerophytes: *Banksia prionotes*, *Conospermum stoechadis* (Proteaceae); *Eucalyptus eudesmoides*, *E. baudiniana*, *Thryptomene baeckeacea* (Myrtaceae); *Acacia spathulata* (Mimosaceae); *Geleznowia verrucosa* (Rutaceae). Hemicryptophyte: *Conostylis candicans* (Haemodoraceae).
Comment: Extensive in the area between the lower Murchison River and Tamala Station, this example really contains two formations. The closed-scrub is often almost a pure stand of *Acacia ligulata*, giving way to open-scrub and heathland of mixed composition on the rises. However, the *Acacia* is often present with the latter also.

3. NORTH OF MURCHISON RIVER (Lat. 27°10'S, Long. 114°35'E; Fig. 7.7)

Landform: Flat or gently undulating plain.
Soil: Red earth with alkaline reaction trend.
Structure: Closed- and open-scrub.
Composition: Phanerophytes: *Eucalyptus oldfieldii, E. jucunda*,

Fig. 7.6. Dirk Hartog Island. Tall closed-heathland with admixture of hummock grassland. Shrubs include *Acacia ligulata, Ptilotus obovatus* and *Scaevola spinescens*.

Fig.7.7. North of Murchison River. Closed-scrub on pale red sand. Species include *Lamarchea hakeifolia, Banksia ashbyi, Eucalyptus* sp., *Callitris huegelii*.

Fig.7.8. North of Murchison River. Low open-heathland in foreground with *Conospermum stoechadis*, *Ecdeiocolea monostachya*, etc. Open-scrub on dune behind, with *Actinostrobus arenarius*, *Banksia sceptrum*.

E. eudesmoides, *Lamarchea hakeifolia* (Myrtaceae); *Acacia neurophylla*, *A. brachystachya* (Mimosaceae); *Bursaria spinosa* (Pittosporaceae); *Banksia ashbyi*, *Hakea stenophylla* (Proteaceae); *Eremophila clarkei* (Myoporaceae).

Comment: Although the tall, dense structure differs from that of the other examples included in this paper, it is related to them in that most of the genera present are typical of the South West Province. Extensive tracts occur between the lower Murchison River and Shark Bay. It lies close to the boundary of the Province, giving way eastwards to *Eucalyptus* woodland and *Acacia* shrubland.

4. NORTH OF MURCHISON RIVER (Lat. 27 29′S, Long. 114 42′E; Fig. 7.8)

Landform: Sand dunes with wide swales.
Soil: Pale yellow siliceous sand.
Structure: Tall shrubland with heathy understorey.
Composition: Phanerophytes: *Banksia sceptrum*, *Grevillea annulifera*, *G. dielsiana*, *Xylomelum angustifolium* (Proteaceae); *Actinostrobus arenarius* (Cupressaceae); *Calothamnus belopharospermus*, *Pileanthus* spp. (Myrtaceae).
Comment: This example occurs in the eastern part of the Kalbarri National Park and extends south eastwards towards Mullewa.

5. KALBARRI NATIONAL PARK (Lat. 27 52′S, Long. 114 30′E)

Landform: Undulating plain.
Soil: White or pale yellow siliceous sand.
Structure: Open-scrub, tall shrubland with heathy understorey tall and low open-heathland.

Composition: Phanerophytes: *Grevillea leucopteris*, *Banksia prionotes*, *Conospermum stoechadis*, *Petrophile ericifolia* (Proteaceae); *Calytrix oldfieldii*, *Verticordia* spp., *Darwinia virescens* (Myrtaceae); *Pityrodia oldfieldii* (Verbenaceae). Hemicryptophyte: *Ecdeiocolea monostachya* (Restionaceae).
Comment: Heathlands characterised by *Grevillea leucopteris* extend from Dongara to the lower Murchison River. The composition of the lower shrubs is somewhat variable.

6. MORESBY RANGE (Lat. 28 29′S, Long. 114 36′E)

Landform: Top and slopes of mesas.
Soil: Lateritic podsol.
Structure: Tall and low closed- and open-heathland.
Composition: Phanerophytes: *Calothamnus homalophyllus*, *Melaleuca* spp., *Verticordia* spp., *Scholtzia* spp. (Myrtaceae); *Acacia* spp. (Mimosaceae); *Halgania sericiflora* (Boraginaceae); *Cryptandra gracilipes* (Rhamnaceae); *Grevillea pinaster* (Proteaceae); *Gastrolobium oxylobioides* (Fabaceae).
Comment: This heathland occurs along the Moresby Range, a series of mesas near Geraldton. It is unusual in the paucity of species of Proteaceae and Epacridaceae.

7. BUNBURY TO DONGARA (Lat. 29–33 S) — COASTAL HEATHLAND ON SAND

Landform: Coastal dunes.
Soil: Calcareous sand.
Structure: Closed- and open-scrub.
Composition: Phanerophytes: *Acacia rostellifera*, *A. cyclops* (Mimosaceae); *Scaevola crassifolia* (Goodeniaceae); *Anthocercis littorea* (Solanaceae); *Xanthorrhoea preissii* (Xanthorrhoeaceae). Climbers: *Clematis microphylla* (Ranunculaceae); *Hardenbergia comptoniana* (Fabaceae). Hemicryptophyte: *Acanthocarpus preissii* (Liliaceae).
Comment: Though the distribution is somewhat disjunct there are some extensive areas of this heathland, especially between Perth and Jurien.

8. BUNBURY TO DONGARA (Lat. 29–33 S) — COASTAL HEATHLANDS ON LIMESTONE

Landform: Low coastal hills.
Soil: Shallow calcareous sand on coastal limestone, rock outcrops frequent.
Structure: Closed-scrub to tall open-heathland.
Composition: Phanerophytes: *Melaleuca huegelii*, *M. acerosa*, *Calothamnus quadrifidus* (Myrtaceae); *Dryandra sessilis*, *Grevillea thelemanniana* (Proteaceae); *Templetonia retusa* (Fabaceae).
Comment: The example is common along this section of coast, though some large disjunctions occur. It extends up to 12 km inland.

9. SOUTH OF ENEABBA (Lat. 29 05′S, Long. 115°18′E)

Landform: Wide, shallow valleys and slopes.
Soil: Deep siliceous sand and lateritic podsol.
Structure: Tall and low open- and closed-heathland.
Composition: Phanerophytes: *Banksia candolleana*, *Strangea cynanchicarpa*, *Conospermum stoechadis*, *Petrophile drummondii* (Proteaceae); *Verticordia grandis*, *Calytrix brachyphylla*,

Fig. 7.9. Near Badgingarra. Low closed-heathland on laterite with emergent *Xanthorrhoea preissii* (left) and *Nuytsia floribunda* (right). Many species of Proteaceae, Myrtaceae, etc.

Eremaea violacea (Myrtaceae); *Leucopogon* spp. (Epacridaceae). Chamaephytes: *Lechenaultia* spp. (Goodeniaceae). Hemicryptophyte: *Anigozanthos pulcherrimus* (Haemodoraceae).
Comment: This example is extremely rich floristically and is extensive in the region from the Moore River to Dongara. This and the following example with many variations form most of the heathlands between the Moore River and Dongara. Many species are endemic to them.

10. NORTH OF BADGINGARRA (Lat. 30°10′S, Long. 115°25′E; Fig. 7.9)

Landform: Low hills and mesas.
Soil: Massive laterite and lateritic podsol.
Structure: Tall and low open- and closed-heathland
Composition: Phanerophytes: *Dryandra* spp., *Banksia sphaerocarpa*, *Hakea conchifolia*, *Lambertia multiflora* (Proteaceae); *Calothamnus sanguineus*, *Hypocalymma linifolium* (Myrtaceae); *Daviesia epiphylla* (Fabaceae); *Astroloma microdonta* (Epacridaceae); *Xanthorrhoea preissii* (Xanthorrhoeaceae). Hemicryptophyte: *Macropidia fuliginosa* (Haemodoraceae).
Comment: Floristically rich and varied, this heath occurs on hills between the Moore River and Eneabba. An outstanding example occurs around Mount Lesueur amd Mount Peron near Jurien, where there are restricted endemic species such as *Banksia tricuspis*, *Hakea neurophylla* (Proteaceae), *Leucopogon plumuliflorus* (Epacridaceae) and *Xanthosia tomentosa* (Apiaceae).

11. NORTH OF REGANS FORD (Lat. 31°20′S, Long. 115°44′E)

Landform: Shallow swale.
Soil: Yellow-brown sand.

Structure: Closed- to open-scrub.
Composition: Phanerophytes: *Acacia blakelyi* (Mimosaceae); *Scholtzia* sp. (Myrtaceae); *Banksia* aff. *sphaerocarpa* (Proteaceae); *Jacksonia* sp. (Fabaceae).
Comments: Dominated by *Acacia blakelyi*, this heath extends from the Moore River to Dongara. In northern areas *Chamelaucium uncinatum* (Myrtaceae) is a common associated species.

12. WUBIN (Lat. 30°03′S, Long. 116°40′S)

Landform: Plain, sometimes gently undulating.
Soil: Sandy loam, sometimes over lateritic gravel.
Structure: Tall closed- and open-heathland.
Composition: Phanerophytes: *Isopogon scabriusculus*, *Grevillea paradoxa*, *Hakea invaginata*, *Persoonia coriacea* (Proteaceae); *Casuarina campestris* (Casuarinaceae); *Acacia neurophylla* (Mimosaceae); *Melaleuca conothamnoides* (Myrtaceae).
Comment: This is an inland heathland, occurring between Morawa and Bonnie Rock, to the southwest of the interzone between the South West and Eremean Botanical Provinces.

13. NORTHEAST OF WONGAN HILLS (Lat. 30°50′S, Long. 116°50′E; Fig. 7.10)

Landform: Undulating plain.
Soil: Pale yellow sand.
Structure: Open-scrub and tall shrubland with heathy understorey
Composition: Phanerophytes: *Xylomelum angustifolium*, *Banksia prionotes*, *Grevillea armigera* (Proteaceae); *Verticordia monadelpha*, *V. brownii*, *Melaleuca cordata* (Myrtaceae); *Actinostrobus arenarius* (Cupressaceae); *Acacia* spp. (Mimosaceae).

Fig.7.10. Northeast of Wongan Hills. Tall shrubland with heathy understorey on yellow sand. Tall shrubs are *Grevillea armigera* and *Actinostrobus arenarius*. Lower shrubs are mostly *Verticordia brownii* and *V. monadelpha*.

Comments: *Banksia prionotes* and/or *Xylomelum angustifolium* dominate this example which occurs as isolated areas between Coorow and Corrigin.

14. WADDOURING (Lat. 30 56′S, Long. 117 49′E)

Landform: Low granitic hill.
Soil: Shallow over granite.
Structure: Open-scrub and tall open-heathland.
Composition: Phanerophytes: *Casuarina campestris* (Casuarinaceae); *Calothamnus quadrifidus*, *Leptospermum roei*, *Baeckea* sp. (Myrtaceae); *Grevillea paniculata* (Proteaceae); *Astroloma serratifolium* (Epacridaceae). Hemicryptophyte: *Borya nitida* (Xanthorrhoeaceae).
Comment: The heathlands and shrublands around granitic outcrops are distinctive floristically due to the different soil and moisture régimes. They themselves also vary greatly in composition from those of the Darling Scarp through the wheatbelt to those inland along the South West-Eremean interzone. Many species are restricted to granitic soils.

15. EAST OF MERREDIN (Lat. 31 23′S, Long. 118 30′E; Fig. 7.11)

Landform: Plain.
Soil: Lateritic podsol.
Structure: Tall closed- and open-heathland.
Composition: Phanerophytes: *Casuarina campestris*, *C. corniculata* (Casuarinaceae); *Hakea scoparia* (Proteaceae); *Baeckea preissiana* (Myrtaceae).
Comment: Known locally as Tamma, this heathland is common in the central-eastern agricultural region of the South West Province. The two species of *Casuarina* are dominant.

Fig.7.11. East of Merredin. Tamma scrub-tall closed-heathland dominated by *Casuarina campestris*.

Fig.7.12. West of Southern Cross. Open-scrub on sand over laterite. *Eucalyptus burracoppinensis*, *Casuarina corniculata*, *Acacia* spp., *Grevillea eriostachya*.

16. WEST OF SOUTHERN CROSS (Lat. 31 22′S, Long. 118 38′E; Fig. 7.12)

Landform: Undulating plains.
Soil: Sandy yellow earth.
Structure: Open-scrub.
Composition: Phanerophytes: *Eucalyptus burracoppinensis*, *E. leptopoda*, *Melaleuca conothamnoides*, *Thryptomene* spp. (Myrtaceae); *Acacia* spp. (Mimosaceae); *Drummondita hassellii*, *Phebalium* spp. (Rutaceae); *Casuarina corniculata* (Casuarinaceae).
Comment: This heathland is common between Merredin and Bullabulling (West of Coolgardie).

17. SOUTH OF SOUTHERN CROSS (Lat. 31 38′S, Long. 119 30′E)

Landform: Undulating plains.
Soil: Yellow sand.
Structure: Tall and low closed- and open-heathland.
Composition: Phanerophytes: *Hakea falcata*, *Grevillea didymobotrya*, *Petrophile* spp. (Proteaceae); *Verticordia* spp., *Beaufortia micrantha*, *Melaleuca* spp., *Baeckea* spp. (Myrtaceae); *Phebalium* spp. (Rutaceae).
Comment: This heathland is common east and southeast from Southern Cross and with similar structure but somewhat different composition southwards towards Ravensthorpe.

18. SOUTH OF SOUTHERN CROSS (Lat. 31 55′S, Long. 119 19′E)

Landform: Undulating plains.
Soil: Pale yellow sand.
Structure: Tall shrubland with heathy understorey and tall open-heathland.

Composition: Phanerophytes: *Grevillea eriostachya, Grevillea* sp. inedit. *Petrophile ericifolia, Hakea falcata* (Proteaceae); *Casuarina campestris* (Casuarinaceae); *Melaleuca cordata, Verticordia chrysantha* (Myrtaceae); *Daviesia croniniana* (Fabaceae). Hemicryptophytes: *Ecdeiocolea monostachya* (Restionaceae), *Lepidosperma* spp. (Cyperaceae).

Comment: Heathland characterised by emergent *Grevillea eriostachya* is widespread between Wongan Hills, Southern Cross and Lake King. The lower shrubs vary in composition.

19. BULLARING (Lat. 32°30'S, Long. 117°44'E)

Landform: Gentle slope.
Soil: Lateritic podsol.
Structure: Tall shrubland with heathy understorey.
Composition: Phanerophytes: *Eucalyptus albida, Leptospermum roei, Verticordia brownii* (Myrtaceae); *Dryandra* spp., *Hakea incrassata, Petrophile striata, Adenanthos argyraeus* (Proteaceae); *Xanthorrhoea nana* (Xanthorrhoeaceae). Hemicryptophytes: *Mesomelaena stygia, Lepidosperma* spp. (Cyperaceae).

Comment: With the mallee *Eucalyptus albida* dominant in the upper storey, this example extends from Lake Grace to Corrigin.

20. SOUTHEAST OF CORRIGIN (Lat. 32°31'S, Long. 117°56'E)

Landform: Undulating plain.
Soil: Lateritic podsol.
Structure: Tall shrubland with heathy understorey.
Composition: Phanerophytes: *Eucalyptus macrocarpa, Melaleuca pungens, Beaufortia micrantha, Verticordia grandiflora* (Myrtaceae); *Dryandra vestita, Hakea gilbertii, Isopogon teretifolius* (Proteaceae); *Hibbertia subvaginata* (Dilleniaceae); *Chloanthes coccinea* (Dicrastylidaceae).

Comment: This example, of distinctive aspect due to the presence of *Eucalyptus macrocarpa*, is now confined to small areas between Wongan Hills and Wickepin. The lower shrub layer varies in composition.

21. NORTH OF TARIN ROCK (Lat. 32°49'S, Long. 118°30'E)

Landform: Higher parts of undulating plains.
Soil: Lateritic podsol with shallow massive laterite.
Structure: Tall shrubland with heathy understorey, tall closed-heathland.
Composition: Phanerophytes: *Casuarina pinaster* (Casuarinaceae); *Eucalyptus falcata, Melaleuca pungens* (Myrtaceae); *Dryandra* spp., *Grevillea* spp., *Hakea ferruginea, Isopogon teretifolius* (Proteaceae); *Physopsis lachnostachya* (Dicrastylidaceae).

Comment: Characterised by *Casuarina pinaster* and numerous Proteaceae, this example extends south and eastwards to Pingrup.

22. STIRLING RANGE NATIONAL PARK (Lat. 34°20'S, Long. 118°E; Fig. 7.13)

Landform: Mountain slopes and some adjacent plains.
Soil: Skeletal.
Structure: Open-scrub, tall closed- and open-heathland.

Fig. 7.13. Stirling Range. Open-scrub with *Xanthorrhoea preissii* and many species of Proteaceae, Myrtaceae and Epacridaceae.

Composition: Phanerophytes: *Eucalyptus decurva, E. preissiana, E. marginata, Beaufortia decussata* (Myrtaceae); *Lambertia ericifolia, Hakea ambigua, Isopogon baxteri* (Proteaceae); *Oxylobium atropurpurem, Gastrolobium velutinum* (Fabaceae); *Andersonia* spp., *Sphenotoma capitatum, Leucopogon* spp. (Epacridaceae); *Kingia australis, Xanthorrhoea preissii* (Xanthorrhoeaceae).

Comment: This heathland is somewhat variable in composition, but is always rich in Proteaceae, Myrtaceae and to a lesser extent Epacridaceae, Fabaceae and Mimosaceae. The higher slopes are characterised by such species as *Isopogon latifolius, Banksia solandri* (Proteaceae), *Sphenotoma dracophylloides, Andersonia axilliflora* (Epacridaceae) and *Casuarina trichodon* (Casuarinaceae).

23. EAST OF MANYPEAKS (Lat. 34°36'S, Long. 118°30'E)

Landform: Undulating plains.
Soil: On rises: lateritic podsol. In swales: deep white sand.
Structure: Closed- and open-scrub, tall closed- and open-heathland.
Composition: On rises — phanerophytes: *Eucalyptus marginata, E. preissiana, Melaleuca exarata,* (Myrtaceae); *Hakea cucullata, H. rubriflora, Isopogon cuneatus, Dryandra baxteri* (Proteaceae). In swales — phanerophytes: *Banksia coccinea, B. baxteri, B. baueri, Lambertia inermis, Adenanthos cuneatus* (Proteaceae).

Comment: These two examples form a close mosaic on the plains southeast of the Stirling Range.

24. ALBANY (Lat. 35°07'S, Long. 117°54'E)

Landform: Coastal hills.
Soil: Shallow sand over granite and limestone.
Structure: Tall and low closed-heathland.
Composition: Phanerophytes: *Scaevola nitida* (Goodeniaceae);

Pimelea ferruginea (Thymelaeaceae); *Dryandra formosa, Hakea trifurcata, Adenanthos sericeus* (Proteaceae); *Xanthorrhoea preissii* (Xanthorrhoeaceae). Hemicryptophyte: *Mesomelaena tetragona* (Cyperaceae).

Comment: This heathland extends from Cape Leeuwin to Esperance though with some variation in composition. It is one of the few coastal formations in which Proteaceae are common.

25. CULHAM INLET, WEST OF HOPETOUN (Lat. 33 56′S, Long. 120 03′E)

Landform: Coastal dunes.
Soil: Deep calcareous sand.
Structure: Tall closed- and open-heathland.
Composition: Phanerophytes: *Acacia rostellifera* (Mimosaceae); *Olearia axillaris* (Asteraceae); *Scaevola nitida* (Goodeniaceae); *Myoporum serratum* (Myoporaceae); *Melaleuca lanceolata, Eucalyptus angulosa* (Myrtaceae).
Comment: This heathland is common between Bremer Bay and Israelite Bay.

26. EAST MOUNT BARREN (Lat. 33 56′S, Long. 120 02′E; Fig. 7.14)

Landform: Hillsides.
Soil: Skelatal with extensive quartzite rocks.
Structure: Tall and low closed-heathland.
Composition: Phanerophytes: *Dryandra quercifolia, Banksia baueri, Adenanthos ellipticus, Hakea victoriae* (Proteaceae); *Eucalyptus coronata, E. lehmannii, Regelia velutina, Agonis* spp., *Melaleuca citrina* (Myrtaceae); *Hibbertia mucronata* (Dilleniaceae). Hemicryptophyte: *Lepidosperma* sp. indet. (Cyperaceae).

Comment: Extremely rich in species including endemic taxa, this heathland occurs on all the quartzite hills of the Fitzgerald River National Park (between Bremer Bay and Hopetoun). A similar type but of different composition occurs on Mt. Ragged, an analogous peak northwest of Israelite Bay.

27. NORTH OF HOPETOUN (Lat. 33 56′S, Long. 120 07′E)

Landform: Consolidated dunes.
Soil: Deep white siliceous sand.
Structure: Open-scrub.
Composition: Phanerophytes: *Banksia speciosa, B. baxteri, B. coccinea, Hakea corymbosa, Petrophile teretifolia* (Proteaceae); *Melaleuca striata, M. thymoides, Calothamnus gracilis* (Myrtaceae). Hemicryptophytes: *Lomandra hastilis* (Xanthorrhoeaceae); *Anigozanthos rufus* (Haemodoraceae); *Caustis dioica* (Cyperaceae).
Comment: *Banksia* spp. are visually dominant in this heathland which extends from Bremer Bay to beyond Israelite Bay. East of Hopetoun, *B. speciosa* is dominant.

28. NORTH OF HOPETOUN (Lat. 33 56′S, Long. 120 08′E; Fig. 7.15)

Landform: Plain.
Soil: Deep white siliceous sand.
Structure: Open-scrub and tall shrubland with heathy understorey.
Composition: Phanerophytes: *Lambertia inermis, Adenanthos cuneatus, Banksia pulchella, B. repens* (Proteaceae); *Melaleuca striata, M. thymoides* (Myrtaceae).
Comment: Though similar to the previous example in structure and lower shrub composition, the dominance of *Lambertia inermis* gives this heathland a distinctive aspect. It extends from Bremer Bay to Israelite Bay.

Fig.7.14. East Mount Barren. Tall closed-heathland on quartzite. Rich mixture especially of Proteaceae, Myrtaceae and Fabaceae. Columnar shrubs are *Hakea victoriae*.

Fig. 7.15. North of Hopetoun. Low open-heathland in foreground, open scrub behind dominated by *Lambertia inermis*.

29. ELVERDTON, SOUTHEAST OF RAVENSTHORPE (Lat. 33 37'S, Long. 120 09'E)

Landform: Low hills.
Soil: Massive laterite.
Structure: Closed- and open-scrub.
Composition: Phanerophytes: *Dryandra quercifolia*, *Banksia lemanniana*, *Hakea obtusa*, *Grevillea patentiloba* (Proteaceae); *Beaufortia orbifolia*, *Agonis parviceps*, *Calothamnus pinifolius*, *Eucalyptus preissiana*, *E. falcata* (Myrtaceae).
Comment: This example occurs chiefly in the Ravensthorpe district but extends west to the lower Fitzgerald River.

30. SOUTH OF LAKE KING (Lat. 33 25'S, Long. 119 55'E; Fig. 7.16)

Landform: Gently undulating plain.
Soil: Lateritic podsol.
Structure: Tall shrubland with heathy understorey.
Composition: Phanerophytes: *Eucalyptus tetragona*, *E. redunca*, *E. incrassata*, *Beaufortia micrantha*, *Melaleuca* spp., *Calothamnus gracilis* (Myrtaceae); *Banksia media*, *Isopogon polycephalus*, *Grevillea nudiflora* (Proteaceae); *Acacia* spp. (Mimosaceae); *Hibbertia* spp. (Dilleniaceae).
Comment: This example, with *Eucalyptus tetragona* as the common species of the tallest layer, is very common between Jerramungup and Israelite Bay, though the composition of the understorey is variable. A visual charge occurs beyond Condingup, east of Esperance, where *E. tetragona* loses it glaucous aspect.

Fig.7.16. South of Lake King. Tall shrubland with heathy understorey. Tall shrubs are the mallee *Eucalyptus tetragona*. Mixed lower storey especially of Myrtaceae, Proteaceae and Fabaceae.

Fig.7.17. South of Lake King. Low closed-heathland. Proteaceae and Myrtaceae predominant.

31. NORTH OF CAPE ARID (Lat. 33 50'S, Long. 123 10'E; Fig. 7.17)

Landform: Plains.
Soil: Grey sand over yellow clay.
Structure: Low closed- and open-heathland.
Composition: Phanerophytes: *Dryandra tenuifolia*, *Isopogon polycephalus*, *Banksia repens*, *Grevillea baxteri* (Proteaceae); *Beaufortia micrantha*, *Melaleuca striata*, *Eucalyptus tetragona* (Myrtaceae); *Hibbertia* spp. (Dilleniaceae); *Acacia* spp. (Mimosaceae); *Xanthorrhoea* sp. (Xanthorrhoeaceae); *Daviesia* spp. (Fabaceae). Hemicryptophytes: many Restionaceae, Cyperaceae.
Comment: Extensive areas of this heathland occur east of Esperance with variation in composition due to soil differences It often contains no *Eucalyptus* species; where *E. tetragona* does occur it is of low stature. Low heathland of somewhat different composition occurs westwards to Lake Grace.

REFERENCES

Beadle, N.C.W., 1966. Soil phosphate and its role in moulding segments of the Australian flora and vegetation, with special reference to xeromorphy and sclerophylly. *Ecology*, 47: 992–1007.

Beard, J.S., 1965. *Descriptive Catalogue of West Australian Plants*. Society for Growing Australian Plants, Sydney, N.S.W., 142 pp.

Beard, J.S., 1969. The vegetation of the Boorabbin and Lake Johnston areas, Western Australia. *Proc. Linn. Soc. N.S.W.*, 93: 239–69.

Beard, J.S., 1973. *Vegetation Survey of Western Australia. The*

Vegetation of the Esperance and Malcolm Areas, Western Australia. Vegmap Publications, Perth, W.A., 41 pp., 1 map.

Beard, J.S., 1975. *Vegetation Survey of Western Australia. The Vegetation of the Nullarbor Area.* University of Western Australia Press, Perth, W.A., 104 pp., 1 map.

Beard, J.S., 1976a. An indigenous term for the Western Australian sandplain and its vegetation. *J. R. Soc. W. Aust.,* 59: 55–57.

Beard, J.S., 1976b. *Vegetation Survey of Western Australia. The Vegetation of the Shark Bay and Edel Area, Western Australia.* Vegmap Publications, Perth, W.A., 20 pp., 1 map.

Beard, J.S. and Burns, A.C., 1976. *Vegetation Survey of Western Australia. The Vegetation of the Geraldton Area, Western Australia.* Vegmap Publications, Perth, W.A., 35 pp., 1 map.

Bettenay, E. and Hingston, F.J., 1961. Soils of the Merredin area, W.A. *CSIRO Soils Land Use Ser.,* No 41: 36 pp.

Bettenay, E., Churchward, H.M. and McArthur, W.M., 1967. *Atlas of Australian Soils. Sheet 6. Meekatharra–Hamersley Range Area, with Explanatory Data.* CSIRO — Melbourne University Press, Melbourne, Vic., 30 pp., 1 map.

Burbidge, N.T., 1960. The phytogeography of the Australian region. *Aust. J. Bot.,* 8: 75–212.

Burbidge, A.A., and George, A.S., 1978. The fauna and flora of Dirk Hartog Island, Western Australia. *J. R. Soc. W.Aust.,* 60: 71–90.

Cain, S.A., 1944. Foundations of Plant Geography. Hafner, New York, N.Y., 556 pp.

Diels, L., 1906. *Die Vegetation der Erde, 7. Die Pflanzenwelt von West-Australien südlich des Wendekreises.* Engelmann, Leipzig, 413 pp.

Erickson, R., George, A.S., Marchant, N.G. and Morcombe, M.K., 1973. *Flowers and Plants of Western Australia.* Reed, Sydney, N.S.W., 216 pp.

Gardner, C.A., 1944. The vegetation of Western Australia with particular reference to the climate and soils. *J. R. Soc. W. Aust.,* 28: 11–87.

Gardner, C.A. and Bennetts, H.W., 1956. *The Toxic Plants of Western Australia.* Western Australian Newspapers Ltd., Perth, W.A., 253 pp.

Hallam, S.J., 1975. Fire and hearth: a study of aboriginal usage and European usurpation in South-Western Australia. *Australian Aboriginal Studies,* No. 58: 158 pp. (Australian Institute of Aboriginal Studies, Canberra, A.C.T.)

Havel, J.J. and Batini, F.E., 1973. Focus on land use conflicts in the northern jarrah forest. *For. Focus,* 11: 3–15 (Forests Department of Western Australia, Perth, W.A.)

Hnatiuk, R.J. and Hopkins, A.J.M. in prep. An ecological survey in shrublands south of Eneabba. Department of Fisheries and Wildlife, Western Australia, Report.

Hooker, J.D., 1860. *The Botany of the Antarctic Voyage of H.M. Discovery Ships "Erebus" and "Terror", in the years 1839–1843. Volume 3. Flora Tasmaniae 1855–1860.* Lovell Reeve, London, 422 pp.

Hopkins, B., 1957. The concept of minimal areas. *J. Ecol.,* 45: 441–449.

Hopkins, A.J.M. and Trudgen, M.E., in prep. The unburnt vegetation. In: A.J.M. Hopkins (Editor), *The Results of*

Studies on Middle Island, Recherche Archipelago. Wildl. Res. Bull., W. Aust.

Hurlburt, S.H., 1971. The nonconcept of species diversity — a critique and alternative parameters. *Ecology,* 52: 577–586.

Johnson, L.A.S. and Briggs, B.G., 1975. On the Proteaceae — the evolution and classification of a southern family. *Bot. J. Linn. Soc.,* 70: 83–185.

Kitchener, D.J., Chapman, A. and Dell, J. 1975. A biological survey of Cape Le Grand National Park. *Rec. W. Aust. Mus., Supp.,* No. 1: 48 pp.

Lamont, B., 1976. *Report on a Biological Survey and Recommendations for Rehabilitating a Portion of Reserve 31030 to be Mined for Heavy Minerals During 1975–81.* WAIT-AID Ltd., Perth, W.A., 62 pp.

McKenzie, N.L., Burbidge, A.A. and Marchant, N.G., 1973. Results of a biological survey of a proposed wildlife sanctuary at Dragon Rocks near Hyden, Western Australia. *Dep. Fish. Wildl., W. Aust., Rep.* No. 12: 27 pp.

Marchant, N.G., 1973. Species diversity in the south-western flora. *J. R. Soc. W. Aust.,* 56: 23–30.

Mueller-Dombois, D. and Ellenberg, E., 1974. *Aims and Methods of Vegetation Ecology.* Wiley, New York, N.Y., 574 pp.

Muir, B.G., 1976. Biological survey of the Western Australian wheatbelt. Part 1. Tarin Rock and North Tarin Rock Reserves. II Vegetation. *Rec. W. Aust. Mus., Suppl.,* No. 2: 21–59.

Muir, B.G., 1977. Biological survey of the Western Australian wheatbelt. Part 2. Vegetation and habitat of Bendering reserve. *Rec. W. Aust. Mus. Suppl.* No. 3: 1–142.

Mulcahy, M.J., 1973. Landforms and soils of south western Australia. *J. R. Soc. W. Aust.,* 56: 16–22.

Nelson, E.C., 1974. Disjunct plant distributions on the south-western Nullarbor Plain, Western Australia. *J. R. Soc. W. Aust.,* 57: 105–117.

Northcote, K.H., 1971. *A Factual Key for the Recognition of Australian Soils.* Rellim Technical Publications, Glenside, 3rd ed., 123 pp.

Northcote, K.H., Bettenay, E., Churchward, H.M. and McArthur, W.M., 1967. *Atlas of Australian Soils. Sheet 5. Perth–Albany Esperance Area, with Explanatory Data.* CSIRO — Melbourne University Press, Melbourne, Vic., 52 pp, 1 map.

Phipps, J.B., and Cullen, J., 1976. Centrew of diversity quantified — A maximum variance approach to a biogeographic problem. *Vegetatio,* 31: 147–159.

Raven, P.H., 1971. The relationship between Mediterranean floras. In: P.H. Davis, P.C. Harper and I.C. Hedge (Editors), *Plant Life of South-West Asia.* Botanical Society, Edinburgh, pp. 119–134.

Smith, F.G., 1972. *Vegetation Survey of Western Australia.* Vegetation Map of Pemberton and Irwin Inlet. Western Australian Department of Agriculture, Perth, W.A., 31 pp., 1 map.

Smith, F.G., 1972. *Vegetation Survey of Western Australia.* Vegetation Map of Busselton and Augusta, Western Australian Department of Agriculture, Perth, W.A., 32 pp., 1 map.

Specht, R.L. and Rayson, P., 1957. Dark Island heath (Ninety Mile Plain, South Australia). I Definition of the ecosystem. *Aust. J. Bot.,* 5: 52–85.

Speck, N.H., 1958. *The Vegetation of the Darling–Irwin Botanical Districts and an Investigation of the Distribution of the Family Proteaceae in South-Western Western Australia.* Thesis, Botany Department, University of Western Australia, Perth, W.A., 637 pp. (unpublished).

UNESCO — FAO, 1963. Ecological study of the Mediterranean Zone. Bioclimatic map of the Mediterranean Zone. *UNESCO Arid Zone Res.*, 21: 2–58.

Whittaker, R.H., 1972. Evolution and measurement of species diversity. *Taxon*, 21: 213–251.

Whittaker, R.H., 1977. Evolution of species diversity in land communities. *Evol. Biol.*, 10: 1–67.

Chapter 8

THE VERTEBRATE FAUNA OF AUSTRALIAN HEATHLANDS — AN EVOLUTIONARY PERSPECTIVE[1]

J. KIKKAWA, G.J. INGRAM and P.D. DWYER

INTRODUCTION

In this chapter the biogeographic and ecological significance of heathlands for the evolution of vertebrate animals will be discussed in the Australian context. For this discussion we regard heathlands as a habitat which comprises plant communities dominated by sclerophyllous shrubs and subshrubs, including shrubland and heathy scrub as well as heathland (Specht, 1970 and also in Chapter 1). Ecological relations between heathlands and other habitats are discussed elsewhere in this volume (Dwyer et al., Chapter 9).

Until recently the direction of evolutionary shift consistently postulated for vertebrate groups has been from Old World tropics to Australia (Darlington, 1957; Keast, 1959a, b; Troughton, 1959; Storr, 1964; Kluge, 1967; Horton, 1972; Heatwole, 1976) and, within Australia, from tropical to temperate regions (Keast, 1961; Horton, 1972; Tyler, 1972) and from wet to dry habitats (Martin, 1967, 1970; Macdonald, 1973; Ford, 1974; Heyer and Liem, 1976; Schodde and McKean, 1976). Migration and subsequent speciation in the southwest of the continent and Tasmania have been interpreted in relation to Quaternary changes of sea level (Main et al., 1958; Serventy and Whittel, 1962; Abbott, 1973; Hope, 1974; Little-john and Martin, 1974; Rawlinson, 1974). In the absence of palaeoecological evidence little attention has been given to the distribution of the autochthonous fauna during the Tertiary (Ride, 1968; Kikkawa and Pearse, 1969; Tyndale-Biscoe, 1973; Rich, 1976). However, information assembled for this volume (Specht, Chapters 1 and 6) and elsewhere (Keast, in press) concerning the chronology of continental drift and the history of the

Australian environment, affords an opportunity for re-examining the problem of past distribution.

As discussed by Specht in Chapter 6 the heathlands in Australia do not occupy vast areas today, though the heathland flora is widely spread over the continent. The heathland vegetation often occurs in association with other vegetation or peculiar substrates to which animals may be primarily adapted. Animals depend ultimately on plants for their energy. Even if there are more obvious associations of animals with abiotic features (e.g. temporary presence of water, rocky or sandy substrate) than with plants, heathlands should be treated as biotic communities. Also, heathy vegetation as a layer in a woodland ·or a forest is considered as forming an integral part of a more complex community in which ecological distribution of animals is influenced by the presence of heathland plants (e.g. as cover). We have therefore considered in our discussion the vertebrates of all those habitats in which heathland vegetation constitutes a prominent layer.

In the biogeographic interpretation of the Australian heathland vertebrates we offer a new perspective, complementing views currently held by other workers and consistent with the following assumptions derived from recent findings on the history of the Australian heathland flora.

(1) The sclerophyllous (heathland) flora originated in Gondwanaland.

(2) Heathland communities developed along with mesophyllous communities.

(3) At some time in the early Tertiary heathlands occupied much of the Australian continent.

---

[1] Manuscript completed November, 1977.

(4) Through the late Tertiary and the Quaternary, heathlands became fragmented so that today they occur in pockets, with the southwest and the east coast of the continent acting as major refuges.

(5) Heathland flora successfully invaded the coastal sand dunes and wind-blown inland dunes which developed in the Pleistocene.

This sequence parallels the presumed history of tropical and subtropical closed-forest (rain forest) in Australia, which today has its major refuge in the northeast of the continent. Under the climatic régime of the Tertiary, heathlands developed on infertile soils while closed-forest occupied fertile land, but subsequently, as the continent became arid and cool, both vegetation types have become fragmented.

The establishment and expansion of heathlands must have entailed adaptations of both autochthonous and allochthonous faunas whereas the contraction and impoverishment of heathlands must have been accompanied by extinction of some taxa, and emancipation of others from obligatory dependence on heathland habitats. These events would have occurred on a continent-wide basis through the Tertiary, and would have resulted in continued speciation and extinction in southern heathlands during the Pleistocene. Fauna specialised for heathland-living may therefore have had the same fate as closed-forest fauna, with many elements becoming extinct and others becoming relicts in isolated pockets. Recent expansion of coastal heathlands would, however, have promoted invasion by open-country species and perhaps led to speciation in some groups.

An examination of endemicity and distribution patterns for heath-adapted vertebrates should reveal some of these hypothesized changes. Differences in antiquity of different vertebrate classes, or in the progress of their radiation on the Australian continent through the Tertiary, imply that different endemicity patterns might emerge for each class. Differences in vagility combined with environmental factors might lead to different patterns of speciation between classes. Contemporaneity is also difficult to establish between similar events postulated for different taxa. Our examination of endemicity patterns is further constrained by differences in the state of taxonomy for vertebrate classes. For example, the

status of some mammals (such as species of *Melomys* and *Pseudomys*) is unclear, and new species of lizards and frogs are still being described. For all these groups, phylogenetic interpretations are in flux. Even birds, which present virtually no systematic problems at the species level and thus are discussed in some detail in this paper, are poorly understood when the affinities of autochthonous elements are examined (Sibley, 1976a).

HISTORY AND GEOGRAPHIC DISTRIBUTION

Vertebrate animals recorded in heathland habitats of various regions of Australia are tabulated in the Appendix, and the number of species recorded for each class is given according to the region in Fig. 8.1.

The information was collected from various sources including unpublished notes of many workers. However, the list is far from complete, and information remains uneven owing to the poor state of distributional knowledge for some groups (e.g. bats) and some regions (e.g. northern Australia). We also found it difficult to determine the distribution within heathlands of those species typical of more arid regions of the continent and of others more typical of forest understorey. We have tried to identify species which are primarily adapted to heathlands and have sought evidence concerning their status and distribution within heathlands.

Amphibians

Savage (1973) regarded the Cretaceous as the most important period for anuran evolution and thought that most existing families had appeared before the Tertiary. Four families (Myobatrachidae, Pelodryadidae, Microhylidae, Ranidae) of frogs and an introduced toad (Bufonidae) constitute the amphibian fauna of Australia. Savage (1973) hypothesized that the ancestors of myobatrachids, microhylids and pelodryadids were represented in the Gondwanaland fauna. In the early Tertiary the myobatrachids dominated southern parts of Australia, whereas the microhylids and the leptodactyline ancestors of the pelodryadids dominated the northern tropical area. Subsequent cooling and

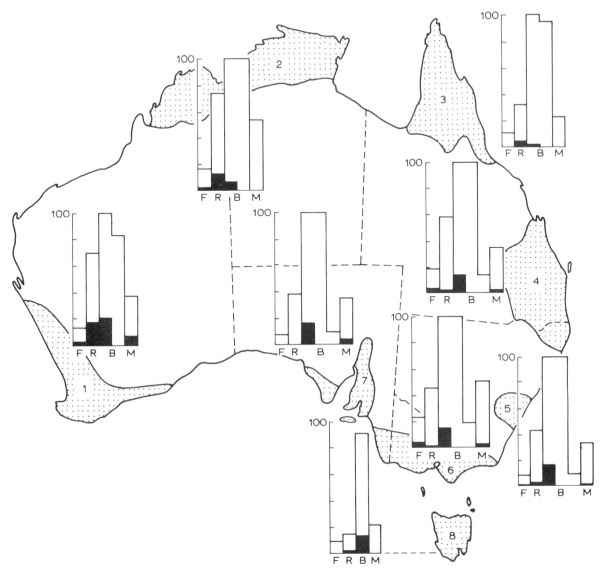

Fig. 8.1. Map showing distribution of eight regions (stippled) within which the areas of heathlands and associated habitats discussed in this paper are located. Histograms show the numbers of indigenous species of vertebrates (F = amphibians, R = reptiles, B = birds, M = mammals) known from heathland areas. For birds the number greater than 100 can be read by summing the columns. The number of species showing special association with heathland habitat of the particular region is indicated by the black part of the column for each vertebrate class. The details of species distribution are given in the Appendix.

drying through the Oligocene forced the two latter groups into smaller areas in the north (Savage, 1973). The myobatrachids would most certainly have radiated into heathlands as these expanded. The dramatic changes of the Quaternary climate have probably caused a subsequent extinction of many heath-adapted myobatrachids. The most striking relict of this group is *Myobatrachus*, a monotypic fossorial genus, adapted to under-

ground living in sandy heathlands of southwestern Australia (Cogger, 1975). Other myobatrachid frogs of heathlands show various degrees of adaptation and phylogenetic isolation: *Philoria*, a monotypic genus, is restricted to one locality in the alpine heathlands of the southeast, with the closest taxon, *Kyarranus*, confined to subtropical rain forest; *Pseudophryne* is well represented in heathlands, with one species (*P. australis*) in the

Hawkesbury Sandstone area, another (*P. cor-roboree*) in *Sphagnum* bogs of southeastern montane heathlands and a large race of a third species (*P. coriacea*) in coastal heathlands of southeastern Queensland (Corben and Ingram, in prep.); *Paracrinia haswelli*, a monotypic genus (Heyer and Liem, 1976), occurs in sandstone and coastal heathlands of the southeast; and *Ranidella* (=*Crinia*) *tinnula* (Straughan and Main, 1966) in northeastern New South Wales and southeastern Queensland is associated with waters of low pH in coastal heathlands. *Heleioporus* shows burrowing adaptations to lateritic soils (Lee, 1967) and has a relict pattern of distribution over the continent, with five species (four in heathlands) in southwestern Australia and one each in Arnhem Land sandstones and southeastern Australia. This suggests that the genus may once have been widespread when heathland was prevalant but subsequently contracted to the continental fringes. It is noteworthy that other primitive burrowing limnodynastine genera, *Neobatrachus* and *Notaden*, are widespread in the arid regions of the continent; these forms may have escaped from the contracting heathlands before the Pleistocene.

The pelodryadid frogs have no early Tertiary relicts restricted to heathlands, but the genus *Litoria* has radiated recently in all wet habitats of the continent and has produced several species restricted to eastern heathlands (Ingram and Corben, 1975). These include a form of *L. ver-reauxii* in subalpine heathlands of the southeast.

Rana daemelii is the only member of the ranid frogs known from Australia, and is confined to the northeast of the continent. It is a recent colonizer, being closely related to the New Guinea member, *R. papua*, (Menzies, 1975), and occurs in streams of closed-forest and heathlands in Cape York Peninsula. The microhylid frogs are considered to be recent recolonizers from New Guinea (Savage, 1973) to northern Australia, where seven species have been described (Cogger, 1975; Zweifel and Parker, 1977). Only one microhylid frog (*Sphenophryne robusta*) is known to penetrate heathlands.

Bufo marinus (Bufonidae) has been introduced to north Queensland and since the 1930s has spread south, west and further north (Covacevich and Archer, 1975). It has colonized heathlands of south-eastern Queensland where it attains high density.

The distribution and endemicity patterns shown by the Amphibia support our thesis that contraction of heathlands through the Tertiary has imposed a relict distribution upon many groups, and has resulted in radiation away from heathland habitats for others. Successful colonization of the Pleistocene heathlands is also evident, the most recent invasion (that of *Bufo marinus*) being a historical event.

Reptiles

Among the reptiles there are genera within eight families (Chelidae, Gekkonidae, Pygopodidae, Agamidae, Varanidae, Scincidae, Typhlopidae, Elapidae) that show close association with the heathlands. Of these the chelid turtles and possibly the gekkonid, pygopodid and agamid lizards in Australia have originated in Gondwanaland (Cracraft, 1974) and radiated during the early Tertiary. The degree of endemism for the chelids, pygopodids and elapids is highest in the southwest of the continent (Storr, 1964). In the family Pygopodidae (legless lizards) six out of the eight genera known (Kluge, 1974) occur in the southwest, where two monotypic genera, *Aclys* and *Pletholax*, are confined to heathlands. Some species (e.g. *Aprasia striolata* and *Delma fraseri*) have disjunct distributions in the south. Of the 26 genera recognized in the Australian Elapidae (Cogger, 1975), the monotypic *Elapognathus* and *Rhinoplocephalus* are restricted to the southwest, the former being found only in heathland. *Hoplocephalus* has three species in the east, one of which (*H. bungaroides*) is confined to heathlands on the Hawkesbury sandstone. *Drysdalia* and *Notechis* exhibit a disjunct pattern of distribution across the south of the continent, each with one species in the southwest and at least one other confined to the southeast, including heathland areas. *Neelaps* has two burrowing species, one confined to southwestern heathland and the other ranging from the southwest to South Australia. *Simoselaps*, with many burrowing species, has one species confined to the southwest heathland. Few other genera of this family have species that show any association with heathlands, though large snakes of the genera *Oxyuranus*, *Pseudechis* and *Pseudonaja* are common in this and other habitats.

The Typhlopidae (blind snakes) are another old

family represented in Australia by a single, wide-spread genus, *Ramphotyphlops* (McDowell, 1974; Stimson et al., 1977), which penetrates heathland habitats in Arnhem Land (Mitchell, 1964) and southeastern Queensland (Covacevich, 1977).

The Gekkonidae (geckos) includes several moderately large or large genera (e.g. *Diplodactylus*, *Gehyra*, *Oedura*), many species of which occur in association with heathlands throughout Australia. This association is often through their attachment to rocky outcrops. Thus two unnamed species of *Gehyra* (Storr and Smith, 1975), *Heteronotia spelea* and *Pseudothecadactylus lindneri* (Cogger, 1975, 1976) occur in northern heathlands developed on sandstone; and *Phyllurus platurus* is confined to the Hawkesbury Sandstone heathlands of the east.

Within the Agamidae (dragon lizards) at least three species of *Amphibolurus* are restricted to the southwestern heathland and one species to the Kimberley heathlands. The widespread, monotypic genus *Chlamydosaurus* is common in eastern heathlands and at least three species of *Diporiphora* appear in heathlands of rock outcrops in the northwest.

In the family Varanidae (goannas) *Varanus glauerti* and another *Varanus* species (Storr and Smith, 1975) are known from the sandstone heathlands of the Kimberleys, and *V. glebopalma* from those of the Kimberleys and Arnhem Land.

Within the Scincidae (skinks) a species of *Omolepida* (Storr, 1976a) is restricted to the Kimberleys, a species of *Anomalopus* to the Hawkesbury–Hunter River area in the east (Cogger, 1975) and a species of *Cryptoblepharus* to Arnhem Land escarpments (H.G. Cogger, pers. comm., 1977). Association with sandstone heathlands is also shown by *Ctenotus mastigura* (Storr, 1975a) in the Kimberleys and Arnhem Land, and by another species of *Ctenotus* in the Cape York Peninsula sandstone area. One species each of *Carlia*, *Ctenotus* and *Lerista* is confined to the coastal heathland of Cape York Peninsula. Several species of *Ctenotis*, *Lerista*, *Egernia* and *Hemiergis* occur in the heathlands of the southwest, the southeast or with disjunct distributions in both areas. *Leiolopisma trilineata* also has a disjunct distribution across the southern part of the continent, whereas a related species, *L. greeni* (Rawlinson, 1974), is a Tasmanian endemic confined to subalpine heathlands. A peculiar pattern of

distribution is shown by some burrowing species (e.g. *Anomalopus truncatus*) and non-burrowing species (e.g. *Anotis graciloides*) which are found only in heathland and closed-forest habitats. The Scincidae is the largest family of Australian lizards, showing an enormous variety of morphology, habits and habitat occupation (Cogger, 1975; Heatwole, 1976). The above-mentioned genera include many species that lack any association with heathlands; the heathland representatives are probably best seen as part of a recent and continent-wide radiation of these genera and others, such as *Saiphos*, *Morethia*, *Sphenomorphus* and *Tiliqua*, whose ranges intrude into heathlands of the east or the southwest.

Within Chelidae (fresh-water turtles), a monotypic genus, *Pseudemydura*, is confined to one locality (Bullsbrook) in the southwest (Cogger, 1975), where it is considered to have been isolated since the Cretaceous (Burbridge et al., 1974), and a new form in the genus *Emydura* is confined to dune lakes of Fraser Island, Queensland (J. Legler, pers. Comm., 1977). Few other chelids penetrate aquatic habitats of heathlands.

With few exceptions (e.g. *Boiga* and *Dendrelaphis* in some southern heathlands; H.G. Cogger, pers. comm., 1977), the remaining reptile families, Boidae, Acrochordidae and Colubridae, show no special association with heathlands.

In the Australian desert, Pianka (1972) recognized that lizards showed great species diversity accompanied by a high degree of habitat specificity, and he argued that the fluctuations of habitats during the Pleistocene led to speciation in a large number of groups. Significantly no genera of the old elements (e.g. Pygopodidae) fit this model.

In heathlands most of the old endemics with relict patterns of distribution are burrowers and occur in the southwest of the continent or in association with rocky outcrops of northern Australia. Their distribution is superimposed by radiants of more recent origin.

Birds

For many birds, the association with heathlands appears to be essentially opportunistic and non-exclusive. Thus, most Australian water birds may be found in heathland waterways when these occur within their respective geographic ranges, but they are seldom abundant there. Open-country species of diurnal and nocturnal birds of prey (e.g. *Elanus, Circus, Haliaeetus, Falco cenchroides, Tyto longimembris*) appear in heathlands opportunistically, as do certain phasianid and turnicid quails (e.g. *Coturnix chinensis, Turnix varia*). Frogmouths, nightjars, swifts and kingfishers and, amongst passerines, many endemic species within Climacteridae, Neosittidae, Cracticidae, Ptilonorhynchinae and Estrildinae, occur in heathlands without any special relationship with this habitat. *Merops ornatus*, the only Australian member of the bee-eaters, is a common bird in heathlands but again it is not restricted to this habitat. Nor are the birds of Gondwanaland origin (the ratites and possibly megapods; Cracraft, 1976) closely associated with heathland, although the emu (*Dromaius novaehollandiae*) occurs in heathlands and one mound-builder (*Leipoa ocellata*) is confined to related mallee habitat in southern Australia.

There are a significant number of species in other groups that show special association with heathlands. Among doves *Geopelia humeralis*, a non-endemic species, is associated with coastal heathland while all three endemic genera of ground-feeding pigeons occur in heathlands, with *Phaps elegans* in the south and *Petrophassa albipennis* in the north being restricted to this habitat. Although many species of parrots, including cockatoos (Cacatuidae) and nectar-feeding lorikeets (Loriidae), have been recorded from heathlands, only one species is restricted to heathland. This is *Pezoporus wallicus*, a primitive, ground-adapted species, endemic to southern and eastern coastal areas. It is likely that *Pezoporus* evolved in the Tertiary heathland and escaped extinction by colonizing the Pleistocene heathlands of coastal dunes. The only surviving member of the contemporaneous radiants would be *Geopsittacus occidentalis*, a near-extinct species of a monotypic genus known from the arid zone. Another parrot with terrestrial habits, *Neophema petrophila*, is a recent radiant to southern and western coastal heathlands from the forest environment (Macdonald, 1973).

The endemic passerine families Menuridae and Atrichornithidae, each represented by two species, are geographic and phylogenetic relicts that are adapted to ground and near-ground feeding in dense vegetation. The lyrebirds (*Menura*) are primarily closed-forest species. One is restricted to subtropical montane forest while the other, found in the southeast, extends its range into eucalypt forest with heathy understorey (Macdonald, 1973; Newsome et al., 1975). One species of scrub-bird (*Atrichornis rufescens*) is restricted to subtropical montane forest (mostly *Nothofagus* with dense understorey) while the other (*A. clamosus*), known currently from a single locality in the southwest, occurs in dense scrub or heathland vegetation, that may or may not have eucalypt overstorey (Webster, 1962; Smith and Robinson, 1976).

The anomalous distribution and habitat association of *Atrichornis* is matched by two other autochthonous genera. These are *Drymodes*, an aberrant group within the turdid–muscicapid complex, and *Psophodes* which has been variously placed within the Timaliidae (Macdonald, 1973) and Orthonychidae (Schodde, 1975). *Drymodes* includes two species: *D. brunneopygia* is confined to mallee-heathland habitats of the south and southeast (Ford, 1971) and *D. superciliaris* occurs in the north and in New Guinea. The Cape York Peninsula race of the latter species is primarily a bird of monsoon rain forest, but does appear in adjacent heathland vegetation. Within *Psophodes*, *P. nigrogularis* is restricted to heathlands in the south and southwest while *P. olivaceus* occurs in closed-forest and associated dense vegetation, including heathland, in the east and northeast. *Sphenostoma*, which is included in *Psophodes* by Schodde (1975), is found only in the arid interior of the continent (Ford, 1971).

Dasyornis is another endemic genus with aberrant features. It is placed in the Acanthizidae and includes three species, all confined to heathland and scrub vegetation. *D. brachypterus* is found in isolated localities in coastal heathlands of eastern New South Wales and mountain heathlands of northeastern New South Wales and southeastern Queensland. The second species, *D. broadbenti*, is represented by three races found respectively in southern and western Victoria, southeastern South Australia, and the southwest corner of Western Australia. The third species, *D. longirostris*, is represented today by small populations in southeastern heathlands (Macdonald, 1973; Smith, 1976).

Among the Australian wrens (Maluridae) and warblers (Acanthizidae) there are many ground-feeding insectivores with heathland representatives. The Maluridae, with three genera, is distributed through all habitats of the continent and Tasmania. Within *Malurus* one species (*M. elegans*) is restricted to the southwestern heathland and another (*M. pulcherrimus*) is represented by populations in the heathlands of southwest and southern South Australia (Serventy, 1951). Some other species of *Malurus* intrude into heathlands in various parts of the continent. Other genera within this family are endemic to Australia. One of the two species of *Stipiturus* (*S. malachurus*) occurs in wet-heathland of the southeast and Tasmania and dry-heathland of the southwest. Three species of *Amytornis* are associated with heathlands by virtue of their very restricted ranges in the sandstone areas of northwestern Kimberleys (*A. housei*), western Arnhem Land (*A. woodwardi*) and the southwestern head of the Gulf of Carpentaria (*A. dorotheae*) (see Gill, 1976; Mason and Schodde, 1976). Within the Acanthizidae associations with heathlands are shown by aberrant members of the family. Two endemic taxa, *Hylacola* and *Calamanthus*, which are now placed under *Sericornis* (Schodde, 1975) show strong associations with heathland and mallee-heathland across the south of the continent. A third member, *Pyrrholaemus*, considered to be derived from a stock of more typical *Sericornis* in wet habitats, lives in arid regions and penetrates heathlands of the south and southwest. the most aberrant member of this family, *Origma*, is a monotypic genus confined to the Hawkesbury sandstone area of New South Wales.

Three other passerine families (Meliphagidae, Ephthianuridae and Artamidae) present other biogeographic patterns that are tied to contraction and expansion of heathlands. They all have brush-tipped tongues (Salomonsen, 1967; McKean,1969; Parker, 1973a), an adaptation to nectar feeding, and hence these birds are important for the pollination of many heathland plants (see Paton and Ford, 1977; Clifford and Drake, Vol. B, Ch. 5).

The Meliphagidae (honey-eaters), with 167 species, have radiated into all habitats in all parts of Australia (67 species), New Guinea (63 species) and islands from Bali to New Zealand (Keast, 1976). The family contributes greatly to the high species diversity of birds in tropical rain forest of northeastern Australia and New Guinea, where most species are sedentary, and in semi-arid formations of Australia where many are considered to be nomadic (Gannon, 1962; Keast, 1968a). Most species take insects, and some take fruit as well as nectar, but few are primarily insectivorous (Officer, 1964; Ford and Paton, 1976). Species that are virtually restricted to heathlands tend to move seasonally within this habitat (Keast, 1968a), and nectar forms a significant part of their diet (Recher and Abbott, 1970a, b; Recher, 1971; Paton and Ford, 1977). *Phylidonyris* and *Trichodere* show the strongest association with heathlands of all Australian meliphagids. Four of the five Australian species of *Phylidonyris* are widely distributed in southern and eastern heathlands, including mountain heathland, and adjoining habitats. In any heathland of the southwest or southeast two or more of these species co-exist. the fifth species (*P. albifrons*) is widespread through the dry interior and reaches dry coastal areas in the south and west where it may be found in heathlands. *Trichodere*, a closely related monotypic genus, is the only bird endemic to Cape York Peninsula, where it is confined to heathlands (Kikkawa, 1976). The distribution pattern shown by this group of honey-eaters suggests that contraction and fragmentation of heathlands late in the Tertiary and early in the Pleistocene may have promoted speciation within isolated areas of heathland, while subsequent expansion of heathlands, especially around the southern coastal fringe, has promoted range extension and overlap for several species.

Among other honey-eaters the endemic genus *Acanthorhynchus*, with two species, exhibits a disjunct distribution: one species (*A. superciliosus*) occurs in heathlands of the southwest, and the other (*A. tenuirostris*) is in the east, where it inhabits heathlands throughout its range but also penetrates subtropical and tropical montane rain forests in the north. Within the genus *Anthochaera*, found in the south of the continent and in Tasmania, *A. chrysoptera* occupies coastal scrub and heathlands in the southwest, southeast and Tasmania, *A. carunculata* is a visitor to mountain and coastal heathlands of the southeast and the south, and *A. paradoxa*, the largest honey-eater and a Tasmanian endemic, is not a heathland bird (Ridpath and Moreau, 1966). Some species within a variety of other meliphagid genera exhibit different degrees of association with heathlands. Thus, all species of *Philemon* may be seasonally abundant in heathlands within their respective ranges, some arid-adapted nomads appear in southern (e.g. *Certhionyx niger*) or northern heathlands (e.g. *Certhionyx pectoralis*), some migratory species appear in southeastern heathlands (e.g. *Melithreptus lunatus, Lichenostomus chrysops* — see Keast, 1968a) and in Cape York Peninsula the rain-forest honey-eaters appear in adjacent heathland and scrub vegetation (e.g. *Xanthotis flaviventer, Meliphaga gracilis*) or in heathlands even far away from closed-forest patches (e.g. *Meliphaga notata*) (Kikkawa, 1976).

The Ephthianuridae (Australian chats) are an endemic family consisting of two genera, *Ephthianura* with four species and the monotypic *Ashbyia*. All are adapted to insect-feeding on or near the ground, but *Ephthianura* is known to take nectar occasionally (Parker, 1973a). One species (*E. albifrons*) is associated with heathland vegetation in the southern parts of its range, while another (*E. tricolor*) has developed long distance nomadism which may be aided by the exploitation of nectar from desert plants [e.g. *Brachysema chambersii* (Podalyrieae); see Parker, 1976]. Sibley (1976b) reported a striking resemblance of the egg-white protein patterns between *Ephthianura* and the meliphagid *Phylidonyris*, and suggested affinity of the two families on this and other bases.

The ten species of Artamidae (wood-swallows) are distributed from India to Fiji and southward through Australia. Four of the six species found in Australia (i.e. *Artamus personatus, A. superciliosus, A. cyanopterus, A. minor*) are endemic, and the family is considered closely related to other endemic families of Australian birds (Sibley, 1976b). Wood-swallows are aerial insectivores, but some species are known to take nectar also (Ingram, 1972; Lowe and Lowe, 1972; Mack, 1973; Parker, 1976).

Both Ephthianuridae and Artamidae may have originated as nectar-feeders within Australian heathlands and subsequently shifted to alternative feeding modes, one to ground-feeding and the other to aerial-feeding, when heathlands contracted and aridity advanced in the interior of the continent.

The foregoing account reveals several patterns of distribution and varying degrees of association with heathlands for a variety of endemic taxa. In many cases it is likely that heathland forms represent relatively recent invasion of this habitat, but in a few important cases they represent relicts of an autochthonous avifauna which radiated within Australian heathlands during the Tertiary. Highly sedentary forms with limited power of flight are amongst the oldest endemics (e.g. *Pezoporus, Atrichornis*). The disjunct distribution of both the old endemics and some recent radiants with similar habits and habitat requirements suggests that the autochthonous groups have repeatedly been subject to extinction since their radiation. The fragmentation of the habitat as a result of increased aridity and cold, the shifting of habitat during the Pleistocene and, more recently, frequent or extensive fire, were probably responsible for this fate. Evolution in heathlands during the early Tertiary probably paralleled that in the subtropical rain forest. Some monotypic genera exist as relicts in rain forest today (Schodde and McKean, 1976). The remarkable fact that some relicts are found only in heathlands and closed-forests, and are separated by thousands of kilometres and long-standing geographical barriers, points to their being contemporaneous radiants rather than unidirectional colonizers from one habitat to another. It is also apparent that speciation has occurred in the heathlands through the Quaternary and is still occurring.

Mammals

Although the pre-Miocene history of Australian marsupials is obscure, most of the modern families appear to have radiated during the early Tertiary [see Keast (1977) and work cited therein].

Habitats occupied by Australian mammals today are often poorly defined. Most species recorded from heathlands occur in other habitats, and there is the suggestion that the heathland is often of marginal importance in providing resources for these species (Dwyer et al., 1979; Posamentier and Recher, in prep.; Newsome and Catling, Chapter 10). The few species of mammals that are closely associated with heathlands fall into four groups.

(1) Species with a small geographic range that are virtually exclusive to heathlands [e.g.

Antechinus apicalis (Morcombe, 1967), *Tarsipes spencerae* (Ride, 1970), *Pseudomys novaehollandiae* (Posamentier and Recher, 1974), *P. albocinereus*]. The honey-possum, *T. spencerae*, is the only member of an endemic family, the Tarsipedidae. It is restricted to the southwest and its relationship with other families of Diprotodonta is unclear. Its tubular mouth and brush-tipped tongue are specializations for collecting nectar and pollen from heathland flowers (Ride, 1970).

(2) Species with a restricted northern distribution that show an association with heathland in areas of rocky outcrops. The monotypic marsupial genera *Wyulda*, *Petropseudes* and *Peradorcas* are examples. *Antechinus macdonnellensis* and the rock rats, *Zyzomys*, have a comparable distribution.

(3) Species with a wide but sometimes fragmented distribution that have close association with heathland in parts of their range [e.g. *Cercartetus concinnus* (Wakefield, 1963), *Sminthopsis crassicaudata*, *Mastacomys fuscus* (Ride, 1970), *Pseudomys gracilicaudatus* (Mahoney and Posamentier, 1975), and *Rattus lutreolus*].

(4) Species having a close association with coastal heathland in parts of their range, where this association arises because either these habitats contain wetlands (e.g. *Melomys littoralis*) or provide an abundance of nectar and pollen (e.g. *Syconycteris australis*, *Pteropus scapulatus*) (Dwyer et al., 1979). Most species within the last two groups, and certainly other species of mammals recorded from heathlands, are members of widely radiated genera.

It is apparent for mammals as a whole that the level of endemicity in heathland is very low. *Tarsipes* stands out as a highly specialized phylogenetic relict, a possible result of radiation at the time of heathland dominance with subsequent extinction of all related forms. The position of the monotypic marsupial genera that are associated with rocky outcrops in northern latitudes is less clear. Undoubtedly they are relicts, but their association appears to be more with the rocky substrate than with heathland vegetation. Similarly the association of some species of the marsupial mouse genera *Antechinus* and *Sminthopsis*, the pigmy possums *Cercartetus*, and the murid rodents *Pseudomys* and *Zyzomys* with heathland, rocky outcrops, or both may be interpreted in terms of radiation either out of or into these habitats. The

differing chronologies of radiation for marsupials and rodents in Australia (Ride, 1970) may imply a different direction for habitat shift in these groups.

Chronology

Several themes emerge from the foregoing survey. First, widely distributed genera or species of heathland animals have a wide range of habitat tolerance or exhibit strong nomadism to locate unpredictably fluctuating supplies of heathland resources (e.g. nectar). Second, heathland species with narrow adaptive zones are local endemics and are often monotypic genera within families whose origin may be in early Tertiary. The continuity of the heathland habitat through time — longer than the evolutionary history of modern birds and mammals — meant that specialization was possible not once but many times as each new group of animals entered this habitat. Speciation, migration and extinction must have accompanied each major shift of the habitat either from the centre to the periphery of the continent, or from sandstone hills to coastal dunes and wind-blown sand-ridges, or along the edge of a moving "arid belt". A large part of the autochthonous vertebrate fauna of Australia must have radiated through the Tertiary (Bartholomai, 1972; Savage, 1973; King and ·King, 1975; Schodde and McKean, 1976). Within heathland habitats this radiation has produced specialized animals that have subsequently been stranded in heathland refuges. The recent great aridity (Beard, 1976) is perhaps responsible for reducing these species to phylogenetic relicts by causing extinction of the majority of related forms. Whether such relicts represent monotypic genera or families depends upon how successful related forms were in their adaptation to currently prevalent habitats. This means that the degree of endemicity does not necessarily indicate the period in which differentiation occurred.

In the following an attempt is made to depict a likely chronology, by grouping species that show gross similarity in the status and distribution within heathlands today.

---

(1) Phylogenetic and geographic relicts showing a high degree of specialization in the heathland habitat and confined to southwestern Western Australia — a possible result of radiation at the time of heathland dominance and subsequent extinction of all or most of the allied forms.

Myobatrachus gouldii (turtle frog): A monotypic genus, specialized in underground termite-feeding.

Pseudemydura umbrina (western swamp turtle): A monotypic genus of chelid turtles with a short neck, found in swamps.

Tarsipes spencerae. (honey-possum). A monotypic family, specialized in nectar-feeding.

(2) Phylogenetic relics showing a high degree of sedentariness and adaptation to ground-living in dense cover, remaining otherwise unspecialized morphologically and having lost links with modern forms.

Philoria frosti (Baw Baw frog): A monotypic genus, restricted to alpine heathlands of Mount Baw Baw, Victoria; the closely allied genus, *Kyarranus*, is restricted to subtropical montane rain forest.

Pezoporus wallicus (swamp parrot): A monotypic, endemic genus, showing geographic radiation at the height of the recent development of coastal wet-heathland, possibly as a result of secondary adaptation and subsequent isolation of populations in fragmented Pleistocene habitats, with some racial differences developing between the eastern and southwestern populations; the closely allied genus, *Geopsittacus*, is a relict in the arid zone.

Atrichornis clamosus (noisy scrub-bird): One of the two species of a primitive passerine family, restricted to heathlands of the southwest; the other species, *A. rufescens* (rufous scrub-bird), is restricted to subtropical montane (temperate) rain forest.

(3) Species showing a relict pattern of distribution in heathlands, as a result of phylogenetic and geographic isolation after radiation into heathlands at the time of heathland dominance.

Paracrinia haswelli (Haswell's frog): A monotypic ground-dwelling genus with disjunct distribution in Hawkesbury Sandstone and coastal heathlands.

Heleioporus inornatus in the southwest and *Heleioporus* sp. (Cogger, 1975) in Arnhem Land — the heathland members of this genus of burrowing frogs; the other members of the genus have wider habitat tolerance but are restricted in geographic distribution.

Phyllurus platurus (southern leaf-tailed gecko): A member of a small forest-dwelling genus, localized in Hawkesbury Sandstone.

Pseudothecadactylus lindneri (giant cave gecko) in western Arnhem Land and Kimberleys, and *P. australis* in Cape York Peninsula — the only known members of a distinctive genus.

Aclys concinna: A monotypic genus of pygopodid lizards with well-developed hindlimb flaps, probably fossorial.

Pletholax gracilis: A monotypic genus of pygopodid lizard with small hindlimb flaps, specialized in fossorial life.

Elapognathus minor (small brown snake): A monotypic genus of elapid snakes with smooth scales.

Neelaps calonotus (western black-stripped snake): A burrowing elapid snake of heathlands in the southwest.

Simoselaps littoralis: A burrowing elapid snake in southwestern heathlands.

Hoplocephalus bungaroides (broad-headed snake): A member of a small forest-dwelling genus, localized in Hawkesbury Sandstone.

Drymodes brunneopygia (southern scrub-robin): The southern member of an autochthonous genus with disjunct distribution;

the northern member is distributed to northern Australia and mid-mountain rain forest of New Guinea.

Psophodes nigrogularis (western whipbird): One of the two species of an endemic genus with disjunct distribution; the other species is a rain-forest bird in the east and northeast.

Dasyornis (bristlebirds): An endemic genus with three species in different parts of continental heathlands, showing further differentiation within heathlands.

Origma solitaria (rock warbler): The sole member of an endemic genus, localized in Hawkesbury Sandstone.

Hylacola (hylacolas): An endemic subgenus with two species showing speciation within southern continental heathlands.

Sericornis (Calamanthus) fuliginosus (fieldwren): A monotypic endemic subgenus in southern heathlands, including Tasmania.

(4) Major taxa showing specialization as nectar-feeders as part of their heathland adaptation and geographic radiation at the time of heathland dominance, but subsequent secondary adaptation to escape from dependence on heathlands; some species of these groups still show association with heathlands without exclusive dependence on this habitat.

Meliphagidae (honey-eaters): An autochthonous family with successful radiation into all land habitats and a wide geographic range; the endemic monotypic *Trichodere* in the northeast and *Phylidonyris* with four species in the south are associated with heathlands; *Acanthorhynchus* is a relict genus with one species in heathlands and sclerophyll forest of the southeast and tropical montane and subtropical closed-forest, and a second endemic to heathlands of the southwest.

Artamidae (wood-swallows): An autochthonous family of secondarily aerial feeders with expanded range.

Ephthianura (Australian chats): An endemic genus of secondarily arid-adapted ground insectivors; *E. albifrons* is associated with heathland in southern parts of its range.

Burramyidae (pigmy possums): An autochthonous family with wide geographic and habitat ranges utilizing insects and nectar; *Cercartetus concinnus* is associated with southern heathlands.

(5) Genera showing radiation during arid periods, with some members inhabiting heathlands.

Anomalopus: A small genus of burrowing skinks with disjunct distribution in eastern Australia, with one species (unnamed) restricted to coastal heathlands of central New South Wales.

Hemiergis: A genus of fossorial skinks, showing a relict pattern of distribution in the southern parts of the continent, with one species (*H. quadrilineatum*) confined to the southwestern heathland.

Lerista: Fossorial skinks, with at least five species restricted to southwestern heathlands and one to those of Cape York Peninsula; the remaining species are widely distributed in the arid region of the continent.

Ctenotus: A large genus of skinks with two species restricted to the southwestern heathlands, two to those of Cape York Peninsula and one to those in the northwest; the remaining species mostly occur in desert regions.

Phaps: P. elegans (brush bronzewing) in eastern, western, southern and Tasmanian heathlands.

Petrophassa: P. albipennis (rock pigeon) with two well-differentiated races (treated as species by some authorities) in northern rocky outcrops.

Stipiturus: *S. malchurus* (southern emu-wren) in eastern, southern, western and Tasmanian heathlands.

Amytornis (grass wrens): Three species in northern and one species in southern heathlands.

Pseudomys (native mice): Several species in southern and eastern heathlands.

Mesembriomys (tree rats): Appearing in northern heathlands.

Leporillus (stick-nest rat): Appearing in southern heathlands.

Notomys (hopping mice): Appearing in northern (*N. aquilo*) and southern (*N. mitchellii*) heathlands.

Zyzomys (rock rats): Found in association with northern rocky outcrops.

Mastacomys: *M. fuscus* (broad-toothed rat), the only species of the genus, appearing in montane heathlands of the southeast.

(6) Genera showing radiation in a wide range of modern habitats including heathlands as an adaptive zone.

Ranidella: *R. tinnula* (wallum froglet) in southeastern Queensland.

Pseudophryne: *P. australis* (red-crowned toadlet) on Hawksebury Sandstone: *P. corroboree* (corroboree frog) in southeastern montane heathlands.

Litoria (tree frogs): A large genus with several heathland species in the east.

Emydura (river turtles): An unnamed species in southeastern Queensland.

Gehyra (dtellas): Two species in northwestern sandstone areas.

Amphibolurus (dragon lizards): A large genus with several heathland species in the southwest and north.

Diporiphora (dragon lizards): *S. superba* in the Kimberleys.

Varanus (monitor lozards): A large genus with several heathland species in the north.

Carlia (rainbow skinks): *C. dogare* in Cape York Peninsula.

Cryptoblepharus (skinks): An unnamed species in Arnhem Land.

Egernia (skinks): A large genus, with one unnamed species having disjunct distribution in southern heathlands.

Leiolopisma (skinks): An eastern humid-zone genus, with one species (*L. greeni*) in subalpine heathlands of Tasmania.

Omolepida (skinks): *O. maxima* in the Kimberleys.

Ramphotyphlops (blind snakes): *R. yirrikalae* in Arnhem Land; an unnamed species in southeastern Queensland.

Neophema: *N. petrophila* (rock parrot) in southern heathlands.

Malurus: *M. pulcherrimus* (blue-breasted wren) in southern heathlands; *M. elegans* (red-winged wren) in the southwest.

Strepera: *S. fuliginosa* (black currawong) in Tasmanian heathlands.

Antechinus: *A. apicalis* (dibbler) in southwestern heathlands; *A. minimus* (swamp antechinus) in southeastern heathlands.

Sminthopsis: *S. murina* (common marsupial-mouse) in southern heathlands.

(7) Species showing recent adaptation to coastal heathlands in parts of the range.

Relict frogs and reptiles showing secondary adaptation to heathlands:

Adelotus brevis (tusked frog): A species of wide habitat tolerance appearing in heathlands of southeastern Queensland.

Anotis graciloides: A relict skink of wet forests appearing in heathlands of southeastern Queensland.

Saiphos equalis: A relict burrowing skink, appearing in heathlands of southeastern Queensland and northeastern New South Wales.

Tropidechis carinatus (rough-scaled snake): A rain-forest relict, appearing in heathlands of southeastern Queensland.

Recent bird colonizers:

Geopelia humeralis (bar-shouldered dove): A woodland species common in northern and eastern heathlands.

Merops ornatus (rainbow bee-eater): A new colonizer to Australia, with a wide breeding range in most continental coastal heathlands.

Loriidae (lorikeets): A group of species in eucalypt forests and woodlands, with brush tongues for nectar- and pollen-feeding; *Trichoglossus haematodus* and *T. chlorolepidotus* in eastern heathlands.

Anthochaera chrysoptera (little wattlebird): A coastal species in eastern and southern heathlands.

Philemon corniculatus (noisy friarbird) and *P. citreogularis* (little friarbird): Woodland species, in eastern heathlands.

Lichmera indistincta (brown honey-eater): A woodland species, in eastern heathlands.

Lichenostomus cratitius (purple-gaped honey-eater): A mallee-heathland species with disjunct distribution, in southern and southwestern heathlands.

Melithreptus brevirostris (brown-headed honey-eater): A woodland species, in southeastern heathlands.

Meliphaga notata (yellow-spotted honey-eater): A rain-forest species, in Cape York Peninsula heathlands.

Recent mammal colonizers:

Rattus lutreolus (swamp rat): A wetland species, in eastern heathlands.

Melomys littoralis (grassland melomys): A grassland species, in coastal heathlands of southeastern Queensland.

Syconycteris australis (Queensland blossom bat): A rain-forest species, occurring in eastern heathlands.

(8) Species showing no obvious association with heathlands.

The majority of vertebrate species recorded from heathlands occur more commonly in other habitats. However, because of peculiar conditions that exist in heathlands, some species, particularly those of low vagility, may be speciating today. *Ranidella insignifera*, for example, has two races parapatrically distributed[1] between heathland and woodland in southwestern Australia (Main, 1969). Polymorphism in a number of species in eastern Australia has also been demonstrated (Straughan and Main, 1966). *Pseudophryne coriacea* has a parapatric form confined to the heathland of Cooloola, southeastern Queensland (Gravatt and Ingram, 1975). It is possible that species of "acid frogs" (see next section) are speciating parapatri-

---

[1] Parapatric taxa have contiguous, non-overlapping distributions.

cally in relation to populations occupying adjoining habitats (cf. Bush, 1975). On the other hand, many species are isolated in fragmented heathlands today and may be undergoing speciation as *Amytornis* and *Dasyornis* appear to have done in the past. Such allopatric speciation has been examined by Keast (1961) for birds in other major habitats. Habitat fluctuation may also lead to local specialization in heathlands as suggested by Pianka (1972) for lizards in Australian deserts.

ECOLOGICAL FACTORS

Australian heathlands as a habitat for land vertebrates are characterized by dense ground cover and seasonally variable types of plant resource. Associated features affecting the distribution of animals include infertile soils, acidic water, seasonal inundation of wet-heathland, substrate characteristics and frequent fire. The restricted nature of most heathland areas means that the nature of neighbouring habitats, and the species supported by those habitats, may importantly influence patterns of abundance and diversity within heathlands. In this section we examine major ecological characteristics of Australian heathlands and their impact upon the associated vertebrate fauna.

Cover

Heathland communities have dense or mid-dense low vegetation cover that may be opened up for periods after fire. Most available information concerning the significance of cover for heathland vertebrates is implicit in studies on the effects of fire. Changes through the pyric succession of small mammal populations and, to a lesser extent, of the populations of some birds, are probably controlled primarily through cover. This aspect will be dealt with in a later section.

Many species of heathland vertebrates, including relict birds such as *Dasyornis*, *Calamanthus*, *Psophodes nigrogularis* and *Atrichornis clamosus*, are closely associated with dense cover created by spreading subshrubs with high orders of branching. These birds tend to be poor flyers but are extremely agile in the dense vegetation and on the ground. For mammals the density of cover may often be a

critical factor determining distribution and influencing behaviour. As in ericoid heathlands of the Palaearctic region where rodents build runways under ground cover (Kikkawa, 1959), areas of very dense heathland in Australia may be criss-crossed by many mammal runways. These runways exposed after a moderate burn in heathlands of southeastern Queensland may lead to the burnt-out ground nests of the murid rodent *Melomys littoralis* or the bandicoot *Isoodon macrourus*, and some of them through dense heathland are well-used routes made by the wallaby *Wallabia bicolor*. *Rattus lutreolus* is also known to utilize runways in heathland; indeed it is difficult to capture this species in traps unless these are placed in runways (cf. Green, 1967). Large mammals use tracks in heathlands (Newsome et al., 1975) and these, especially of *W. bicolor*, and narrow walking tracks made by humans, may act as minor firebreaks during light or moderate burns (P.D. Dwyer, personal observation, 1977).

The density of heath vegetation may preclude effective movement of many mammals and some birds (e.g. water rail, *Rallus pectoralis*) without a runway system. For large species (e.g. kangeroos) it may render passage extremely difficult and discourage utilization of the habitat. Dense ground cover may, however, allow increased levels of daylight activity, without greatly increasing the risk of predation, in species that are generally nocturnal. On a grid of 6.25 ha in heathland at Beerwah, southeastern Queensland, 32% of *Rattus lutreolus* captures, 19% of *Antechinus flavipes* captures, and 17% of *Isoodon macrourus* captures were in daylight (Dwyer and Willmer, unpublished). The trapping procedure was such that these values understated the level of diurnal activity for all species, but particularly for the bandicoot *I. macrourus*. *R. lutreolus* is well known for high levels of diurnal activity, but this behaviour is not typical for the other two species. There are also a few diurnal species penetrating heathlands in the southeast (e.g. *Antechinus swainsonii*) and in the southwest (e.g. *Myrmecobius fasciatus*).

Both density of cover *per se* and quality of cover may be important for some species. On a grid of 6.25 ha in low heathland of the Noosa Plain at Cooloola in southeastern Queensland, the vegetation at trap sites was classified on a five point scale from (1) very dense, tall cover of 1.5–2.5 m to

(5) moderately dense low cover up to 1 m high. For the 96 sites classified in this way 361 captures of 73 *Melomys littoralis* were distributed as follows: (1) 7.2 captures per site ($n=6$ sites), (2) 5.0 captures per site ($n=8$), (3) 4.5 captures per site ($n=44$), (4) 3.1 captures per site ($n=15$), (5) 1.4 captures per site ($n=23$), (Dwyer and Willmer, unpublished). Sites in category 5 were dominated by beds of erect restiads (*Restio*) or included erect restiads that penetrated a light tangle of other restiad species (*Lepyrodia*). Category 3 included sites where *Banksia robur* and clumps of the sedge *Gahnia* predominated above dense tangles of *Lepyrodia*, while category 1 was complex in having *B. robur* mixed with restiads, fern and low shrubs (*Leptospermum* and *Melaleuca*). The wettest sites were those in categories 4 and 5 where surface water was usually present throughout the year, and those in categories 1 and 2 where minor drainage channels ran across the plain. Sites in category 3, which produced intermediate numbers of *M. littoralis*, were subject to seasonal inundation. In this preliminary analysis it is apparent that this species favours areas of very dense cover especially those areas where plant species diversity is high.

Substrates

The heathland flora which had developed on moist but infertile subtropical soils of the Tertiary now survives best in southwestern Australia. It also remains on sandstone and granite outcrops of northern and southeastern Australia, and has spread to Quaternary siliceous sands around the coast (Specht, Chapter 6). Many animals found in heathlands are present by virtue of their association with lateritic soils, sands or rocky outcrops.

The sandy substrate seems to have played a significant role in the evolution of heathland frogs and reptiles. *Myobatrachus gouldii* provides the most striking example of this relationship. It is the only truly fossorial frog, and is a phylogenetic relict confined to heathlands of the southwest. For reptiles it is significant that 36% of all heathland-adapted species, compared with 18% of the rest, are fossorial. Indeed Storr (1964) suggests that the distribution of reptiles is influenced more by the nature of the substrate than by vegetation. Given the sedentary nature of reptiles the fossorial habit is also an important survival factor in the fire-

prone heathlands. Although the sandy substrate is not a sufficient condition for survival, many hollow-nesting birds, such as *Merops ornatus*, *Cheramoeca leucosternum*, *Cecropis ariel* and *Pardalotus*, and burrowing mammals, such as *Tachyglossus*, *Vombatus* and introduced rabbits, live in parts of sandy heathlands.

Of animals associated with rocky substrates the giant cave gecko (*Pseudothecadactylus lindneri*), dtellas (*Gehyra*) and other reptiles are found in sandstone areas of Arnhem Land and the Kimberleys while the pigeons *Ptilinopus cinctus* and *Petrophassa albipennis*, the rock warbler (*Origma solitaria*), grass wrens (*Amytornis*), the sandstone shrike-thrush (*Colluricincla woodwardi*), the white-lined honey-eater (*Meliphaga albilineate*), the rock wallabies (*Petrogale* and *Peradorcas*) and the rock rats (*Zyzomys*) are all rock-adapted forms which, at least in parts of their range, are associated with heathland flora. Some of these species qualify as phylogenetic relicts (see pp. 239–241) and therein support the view that sandstone heathlands have provided important refuge areas for a diminishing heathland fauna.

Because areas of rocky outcrops are not contiguous, isolation has led to population differentiation for some of these groups. Isolation is pronounced between the Arnhem Land escarpment and the Kimberley sandstones of northern Australia. Here *Amytornis* is represented by different species in the two regions, and *Petrophassa albipennis* and *Peradorcas concinna* have differentiated at least to the level of subspecies. Differentiation within this habitat is also shown by a few wide-ranging species. Thus, in Arnhem Land the sandstone shrike-thrush *Colluricincla woodwardi* of the escarpment is replaced by *C. harmonica brunnea* on adjoining lowlands (Hall, 1974) and, in Cape York Peninsula, pied currawongs are represented by *Strepera graculina graculina* in closed-forest and *S. g. magnirostris* in sandstone hills (Kikkawa, 1976).

Aquatic habitats

Sandstone heathlands have very little surface water; their soils have low water-retaining capacity and are well drained. While the same properties apply for the Quaternary sand soils, these regions may develop peaty soil with high water-retaining capacity, and the water table may be high, promoting development of high sedge and shrub swamps (cf. Pidgeon, 1938). In coastal sand dunes the accumulation of humic material and the burying of peat beds contribute to development of "perched" lakes[1] (see Bayly, 1975; Laycock, 1975) and, where erosion occurs in association with spring sapping, "window" lakes[1] may be formed (Bensink and Burton, 1975). In southeastern Queensland, northeastern New South Wales, and Gippsland in Victoria, these waters are dystrophic, with low pH and a high concentration of allochthonous organic matter (Coaldrake, 1961; Timms, 1973; Bayly et al., 1975). Window lakes are less acid and more transparent than perched lakes, but the productivity of both types is very low (Bayly, 1967). Animals associated with them depend primarily on saprophytic food chains for nutrition and the benthos are very unproductive (Timms, 1973; Bayly, 1975; Bensink and Burton, 1975). The vertebrate fauna of this habitat is restricted, although tortoises (*Emydura* sp.) are abundant in some lakes of Fraser Island, southeastern Queensland.

Some of the dune systems apparently date back to pre-Pleistocene times (Coaldrake, 1962), and these contain endemic species of animals that exhibit adaptation to the peculiar acidic conditions of included lakes and swamps. Thus the sunfish *Rhadinocentrus ornatus* (Melanotaeniidae) utilizes wind-blown insects and pollen as a major food source (Bayly et al., 1975). There are at least five species of frog (*Ranidella tinnula*, *Litoria freycineti*, *L. cooloolensis*, *L. olongburensis Heleioporus inornatus*) confined to this habitat, and the first four listed have eggs and tadpoles known to tolerate exceptionally low pH (4.3–5.2). Ingram and Corben (1975) called these species "acid frogs". *Litoria olongburensis* and *L. cooloolensis* are arboreal species that disperse from the aquatic habitat after the breeding season, while the others are ground-dwelling species associated with the sandy substrate of the dune system. Most of the "acid frogs" have sibling species in neighbouring habitats, and it is probable that they reflect recent radiation within the heathland communities. Ingram and Corben (1975) suggested that parapatry between "acid frogs" and their siblings is maintained by habitat selection of adults, competition between adults, and the tolerance of larvae

to different pH ranges. There is evidence that a low pH of 3.6–5.2 restricts the anuran fauna in sphagnaceous barren habitats of southern New Jersey, U.S.A. and that the embryos of two endemic species found there (*Rana virigatipes* and *Hyla andersoni*) tolerate the acidity (Gosner and Black, 1957; Means and Longden, 1976). Acidity affects amphibian eggs as does salinity by modifying osmotic pressures and preventing the eggs from absorbing enough water for normal development (Goin and Goin, 1962).

Another mode of adaptation is shown by *Philoria frosti*, which is an endemic frog restricted to *Sphagnum* bogs in subalpine heathlands of the Australian Alps. It lays large eggs in waterlogged burrows, and the larvae do not feed although they may have a brief free-swimming phase (Littlejohn, 1963). *Myobatrachus gouldii* of the southwestern Australian heathlands also has large eggs, and it is possible that these develop entirely out of water (Main et al., 1959). Emancipation from aquatic dependence is well recognized in Australian frogs, and has generally been considered an arid adaptation (Martin, 1967). The fact that many old endemics, including species of mesic habitats (e.g. *Kyarranus*), show this behaviour is significant. It may imply adaptation to excessively well-drained soils in moist regions. The fact that heathland representatives of *Heleioporus* and *Pseudophryne* breed in burrows (e.g. *H. inornatus* in the southwest and *P. corroboree* in alpine areas of the southeast) or in temporary creeks and gutters of sandstone areas (e.g. *P. australis* on the Hawkesbury Sandstone and *Heleioporus* sp. in the western escarpment of Arnhem Land) is in marked contrast to the recent adaptation of the aquatic "acid frogs". In southeastern Queensland the capacity of the eggs of the "acid frogs" to tolerate low pH contrasts with the behaviour of *Pseudophryne coriacea* with which they are sympatric. In wet-heathlands the latter species lays terrestrially and the eggs hatch within minutes when local flooding occurs.

---

[1] A "perched" lake is formed as a result of the accumulation of fine hill wash and organic matter in dune depressions, rendering these relatively less permeable than the surrounding dunes; a "window" lake is an exposure of water table formed by ground-water springs eroding surrounding dunes (Bensink and Burton, 1975). (See Fig. 6.12 in Chapter 6, and also Volume B, Chapter 10.)

Aquatic habitats of Australian heathlands are of limited importance to reptiles, birds and mammals. With few exceptions most species seem to avoid very wet heathland. Only five reptile species are known from wet-heathland at Cooloola, southeastern Queensland, whereas adjacent dry-heathlands may have eighteen species (see Chapter 9, Fig. 9.1). A few birds are associated with very wet areas in heathland (e.g. water rail, *Rallus pectoralis*), but the food resources of the aquatic habitat are limited, and consequently there are seldom large concentrations of waterfowl. The waterfowl breeding in permanent waters of heathlands are the black swan (*Cygnus atratus*), black duck (*Anas superciliosa*), grey teal (*A. gibberifrons*), wood duck (*Chenonetta jubata*), blue-billed duck (*Oxyura australis*) and musk duck (*Biziura lobata*) in the south, and the pied goose (*Anseranas semipalmata*), whistling tree-duck (*Dendrocygna arcuata*), plumed tree-duck (*D. eytoni*) and black duck in the north (Frith, 1967). The highly saline dune lakes of southern Australia, probably caused by marine flooding and subsequent evaporation in Recent times (Bayly, 1967), provide few breeding habitats for waterfowl, but brackish tea-tree swamps (*Leptospermum–Melaleuca*) of southern Victoria and South Australia are said to have been occupied once by large numbers of chestnut teal (*Anas castanea*) (Frith, 1967). Heathland near the water may become a nesting site of other ducks (e.g. shoveller, *Anas rhynchotis*), and along the southern beaches may support colonies of seals (e.g. *Neophoca cinera* and *Arctocephalus doriferus*). Amongst mammals both *Hydromys chrysogaster* and *Xeromys myoides* are present in wet-heathland and associated waterways in southeastern Queensland. However, *Hydromys* is more common outside heathland. The murid rodents *Rattus lutreolus* and *Melomys littoralis* are both more abundant in wet-heathland than in other heathland habitats in southeastern Queensland (see Dwyer et al., Chapter 9).

Food resources

Despite the relatively high floristic diversity of heathlands, productivity is low due to the low nutrient status of the substrate (Specht, Chapter 6). The total energy available for animals is limited compared with habitats such as grassland where plant biomass is similar. This implies that the abundance and diversity of heathland animals will be low and, in Australia, low abundance and diversity has been reported for foliage insects and litter fauna (Edmonds and Specht, Vol. B, Ch. 3), birds (Bell, 1966; Recher, 1969) and mammals (Dwyer et al., 1979; Posamentier and Recher, in prep.). Abbott (1976) compared habitat structure, plant species diversity and the diversity of arthropods and passerine birds for *Acacia* scrub and dune scrub in Western Australia and found that the abundance of arthropods and the number of resident passerine species were lowest in the dune scrub where plant species diversity was greatest.

Because nutrient levels are low in heathland communities plants must, of necessity, be efficient at obtaining these nutrients and retaining them within living tissue. Herbivory could, therefore, be more hazardous for heathland plants than for those growing in nutrient-rich environments. Janzen (1974) suggests that exceptionally high concentrations of tannins and other phenols in plants that grow on nutrient-poor sandy soils will be debilitating to the animal communities supported by those plants. He argues that these characteristics of plants provide chemical defence against herbivory. The generally poor representation of large herbivorous mammals in Australian heathlands can be understood in these terms. If such mammals are to utilize heathland plants they must be able to detoxify any inimical chemicals encountered. There is evidence that some Australian mammals have this capacity. Oliver et al. (1977) have found that *Macropus fuliginosus*, *Trichosurus vulpecula* and *Rattus fuscipes* from Western Australia have substantially higher tolerance of sodium monofluoroacetate (1080 pesticide) than their eastern counterparts. In Western Australia these species are found within the range of occurrence of fluoroacetate-bearing species of *Gastrolobium* and *Oxylobium* (heathland genera of the tribe Podalyrieae of the family Fabaceae). For herbivores additional stresses may arise as a direct consequence of nutrient deficiency. Both native and introduced herbivorous mammals are known to be adversely affected by the low levels of phosphorus and copper in coastal heathlands (Coaldrake, 1961), nitrogen in the arid zone (Ealey and Main, 1967), and sodium in montane habitats (Blair-West et al., 1968).

Nectar may be the most important energy source in heathlands; it is readily available though subject to seasonal variation (Baker and Baker, 1975; Carpenter and Recher, in press). Flowers of at least twenty genera belonging to the Epacridaceae, Proteaceae (Grevilleoideae), Myrtaceae (Leptospermoideae), Fabaceae, Haemodoraceae, Rutaceae and Xanthorrhoeaceae are visited by birds in heathlands (Sargent, 1928; Paton and Ford, 1977; Recher, 1977; see also Clifford and Drake, Vol. B, Ch. 5). However, it seems that nectar is not available in sufficient quantities in all months to permit complete nectarivory or to support large sedentary populations of nectar-feeders at any one locality (cf. Edmonds and Specht, Vol. B, Ch. 3). Thus most honey-eaters (Meliphagidae) feed on insects as well as nectar and move between regions to take advantage of asynchronous flowering régimes (Keast, 1968a). The heathland specialists (*Phylidonyris*) do not perform long-distance migrations but appear to be local nomads that exploit many different species of plant. For *P. nigra* as much as 97.5% of foraging time may be spent visiting flowers where insects as well as nectar are obtained (Recher and Abbott, 1970b). Short-beaked species of honey-eaters may spend only 11 to 38% of foraging time in nectar-feeding (Ford and Paton, 1976).

Pollen grains also provide a potential food source (West and Todd, 1955). Recher (1977) considers pollen of *Banksia* and *Grevillea* to be an important food item for at least *Phylidonyris nigra* and *P. novaehollandiae*. However, Christensen (1971) suggests that the limited pollen intake of wattlebirds (*Anthochaera*) is unaffected by digestive processes. Certainly, the pollen of eucalypts is ingested by the purple-crowned lorikeet (*Glossopsitta porphyrocephala*). This species is able to collect pollen grains with its brush tongue, without causing significant damage to flowers or reducing seed set (Churchill and Christensen, 1970; Christensen, 1971). The rainbow lorikeet (*Trichoglossus haematodus*) is frequently observed at the inflorescence of *Xanthorrhoea*, but the extent to which lorikeets consume pollen grains from heathland flowers is not known.

Mammals known to utilize heathland flowers include *Acrobates* (Troughton, 1965), *Cercartetus* (Wakefield, 1963), *Tarsipes* (Vose, 1973), *Antechinus apicalis* (Morcombe, 1967), *Rattus*

fuscipes (Morcombe, 1967; Carpenter, 1978) and *Nyctimene* and *Syconycteris* (Nelson, 1964; Troughton, 1965). The flying fox (*Pteropus scapulatus*) is primarily dependent upon eucalypt blossoms but, in coastal southeastern Queensland, will also feed upon *Banksia aemula* (=*B. serratifolia*). Carpenter (1978) contends that the hooked styles of some fifteen species of *Banksia* (e.g. *B. ericifolia*) are adaptations for mammal pollination. She suggests that these species produce more nectar at dusk and at night than they do through the day. The strongly scented nectar of *B. ericifolia* apparently runs down longitudinal troughs within the inflorescence structure and reaches the ground.

In general the associations recognized between animals and heathland plants appear to be broadly based. This makes possible alternative pollinators for the plants and alternative food sources for the animals. Although co-existing species of honey-eaters show pronounced ecological differences (Keast, 1968b; Recher, 1971), when food supplies are locally abundant they may overlap considerably in the sources they exploit (Paton and Ford, 1977; Recher, 1977). Species-specific pollination, such as the wasp *Lissopimpla* pollinating the orchid *Cryptostylis leptochila* (Coleman, 1928), does not seem to exist for vertebrates. The closest association involving vertebrates is seen between the western spinebill (*Acanthorhynchus superciliosus*) and the kangaroo paw (*Anigozanthos manglesii*, Haemodoraceae) (Mees, 1967). *Acanthorhynchus* has the longest bill (25 mm) among the small honeyeaters and feeds on nectar, particularly from the flowers of Proteaceae and Epacridaceae. When the western spinebill forces its bill into the basal slit of the flower tube of the kangaroo paw the flower is bent downwards so that the pistil and anthers contact the bird's back. No other ornithophilous flowers of Western Australia are known to have pollen transported in this way. Most birds that visit flowers carry pollen on their foreheads, chins, faces and bills, and if flowers of different species are visited in close succession pollination is unlikely to be very efficient. It is possible that nectar and pollen were more important as food sources for heathland birds and mammals in the past and that the generalist orientation of current relationships between heathland plants and animals reflects recent adaptation within the heathland ecosystems.

The fruits and seeds of heathland plants provide another food source that may be exploited by birds which can then act as vectors for seed dispersal (Clifford and Drake, Vol. B, Ch. 5). The mistletoe bird (*Dicaeum hirundinaceum*) and the painted honey-eater (*Grantiella picta*) utilize fruit in heathlands (Gannon, 1966) and berries of heathland plants may be important to some honey-eaters (e.g. *Acanthagenys rufogularis*, *Lichenostomus vitescens*) and to flocks of silvereyes (*Zosterops*) that migrate in the non-breeding season. Emus may encourage seed germination by removing the pericarp through ingestion (Noble, 1975) and may also help dispersal within heathland communities (cf. Davies, 1963; Edmonds and Specht, Vol. B, Ch. 3). A variety of pigeons and doves (e.g. *Phaps elegans*, *Geopelia humeralis*) and a few parrots are the major seed-eating vertebrates of heathlands. In southeastern Queensland the yellow-tailed black cockatoo (*Calyptorhynchus funereus*) eats seeds from *Banksia* species and *Hakea gibbosa* after the seed capsules have been opened by fire (Gravatt, 1974). Small seeds are eaten by bristlebirds (*Dasyornis*) and some grass-finches (*Estrildinae*). For other vertebrate classes seed-eating is not an important foraging mode in heathland. The New Holland mouse (*Pseudomys novaehollandiae*) consumes seeds of legumes in forest habitats (Keith and Calaby, 1968) but does not appear to do so in coastal heathlands (Posamentier and Recher, 1974). Certainly seeds form part of the diet of some murid rodents [e.g. *Pseudomys gracilicaudatus* (Mahoney and Posamentier, 1975), *Rattus lutreolus*], of bandicoots (*Isoodon macrourus*) and perhaps some skinks [e.g. *Egernia* and *Tiliqua* (Cogger, 1975)], but in Australian heathlands it seems that ants are the major exploiters of seeds (cf. Berg, 1975, and Vol. B, Ch. 6).

Insectivory is a common mode of feeding for heathland vertebrates. Some birds (e.g. *Merops*) and insectivorous bats utilize insects attracted to flowering plants. Swifts, swallows and woodswallows also forage frequently above heathland. For frogs, most reptiles, some terrestrial mammals and many birds, insects predominate in the diet. The extremely high densities (often greater than 50 per ha) attained by the introduced cane toad *Bufo marinus* in southeastern Queensland attest to the potential of insects as a food source — although, in fact, the density of native insectivores is generally

low. The specialist insectivores concentrate on abundant insects. For example, the burrowing frog *Myobatrachus* utilizes termites (Calaby, 1956), while some lizards (e.g. *Amphibolurus*), snakes (e.g. *Ramphotyphlops*), birds (e.g. *Climacteris*) and mammals (e.g. *Tachyglossus*) feed on ants.

Insectivory and nectarivory are the major feeding strategies available for heathland vertebrates. The generally low densities of primary consumers and decomposers as well as their seasonality and patchiness (Edmonds and Specht, Vol. B, Ch. 3) impose major constraints upon the abundance of terrestrial insectivores. The asynchronous flowering régimes of heathland plants may support many insects that are available to aerially foraging species, and nectarivory is possible where it is combined with migratory behaviour or local nomadism. Low densities of insectivores and herbivores are further reflected in reduced predator abundance and diversity. Where grazing occurs, dingoes (*Canis familiaris*) and feral cats (*Felis catus*) may depend on herbivores, but for large predators dense vegetation is not conducive to hunting. On the other hand, there are many predatory reptiles which hunt for frogs and skinks. Major food chains involving vertebrates take their origin either in the litter or in nectar, but in both cases they are reduced; community structure is relatively simple and opportunism in foraging mode is common. Specializations in food niches are few and restricted to lower trophic levels.

Fire

The fire history of Australian heathlands is poorly documented. It is generally accepted that sclerophyll vegetation was burnt by aborigines before European settlement (Merrilees, 1968), and lightning can be an important cause of wildfire (Wallace, 1966). There is evidence that an intense fire as early as 7340 B.C. caused severe damage to sandy heathlands and truncated peat in Western Australia (Churchill, 1968), but it is likely that in pre-European times, and across much of Australia, fires were not frequent (cf. Gill, 1977). Through the last century, and the early decades of this century, burning was widespread and subject to little control. A period of complete protection from fire followed (except in northern Australia) until intense wildfires destroyed large areas of forest country. As a result of these fires, various régimes of prescribed burning were implemented through much of the continent; these were designed to reduce the accumulation of fuels on the forest floor (Gill, 1975, 1977). Today most heathlands are burnt on a regular and often rotational basis, with burning programmes spaced yearly in the north (Stocker, 1966; Pedley and Isbell, 1971) and from two to ten years elsewhere (I. Fergus and J. Walker, pers. comm., 1977; Gill and Groves, Vol. B, Ch. 7).

It is probable that all heathland plants incorporate mechanisms that promote survival in fire-prone environments (Gardner, 1957: Gill, 1975). Experimental work in dry sclerophyll forest with heathy understorey near Canberra showed that the floristic composition of the post-fire community was based on the species present at the time of the fire either as living plants or as seeds in the area (Purdie and Slatyer, 1976). Near Dark Island Soak, in South Australia, succession following fire in heathland was characterized by suppression of early dominants, usually herb and subshrub species, and increasing biomass of long-lived, larger heathland species (Specht et al., 1958). The latter included many plants with nectariferous inflorescences (e.g. *Banksia ornata*).

The effects of fire on heathland vertebrates will vary with the frequency and intensity of the fire, with its patchiness, with the type of heathland burnt and the time of the year when the fire occurs. Different species will respond in different ways to these impacts, and habitat changes associated with the pyric succession will be reflected in corresponding changes in faunal composition and abundance. The immediate visual impact following fire is of the destruction of ground vegetation. Presumably this entails destruction of some associated fauna; and certainly, where wildfires occur, the charred bodies of vertebrates are often taken as dramatic evidence for such destruction (cf. Newsome et al., 1975). However, many vertebrates do escape fire (Komarek, 1969; Main, Vol. B, Ch. 8). A wildfire at Nadgee Nature Reserve, southeastern New South Wales, that occurred in December 1972 after forty years' protection from fire did not cause immediate extermination of vertebrate fauna (Newsome et al., 1975; Recher et al., in prep). In heathland in southeastern Queensland a trapping study of small mammals on a grid of 6.25 ha showed that at least 38 (5 species) of 45 individuals (7 species) survived

a prescribed burn that removed 65% of ground vegetation (Dwyer and Willmer, unpublished). One of the species (*Melomys littoralis*) for which survival was high (13 to 16 individuals) builds nests from ground level to about one metre above the ground.

Many birds are known to be attracted to fire and smoke. Species of *Ardeola*, *Dicrurus*, *Eurystomus*, *Merops*, *Milvus* and *Threskiornis* are attracted to fire in African savannah (Komarek, 1969); these genera all have Australian representatives. Fire may temporarily increase the availability of insects or seeds and, in Australian heathlands, wood-swallows (*Artamidae*), some flycatchers (e.g. *Microeca leucophaea*), rosellas (*Platycercus*), cockatoos (*Calyptorhynchus*), kookaburras (*Dacelo*), lyrebirds (*Menura*), magpie larks (*Grallina*), butcherbirds (*Cracticus*), magpies (*Gymnorhina*), currawongs (*Strepera*) and ravens and crows (*Corvus*) all visit burnt heathland.

The impact of the Nadgee wildfire upon mammal populations has been followed in some detail. Newsome et al. (1975) found that bandicoots (*Isoodon obesulus* and *Perameles nasuta*), possums (*Trichosurus vulpecula* and *Pseudocheirus peregrinus*), wallabies and young kangaroos (*Wallabia bicolor*, *Macropus rufogriseus* and *M. giganteus*), foxes (*Vulpes vulpes*) and feral cats (*Felis catus*) all suffered mortality as a result of the fire. The abundance of wallabies, foxes and cats was still much lower one year after the fire than it was before the fire. However, bandicoots and possums increased in abundance over the same period. Wombats (*Vombatus ursinus*), rabbits (*Oryctolagus cuniculus*), dingoes (*Canis familiaris*) and adult grey kangaroos (*Macropus giganteus*) did not decline in numbers immediately after the fire. Of these species, the first three increased significantly in the following year, while adult grey kangaroos showed a slight decrease in abundance. Neither the potoroo (*Potorous tridactylus*) nor the superb lyrebird (*Menura novaehollandiae*), which was the only bird studied, showed significant change in abundance through the period of study. However, records for one reptile, the lace monitor (*Varanus varius*), indicated a reduction in population size. Newsome et al. (1975) report that dingoes switched their prey from small mammals before the fire to wallabies and kangaroos after the fire; they suggest that the fox and feral cat may depend on small mammals for their food. The small mammal population as a whole declined markedly after the fire (Recher et al., in prep). Those that survived in burnt areas had reduced mean weights (Newsome et al., 1975). Poor conditions resulting from reduced food intake combined with increased predation have been implicated by the above authors as possible factors inducing population decline. Local extinction occurred for some small mammal species (e.g. *Rattus lutreolus*, *R. fuscipes*, *Antechinus stuartii* and *A. swainsonii*). Disappearance after fire has also been reported for populations of *R. fuscipes* living in the dense understorey of sclerophyll forest in Western Australia (Christensen and Kimber, 1975).

Roberts (1970) has reported on the impact of fire upon birds in heath vegetation of the Hawkesbury Sandstones. The area concerned had not been burnt for fifteen years when a moderate fire destroyed most of the understorey vegetation. Of nine species resident before the fire only one, the grey shrike-thrush (*Colluricincla harmonica*), was present after the fire. The brown thornbill (*Acanthiza pusilla*) and two honey-eaters, *Anthochaera chrysoptera* and *Acanthorhynchus tenuirostris*, moved to non-heath vegetation. The chestnut-rumped hylacola (*Sericornis pyrrhopygius*) re-appeared as a resident within a year of the fire but three species, the eastern whipbird (*Psophodes olivaceus*), variegated wren (*Malurus lamberti*) and red-browed firetail (*Emblema temporalis*), all associated with ground cover, had not recolonized two and a half years after the fire. In this period the white-cheeked honey-eater (*Phylidonyris nigra*) which, among the honey-eaters, had the strongest association with heathland re-appeared only as a visitor while the white-eared honey-eater (*Lichenostomus leucotis*), which was not in the area at the time of the fire, nested there one year later. Roberts (1970) observed that *Banksia ericifolia* which normally attracts honey-eaters had not flowered for at least three years after the fire.

Kimber (in prep) studied the effect of intense fire on the avifauna of tall dry sclerophyll forest with heathy understorey in southwestern Australia. Passerine species associated with the ground vegetation and the lower portions of the forest understorey decreased in abundance immediately after burning but increased beyond pre-fire levels within two years. Indeed after this period the total abundance of this assemblage of birds was greater than in an area of comparable habitat that had been protected from fire for forty years.

The phase of the pyric succession may significantly influence abundance and diversity of bird species, and the frequency with which fires interrupt this succession may contribute to longer-term habitat changes that have consequences for vertebrate populations. With reference to southeastern Queensland Gravett (1974) remarks that, where open-heathland and swamp communities are protected from fire for a long time, nectar-producing species, such as *Banksia robur*, may be replaced by rushes and sedges. This would clearly affect local populations of honey-eaters (e.g. *Phylidonyris*). Where sclerophyll forest is frequently burnt the shrub layer may be suppressed and grasses may become important in the understorey. In southeastern Queensland this may promote increase of the double-barred finch (*Poephila bichenovii*) at the expense of the red-browed firetail (*Emblema temporalis*), or the replacement of the bar-shouldered dove (*Geopelia humeralis*) and bronzewing pigeons (*Phaps* spp.) by the peaceful dove (*Geopelia striata*) (Gravatt, 1974). It may also encourage utilization of heathland areas by grey kangeroos (Dwyer et al., 1979).

Among mammals the New Holland mouse (*Pseudomys novaehollandiae*) and the feral house mouse (*Mus musculus*) may increase in numbers in the early stages of succession in heathland. Posamentier and Recher (1974, in prep) found *P. novaehollandiae* to be most common in disturbed areas supporting xeric vegetation; in New South Wales it responds to post-fire conditions in coastal heathlands where its local distribution is associated with sparse cover. The house mouse may invade heathlands within a year of burning and may be common in fire-deflected heathlands (Posamentier and Recher, 1974; Christensen and Kimber, 1975; Newsome et al., 1975); in these habitats its requirements are apparently similar to those of *P. novaehollandiae* (Posamentier and Recher, in prep). The swamp parrot (*Pezoporus wallicus*) also appears to favour heathlands in early stages of pyric succession (Forshaw, 1969; Gravatt, 1974; Roberts and Ingram, 1976).

The limited data available on the effects of fire

upon vertebrates of heathlands in Australia suggest that different fire régimes will have diverse consequences for the fauna. In southern North American forests prescribed burning is known to increase protein and phosphorus content in the new growth (DeWitt and Derby, 1955; Lay, 1957); and, after a chaparral fire, deer were found to concentrate on new growth (Hendricks, 1968). These effects upon nutritional status apparently last only one or two years and have led to recommendations that burning should be on a cycle of three years or less for wildlife (Crow, 1973). In Scotland the red grouse (*Lagopus lagopus*) depends on heather (*Calluna vulgaris*) for food (Jenkins et al., 1963) and heather is burnt approximately every 12 to 15 years in patches not more than 30 m wide to maintain optimum conditions for grouse management (Watson and Miller, 1976). In the southeastern United States marsh burning is an accepted management practice for waterfowl refuges (Givens, 1962). Fire removes dense marsh grass and exposes seed-bearing plants. It also encourages growth of succulent sprouts suitable for browsing by waterfowl. In southwestern Australia kangeroo populations may increase for a year or two after moderate fires, while lush feed is available in the early stages of regeneration (Christensen and Kimber, 1975). It is possible that changes in the nutrient status of food sources is involved here; certainly in parts of southern Australia trace elements, such as zinc and copper, are released after fire, and these too will be incorporated in new plant growth (Specht et al., 1958).

We suggest that the disappearing heath-specialized vertebrates were adapted to the conditions of soil and vegetation characteristic of a long fire-cycle and a large-scale mosaic of habitat. Such conditions include availability of nectar throughout the year as a major source of food. One relict species of marsupial mammal, *Tarsipes*, confined to southwestern Australia, is considered to require a continued supply of nectar for survival (Vose, 1973). Another rare heathland mammal, *Antechinus apicalis*, has been found in an area which Ride (1970, p. 22) described as "the waist-high coastal sandplain with its stunted banksias, Christmas tree, dryandras and other flamboyant blossoms" and "incredibly rich in small mammals". These post-fire conditions exist in parts of southwest Western Australia, where there are more

species of heathland-specialized reptiles, birds and mammals than in any other heathland region of Australia (see Fig. 8.1).

In the foregoing analysis climatic factors were not considered. For example, low temperatures in winter in southern and subalpine heathlands induce torpidity in frogs, reptiles, pigmy possums and bats, whereas dry, hot conditions in northern heathlands effect aestivation. Similarly, the breeding season of birds is restricted by cold climate in the south and dry conditions in the north. However, these factors are not unique to heathlands. We have shown that cover, soil nutrients, nature of substrates, aquatic conditions, food resources and fire combine as ecological factors of heathlands to create an adaptive zone ranking equal to any other major habitat in Australia.

CONCLUSION

Within Australia, heathland habitats show a fragmented distribution. They support a formerly widespread flora that through the Tertiary became increasingly restricted to coastal or subcoastal refuges or, during Quaternary times, invaded coastal sand dunes and wind-blown inland dunes. This sequence was paralleled by changes in the extent of tropical and sub-tropical closed-forests. Both habitat types must count as major adaptive zones in any consideration of the biogeography of Australian vertebrates. For heathland habitats the density of ground cover, the generally low status of soil nutrients, the sandy or rocky nature of the substrate, the restricted or special nature of food resources, and the important role of fire have interacted to produce a unique ecosystem. Patterns of vertebrate adaptation to this system have been in flux as the habitat itself has undergone change. Specialization to heathland living must have occurred many times and with differing chronologies for different major vertebrate taxa. Notable specializations include production of large eggs and reduced dependence upon water for breeding in frogs, burrowing in frogs and reptiles, and adaptations to nectar- and pollen-feeding in birds and mammals. Toxic subtances in plant tissues make herbivory hazardous; the few browsing and grazing vertebrates tend to utilize seasonal or post-fire growth of young foliage. Contraction of heathlands would have caused widespread extinction of forms or encouraged radiation away from heathlands for some groups. A relict pattern of distribution resulted for certain surviving forms. Comparable biogeographic patterns have been inferred for some closed-forest vertebrates and it is now clear that in several cases related types have survived as relicts in both closed-forest and heathland — sometimes at "opposite" ends of the continent.

Processes of radiation into or specialization within Australian heathlands continue to the present, and are well illustrated by the "acid-adapted" frogs and the *Phylidonyris* honey-eaters of eastern Australia. Thus specialization to heathland living is not restricted merely to relict species, but should be seen as a continual process involving taxa from the oldest to the latest heathland colonizers. Because, in eastern Australia, much of the heathland community is developed on recent Quaternary landscapes, the assemblages of vertebrates found there often include many invaders from adjoining non-heath open-country habitats.

Distribution of vertebrates within heathland is often patchy in consequence of fire history and asynchronous flowering régimes that characterize nectariferous plants of different regions. Abundance of animals is frequently low, and food chains are reduced and simplified in consequence of low nutrient status of soils and low primary production in heathland.

These biogeographic and ecological considerations reveal a unique process of community evolution in the heathland ecosystem of Australia.

ACKNOWLEDGEMENTS

We thank Professor Ray Specht who inspired us to attempt this review and Professor Allen Keast of Queens University, Canada, Dr. Harold Cogger of the Australian Museum, and Drs. Harold Heatwole and Hugh Ford of the University of New England, who made critical comments and provided useful information. Dr. Harry Recher and Heimo Posamentier of the Australian Museum and Dr. Lynn Carpenter of the University of California at Irvine permitted us to cite unpublished manuscripts. Many unpublished sources were used to compile the species lists in the Appendix and we are grateful to the following, who helped to update information: Dr. Glen Storr and Alex Baynes of the Western Australian Museum, Shane Parker of the South Australian Museum, John Coventry of the National Museum of Victoria, Basil Marlow of the Australian Museum, Dr. John Winter and Dr. Greg Gordon of Queensland National Parks and Wildlife Service, Dr. Michael Archer and Jeanette Covacevich of the Queensland Museum, Kath Shurcliff and other members of the South Australian Naturalists' Mammal Club, Marion Anstis, Peter Rankin and Chris Corben. David Hancock helped with listing of species.

The original work of the authors, from which information was extracted for this review, has been supported by the Interim Council of the Australian Biological Resources Study to Peter D. Dwyer (southeastern Queensland) and Jiro Kikkawa (Cape York Peninsula).

APPENDIX I: VERTEBRATES OF AUSTRALIAN HEATHLANDS

Fishes, seabirds and marine mammals are excluded.

Legend: + = presence of species in heathlands within the region shown at the top of the column; ‡ = species showing special association with the heathland habitat of that region; in = presence due to local introduction; ex = recorded only in the past; * = species introduced into Australia.

Geographic grouping of heathlands into eight regions is indicated in Fig. 8.1.

The nomenclature follows Cogger (1975) for amphibians and reptiles except for additions and revisions made by Greer (1974), McDowell (1974, 1975), Rawlinson (1975), Storr (1975a, b, 1976a, b, c, 1977b), Stimson et al. (1977); Condon (1975) and Schodde (1975) for birds; and Ride (1970) for mammals. The list was compiled from the following sources.

Amphibians and reptiles: Main (1957, 1963), Mitchell (1964), Lee (1967), Littlejohn (1967), Kluge (1974), Littlejohn and Martin (1974), Rawlinson (1974, 1975), Cogger (1975 and pers. comm., 1977), Covacevich and Ingram (1975a, b), Ingram and Corben (1975), Storr (1975a, b, 1976a, b, c, pers. comm., 1977), Storr and Smith (1975), Marion Anstis (pers. comm., 1977), Covacevich (1977), J. Coventry (pers comm., 1977), Hugh Ford (pers. comm., 1977), P. Rankin (pers. comm., 1977), Ingram (unpubl.).

Birds: Bryant (1930), McGill and Lane (1955), Officer (1958), Serventy and Whittell (1962), Deignan (1964), Hore-Lacy (1964), Bell, (1966), Gannon (1966), Ridpath and Moreau (1966), Storr (1967, 1977a), McEvey and Middleton (1968), Kikkawa (1970, 1976, unpubl.), Ford (1971), Abbot (1974), Miles and Burbidge (1975), Recher (1975), Vernon and Martin (1975), Roberts and Ingram (1976), S. Parker (pers. comm., 1977).

Mammals: Wood Jones (1923–25), Tate (1953), Anonymous (1961), Johnson (1964), Aitken (1970), Ride (1970), Frankenberg (1971), Land Conservation Council of Victoria (1972a, b, 1973, 1974a, b, 1976), Parker (1973b), Taylor and Horner (1973), Green (1974), Martin (1975), Miles and Burbidge (1975), Newsome et al. (1975), M. Archer (pers. comm., 1977), Barry and Campbell (1977), Baynes (pers. comm., 1977), P. Campbell (pers. comm., 1977), G. Gordon (pers. comm., 1977), B. Marlow (pers. comm., 1977), H. Recher (pers. comm., 1977), Kath Shurcliff (pers. comm., 1977), South Australian Field Naturalists Mammal Club (pers. comm., 1977), Dwyer et al. (1979), Posamentier and Recher (in prep.), J. Winter (pers. comm., 1977).

TABLE 8.1

Frogs of Australian heathlands

	1	2	3	4	5	6	7	8
MYOBATRACHIDAE								
Adelotus brevis	.	.	.	+	.	.	.	.
Australocrinia tasmaniensis	.	.	.	.	.	.	.	+
Geocrinia laevis	.	.	.	.	.	+	.	+
G. rosea	+	.	.	.	.	.	.	.
G. victoriana	.	.	.	.	.	+	.	.
Heleioporus albopunctatus	+	.	.	.	.	.	.	.
H. australiacus	.	.	.	.	+	+	.	.
H. eyrei	+	.	.	.	.	.	.	.
H. inornatus	‡	.	.	.	.	.	.	.
H. psammophilus	+	.	.	.	.	.	.	.
H. sp.	.	‡	.	.	.	.	.	.
Limnodynastes depressus	.	+	.	.	.	.	.	.
L. peroni	.	.	.	+	+	+	.	.
L. tasmaniensis	.	.	.	.	.	+	+	+
Myobatrachus gouldii	‡	.	.	.	.	.	.	.
Neobatrachus centralis	.	.	.	.	.	+	+	.
Paracrinia haswelli	.	.	.	.	‡	‡	.	.
Philoria frosti	.	.	.	.	.	‡	.	.
Platyplectron dorsalis	+	.	.	.	.	.	.	.
P. dumerilii	.	.	.	+	.	+	+	+
P. ornatus	.	.	.	.	+	.	.	.
P. terraereginae	.	.	.	+	.	.	.	.
Pseudophryne australis	.	.	.	.	‡	.	.	.
P. bibroni	.	.	.	.	.	+	+	.
P. coriacea	.	.	.	+	.	+	.	.
P. corroborree	.	.	.	.	.	‡	.	.
P. dendyi	.	.	.	.	.	+	.	.
P. guentheri	+	.	.	.	.	.	.	.
P. semimarmorata	.	.	.	.	.	+	.	+
Ranidella insignifera	+	.	.	.	.	.	.	.
R. signifera	.	.	.	+	+	+	+	+
R. subinsignifera	+	.	.	.	.	.	.	.
R. tinnula	.	.	.	‡	.	.	.	.
R. sp.	.	+	.	.	.	.	.	.
Uperoleia laevigata	.	.	.	+	.	.	.	.
U.sp. A	.	+	.	.	.	.	.	.
U. sp. B	.	.	+	.	.	.	.	.

TABLE 8.1 *(continued)*

Frogs of Australian heathlands

	1	2	3	4	5	6	7	8
PELODRYADIDAE								
Litoria adelaidensis	+	·	·	·	·	·	·	·
L. bicolor	·	+	+	·	·	·	·	·
L. burrowsi	·	·	·	·	·	·	·	+
L. caerulea	·	+	+	·	+	·	·	·
L. cooloolensis	·	·	·	‡	·	·	·	·
L. coplandi	·	+	·	·	·	·	·	·
L. cyclorhynchus	+	·	·	·	·	·	·	·
L. dentata	·	·	·	+	·	+	·	·
L. ewingi	·	·	·	+	·	+	+	+
L. fallax	·	·	·	+	·	+	·	·
L. freycineti	·	·	·	‡	·	·	·	·
L. gracilenta	·	·	+	·	·	·	·	·
L. inermis	·	+	+	·	·	·	·	·
L. jervisiensis	·	·	·	·	+	+	·	·
L. meiriana	·	+	·	·	·	·	·	·
L. microbelos	·	+	+	·	·	·	·	·
L. moorei	+	·	·	·	·	·	·	·
L. nasuta	·	+	+	+	·	·	·	·
L. olongburensis	·	·	·	‡	·	·	·	·
L. peronii	·	·	+	+	+	+	+	·
L. raniformis	·	·	·	·	·	+	+	+
L. rothi	·	+	+	·	·	·	·	·
L. rubella	·	+	+	+	·	·	·	·
L. tornieri	·	+	·	·	·	·	·	·
L. verreauxii	·	·	·	·	·	+	·	·
L. wotjulumensis	·	+	·	·	·	·	·	·
L. sp. A	·	·	·	+	·	·	·	·
L. sp. B	·	·	·	+	·	·	·	·
BUFONIDAE								
*Bufo marinus**	·	·	·	+	·	·	·	·
MICROHYLIDAE								
Sphenophryne robusta	·	+	+	·	·	·	·	·
RANIDAE								
Rana daemeli	·	·	+	·	·	·	·	·

TABLE 8.2

Reptiles of Australian heathlands

	1	2	3	4	5	6	7	8
CROCODYLIDAE								
Crocodylus johnstoni	·	+	+	·	·	·	·	·
C. prosus	·	·	+	·	·	·	·	·
CHELIDAE								
Chelodina longicollis	·	·	·	+	+	+	·	·
C. novaeguineae	·	·	+	·	·	·	·	·
C. rugosa	·	+	·	·	·	·	·	·
Elseya dentata	·	+	·	·	·	·	·	·
E. latisternum	·	·	+	·	·	·	·	·
Emydura sp.	·	·	·	‡	·	·	·	·
Pseudemydura umbrina	‡	·	·	·	·	·	·	·
GEKKONIDAE								
Diplodactylus alboguttatus	+	·	·	·	·	·	·	·
D. intermedius	·	·	·	·	·	+	+	·
D. maini	+	·	·	·	·	·	·	·
D. michaelseni	+	+	·	·	·	·	·	·
D. spinigerus	+	·	·	·	·	·	·	·
D. stenodactylus	·	+	·	·	·	·	·	·
D. vittatus	·	·	·	+	+	+	+	·
Gehyra australis	·	+	·	·	·	·	·	·
G. punctata	·	+	·	·	·	·	·	·
G. sp. A	·	‡	·	·	·	·	·	·
G. sp. B	·	‡	·	·	·	·	·	·
Heteronotia binoei	·	+	·	·	·	·	·	·
H. spelea	·	+	·	·	·	·	·	·
Nephrurus asper	·	+	·	·	·	·	·	·
N. stellatus	+	·	·	·	·	·	·	·
Oedura lesueurii	·	·	·	·	+	·	·	·
O. marmorata	·	+	·	·	·	·	·	·
O. reticulata	+	·	·	·	·	·	·	·
O. rhombifer	·	+	·	·	·	·	·	·
O. sp.	·	·	·	+	·	·	·	·
Phyllodactylus marmoratus	+	·	·	·	·	+	+	·
Phyllurus platurus	·	·	·	·	‡	·	·	·
Pseudothecadactylus australis	·	·	‡	·	·	·	·	·
P. lindneri	·	‡	·	·	·	·	·	·
Underwoodisaurus milii	·	·	·	+	+	+	+	·
PYGOPODIDAE								
Aclys concinna	‡	·	·	·	·	·	·	·
Aprasia pulchella	+	·	·	·	·	·	·	·
A. repens	+	·	·	·	·	·	·	·
A. striolata	+	·	·	·	·	·	+	·
Delma australis	·	·	·	·	·	·	+	·
D. borea	·	+	·	·	·	·	·	·
D. fraseri	+	·	·	·	·	·	+	·
D. grayii	+	·	·	·	·	·	·	·
D. inornata	·	·	·	·	·	·	+	·
D. tincta	·	+	·	·	·	·	·	·

TABLE 8.2 *(continued)*

Reptiles of Australian heathlands

	1	2	3	4	5	6	7	8
PYGOPODIDAE *(continued)*								
Lialis burtonis	+	+	+	+	+	+	+	.
Pletholax gracilis	‡	.	.	.	.	.	.	.
Pygopus lepidopodus	+	.	.	+	+	+	+	.
AGAMIDAE								
Amphibolurus adelaidensis	‡	.	.	.	.	.	.	.
A. barbatus	.	.	.	+	+	+	+	.
A. caudicinctus	.	+	.	.	.	.	.	.
A. diemensis	.	.	.	.	+	+	.	+
A. maculatus	‡	.	.	.	.	.	.	.
A. microlepidotus	.	‡	.	.	.	.	.	.
A. minimus	+	.	.	.	.	.	.	.
A. muricatus	.	.	.	+	+	+	+	.
A. parviceps	‡	.	.	.	.	.	.	.
Chelosania brunea	.	+	.	.	.	.	.	.
Chlamydosaurus kingii	.	+	+	+	.	.	.	.
Diporiphora albilabris	.	+	.	.	.	.	.	.
D. australis	.	.	+	.	.	.	.	.
D. bennettii	.	+	.	.	.	.	.	.
D. bilineata	.	+	.	.	.	.	.	.
D. superba	.	‡	.	.	.	.	.	.
D. sp.	.	.	.	+	.	.	.	.
Lophognathus gilberti	.	+	+	.	.	.	.	.
Physignathus lesueurii	.	.	.	+	+	+	.	.
Tympanocryptis lineatus	.	.	.	.	.	.	+	.
VARANIDAE								
Varanus acanthurus	.	+	.	.	.	.	.	.
V. glauerti	.	‡	.	.	.	.	.	.
V. glebopalma	.	‡	.	.	.	.	.	.
V. gouldii	+	+	+	+	+	.	+	.
V. mertensi	.	+	.	.	.	.	.	.
V. mitchelli	.	+	.	.	.	.	.	.
V. timorensis	.	+	+	.	.	.	.	.
V. tristis	.	+	.	.	.	.	.	.
V. varius	.	.	.	+	+	+	+	.
V. sp.	.	‡	.	.	.	.	.	.
SCINCIDAE								
Anomalopus ophioscincus	.	.	.	+	.	.	.	.
A. reticulatus	.	.	.	+	.	.	.	.
A. truncatus	.	.	.	+	.	.	.	.
A, verreauxii	.	.	.	+	.	.	.	.
A. sp.	.	.	.	.	‡	.	.	.
Anotis graciloides	.	.	.	+	.	.	.	.
A. sp.	.	.	.	+	.	.	.	.
Carlia amax	.	+	.	.	.	.	.	.
C. burnetti	.	.	+	.	+	.	.	.
C. dogare	.	.	‡	.	.	.	.	.
C. foliorum	.	+	.	.	.	.	.	.

TABLE 8.2 *(continued)*

Reptiles of Australian heathlands

	1	2	3	4	5	6	7	8
SCINCIDAE *(continued)*								
C. jarnoldae	·	·	+	·	·	·	·	·
C. johnstonei	·	+	·	·	·	·	·	·
C. pectoralis	·	·	·	+	·	·	·	·
C. schmeltzii	·	·	·	+	·	·	·	·
C. tricantha	·	+	·	·	·	·	·	·
C. vivax	·	·	·	+	·	·	·	·
Crytoblepharus megastictus	·	+	·	·	·	·	·	·
C. plagiocephalus	·	+	·	·	·	·	·	·
C. virgatus	·	·	+	+	·	·	·	·
C. sp.	·	‡	·	·	·	·	·	·
Ctenotus catenifer	+	·	·	·	·	·	·	·
C. essingtonii	·	+	·	·	·	·	·	·
C. delli	+	·	·	·	·	·	·	·
C. fallens	+	·	·	·	·	·	·	·
C. gemmula	+	·	·	·	·	·	·	·
C. impar	+	·	·	·	·	·	·	·
C. inornatus	·	+	·	·	·	·	·	·
C. lesueurii	‡	·	·	·	·	·	·	·
C. mastigura	·	‡	·	·	·	·	·	·
C. rubustus	·	+	+	+	+	+	+	·
C. spaldingi	·	+	+	·	·	·	·	·
C. taeniolatus	·	·	·	+	+	·	·	·
C. sp. A.	·	·	‡	·	·	·	·	·
C. sp. B.	·	·	‡	·	·	·	·	·
Egernia cunninghami	·	·	·	·	+	·	·	·
E. frerei	·	+	+	+	·	·	·	·
E. kingii	+	·	·	·	·	·	·	·
E. multiscutata	+	·	·	·	·	+	+	·
E. napoleonis	+	·	·	·	·	·	·	·
E. pulchra	+	·	·	·	·	·	·	·
E. richardi	·	+	·	·	·	·	·	·
E. whitii	·	·	·	·	+	+	+	+
E. sp.	·	·	·	·	·	‡	·	·
Hemiergis decresiensis	·	·	·	·	·	·	+	·
H. peronii	+	·	·	·	·	+	+	·
H. quadrilineatum	‡	·	·	·	·	·	·	·
Lampropholis delicata	·	·	·	+	+	+	+	·
L. guichenoti	·	·	·	+	+	+	+	·
L. mustelina	·	·	·	+	+	+	·	·
Leiolopisma entrecasteauxii	·	·	·	·	·	+	+	+
L. greeni	·	·	·	·	·	·	·	‡
L. metallica	·	·	·	·	·	+	·	+
L. ocellata	·	·	·	·	·	·	·	+
L. platynota	·	·	·	·	+	·	·	·
L. pretiosa	·	·	·	·	·	·	·	+
L. trilineata	+	·	·	·	+	+	+	+
Lerista bougainvillii	·	·	·	·	·	+	+	+
L. connivens	‡	·	·	·	·	·	·	·
L. distinguenda	+	·	·	·	·	·	·	·
l. elegans	+	·	·	·	·	·	·	·
L. karlschmidti	·	+	·	·	·	·	·	·

TABLE 8.2 *(continued)*

Reptiles of Australian heathlands

	1	2	3	4	5	6	7	8
SCINCIDAE *(continued)*								
L. lineata	‡	.	.	.	.	.	.	.
L. lineopunctulata	‡	.	.	.	.	.	.	.
L. microtis	+	.	.	.	.	.	.	.
L. nichollsi	+	.	.	.	.	.	.	.
L. planiventralis	‡	.	.	.	.	.	.	.
L. praepedita	‡	.	.	.	.	.	.	.
L. stylis	.	+	.	.	.	.	.	.
L. walkeri	.	+	.	.	.	.	.	.
L. sp.	.	.	‡	.	.	.	.	.
Menetia greyii	+	.	.	.	.	.	.	.
Morethia adelaidensis	.	.	.	.	.	.	.	+
M. boulengeri	.	.	.	+	.	.	.	.
M. lineoocellata	+	.	.	.	.	.	.	.
M. obscura	+	.	.	.	.	+	+	.
M. taeniopleura	.	+	.	+	.	.	.	.
Notoscincus ornatus	.	+	.	.	.	.	.	.
Omolepida branchialis	+	.	.	.	.	.	.	.
O. casuarinae	.	.	.	.	+	+	.	+
O. maxima	.	‡	.	.	.	.	.	.
Proablepharus tenuis	.	+	.	.	.	.	.	.
Saiphos equalis	.	.	.	+	+	.	.	.
Sphenomorphus australis	+	.	.	.	.	.	.	.
S. crassicaudatus	.	+	.	.	.	.	.	.
S. douglasi	.	+	.	.	.	.	.	.
S. isolepis	.	+	.	.	.	.	.	.
S. kosciuskoi	.	.	.	.	.	+	.	.
S. quoyii	.	.	.	+	+	.	.	.
S. scutirostrum	.	.	.	+	.	.	.	.
S. tympanum	.	.	.	.	.	+	.	.
S. sp.	.	.	.	+	+	.	.	.
Tiliqua nigrolutea	.	.	.	.	.	+	.	+
T. occipitalis	+	.	.	.	.	.	+	.
T. rugosa	+	.	.	.	.	.	+	.
T. scincoides	.	+	+	+	+	+	+	.
TYPHLOPIDAE								
Ramphotyphlops australis	+	.	.	.	.	.	+	.
R. nigrescens	.	.	.	+	+	.	.	.
R. pinguis	+	.	.	.	.	.	.	.
R. yirrikalae	.	+	.	.	.	.	.	.
R. sp.	.	.	.	‡	.	.	.	.
BOIDAE								
Aspidites melanocephalus	.	+	+	.	.	.	.	.
Liasis olivaceus	.	+	.	.	.	.	.	.
Python spilotus	.	+	+	+	+	+	.	.
COLUBRIDAE								
Amphiesma mairii	.	+	+	+	.	.	.	.
Boiga irregularis	.	+	.	+	+	+	.	.

TABLE 8.2 *(continued)*

Reptiles of Australian heathlands

	1	2	3	4	5	6	7	8
COLUBRIDAE *(continued)*								
Dendrelaphis calligaster	.	.	+	.	.	.	.	.
D. punctulatus	.	+	+	+	+	+	.	.
Enhydris polylepis	.	+	+	.	.	.	.	.
E. punctata	.	+	.	.	.	.	.	.
Stegonotus cucullatus	.	.	+	.	.	.	.	.
ELAPIDAE								
Acanthopis antarticus	.	+	+	+	+	+	.	.
Austrelaps superbus	.	.	.	.	.	+	+	+
Cacophis harriettae	.	.	.	+	.	.	.	.
C. krefftii	.	.	.	+	.	.	.	.
C. squamulosus	.	.	.	+	+	+	.	.
Cryptophis nigrescens	.	.	.	+	+	+	.	.
C. pallidiceps	.	+	.	.	.	.	.	.
Demansia atra	.	+	+	+	.	.	.	.
D. olivacea	.	+	.	.	.	.	.	.
D. psammophis	.	.	.	+	+	+	+	.
D. reticulata	+	.	.	.	.	.	.	.
D. torquata	.	.	+	.	.	.	.	.
Drysdalia coronata	+	.	.	.	.	.	.	.
D. coronoides	.	.	.	.	.	+	.	+
Echiopsis curta	+	.	.	.	.	+	+	.
Elapognathus minor	‡	.	.	.	.	.	.	.
Furina diadema	.	.	.	+	+	.	.	.
Hemiaspis signata	.	.	.	+	+	.	.	.
Hoplocephalus bungaroides	.	.	.	.	‡	.	.	.
Neelaps bimaculatus	+	.	.	.	.	.	.	.
N. calonotus	‡	.	.	.	.	.	.	.
Notechis ater	+	.	.	.	.	.	+	+
N. scutatus	.	.	.	+	+	+	.	.
Oxyuranus scutellatus	.	+	+	+	.	.	.	.
Pseudechis australis	+	+	.	.	.	.	.	.
P. porphyriacus	.	.	.	+	+	+	.	.
Pseudonaja affinis	+	.	.	.	.	.	.	.
P. textilis	+	.	.	+	+	+	+	.
P. nuchalis	+	.	.	.	.	.	.	.
Rhinoplocephalus bicolor	+	.	.	.	.	.	.	.
Simoselaps bertholdi	+	.	.	.	.	.	.	.
S. fasciolatus	+	.	.	.	.	.	.	.
S. littoralis	‡	.	.	.	.	.	.	.
S. semifasciatus	+	.	.	.	.	.	.	.
Tropidechis carinatus	.	.	.	+	.	.	.	.
Unechis brevicaudus	.	.	.	.	.	.	+	.
U. flagellum	.	.	.	.	.	+	+	.
U. gouldii	+	.	.	.	.	.	.	.
Vermicella annulata	.	.	.	+	.	.	.	.

TABLE 8.3

Birds of Australian heathlands

	1	2	3	4	5	6	7	8
DROMAIIDAE								
Dromaius novaehollandiae	+	+	+	+	+	+	+	.
PODICIPEDIDAE								
Podiceps cristatus	+	.	.	+	+	+	.	.
Poliocephalus poliocephalus	+	.	.	.	+	+	+	.
Tachybaptus novaehollandiae	+	+	+	+	+	+	+	.
PELECANIDAE								
Pelecanus conspicillatus	+	+	+	+	+	+	+	.
ANHINGIDAE								
Anhinga melanogaster	+	+	+	+	+	+	.	.
PHALACROCORACIDAE								
Phalacrocorax carbo	+	.	+	+	+	+	+	+
P. melanoleucos	+	+	+	+	+	+	+	.
P. sulcirostris	+	+	+	+	+	+	+	.
P. varius	.	.	+	+	+	+	+	.
ARDEIDAE								
Ardea novaehollandiae	+	+	+	+	+	+	+	+
A. pacifica	+	+	+	+	+	+	+	.
A. picata	.	+	+	.	.	.	.	.
Ardeola ibis	+	+	.	+	.	.	.	.
Botaurus poiciloptilus	+	.	.	+	+	+	.	.
Dupetor flavicollis	+	+	+	+	+	.	.	.
Egretta alba	+	+	+	+	+	+	+	.
E. garzetta	.	+	+	+	+	.	.	.
E. intermedia	.	+	+	+	+	+	.	.
Ixobrychus minutus	+	.	+	+	+	.	.	.
Nycticorax caledonicus	+	+	+	+	+	+	+	.
CICONIIDAE								
Xenorhynchus asiaticus	.	+	+	+	.	.	.	.
PLATALEIDAE								
Platalea flavipes	.	+	.	+	+	+	+	.
P. regia	+	+	+	+	+	+	+	.
Plegadis falcinellus	.	+	.	+	.	.	.	.
Threskiornis moluccus	.	+	+	+	+	+	+	.
T. spinicollis	+	+	+	+	+	+	+	.
ANATIDAE								
Anas castaneae	+	.	.	+	+	+	+	+
A. gibberifrons	+	+	+	+	+	+	+	+
A. rhynchotis	+	.	.	.	.	+	+	+
A. superciliosa	+	+	+	+	+	+	+	+

TABLE 8.3 *(continued)*

Birds of Australian heathlands

	1	2	3	4	5	6	7	8
ANATIDAE *(continued)*								
Anseranas semipalmata	·	+	+	·	·	·	·	·
Aythya australis	+	+	·	+	+	+	+	·
Biziura lobata	+	·	·	+	+	+	+	+
Cereopsis novaehollandiae	·	·	·	·	·	+	+	·
Chenonetta jubata	+	·	·	+	+	+	+	·
Cygnus atratus	+	·	·	+	+	+	+	·
Dendrocygna arcuata	·	+	+	·	·	·	·	·
D. eytoni	·	+	+	+	·	·	·	·
Malacorhynchus membranaceus	+	+	·	·	+	+	+	·
Nettapus pulchellus	·	+	+	·	·	·	·	·
Oxyura australis	+	·	·	·	·	+	·	·
Stictonetta naevosa	+	·	·	·	·	+	·	·
Tadorna radjah	·	+	+	·	·	·	·	·
T. tadornoides	+	·	·	·	·	+	+	+
PANDIONIDAE								
Pandion haliaetus	+	+	+	+	+	+	+	+
ACCIPITRIDAE								
Accipiter cirrocephalus	+	+	+	+	+	+	+	+
A. fasciatus	+	+	+	+	+	+	+	+
A. novaehollandiae	·	+	+	+	+	+	·	+
Aquila audax	+	+	+	+	+	+	+	+
Aviceda subcristata	·	·	+	+	+	·	·	·
Circus aeruginosus	+	+	+	+	+	+	+	+
C. assimilis	·	+	+	·	·	·	+	·
Elanus notatus	+	+	+	+	+	+	+	·
Haliaeetus leucogaster	·	+	+	+	+	+	+	+
Haliastur indus	·	+	+	+	·	·	·	·
H,. sphenurus	+	+	+	+	+	+	+	·
Hamirostra melanosternon	·	+	+	·	·	·	+	·
Hieraaetus morphnoides	+	+	+	+	+	+	·	·
Lophoictinia isura	+	+	+	·	·	·	+	·
Milvus migrans	·	+	+	+	·	+	+	·
FALCONIDAE								
Falco berigora	+	+	+	+	+	+	+	+
F. cenchroides	+	+	+	+	+	+	+	·
F. longipennis	+	+	+	·	+	+	+	·
F. peregrinus	+	+	+	+	+	+	+	+
MEGAPODIIDAE								
Alectura lathami	·	·	+	+	·	·	in	·
Leipoa ocellata	+	·	·	·	·	·	+	·
Megapodius freycinet	·	+	+	·	·	·	·	·
PHASIANIDAE								
Coturnix australis	+	+	+	+	+	+	·	+
C. chinensis	·	+	+	+	+	+	·	·

TABLE 8.3 *(continued)*

Birds of Australian heathlands

	1	2	3	4	5	6	7	8
PHASIANIDAE *(continued)*								
C. pectoralis	+	.	.	.	+	+	+	.
TURNICIDAE								
Turnix castanota	.	+	+	.	.	.	.	.
T. maculosa	.	+	+	.	.	.	.	.
T. pyrrhothorax	.	+	.	.	.	.	.	.
T. varia	+	.	+	+	+	+	+	+
T. velox	+	+	.	.	.	.	+	.
RALLIDAE								
Fulica atra	+	.	.	+	+	+	+	+
Gallinula mortierii	.	.	.	.	.	.	.	+
G. olivacea	.	.	+	+	.	.	.	.
G. tenebrosa	+	.	.	+	+	+	+	+
G. ventralis	+	.	.	.	.	+	+	.
Porphyrio porphyrio	+	.	.	+	+	+	+	+
Porzana fluminea	+	.	.	+	+	+	+	+
P. pusilla	+	.	.	.	+	+	+	+
P. tabuensis	+	.	.	.	.	.	.	+
Rallus pectoralis	+	.	.	+	+	+	.	+
R. philippensis	+	+	+	+	+	+	.	.
GRUIDAE								
Grus rubicundus	.	+	+	+	.	.	.	.
OTIDIDAE								
Ardeotis australis	+	+	+	.	.	.	.	.
JACANIDAE								
Irediparra gallinacea	.	+	+	+	.	.	.	.
BURHINIDAE								
Burhinus magnirostris	+	+	+	+	.	.	.	.
ROSTRATULIDAE								
Rostratula benghalensis	.	.	.	+	+	.	.	.
CHARADRIIDAE								
Charadrius melanops	+	+	+	+	+	+ .	+	.
C. rubricollis	+	.	.	.	.	+	.	+
C. ruficapillus	+	+	+	+	+	+	+	.
Erythrogonys cinctus	+	+	.	.	.	+	+	.
Peltohyas australis	.	.	.	.	.	.	+	.
Vanellus miles	+	+	+	+	+	+	+	+
V. tricolor	+	.	.	+	+	+	+	.

TABLE 8.3 *(continued)*

Birds of Australian heathlands

	1	2	3	4	5	6	7	8
RECURVIROSTRIDAE								
Cladorhynchus leucocephalus	+	·	·	·	·	·	+	·
Himantopus himantopus	·	·	+	+	+	+	+	·
Recurvirostra novaehollandiae	·	+	+	·	·	·	+	·
SCOLOPACIDAE								
Calidris acuminata	+	+	+	+	+	+	·	·
C. ruficollis	+	+	+	+	+	+	·	+
Gallinago hardwickii	·	·	+	+	+	+	·	+
Numenius minutus	·	+	+	·	·	·	·	·
Tringa glareola	·	+	·	+	+	·	·	·
T. hypoleucos	+	+	+	+	+	+	·	·
T. nebularia	·	+	+	+	+	+	·	·
GLAREOLIDAE								
Stiltia isabella	·	+	+	·	·	·	+	·
LARIDAE								
Chlidonias hybrida	+	+	·	+	+	+	+	·
C. leucoptera	·	+	+	+	·	·	·	·
Gelochelidon nilotica	+	+	+	+	+	+	+	·
Larus novaehollandiae	+	+	+	+	+	+	+	+
COLUMBIDAE								
Chalcophaps indica	·	+	+	+	+	·	·	·
Ducula spilorrhoa	·	+	+	·	·	·	·	·
Geopelia cuneata	·	+	+	·	·	·	+	·
G. humeralis	·	‡	‡	‡	‡	·	·	·
G. striata	·	+	+	+	+	+	+	·
Ocyphaps lophotes	·	+	+	+	·	·	+	·
Petrophassa albipennis	·	‡	·	·	·	·	·	·
P. scripta	·	·	+	·	·	·	·	·
P. smithii	·	+	·	·	·	·	·	·
Phaps chalcoptera	+	+	·	+	+	+	‡	‡
P. elegans	‡	·	·	‡	‡	‡	‡	‡
Ptilinopus cinctus	·	+	·	·	·	·	·	·
P. regina	·	+	+	·	·	·	·	·
CACATUIDAE								
Cacatua galerita	·	+	+	+	+	+	+	·
C. leadbeateri	·	·	·	·	·	+	+	·
C. roseicapilla	+	·	+	+	+	+	+	·
C. sanguinea	·	+	+	·	·	+	+	·
Callocephalon fimbriatum	·	·	·	·	+	+	in	·
Calyptorhynchus baudinii	+	·	·	·	·	·	·	·
C. funereus	·	·	·	+	+	+	+	+
C. lathami	·	·	·	+	+	·	+	·
C. magnificus	+	+	+	+	+	·	+	·

TABLE 8.3 *(continued)*

Birds of Australian heathlands

	1	2	3	4	5	6	7	8
LORIIDAE								
Glossopsitta concinna	.	.	.	.	+	+	+	.
G. porphyrocephala	+	.	.	.	.	+	+	.
G. pusilla	.	.	.	+	+	+	.	+
Psitteuteles versicolor	.	+	+	.	.	.	.	.
Trichoglossus chlorolepidotus	.	.	.	+	.	.	.	.
T. haematodus	.	.	+	+	+	+	+	.
T. rubritorquis	.	+	.	.	.	.	.	.
POLYTELIDAE								
Aprosmictus erythropterus	.	+	+	+	.	.	.	.
Nymphicus hollandicus	.	.	.	.	.	.	+	.
Polytelis anthopeplus	+	.	.	.	.	.	+	.
PLATYCERCIDAE								
Barnardius barnardi	.	.	.	.	.	.	+	.
B. zonarius	+	.	.	.	.	.	+	.
Lathamus discolor	.	.	.	.	.	+	.	+
Melopsittacus undulatus	.	.	.	.	.	.	+	.
Neophema chrysogaster	.	.	.	.	.	+	.	+
N. chrysostoma	.	.	.	.	.	+	+	+
N. elegans	+	.	.	.	.	+	+	.
N. petrophila	‡	.	.	.	.	.	‡	.
N. pulchella	.	.	.	+	+	.	.	.
Northiella haematogaster	.	.	.	.	.	+	+	.
Pezoporus wallicus	‡	.	.	‡	‡	‡	ex	‡
Platycercus adscitus	.	.	+	+	.	.	.	.
P. caledonicus	.	.	.	.	.	.	.	+
P. elegans	.	.	.	.	+	+	+	.
P. eximius	.	.	.	.	+	+	.	.
P. icterotis	+	.	.	.	.	.	.	.
P. venustus	.	+	.	.	.	.	.	.
Psephotus haematonotus	.	.	.	.	.	+	+	.
P. varius	+	.	.	.	.	.	+	.
Purpureicephalus spurius	+	.	.	.	.	.	.	.
CUCULIDAE								
Centropus phasianinus	.	+	+	+	+	.	.	.
Chrysococcyx basalis	+	+	+	+	+	+	+	+
C. lucidus	+	.	+	+	+	+	+	+
C. osculans	.	.	.	.	.	+	+	.
C. russatus	.	.	+	+	.	.	.	.
Cuculus pallidus	+	+	+	+	+	+	+	+
C. pyrrhophanus	+	.	+	+	+	+	+	+
C. saturatus	.	+	+	.	.	.	.	.
C. variolosus	.	+	+	+	+	+	.	.
Eudynamys scolopacea	.	+	+	+	+	.	.	.
Scythrops novaehollandiae	.	.	+	+	+	.	.	.

TABLE 8.3 *(continued)*

Birds of Australian heathlands

	1	2	3	4	5	6	7	8
STRIGIDAE								
Ninox connivens	+	+	+	+	+	+	.	.
N. novaeseelandiae	+	+	+	+	+	+	+	+
TYTONIDAE								
Tyto alba	+	+	+	+	+	+	+	+
T. longimembris	.	.	.	+	.	.	.	.
T. novaehollandiae	+	+	.	+	+	+	.	+
PODARGIDAE								
Podargus strigoides	+	+	+	+	+	+	+	+
AEGOTHELIDAE								
Aegotheles cristatus	+	+	.	+	+	+	+	.
CAPRIMULGIDAE								
Caprimulgus guttatus	+	+	+	.	.	.	+	.
C. macrurus	.	+	+	.	.	.	.	.
C. mystacalis	.	.	+	+	+	+	.	.
APODIDAE								
Apus pacificus	.	+	+	+	+	+	.	.
Hirundapus caudacutus	.	.	+	+	+	+	+	.
ALCEDINIDAE								
Ceyx azureus	.	+	+	+	+	+	.	+
Dacelo leachii	.	+	+	+	.	.	.	.
D. novaeguineae	+	.	+	+	+	+	+	.
Halcyon macleayii	.	+	+	+	.	.	.	.
H. pyrrhopygia	.	+	+	.	.	.	+	.
H. sancta	+	+	+	+	+	+	+	+
MEROPIDAE								
Merops ornatus	+	+	+	+	+	+	+	.
CORACIIDAE								
Eurystomus orientalis	.	+	+	+	+	.	.	.
MENURIDAE								
Menura novaehollandiae	.	.	.	.	+	+	.	.
ATRICHORNITHIDAE								
Atrichornis clamosus	‡	.	.	.	.	.	.
ALAUDIDAE								
Mirafra javanica	.	+	.	+	+	+	+	.

TABLE 8.3 *(continued)*

Birds of Australian heathlands

	1	2	3	4	5	6	7	8
HIRUNDINIDAE								
Cecropis ariel	·	+	+	+	+	+	+	·
C. nigricans	+	+	+	+	+	+	+	+
Cheramoeca leucosternum	+	·	·	+	·	+	+	·
Hirundo neoxena	+	·	·	+	+	+	+	+
MOTACILLIDAE								
Anthus novaeseelandiae	+	+	+	+	+	+	+	+
CAMPEPHAGIDAE								
Coracina novaehollandiae	+	+	+	+	+	+	+	+
C. papuensis	·	+	+	+	+	+	·	·
C. tenuirostris	·	·	+	+	+	·	·	·
Lalage leucomela	·	+	+	+	·	·	·	·
L. sueurii	+	+	+	+	+	+	+	·
MUSCICAPIDAE								
Arses telescophthalmus	·	·	+	·	·	·	·	·
Colluricincla harmonica	+	+	+	+	+	+	+	+
C. megarhyncha	·	+	+	+	·	·	·	·
C. woodwardi	·	+	·	·	·	·	·	·
Drymodes brunneopygia	‡	·	·	·	·	·	‡	·
D. superciliaris	·	·	+	·	·	·	·	·
Eopsaltria australis	·	·	·	+	+	+	·	·
E. georgiana	+	·	·	·	·	·	·	·
E. griseogularis	+	·	·	·	·	·	+	·
Falcunculus frontatus	+	·	·	·	+	+	·	·
Melanodryas cucullata	+	+	·	·	+	+	+	·
M. vittata	·	·	·	·	·	·	·	+
Microeca flavigaster	·	+	+	·	·	·	·	·
M. leucophaea	+	+	+	+	+	+	+	·
Monarcha leucotis	·	·	+	+	·	·	·	·
M. melanopsis	·	·	+	+	·	·	·	·
Myiagra alecto	·	+	+	+	·	·	·	·
M. cyanoleuca	·	·	·	+	+	+	·	·
M. inquieta	+	+	·	+	+	+	+	·
M. rubecula	·	+	+	+	+	+	·	·
Oreoica gutturalis	+	·	·	·	·	·	+	·
Pachycephala inornata	+	·	·	·	·	·	+	·
P. olivacea	·	·	·	·	·	+	·	+
P. pectoralis	+	·	·	+	+	+	+	+
P. rufiventris	+	+	+	+	+	+	+	·
P. rufogularis	·	·	·	·	·	·	+	·
Petroica goodenovii	+	·	·	·	+	+	+	·
P. multicolor	+	·	·	·	+	+	+	+
P. phoenicea	·	·	·	·	+	+	+	+
P. rodinogaster	·	·	·	·	·	+	·	·
P. rosea	·	·	·	+	+	+	·	·
Rhipidura fuliginosa	+	+	+	+	+	+	+	+
R. leucophrys	+	+	·	+	+	+	+	·
R. rufifrons	·	+	+	+	+	+	·	·

TABLE 8.3 *(continued)*

Birds of Australian heathlands

	1	2	3	4	5	6	7	8
MUSCICAPIDAE *(continued)*								
R. rufiventris	·	+	+	·	·	·	·	·
Zoothera dauma	·	·	·	·	·	+	·	+
ORTHONYCHIDAE								
Cinclosoma castanotum	+	·	·	·	·	·	+	·
C. punctatum	·	·	·	·	+	+	·	+
Psophodes nigrogularis	‡	·	·	·	·	·	‡	·
P. olivaceus	·	·	·	+	+	+	·	·
TIMALIIDAE								
Pomatostomus ruficeps	·	·	·	·	·	·	+	·
P. superciliosus	+	·	·	·	·	+	+	·
P. temporalis	·	+	+	+	+	+	·	·
SYLVIIDAE								
Acrocephalus stentoreus	+	·	·	+	+	+	+	+
Cinclorhamphus cruralis	+	+	·	·	·	+	+	·
C. mathewsi	+	+	+	+	+	+	+	·
Cisticola exilis	·	+	+	+	+	+	+	+
Megalurus gramineus	+	·	·	+	+	+	+	+
M. timoriensis	·	+	+	+	+	·	·	·
MALURIDAE								
Amytornis dorotheae	·	‡	·	·	·	·	·	·
A. housei	·	‡	·	·	·	·	·	·
A. striatus	·	·	·	·	·	·	+	·
A. textilis	+	·	·	·	·	·	+	·
A. woodwardi	·	‡	·	·	·	·	·	·
Malurus coronatus	·	+	·	·	·	·	·	·
M. cyaneus	·	·	·	+	+	+	+	+
M. elegans	‡	·	·	·	·	·	·	·
M. lamberti	+	+	+	+	+	·	+	·
M. leucopterus	+	·	·	·	·	·	+	·
M. melanocephalus	·	+	+	+	·	·	·	·
M. pulcherrimus	‡	·	·	·	·	·	‡	·
M. splendens	+	·	·	·	·	·	+	·
Stipiturus malachurus	‡	·	·	‡	‡	‡	·	‡
S. ruficeps	·	+	·	·	·	·	+	·
ACANTHIZIDAE								
Acanthiza apicalis	+	·	·	·	·	·	+	·
A. ewingii	·	·	·	·	·	·	·	+
A. chrysorrhoa	·	·	·	+	+	+	+	·
A. inornata	+	·	·	·	·	·	·	·
A. iredalei	·	·	·	·	·	·	+	·
A. lineata	·	·	·	+	+	+	·	·
A. nana	·	·	·	+	+	+	+	·
A. pusilla	·	·	·	+	+	+	+	+
A. reguloides	·	·	·	+	+	+	+	·

TABLE 8.3 *(continued)*

Birds of Australian heathlands

	1	2	3	4	5	6	7	8
ACANTHIZIDAE *(continued)*								
A. uropygialis	+	.	.	.	.	.	+	.
Aphelocephala leucopsis	.	.	.	.	+	+	+	.
Dasyornis brachypterus	.	.	.	‡	‡	.	.	.
D. broadbenti	.	.	.	.	.	‡	.	.
D. longirostris	‡	.	.	.	.	.	.	.
Gerygone chloronota	.	+	.	.	.	.	.	.
G. fusca	+	.	.	.	.	.	+	.
G. olivacea	.	+	+	+	+	+	.	.
G. palpebrosa	.	.	+	.	.	.	.	.
Origma solitaria	.	.	.	.	‡	.	.	.
Sericornis brunneus	+	.	.	.	.	.	+	.
S. cautus	‡	.	.	.	.	.	‡	.
S. frontalis	+	.	.	+	+	+	+	+
S. fuliginosus	‡	.	.	.	.	‡	‡	‡
S. magnus	.	.	.	.	.	.	.	+
S. pyrrhopygius	.	.	.	.	‡	‡	+	.
S. sagittatus	.	.	.	+	+	+	.	.
Smicrornis brevirostris	+	+	+	+	+	+	+	.
NEOSITTIDAE								
Daphoenositta chrysoptera	+	+	+	+	+	+	+	.
CLIMACTERIDAE								
Climacteris leucophaea	.	.	.	+	+	+	.	.
C. melanura	.	+	.	.	.	.	.	.
C. picumnus	.	.	.	+	+	+	+	.
C. rufa	+	.	.	.	.	.	.	.
MELIPHAGIDAE								
Acanthagenys rufogularis	+	.	.	.	.	+	+	.
Acanthorhynchus superciliosus	‡	.	.	.	.	.	.	.
A. tenuirostris	.	.	.	‡	‡	‡	‡	‡
Anthochaera carunculata	+	.	.	.	+	+	+	.
A. chrysoptera	‡	.	.	‡	‡	‡	‡	‡
A. paradoxa	.	.	.	.	.	.	.	+
Certhionyx niger	+	.	.	.	.	+	+	.
C. pectoralis	.	+	+	.	.	.	.	.
C. variegatus	+	.	.	.	.	.	+	.
Conopophila albogularis	.	+	+	.	.	.	.	.
C. rufogularis	.	+	+	.	.	.	.	.
Entomyzon cyanotis	.	+	+	+	.	.	+	.
Grantiella picta	.	+	.	.	+	+	.	.
Lichenostomus chrysops	.	.	.	‡	‡	+	+	.
L. cratitius	‡	.	.	.	.	+	‡	.
L. fasciogualris	.	.	.	+	.	.	.	.
L. flavescens	.	+	.	.	.	.	.	.
L. flavicollis	.	.	.	.	.	.	.	‡
L. flavus	.	.	+	.	.	.	.	.
L. fuscus	.	.	.	+	+	+	+	.
L. leucotis	+	.	.	.	‡	+	+	.

TABLE 8.3 *(continued)*

Birds of Australian heathlands

	1	2	3	4	5	6	7	8
MELIPHAGIDAE *(continued)*								
L. melanops	·	·	·	+	+	+	+	·
L. ornatus	+	·	·	·	·	+	‡	·
L. penicillatus	·	·	·	·	+	+	+	·
L. plumulus	·	·	·	·	·	·	+	·
L. unicolor	·	+	·	·	·	·	·	·
L. versicolor	·	·	+	·	·	·	·	·
L. virescens	+	+	·	·	·	+	+	·
Lichmera indistincta	+	‡	‡	‡	+	·	·	·
Manorina flavigula	+	+	·	·	·	·	+	·
M. melanocephala	·	·	·	+	+	+	+	+
M. melanotis	·	·	·	·	·	·	+	·
Meliphaga albilineata	·	+	·	·	·	·	·	·
M. gracilis	·	·	+	·	·	·	·	·
M. lewinii	·	·	·	+	·	·	·	·
M. notata	·	·	+	·	·	·	·	·
Melithreptus affinis	·	·	·	·	·	·	·	+
M. albogularis	·	+	+	+	·	·	·	·
M. brevirostris	+	·	·	·	‡	+	+	·
M. gularis	·	+	+	·	+	+	+	·
M. lunatus	+	·	·	+	+	+	+	·
M. validirostris	·	·	·	·	·	·	·	+
Myzomela obscura	·	+	+	·	·	·	·	·
M. sanguinolenta	·	·	+	+	+	+	·	·
Philemon argenticeps	·	+	+	·	·	·	·	·
P. buceroides	·	+	+	·	·	·	·	·
P. citreogularis	·	+	+	‡	·	‡	+	·
P. corniculatus	·	·	·	‡	+	+	·	·
Phylidonyris albifrons	‡	·	·	·	·	‡	‡	·
P. melanops	‡	·	·	·	‡	‡	‡	‡
P. nigra	‡	·	·	‡	‡	·	·	·
P. novaehollandiae	‡	·	·	‡	‡	‡	‡	‡
P. pyrrhoptera	·	·	·	·	‡	‡	‡	‡
Plectorhyncha lanceolata	·	·	·	+	·	·	+	·
Ramsayornis fasciatus	·	+	+	·	·	·	·	·
R. modestus	·	·	+	·	·	·	·	·
Trichodere cockerelli	·	·	‡	·	·	·	·	·
Xanthomyza phrygia	·	·	·	+	+	+	+	·
Xanthotis flaviventer	·	·	+	·	·	·	·	·
EPHTHIANURIDAE								
Ephthianura albifrons	‡	·	·	·	+	+	‡	‡
E. crocea	·	+	·	·	·	·	·	·
E. tricolor	·	·	·	·	·	·	+	·
NECTARINIIDAE								
Nectarinia jugularis	·	·	+	·	·	·	·	·
DICAEIDAE								
Dicaeum hirundinaceum	+	+	+	+	+	+	+	·

TABLE 8.3 *(continued)*

Birds of Australian heathlands

	1	2	3	4	5	6	7	8
PARDALOTIDAE								
Pardalotus punctatus	+	·	·	+	+	+	·	·
P. striatus	+	+	+	+	+	+	+	+
P. xanthopygus	+	·	·	·	·	·	+	·
ZOSTEROPIDAE								
Zosterops lateralis	+	·	·	+	+	+	+	+
PLOCEIDAE								
Emblema bella	·	·	·	·	+	+	+	+
E. guttata	·	·	·	·	+	+	+	·
E. oculata	+	·	·	·	·	·	·	·
E. temporalis	·	·	+	+	+	+	+	·
Erythrura gouldiae	·	+	·	·	·	·	·	·
Lonchura castaneothorax	·	+	+	+	·	·	·	·
L. flaviprymna	·	+	·	·	·	·	·	·
L. pectoralis	·	+	·	·	·	·	·	·
Neochmia phaeton	·	+	·	·	·	·	·	·
Poephila acuticauda	·	+	·	·	·	·	·	·
P. bichenovii	·	+	+	+.	+	·	·	·
P. cincta	·	·	+	·	·	·	·	·
P. personata	·	+	+	·	·	·	·	·
P. guttata	·	·	·	·	·	+	+	·
ORIOLIDAE								
Oriolus flavocinctus	·	+	+	·	·	·	·	·
O. sagittatus	·	+	+	+	+	+	·	·
Sphecotheres viridis	·	+	+	+	·	·	·	·
DICRURIDAE								
Dicrurus hottentottus	·	+	+	+	+	·	·	·
PARADISAEIDAE								
Chlamydera cerviniventris	·	·	+	·	·	·	·	·
C. nuchalis	·	+	+	·	·	·	·	·
CORCORACIDAE								
Corcorax melanorhamphos	·	·	·	·	+	+	+	·
GRALLINIDAE								
Grallina cyanoleuca	+	+	+	+	+	+	+	·
ARTAMIDAE								
Artamus cinereus	+	+	+	·	·	·	+	·
A. cyanopterus	+	·	·	+	+	+	+	+
A. leucorhynchus	·	+	+	+	·	·	+	·
A. minor	·	+	+	+	·	·	·	·
A. personatus	·	+	·	·	+	+	+	·
A. superciliosus	·	+	·	+	+	+	+	·

TABLE 8.3 *(continued)*

Birds of Australian heathlands

	1	2	3	4	5	6	7	8
CRACTICIDAE								
Cracticus mentalis	·	·	+	·	·	·	·	·
C. nigrogularis	·	+	+	+	+	·	+	·
C. quoyi	·	+	+	·	·	·	·	·
C. torquatus	+	+	·	+	+	+	+	·
Gymnorhina tibicen	+	·	+	+	+	+	+	·
Strepera fuliginosa	·	·	·	·	·	·	·	‡
S. graculina	·	·	+	+	+	+	·	·
S. versicolor	+	·	·	·	+	+	+	·
CORVIDAE								
Corvus bennetti	·	·	·	·	·	·	+	·
C. coronoides	+	·	·	·	+	+	+	+
C. mellori	·	·	·	·	·	+	+	·
C. orru	·	+	+	+	·	·	·	·

TABLE 8.4

Mammals of Australian heathlands

	1	2	3	4	5	6	7	8
MACROPODIDAE								
Aepyprymnus rufescens	·	·	·	+	·	·	·	·
Bettongia lesueur	+	·	·	·	·	·	·	·
B. penicillata	+	·	·	·	·	·	ex	·
Lagorchestes conspicillatus	·	+	·	·	·	·	·	·
L. hirsutus	+	·	·	·	·	·	·	·
Lagostrophus fasciatus	+	·	·	·	·	·	·	·
Macropus agilis	·	+	+	+	·	·	·	·
M. antilopinus	·	+	·	·	·	·	·	·
M. eugenii	+	·	·	·	·	·	+	·
M. fuliginosus	+	·	·	·	·	+	+	·
M. giganteus	·	·	·	+	+	+	+	+
M. irma	+	·	·	·	·	·	·	·
M. robustus	·	+	·	·	+	+	·	·
M. rufogriseus	·	·	·	·	+	+	+	+
Peradorcas concinna	·	+	·	·	·	·	·	·
P. brachyotis	·	+	·	·	·	·	·	·
Petrogale penicillata	+	+	·	·	·	+	+	·
Potorous apicalis	·	·	·	·	·	+	·	+
P. platyops	+	·	·	·	·	·	·	·
P. tridactylus	·	·	·	+	+	+	ex	·
Setonix brachyurus	+	·	·	·	·	·	·	·
Thylogale billardierii	·	·	·	·	·	+	·	+
Wallabia bicolor	·	·	·	+	+	+	·	·

TABLE 8.4 *(continued)*

Mammals of Australian heathlands

	1	2	3	4	5	6	7	8
PHALANGERIDAE								
Trichosurus arnhemensis	·	+	·	·	·	·	·	·
T. vulpecula	·	·	·	+	+	+	+	+
Wyulda squamicaudata	·	+	·	·	·	·	·	·
PETAURIDAE								
Petaurus australis	·	·	·	+	+	+	·	·
P. breviceps	·	+	+	+	+	+	+	+
P. norfolcensis	·	·	·	+	+	+	·	·
Petropseudes dahli	·	+	·	·	·	·	·	·
Pseudocheirus peregrinus	·	·	·	+	+	+	+	+
Schoinobates volans	·	·	·	+	·	·	·	·
BURRAMYIDAE								
Acrobates pygmaeus	·	·	·	+	+	+	+	·
Burramys parvus	·	·	·	·	·	+	·	·
Cercartetus concinnus	+	·	·	·	·	·	‡	·
C. lepidus	·	·	·	·	·	·	+	+
C. nanus	·	·	·	·	·	+	·	·
TARSIPEDIDAE								
Tarsipes spencerae	‡	·	·	·	·	·	·	·
PHASCOLARCTIDAE								
Phascolarctos cinereus	·	·	·	+	+	+	in	·
VOMBATIDAE								
Vombatus ursinus	·	·	·	·	·	+	+	+
PERAMELIDAE								
Chaeropus ecaudatus	+	·	·	·	·	·	·	·
Echymipera rufescens	·	·	+	·	·	·	·	·
Isoodon macrourus	·	+	+	+	+	·	·	·
I. obesulus	+	·	+	·	·	+	+	+
Parameles bougainville	‡	·	·	·	·	·	+	·
P. gunnii	·	·	·	·	·	+	·	·
P. nasuta	·	·	·	+	+	+	·	·
DASYURIDAE								
Antechinus apicalis	‡	·	·	·	·	·	·	·
A. flavipes	·	·	·	+	+	·	+	·
A. macdonnellensis	·	+	·	·	·	·	·	·
A. maculatus	·	·	·	+	+	·	·	·
A. minimus	·	·	·	·	·	‡	·	·
A. stuartii	·	·	·	·	+	+	·	·
A. swainsonii	·	·	·	·	+	+	·	+
Dasycercus cristicauda	+	·	·	·	·	·	·	·
Dasyurus geoffroii	+	·	·	·	·	·	ex	·

TABLE 8.4 *(continued)*

Mammals of Australian heathlands

	1	2	3	4	5	6	7	8
DASYURIDAE *(continued)*								
D. hallucatus	·	+	+	·	·	·	·	·
D. maculatus	·	·	·	·	·	+	ex	·
D. viverrinus	·	·	·	·	·	ex in	ex	+
Myrmecobius fasciatus	+	·	·	·	·	·	·	·
Phascogale calura	+	·	·	·	·	·	·	·
P. tapoatafa	·	+	·	·	·	·	·	·
Sarcophilus harrisii	·	·	·	·	·	·	·	+
Sminthopsis crassicaudata	+	·	·	·	·	·	·	·
S. granulipes	‡	·	·	·	·	·	·	·
S. hirtipes	+	·	·	·	·	·	·	·
S. leucopus	·	·	·	·	·	+	·	+
S. murina	+	·	·	+	+	+	‡	·
S. nitela	·	+	·	·	·	·	·	·
S. rufigenus	·	+	·	·	·	·	·	·
THYLACINIDAE								
Thylacinus cynocephalus	·	·	·	·	·	·	·	+
MURIDAE								
Conilurus penicillatus	·	+	·	·	·	·	·	·
Hydromys chrysogaster	·	+	+	+	+	+	+	+
Leporillus conditor	·	·	·	·	·	·	+	·
Mastacomys fuscus	·	·	·	·	·	+	·	+
Melomys cervinipes	·	·	·	+	·	·	·	·
M. littoralis	·	·	·	+	·	·	·	·
M. lutillus	·	·	+	·	·	·	·	·
M. spp.	·	+	+	·	·	·	·	·
Mesembriomys gouldii	·	+	·	·	·	·	·	·
M. macrurus	·	+	·	·	·	·	·	·
Mus musculus*	+	+	·	+	+	+	+	+
Notomys aquilo	·	+	·	·	·	·	·	·
N. mitchellii	‡	·	·	·	·	·	‡	·
Pseudomys albocinereus	‡	·	·	·	·	+	‡	·
P. australis	·	·	·	·	·	+	·	·
P. delicatulus	·	+	·	+	·	·	·	·
P. fumeus	·	·	·	·	·	+	·	·
P. gracilicaudatus	·	+	·	·	‡	·	·	·
P. hermannsburgensis	·	·	·	·	·	·	+	·
P. nanus	·	+	·	·	·	·	·	·
P. novaehollandiae	·	·	·	·	‡	‡	·	·
P. occidentalis	+	·	·	·	·	·	·	·
P. praeconis	+	·	·	·	·	·	·	·
P. shortridgei	+	·	·	·	·	‡	·	·
Rattus fuscipes	+	·	·	+	+	+	+	·
R. lutreolus	·	·	·	+	+	+	+	+
R. rattus*	·	·	·	+	+	+	·	+
R. sordidus	·	+	+	·	·	·	·	·
R. tunneyi	·	+	+	+	·	·	·	·

TABLE 8.4 *(continued)*

Mammals of Australian heathlands

	1	2	3	4	5	6	7	8
MURIDAE *(continued)*								
Xeromys myoides	.	+	.	+	.	.	.	.
Zyzomys argurus	.	+	.	.	.	.	.	.
Z. woodwardi	.	+	.	.	.	.	.	.
TACHYGLOSSIDAE								
Tachyglossus aculeatus	+	+	+	+	+	+	+	+
ORNITHORHYNCHIDAE								
Ornithorhynchus anatinus	.	.	.	.	.	+	.	.
MEGADERMATIDAE								
Macroderma gigas	+	+	.	.	.	.	.	.
VESPERTILIONIDAE								
Chalinolobus gouldii	+	+	.	+	+	+	+	+
C. morio	.	.	.	.	.	+	+	.
C. rogersi	.	+	.	.	.	.	.	.
C. sp. (cf. *dwyeri*)	.	.	.	+	.	.	.	.
Eptesicus pumilus	+	+	+	.	.	+	+	+
Miniopterus australis	.	.	.	+	.	.	.	.
M. schreibersii	.	+	+	.	+	+	+	.
Myotis adversus	.	.	.	+	.	+	.	.
M. australis	.	.	.	.	+	.	.	`
Nycticeius greyi	.	+	.	.	+	+	.	.
N. rueppellii	.	.	.	.	+	.	.	.
Nyctophilus arnhemensis	.	+	.	.	.	.	.	.
N. bifax	.	+	.	.	.	.	.	.
N. geoffroyi	+	.	.	+	.	+	+	+
N. timoriensis	.	.	.	+	+	+	.	.
Pipistrellus tasmaniensis	.	.	.	.	.	+	.	.
RHINOLOPHIDAE								
Rhinolophus megaphyllus	.	.	+	.	.	.	.	.
HIPPOSIDERIDAE								
Hipposideros ater	.	+	.	.	.	.	.	.
H. bicolor	.	+	.	.	.	.	.	.
H. diadema	.	+	.	.	.	.	.	.
H. galeritus	.	+	.	.	.	.	.	.
H. stenotis	.	+	.	.	.	.	.	.
Rhinonicteris aurantius	.	+	.	.	.	.	.	.
MOLOSSIDAE								
Tadarida australis	+	.	.	.	.	+	+	.
T. planiceps	+	.	.	.	.	.	.	.
T. sp.	.	.	.	+	.	.	.	.

TABLE 8.4 *(continued)*

Mammals of Australian heathlands

	1	2	3	4	5	6	7	8
EMBALLONURIDAE								
Taphozous australis	.	+	+	.	.	.	.	.
T. flaviventris	.	+	.	.	.	.	.	.
T. georgianus	.	+	.	.	.	.	.	.
PTEROPODIDAE								
Dobsonia moluccense	.	.	+	.	.	.	.	.
Macroglossus lagochilus	.	+	+	.	.	.	.	.
Pteropus alecto	.	+	+	.	.	.	.	.
P. poliocephalus	.	.	.	+	+	.	.	.
P. scapulatus	.	+	+	+	.	.	.	.
Nyctimene robinsoni	.	.	+	+	.	.	.	.
Syconycteris australis	.	.	.	+	.	.	.	.
LEPORIDAE								
*Leporus europaeus**	.	.	.	+	.	+	.	.
*Oryctolagus cuniculus**	+	.	.	.	.	+	+	+
CANIDAE								
Canis familiaris dingo	+	+	+	+	+	+	.	.
*Vulpes vulpes**	+	.	.	+	.	+	+	.
FELIDAE								
*Felis catus**	+	+	.	+	+	+	+	+
CERVIDAE								
*Axis porcinus**	.	.	.	.	.	+	.	.
*Cervus elaphus**	+	.	.	.	.	+	.	.
*C. unicolor**	.	.	.	.	.	+	.	.
EQUIDAE								
*Equus asinus**	.	+	.	.	.	.	.	.
*E. caballus**	.	.	.	+	.	+	.	.
SUIDAE								
*Sus scrofa**	+	+	+	+	.	.	+	.
BOVIDAE								
*Bos taurus**	.	.	.	+	.	.	.	.
*Bubalus bubalis**	.	+	.	.	.	.	.	.
*Capra hircus**	+	.	.	+	.	.	+	.

REFERENCES

Abbott, I.J., 1973. Birds of Bass Strait. Evolution and ecology of the avifauna of some Bass Strait islands, and comparisons with those of Tasmania and Victoria. *Proc. R. Soc. Vic.*, 85: 197–223.

Abbott, I.J., 1974. The avifauna of Kangaroo Island and causes of its impoverishment. *Emu*, 74: 124–134.

Abbott, I.J., 1976. Comparisons of habitat structure and plant, arthropod and bird diversity between mainland and island sites near Perth, Western Australia. *Aust. J. Ecol.*, 1: 275–280.

Aitken, P.F., 1970. South Australian mammals. In: *South Australian Yearbook, 1970.* Government Printer, Adelaide, S.A., pp. 42–49.

Anonymous, 1961. Tiger cats on Wilson's Promontory. *Vic. Nat.*, 77: 355.

Baker, H.G. and Baker, I., 1975. Species of nectar-constitution and pollinator–plant coevolution. In: L.E. Gilbert and P.H. Raven (Editors), *Coevolution of Animals and Plants.* University of Texas Press, Austin, Texas, pp. 100–140.

Barry, D.H. and Campbell, P.R., 1977. A survey of the mammals and herptiles of Fraser Island, with comments on the Cooloola Peninsula, North Stradbroke, Moreton and Bribie Islands. In: P.K. Lauer (Editor), *Fraser Island.* Occasional Papers in Anthropology, No. 8. Anthropology Museum, University of Queensland, Brisbane, pp. 147–177.

Bartholomai, A., 1972. Aspects of the evolution of the Australian marsupials. *Proc. R. Soc. Qld.*, 82: v–xviii.

Bayly, I.A.E., 1967. The general biological classification of aquatic environments with special reference to those of Australia. In: A.H. Weatherley (Editor), *Australian Inland Waters and their Fauna.* Australian National University Press, Canberra, A.C.T., pp. 78–104.

Bayly, I.A.E., 1975. Interactions between terrestrial and aquatic environments: Queensland's coastal lakes. *Proc. Ecol. Soc. Aust.*, 9: 325–328.

Bayly, I.A.E., Ebsworth, E.P., and Wan, H.F., 1975. Studies on the lakes of Fraser Island, Queensland. *Aust. J. Mar. Freshw. Res.*, 26: 1–13.

Beard, J.S., 1976. The evolution of Australian desert plants. In: D.W. Goodall (Editor), *Evolution of Desert Biota.* University of Texas Press, Austin, Texas, pp. 51–63.

Bell, H.L., 1966. A population study of heathland birds. *Emu*, 65: 295–304.

Bensink, H.A. and Burton, H., 1975. North Stradbroke Island — A place for freshwater invertebrates. *Proc. R. Soc. Qld.*, 86: 29–45.

Berg, R.Y., 1975. Myrmecochorous plants in Australia and their dispersal by ans. *Aust. J. Bot.*, 23: 475–508.

Blair-West, J.R., Coghlan, J.P., Denton, D.A., Nelson, J.F., Orchard, E., Scoggins, B.A., Wright, R.D., Myers, K. and Junqueira, C.L., 1968. Physiological, morphological and behavioural adaptation to a sodium deficient environment by wild native Australian and introduced species of animals. *Nature, Lond.*, 217: 922–928.

Bryant, C., 1930. Birds notes of a recent trip to Wilson's Promontory, Victoria. *Emu*, 29: 297 301.

Burbidge, A.A., Kirsch, A.W. and Main, A.R., 1974. Relation-

ships within the Chelidae (Testudines: Pleurodira) of Australia and New Guinea. *Copeia*, 1974: 392–409.

Bush, G.L., 1975. Modes of animal speciation. *Annu. Rev. Ecol. Syst.*, 6: 339–364.

Calaby, J.H., 1956. The food habits of the frog *Myobatrachus gouldii* (Gray). *W. Aust. Nat.*, 5: 93–96.

Carpenter, F.L., 1978. Hooks for mammal pollination? *Oecologia*, 35: 123–132.

Carpenter, F.L. and Recher, H.F., in press. Pollination, reproduction, and fire. *Am. Nat.*

Christensen, P., 1971. The purple-crowned lorikeet and eucalypt pollination. *Aust. For.*, 35: 263–270.

Christensen, P.E. and Kimber, P.C., 1975. Effect of prescribed burning on the flora and fauna of south-west Australian forests. *Proc. Ecol. Soc. Aust.*, 9: 85–106.

Churchill, D.M., 1968. The distribution and prehistory of *Eucalyptus diversicolor* F. Muell., *E. marginata* Donn ex Sm., and *E. calophylla* R. Br. in relation to rainfall. *Aust. J. Bot.*, 16: 125–151.

Churchill, D.M. and Christensen, P., 1970. Observations on pollen harvesting by brush-tongued lorikeets. *Aust. J. Zool.*, 18: 427–437.

Coaldrake, J.E., 1961. The ecosystem of the coastal lowlands ("wallum") of southern Queensland. *C.S.I.R.O. Aust. Bull.*, No. 283: 138 pp.

Coaldrake, J.E., 1962. The coastal sand dunes of southern Queensland. *Proc. R. Soc. Qld.*, 62: 101–116.

Coaldrake, J.E., 1975. The natural history of Cooloola. *Proc. Ecol. Soc. Aust.*, 9: 308–313.

Cogger, H.G., 1975. *Reptiles and Amphibians of Australia.* A.H. and A.W. Reed, Sydney, N.S.W., 584 pp.

Cogger, H.G., 1976. New lizards of the genus *Pseudothecadactylus* (Lacertilia: Gekkonidae) from Arnhem Land and northwestern Australia. *Rec. Aust. Mus.*, 30: 87–97.

Coleman, E., 1928. Pollination of *Cryptostylis leptochila* F.v.M. *Vic. Nat.*, 49: 179–186.

Condon, H.T., 1975. *Checklist of the Birds of Australia. Part 1. Non-Passerines.* R. Aust. Ornithol. Union, Melbourne, Vic., 311 pp.

Covacevich, J. (Editor), 1977. *Fauna of Eastern Australian Rainforests. II.* Queensland Museum, Brisbane, Qld., 99 pp.

Covacevich, J., and Archer, M., 1975. The distribution of the cane toad, *Bufo marinus*, in Australia and its effects on indigenous vertebrates. *Mem. Qld. Mus.*, 17: 305–310.

Covacevich, J. and Ingram, G.J., 1975a. Three new species of rainbow skinks of the genus *Carlia* from northern Queensland. *Vic. Nat.*, 92: 19–22.

Covacevich, J. and Ingram, G.J., 1975b. The reptiles of Stradbroke Island. *Proc. R. Soc. Qld.* 86: 55–60.

Cracraft, J., 1974. Continental drift and vertebrate distribution. *Annu. Rev. Ecol. Syst.*, 5: 215–261.

Cracraft, J., 1976. Avian evolution on southern continents: influences of palaeogeography and palaeoclimatology. *Proc. 16th Int. Ornithol. Congr.*, pp. 40–52.

Crow, A.B., 1973. Use of fire in southern forests. *J. For.*, 71: 629–632.

Darlington, P.J., 1957. *Zoogeography: the Geographical Distribution of Animals.* Wiley, New York, N.Y., 675 pp.

Davies, S.J.J.F., 1963. Emus. *Aust. Nat. Hist.*, 14: 225–229.

Deignan, H.G., 1964. Birds of the Arnhem Land expedition. In: R.L. Specht (Editor), *Records of the American–Australian Scientific Expedition to Arnhem Land. 4. Zoology.* Melbourne University Press, Melbourne, Vic., pp. 345–425.

DeWitt, J.B. and Derby, J.V., 1955. Changes in nutritive value of browse plants following forest fires. *J. Wildl. Manage.,* 19: 65–70.

Dwyer, P., Hockings, M. and Willmer, J., 1979. Mammals of Cooloola and Beerwah. *Proc. R. Soc. Qld.,* 90: 65–84.

Ealey, E.H.M. and Main, A.R., 1967. Ecology of the euro, *Macropus robustus* (Gould), in north-western Australia. III. Seasonal changes in nutrition. *C.S.I.R.O. Wildl. Res.,* 12: 53–65.

Ford, H.A. and Paton, D.C., 1976. Resource partitioning and competition in honeyeaters of the genus *Meliphaga. Aust. J. Ecol.,* 1: 281–287.

Ford, J., 1971. Distribution, ecology and taxonomy of some Western Australian passerine birds. *Emu,* 71: 103–120.

Ford, J., 1974. Speciation in Australian birds adapted to arid habitats. *Emu,* 74: 161–168.

Forshaw, J.M., 1969. *Australian Parrots.* Lansdowne, Melbourne, Vic., 306 pp.

Frankenberg, J., 1971. *Nature Conservation in Victoria.* Victorian National Parks Association, Melbourne, Vic., 145 pp.

Frith, H.J., 1967. *Waterfowl in Australia.* Angus and Robertson, Sydney, 328 pp.

Gannon, G.R., 1962. Distribution of the Australian honeyeaters. *Emu,* 62: 145–166.

Gannon, G.R., 1966. The influence of habitat on the distribution of Australian birds. *Emu,* 65: 241–253.

Gardner, C.A., 1957. The fire factor in relation to the vegetation of Western Australia. *W. Aust. Nat.,* 5: 166–173.

Gill, A.M., 1975. Fire and the Australian flora: a review. *Aust. For.,* 38: 4–25.

Gill, A.M., 1977. Management of fire-prone vegetation for plant species conservation in Australia. *Search,* 8: 20–26.

Gill, H.B., 1976. Black grass wren *Amytornis housei..* In: H.J. Frith (Editor), *Complete Book of Australian Birds.* Reader's Digest Services, Sydney, N.S.W., p. 419.

Givens, L.S., 1962. Use of fire on southeastern wildlife refuges. *Proc. First Annu. Tall Timbers Fire Ecol. Conf.,* Tallahassee, Fla., pp. 121–126.

Goin, C.J. and Goin, O.B., 1962. *Introduction to Herpetology.* W.H. Freeman, San Francisco, Calif., 353 pp.

Gosner, K.L. and Black, I.H., 1957. The effects of acidity on the development and hatching of New Jersey frogs. *Ecology,* 38: 256–262.

Gravatt, D., 1974. Birds that eat plant products in coastal south Queensland. *Inst. For. Aust. Seventh Triennial Conf.,* Caloundra, Qld., Vol. 1, Working Papers, pp. 339–348.

Gravatt, D.J. and Ingram, G.J., 1975. Comments on the land vertebrates of Cooloola. *Proc. Ecol. Soc. Aust.,* 9: 321–325.

Green, R.H., 1967. The murids and small dayurids in Tasmania, Parts 1 and 2. *Rec. Queen Victoria Mus.,* Launceston, 28: 1–19.

Green, R.H., 1974. Mammals. In: W.D. Williams (Editor), *Biogeography and Ecology in Tasmania.* W. Junk, The Hague, pp. 367–396.

Greer, A.E., 1974. The generic relationships of the scincid lizard

genus *Leiolopisma* and its relatives. *Aust. J. Zool. Suppl. Ser.,* No. 31: 1–67.

Hall, B.P. (Editor), 1974. *Birds of the Harold Hall Australian Expeditions 1962–70.* Trustees of the British Museum (Natural History), London, 396 pp.

Hawkins, P.J., 1975. Forest management of Cooloola State Forest. *Proc. Ecol. Soc. Aust.,* 9: 328–333.

Heatwole, H., 1976. *Reptile Ecology.* University of Queensland Press, St. Lucia, Qld., 178 pp.

Hendricks, J.H., 1968. Control burning for deer management in chaparral in California. *Proc. Annu. Tall Timbers Fire Ecol. Conf.,* Tallahassee, Fla., No. 8: 219–233.

Heyer, W.R. and Liem, D.S., 1976. Analysis of the intergeneric relationships of the Australian frog family Myobatrachidae. *Smithsonian Contrib. Zool.,* No. 233: 1–29.

Hope, J.H., 1974. The biogeography of the mammals of the islands of Bass Strait. In: W.D. Williams (Editor), *Biogeography and Ecology in Tasmania.* W. Junk, The Hague, pp. 397–415.

Hore-Lacy, I., 1964. Birds of the You Yangs, Victoria: habitat selection and status. *Emu,* 64: 28–32.

Horton, D.R., 1972. Speciation of birds in Australia, New Guinea and the south-western Pacific islands. *Emu,* 72: 91–109.

Ingram, G., 1972. Notes on the feeding of white-browed wood-swallows (*Artamus superciliosus*). *Sunbird,* 3: 64–65.

Ingram, G.J. and Corben, C.J., 1975. The frog fauna of North Stradbroke Island, with comments on the 'acid' frogs of the wallum. *Proc. R. Soc. Qld.,* 86: 49–54.

Janzen, D.H., 1974. Tropical blackwater rivers, animals, and mast fruiting by the Dipterocarpaceae. *Biotropica,* 6: 69–103.

Jenkins, D., Watson, A. and Miller, G.R., 1963. Population studies on red grouse, *Lagopus lagopus scoticus* (Lath.) in north-east Scotland. *J. Anim. Ecol.,* 32: 317–376.

Johnson, D.H., 1964. Mammals of the Arnhem Land Expedition. In: R.L. Specht (Editor), *Records of the American–Australian Scientific Expedition to Arnhem Land.. 4. Zoology.* Melbourne University Press, Melbourne, Vic., pp. 427–515.

Keast, A., 1959a. Australian birds: their zoogeography and adaptations to an arid continent. In: A. Keast, R.L. Crocker and C.S. Christian (Editors), *Biogeography and Ecology in Australia.* W. Junk, The Hague, pp. 89–114.

Keast, A., 1959b. The reptiles of Australia. In: A. Keast, R.L. Crocker and C.S. Christian (Editors), *Biogeography and Ecology in Australia.* W. Junk, The Hague, pp. 115–135.

Keast, A., 1961. Bird speciation on the Australian continent. *Bull. Mus. Comp. Zool. Harvard,* 123: 305–495.

Keast, A., 1968a. Seasonal movements in the Australian honeyeaters (Meliphagidae) and their ecological significance. *Emu,* 67: 159–209.

Keast, A., 1968b. Competitive interactions and the evolution of ecological niches as illustrated by the Australian honeyeater genus *Melithreptus* (Meliphagidae). *Evolution,* 22: 762–784.

Keast, A., 1976. The origin of adaptive zone utilizations and adaptive radiations, as illustrated by the Australian Meliphagidae.*Proc. 16th Int. Ornithol. Congr.,* Canberra, A.C.T., pp. 71–82.

Keast, A., 1977. Historical biogeography of the marsupials. In: B. Stonehouse and D. Gilmore (Editors), *The Biology of Marsupials.* Macmillan Press, Melbourne, Vic., pp. 69–95.

Keast, A. (Editor), in press. *Ecological Biogeography in Australia.* W. Junk, The Hague.

Keith, K. and Calaby, J.H., 1968. The New Holland mouse, *Pseudomys novaehollandiae* (Waterhouse), in the Port Stephens District, New South Wales. *C.S.I.R.O. Wildl. Res.,* 13: 45–58.

Kikkawa, J., 1959. Habitats of the field mouse on Fair Isle in spring, 1956. *Glasgow Nat.,* 18: 65–77.

Kikkawa, J., 1970. The birds of the Richmond valley. In: F. Bitmead (Editor), *Richmond Valley.* University of New England Department of University Extension Regional Office, Lismore, N.S.W., pp. 59–75.

Kikkawa, J., 1976. The birds of Cape York Peninsula. *Sunbird,* 7: 25–41; 81–106.

Kikkawa, J. and Pearse, K., 1969. Geographical distribution of land birds in Australia — a numerical analysis. *Aust. J. Zool.,* 17: 821–840.

Kimber, P.C., in prep. Some effects of prescribed burning on jarrah forest birds. *Third Fire Ecol. Symp. Pap.,* Monash University, Vic.

King, M. and King, D., 1975. Chromosomal evolution in the lizard genus *Varanus* (Reptilia). *Aust. J. Biol. Sci.,* 28: 89–108.

Kluge, A.G., 1967. Systematics, phylogeny, and zoogeography of the lizard genus *Diplodactylus* Gray (Gekkonidae). *Aust. J. Zool.,* 15: 1007–1108.

Kluge, A.G., 1974. A taxonomic revision of the lizard family Pygopodidae. *Misc. Publ. Mus. Zool. Univ. Mich.,* 147: 1–221.

Komarek, E.V., 1969. Fire and animal behavior. *Proc. Annu. Tall Timbers Fire Ecol. Conf.,* Tallahassee, Fla., No. 9: 161–207.

Land Conservation Council, Victoria, 1972a. *Report on the South Western Study Area* (district 1). L.C.C., Melbourne, Vic., 254 pp.

Land Conservation Council, Victoria, 1972b. *Report on the South Gippsland Study Area* (district 1). L.C.C., Melbourne, Vic., 123 pp.

Land Conservation Council, Victoria, 1973. *Report on the Melbourne Study Area.* L.C.C., Melbourne, Vic., 444 pp.

Land Conservation Council, Victoria, 1974a. *Report on the Mallee Study Area.* L.C.C., Melbourne, Vic., 263 pp.

Land Conservation Council, Victoria, 1974b. *Report on the East Gippsland Study Area.* L.C.C., Melbourne, Vic., 236 pp.

Land Conservation Council, Victoria, 1976. *Report on the Corangamite Study Area.* L.C.C., Melbourne, Vic., 310 pp.

Lay, D.W., 1957. Browse quality and the effects of prescribed burning in southern pine forests. *J. For.,* 55: 342–347.

Laycock, J.W., 1975. Hydrogeology of North Stradbroke Island. *Proc. R. Soc. Qld.,* 86: 15–19.

Lee, A.K., 1967. Studies in Australian Amphibia. 2. Taxonomy, ecology and evolution of the genus *Heleioporus* Gray (Anura: Leptodactylidae). *Aust. J. Zool.,* 15: 367–439.

Littlejohn, M.J., 1963. The breeding biology of the baw baw frog *Philoria frosti* Spencer. *Proc. Linn. Soc. N.S.W.,* 88: 273–276.

Littlejohn, M.J., 1967. Patterns of zoogeography and speciation in south-eastern Australian Amphibia. In: A.H. Weatherley (Editor), *Australian Inland Waters and Their Fauna.* Australian National University Press, Canberra, A.C.T., pp. 150–174.

Littlejohn, M.J. and Martin, A.A., 1974. The amphibia of Tasmania. In: W.D. Williams (Editor), *Biogeography and Ecology in Tasmania.* W. Junk, The Hague, pp. 251–289.

Lowe, V.T. and Lowe, T.G., 1972. Wood-swallows in mid-northern Victoria. *Aust. Bird Watcher,* 4: 205–206; 216.

Macdonald, J.D., 1973. *Birds of Australia.* A.H. and A.W. Reed, Sydney, N.S.W., 552 pp.

McDowell, S.B., 1974. A catalogue of the snakes of New Guinea and the Solomons, with special reference to those in the Bernice P. Bishop Museum. Part 1. Scolecophidia. *J. Herpetol.,* 8: 1–57.

McDowell, S.B., 1975. A catalogue of the snakes of New Guinea and the Solomons, with special reference to those in the Bernice P. Bishop Museum. Part 2. Anilioidea, Pythoninae. *J. Herpetol.,* 9: 1–79.

McEvey, A.R. and Middleton, W.G., 1968. Birds and vegetation between Perth and Adelaide. *Emu,* 68: 161–212.

McGill, A.R. and Lane, S.G., 1955. The Mt Keira camp-out. *Emu,* 55: 49–71.

Mack, K.J., 1973. Wood-swallows feeding on Sturt peas. *S. Aust. Ornithol..,* 26: 90.

McKean, J.L., 1969. The brush tongue of Artamidae. *Bull. Br. Ornithol. Club,* 89: 129–130.

Mahoney, J.A. and Posamentier, H., 1975. The occurrence of the native rodent *Pseudomys gracilicaudatus* (Gould, 1845) (Rodentia: Muridae) in New South Wales. *J. Aust. Mammal. Soc.,* 1: 333–346.

Main, A.R., 1957. Studies in Australian Amphibia. I. The genus *Crinia* in south-western Australia and some species from south-eastern Australia. *Aust. J. Zool.,* 5: 30–55.

Main, A.R., 1963. A new species of *Crinia* (Anura: Leptodactylidae) from National Park, Nornlup. *W. Aust. Nat.,* 6: 143–144.

Main, A.R., 1969. Ecology, systematics and evolution of Australian frogs. *Adv. Ecol. Res.,* 5: 37–86.

Main, A.R., Lee, A.K. and Littlejohn, M.J., 1958. Evolution in three genera of Australian frogs. *Evolution,* 12: 224–233.

Main, A.R., Littlejohn, M.J. and Lee, A.K., 1959. Ecology of Australian frogs. In: A. Keast, R.L. Crocker and C.S. Christian (Editors), *Biogeography and Ecology in Australia.* W. Junk, The Hague, pp. 396–411.

Martin, A.A., 1967. Australian anuran life histories: some evolutionary and ecological aspects. In: A.H. Weatherley (Editor), *Australian Inland Waters and Their Fauna.* Australian National University Press, Canberra, A.C.T., pp. 175–191.

Martin, A.A., 1970. Parallel evolution in the adaptive ecology of leptodactylid frogs of South America and Australia. *Evolution,* 24: 643–644.

Martin, J.H.D., 1975. A list of mammals from Stradbroke Island. *Proc. R. Soc. Qld.,* 86: 73–76.

Mason, I.J. and Schodde, R., 1976. White-throated grass wren, *Amytornis woodwardi,* Carpentarian grass wren, *Amytornis dorotheae.* In: H.J. Frith (Editor), *Complete Book of Australian Birds.* Reader's Digest Services, Sydney, N.S.W., pp. 419–420.

Means, D.B. and Longden, C.J., 1976. Aspects of the biology and zoogeography of the Pine Barrens treefrog (*Hyla andersonii*) in northern Florida. *Herpetologica*, 32: 117–130.

Mees, G.F., 1967. A note on the pollination of the kangaroo paw *Anigozanthos manglesii*. *W. Aust. Nat.*, 10: 149–151.

Menzies, J.I., 1975. *Handbook of Common New Guinea Frogs*. Wau Ecology Institute, Wau, Papua New Guinea, 74 pp.

Merrilees, D., 1968. Man the destroyer: late Quaternary changes in the Australian marsupial fauna. *J. R. Soc. W. Aust.*, 51: 1–24.

Miles, J.M. and Burbidge, A.A. (Editors), 1975. A biological survey of the Prince Regent River Reserve, North-West Kimberley, Western Australia. *Wildl. Res. Bull. W. Aust.*, 3: 1–116.

Mitchell, F.J., 1964. Reptiles and amphibians of Arnhem Land. In: R.L. Specht (Editor), *Records of the American–Australian Scientific Expedition to Arnhem Land. 4. Zoology*. Melbourne University Press, Melbourne, Vic., pp. 309–343.

Morcombe, M.K., 1967. The rediscovery after 83 years of the dibbler *Antechinus apicalis* (Marsupialia, Dasyuridae). *W. Aust. Nat.*, 10: 103–111.

Nelson, J.E., 1964. Notes on *Syconycteris australis* Peters, 1867 (Megachiroptera). *Mammalia*, 28: 429–432.

Newsome, A.E., McIlroy, J. and Catling, P., 1975. The effects of an extensive wildfire on populations of twenty ground vertebrates in south-east Australia. *Proc. Ecol. Soc. Aust.*, 9: 107–123.

Noble, J., 1975. The effects of emu (*Dromaius novaehollandiae* Latham) on the distribution of the nitre bush (*Nitraria billardieri* DC.). *J. Ecol.*, 63: 979–984.

Officer, H.R., 1958. Birds of the R.A.O.U. 1957 camp-out — Portland District, Victoria. *Emu*, 58: 375–390.

Officer, H.R., 1964. *Australian Honeyeaters*. The Bird Observers Club, Melbourne, Vic., 86 pp.

Oliver, A.J., King, D.R. and Mead, R.J., 1977. The evolution of resistance to fluoroacetate intoxication in mammals. *Search*, 8: 130–132.

Parker, S.A., 1973a. The tongues of *Ephthianura* and *Ashbyia*. *Emu*, 73: 19–20.

Parker, S.A., 1973b. An annotated checklist of the native land mammals of the Northern Territory. *Rec. S. Aust. Mus.*, 16(11): 1–57.

Parker, S.A., 1976. Crimson chat *Ephthianura tricolor*. In: H.J. Frith (Editor), *Complete Book of Australian Birds*. Reader's Digest Services, Sydney, N.S.W., p. 510.

Paton, D.C. and Ford, H.A., 1977. Pollination by birds of native plants in South Australia. *Emu*, 77: 73–85.

Pedley, L. and Isbell, R.F., 1971. Plant communities of Cape York Peninsula. *Proc. R. Soc. Qld.*, 82: 51–74.

Pianka, E.R., 1972. Zoogeography and speciation of Australian desert lizards: an ecological perspective. *Copeia*, 1972: 127–145.

Pidgeon, I.M., 1938. The ecology of the central coastal area of New South Wales. II. Plant succession on the Hawkesbury Sandstone. *Proc. Linn. Soc. N.S.W.*, 63: 1–26.

Posamentier, H. and Recher, H.F., 1974. The status of *Pseudomys novaehollandiae* (the new holland mouse). *Aust. Zool.*, 18: 66–71.

Posamentier, H.G. and Recher, H.F., in prep. Habitat and small mammals in coastal heathlands of New South Wales.

Purdie, R.W. and Slatyer, R.O., 1976. Vegetation succession after fire in sclerophyll woodland communities in south-eastern Australia. *Aust. J. Ecol.*, 1: 223–236.

Rawlinson, P.A., 1974. Biogeography and ecology of the reptiles of Tasmania and the Bass Strait area. In: W.D. Williams (Editor), *Biogeography and Ecology in Tasmania*. W. Junk, The Hague, pp. 291–338.

Rawlinson, P.A., 1975. Two new lizard species from the genus *Leiolopisma* (Scincidae: Lygosominae) in south-eastern Australia and Tasmania. *Mem. Natl. Mus. Vic.*, 36: 1–15.

Recher, H.F., 1969. Bird species diversity and habitat diversity in Australia and North America. *Am. Nat.*, 103: 75–80.

Recher, H.F., 1971. Sharing of habitat by three congeneric honeyeaters. *Emu*, 71: 147–152.

Recher, H.F., 1975. Survey of the avifauna of Myall Lakes, N.S.W.: report of the 1972 RAOU field-outing. *Emu*, 75: 213–225.

Recher, H.F., 1977. Ecology of co-existing white-cheeked and new holland honeyeaters. *Emu*, 77: 136–142.

Recher, H.F. and Abbott, I.J., 1970a. The possible ecological significance of hawking by honeyeaters and its relation to nectar feeding. *Emu*, 70: 90.

Recher, H.F. and Abbott, I.J., 1970b. Some differences in use of habitat by white-eared and white-cheeked honeyeaters. *Emu*, 70: 117–125.

Recher, H.F., Lunney, D. and Posamentier, H., in prep. Effects of wildfire on small mammals at Nadgee Nature Reserve, N.S.W. *Third Fire Ecol. Symp. Pap.*, Monash University, Vic.

Rich, P.V., 1976. The history of birds on the island continent Australia. *Proc. 16th Int. Ornithol. Congr.*, Canberra, A.C.T., pp. 53–65.

Ride, W.D.L., 1968. On the past, present and future of Australian mammals. *Aust. J. Sci.*, 31: 1–11.

Ride, W.D.L., 1970. *A Guide to the Native Mammals of Australia*. Oxford University Press, London, 249 pp.

Ridpath, M.G. and Moreau, R.E., 1966. The birds of Tasmania: ecology and evolution. *Ibis*, 108: 348–393.

Roberts, G.J. and Ingram, G.J., 1976. An annotated list of the land birds of Cooloola. *Sunbird*, 7: 1–20.

Roberts, P.E., 1970. Some effects of a bushfire on heathland bird-life. *Proc. R. Zool. Soc. N.S.W.*, 1968–69: 40–43.

Salomonsen, F., 1967. Family Meliphagidae. In: R.A. Paynter (Editor), *Check-list of Birds of the World. XII.* Mus. Comp. Zool., Cambridge, Mass., pp. 338–450.

Sargent, O.H., 1928. Reactions between birds and plants. *Emu*, 27: 185–192.

Savage, J.M., 1973. The geographic distribution of frogs: patterns and predictions. In: J. Vial (Editor), *Evolutionary Biology of the Anurans*. University of Missouri Press, Columbia, Mo., pp. 351–445.

Schodde, R., 1975. *Interim List of Australian Songbirds, Passerines*. R. Aust. Ornithol. Union, Melbourne, Vic., 46 pp.

Schodde, R. and McKean, J.L., 1976. The relations of some monotypic genera of Australian oscines. *Proc. 16th Int. Ornithol. Congr.*, Canberra, A.C.T., pp. 530–541.

Serventy, D.L., 1951. The evolution of the chestnut-shouldered wrens (*Malurus*). *Emu*, 51: 113–120.

Serventy, D.L. and Whittell, H.M., 1962. *Birds of Western Australia*. Paterson Brokensha, Perth, W.A., 3rd ed., 427 pp.

Sibley, C.G., 1976a. Protein evidence of the origin of certain Australian birds. *Proc. 16th Int. Ornithol. Congr.*, Canberra, A.C.T., pp 64–70.

Sibley, C.G., 1976b. Protein evidence of the relationships of some Australian passerine birds. *Proc. 16th Int. Ornithol. Congr.*, Canberra, A.C.T., pp. 557–570.

Smith, G.T., 1976. Western bristlebird *Dasyornis longirostris*. In: H.J. Frith (Editor), *Complete Book of Australian Birds*. Reader's Digest Services, Sydney, N.S.W., p. 422.

Smith, G.T. and Robinson, F.N., 1976. The noisy scrub-bird: an interim report. *Emu*, 76: 37–42.

Specht, R.L., 1970. Vegetation. In: G.W. Leeper (Editor), *The Australian Environment*. C.S.I.R.O. — Melbourne University Press, Melbourne, Vic., 4th ed., pp. 44–67.

Specht, R.L., Rayson,, P. and Jackman, M.E., 1958. Dark Island heath (Ninety-Mile Plain, South Australia). VI. Pyric succession: Changes in composition, coverage, dry weight, and mineral nutrient status. *Aust. J. Bot.*, 6: 59–88.

Stimson, A.S., Robb, J. and Underwood, G., 1977. *Leptotyphlops* and *Rhamphotyphlops* Fitzinger, 1843 (Reptilia, Serpentes); proposed conservation under the plenary powers. *Bull. Zool. Nomencl.*, 33: 204–207.

Stocker, G.C., 1966. Effects of fires on vegetation in the Northern Territory. *Aust. For.*, 30: 223–230.

Storr, G.M., 1964. Some aspects of the geography of Australian reptiles. *Senckenbergiana Biol.*, 45: 577–589.

Storr, G.M., 1967. List of Northern Territory birds. *Spec. Publ. W. Aust. Mus.*, 4: 1–90.

Storr, G.M., 1975a. The genus *Ctenotus* (Lacertilia, Scincidae) in the Kimberley and north-west divisions of Western Australia, *Rec. W. Aust. Mus.*, 3: 209–243.

Storr, G.M., 1975b. The genus *Hemiergis* (Lacertilia, Scincidae) in Western Australia. *Rec. W. Aust. Mus.*, 3: 251–260.

Storr, G.M., 1976a. The genus *Omolepida* (Lacertilia, Scincidae) in Western Australia. *Rec. W. Aust. Mus.*, 4: 163–170.

Storr, G.M., 1976b. Revisionary notes on the *Lerista* (Lacertilia: Scincidae) of Western Australia. *Rec. W. Aust. Mus.*, 4: 241–256.

Storr, G.M., 1976c. The genus *Menetia* (Lacertilia, Scincidae) in Western Australia. *Rec. W. Aust. Mus.*, 4: 189–200.

Storr, G.M, 1977a. Birds of the Northern Territory. *Spec. Publ. W. Aust. Mus.*, 7: 1–130.

Storr, G.M., 1977b. *Amphibolurus adelaidensis* species group (Lacertilia, Agamidae) in Western Australia. *Rec. W. Aust. Mus.*, 5: 73–81.

Storr, G.M. and Smith, L.A., 1975. Amphibians and reptiles of the Prince Regent River Reserve, north-western Australia. In: J.M. Miles and A.A. Burbidge (Editors), *A Biological Survey of the Prince Regent River Reserve, North-West Kimberley, Western Australia. Wildl. Res. Bull. W. Aust.*, 3: 85–88.

Straughan, I.R. and Main, A.R., 1966. Speciation and polymorphism in the genus *Crinia* Tschudi (Anura, Leptodactylidae) in Queensland. *Proc. R. Soc. Qld.*, 78: 11–28.

Tate, G.H.H., 1953. Notes on the mammals of Cape York Peninsula. In: L.J. Brass, Results of the Archbold Expeditions. No. 68. Summary of the 1948 Cape York (Australia) expedition. *Bull. Am. Nat. Hist.*, 102: 199–203.

Taylor, J.M. and Horner, B., 1973. Results of the Archbold Expeditions. No. 98. Systematics of native Australian *Rattus* (Rodentia, Muridae). *Bull. Am. Mus. Nat. Hist.*, 150: 1–130.

Timms, B.V., 1973. A limnological survey of the freshwater coastal lakes of East Gippsland, Victoria. *Aust. J. Mar. Freshwater Res.*, 24: 1–20.

Troughton, E. Le G., 1959. The marsupial fauna: its origin and radiation. In: A. Keast, R.L. Crocker and C.S. Christian (Editors), *Biogeography and Ecology in Australia*. W. Junk, The Hague, pp. 69–88.

Troughton, E., 1965. *Furred Animals of Australia*. Angus and Robertson, Sydney, N.S.W., 8th ed., 376 pp.

Tyler, M.J., 1972. An analysis of the lower vertebrate faunal relationships of Australia and New Guinea. In: D. Walker (Editor), *Bridge and Barrier: The Natural and Cultural History of Torres Strait*. Australian National University, Canberra, pp. 231–256.

Tyndale-Biscoe, H., 1973. *Life of Marsupials*. Edward Arnold, London, 254 pp.

Vernon, D.P. and Martin, J.H.D., 1975. Birds of Stradbroke Island. *Proc. R. Soc. Qld.*, 86: 61–72.

Vose, H.M., 1973. Feeding habits of the Western Australian honey possum, *Tarsipes spenserae*. *J. Mammal.*, 54: 245–247.

Wakefield, N.A., 1963. The Australian pigmy-possums. *Vic. Nat.*, 80: 99–116.

Wallace, W.R., 1966. Fire in the jarrah forest environment. *J. R. Soc. W. Aust.*, 49: 33–44.

Watson, A. and Miller, G.R., 1976. *Grouse Management*. Booklet 12. The Game Conservancy, Fordingbridge, Hamps., 78 pp.

Webster, H.O., 1962. Re-discovery of the noisy scrub-bird, *Atrichornis clamosus. W. Aust. Nat.*, 8: 57–59; 81–84.

West, E.S. and Todd, W.R., 1955. *Textbook of Biochemistry*. Macmillan, New York, N.Y., 2nd ed., 1356 pp.

Wood Jones, F., 1923–25. *The Mammals of South Australia. I–III*. 1968 reprint. Government Printer, Adelaide, S.A., 458 pp.

Zweifel, R.G. and Parker, F., 1977. A new species of frog from Australia. *Am. Mus. Novit.*, 2614: 1–10.

Chapter 9

HABITAT RELATIONS OF VERTEBRATES IN SUBTROPICAL
HEATHLANDS OF COASTAL SOUTHEASTERN QUEENSLAND[1]

P.D. DWYER, J. KIKKAWA and G.J. INGRAM

INTRODUCTION

Heathlands in the subtropical region of Australia are smaller and floristically less diverse than southern heathlands, and appear in closer association with other communities than their southern counterparts. This chapter discusses habitat relations of vertebrate animals found on coastal Quaternary sands of southeastern Queensland. Here, adaptive processes are occurring in a comparatively new environment, of which heathland vegetation is a significant part. The habitats concerned are diverse, ranging from closed-forest and tall open-forest to open-forest (including swamp forest), woodland, scrub, heathland, herbland and sedgeland. The understorey of scrub and woodland is typically of sclerophyllous (heathy) shrubs and that of open forest may be either of heathland or grassland species. Heathland elements may be conspicuous even in tall open-forest. The distribution of these structural features and plant associations can be correlated with "soil moisture, soil aeration, soil minerals, nitrogen accretion, topography and history" (Webb and Tracey, 1975). These produce a mosaic of vegetation types that may, in different regions, be either finely or coarsely interspersed. Broad geomorphological correlates of these types are frequently confounded by logging and fire history, so that different types may often be viewed as successional stages (Coaldrake, 1961, 1962, 1975).

The work considered here is in progress and most of it is unpublished. It is based upon extensive studies of all vertebrates, except fish, at Cooloola (lat. 25°49′ to 26°14′S, long. 152°51′ to 153°11′E), and intensive studies of small mammals on 600 ha of heathland at Beerwah (lat. 26°51′S, long. 152°58′E).

HABITAT RELATIONS OF VERTEBRATES

The various types of vegetation represented are arranged in Fig. 9.1 along an idealized transect of about 15 km across the Cooloola study area from the Como Escarpment in the west to the Pacific Ocean in the east (habitat categories A to L). The major relations we have detected between distribution and diversity patterns for vertebrate animals and plant formations are summarized in Fig. 9.1. Details for each vertebrate group are given in the Appendix. The three landscapes included in the figure are described by Coaldrake (1961); they differ in details of parent material, topography, soil type and plant associations represented. In the west, Mesozoic sandstones dominate the parent material of the Womalah landscape. Here soils are mainly shallow podzolics, but gleys and gleyed podzolics are also found in drainage lines and sometimes on flats. Open forest (A), typically with grassy understorey, is common on slopes, while woodland (B) with grass and heathy understorey occupies lower parts. Small pockets of closed-forest (rain forest), not shown in the figure, may interrupt these formations. The Noosa Landscape comprises low-lying land adjoining the Noosa River (C) and its major tributaries. Soils here are humus podzols throughout the sandplain with some development of humic gleys and peaty podzols. Heathland and sedgeland are the dominant vegetation formations (E) but woodland and open-forest with well-developed heathy understorey may intrude on low rises across the Noosa Plains and along the banks (D north) of the Noosa

---

[1] Manuscript completed November, 1977.

281

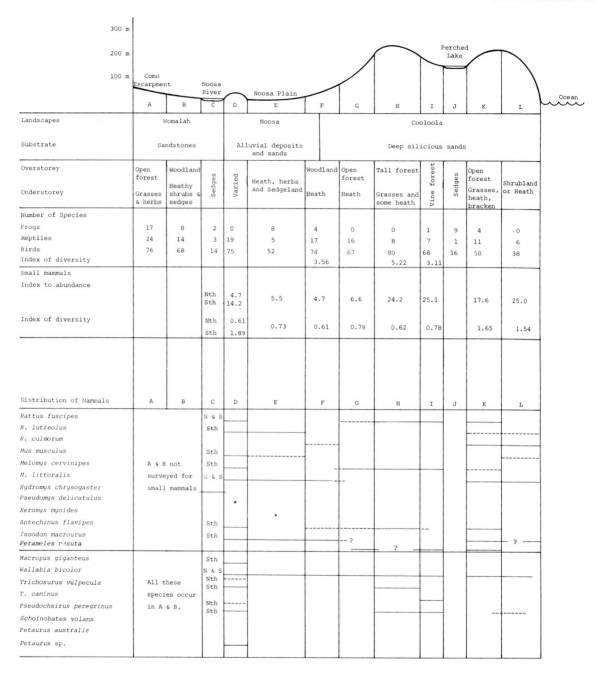

Fig. 9.1. An idealized transect across Cooloola, southeastern Queensland, showing landscapes (after Coaldrake, 1961), major substrate and vegetation characteristics (see text for details) and characteristics of the vertebrate fauna. Habitat category D varies in the north and south of the region; in the north vegetation is woodland with heathy understorey (cf. category F), in the south a complex of habitats are present. For mammals differences in the north and south are indicated on the figure; for other vertebrate groups the data given refer to northern habitats only. For frogs, species number for each habitat category refers to species breeding in that habitat. The index of diversity (Williams, 1947) for birds was calculated from netted samples over two seasons for three habitat categories (D.J. Gravatt supplied the netting data for the category F). For small mammals the number of species and the index of abundance (per cent capture rate) are based on trapping returns and the Williams index of diversity is based on total captures.

River. In the south of the Coolola study area the Noosa River gives way to an extensive system of freshwater and brackish lakes. Here alluvial deposits are better developed, sandstone outcrops in places, and vegetation is more diverse (D south). Swamp forest of *Casuarina* and *Melaleuca* may be closely interspersed with sedgelands, with pockets of grassy open-forest or fringing closed-forest. The Coolola Landscape to the east comprises a high dune system where the main soils are relatively thick podzols (Thompson, 1975). Open-forest on slopes (G, K) and tall open-forest with blackbutt (*Eucalyptus pilularis*) on ridges (H) are common. On lower slopes, however, woodland (F) with *Banksia aemula* is found on groundwater podzols, closed-forest (I) is present in valleys between sandhills or on smaller sandhills and is accompanied by development of humus podzols. Perched lakes (J) are supported by an organic hard-pan layer in pockets between dunes.

From the Noosa River eastwards the heathland component in the understorey of these forests tends to decline except along the eastern seaboard slopes, where wind-swept dense heathland or shrubland (L) appears. Throughout the high dune formations heathland development in the understorey of forests is strongly influenced by fire and logging history, fire increasing the grass component in more open-forests (Coaldrake, 1961) and logging of open-forest promoting growth of heathy elements (Hawkins, 1975).

The habitats along the transect contain a wide variety of vertebrates (Gravatt and Ingram, 1975). However, not all species available in the region are represented. For example, comparisons of the mammal fauna of Coolola with those of the Warwick district (Kirkpatrick, 1966) to the southwest, the Upper Clarence and Richmond Rivers district (Calaby, 1966) to the south, and the Moreton and Wide Bay/Burnett Regions (Queensland Museum, 1976) including Coolola, give 30, 52, 55 and 58 species respectively. If bats, which have been incompletely surveyed at Coolola, and introduced mammals are excluded from the comparison the numbers of species are reduced to 22, 34, 34 and 43 respectively. For birds excluding sea birds, migratory waders and introduced species, the Warwick district has 219 species (Kirkpatrick, 1967), and the lower Richmond Valley (with size and habitat range

comparable to Coolola) has 177 species (Kikkawa, 1970) compared with 153 in Coolola. The Warwick district has 51 species of reptiles (Kirkpatrick, 1968) and the Moreton district 95, compared with 39 in Coolola. Forty-six species of frogs have been recorded from the Moreton district compared with 24 species in Coolola.

The heathland of Coolola lacks 16 species of mammal (7 introduced), 73 species of land birds, 23 species of reptiles and 3 species of frogs that have been recorded from other heathlands within the region of southeastern Queensland and northeastern New South Wales (see Kikkawa et al., in Chapter 8, Fig. 8.1, and the Appendix to that chapter). On the other hand, there are species within the Coolola area that are absent from heathland vegetation. These are 2 species of mammal, 15 species of bird, 2 species of lizard and 1 species of frog confined to closed-forest, and 2 species of lizard and 7 species of frog confined to grassy open-forest.

Within the Coolola area there are also a few representatives in each class of vertebrates that are either confined to or have strong associations with the heathland communities. Among mammals the murid rodents *Xeromys myoides* and *Melomys littoralis* fall into this category although, in north Queensland and in the Northern Territory, these species are found in other habitat types (Redhead and McKean, 1975; Magnussen et al., 1976). Among birds a significant number of heathland specialists have their northernmost distributions in this region. Thus *Phaps elegans*, *Pezoporus wallicus*, *Stipiturus malachurus* and *Anthochaera chrysoptera* do not occur further north, while *Dasyornis brachypterus*, *Phylidonyris novaehollandiae* and *P. melanops* taper out south of Coolola. Other species associated with heathlands are mostly honey-eaters and are not restricted to this habitat. Among reptiles, only one undescribed species of blind snake (*Ramphotyphlops*) is restricted to the heathland of Coolola. Compared with other heathlands, frogs have more heathland specialists in Coolola. These are the "acid" frogs (Ingram and Corben, 1975) *Ranidella tinnula*, *Litoria cooloolensis*, *L. freycineti* and *L. olongburensis*. The first three mentioned have parapatric siblings in open-forest and woodland: *Ranidella parinsignifera*, *Litoria fallax* and *L. latopalmata* respectively.

Accepting that the vertebrate fauna of Cooloola is a sample of the pool of species available in southeastern Queensland, a question arises as to whether this sample is biased with respect to either taxonomic or ecological groups. Among mammals, low species number is due primarily to the poor representation of larger herbivorous species, especially macropods. Only two macropods, *Macropus giganteus* and *Wallabia bicolor*, are known from Cooloola though two smaller species (*Aepyprymnus rufescens* and *Potorous tridactylus*) have been recorded from comparable habitats immediately to the south. Eleven species of macropods may co-occur in areas to the southwest and further south. The dasyurid fauna of Cooloola is also depauperate relative to inland areas. Only 2 species (*Antechinus flavipes* and *Sminthopsis murina*) have been recorded from Cooloola compared to 5 from the Warwick district and 6 from the Upper Clarence and Richmond Rivers district. Outside Cooloola there are records of one other small dasyurid from the coastal lowland communities (Dwyer et al., 1979), but larger species (e.g. *Phascogale*, *Dasyurus*) are certainly missing from the coastal lowlands. This depauperization of the mammal fauna is in part compensated by an increased number of murid rodents. Among birds 42 species of water birds known to appear in heathlands of this region are missing from the Cooloola list. However, many of them occur in neighbouring areas south of Cooloola. Of land birds, some elements of sclerophyll-forest and rain forest are absent from Cooloola, but no particular group is underrepresented in relation to others. Incomplete survey may be partly responsible for the absence from Cooloola records of 8 species of lizard and of 13 species of snake which have been recorded from other heathlands of the region. Among lizards, for example, *Anomalopus truncatus*, which occurs in similar habitats on North Stradbroke Island and in closed-forest on the mainland (Covacevich and Ingram, 1975), has not been recorded from Cooloola. There are no freshwater turtles and very few geckos, legless lizards or snakes recorded from Cooloola. Many sclerophyll-woodland and closed-forest species of lizards, snakes and frogs known from the Moreton district (Queensland Museum, 1976) are missing from Cooloola.

The twelve habitat categories depicted in Fig. 9.1

support widely varying numbers of species. For mammals, reptiles and frogs the species number west of the Noosa River (habitats A and B) is higher than for habitats to the east. This relationship does not hold for birds, which are equally diverse in woodlands east of the river. Especially high diversity in tall open-forest (H) is due to the presence of some closed-forest birds in this habitat. Similar habitats to the west of the Como Escarpment, however, support more species than the tall open-forest east of the river. The number of frog species increases near still water, showing association with perched lakes (J) and wet heathland on the Noosa Plain (E).

Despite the high level of species overlap (see Table 9.1) some patterns of habitat selection are evident. For mammals these patterns are expressed at two levels. First, while macropods and phalangerines are well represented west of the river in terms of species present, they are poorly represented to the east. Indeed the distribution of these species, with the notable exception of *Wallabia bicolor*, is essentially peripheral to heathland. Thus *Macropus giganteus* is relatively common in open-forests with grassy understorey and may also utilize forest or woodland with well-developed heathy understorey if it is invaded by grass. In areas south of Cooloola *M. giganteus* may become locally common where firebreaks have been created between heathland and exotic pine plantations. They shelter in adjoining plantations and native forest by day. Similar considerations apply for phalangerines. Most phalangerine species recorded from Cooloola occur in closed-forest pockets (i.e. *Trichosurus caninus*, *Pseudocheirus peregrinus*) or tall open-forest (i.e. *T. vulpecula*, *P. peregrinus*) within or near heath-dominated habitats. The last two species may use pockets and corridors of suitable habitat that run into heathland but within these abundance is always low. Nor have gliders (*Schoinobates* and *Petaurus*) been located within heath-dominated habitats. Secondly, habitat segregation for mammals is shown by the fact that related species are generally found in different habitat categories. With the exception of habitats D and K this applies for the three species of *Rattus*, the two species of *Melomys*, the two bandicoots (*Isoodon* and *Perameles*) and the two species of *Trichosurus*. Within each of these taxa one species (i.e. *R. fuscipes*, *M. cervinipes*, *P. nasuta* and *T.*

caninus) occurs in closed-forest. *T. caninus* is restricted to this habitat and is replaced by *T. vulpecula* in tall open-forest. The two murid rodents and *P. nasuta* occur through other forest types (e.g. G and H), but are replaced by related species where vegetation changes to woodland or heathland. The overlapping distributions recorded for the habitat D in the south of the Cooloola area can be understood in terms of the complex interspersion of many habitat types here. This explanation is, however, not sufficient to account for the overlaps observed in the habitat K. The survey area within this habitat has been subject to extensive disturbance from wind-blast and fire, and it is possible that populations of mammals in the area are in a state of flux. If this is true, the co-occurrence of related species within this habitat may reflect the dynamics of seral change in the vegetation.

In accordance with the limited number of species and their association with restricted habitats, frogs show little habitat overlap (Table 9.1A). Along the Noosa River only two species have been recorded, both characteristic of heathland but appearing in adjoining woodland. In contrast the wet-heathland of the Noosa Plain (E) and the perched lake in high dunes (J) have 8 and 9 species respectively (see Fig. 9.1). Of these, only 4 are common to both. The peculiarity of distribution is shown by the species in high dune lakes. *Litoria cooloolensis*, for example, is known only from dune lakes of Cooloola, Stradbroke Island and Fraser Island.

The overlap pattern for reptiles (Table 9.1B) shows attenuation of the dune fauna and some

TABLE 9.1

Overlap of species between different habitats along the Cooloola transect (see Fig. 9.1). The numbers of species shared between the habitats (A–L) are shown to the right of the diagonal line and their percentages out of the total for two habitats are shown to the left of the diagonal line. For mammals the aquatic habitats (C and J) and habitat category A are excluded; category D is subdivided as in the Appendix. Bats are not included in the analysis.

A. FROGS

	A	B	C	D	E	F	G	H	I	J	K	L
A	\	6			4	2				4	1	
B	32	\	2		6	3				4	1	
C		25	\		2	2				1	1	
D				\								
E	19	60	25		\	3				4	1	
F	11	33	50		33	\				2	2	
G							\					
H								\				
I									\			
J	18	31	10		31	18				\		4
K	5	9	20		9	33				44	\	
L												\
No. of species	17	8	2	0	8	4	0	0	1	9	4	0

B. REPTILES

	A	B	C	D	E	F	G	H	I	J	K	L
A	\	13	2	14	4	13	11	5	1	1	12	6
B	52	\	1	10	3	11	8	4	1	1	9	5
C	8	6	\	3	1				1			
D	42	42	15	\	4	16	8	4	1	1	11	5
E	16	19	14	19	\	4	3		1		3	2
F	45	52		73	21	\	9	4			12	6
G	38	36		29	17	36	\	5	1		10	6
H	18	21		16		17	25	\	4		5	2
I	3	5		4		5	33		\		2	
J	4	7	33	5	20					\		
K	44	45		46	18	57	48	26	10		\	6
L	25	33		24	22	33	38	15			40	\
No. of species	24	14	3	20	5	18	16	9	7	1	15	6

TABLE 9.1 *(continued)*

C. BIRDS

	A	B	C	D	E	F	G	H	I	J	K	L
A		62	7	60	25	59	59	46	33	10	48	31
B	76		8	61	24	57	54	45	32	11	47	33
C	8	11		10	9	8	7	5	6	8	7	10
D	66	74	13		32	64	56	49	33	9	48	36
E	24	25	16	34		32	23	17	10	9	19	18
F	65	67	10	75	34		55	46	34	8	46	33
G	70	67	9	65	24	64		53	37	9	53	31
H	42	44	6	46	15	43	56		56	7	47	27
I	30	31	8	30	9	31	38	61		6	33	18
J	12	15	36	11	15	10	12	8	8		9	8
K	55	58	9	55	20	52	72	51	35	13		32
L	37	45	24	47	25	42	42	30	20	17	50	
No. of species	76	68	14	75	52	74	67	80	68	16	60	38

D. MAMMALS (without bats)

	B	D(S)	D(N)	E	F	G	H	I	K	L
B		11	6	5	5	5	6	5	9	7
D(S)	69		5	5	4	6	7	6	10	8
D(N)	40	36		3	3	3	4	3	4	4
E	31	36	27		3	2	1	1	4	5
F	31	27	27	27		4	3	3	6	6
G	31	46	27	17	40		6	6	7	4
H	38	54	36	7	25	67		7	7	4
I	29	43	25	7	25	67	78		7	4
K	50	67	25	25	43	54	50	50		8
L	41	57	31	42	55	31	29	29	53	
No. of species	14	13	7	7	7	7	8	8	13	10

closed-forest species (i.e. *Anotis graciloides, Tiliqua gerrardii*) restricted in distribution. *Anomalopus reticulatus*, a species found in closed-forest elsewhere, is found in closed-forest (I) and in neighbouring tall open-forest (H). *Ramphotyphlops* sp., the only species endemic to Cooloola, occurs in the high dune forests of G, H and I.

Apart from water birds in the habitats C and J, birds reveal two patterns of overlap in their habitat distribution (Table 9.1C). One is that most open-forest and woodland species have wide tolerance over the range of the Cooloola habitats. This is shown by high percentages of overlap among the habitats A, B, D, F and G. The habitat K also belongs to this series, but the number of species is reduced here. Differences between the west and the east of the Noosa Plain are not a simple result of the attenuation of the escarpment fauna in the dune habitats. There are coastal elements and inland elements augmenting the impoverished fauna of open-forests and woodlands on dunes.

The coastal elements are few (e.g. *Pandion haliaetus, Aviceda subcristata*), but the presence of inland semi-arid adapted species in the dune habitats is conspicuous. They include *Dromaius novaehollandiae, Aprosmictus erythropterus, Neophema pulchella, Artamus superciliousus* and *A. cyanopterus*. Another pattern of overlap is shown in the restricted habitat range of birds of closed-forest (habitat I). Between this and the neighbouring tall open-forest (H) 61% of the species are shared, increasing the species number of the latter habitat to 80, the largest along the transect (see Fig. 9.1). Typical closed-forest birds show a high degree of association (Kikkawa, 1974), and at Cooloola they stay within these two habitat categories.

Overlap patterns for mammals (Table 9.1D) are in large part influenced by species complement, and the attenuation of fauna in heath-dominated and forest habitats. Thus high overlap occurs for habitat categories B, D (South), K and L where species number is greatest, and for the forest categories G, H and I. The seasonally wet-heathland (E) overlaps more with coastal heathland (L) than with other habitat categories.

Thus it appears that the diversity of habitat types found on the coastal lowlands largely accounts for the range of vertebrate species known to occur in the Cooloola area. However, within each habitat category species diversity is low.

Extensive studies of small mammal species based upon trapping have revealed very low within-habitat diversity for all habitat categories examined at Cooloola. Estimates of the α index of diversity (Williams, 1947) for trapped mammals showed relatively high diversities in coastal heathland, in open-forest of the foredunes, and in the complex of habitats fringing the southern reaches of the Noosa River (habitat D south, K and L); but elsewhere diversity is consistently very low (Fig. 9.1). With the exception of tall open-forest and closed-forest (H and I), where diversity is low and relative abundance of small mammals is high, the abundance ratings given in Fig. 9.1 mirror the trend in the diversity values. Dwyer et al. (1979) suggest that mammal abundance in habitats of the coastal lowlands is correlated with soil nutrient status. They argue that mammal abundance will be higher in the Womalah Landscape than in the Noosa and Cooloola Landscapes because the soils of the former maintain high nutrient status owing to rich parent material, and that within the latter landscapes high abundance ratings occur where soil material is enriched by alluvium, organic hard pan or ocean spray.

Density estimates obtained at Cooloola and from similar habitats to the south give maximum values of 3.5 per ha for *Antechinus flavipes*, 1.5 per ha for *Isoodon macrourus*, 11.8 per ha for *Rattus fuscipes*, 2.9 per ha for *R. lutreolus*, 4.25 per ha for *R. tunneyi culmorum*, 5 per ha for *Melomys cervinipes* and 6.1 per ha for *M. littoralis*. For *R. fuscipes* the estimate is derived from trapping in closed-forest, but for other species the estimates come from open-forest and woodland formations with heathy understorey or from heathland and sedgeland. Even if all species are combined for each area the density is low. Results from four grids have indicated maximum densities of 7.0 per ha (5 species) and 8.3 per ha (4 species) for two grids of 6.25 ha comprising nearly equal portions of

heathland and open-forest; 7.0 per ha (4 species) for a grid of the same size in mixed heathland and sedgeland (habitat E); and 9.25 per ha (6 species) for a grid of 4.0 ha in open-forest (habitat K). These data, combined with relatively high levels of diurnal activity exhibited by some small mammals, imply that the resource base exploited by these species is relatively poor.

The density of birds, as indicated by the number netted in two seasons, is lowest in closed-forest (I) and highest in tall open-forest (H). In all sclerophyll habitats the number of birds fluctuates greatly depending on the flowering of *Banksia*. As in mammals the density and the diversity (α index) of netted birds increased together (Fig. 9.1).

Seasonal movements are the most important aspect of habitat relations for birds. Both migratory and nomadic honey-eaters congregate and breed in habitats B, D and F during the flowering seasons of shrubs, which may occur at any time of the year but particularly in winter months. Consequently there are more honey-eaters in the area in winter than in summer. In contrast there are more insectivorous birds in summer than in winter. Typical summer visitors are *Eudynamys scolopacea*, *Coracina tenuirostris*, *Dicrurus hottentottus* and *Monarcha trivirgatus*, the last mentioned being the commonest insectivorous bird of closed-forest at Cooloola. There is only one common winter visitor among the insectivorous birds in the area (*Rhipidura fuliginosa*). The greatest fluctuation of number between summer and winter is seen in *Zosterops lateralis* which appears in large numbers during winter months. Their density in the shrub layers of tall open-forest (H) may be as high as 100 birds per ha.

No data are available for the abundance of frogs and reptiles at Cooloola.

The heathland communities of Cooloola developed on siliceous sands of exceptionally low nutrient status contain many plants rich in toxic compounds, and runoff from the area produces black-water creeks. The scarcity of herbivorous mammals within these communities may be understood in terms of the unfavourable diet offered (see Kikkawa et al., in Chapter 8). The relatively high level of toxic compounds in litter should also lead to litter and soil communities of low biomass and diversity. Food chains supported by such decomposer communities are therefore likely to be reduced and simplified. In combination, these attributes of the Cooloola ecosystem will reduce abundance and diversity of insectivorous species. This means that mammals with a relatively generalized diet, or those that can take advantage of the asynchronous flowering régime prevailing in these communities, are likely to be best represented. Murid rodents are well represented in the Cooloola fauna but it is noteworthy that, within this group, specialist seed-eaters (e.g. *Pseudomys*) are poorly represented. Blossom- and nectar-eating bats are also an important component of the fauna; large numbers of *Pteropus scapulatus* feed in *Banksia* and *Eucalyptus* through late summer and autumn, and the small *Syconycteris australis* is present throughout the year presumably exploiting a wider range of heathland species whose flowering is scattered throughout the year. Such considerations also explain the resource-utilization pattern of birds, and the low species diversity of frogs, lizards and snakes in the area.

Generally speaking, habitat selection by vertebrate animals may be a response to the structural forms of vegetation (e.g. many birds), the density of cover near the ground (e.g. many mammals) or to substrate characteristics (e.g. many reptiles and frogs). On the coastal lowlands of southeastern Queensland these variables interact in subtle ways; hence, an appreciation of the dynamics of vertebrate populations or species diversity within a single plant formation demands understanding of relationships between adjoining habitats. This is particularly true in the case of heathlands without an overstorey, for these often occupy small pockets within woodland, or relatively minor topographic zones interposed between sedgeland and woodland or open-forest. This habitat may be subject to seasonal flooding in low-lying areas or may be subject to (and perhaps maintained by) wind-blast off the sea. Similarly the extent to which a heathy understorey is developed in woodland and forest formations may be directly related to soil nutrient status, topography, fire history or disturbance from logging operations. Depending on the size of local heathland areas and their proximity to other habitats, these factors significantly influence the occurrence and diversity of vertebrate species.

A detailed study of mammals in heath vegetation at Beerwah and Cooloola in southeastern Queensland further elucidated these effects. At Beerwah trapping has been carried out within an area of 600 ha where many habitat types are closely interspersed; at Cooloola (approximately 56,600 ha) a wider range of habitats shows interspersion on a broader scale. Areas of equal size and comparable habitat support more species at Beerwah than at Cooloola. Two grids of 6.25 ha at Beerwah yielded 8 species and 10 species, in the first 150 individuals trapped. These grids respectively included 3.88 ha and 3.13 ha of low heathland, the remaining area comprising *Eucalyptus* open-forest and *Banksia* woodland both with a heathy understorey. In the period of observations 4 species too large to be trapped were recorded from the first grid, with 5 additional species from the second grid. At Cooloola the first 150 individuals trapped on a grid of the same size (4.81 ha heathland, 1.44 ha sedgeland) comprised only 5 species, and no additional species were recorded. The greater number of species at Beerwah is attributable to the presence of many habitat types in close proximity.

The scale of habitat interspersion can influence both seasonal patterns of habitat utilization and recovery of mammal populations after fire within particular habitats. At Beerwah many males of the marsupial mouse *Antechinus flavipes* disperse from woodlands and open-forests (where they are born) to contiguous heathland. Those individuals entering heathland may attain larger size than individuals remaining in the habitat of their birth. Since most of the surviving dispersers return to the original habitats before mating, they may be at some advantage with respect to reproduction. This seasonal shift in habitat for male *A. flavipes* has not been observed in the more extensive heathland tracts found at Cooloola. In this regard insects found in Cooloola heathlands are essentially unavailable to local populations of *A. flavipes*. A similar seasonal shift has been observed for *Rattus fuscipes*. In this case a post-breeding influx of individuals occurred into the heathy understorey of open-forest that adjoined closed-forest (M. Hockings, pers. comm., 1977). The failure of this open-forest to sustain a viable population of *R. fuscipes* through the year may be related to the recency of prescribed burning, which creates a relatively open understorey. Pockets of closed-forest on the Cooloola sandmass may provide important refuges from fire for *R. fuscipes*. The Beerwah study area does not support populations of *R. fuscipes* even where habitat appears suitable. The area is more frequently and evenly burnt than Cooloola and has no closed-forest in it.

The distribution and abundance of the murid rodent *Melomys littoralis* within different habitats is also affected differentially by fire according to the scale of habitat interspersion. This species is most abundant in wet-heathland but also occurs in dry-heathland and the heathy understorey of woodland and open-forest. It has been found in the grassy understorey of woodlands and open-forests that are close to heathland areas and, occasionally, in tall open-forest and closed-forest that have been heavily logged. Data from Beerwah show that the species may disappear from woodland and open-forest areas following fire, but that survival may be relatively high within small pockets of wet-heathland that escape burning or that recover rapidly after burning. Where habitats are closely interspersed these wet refuges provide foci for colonization of adjoining dry areas where burning tends to be more complete and regrowth is suppressed. Where habitat interspersion is coarse, population recovery in drier habitats may require more time. Adverse effects of flooding upon juvenile recruitment of *M. littoralis* have been detected in low-lying heathland. These effects may be alleviated if higher land supporting dry-heathland or woodland exists nearby. Some individuals of *Rattus lutreolus* have made shifts from wet to dry habitats during periods of heavy rain, while the bandicoot *Isoodon macrourus* has utilized areas of wet-heathland when these have dried out sufficiently through the winter months.

Changes in the utilization of heathland after fire have also been observed for the introduced cane toad, *Bufo marinus*. The toad frequently enters traps set for mammals, and on one grid at Beerwah the proportion of captures made in heathland rose from 14.5% to 25.4% the next summer, following a winter burn. This increase in utilization of heathland after fire appears to result from a reduction in density of vegetation near the ground. The toad typically feeds in relatively open situations and has difficulty in moving through dense vegetation.

CONCLUSIONS

The foregoing analysis clearly indicates that the dynamics of populations, the patterns of diversity,

distribution and seasonal behaviour of species, and their responses to fire and inundation of habitat, can only be understood if heathland is recognized as part of a series of associated habitats. In southeastern Queensland, areas of heathland are seldom extensive. They occur as patches of various sizes that give way to sedgelands or to woodlands and open-forests. These latter formations grow on similar soils and often share many plant species with heathland. Fire and changes in water table tend to be the primary determinants of vegetation structure.

The resources exploited by heath-dwelling vertebrates (cover, animal or plant foods, or attributes of substrate or the aquatic environment) may be available to different species in several distinct habitats which are interspersed at varying scales. The pattern of interspersion of these habitats, combined with the seasonality of many food sources (especially of nectar), impose major organizational constraints — both in space and in time — upon the associated vertebrate community. Vertebrates utilizing nectar may be locally abundant in patches of heathland through the flowering season, whereas most species utilizing other food sources are characterized by their low abundance in heathland. Both abundance and diversity of vertebrates fluctuate in response to food supply or community succession following fire.

The theme that emerges is one of a high level of opportunism superimposed upon a low degree of specialization, the latter being evidenced only in frogs and a few birds. Given the recency of these southeastern Queensland coastal heathlands, their relatively meagre complement of heathland flora (Specht, Chapter 6), their limited extent and their occurrence within a zone of highly variable rainfall, it seems likely that conditions may be more stressful for heathland-adapted fauna than those encountered in other Australian heathlands. The prevailing theme of opportunism may reflect one end of a continuum of patterns of community organization for Australian heathland vertebrates.

ACKNOWLEDGEMENTS

We thank Paul Campbell, David Gravatt, Marc Hockings and Jeffrey Willmer of the University of Queensland Zoology Department for information.

The Queensland and National Parks and Wildlife Service and Queensland Forestry Department provided permits to work at Cooloola and Beerwah. Our work has been supported by the Australian Research Grants Committee to Jiro Kikkawa and by the Interim Council of the Australian Biological Resources Study to Peter D. Dwyer.

APPENDIX I

Ecological distribution of vertebrates across the Cooloola transect shown in Fig. 9.1.

The frog data are based on censuses made at night at the height of the breeding season. The lists for reptiles and birds are compiled from Gravatt and Ingram (1975), Roberts and Ingram (1976), Covacevich (1977), and unpublished records kept by C.J. Corben, D.J. Gravatt, and T. Low, and by the authors. Mammal lists are based on trap and track records.

Legend: ‡=relatively abundant; +=present but not common; *=a single record; ?=tracks observed, but species identification not confirmed; 0=absent.

Habitat A has not been surveyed for small mammals and the data for habitat B are incomplete. Habitat D is divided into south (S) and north (N).

TABLE 9.2

Frogs

	A	B	C	D	E	F	G	H	I	J	K	L
MYOBATRACHIDAE												
Adelotus brevis	+	+	0	0	0	0	0	0	0	0	0	0
Limnodynastes peroni	+	+	0	0	+	0	0	0	0	+	0	0
Mixophyes fasciolatus	0	0	0	0	0	0	0	0	+	0	0	0
Platyplectron ornatus	+	0	0	0	0	0	0	0	0	0	0	0
P. terraereginae	0	0	0	0	0	0	0	0	0	+	+	0
Pseudophryne bibroni	+	0	0	0	0	0	0	0	0	0	0	0
P. coriacea	+	0	0	0	0	+	0	0	0	+	+	0
Ranidella parinsignifera	+	0	0	0	0	0	0	0	0	0	0	0
R. signifera	+	0	0	0	0	0	0	0	0	0	0	0
R. tinnula	0	+	+	0	+	+	0	0	0	+	+	0
Uperoleia laevigata	0	0	0	0	0	0	0	0	0	+	+	0
PELODRYADIDAE												
Litoria caerulea	+	0	0	0	0	0	0	0	0	0	0	0
L. cooloolensis	0	0	0	0	0	0	0	0	0	+	0	0
L. dentata	+	0	0	0	0	0	0	0	0	0	0	0
L. fallax	+	0	0	0	0	0	0	0	0	0	0	0
L. freycineti	0	0	0	0	+	0	0	0	0	+	0	0
L. gracilenta	+	0	0	0	0	0	0	0	0	0	0	0
L. latopalmata	+	0	0	0	0	0	0	0	0	0	0	0
L. nasuta	+	+	0	0	+	0	0	0	0	+	0	0
L. olongburensis	0	+	+	0	+	+	0	0	0	0	0	0
L. peronii	+	+	0	0	0	0	0	0	0	+	0	0
L. rothi	+	0	0	0	0	0	0	0	0	0	0	0
L. rubella	+	+	0	0	+	0	0	0	0	0	0	0
L. sp. A	0	0	0	0	+	0	0	0	0	0	0	0
BUFONIDAE												
Bufo marinus (introduced)	+	+	0	0	+	+	0	0	0	0	0	0

TABLE 9.3

Reptiles

	A	B	C	D	E	F	G	H	I	J	K	L
GEKKONIDAE												
Diplodactylus vittatus	+	0	0	+	0	+	0	0	0	0	0	
Oedura sp.	0	0	0	0	0	0	+	0	0	0	0	0
PYGOPODIDAE												
Lialis burtonis	+	+	0	+	0	+	0	0	0	0	+	0
AGAMIDAE												
Amphibolurus barbatus	+	+	0	0	0	+	0	0	0	0	0	0
Chlamydosaurus kingii	+	+	0	+	+	+	+	0	0	0	+	0
Diporiphora australis	+	0	0	0	0	0	0	0	0	0	0	0
D. sp.	0	0	0	+	+	+	0	0	0	0	0	0
Physignathus lesueurii	+	0	+	+	0	0	0	0	0	0	0	0
VARANIDAE												
Varanus varius	+	+	0	+	0	+	+	+	0	0	+	0
SCINCIDAE												
Anomalopus ophioscincus	0	0	0	0	0	0	0	0	+	0	+	0
A. reticulatus	0	0	0	0	0	0	0	+	+	0	0	0
A. verreauxii	+	+	0	+	0	+	0	0	0	0	0	0
Anotis graciloides	0	0	0	0	0	0	0	0	+	0	0	0
A. sp.	0	0	0	0	0	0	+	0	0	0	0	0
Carlia burnetti	+	0	0	0	0	0	0	0	0	0	0	0
C. pectoralis	0	+	0	+	0	+	0	0	0	0	0	0
C. vivax	+	+	0	+	0	+	+	0	0	0	+	+
Cryptoblepharus virgatus	+	+	0	+	0	+	+	0	0	0	+	+
Ctenotus robustus	+	0	0	+	+	+	+	0	0	0	+	+
C. taeniolatus	+	+	0	0	+	+	+	0	0	0	+	+
Egernia frerei	0	0	0	+	0	+	0	0	0	0	+	0
Lampropholis delicata	+	+	0	0	0	0	0	+	+	0	+	0
L. guichenoti	+	0	0	0	0	0	0	0	0	0	0	0
Morethia boulengeri	0	0	0	+	0	+	0	0	0	0	0	0
M. taeniopleura	0	0	0	+	0	+	+	0	0	0	+	0
Sphenomorphus quoyii	0	0	+	+	0	0	0	0	0	0	0	0
S. scutirostrum	0	0	0	0	0	0	+	+	0	0	0	0
S. tenuis	0	0	0	0	0	0	0	+	+	0	0	0
S. sp.	+	0	0	0	0	0	+	0	0	0	+	0
Tiliqua gerrardii	0	0	0	+	0	0	0	0	+	0	0	0
T. scincoides	+	0	0	0	0	0	+	0	0	0	0	0
TYPHLOPIDAE												
Ramphotyphlops sp.	0	0	0	0	0	0	+	+	+	0	0	0
BOIDAE												
Python spilotus	+	+	0	0	0	0	+	0	0	0	0	0
COLUBRIDAE												
Amphiesma mairii	+	+	+	+	+	0	0	0	0	+	0	0
Dendrelaphis punctulatus	+	+	0	+	0	+	+	+	0	0	+	+

TABLE 9.3 *(continued)*

Reptiles

	A	B	C	D	E	F	G	H	I	J	K	L
ELAPIDAE												
Cryptophis nigrescens	+	0	0	+	0	+	0	+	0	0	+	0
Demansia psammophis	+	+	0	+	0	+	+	+	0	0	+	+
Pseudechis porphyriacus	+	0	0	0	0	0	0	0	0	0	0	0
Pseudonaja textilis	+	0	0	0	0	0	0	0	0	0	0	0

TABLE 9.4

Birds

	A	B	C	D	E	F	G	H	I	J	K	L
DROMAIIDAE												
Dromaius novaehollandiae	0	0	0	0	+	+	+	0	0	0	+	0
PODICIPEDIDAE												
Tachybaptus novaehollandiae	0	0	0	0	0	0	0	0	0	+	0	0
PHALACROCORACIDAE												
Phalacrocorax melanoleucos	0	0	+	0	0	0	0	0	0	0	0	0
ARDEIDAE												
Ardea novaehollandiae	0	0	0	0	+	0	0	0	0	0	0	0
Egretta alba	0	0	0	0	0	0	0	0	0	+	0	0
CICONIIDAE												
Xenorhynchus asiaticus	0	0	0	0	+	0	0	0	0	0	0	0
PLATALEIDAE												
Threskiornis moluccus	0	0	0	0	+	0	0	0	0	0	0	0
ANATIDAE												
Anas superciliosa	0	0	0	0	0	0	0	0	0	+	0	0
PANDIONIDAE												
Pandion haliaetus	0	0	+	+	+	0	0	0	0	0	0	+
ACCIPITRIDAE												
Accipiter cirrocephalus	+	+	0	0	0	+	+	0	+	0	0	0
A. novaehollandiae	0	0	0	0	0	+	0	+	+	0	0	0
Aquila audax	+	0	0	+	0	+	0	0	0	0	0	0
Aviceda subcristata	0	0	0	0	0	0	+	+	0	0	+	0
Circus aeruginosus	0	0	0	0	+	0	0	0	0	0	0	0
Elanus notatus	0	0	0	0	+	+	0	0	0	0	0	0
Haliaeetus leucogaster	0	0	+	0	0	0	0	0	0	0	0	+
Haliastur indus	0	0	+	+	0	+	0	0	0	0	0	+
H. sphenurus	+	+	+	+	+	+	+	0	+	+	+	+
Hieraaetus morphnoides	0	0	0	0	0	0	0	+	+	0	0	0

TABLE 9.4 *(continued)*

Birds

	A	B	C	D	E	F	G	H	I	J	K	L
FALCONIDAE												
Falco berigora	+	+	0	+	+	+	+	0	0	0	0	0
F. cenchroides	+	0	0	+	+	+	0	0	0	0	0	0
F. peregrinus	+	+	0	+	+	+	+	+	+	0	0	0
MEGAPODIIDAE												
Alectura lathami	0	0	0	+	0	+	0	+	+	0	0	0
PHASIANIDAE												
Coturnix australis	0	0	0	+	+	+	0	0	0	0	0	0
C. chinensis	0	0	0	0	+	0	0	0	0	0	0	0
TURNICIDAE												
Turnix varia	+	+	0	+	+	+	0	0	0	0	0	0
RALLIDAE												
Gallinula olivacea	+	+	0	0	0	0	0	0	0	+	0	0
Porphyrio porphyrio	0	0	+	0	0	0	0	0	0	0	0	0
Rallus philippensis	+	0	0	0	0	0	0	0	0	0	0	0
GRUIDAE												
Grus rubicundus	0	0	0	0	+	0	0	0	0	0	0	0
BURHINIDAE												
Burhinus magnirostris	+	+	0	+	+	+	0	0	0	0	0	0
CHARADRIIDAE												
Vanellus miles	0	0	0	0	+	0	0	0	0	0	0	0
SCOLOPACIDAE												
Gallinago hardwickii	0	0	0	0	+	0	0	0	0	0	0	0
COLUMBIDAE												
Columba leucomela	0	0	0	0	0	0	0	0	+	0	0	0
Chalcophaps indica	0	0	0	0	0	+	0	+	+	0	+	0
Geopelia humeralis	+	+	0	+	0	+	+	+	0	0	0	+
G. striata	+	+	0	0	0	+	0	0	0	0	0	0
Lopholaimus antarcticus	0	0	0	0	0	0	0	+	+	0	0	0
Macropygia amboinensis	0	0	0	0	0	0	0	+	+	0	0	0
Ocyphaps lophotes	+	+	0	0	0	0	0	0	0	0	0	0
Phaps chalcoptera	+	+	0	+	0	+	+	0	0	0	0	0
P. elegans	0	0	0	0	0	+	0	0	0	0	0	0
Ptilinopus magnificus	0	0	0	0	0	0	0	+	+	0	0	0
P. regina	0	0	0	0	0	0	0	0	+	0	0	0

TABLE 9.4 *(continued)*

Birds

	A	B	C	D	E	F	G	H	I	J	K	L
CACATUIDAE												
Cacatua galerita	+	0	0	0	0	0	+	+	+	0	+	0
Calyptorhynchus funereus	+	+	0	+	+	+	+	+	+	0	+	+
C. lathami	0	0	0	+	0	0	+	+	0	0	0	0
LORIIDAE												
Trichoglossus chlorolepidotus	+	+	0	+	0	+	+	+	+	0	+	0
T. haematodus	+	+	0	+	0	+	+	+	+	0	+	0
POLYTELIDAE												
Alisterus scapularis	0	0	0	0	0	0	0	+	+	0	0	0
Aprosmictus erythropterus	+	+	0	+	+	+	0	0	0	0	0	0
PLATYCERCIDAE												
Neophema pulchella	0	0	0	0	0	+	0	0	0	0	0	0
Pezoporus wallicus	0	0	0	0	+	0	0	0	0	0	0	0
Platycercus adscitus	+	+	0	+	+	+	+	+	0	0	+	0
CUCULIDAE												
Centropus phasianinus	+	+	0	+	+	+	+	0	0	0	0	0
Chrysococcyx basalis	0	0	0	+	+	+	0	0	0	0	0	0
C. lucidus	+	+	0	+	0	+	+	+	+	0	+	0
Cuculus pallidus	+	+	0	+	+	+	+	+	0	0	0	0
C. pyrrhophanus	0	0	0	+	0	0	+	+	+	0	+	0
C. variolosus	+	0	0	+	0	0	+	+	+	0	0	0
Eudynamys scolopacea	+	0	0	0	0	0	+	+	+	0	0	0
Scythrops novaehollandiae	+	0	0	+	0	+	+	0	0	0	0	0
STRIGIDAE												
Ninox connivens	0	0	0	0	+	0	0	+	+	0	+	0
N. novaeseelandiae	+	+	0	+	0	+	+	+	+	0	+	+
N. strenua	0	0	0	0	0	0	0	+	+	0	0	0
TYTONIDAE												
Tyto alba	0	0	0	0	0	0	0	0	0	0	+	0
T. longimembris	0	0	0	0	+	0	0	0	0	0	0	0
PODARGIDAE												
Podargus strigoides	+	+	0	+	0	+	+	+	+	0	+	0
AEGOTHELIDAE												
Aegotheles cristatus	+	+	0	+	0	+	+	+	+	0	+	0
CAPRIMULGIDAE												
Caprimulgus mystacalis	+	+	0	+	+	+	+	+	0	0	+	0

TABLE 9.4 *(continued)*

Birds

	A	B	C	D	E	F	G	H	I	J	K	L
APODIDAE												
Apus pacificus	+	+	+	+	+	+	+	+	+	+	+	+
Hirundapus caudacutus	+	+	+	+	+	+	+	+	+	+	+	+
ALCEDINIDAE												
Ceyx azureus	0	+	+	+	+	0	0	0	0	+	0	0
Dacelo novaeguineae	+	+	0	+	0	+	+	+	+	+	+	0
Halcyon macleayii	+	+	0	0	0	0	+	+	+	+	+	0
H. sancta	+	+	+	+	0	+	+	0	0	+	+	+
MEROPIDAE												
Merops ornatus	+	+	+	+	+	+	+	+	+	+	+	+
CORACIIDAE												
Eurystomus orientalis	+	0	0	+	0	+	+	+	0	0	+	0
PITTIDAE												
Pitta versicolor	0	0	0	0	0	0	0	+	+	0	0	0
HIRUNDINIDAE												
Cecropis ariel	0	0	+	0	+	0	0	0	0	0	0	0
C. nigricans	+	+	+	+	+	+	+	+	+	+	+	+
Hirundo neoxena	+	+	+	+	+	+	+	+	+	+	+	+
MOTACILLIDAE												
Anthus novaeseelandiae	0	0	0	0	+	0	0	0	0	0	0	0
CAMPEPHAGIDAE												
Coracina novaehollandiae	+	+	0	+	0	+	+	+	+	0	+	+
C. papuensis	+	0	0	0	0	0	0	0	0	0	0	0
C. tenuirostris	+	+	0	+	0	+	+	+	+	0	+	0
Lalage leucomela	0	0	0	0	0	0	0	+	+	0	0	0
L. sueurii	+	+	0	+	0	+	+	0	0	0	0	0
MUSCICAPIDAE												
Colluricincla harmonica	+	+	0	+	0	+	+	+	+	0	+	+
C. megarhyncha	0	0	0	0	0	0	0	+	+	0	0	0
Eopsaltria australis	0	0	0	0	0	0	0	+	+	0	0	0
Microeca leucophaea	+	+	0	0	0	0	0	0	0	0	0	0
Monarcha leucotis	0	0	0	0	0	0	0	0	+	0	0	0
M. melanopsis	0	0	0	0	0	0	0	+	+	0	0	0
M. trivirgatus	0	0	0	0	0	0	0	+	+	0	0	0
Myiagra inquieta	0	+	0	+	+	0	0	+	+	0	0	0
M. rubecula	+	+	0	+	0	+	+	+	+	0	+	+
Pachycephala pectoralis	0	0	0	0	0	0	0	+	+	0	0	0
P. rufiventris	+	+	0	+	0	+	+	+	0	0	+	+
Petroica rosea	0	+	0	+	0	0	0	+	0	0	0	0
Rhipidura fuliginosa	+	+	0	+	0	+	+	+	+	0	+	+
R. leucophrys	0	0	0	+	0	0	0	0	0	0	+	+

TABLE 9.4 *(continued)*

Birds

	A	B	C	D	E	F	G	H	I	J	K	L
MUSCICAPIDAE *(continued)*												
R. rufifrons	0	0	0	0	0	0	0	+	+	0	0	0
Tregellasia capito	0	0	0	0	0	0	0	0	+	0	0	0
Zoothera dauma	0	0	0	0	0	0	0	0	+	0	0	0
ORTHONYCHIDAE												
Psophodes olivaceus	0	0	0	0	0	0	0	+	+	0	0	0
SYLVIIDAE												
Cisticola exilis	0	0	0	0	+	0	0	0	0	+	0	+
Megalurus timoriensis	0	0	0	0	+	0	0	0	0	+	0	0
MALURIDAE												
Malurus lamberti	+	+	0	+	0	+	+	+	0	0	+	0
M. melanocephalus	+	+	0	+	+	+	+	0	0	0	+	+
Stipiturus malachurus	0	0	0	0	+	0	0	0	0	0	0	0
ACANTHIZIDAE												
Acanthiza pusilla	+	+	0	+	0	+	+	+	+	0	+	+
Gerygone mouki	0	0	0	0	0	0	0	0	+	0	0	0
G. olivacea	+	0	0	0	0	0	0	0	0	0	0	0
Sericornis frontalis	0	0	0	0	0	+	+	+	+	0	0	0
S. magninostris	0	0	0	0	0	0	0	+	+	0	0	0
Smicrornis brevirostris	+	0	0	0	0	0	0	0	0	0	0	0
NEOSITTIDAE												
Daphoenositta chrysoptera	0	0	0	0	0	0	+	+	0	0	+	0
CLIMACTERIDAE												
Climacteris leucophaea	+	+	0	+	0	+	+	+	+	0	+	0
MELIPHAGIDAE												
Acanthorhynchus tenuirostris	+	+	0	+	0	+	+	+	+	0	0	0
Anthochaera chrysoptera	+	+	0	+	+	+	+	+	0	0	+	+
Entomyzon cyanotis	+	+	0	+	0	+	+	0	0	0	+	0
Lichenostomus chrysops	+	+	0	+	0	+	+	+	0	0	+	+
Lichmera indistincta	+	+	0	+	+	+	+	0	0	0	+	+
Manorina melanocephala	+	0	0	0	0	0	0	0	0	0	0	0
Meliphaga lewinii	0	0	0	0	0	0	0	+	+	0	0	0
Melithreptus albogularis	+	+	0	+	0	0	+	+	0	0	+	+
M. lunatus	0	+	0	+	0	0	0	0	0	0	+	0
Myzomela sanguinolenta	+	+	0	+	+	+	+	+	+	0	+	+
Philemon citreogularis	+	+	0	+	+	+	+	+	0	0	+	+
P. corniculatus	+	+	0	+	+	+	+	+	0	0	+	+
Phylidonyris nigra	0	+	0	+	+	+	0	+	0	0	0	+
DICAEIDAE												
Dicaeum hirundinaceum	+	+	0	+	0	+	+	+	+	0	+	+

TABLE 9.4 *(continued)*

Birds

	A	B	C	D	E	F	G	H	I	J	K	L
PARDALOTIDAE												
Pardalotus punctatus	+	+	0	+	0	+	+	+	+	0	+	0
P. striatus	+	+	0	+	0	+	+	+	0	0	+	+
ZOSTEROPIDAE												
Zosterops lateralis	+	+	0	+	0	+	+	+	+	0	+	+
PLOCEIDAE												
Emblema temporalis	0	+	0	+	0	+	0	+	+	0	+	+
Lonchura castaneothorax	0	0	0	0	+	0	0	0	0	0	0	0
Poephila bichenovii	+	0	0	0	0	0	0	0	0	0	+	0
ORIOLIDAE												
Oriolus sagittatuts	0	0	0	0	0	0	+	+	0	0	+	0
Sphecotheres viridis	0	0	0	0	0	0	0	+	+	0	0	0
DICRURIDAE												
Dicrurus hottentottus	+	+	0	+	0	+	+	+	+	0	+	+
PARADISAEIDAE												
Ailuroedus crassirostris	0	0	0	0	0	0	0	0	+	0	0	0
Sericulus chrysocephalus	0	0	0	0	0	0	0	+	+	0	0	0
GRALLINIDAE												
Grallina cyanoleuca	+	+	0	+	0	0	0	0	0	0	0	0
ARTAMIDAE												
Artamus cyanopterus	0	0	0	0	+	+	0	0	0	0	0	0
A. leucorhynchus	+	+	0	+	+	+	+	0	0	0	+	+
A. superciliosus	0	0	0	+	+	+	0	0	0	0	0	0
CRACTICIDAE												
Cracticus nigrogularis	+	+	0	+	+	+	+	0	0	0	+	+
C. torquatus	+	+	0	0	0	0	+	+	0	0	+	0
Gymnorhina tibicen	+	0	0	0	0	0	0	0	0	0	0	0
Strepera graculina	0	0	0	0	0	0	+	+	+	0	+	0
CORVIDAE												
Corvus orru	+	+	0	+	0	+	+	+	0	0	+	+

TABLE 9.5

Mammals

	A	B	C	D(S)	D(N)	E	F	G	H	I	J	K	L
TACHYGLOSSIDAE													
Tachyglossus aculeatus	·	·	0	0	0	0	*	0	0	0	0	*	0
DASYURIDAE													
Antechinus flavipes	·	‡	0	‡	0	0	+	+	‡	+	0	‡	+
Sminthopsis murina	·	·	0	0	0	0	0	0	0	0	0	0	*
PERAMELIDAE													
Isoodon macrourus	·	‡	0	‡	0	+	‡	?	0	0	0	‡	?
Perameles nasuta	·	·	0	0	0	0	0	+	?	‡	0	+	0
PETAURIDAE													
Petaurus australis	+	+	0	0	0	0	0	0	0	0	0	0	0
P. sp.	+	+	0	+	0	0	0	0	0	0	0	+	0
Pseudocheirus peregrinus	‡	‡	0	‡	+	0	0	0	‡	‡	0	+	+
Schoinobates volans	+	+	0	0	0	0	0	0	0	0	0	0	0
PHALANGERIDAE													
Trichosurus caninus	0	0	0	0	0	0	0	0	0	‡	0	0	0
T. vulpecula	‡	‡	0	‡	+	0	0	0	+	0	0	0	0
MACROPODIDAE													
Macropus giganteus	‡	‡	0	+	0	0	0	0	0	0	0	0	0
Wallabia bicolor	‡	‡	0	‡	+	+	+	+	+	+	0	+	+
MURIDAE													
Hydromys chrysogaster	·	0	‡	0	0	0	0	0	0	0	0	0	0
Melomys cervinipes	·	·	0	‡	0	0	0	‡	‡	‡	0	+	0
M. littoralis	·	‡	0	‡	‡	‡	‡	0	0	0	0	‡	‡
Mus musculus	·	·	0	‡	0	+	0	0	0	0	0	0	+
Pseudomys delicatulus	·	·	0	0	*	0	0	0	0	0	0	0	0
Rattus fuscipes	·	+	0	‡	0	0	0	+	‡	‡	0	‡	0
R. lutreolus	·	+	0	‡	0	‡	0	0	0	0	0	+	+
R. tunneyi culmorum	·	·	0	0	0	0	+	0	0	0	0	‡	‡
Xeromys myoides	·	·	0	0	0	*	0	0	0	0	0	0	0
CANIDAE													
Canis familiaris	·	+	0	+	+	0	+	+	+	+	0	+	+
Vulpes vulpes	·	+	0	0	+	+	0	0	0	0	0	0	0
VESPERTILIONIDAE													
Miniopterus australis	·	·	0	0	0	0	*.	0	0	0	0	0	0
Nyctophilus timoriensis	·	·	0	0	0	0	+	0	0	‡	0	+	0
PTEROPODIDAE													
Nyctimene robinsoni	·	·	0	0	0	0	0	0	0	+	0	0	0
Pteropus scapulatus	·	‡	0	‡	‡	0	‡	‡	‡	‡	0	‡	0
Syconycteris australis	·	·	0	0	‡	0	‡	‡	‡	‡	0	+	+

REFERENCES

Calaby, J.H., 1966. Mammals of the upper Richmond and Clarence Rivers, New South Wales. *Tech. Pap. Div. Wildl. Res. C.S.I.R.O., Aust.*, No. 10: 55 pp.

Coaldrake, J.E., 1961. The ecosystem of the coastal lowlands ("wallum") of southern Queensland. *C.S.I.R.O. Aust. Bull.* No. 283: 138 pp.

Coaldrake, J.E., 1962. The coastal sand dunes of southern Queensland. *Proc. R. Soc. Qld.*, 62: 101–116.

Coaldrake, J.E., 1975. The natural history of Cooloola. *Proc. Ecol. Soc. Aust.*, 9: 308–313.

Covacevich, J. (Editor), 1977. *Fauna of Eastern Australian Rainforests II.* Queensland Museum, Brisbane, Qld., 99 pp.

Covacevich, J. and Ingram, G.J., 1975. The reptiles of Stradbroke Island. *Proc. R. Soc. Qld.*, 86: 55–60.

Dwyer, P., Hockings, M. and Willmer, J., 1979. Mammals of Cooloola and Beerwah. *Proc. R. Soc. Qld.*, 90: 65–84.

Gravatt, D.J. and Ingram, G.J., 1975. Comments on the land vertebrates of Cooloola. *Proc. Ecol. Soc. Aust.*, 9: 321–325.

Hawkins, P.J., 1975. Forest management of Cooloola State Forest. *Proc. Ecol. Soc. Aust.*, 9: 328–333.

Ingram, G.J. and Corben, C.J., 1975. The frog fauna of North Stradbroke Island, with comments on the 'acid' frogs of the wallum. *Proc. R. Soc. Qld.*, 86: 49–54.

Kikkawa, J., 1970. The birds of the Richmond valley. In: F. Bitmead (Editor), *Richmond Valley.* University of New England Department of University Extension Regional Office, Lismore, N.S.W., pp. 59–75.

Kikkawa, J., 1974. Comparison of avian communities between wet and semiarid habitats in eastern Australia. *Aust. Wildl. Res.*, 1: 107–116.

Kirkpatrick, T.H., 1966. Mammals, birds and reptiles of the Warwick District, Queensland. 1. Introduction and mammals. *Qld. J. Agric. Anim. Sci.*, 23: 591–604.

Kirkpatrick, T.H., 1967. Mammals, birds and reptiles of the Warwick District, Queensland. 2. Birds. *Qld. J. Agric. Anim. Sci.*, 24: 81–91.

Kirkpatrick, T.H., 1968. Mammals, birds and reptiles of the Warwick District, Queensland. 3. Reptiles and general conclusions. *Qld. J. Agric. Anim. Sci.*, 25: 235–241.

Magnussen, W.E., Webb, G.J.W. and Taylor, J.A., 1976. Two new locality records, a new habitat and a nest description for *Xeromys myoides* Thomas (Rodentia: Muridae). *Aust. Wildl. Res.*, 3: 153–157.

Queensland Museum, 1976. *The National Estate in the Moreton and Wide Bay–Burnett Regions, South-East Queensland and Recommendations on Its Managemrnts.* Queensland Museum, Brisbane, Qld., revised ed., 272 pp.

Redhead, T.D. and McKean, J.L., 1975. A new record of the false water-rat, *Xeromys myoides*, from the Northern Territory of Australia. *Aust. Mammal.*, 1: 347–354.

Roberts, G.J. and Ingram, G.J., 1976. An annotated list of the land birds of Cooloola. *Sunbird*, 7: 1–20.

Thompson, C.H., 1975. Coastal areas of southern Queensland, some land-use conflicts. *Proc. R. Soc. Qld.*, 86: 109–120.

Webb, L.J. amd Tracey, J.G., 1975. The Cooloola rain forests. *Proc. Ecol. Soc. Aust.*, 9: 317–321.

Williams, C.B., 1947. The logarithmic series and its application to biological problems. *J. Ecol.*, 34: 253–272.

Chapter 10

HABITAT PREFERENCES OF MAMMALS INHABITING HEATHLANDS OF WARM TEMPERATE COASTAL, MONTANE AND ALPINE REGIONS OF SOUTHEASTERN AUSTRALIA[1]

A.E. NEWSOME and P.C. CATLING

INTRODUCTION

Much of southeastern Australia is covered by an open-forest dominated by a number of species of *Eucalyptus*. The open-forest ranges in height from 10 to 30 m but, in wetter areas, may be even taller (up to 100 m in a tall open-forest). An intergrading range of understoreys — from "heathy" on the most infertile soils, to "grassy" (*Poa, Themeda, Danthonia*, etc.) on the more fertile soils, or "ferny" (sometimes including tree ferns) in the wetter tall open-forests — occurs as a mosaic under much of the forest. Terms such as "dry sclerophyll forest" (including both "heathy open-forest" and "grassy open-forest") and "wet sclerophyll forest" (tall open-forest, with or without tree ferns) have been used to describe the vegetation. In extreme habitats from lowland to alpine areas, the eucalypt trees may become stunted or fail to survive. The understoreys of "heathy" or "grassy" species found in the open-forest are then dominant as a heathland or *Poa* grassland, sometimes with a few eucalypt trees scattered across the community. These open areas of heathland or grassland are relatively small in area compared with the considerable expanse of open-forest. The purpose of this paper is to assess the habitat preferences of mammals in the range of heathy and grassy communities (with or without trees) which exist from lowland to alpine sites in southeastern Australia.

Southeastern Australia, Tasmania included, has one of the richest mammalian faunas in the land. Tyndale-Biscoe and Calaby (1975) attributed that fact to the prevalence of *Eucalyptus* forests there. Of the 57 species of mammal indigenous to the region, they see 10 to 20 as totally dependent on the forests, and the rest use them for shelter. Examples of totally dependent mammals are the koala (*Phascolarctos cinereus*) and the possums (*Acrobates, Gymnobelideus, Petaurus* spp., *Schoinobates* and *Trichosurus caninus*). One small insectivorous marsupial, *Sminthopsis leucopus*, uses heathland as its optimum habitat though three more, the two large marsupial predators in Tasmania *Sarcophilus harrisii* and *Thylacinus cynocephalus* (believed extinct) and a pygmy possum, *Cercartetus nanus*, use heathlands as well as forests.

There are, however, few systematically collected data on the distribution of the fauna among the various habitats of the region. This paper reports results of trapping (47 sites) and tracking (217 sites) of vertebrates in major habitats in the southeast corner of the Australian mainland, in the coastal Nadgee Nature Reserve, in the montane to alpine Kosciusko National Park, and on Mount Tinderry, a high montane outlier further north on the Great Dividing Range. The habitats are broadly mapped in Fig. 10.1. Heathlands comprise less than 1.5% of the Australian land mass (Williams, 1955; Newsome, 1973), but are well represented in Nadgee and Kosciusko (Williams, 1955). Small pockets, too small or scattered to map in Fig. 10.1, grow along the mountain streams in bands a few metres wide. Some species of heathy shrubs grow not only in the open valleys but in the adjacent wooded habitats as well. Trapping sites were located to take advantage of these comparisons. An extensive heathland of *Casuarina nana* was remote from sites used in this study and so was not sampled.

---

[1] Manuscript completed March, 1977.

Fire, probably long a characteristic of the Australian environment and certainly of some heathlands (Specht et al., 1958), destroyed around 95% of Nadgee Nature Reserve during our study; its effect on the vertebrates was also studied (Newsome et al., 1975).

ENVIRONMENTS

Nadgee Nature Reserve

Nadgee is a wilderness area of about 15 000 ha. It stretches about 22 km north from Cape Howe, the most southeasterly point of Australia, and 6 to 8 km inland from the sea to a coastal range 500 to 600 m high (Fig. 10.1).

The climate (Fox, 1970) is temperate and mild marine. Maximum and minimum temperatures are 23°C and 14°C in January and 14°C and 7°C in July. There are about 60 frosty days per year and about 150 wet days (mists included). Average rainfall since 1966 in the north of the reserve has been 972 mm (range: 750 to 1226 mm). Heavy, irregular thunderstorms yield 250 to 400 mm of rain in 24 hours, flooding the streams. Evaporation per year is about 900 mm. The years 1971 and 1972 were droughty, but 1973, 1974 and 1975 were wet. Rainfall is highest on the coastal range.

The feature along the coast is the heathlands, which are perched behind sea cliffs about 20 to 50 m high. The broad vegetation map (Scobie and Fox, 1972) indicates that heaths comprise about 7.5% of the area (Fig. 10.1). The shrubs are sparse and low. There are three major communities, each about 2 km² in size, with their dominant shrubs differing from north to south, due probably to the soils and to fires (Fox, 1970, and pers. comm., 1972). Posamentier (1975) provides an extensive species list for parts of the two northern coastal heathlands shown on Fig. 10.1, and A. Fox (pers. comm., 1972) has data for smaller coastal ones beyond our study area. The heathland on top of the range is about twice the size of any on the coast, and more complex, with rocky outcrops, moist peaty soils and different vegetation (Table 10.1).

Dry sclerophyll forest of *Eucalyptus sieberi* and *E. muellerana* predominate between the coastal and upland heathlands. Small areas of wet sclerophyll forest (including *E. longifolia*, *Melaleuca squarrosa*

and lianas) extend along watercourses and in wet gullies. *Leptospermum* spp. and *Melaleuca* spp. grow in thickets on the dunes and in the swales behind the beaches. There are many swamps, some of them quite large. Ground vertebrates have been recorded in all of these habitats (see below).

Kosciusko National Park

This area lies about 175 km to the northwest of Nadgee. It is over 600 000 ha in size and straddles the main spine of the Australian Alps, the highest part of the Great Dividing Range (Fig. 10.1). Our study areas were along mountain trails connecting the eastern side of the Park at an altitude of about 1000 m, with Mount Kosciusko, Australia's highest peak, at 2227 m (Fig. 10.1). The topography of the region is most complex, with rugged, steep, heavily wooded mountain ridges deeply dissected by valleys. Snow lies on the ground above 1800 m for at least four months each year, and may be on the taller peaks for much of the year. Tree line is at about 1800 m.

Weather data are scanty, and non-existent for our main study areas. At Thredbo (1365 m) near the subalpine region, rainfall averages 1950 mm per year and maximum and minimum temperatures average 20.4°C and 7.4°C in January, and 4.5°C and −3.6°C in July. Just below the alpine region at Spencer's Creek (1767 m), the annual rainfall is 1800 mm, and temperatures are 15.3°C and 5.7°C in January, and 1.0°C and −4.6°C in July. Snow falls on an average of 104 days there. The eastern side of the Park where we mostly worked is in a partial rain shadow but inaccessible in winter due to snow and trees brought down by storms.

The map of the dominant vegetation types (Fig. 10.1) was adapted from Costin (1954). Sclerophyll forest and woodland predominate, the dominant trees being all eucalypts regardless of altitude. In the lower altitudes of the dry eastern side of the Park, the main species are *Eucalyptus pauciflora* (white sally), *E. dives* and *E. rubida*. At montane elevations of about 1200–1500 m altitude, mountain gum (*E. dalrympleana*) and white sally form woodlands and open-forests. On the moist southern slopes, wet sclerophyll forests of tall alpine ash, *E. delegatensis*, form large stands. At higher altitudes, both these communities give way

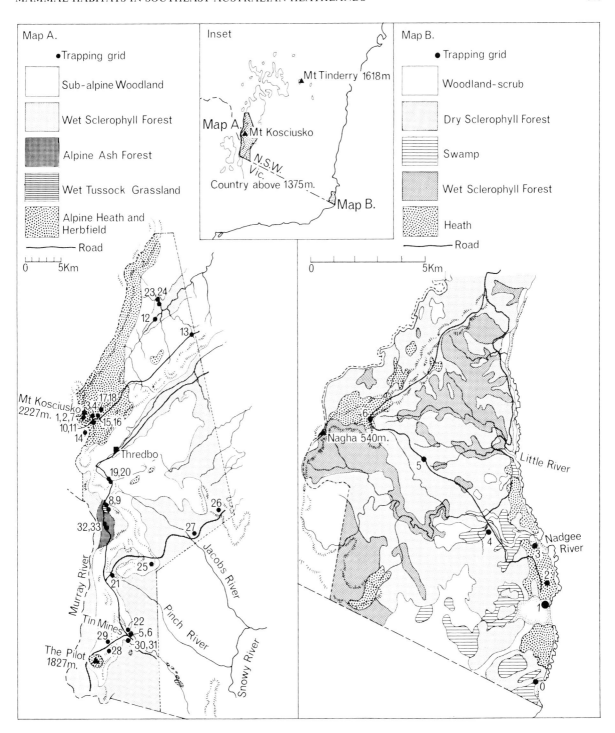

Fig. 10.1. Vegetation maps of the study areas.

to subalpine woodlands of snowgum, *E. pauciflora* spp. *niphophila*. Above are extensive alpine herb fields and heathlands. Valleys in these wooded areas are open and grassy with no trees due to pooling of cold air and water.

There are three major kinds of heathlands: those along streams and bogs, those in the shelter of rocky outcrops and boulders, and short heathlands on exposed, well-drained slopes (Wimbush and Costin, 1973). The shrubs along the water drainages are mainly *Epacris*, *Richea* and *Baeckea*. *Kunzea* and *Bossiaea* grow on well-drained slopes. Dense tussock grasslands (*Poa* spp.) grow across the lower valleys and up the slopes to the lower tree line. In the forests and woodlands, there is often a dense understorey of shrubs, mostly of *Pultenaea*, *Bossiaea* and *Daviesia*. The species of shrubs on particular trapping grids are listed in Table 10.2 below. Soils are granitic and usually shallow on the mountains, ranges and slopes, but are wet and peatty in the valleys, where bogs and fens are common.

Mount Tinderry

No climatic records exist for this outlier (1618 m). Rainfall is lower and temperatures higher than in regions of similar altitude in Kosciusko because there are no other high mountains. Snow falls infrequently and remains on the ground only a few days. Sclerophyll forest and woodland predominate, but there are heathlands on the exposed, higher slopes.

METHODS

Small mammals

Seven major habitats were trapped at Nadgee Nature Reserve between April 1972 and August 1976 (Fig. 10.1, Table 10.1). Small mammals were caught in Elliott traps set on grids 0.09 ha in size. Traps were in two lines of ten traps each and set 7 m apart from one another. Each captive was sexed, weighed, toe-clipped, reproductively classified and released. Trapping was for three consecutive nights every month for the first three years, and every one to three months subsequently depending on population trends, which also dic-

tated whether more than one trap was set per station (Newsome et al., 1975).

Traps were set similarly in Kosciusko National Park, at 25 different sites in the major habitats (Fig. 10.1, Table 10.1). Six grids were trapped for five to nine consecutive nights in February and May 1975, and in February 1976, at times of the year small mammals tend to be most abundant. Thus, marsupial mice (*Antechinus*) breed in August, and young are independent by January or February (Woolley, 1966; Wood, 1970). The native rats (*Rattus*, *Mastacomys*) breed from spring to summer (J.H. Calaby, pers. comm., 1975; Newsome et al., 1975). Eleven extra grids were trapped for five to nine consecutive nights between February and April 1976 to generate a minimum of 120 trap nights per site. Some habitats had two grids each. Results from another eight grids in the alpine and subalpine regions were obtained from other original data of Mr H. Dimpel (H. Dimpel and J.H. Calaby, pers. comm., 1976).

There were seven sites at Mount Tinderry (see Fig. 10.1) which were trapped irregularly between May 1967 and October 1968, generating about fifty trap nights per site. All habitats trapped are briefly described in Tables 10.2 and 10.4 below, and also placed in Specht's classification of structural formations (Specht et al., 1974).

Large vertebrates

Tracks of large vertebrates were detected on plots of soil about 1 m wide dug up across thoroughfares like beaches and vehicle tracks (Fig. 10.1) (Newsome et al., 1975). At Nadgee there were 112 plots about 0.4 km apart, and in Kosciusko 105 plots about 0.8 km apart. Plots were tilled one day and read the next for three to five consecutive days depending on weather. Tracks on each plot were identified and the direction and number of every set of tracks noted. It was sometimes clear that more than one individual of a species had crossed, for instance, when tracks were of different sizes. The plots were not readable after heavy rain or snow, or, on beaches, after high tides and winds. In Kosciusko in winter most plots were frozen or under snow and therefore useless. Accordingly, a transect 2.5 km long was skied in the subalpine region, and transects from 3 to 10 km long walked in other habitats, noting sign (tracks scratchings, or fresh dung) every 30 m approximately. Each interval was regarded as a plot. Winter and summer plots are different and therefore cannot be compared; but changes in the relative frequences in the different habitats indicate differential usage by the animals in the two seasons. Sampling in summer was more representative.

RESULTS

Small mammals

Coastal habitats: Nadgee Nature Reserve
Before fire. Five species of small mammals were caught commonly in the study: the marsupials *Antechinus stuartii* and *A. swainsonii*, the native rats *Rattus fuscipes* and *R. lutreolus*, and the introduced *Mus musculus*. Their relative abundance in different habitats is presented in Table 10.1. Three other small mammals have been caught, one specimen of the pigmy possum (*Cercartetus nanus*) and the introduced rat (*Rattus rattus*) in our traps, and a marsupial mouse, (*Sminthopsis murina*) in another study (J. McIlroy, pers. comm., 1976).

The most noteworthy result was that the coastal (lowland) and upland heathlands were the least and most favoured habitats respectively. No native mammal was caught in the former, but all four species were caught in the latter (Table 10.1) at high and relatively even frequencies ranging from 5.0 to 8.9 per thousand trap nights, with a total of 28.5. The grid in the lower heathland was placed purposely in the middle to avoid edge effects from the swamp on the southern edge or from the backing sclerophyll forest. Posamentier (1975), however, trapped in a small gully in the northern edge of this moor and near the southern swamp, catching five *R. fuscipes* and 22 *R. lutreolus* in 1920 trap nights. Results from other habitats in our study were all similar, with one or sometimes two species predominant.

The captures indicated the preferences of different species for different habitats. For example, 65% of *Rattus lutreolus* were from the swamp, and 86% came from the swamp plus the moist upland heathland. The majority (76%) of *Antechinus swainsonii*, also moisture-loving, were from the wet upland heathland and the swamp as well. *R. fuscipes*, regarded as an animal of forests, was relatively evenly spread throughout habitats excepting the lower moor. *A. stuartii* is an animal of the forests, where 60% of them were caught. The upland heathland and upland sclerophyll forest were the most favoured habitats.

After fire. The wildfire of December 1972 devastated 95% of the Reserve, especially the heathland habitats. Populations of small mammals were also devastated (Fig. 10.2), most noticeably on the upland heathland where populations of native species were almost exterminated by the fire. All species survived in the habitats, however, though in scattered remnants. Species survived best in sclerophyll forest, especially *Antechinus stuartii*. Relative abundances of all native species in various habitats changed. Whereas 22% of captures were in sclerophyll forest before the fire, this proportion increased to 55% in the first year afterwards. Captures on the upland heathland fell from 42% of the total before the fire to 19% afterwards.

There was an invasion of the alien *Mus musculus* after the fire. Unseen beforehand, it first appeared on the upland heathland (Fig. 10.2). Though never abundant, *Mus* was the only species caught there the second year after the fire. *Mus* was also uncommon in the sclerophyll forest. It became quite abundant in the moist sandy habitats, however, more so than any other small mammal at any time during the study. At the end of the period of the results presented here, *Mus* appeared to be in decline as native species were increasing in abundance once more.

Montane to alpine habitats: Kosciusko National Park and Mount Tinderry
With the exception of Grid 27 (Table 10.2) there is no recent history of extensive fire in any habitats trapped by us in Kosciusko National Park (A. Jelinek, pers. comm., 1975), nor any botanical indication of it (M. Parris, pers. comm., 1976). Communities have developed largely undisturbed since the summer grazing of livestock ceased in the high country in 1966. Mount Tinderry is grazed by cattle.

At our Kosciusko sites, a total of 280 individuals of five native and one alien species of small mammals were caught in 5890 trap nights. There

TABLE 10.1

Distribution of small mammals in coastal habitats (Number of individuals caught, Nadgee Nature Reserve 1972–1976)

Habitats	Structural formations of vegetation (Specht et al., 1974)	Trapping grid	Antechinus stuartii	Antechinus swansonii	Rattus fuscipes	Rattus lutreolus	Rattus rattus[1]	Mus musculus	Totals	Trap nights	Predominant vegetation, etc.
Heathlands 1. Lowland	open graminoid heathland	2	0	0	0	0	0	21	21	2621	*Epacris impressa, Grevillea lanigera, Casuarina distyla; Stipa, Aristida, Themeda, Enneapogon* vegetation fire-dependent; shrubs scattered; soils very hard groundwater podzols
2. Upland	Tall open-heathland	6	18	31	20	30	0	16	115	3478	*Leptospermum attenuatum, Banksia serrata, Leptocarpus tenax, Lomandra* spp., *Poa australis* shrubs and sedges common; peaty moist soils
Coastal Thicket	Closed-scrub	3	16	3	26	6	0	68	119	2669	*Leptospermum laevigatum, Melaleuca armarallis* beach succession in dune; wet, sandy soils
		0	1	2	41	2	0	44	90	1200	*Melaleuca armarallis, Banksia marginata; Spinifex hirsutus* dune-beach ecotone; moist, sandy soils
Swamp	Closed graminoid heathland	1	5	7	29	94	1	124	260	3614	*Persoonia, Epacris, Banksia serrata; Lomandra, Poa poiformis* small swamp; graminoids abundant; black soils; fills after rains
Sclerophyll Forest 1. Lowland	open-forest with heathy understorey	4	33	3	15	1	0	1	53	2720	*Eucalyptus sieberi, E. muellerana; Hakea, Grevillea, Banksia, Leptospermum* mixed grasses; dry sclerophyll forest; coastal
2. Upland	open-forest	5	26	4	24	11	0	5	70	2484	*Eucalyptus sieberi, E. muellerana; Poa poiformis.* dry sclerophyll forest on coastal range; some rocky outcrops
Totals		7	99	50	155	144	1	279	728	18 786	
Relative abundance (%)			13.6	6.9	21.3	19.8	0.1	38.3			

[1] From nearby *Banksia* scrub.

were 79 *Antechinus stuartii*, 55 *A. swansonii*, 116 *Rattus fuscipes*, 6 *Mus* (the same species as in coastal habitats), plus 14 *Mastacomys fuscus*, and 10 mountain pygmy possums, *Burramys parvus*. The last two are rare species found mostly in subalpine and alpine regions (Calaby and Wimbush, 1964). The lowest altitude for *Mastacomys* was about 1280 m, and the lowest for *Burramys* was about 1480 m, with the highest at 2227 m, right on the top of Mount Kosciusko (Dimpel, 1976).

Habitat preferences of the small mammals were

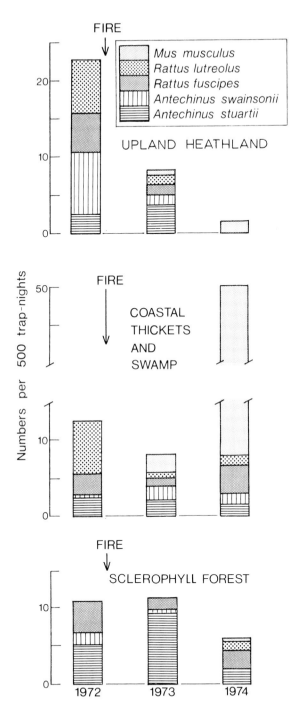

Fig. 10.2. Gross changes after fire in numbers and species-diversity of small mammals in habitats at Nadgee.

grids in the alpine and subalpine zones which included clusters of granite boulders or tors. Captures of *Mastacomys* were all along the watercourses. Most captures of *R. fuscipes* (75.9%) were in the forests and woodlands. The remainder came from localities with dense shrubby or grassy vegetation, twenty from along watercourses and eight from the moist hillside mentioned above. The alien *Mus* was rarely caught; four of the five captures were from altitudes below 1400 m, and three from along streams.

Capture rates increased with increasing complexity of the habitat. Thus, in the simply structured alpine herbfields and grasslands the capture rate was 1 per 100 trapnights, increasing to 3.3 in heathlands, 6.4 in woodlands and 7.5 in forests.

In an attempt to formalise these trends in the fauna with increasing complexity of the habitat, habitat-complexity scores were calculated for every trapping site. Four factors were considered: tree canopy, the cover and height of the understorey and the ground cover, the prevalence of boulders, fallen trees, debris etc., and moistness of the soil. Each factor was given a score from 0 to 3 (Table 10.3) and the total scores calculated for each site (Table 10.4). When the number of species caught was plotted against the total of these scores, a clear trend was shown (Fig. 10.3). Total density of small mammals (number of captures per 100 trap nights) did not show such a clear trend (Fig. 10.4). Results from other study areas are similarly presented in Figs. 10.3 and 10.4.

Large vertebrates

Nadgee Nature Reserve

Twelve kinds of tracks were recognised on the soil plots. They were caused by fifteen species of large vertebrates (Newsome et al., 1975): a large goanna (*Varanus varius*), lyrebird (*Menura novaehollandiae*), two bandicoots (*Isoodon obesulus*, *Perameles nasuta*), wombat (*Vombatus ursinus*), two possums (*Trichosurus vulpecula*, *Pseudocheirus peregrinus*), four macropodids [the potoroo (*Potorous tridactylus*), two large wallabies (*Wallabia bicolor*, *Macropus rufogriseus*) and the grey kangaroo (*Macropus giganteus*)], the dingo (*Canis familiaris dingo*), and the European rabbit (*Oryctolagus cuniculus*), fox (*Vulpes vulpes*) and cat (*Felis catus*). The tracks of the two bandicoots were

clear-cut. Amongst the marsupials, *A. stuartii* was caught only in forests and woodlands, most (87%) individuals of *A. swainsonii* were caught along creeks, and all of the *Burramys* possums came from

TABLE 10.2

Distribution of small mammals in mountainous habitats (Number of individuals caught, Kosciusko National Park 1975–1976, and on Mount Tinderry[1] 1967–1968)

Habitats	Structural formations of vegetation (Specht et al., 1974)	Trapping grids	Predominant vegetation	Altitude (m)	*Antechinus stuartii*	*Antechinus swainsonii*	*Burramys parvus*	*Rattus fuscipes*	*Mastacomys fuscus*	*Mus musculus*	Totals	No. trap nights	Comments
Alpine herbfields and grassland	open-herbland	1	*Poa* spp., *Celmisia longifolia*, *Epacris* spp; *Aciphylla glacialis*	2200	0	0	1	0	0	0	1	20	rock scree and feldmark; Mount Kosciusko
		7	*Poa* spp., *Celmisia longifolia*, *Epacris* spp.	2227	0	0	6	0	0	0	6	60	Huge boulders and feldmark; Mount Kosciusko
	closed-herbland	2	*Sphagnum cristatum*	2100	0	0	0	3	1	0	4	80	small watercourse
		3,4	*Celmisia longifolia/Poa* spp.	2000	0	0	0	0	0	0	0	240	exposed plain
	closed-grassland	5,6	*Poa* spp; *Helichrysum* sp. *Ranunculus* spp.	1240	0	0	0	0	0	0	0	160	cold valley floor
Heathlands	low open-heathland	8,9	*Hovea purpurea* var. *montana, Bossiaea foliosa* (dwarf)	1480	0	0	0	0	0	0	0	320	shaded valley slope
		10,11	*Phebalium ovatifolium*	2040	0	7	0	8	0	0	15	240	granite tors, moist
Heathy shrubs along streams	low open-heath/grassland	14	*Richea continentis; Chionochloa frigida*	2060	0	0	0	2	2	0	4	120	lake outflow, boulders
		15,16	*Prostanthera cuneata; Chionochloa frigida*	1920	0	10	0	5	2	0	17	240	tall grass on small watercourse
	low closed-heathland	13	*Richea continentis*	1975	0	3	0	10	2	0	15	80	creek and extensive adjacent bog
		17,18	*Kunzea ericifolia*	1750	0	6	0	0	1	0	7	240	creek, on steep bank
		19	*Epacris glacialis; E. microphylla; Richea continentis*	1600	0	4	0	0	0	0	4	160	creek flat and small fen
		20	*Richea continentis*	1600	0	2	0	0	0	0	2	160	creek and adjacent bog
	tall open-heathland	21	*Hakea microcarpa; Baeckea* spp; *Epacris* spp.	1280	0	4	0	0	0	1	5	440	creek and bog
	tall closed-heathland	22	*Hakea microcarpa; Baeckea* spp; *Epacris* spp.	1280	0	0	0	0	4	2	6	400	creek and bog
		34	*Baeckea utilis, Hakea microcarpa, Leptospermum lanigerum*	1370	0	0	0	1	0	0	1	57	swamp on creek; few *Eucalyptus stellulata*
Sclerophyll woodland with heathy understorey	woodland with heathland	26	*Eucalyptus dalrympleana; Pultenaea juniperina* var; *Leucopogon* sp.	1160	15	0	0	0	0	0	15	240	mountain gum woodland, few rotting logs
		27	*Eucalyptus rubida; Pultenaea juniperina* var., *Leptospermum myrtifolium*	920	0	0	0	0	0	0	0	240	candle-bark woodland; recent fire

TABLE 10.2 *(continued)*

Habitats	Structural formations of vegetation (Specht et al., 1974)	Trapping grids	Predominant vegetation	Altitude (m)	*Antechinus stuartii*	*Antechinus swainsonii*	*Burramys parvus*	*Rattus fuscipes*	*Mastacomys fuscus*	*Mus musculus*	Totals	No. trap nights	Comments
		30, 31	*Eucalyptus dalrympleana; Bossiaea foliosa* (dwarf); *Leucopogon* sp.	1260	5	0	0	0	0	0	5	160	mountain gum woodland
	low open-woodland with heathy understorey	12	*Eucalyptus pauciflora* subsp. *niphophila; Leptospermum obovatum, Baeckea gunniana*	1480	1	7	1	6	0	0	15	150	snowgum woodland; creek nearby; boulders
		23	*Eucalyptus pauciflora* subsp.; *Bossiaea foliosa* (tall)	1760	0	11	1	7	2	1	22	240	snowgum woodland; shrubs; creek; boulders
		24	*Eucalyptus pauciflora* subsp.; *Bossiaea foliosa* (tall)	1520	0	0	1	20	0	0	21	170	snowgum; shrubs; creek nearby; boulders
		25	*Eucalyptus pauciflora* subsp.; *Bossiaea foliosa* (tall)	1640	3	0	0	6	0	0	9	160	snowgum woodland
		37	*Eucalyptus dives, E. bridgesiana; Cassinia longifolia; Poa* spp.	1150	0	0	0	0	0	0	0	50	dry sclerophyll; bare stony slope
Sclerophyll forest with heathy understorey	open-forest	28	*Eucalyptus aatrympleana; Pultenaea juniperina* var.	1400	8	0	0	14	0	0	22	360	mountain gum; many rotting logs
		29	*Eucalyptus dalrympleana; Pultenaea juniperina* var.	1400	29	0	0	1	0	2	32	360	mountain gum; many rotting logs
		35	*Eucalyptus dalrympleana; Daviesia mimosoides*	1370	7	0	0	12	0	0	19	129	mountain gum; boulders; abundant litter
		36	*Eucalyptus dalrympleana; Daviesa mimosoides; Poa* spp.	1370	4	0	0	3	0	0	7	148	mountain gum
		38	*Eucalyptus dalrympleana; Daviesia mimosoides*	1370	0	0	0	1	0	0	1	50	mountain gum
		39	*Eucalyptus dalrympleana, Acacia dealbata; Pteridium esculentum*	1350	1	0	0	0	0	0	1	48	bracken gully in mountain gum
		32,33	*Eucalyptus delegatensis; Bossiaea foliosa* (tall)	1320	6	1	0	17	0	0	24	320	wet alpine ash; much litter and rotting logs; creek nearby
Sclerophyll forest over grassland		40	*Eucalyptus rubida; Poa* spp.	1150	0	0	0	0	0	0	0	48	candle-bark forest
Totals		40		920–2227	79	55	10	116	14	6	280	5890	

[1] Trapping grids Nos. 34–40 (Mount Tinderry) not mapped in Fig. 10.1.

TABLE 10.3

Visual method for scoring complexity of the habitat at trap-sites

Complex	Score			
	0	1	2	3
Tree canopy (%)	0	<30	30–70	>70
Shrub canopy (%)	0	<30	30–70	>70
Ground herbage	sparse <0.5 m	sparse >0.5 m	dense <0.5 m	dense >0.5 m
Logs, rocks, debris, etc. (ground-cover %)	0	<30	30–70	>70
Moisture in the soil	dry	moist	perma- nent water adjacent	water- logged

TABLE 10.4

Habitat-complexity scores for trapping grids

	Nadgee Nature Reserve	Kosciusko National Park	Mount Tinderry
Totals			
2	2	5,6	
3		3,4	
4		1,8,9	37
5	5,0	2,19,27	40
6	3	7,10,11,17,18,30,31	38
7	1,6,4	21,22,25,26	
8		13,15,16,20,29	36
9		14	35,39
10		12,23,24,28	34
11			
12		32,33	

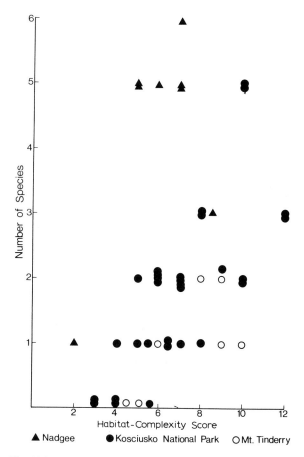

Fig. 10.3. Number of species of small mammals and habitat diversity at each trapping grid. The identity of each grid is obtainable from Table 10.4, the habitats from Tables 10.1 and 10.2, and the method of scoring habitats from Table 10.3.

indistinguishable from each other as were those of the two possums. In general, the tracks of the two wallabies and of small kangaroos could not be distinguished from each other, though tracks of large kangaroos were clearly separable. Since trends in numbers of tracks of the latter resembled those for the smaller macropodids, their results have been lumped. All results for the years 1972 to 1976 are presented in Table 10.5, sorted on the major habitats.

Few species were detected commonly. Except for the rabbit and dingo on the beaches, tracks were most abundant in the sclerophyll forests and coastal thickets. Wallabies were most abundant in the latter habitat (which was too narrow to map in Fig. 10.1). Rabbits were also prevalent there, and so was the fox. Tracks of lyrebird and potoroo (all found coastally) were rare and were found near moist overgrown gullies; the potoroo was also near swampy overgrown areas including patches in the coastal heathlands (Table 10.5). The wombat mostly favoured the upland schlerophyll forest.

Tracks of all kinds were least common on the heathlands, both coastal and on the range, though the macropodids, the dingo and the fox were moderately common there. The records for the dingo on the elevated heathland were mostly the result of pups being whelped there in 1974.

The effects of the bushfire have been reported for

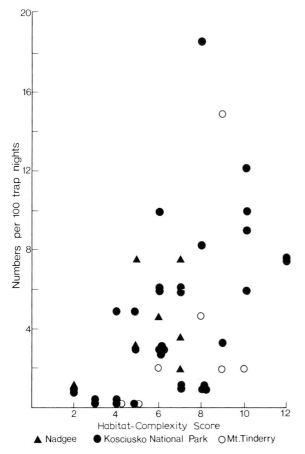

Fig. 10.4. Capture rate of small mammals and diversity of habitat at each trapping grid. Grid identity as for Fig. 10.3.

▲ Nadgee   ● Kosciusko National Park   ○ Mt.Tinderry

dicoots and the wombat on heathlands, but increases in the wooded habitats. Overall increases were recorded for the other burrower, the rabbit; it was mostly confined to the low dense scrub on dunes behind the beaches before the fire, but afterwards it was recorded on adjacent plots with the scrub all burnt. The other herbivores, the wallabies and kangaroos, declined in all habitats, the decline probably being aided by selective predation by dingoes. The introduced predators, fox and cat, provide a contrast with the dingo, as they declined in general, probably on account of lack of small mammals (Newsome et al., 1975).

Kosciusko National Park

Tracks of thirteen species were recorded: lyrebird (*Menura novaehollandiae*), emu (*Dromaius novaehollandiae*), wombat (*Vombatus ursinus*), two possums (*Trichosurus vulpecula*, *Pseudocheirus peregrinus*), the same three macropodids (*Wallabia bicolor*, *Macropus rufogriseus* and *Macropus giganteus*) and dingo (*Canis familiaris dingo*) and the introduced hare (*Lepus capensis*), rabbit (*Oryctolagus cuniculus*), horse (*Equus caballus*), fox (*Vulpes vulpes*) and cat (*Felis catus*). Records of their tracks for various habitats are presented in Table 10.7. It was clear from the groups of tracks that emus, dingoes, horses and foxes sometimes moved long distances, especially dingoes and horses. Hence, estimates within and between habitats (Table 10.7) were not always independent. It is however assumed that results in a given season can be compared between species.

Lyrebirds, possums, macropodids and cats were not found in the subalpine zones, only at lower altitudes. In contrast, the hare was found only in the subalpine zone, near open grassy areas. The most favoured habitat for the fauna was the ecotone between the grassy valley and mountain gum woodlands. The rabbit, horse and dingo were commonest there, the fox and wombat also being prevalent. The dingo was uncommon in the alpine ash complex where the fox was commonest. Both dingoes and foxes were common at the low altitudes (900 m) of the dry sclerophyll forest. Kangaroos and wallabies were most abundant there, and wombats and rabbits were common. Emus moved into the subalpine regions in summer only and were often accompanied by chicks about quarter-grown.

the Reserve as a whole (Newsome et al., 1975). Briefly, tracks of possums and bandicoots first declined but then increased a year after the fire. Tracks of goannas, wallabies, kangaroo, fox and cat continued to decline throughout that first year. Only the wombat and rabbit, both burrowers, increased. Of the predators, only the dingo held its own, probably because it altered its diet from small to large mammals after the fire. The macropodids were slowest to recover. As mentioned above, these changes were reported for Nadgee as a whole. Results for different habitats are presented (Table 10.6) for species with significant fluctuations. There was a contrast between data from the open habitats (beaches and heathlands) on the one hand, and those from the wooded habitats on the other. Declines were recorded for the ban-

TABLE 10.5

Distribution of large vertebrates in coastal habitats (Nadgee Nature Reserve, 1972–1976)

Habitats[1]	Plots with tracks (%)											Total plot-nights
	Goanna	Lyrebird	Two bandicoots	Wombat	Two possums	Potoroo	Two wallabies and kangaroos	Dingo	Rabbit	Fox	Cat	
Heathlands												
1. Lowland	0.4	0	1.8	0.2	0.5	0.2	5.8	25.9	8.0	18.9	0.2	1937
2. Upland	0.2	0	2.6	1.2	1.4	0	14.5	18.2	1.1	16.2	2.2	585
Beach	1.2	0	0.3	0.2	1.2	0	2.7	39.5	60.6	20.9	0.1	1496
Coastal thickets	1.6	0.1	4.1	1.6	2.6	0.3	34.3	9.2	19.4	40.0	6.3	772
Sclerophyll forests												
1. Lowland	2.1	1.3	4.9	6.7	6.1	0.4	16.2	13.3	3.0	31.3	11.3	3204
2. Upland	1.3	2.8	5.6	13.3	3.0	0	16.9	18.5	0.4	32.0	7.8	1806

[1] Structural formations of vegetation as in Table 10.1.

TABLE 10.6

Changes in abundance of tracks (%) of large mammals in various coastal habitats due to the fire (Nadgee Nature Reserve)

Habitats	Plots with tracks (%)															
	bandicoots		wombats		macropodids		dingo		rabbit		fox		cat		plot-nights	
	1972	1973	1972	1973	1972	1973	1972	1973	1972	1973	1972	1973	1972	1973	1972	1973
Heathlands	4.2***	1.2	0	0.4	11.7***	5.8	18.6**	24.6	2.5**	6.0	19.0***	7.3	1.2	0.6	521	949
Beach	0.5	0.5	0.5	0.3	7.4***	2.0	40.6	34.2	68.6**	58.5	39.6***	16.3	0.5	0	217	646
Forests and thickets	1.8**	3.6	5.0***	10.2	23.0***	17.4	12.5***	17.5	1.8***	3.8	34.4***	24.6	12.0***	5.3	1214	2182

** $P < 0.01$; *** $P < 0.001$.

TABLE 10.7

Distribution of large vertebrates in mountainous habitats (Kosciusko National Park 1975–1976)

	Plots with tracks (%) (whole numbers)									
	Sclerophyll woodlands and forests						Open grassy, heathy valleys and woodland ecotones			
Habitats	dry montane		open to dense montane		dense wet montane		montane		subalpine (deep snow in winter)	
Altitudes (m)	800–1100		1200–1600		1400–1600		1200–1300		1500–1800	
Season	summer	winter	summer	winter	summer	winter	summer	winter	summer	winter
Lyrebird	0	–	1	0*	0	0	0	0	0	0
Emu	4	–	1	0	2	0	1	0	4	0
Wombat	13	–	11	64	21	52	16	62	11	16
Possums	1	–	3	2*	1	0	1	0	0	0
Kangaroos and wallabies	18	–	1	0	2	0	1	4	0	0
Dingo	20	–	13	21	2	15	26	52	9	0
Hare	0	–	0	0	0	0	0	0	7 }	24
Rabbit	40	–	18	24	11	0	44	42	24 }	
Horse	0	–	3	30	2	60	9	75	1	0
Fox	51	–	38	34	55	12	24	14	34	22
Cat	11	–	3	0*	4	0	3	0	0	0
Total plot nights	205	0	643	238	242	158	339	302	102	100
Structural formation of vegetation (Specht et al., 1974)	woodland with heathy understorey		woodland to open-forest or dense open-forest		open-forest with heathy understorey		woodland with heathy understorey and low open graminoid heathland		low open-woodland with heathy understorey and low open graminoid heathland	
Predominant vegetation	*Eucalyptus dalrympleana*; *E. rubida, E. dives; Pultenaea juniperina* var.; *Poa* spp. *Themeda australis*		*Eucalyptus delegatensis*; *E. dalrympleana*; *E. pauciflora* subsp.; *Pultenaea juniperina* var.; *Daviesia latifolia; Poa* spp.; *Bossiaea foliosa*		*Eucalyptus delegatensis*; *Bossiaea foliosa*; *Daviesia latifolia*		*Eucalyptus dalrympleana*; *Bossiaea foliosa*; *Daviesia ulicifolia; Poa* spp.		*Eucalyptus pauciflora* subsp. *niphophila*; *Richea continentis*; *Epacris glacialis*; *Poa* spp.	

* 95 Plot-night only.

N.B. Plots in summer and winter were different (see text).

DISCUSSION

Pyric succession

The wildfire at Nadgee Nature Reserve in summer 1972 was one of the fiercest for decades (Newsome et al., 1975). The destruction of habitat seemed absolute except in moist gullies. Populations of small mammals collapsed in all habitats but species survived best in the *Eucalyptus* sclerophyll forests (=eucalypt open-forest with heathy understorey). Most of the charred and leafless trees sprouted adventitiously during the first year after the fire, and a great tangle of undergrowth also grew. The heathlands were bared to the soil; but they too have regrown remarkably so that now they would appear to the casual observer never to have been burnt. Since the small mammals have not recovered as quickly as the vegetation, there may be hysteresis in the relationship between numbers and diversity of the fauna and the habitat score.

Populations of native small mammals, though resurging now, do not resemble their state five years ago before the fire. Of course they may be expected to have a long pyric succession just as for the vegetation. Specht et al. (1958) studied plants of various ages since the last fire, in a dry-heathland in South Australia at about the same latitude as Nadgee, and were able to reconstruct a succession covering at least fifty years. That the succession for the rodents may be short is indicated by the rapidly declining population of the invader, *Mus musculus*, and the recolonization of habitats by native species. *Antechinus stuartii* reappeared after an absence of three years, but *A. swainsonii* has not been caught yet.

Except for the burrowers and the dingo, the response of the larger fauna to the fire was firstly to decline in abundance and then to increase. Indeed, signs of some species are now more abundant than before the fire, for instance the bandicoots. The refuge for these animals, as for the small ones, was the eucalypt sclerophyll forest. No large vertebrate was characteristic of the heathlands either before or after the fire. The birds are the best studied fauna in Australia including the regions discussed here. Four species of birds are confined to the coastal heathland at Nadgee, the ground parrot (*Pezoporus wallicus*), striated field wren (*Calamanthus fuliginosus*), emu wren (*Stipiturus malachurus*), and tawny-crowned honey-eater (*Phylidonyris melanops*) (M. Ridpath, in Fox, 1970). The ground parrot is seen more commonly now than before the fire despite the bushiness of the vegetation.

It seems probable that wildfire may have reset the sere for both plants and animals, i.e. that the entire biota may be adapted to frequent fires[1]. It is curious, however, that no example of the New Holland mouse (*Pseudomys novaehollandiae*), which is dependent on pyric succession in heathlands (Keith and Calaby, 1968; Posamentier and Recher, 1974), has been caught at Nadgee. This omission is perhaps even more surprising considering that one habitat, the coastal heathland, seemed devoid of mammals except at the edges. Competitive interaction with *Mus* may be a factor as has been suggested for other *Pseudomyd* species in central Australia (Newsome and Corbett, 1975).

Habitat preferences

Over a period of four years, changes in the populations of small and large mammals were studied, by trapping and tracking, in a series of sclerophyllous communities (with or without trees) located in the following regions:

Nadgee Nature Reserve (a lowland area)	7 trapping grids
	112 tracking plots
Kosciusko National Park (a montane to alpine area)	33 trapping grids
	105 tracking plots
Mount Tinderry (a high montane outlier)	7 trapping grids

Two general conclusions may be derived from these observations.

(1) The first conclusion is the remarkable fact that many of the same species of mammals, large and small, occupy the habitats from the coast to the snow line in southeastern Australia. At higher altitudes, *Rattus lutreolus* disappears in moist habitats and *Mastacomys fuscus* appears, and the mountain pigmy possum (*Burramys parvus*; a recent discovery, Seebeck, 1967), is added to the fauna. This problem is central to understanding the ecology of mammals, and possibly other animals, in Australia. The droughty climate and frequency of fires have been suggested as contributing to the low species diversity of mammals in Australia (Newsome et al., 1975; Recher et al., 1975). Paine's (1966) hypothesis that low rates of predation may promote competitive displacement of potential prey species may be relevant in that generalists rather than specialists are promoted.

Populations of small mammals were greater for a given habitat-complexity score near the coast than in the montane to alpine environments, probably because of the milder climate there. Mount Tindery was the least favourable area probably because it is drier than at Kosciusko National Park, and cattle graze there. Populations declined greatly at Nadgee after the fire, most notably in the heathlands, but in other habitats also as habitat diversity declined.

---

[1] In today's Australia, however, much of the forests have been cleared or are managed to prevent wildfire. The ultimate effects on the dependent fauna of being confined to archipelagos of bushland is unknown.

For example, the habitat-complexity scores for the coastal heathland declined from 2 to 0, and that for the upland heathland from 7 to 1.

(2) The second conclusion is that the vertebrate fauna was associated most with the forested environments, often with heathy understorey, or with their grassland ecotones. The short open herb fields and grasslands, and the sparse heathlands, were the least favourable habitats. Dense, tall stands of heathy shrubs, especially those on the coastal uplands, supported intermediate populations. Before destruction in the fire, this last habitat was one of the most favourable for small mammals. Though open, its physical and botanical structures were complicated by the presence of rocky outcrops and seepages.

The simple grouping of environments into "habitats" seemed more satisfactory in predicting species diversity than when the structural formations as distinguished in the vegetation by Specht et al. (1974; see also Chapter 1) were used. The reason was probably that other elements of the environment — creeks, swamps, and large boulders, etc. — were considered as part of the "habitat". We wish to make another point about the use of structural formations as reference points for mammalian inhabitants. We were struck with the fact that the heathy shrubs in our study-areas grew tallest, most densely and in broadest expanse in the woodlands and forests. Structural formations (using the upper stratum only) allowed for several differences if the upper canopy were shrubs but not if the shrubs were an understorey. This approach may consequently be deficient as a tool for understanding of the biota as a whole, because the small mammals were at their best in shrubby forested habitats.

It would be expected that destruction of the understorey through 'control' burning of forests, or clear-felling and strip-felling the forests, should menace the fauna. Such was the case for the marsupial greater glider (*Schoinobates volans*) studied by Tyndale-Biscoe and Smith (1969) and for the mountain possum (*Trichosurus caninus*) studied by How (1972). The birds which feed on nectar in forest flowers decline similarly (Gravatt, 1974).

There is a way in which partial clearing of the dry sclerophyll forest can aid some mammals, however, through the creation of edge effects at the newly created forest/grassland ecotones. Examples recorded are for the grey kangaroo (*Macropus giganteus*), and whiptail wallaby (*M. parryi*) (Calaby, 1966), and the wombat (*Vombatus ursinus* (McIlroy, 1973). It may be noted that the forest/grassland ecotone was most favoured by the wildlife at Kosciusko. The principle is the same as for the two desert kangaroos (*M. robustus*, *M. rufus*) whose food supplies were increased as a result of ruminants grazing their natural habitats (Newsome, 1962, 1975).

The overall conclusion of this study is that the most favourable habitats for mammals (probably also birds) in southeastern Australia between the coast and range are the sclerophyll (heathy) woodlands and forests with dense heathy understoreys. Dense, tall stands of heathy shrubs by themselves are less favourable. All other habitats, though occupied variously and sometimes well, may be ancillary. The exceptions appear in the subalpine valleys and the alpine zone where there are no trees. There the stream frontages harbour the distinctive *Mastacomys*, and the tor piles, *Burramys*.

ACKNOWLEDGEMENTS

There are many people to thank: Mrs. M. Parris particularly, Mr. L. Adams and Mr. A. Fox on botany; Mr. H. Dimpel and Mr. J. Calaby for unpublished data; Mrs M. Parris and Mr. J. Calaby for reading the manuscript; Messrs L. Corbett, R. Burt, P. Hanisch, J. Lemon and Ms. R. Perry for technical assistance; Mr. F. Knight for illustrations; New South Wales National Parks and Wildlife Service for assistance and permission to work in their Parks; and the Australian Meat Board for finance, project CS8S, on dingoes.

REFERENCES

Calaby, J.H., 1966. Mammals of the Upper Richmond and Clarence Rivers, New South Wales. *C.S.I.R.O. Div. Wildl. Res. Tech. Pap.*, No. 10: 55 pp.

Calaby, J.H. and Wimbush, D.J., 1964. Observations on the broad-toothed rat, *Mastacomys fuscus* (Thomas). *C.S.I.R.O. Wildl. Res.*, 9: 123–133.

Costin, A.B., 1954. *A Study of the Ecosystems of the Monaro Region of New South Wales*. A.H. Pettifer Government Printer, Sydney, N.S.W., 860 pp.

Dimpel, H., 1976. Research on some of the endangered marsupials. *Burramys* on Mount Kosciusko. *Parks Wildl.*, 1(5): 156–158.

Ealey, E.H.M., 1967. Ecology of the euro, *Macropus robustus* (Gould), in north-western Australia. I. The environment and changes in euro and sheep populations. *C.S.I.R.O. Wildl. Res.*, 12: 9–25.

Fox, A., 1970. *Development plan, 1970–75: Nature Reserve No. 6 — Nadgee*. N.S.W. National Parks and Wildlife Service Report, 50 pp.

Gravatt, D., 1974. Birds that eat plant products in coastal South Queensland. *Proc. 7th Triennial Conf., Institute of Foresters of Australia, Caloundra*, 1: 339–348.

How, R.A., 1972. *The Ecology and Management of Trichosurus spp. Populations in Northern New South Wales*. Thesis, University of New England, Armidale, N.S.W., 212 pp.

Keith, K. and Calaby, J.H., 1968. The New Holland mouse, *Pseudomys novaehollandiae* (Waterhouse), in the Port Stephens District, N.S.W., *C.S.I.R.O. Wildl. Res.*, 13: 45–58.

McIlroy, J., 1973. *Aspects of the Ecology of the Common Wombat, Vombatus ursinus (Shaw 1800)*. Thesis, Australian National University, Canberra, A.C.T., 285 pp.

Newsome, A.E., 1962. *The Biology of the Red Kangaroo, Macropus rufus (Desmarest) in Central Australia*. Thesis, University of Adelaide, Adelaide, S.A., 111 pp.

Newsome, A.E., 1973. The adequacy and limitations of flora conservation for fauna conservation in Australia and New Zealand. In: A.B. Costin and R.H. Groves (Editors), *Nature Conservation in the Pacific*. ANU Press, Canberra, A.C.T., pp. 93–110.

Newsome, A.E. 1975. An ecological comparison of the two arid-zone kangaroos of Australia, and their anomalous prosperity since the introduction of ruminant stock to their environment. *Q. Rev. Biol.*, 50: 389–424.

Newsome, A.E., and Corbett, L.K., 1975. Outbreaks of rodents in semi-arid and arid Australia: cause, preventions, and evolutionary considerations. In: I. Prakash and P.K. Ghosh (Editors), *Rodents in Desert Environments, I*. Junk, The Hague, pp. 117–153.

Newsome, A.E., McIlroy, J. and Catling, P.C., 1975. The effects of an extensive wildfire on populations of twenty ground vertebrates in south-east Australia. *Proc. Ecol. Soc. Aust.*, 9: 107–123.

Paine, R.T., 1966. Food web complexity and species diversity. *Am. Nat.*, 100: 65–75.

Posamentier, H., 1975. *Habitat Requirements of Small Mammals in Coastal Heathlands of New South Wales*. Thesis, University of Sydney, Sydney, N.S.W., 133 pp.

Posamentier, H. and Recher, H.F., 1974. The status of *Pseudomys novaehollandiae* (the New Holland Mouse). *Aust. Zool.*, 18(2): 66–71.

Recher, H., Lunny, D. and Posamentier, H., 1975. A grand natural experiment. The Nadgee wildfire. *Aust. Nat. Hist.*, 18: 153–163.

Scobie, P. and Fox, A., 1972. *A Map of the Vegetation of Nadgee Nature Reserve, N.S.W.* National Parks and Wildlife Service, Sydney, N.S.W.

Seebeck, J., 1967. Rediscovery of two 'extinct' marsupials. *Animals*, 10: 271–272.

Specht, R.L., Rayson, P. and Jackman, M.E., 1958. Dark Island heath (Ninety-Mile Plain, South Australia). VI. Pyric succession: changes in composition, coverage, dry weight, and mineral nutrient status. *Aust. J. Bot.*, 6: 59–88.

Specht, R.L., Roe, E.M. and Boughton, V.H., 1974. Conservation of major plant communities in Australia and Papua New Guinea. *Aust. J. Bot., Suppl. Ser.*, No. 7: 667 pp.

Tyndale-Biscoe, C.H. and Calaby, J.H., 1975. Eucalypt forests as refuge for wildlife. *Aust. For.*, 38: 117–133.

Tyndale-Biscoe, C.H. and Smith, R.F.C., 1969. Studies on the marsupial glider, *Schoinobates volans* (Kerr). III. Response to habitat destruction. *J. Anim. Ecol.*, 38: 651–659.

Williams, O.B., 1955. Studies on the ecology of the riverine plain. I. The gilgai microrelief and associated flora. *Aust. J. Bot.*, 3: 99–112.

Wimbush, D.J. and Costin, A.B., 1973. Vegetation mapping in relation to ecological interpretation and management in the Kosciusko alpine area. *C.S.I.R.O. Div. Plant Ind., Tech. Pap.*, No. 32: 22 pp.

Wood, D.H., 1970. An ecological study of *Antechinus stuartii* (Marsupialia) in a south-east Queensland rainforest. *Aust. J. Zool.*, 18: 185–207.

Woolley, P., 1966. Reproduction in *Antechinus* spp. and other dasyurid marsupials. *Symp. Zool. Soc. Lond.*, 15: 281–294.

Chapter 11

"MAQUIS" VEGETATION OF NEW CALEDONIA[1]

R.L. SPECHT (Compiler)

GENERAL

New Caledonia is a mountainous, mineral-rich, and geologically complex island. The main island, la Grande-Terre, is about 400 km long by an average of 50 km wide. Oriented along a NNW–SSE axis, it lies between 20°0' and 22°25'S latitude, and between 163°56' and 167°03'E longitude.

The island has been greatly disturbed geologically with many, possibly Cretaceous or Eocene, metamorphics. It has apparently been very unstable with much mountain building, peneplantation, lateritization, and vertical oscillation until relatively recent times. No volcanic activity has been recorded since the Oligocene. Extensive areas of serpentines and peridotites bear a rich, endemic flora of sparse, sclerophyllous scrubby vegetation, termed "marquis" (Thorne, 1965)[2]. Areas of closed-forest (containing *Nothofagus*, *Agathis*, *Araucaria*, etc.) and savannah woodland (dominated by the paperbark, *Melaleuca quinquenervia*) are found on more fertile soils. Small pockets of mangrove and coastal dune vegetation occur around the coast.

[1] Manuscript completed July, 1977.

[2] This sclerophyllous vegetation, found on oligotrophic soils, should not be equated with Mediterranean maquis but with heathland and shrubland communities, as defined in Chapter 1 of this volume.

NEW CALEDONIA

Diagrammatic cross-section (Le Borgne, 1964):

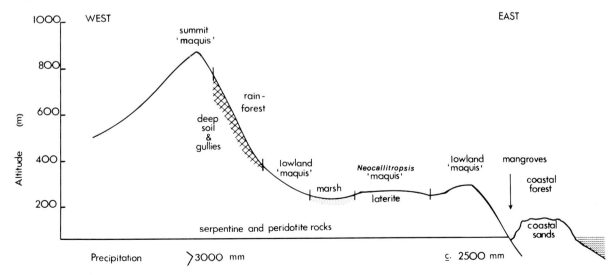

Fig. 11.1. Diagrammatic cross-section, New Caledonia.

NEW CALEDONIA *(continued)*

Climate: Af (Köppen, 1923)

Climatic data[1]	Summer (Dec–Feb)	Autumn (March–May)	Winter (June–Aug)	Spring (Sept–Nov)	Year (Jan–Dec)
Air temperature (°C)					
Mean daily maximum	29.8	27.9	24.6	26.9	27.3
Mean daily minimum	22.0	20.7	16.9	18.5	19.5
Precipitation (mm)	290	389	251	175	1105
Precipitation (days)	28	44	38	22	132
Relative humidity (%)					
— at 0900 hours	72	76	75	69	73
— at 1500 hours	70	73	69	67	70
Cyclones (per 100 years	Dec–Apr (most in March)		0	0	*c.* 30

[1] Climate station: Nouméa (22°16′S, 166°27′E, 10 m).

The climate at Montagne des Sources, the highest elevation (1050 m) near Nouméa, is cooler (*c.* 13°C mean annual temperature) and wetter (more than 3000 mm precipitation per annum).

Dominant soils: Extensive areas of lateritised red soils developed on serpentine and peridotite rocks — often considerably eroded. The shallow soils developed on the exposed rock are poor in essential plant nutrients, especially phosphorus and potassium.

Species composition: "maquis" and allied sclerophyllous, scrubby vegetation (Guillaumin et al., 1965; Thorné, 1965)

Plant communities: "maquis" (m); thicket (th); exposed slopes, plateaux and summits (s)

FERNS

Lycopodiaceae   *Lycopodium cernuum* (s), *L. clavatum* (s), *L. deuterodensum* (s), *L. scariosum* (s), *L. squarrosum* (s)

Schizaeaceae   *Schizaea dichotoma* (s), *S. fistulosa* (s)

Gleicheniaceae   *Gleichenia dicarpa* (s), *Sticherus flabellatus* (s), *Stromatopteris moniliformis* (s)

Pteridaceae   *Tapeinidium flavicans* (s)

GYMNOSPERMS

Araucariaceae   *Araucaria montana* (s), *A. muelleri* (m)

Cupressaceae   *Neocallitropsis araucarioides* (m)

Podocarpaceae   *Dacrydium araucarioides* (m), *Podocarpus gnidioides* (s)

MONOCOTYLEDONS

Flagellariaceae   *Flagellaria indica* var. *minor* (m)

Liliaceae   *Astelia neo-caledonica* (s), *Xeronema moorei* (s)

Xanthorrhoeaceae   *Lomandra banksii* forma *neo-caledonica* (m)

DICOTYLEDONS

Annonaceae   *Xylopia pancheri* (m)

Apocynaceae   *Alstonia balansae* (m), *A. deplanchei* (m), *A. legouixiae* (s), *Alyxia caletioides* (m), *A. disphaerocarpa* (m), *A. doliolifera* (m), *A. leucogyne* (s), *Alyxia* sp. aff. *A. sarasinii* (m), *Melodinus* sp. (m), *Parsonsia carnea* (m), *Rauwolfia semperflorens* (m), *R. viridis* (m)

Araliaceae   *Delarbrea collina* (m), *Dyzygotheca elegantissima* (m), *Meryta coriacea* (s), *Myodocarpus fraxinifolius* (m), *M. involucratus* (s), *Tieghemopanax austrocaledonicus* (m), *T. calophyllus* (m), *T. dioicus* (m), *T. pulchellus* (m), *T. schlechteri* (m)

Asteraceae   *Helichrysum cinereum* (m)

Bombacaceae   *Maxwellia lepidota* (m)

Caesalpiniaceae   *Storkiella pancheri* (m)

Capparidaceae   *Oceanopapaver neo-caledonicum* (th)

Casuarinaceae   *Casuarina deplancheana* (m), *C. leucodon* (m), *C. poissoniana* (m)

Celastraceae   *Maytenus bureaviana* (m), *M. sebertiana* (th), *Peripterygia marginata* (m)

Clusiaceae   *Garcinia amplexicaulis* (m), *G. neglecta* (m), *Montrouziera rhodoneura* (m), *M. sphaeroidea* (s)

Cunoniaceae   *Codia albifrons* (s), *Cunonia lenormandii* (s), *C. pterophylla* (s), *Pancheria alaternoides* (m), *P. brunhesii* (s), *P. confusa* (s), *P. engleriana* (s), *P. robusta* (s)

DICOTYLEDONS *(continued)*

Dilleniaceae	*Hibbertia altigena* (m), *H. baudouinii* (s), *H. brongniartii* (m), *H. coriacea* (m), *H. deplancheana* (m), *H. patula* (s), *H. scabra* (s), *H. vieillardii* (m), *H. virotii* (s), *H. wagapii* (m)
Droseraceae	*Drosera neo-caledonica* (s)
Ebenaceae	*Maba rosea* (m)
Elaeocarpaceae	*Dubouzetia confusa* (m), *Sloanea montana* (m)
Epacridaceae	*Cyathopsis floribunda* (m), *Dracophyllum ramosum* (m), *D. verticillatum* (m), *Leucopogon albicans* (m), *L. cymbulae* (m), *L. dammarifolius* (s), *L. enervis* (m), *L. pancheri* (m), *L. violaceo-spicatus* (s)
Erythroxylaceae	*Erythroxylon novo-caledonicum* (th)
Euphorbiaceae	*Baloghia bureavii* (m), *B. drimiflora* (m), *Bocquillonia castaneaefolia* (m), *Bureavia carunculata* (m), *B. rubiginosa* (m), *Cleistanthus stipitatus* (th), *Croton insularis* (m), *Longetia buxoides* (m), *Phyllanthus chrysanthus* (m), *P. induratus* (m), *P. koumacensis* (m), *P. peltatus* (m)
Flacourtiaceae	*Casearia melistaurum* (m)
Goodeniaceae	*Scaevola beckii* (m), *S. cylindrica* (m), *S. montana* (s)
Icacinaceae	*Sarcanthidion sarmentosum* (s)
Loganiaceae	*Geniostoma balansaeanum* (s), *G. densiflorum* (m), *G. oleifolium* (m)
Loranthaceae	*Korthelsella dichotoma* (m)
Malphigiaceae	*Acridocarpus austro-caledonicus* (m)
Meliaceae	*Dysoxylum lessertianum* (m)
Mimosaceae	*Acacia spirorbis* (m), *Albizia callistemon* (th)
Myrsinaceae	*Rapanea lecardii* (m), *Tapeinosperma grandiflorum* (s)
Myrtaceae	*Baeckea ericoides* (m), *B. pinifolia* (th), *B. virgata* (s), *Callistemon buseanus* (s), *C. suberous* (s), *Caryophyllus* sp. cf. *C. multipetalus* (m), *Cloëzia angustifolia* (m), *C. artensis* (m), *C. canescens* (m), *Eugenia stricta* (m), *Melaleuca brongniartii* (m), *Metrosideros demonstrans* (s), *M. engleriana* (s), *Myrtus rufo-punctata* (s), *Piliocalyx baudouinii* (m), *Rhodamnia andromedoides* (m), *Syzygium pancheri* (m), *Tristania callobuxus* (m), *T. glauca* (m), *T. guillaumii* (m), *Xanthostemon aurantiacus* (m), *X. laurinus* (m), *X. longipes* (m), *X. macrophyllus* (m), *Xanthostemon* sp. aff. *X. myrtifolius* (m)
Nepenthaceae	*Nepenthes vieillardii* (s)
Oleaceae	*Jasminum artense* (m), *J. leratii* (m), *Olea paniculata* (m), *Osmanthus austro-caledonica* (m)
Oxalidaceae	*Oxalis neo-caledonica* (m)
Pittosporaceae	*Pittosporum gracile* (m), *P. koghiense* (s), *P. poumense* (m), *Pittosporum* sp. cf. *P. echinatum* (m)
Proteaceae	*Beauprea spathulaefolia* (s), *Beauprea* sp. (m), *Cenarrhenes paniculata* (s), *Grevillea exul* (th), *G. gillivrayi* (th), *G. meisneri* (th), *G. rhododesmia* (th), *G. rubiginosa* (s), *Stenocarpus intermedius* (th), *S. milnei* (m), *S. trinervis* (m), *S. umbelliferus* (s)
Rhamnaceae	*Alphitonia neo-caledonica* (m)
Rubiaceae	*Coelospermum monticolum* (m), *Ecremocarpus rupicolus* (m), *Gardenia oudiepe* (m), *G. urvillei* (m), *Morinda candollei* (m), *M. collina* (s), *M. deplanchei* (m), *M. glaucescens* (s), *M. neo-caledonica* (m), *Normandia neo-caledonica* (s), *Psychotria calorhamnus* (m), *P. collina* (m), *P. phyllanthoides* (m)
Rutaceae	*Acronychia laevis* (m), *Boronella pancheri* (m), *B. verticillata* (m), *Halfordia kendack* (m), *Myrtopsis deplanchei* (s), *M. novae-caledoniae* (m), *M. sellingii* (s), *Zieria chevalieri* (s)
Sapindaceae	*Dodonaea viscosa* (m), *Gongrodiscus parvifolius* (s), *Guioa crenulata* (m), *G. gracilis* (m), *G. villosa* (m), *Loxodiscus coriaceus* (s), *Podonephelium homei* (m), *Storthocalyx chryseus* (th)
Sapotaceae	*Madhuca* (?) (m), *Planchonella crebrifolia* (m), *P. sebertii* (m)
Saxifragaceae (Escalloniaceae)	*Argophyllum brevistylum* (m), *A. montanum* (m), *A. vernicosum* (m)
Simaroubaceae	*Soulamea cardioptera* (m), *S. cycloptera* (th), *S. pancheri* (m)
Solanaceae	*Duboisia myoporoides* (m)
Thymelaeaceae	*Microsemma salicifolia* (m), *M. thornei* (m), *Solmsia calophylla* (m), *Wikstroemia indica* (m)
Verbenaceae	*Vitex rapinii* (s)
Violaceae	*Agatea longipedicellata* (m), *A. pancheri* (m), *Hybanthus caledonicus* (m)

REFERENCES

Guillaumin, A., Thorne, R.F. and Virot, R., 1965. Vascular plants collected by R.F. Thorne in New Caledonia in 1959. *Univ. Iowa Stud. Nat. Hist.*, 20(7): 15–65.

Köppen, W., 1923. *Die Klimate der Erde.* Bornträger, Berlin, 369 pp.

Le Borgne, J., 1964. *Géographie de la Nouvelle-Calédonie et des Iles Loyauté.* Ministère de l'Éducation, Nouméa, Nouvelle-Calédonie, 308 pp.

Thorne, R.F., 1965. Floristic relationships of New Caledonia. *Univ. Iowa Stud. Nat. Hist.*, 20(7): 1–14.

Chapter 12

HEATHLANDS AND RELATED SHRUBLANDS OF MALESIA (WITH PARTICULAR REFERENCE TO BORNEO AND NEW GUINEA)[1]

R.L. SPECHT and J.S. WOMERSLEY[2] (compilers)

INTRODUCTION

In the introductory chapter to the *Flora Malesiana*, Van Steenis (1948) considered that the natural demarcation lines of the Malesian (Malaysian) flora extend from the Isthmus of Kra, between the Philippines and Taiwan, through Torres Strait, and include the Louisiades and the Bismarck Archipelago. Thus, Malesia includes the Malay Peninsula, Sumatra, Java, Timor, Borneo, the Philippines, Celebes, the Moluccas, the whole of New Guinea, and the islands of the Bismarck Archipelago. In the northwest, quite a number of typical Malesian genera do not extend northwards into the Indochinese Peninsula. The Philippines possess an essentially Malesian flora, in contrast to the Japano-Chinese flora found on Taiwan. Much of New Guinea is covered by tropical rain forest with strong Malesian affinities; in contrast, open-communities (in the drier monsoonal areas of the south coast, some inland valleys and in subalpine localities) include plants closely related to or identical with the Australian flora.

The Malesian region, as defined above, was actually composed of two distinct parts. During the Mesozoic, the region to the northwest of New Guinea was closely associated with the large Eurasian tectonic plate; New Guinea was part of the Australasian plate, 35° to 40° of latitude to the south (see Specht's Fig. 1.2, in Chapter 1). At least since the middle Cretaceous, Borneo (West Kalimantan, Sarawak and Sabah) and the Malay Peninsula have behaved as a unit, located on the Equator, associated with, but apparently independent of, the mainland of Asia (Haile et al., 1977). Some break apparently existed between the Malay Archipelago and Eurasia, as it was only

during the Tertiary that the Malay Archipelago rotated anticlockwise through *c.* 50°, thus bringing the Malay Peninsula into conjunction with Southeast Asia (Haile et al., 1977).

Over the last 70 million years, the Australasian plate, after breaking away from Antarctica, drifted northwards at an average rate of 66 ± 5 mm per year (Wellman and McDougall, 1974). By the late Miocene, the western end of New Guinea (on the Australasian plate) apparently came in contact with the Indonesian extremity of Eurasia (Smith and Briden, 1977). Considerable intermingling of the floras of the Malay Archipelago and Australasia was thus possible; heathland elements, among others, were able to migrate from Australasia into the Archipelago. However, contact between New Guinea and Malay Archipelago was short-lived; the northward drift of the Australian–New Guinea plate continued until today, thus severing the temporary continuity between the two land masses (Smith and Briden, 1977).

Until the late Miocene, the distance between the Malay Archipelago and the New Guinean section of Australasia was too great for intermingling of the floras. Many birds, evolving in Asia during the early Tertiary, established migration routes from northeastern Siberia and Japan which terminated in the northern Philippines or the Malay Peninsula; a few species penetrated into Borneo and Java. Only a very small number of species now migrate over Wallace's Line to winter in New Guinea or Australia (McClure, 1973). This restricted bird

---

[1] Manuscript completed July, 1977.
[2] Former Address: Department of Forests, Division of Botany, P.O. Box 314, Lae, Papua New Guinea.

migration in Southeast Asia contrasts markedly with that observed from North to South America and from Eurasia to Africa, where breeding birds migrate as far south as latitude 10°S.

During the Tertiary, Australia and New Guinea drifted northwards experiencing a latitudinal change of 35° since the middle Cretaceous (Wellman and McDougall, 1974). Similarly, the Indian tectonic plate broke away from Antarctica and drifted northward to impinge on Asia. The drift history has therefore involved the convergence of both Australasia and the Indian subcontinent on Southeast Asia. Intermingling of floras of the four areas — India; Southeast Asia; the Malay Archipelago; New Guinea associated with Australia — occurred, the extent depending on the dispersal mechanisms of the flora. The humid, tropical climate over most of Malesia would have favoured the flora of closed- (rain-) forest communities at the expense of the open-communities characteristic of the drier and more seasonal climates of Australia. The diversity of genera and species in closed-communities is highest in New Guinea and decreases as one progresses down the eastern Australian coast from tropical to subtropical to warm temperate regions. Conversely, the heathland elements prominent in southern Australia decline in diversity in subtropical and tropical Australia, and survive in only a few restricted habitats in New Guinea.

HEATHLAND ELEMENTS IN MALESIA

It is probable that the forerunners of Malesian heathland elements had already evolved on the supercontinent of Gondwanaland before it began to break up into tectonic plates. Two tectonic plates — the Indian plate, and the Australasian plate — would have carried heathland (and other) elements northward. After 50 to 100 million years of continental drift, these two tectonic plates eventually came again in relative close proximity with each other and with the southeast corner of the Eurasian plate (including the Malay Archipelago). Movement of wind- and bird-dispersed floral elements between the tectonic plates again became feasible. Land connexions between several of the plates (especially when their continental shelves were exposed during the

Pleistocene fluctuations in sea level) facilitated the mingling of floral elements even further.

In most parts of the world the two angiosperm families, Ericaceae and Epacridaceae, are regarded as indicative of the heathland communities, though by no means confined to these communities. Twelve genera of Ericaceae (*Agapetes, Andresia* syn. *Wirtgenia, Costera, Dimorphanthera, Diplycosia, Gaultheria, Lyonia, Monotropastrum, Pernettyopsis, Pyrola, Rhododendron,* and *Vaccinium*) and three genera of Epacridaceae (*Decatoca, Styphelia* — including *Cyathodes* and *Leucopogon,* and *Trochocarpa*) have been recorded in Malesia (Sleumer, 1964, 1966–67).

The genera of Ericaceae and Epacridaceae occupy sites in exposed parts of the interface between rain forest and forest, in tree crowns, as epiphytes, and in exposed areas brought about by windfall or death of old large trees or landslides. They may be regarded as pioneer plants which can occupy breaks in the rain-forest environment.

Quite a number of heathland species are found above the timberline, mainly as dwarf shrubs in subalpine grassland or even in true alpine heathland and shrubland, where they ascend to 4000 to 4400 m on Mount Kinabalu in Borneo, and in the Main Range of New Guinea (or even to 4700 m on Mount Carstensz and Mount Wilhelm). In the mountains, they are found mostly in open or rather open places, or again as epiphytes on the sunlit branches of the tallest trees of the rain forest. A few live at low altitudes on podzolized sands on coastal plains and lowland hills. Some species are even found on coastal sand dunes, or epiphytic on mangrove trees. Two genera (*Andresia* and *Monotropastrum*) lack chlorophyll and are saprophytic, growing in shady places on the forest floor of lower montane forests.

Almost all the species of Ericaceae and Epacridaceae are found on acid, oligotrophic, sandy or peaty soil; the amount of mineral nutrients available to epiphytic plants must be just as poor as that available to terrestrial plants on oligotrophic soil.

Heathland species should typically possess sclerophyllous leaves, usually relatively small in size (leptophyll to nanophyll); lignotubers, from which several stems arise, are often characteristic. In exposed habitats on oligotrophic soil, both ericaceous and epacridaceous species may possess these characteristics, but in more protected habitats leaves tend to be larger and coriaceous rather than sclerophyllous. The stem base and upper part of the roots, especially of epiphytic species of *Vaccinium* (and of the related genera *Agapetes*, *Costera*, and *Dimorphanthera*) are often swollen and superficially resemble a lignotuber. Many terrestrial species of these genera have a very short and perhaps swollen trunk from which the scandent branches, often numerous, arise. The transition of such species from terrestrial habitats in crevices of rocks, poor in plant nutrients, to their epiphytic position in the branches of upper-montane trees (or even vice versa) is not hard to visualize. In Papua-New Guinea, all species of *Rhododendron* known to grow as epiphytes also occur as terrestrial plants.

The probable three centres of development of the heathland species are obvious from the distribution maps of representative ericaceous and epacridaceous species in Malesia (Fig. 12.1 and 12.2). The statistics presented by Van Steenis (1949) on the distribution of endemic genera of phanerogams showed the same three centres of evolution.

Rhododendron

The genus *Rhododendron* (Fig. 12.1A) includes approximately 850 species, about 525 of which occur in eastern and southeastern Asia; about 25 species are found in North America and 9 species in Europe. Almost all of the 288 species found in Malesia belong to the lepidote Section *Vireya* of the genus, only five elepidote species from southeastern Asia extending into the Philippines, the Malay Peninsula and Sumatra.

The elepidote subgenera *Hymenanthes*, *Pentanthera*, *Tsutsutsi*, and *Azaleastrum* of the genus *Rhododendron* apparently developed in the southern part of Eurasia when sections of Gondwanaland collided with the northern continent. As the light, winged or tailed seed of *Rhododendron* spp. is easily carried by wind — and, in fact, the genus is a pioneer on disturbed sites — the genus may be rapidly dispersed. It is here suggested that the ancient *Rhododendron* flora of the Indian plate rapidly expanded into South-east Asia and extended northward towards Japan; only recently did it have the opportunity to invade the Malay Archipelago.

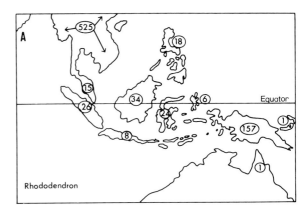

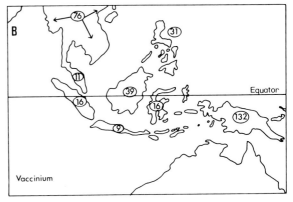

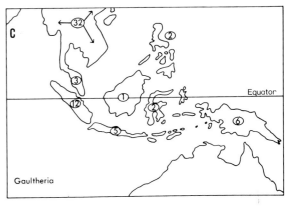

Fig. 12.1. Distribution of species of three representative ericaceous genera throughout Malesia: A, *Rhododendron*; B, *Vaccinium*; C, *Gaultheria*.

About half of the Malesian lepidote species of *Rhododendron* are found in New Guinea (one in Australia). The lepidote subgenus *Rhododendron* Section *Vireya* must have developed on the Australasian tectonic plate and then migrated into the Malay Archipelago when the two areas became contiguous in the late Miocene (see Introduction

324

above). Five subsections — *Pseudovireya, Albovireya, Solenovireya, Euvireya,* and *Phaeovireya* — are common to New Guinea and the islands to the west, while Subsection *Siphonovireya* is confined to New Guinea and Subsection *Malayovireya* to the western part of Malesia. In spite of easy wind-dispersal of *Rhododendron* seeds, the isolation of the montane habitats must have ensured independent evolution. Little migration of species occurred between Malesian islands and only a few species of Subsection *Pseudovireya* penetrated northward from Malesia into southeastern Asia — just as limited a migration northward as has occurred southward from the southeastern Asian elepidote species of *Rhododendron.*

Vaccinium

The genus *Vaccinium* (Fig. 12.1B) includes about 450 species distributed as follows:

Approximate species number	Region
6	Europe
65	temperate North America
30	tropical America
5	southeastern Africa and Madagascar
76	tropical southern and southeastern Asia
22	Japan
11	Pacific
240	Malesia (more than half restricted to New Guinea)

The fruits of most species are juicy and soft or, in some species, somewhat drier and rather hard; birds, rodents, and small mammals are known to eat the fruit and disperse the seeds widely. In spite of this ease in dispersal, Sections *Pachyanthum* (5 species) and *Neojunghunia* (13 species) are confined to New Guinea, Section *Rigiolepis* (23 species) to western Malesia. The Section *Oarianthe* (46 species) is found all over Malesia with one species extending to the New Hebrides. Section *Bracteata* (about 163 species) is by far the largest group, with essentially the same wide distribution as Section *Oarianthe,* but with ten species extending into southeastern Asia as well. Only one species of the Section *Galeopetalum* (20 species) in Southeast Asia has extended down the Malay Peninsula into

Sumatra and Java — but other sections derived from the ancient taxa on the Indian Plate have extended from Southeast Asia into Japan.

Gaultheria

The genus *Gaultheria* (Fig. 12.1C) includes about 150 species, distributed as follows:

Approximate species number	Region
5	North America
85	Central and South America
32	eastern Asia including Japan
3	southeastern Australia
8	New Zealand
26	Malesia

Of the 26 species recorded in Malesia, 6 are endemic in New Guinea, 18 are found only in the western section of Malesia, while 2 (*G. leucocarpa* and *G. nummularioides*) are Southeast Asian species which have extended their range into the western section of Malesia. The dry capsular fruits are usually completely enveloped by the enlarged, more-or-less succulent calyx which makes the fruit attractive to birds and small rodents. Dispersal should be assured, but the genus shows centres of development noted above for *Rhododendron* and *Vaccinium* — southeastern Asia, western Malesia, and New Guinea.

Agapetes

The genus *Agapetes* includes about 95 species, 80 of which are endemic in southeastern Asia with one species in the Malay Peninsula (Fig. 12.2A). Unlike the genera discussed above, no representative of the genus has been found in the western section of Malesia. Ten species are found in the eastern half of New Guinea, two in northeastern Australia. One species each is found in New Caledonia and Fiji.

The fleshy, purplish or blackish berries of this genus are eaten by birds, which should facilitate dispersal.

Family Epacridaceae (*Decatoca, Styphelia, Trochocarpa*)

In contrast to the four genera of the family Ericaceae (discussed above) which have three

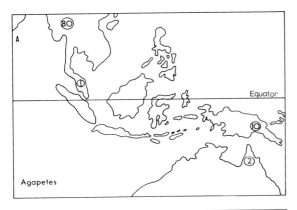

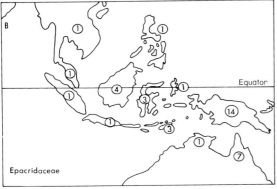

Fig. 12.2. Distribution of species of: A, Family Ericaceae —
Agapetes; and B, Family Epacridaceae — *Decatoca*, *Styphelia*,
Trochocarpa; throughout Malesia.

savannah communities, usually on somewhat
infertile soils. For example, the monocotyledonous
families Restionaceae, Centrolepidaceae, and
Xyridaceae, common in wet-heathlands in
southern Australia, extend onto seasonally-
waterlogged sandy soils (to form reed-swamps or
swampy woodlands) in northern Australia and
Malesia. *Astelia* (Liliaceae), a heathland plant of
the high altitudes in Papua New Guinea, includes
species in New Zealand which are epiphytes. Some
dicotyledonous "heathland" families (e.g.
Droseraceae, Haloragaceae, Stylidiaceae) show the
same versatility.

No comprehensive survey of Malesian heathland
species can be presented at this stage. Nevertheless,
the monographic studies published in *Flora
Malesiana* to date (1976) have not only covered
several major heathland families, but have included
ecological information on the distribution of
species. The following lists of heathland species
recorded in lowland and highland heathlands and
sclerophyllous scrubs in Borneo, Java, Malay
Peninsula and Sumatra have been abstracted from
these monographs.

centres of development, the family Epacridaceae is
strongly developed in Australia and New Guinea,
and to a lesser extent in Borneo and Celebes (Fig.
12.2B). One fleshy-fruited species, *Styphelia ma-
layana*, extends into southern Thailand, lower
Burma and southern Indo-China. It seems that
epacridaceous taxa had either not evolved or had
not extended their range onto the Indian tectonic
plate before it broke asunder from Gondwanaland
and drifted northwards.

Other "heathland" families

In South Africa and Australia many other
angiosperm families coexist with the families
Ericaceae and Epacridaceae to form heathlands
and related communities. These families (or particular "heathland" genera) prominent on the
Australian–New Guinea tectonic plate, are present
in isolated areas of western Malesia, but few extend
into Southeast Asia. The genera are not necessarily
confined to heathlands but extend into open

BORNEO

1. Lowland heathland species

Casuarinaceae	*Casuarina sumatrana*
Cyperaceae	*Fimbristylis fuscoides*
	F. insignis
	Schoenus calostachyus
	Tetraria borneensis
	Tricostularia undulata
Epacridaceae	*Styphelia abnormis*
	S. malayana
Ericaceae	*Rhododendron longiflorum*
	R. malayanum var. *axillare*
	R. nieuwenhuisii
	R. orbiculatum
	Vaccinium bancanum
	V. borneense
	V. bracteatum
	V. clementis
	V. costerifolium
	V. laurifolium var. *glanduligerum*
	V. moultonii
	V. stenanthum
Myrtaceae	*Baeckea frutescens*
	Leptospermum flavescens
Xyridaceae	*Xyris borneensis*
	X. complanata
	X. pauciflora

BORNEO *(continued)*

2. Highland heathland species

Apiaceae	*Oreomyrrhis andicola*
Casuarinaceae	*Casuarina sumatrana*
Cyperaceae	*Gahnia javanica*
	Lepidosperma chinense
	Schoenus longibracteatus
	S. melanostachys
	Scirpus subcapitatus
Droseraceae	*Drosera spathulata*
Epacridaceae	*Styphelia malayana*
	S. suaveolens
	Trochocarpa celebica
Ericaceae	*Gaultheria borneensis*
	Rhododendron bagobonum
	R. buxifolium
	R. ericoides
	R. fuchsii
	R. malayanum var. *malayanum*
	R. quadrasianum
	Vaccinium monanthum
Myrtaceae	*Baeckea* sp.
	Leptospermum flavescens
	Tristania anomala
	Tristania sp.
Thymelaeaceae	*Drapetes ericoides*

JAVA

1. Lowland heathland species

Cyperaceae	*Fimbristylis ovata*
Ericaceae	*Vaccinium bancanum*
	V. bracteatum
	V. laurifolium var. *glanduligerum*
Xyridaceae	*Xyris pauciflora*

2. Highland heathland species

Cyperaceae	*Carex* spp.
	Gahnia javanica
Droseraceae	*Drosera peltata*
Ericaceae	*Gaultheria nummularioides*
	G. punctata
	G. solitaria
	Rhododendron javanicum
	R. malayanum
	R. retusum
	Vaccinium laurifolium
	V. varingiaefolium
Epacridaceae	*Styphelia javanica*
Liliaceae	*Dianella javanica* (= *Rhuacophila javanica*)

MALAY PENINSULA

1. Lowland heathland species

Cyperaceae	*Fimbristylis fuscoides*
	F. insignis

	F. ovata
	F. tristachya
	Schoenus calostachyus
	Scleria neesii
	Tricostularia undulata
Ericaceae	*Rhododendron longiflorum*
	Vaccinium bancanum
	V. bracteatum
Epacridaceae	*Styphelia malayana*
Xyridaceae	*Xyris complanata*

2. Highland heathland species

Cyperaceae	*Gahnia javanica*
	Lepidosperma chinense
	Oreobolus kükenthalii
	Scirpus subcapitatus
Droseraceae	*Drosera spathulata*
Ericaceae	*Agapetes scortechinii*
	Rhododendron javanicum
	R. malayanum
	R. wrayi
	Vaccinium varingiaefolium
	V. viscifolium
Epacridaceae	*Styphelia malayana*

SUMATRA

1. Lowland heathland species

Campanulaceae	*Wahlenbergia erecta*
Cyperaceae	*Fimbristylis tristachya*
	Schoenus calostachyus
	Tricostularia undulata
Epacridaceae	*Styphelia malayana*
Ericaceae	*Rhododendron longiflorum*
	Vaccinium bancanum
	V. bracteatum
	V. laurifolium var. *glanduligerum*
Xyridaceae	*Xyris bancana*
	X. borneensis
	X. complanata
	X. pauciflora

2. Highland heathland species

Campanulaceae	*Lobelia sumatrana*
Centrolepidaceae	*Centrolepis fascicularis*
Cyperaceae	*Gahnia javanica*
	Kobresia kobresioidea
	Lepidosperma chinense
	Oreobolus kükenthalii
	Scirpus junghuhnii
	S. subcapitatus
Droseraceae	*Drosera spathulata*
Epacridaceae	*Styphelia malayana*
Ericaceae	*Gaultheria abbreviata*
	G. atjehensis
	G. kemiriensis
	G. nummularioides
	G. pernettyoides

SUMATRA *(continued)*

	G. punctata
	Rhododendron aequabile
	R. adinophyllum
	R. citrinum var. *discoloratum*
	R. javanicum
	R. malayanum
	R. retusum
	R. sumatranum
	R. vanderbiltianum
	R. vinicolor
	Vaccinium bartlettii
	V. laurifolium
	V. rigidifolium
	V. varingiaefolium
Haloragaceae	*Haloragis philippinensis*
Liliaceae	*Dianella javanica* (= *Rhuacophila javanica*)
Xyridaceae	*Xyris flabellata*

ECOLOGY OF HEATHLAND ELEMENTS IN MALESIA

Remnants of the Gondwanaland heathland flora are seen throughout Malesia in two main environments:

(1) *Padang vegetation on lowland sites* on very infertile, sandy podzolic soils, developed from quartzitic and sandstone rocks.

(2) *Subalpine heathlands on highland sites*, particularly on infertile soils, amongst depauperate subalpine shrublands and meadows.

It must be stressed that these habitats are very restricted — most of Malesia is (or was) clothed in tropical rain forest. However, the heathland elements do penetrate the tropical rain forests. Several heathland genera, together with sclerophyllous, gymnospermous trees, are characteristic of the most unusual *kerangas* closed-forest (*Heidewald*) growing on infertile, sandy podzolic soils in Sarawak. As well, a number of heathland genera may also be found in dipterocarp rain forest, typical of more fertile soils — but now as epiphytic plants in the sunlit branches of emergent trees. A number of these epiphytic species characteristic of the rain forest are terrestrial in habit in shrublands and heathlands of the subalpine zone (Sleumer, 1966–67). In spite of the comparatively fertile environment on which the rain forest flourishes the mineral nutrient supply to the epiphytic plants must be minimal — ions in rainfall, leached out of overhanging leaves and bark, and in decomposing debris trapped at the base of the epiphyte. In fact, the epiphytes must grow in a habitat just as oligotrophic as that found in infertile sandy podzolic soils.

Heathland in New Guinea occurs in two environments. Throughout the lowlands, where in general the nutrient and moisture status of the soils is at least moderate and not limiting, locations with soils of low nutrient status do occur. This is due to the underlying parent material: the sandstone formations on Normanby Island or the deep podzolized soil of Green River, Kunjengini and Telefomin in the Sepik Basin. While the soil of the heathland on Normanby Island is well drained, impeded drainage due to subterranean clay, with consequent crab-hole development, is the feature of the Sepik occurrences. At higher altitudes, heathland formations occur in the subalpine zones as shrublands in which Ericaceae and Epacridaceae predominate as woody plants; in alpine meadows such families as Burmanniaceae, Droseraceae, Cyperaceae, Xyridaceae, and Eriocaulaceae are common. The impression is that many of the plants occupying these sites are there because no other species can survive in this oligotrophic environment.

The heathland elements of Malesia thus survive in habitats, similar to those occupied by heathland elements in other parts of the world — exposed, oligotrophic environments.

"Heath-forests" of Borneo

In Borneo, the primary lowland rain forest is dominated by many species of the family Dipterocarpaceae. Pockets and belts of this mixed dipterocarp forest are found on sandy loam soils (usually yellow podzolic latosols with transitions to grey hydromorphic soils or humus podzols) on slopes and hills up to 1150 m. The canopy is dense and the undergrowth thin, but with a ground flora moderately rich in species. Many trees are buttressed at the base of the trunks.

On poor, sandy soils (strongly podzolized or even seasonally-waterlogged to form humus podzols, humus iron podzols, ground-water humus podzols) the mixed dipterocarp forest changes, often abruptly, into a peculiar sub-xerophilous closed-forest called *Heidewald* (Winkler, 1914) or "heath-forest" (Richards, 1936), but locally called *kerangas* (Brünig, 1961a and b, 1968). The kerangas forest varies considerably, depending on soil and site. Generally the dipterocarps, though still present, yield their dominance to other families. The structure, texture and whole colour of the kerangas forest is in sharp contrast to the lowland dipterocarp forest. The lowland dipterocarp forest is a rather loose tangle of integrated storeys of which none is clearly separated or dominating the others; the whole growing space is loosely and evenly filled with green foliage and the general impression is of a sombre but fresh green. In the kerangas forest, the storey formed by large saplings and small poles dominates the stand; the lower storeys are dense and composed of straight stems of saplings and poles forming a tidy and orderly but forbidding phalanx which is often difficult to penetrate. The main canopy is low (top height *c.* 40 m), uniform and usually densely packed; brown and reddish colours prevail in the foliage of the canopy and the sun fills the forest with a rather bright light of reddish-brown hue — considerably brighter than the light in the lowland dipterocarp forest (Brünig, 1961a). From the air, the canopy of the kerangas forest is much more even than the lowland dipterocarp forest, being composed of smaller crowns (Brünig, 1970). The average leaf size-class falls into the category of notophyll (2025–4500 mm²), much smaller than the mesophyll category (4500–18 225 mm²), characteristic of the dipterocarp forests (Brünig, 1968). The general composition of the "heath-forest" is compared with mixed dipterocarp forest, peat swamp forest and *padang* vegetation in Appendix Table 12.1. Many species of the mixed dipterocarp forests are found in the heath-forests but not in the peat-swamp forests. The peat-swamp forests and the heath-forests have a close floristic affinity with a large number of species in common. Few of the species extend into the dipterocarp forests.

The term "heath-forest" is somewhat misleading. Few characteristic "heathland" species are present, though sub-xerophyllous, leptophyllous species — *Gymnostoma nobile* (formerly under *Casuarina* sp.), *Dacrydium beccarii* — become dominant on shallow sands and clays with extremely fluctuating water tables, while microphyllous species (such as *Agathis borneensis*) are gregarious on deep, well-drained sands with water tables within reach of deep-rooting species (Brünig, 1970). Use of the term "heath-forest" puts emphasis on the characteristic of the community as an edaphic climax rain forest developed on very infertile soils, similar to the environment on which heathlands of other parts of the world are found.

The structure (in all its aspects) of the Bornean "heath-forest" is remarkably similar to that of the *wallaba* forest of Guyana (formerly British Guiana), described by Davis and Richards (1933–34). The differences between the "heath-forest" and mixed dipterocarp forest of Mount Dulit, Sarawak, are almost all analogous to the differences between the wallaba and mixed-forest types of Moraballi Creek in Guyana (Richards, 1936).

In Malesia, "heath-forests" have been recorded from all over Borneo, and on several islands in the South China Sea (Karimata, Banka, Billiton, Anambas and Natuna Islands). It is probably found in some parts of southern Sumatra and the eastern part of the Malay Peninsula. Localized stands have been recorded in central Celebes and West Irian (Van Steenis, 1957).

Padang heathlands of Borneo

With increasing severity of site conditions the kerangas forest on sandy soils is gradually or abruptly replaced by an open-scrubland or *padang*. These open-scrubs seem to be partly primary due to the direct result of poor site conditions, partly secondary resulting from poor site conditions combined with fire (Brünig, 1961a).

The vegetation is a very low and open-scrub of stunted trees and shrubs (mean height 2 to 3 m but ranging from 0.5 to 7 m), among which the most noticeable are *Baeckea frutescens*, *Cratoxylon glaucum*, *Calophyllum canum*, *Dacrydium beccarii* var. *subelatum*, *Eugenia tetraptera* and *Ploiarium alternifolium*. The ground vegetation consists of sedges, more or less dense, with a few weakly developed individuals of *Gleichenia linearis* var. *normalis* and *Lycopodium cernuum*. *Nepenthes* species and *Flagellaria indica* occur on the ground and as climbers. *Drosera spathulata* and *Burmannia coelestis* favour open but marshy spots whereas *Burmannia disticha* and *Utricularia caerulea* occur throughout. The ground orchid (*Bromheadia findlaysoniana*) grows in fairly dry, well-drained places. A complete list of species recorded in the primary padangs of Bako National Park, Sarawak is presented in Appendix, Table 12.2.

The primary padang can be expected: (1) on very wet, waterlogged sites with a rapidly fluctuating water table; (2) on shallow, heavy clays which can probably be regarded as physiologically dry; (3) on sites with shallow, sandy soil on rocks with excessive drainage, which are extremely dry throughout the year except during actual rainfall. The ecological relationships of the kerangas forest and the primary padang are shown diagrammatically in Fig. 12.3 (after Brünig, 1961a and b).

A padang on more favourable sites may be suspected of being of secondary origin, usually the result of fire. The physiognomy of the fire padang is very much like those of a primary padang, but is usually more luxuriant, denser, and richer in species than the natural padang vegetation. The differences between fire padang and primary padang communities are presented in Appendix Table 12.2.

Padang heathlands have been described from Billiton and Anabas Islands, as well as from Borneo (Van Steenis, 1957, 1958). In structure and physiognomy, the padang vegetation is closely analogous to the *muri* vegetation of Guyana, a kind of scrub dominated by *Houmiri floribunda* var. *guianensis* (Richards, 1957). Muri scrub appears to be an edaphic climax related to the wallaba forest of the same region in much the same way as padang heath is related to "heath-forest". (See also Chapter 21 by Cooper and Chapter 22 by Klinge and Medina).

Lowland heathlands in Papua-New Guinea

Heathland on Normanby Island is developed on hillsides with surface covering of weathered sandstone. Scattered trees of *kasi kasi* (*Xanthostemon* sp., Myrtaceae) dominate the scene but there is a rather open shrubland of *Baeckea frutescens*, *Myrtella beccarii* and occasional shrubby plants of *Xanthomyrtus* sp. (all Myrtaceae). *Nepenthes* sp. (probably *N. mirabilis*) is very abundant and *Melastoma malabathricum* and the related dwarf shrub *Osbeckia chinensis* are common. The ground flora is sparse between medium to large sandstone blocks. Here *Drosera* sp. and several sedges are found. Perched as an epiphyte exposed to the sun and drying sea breezes is a white-flowered *Dendrobium* (sect. *Ceratobium*). This orchid is frequently very abundant, and is associated with the curious 'ant plants' (*Hydnophytum* sp. and *Myrmecodia* sp.) also growing as epiphytes or sometimes perched on large rocks. The whole community is very open and exposed to the sun. Although the floristics are different there is a degree of similarity with the kerangas formations of the sandstones of Sarawak. Trees, when occurring, tend to have a sparse crown.

Elsewhere in Papua-New Guinea there are lowland heath-like communities of some extent in the Sepik Basin. Of the low-altitude communities those at Kunjengini and Green River are known in some detail, but both require detailed pedological and vegetational study. The Kunjengini grasslands occupy in part the Yambi land system (Haantjens, 1968). The vegetation is of the nature of a herbaceous heathland. Robbins (in Haantjens, 1968) regards the community as disclimax developed on poor podzolic soils with impeded drainage, and poorly drained alluvial black clays. Ribbons of forest traverse the grassland along the line of slope toward the Sepik River. The composition of these forest ribbons is not markedly different from that elsewhere in Papua-New Guinea lowlands. Grass genera include *Themeda*, *Ischaemum*, *Arundinella*, and *Apluda*. Sedges include *Cyperus*, *Eleocharis*, *Fimbristylis*, *Fuirena*, *Rhynchospora*, *Scleria* and *Thoracostachyum*. More significant and indicative of heathland are the dwarf shrubs, which include *Buchnera*, *Commelina*, *Desmodium*, *Dianella*, *Eriocaulon*, *Heliotropium*, *Hibiscus abelmoschus*, *Ipomoea*, *Lindernia*, *Nepenthes mirabilis*, *Phyllanthus*, *Pimelea*, *Polygala*, *Portulaca*, *Pycnospora lutescens*, *Salvia*, *Stackhousia intermedia* and *Osbeckia*. The absence of larger perennial woody plants is probably due to the periodic, probably annual, burning of the grass by local inhabitants.

The heathland vegetation at Green River in the upper Sepik Basin is more characteristic, including woody shrubs. Unlike Kunjengini, the narrow forest ribbons fringing drainage channels include species not found in the usual lowland forest. The following list is of collections made at Green River in 1957:

Herbs and low perennials

Lycopodiaceae	*Lycopodium carolinianum*
Burmanniaceae	*Burmannia* sp.
Cyperaceae	*Cladium undulatum*, *Cyperus* sp.,

Herbs and low perennials *(continued)*

	Rhynchospora rugosa, *Scleria* sp., *Tricostularia undulata*
Droseraceae	*Drosera burmannii*
Eriocaulaceae	*Eriocaulon australe*
Flagellariaceae	*Flagellaria* sp.
Nepenthaceae	*Nepenthes* sp.
Orchidaceae	*Spiranthes* sp.
Poaceae	*Eriachne pallescens*, *E. triseta*, *Garnotia mezii*, *Isachne confusa*, *I. globosa*, *I. myosotis*, *Ischaemum barbatum*, *Leersia hexandra*
Polygalaceae	*Salomonia* sp.
Xyridaceae	*Xyris complanata*, *X. papuana*, *Xyris* sp.

Woody plants, shrubs or small trees within or marginal to the open heathland communities

Elaeocarpaceae	*Elaeocarpus sepikanus*
Ericaceae	*Rhododendron zoelleri*
Lauraceae	*Litsea* sp.
Melastomataceae	*Kibessia galeata*
Myrtaceae	*Baeckea frutescens*, *Metrosideros eugenioides*, *Xanthostemon* sp.
Orchidaceae	*Dipodium pandanum*
Rubiaceae	*Mussaenda ferruginea*
Rutaceae	*Euodia* sp.
Sterculiaceae	*Sterculia lepido-stellata*
Verbenaceae	*Gmelina dalrympleana* var. *schlechteri*

High altitude heath-like vegetation, Papua-New Guinea

At altitudes in excess of 3000 m, forest trees become much reduced in stature, forming mixed communities with true shrubs. Species which at lower altitudes develop tree form include *Xanthomyrtus* spp., *Papuacedrus papuanus*, *Phyllocladus hypophyllus* and *Pittosporum pullifolium*. The true shrubs include *Coprosma divergens* and numerous species of Epacridaceae and Ericaceae. On valley flats, scree slopes, and at altitudes above 3500 m, the shrubland becomes more open, developing a heath-like appearance with scattered shrubs not exceeding 1 m in height. Common among these is *Styphelia suaveolens*. Of local occurrence being confined to Mount Wilhelm and the Salawaket Range is *Detzneria tubata* (Scrophulariaceae). Several *Rhododendron* species, including *R. gaultheriifolium*, *R. culminicolum* and *R. womersleyi*, are abundant. Lower or prostrate shrubs cover exposed rocks or stream banks; *Gaultheria mundula* and *Vaccinium* spp. are found here.

Open valleys above 3000 m frequently have a grass flora with scattered shrubs and dispersed, or sometimes closely aggregated, tree ferns of the genus *Cyathea*. Grass species are usually tussock-forming. Paijmans (1975) considered this community to be secondary, following destruction of the former forest cover by fire. This oversimplifies the picture, as these specialized species of tree ferns have their own niche apart from the pyrogenous communities. Throughout the forest-clad, high-altitude topography of Papua-New Guinea there are numerous areas where cold air drains from higher slopes into depressions, and there stands in quiet isolation for long periods. Frosts are frequent and the vegetation of these sites is best described as frost-induced heathland. Grass species include *Deschampsia*, *Poa*, *Danthonia*, *Deyeuxia* and *Festuca*. Among these tussocks are *Hypericum macgregorii*, *Styphelia suaveolens*, *Vaccinium amblyandrum*, etc. Tree ferns frequent forest margins, and some species occur as scattered individuals in the centre of these depressions. *Cyathea percrassa*, *C. atrox*, and *C. muelleri* are typical of these sites. It is from these climatically controlled communities that spore and seed material, which invades pyrogenous communities, arises.

The ground flora of these heath-like communities includes *Gentiana*, *Ranunculus*, *Rubus*, numerous cyperaceous species, *Eriocaulon*, *Juncus*, *Astelia*, etc. Elsewhere, as for example on Mount Giluwe, the dwarf *Rhododendron saxifragoides* is abundant. This plant forms a heath-like shrub submerged in a covering of peat and moss through which the rosettes of small leaves appear. No woody branches protrude above the level of the moss and peat. From the leafy rosettes arise slender inflorescences each carrying a bright orange-red flower. Another rare, but locally common, unusual heathland plant is the fern *Papuapteris linearis*.

Detailed plant lists for Mount Wilhelm are given by Johns and Stevens (1974). An abridged listing of the plants of the heath-like vegetation occurring at altitudes greater than 3500 m follows.

SPECIES OF THE HEATHLAND FLORA OF MOUNT WILHELM

FERNS AND FERN ALLIES

Aspidiaceae	*Papuapteris linearis*
Athyriaceae	*Cystopteris* sp.

FERNS AND FERN ALLIES *(continued)*

Cyatheaceae	*Cyathea atrox, C. gleichenioides, C. macgregorii, C. muelleri, C. percrassa, C. vandeusenii*
Equisetaceae	*Equisetum debile*
Gleicheniaceae	*Gleichenia bolanica, G. vulcanica*
Grammitidaceae	*Grammitis ornatissima, Grammitis* sp., *Oreogrammitis* sp.
Hymenophyllaceae	*Hymenophyllum foersteri, H. ooides*
Lomariopsidaceae	*Elaphoglossum angulatum*
Lycopodiaceae	*Lycopodium scariosum, L. selago, Lycopodium* sp.
Polypodiaceae	*Loxogramme* sp.
Pteridaceae	*Pteris* sp.
Schizaeaceae	*Schizaea fistulosa*

Gymnosperms

Podocarpaceae	*Dacrycarpus compactus, Podocarpus brassii*

MONOCOTYLEDONS

Centrolepidaceae	*Centrolepis philippinensis, Gaimardia setacea*
Cyperaceae	*Carex cappillacea, C. celebica, C. gaudichaudiana, C. perciliata, Carpha alpina, Oreobolus pumilio, Schoenus curvulus, S. maschalinus, Scirpus aucklandicus, S. crassiusculus, S. subcapitatus, S. subtilissimus, Uncinia riparia*
Eriocaulaceae	*Eriocaulon montanum*
Juncaceae	*Juncus* spp.
Liliaceae	*Astelia papuana*
Poaceae	*Agrostis reinwardtii, Brachypodium sylvaticum* var. *luzoniense, Danthonia archboldii, D. vestita, Danthonia* sp., *Deschampsia klossii, Deyeuxia brassii, Festuca papuana, Hierochloe redolens, Monostachya oreoboloides, Poa callosa, P. crassicaulis, P. epileuca, Poa* sp.

DICOTYLEDONS

Apiaceae	*Oreomyrrhis pumila, O. linearis*
Asteraceae	*Abrotanella papuana, Anaphalis mariae, Gnaphalium breviscapum, G. clemensiae, Ischnea* sp., *Keysseria radicans, Lactuca* sp., *Senecio* sp., *Tetramolopium alinae, T. macrum*
Boraginaceae	*Trigonotis papuana, T.* sp.
Campanulaceae	*Lobelia archboldiana*
Caryophyllaceae	*Cerastium keysseri, C. papuanum, Sagina papuana, Scleranthus singuliflorus*
Epacridaceae	*Styphelia suaveolens, Trochocarpa dekockii, T. dispersa*
Ericaceae	*Gaultheria mundula, Rhododendron commonae, R. culminicolum, R. gaultheriifolium, R. womersleyi, R. yelliottii*
Gentianaceae	*Gentiana ettinghausenii, G. cruttwellii, G. piundensis*

Geraniaceae	*Geranium monticola, G. potentilloides*
Haloragaceae	*Haloragis halconensis*
Hydrocotylaceae	*Hydrocotyle sibthorpioides, Trachymene tripartita*
Hypericaceae	*Hypericum macgregorii*
Myrtaceae	*Xanthomyrtus* sp.
Onagraceae	*Epilobium detznerianum, E. hooglandii, E. keysseri*
Plantaginaceae	*Plantago aundensis*
Ranunculaceae	*Ranunculus pseudolowii, R. saruwagedicus, Ranunculus* sp.
Rosaceae	*Potentilla forsteriana, P. parvula, Rubus archboldianus, R. montis-wilhelmi*
Rubiaceae	*Amaracarpus* sp., *Coprosma divergens, C. papuensis, Nertera granadensis, Nertera* sp.
Saxifragaceae (Escalloniaceae)	*Quintinia* sp.
Scrophulariaceae	*Detzneria tubata, Euphrasia mirabilis, Parahebe ciliata, P. tenuis*
Theaceae	*Eurya brassii*
Thymelaeaceae	*Drapetes ericoides*
Vacciniaceae	*Vaccinium amblyandrum, V. amplifolium*
Violaceae	*Viola kjellbergii*
Winteraceae	*Drimys piperita*

CONCLUSION

The heathland species of Malesia appear to have been derived from the ancient flora which existed on the northeast section of Gondwanaland. Two distinct subgroups resulted when Gondwanaland spilt asunder — one on the Indian plate, the second on the Australian–New Guinean plate. These two tectonic plates eventually impinged on the southeastern corner of the Eurasian plate (including the Malay Archipelago) in the late Miocene. Heathland (and other) floral elements then invaded southeastern Asia from the Indian plate, and the Malay Archipelago from the Australian–New Guinean plate and there continued to evolve. Some discontinuity apparently existed between southeastern Asia and the Malay Archipelago and this acted as a barrier to the dispersal of even the most mobile wind- and bird-distributed heathland species.

The climate of Malesia generally favours tropical rain forest and mossy rain forest (above about 1200 m). On very infertile sandy soils in lowland regions, a peculiar sub-xeromorphic rain forest, termed "heath-forest" or kerangas forest, has developed. On extreme sites, the kerangas forest has degenerated to padang (open-scrub) vegetation with strong heathland affinities.

Even in the lowland dipterocarp rain forest and the montane mossy rain forest, heathland species (with coriaceous leaves and pseudo-lignotubers) may be found as epiphytic plants in the sunlit branches of the tallest trees — an oligotrophic environment. A number of these epiphytic species become terrestrial in subalpine conditions and, on infertile soils and rocks, form a subalpine heathland amongst shrubby vegetation and alpine meadows.

In general, "heath" species are a minor component of the Malesian vegetation, but their ecological distribution is identical with the worldwide pattern on sunny, oligotrophic environments from lowland to subalpine conditions. The emergence of "heath" species from the primary tropical and lower montane rain forest appears to hold clues to the development of heathland on oligotrophic soils throughout the world.

APPENDIX I

TABLE 12.1

Floristic composition of "heath-forest", peat-swamp forest, dipterocarp forest, and padang in Sarawak and Brunei, North Borneo (Brünig, 1961a and b, 1968).

Family	"Heath-forest"			Peat-swamp forest		Dipterocarp forest		Padang		
	total genera	total species	tree species	tree species	tree species common with "heath-forest"	tree species	tree species common with "heath-forest"	total genera	total species	total species common with "heath forest"
PTERIDOPHYTA										
Eight families	31	45	2	–	–	–	–	3	3	1
GYMNOSPERMAE										
Gnetaceae	1	3	1	–	–	–	–	–	–	–
Pinaceae	1	2	2	–	–	–	–	–	–	–
Podocarpaceae	1	3	3	1	1	–	–	1	1	1
Taxaceae	2	6	6	1	1	–	–	–	–	–
DICOTYLEDONS										
Acanthaceae	2	3	–	–	–	–	–	–	–	–
Anacardiaceae	11	24	24	11	9	30	5	–	–	–
Anisophylleaceae	–	–	–	–	–	–	–	2	2	–
Annonaceae	11	28	27	9	9	18	7	–	–	–
Apocynaceae	5	10	3	3	1	4	–	–	–	–
Aquifoliaceae	1	5	5	2	1	3	–	–	–	–
Araliaceae	2	11	8	2	2	–	–	–	–	–
Asclepiadaceae	3	8	–	–	–	–	–	–	–	–
Asteraceae	1	1	1	–	–	–	–	–	–	–
Barringtoniaceae	1	1	–	–	–	–	–	1	1	?
Bignoniaceae	1	1	1	–	–	–	–	–	–	–
Bombacaceae	3	6	6	2	2	9	3	–	–	–
Bonnetiaceae	1	1	1	–	–	–	–	1	1	–
Burseraceae	3	15	15	6	4	21	4	–	–	–
Casuarinaceae	2	2	2	1	1	–	–	1	1	1
Celastraceae	4	12	12	4	3	6	3	–	–	–
Chrysobalanaceae	4	12	12	3	1	11	3	1	1	1
Clusiaceae	4	59	59	16	11	34	11	3	5	5
Connaraceae	2	2	1	1	–	–	–	–	–	–
Convolvulaceae	1	1	–	–	–	–	–	–	–	–

TABLE 12.1 *(continued)*

Family	"Heath-forest"			Peat-swamp forest		Dipterocarp forest		Padang		
	total genera	total species	tree species	tree species	tree species common with "heath-forest"	tree species	tree species common with "heath-forest"	total genera	total species	total species common with "heath-forest"
DYCOTYLEDONS *(continued)*										
Cornaceae	1	1	1	1	–	–	–	–	–	–
Ctenolophonaceae	1	1	–	–	–	–	–	–	–	–
Cunoniaceae	1	3	2	–	–	–	–	–	–	–
Dilleniaceae	2	5	4	1	1	7	3	1	2	1
Dipterocarpaceae	8	86	86	15	11 (13)	97	42	3	5	4
Droseraceae	–	–	–	–	–	–	–	1	1	–
Ebenaceae	1	25	25	4	4	22	5	–	–	–
Elaeocarpaceae	1	2	–	–	–	–	–	1	1	?
Epacridaceae	1	1	–	–	–	–	–	1	1	1
Ericaceae	4	30	5	–	–	–	–	1	1	?
Erythroxylaceae	1	1	1	–	–	–	–	–	–	–
Euphorbiaceae	24	63	59	16	10	72	19	5	5	2
Fabaceae, etc.	17	27	18	7	4	36	8	1	1	1
Fagaceae	3	21	21	5	3	14	2	–	–	–
Flacourtiaceae	5	13	13	2	2	7	1	–	–	–
Gesneraceae	1	1	–	–	–	–	–	–	–	–
Hypericaceae	1	3	3	2	2	3	1	–	–	–
Icacinaceae	4	9	9	2	2	4	3	1	1	1
Ixonanthaceae	1	1	1	–	–	–	–	1	1	1
Juglandaceae	1	1	1	–	–	–	–	–	–	–
Lauraceae	11	36	36	18	8	29	3(?)	–	–	–
Lecythidaceae	1	4	4	–	–	6	?	–	–	–
Lentibulariaceae	–	–	–	–	–	–	–	1	2	–
Linaceae	2	4	4	2	2	–	–	–	–	–
Loganiaceae	2	6	4	1	–	3	1	–	–	–
Loranthaceae	7	12	–	–	–	–	–	–	–	–
Magnoliaceae	3	3	3	1	1	1	–	–	–	–
Melastomataceae	14	36	16	2	2	11	7	2	2	–
Meliaceae	7	12	12	3	2	24	2	–	–	–
Memecylaceae	1	1	–	–	–	–	–	1	1	?
Menispermaceae	1	1	1	–	–	–	–	–	–	–
Moraceae	4	28	7	3	2	13	5	–	–	–
Myricaceae	1	1	1	–	–	–	–	1	1	–
Myristicaceae	4	26	26	7	4	31	14	–	–	–
Myrsinaceae	7	21	7	4	–	2	1	–	–	–
Myrtaceae	8	58	54	16	6	32	20	5	7	6
Nepenthaceae	1	11	–	–	–	–	–	1	5	2
Ochnaceae	6	12	7	2	1	2	1	2	2	1
Olacaceae	1	1	1	3	1	4	2	–	–	–
Oleaceae	1	8	8	3	1	3	2	–	–	–
Passifloraceae	1	1	–	–	–	–	–	–	–	–
Piperaceae	1	1	–	–	–	–	–	–	–	–
Pittosporaceae	1	1	1	–	–	–	–	–	–	–
Polygalaceae	3	16	15	4	2	12	3	–	–	–
Potaliaceae	1	1	1	–	–	–	–	1	1	–
Proteaceae	1	2	2	–	–	2	1	–	–	–

TABLE 12.1 *(continued)*

Family	"Heath-forest"			Peat-swamp forest		Dipterocarp forest		Padang		
	total genera	total species	tree species	tree species	tree species common with "heath-forest"	tree species	tree species common with "heath-forest"	total genera	total species	total species common with "heath-forest"
DYCOTYLEDONS *(continued)*										
Rhamnaceae	1	1	1	–	–	–	–	–	–	–
Rhizophoraceae	4	8	8	2	2	6	5	–	–	–
Rubiaceae	32	67	28	6	2	30	7	4	4	3
Rutaceae	5	8	7	3	1	3	1	1	1	–
Santalaceae	2	4	2	–	–	–	–	–	–	–
Sapindaceae	8	15	15	4	3	13	3	–	–	–
Sapotaceae	7	30	30	12	8	24	6	–	–	–
Saxifragaceae	1	3	1	–	–	1	–	–	–	–
Simaroubaceae	3	3	3	2	1	–	–	–	–	–
Sterculiaceae	3	8	8	4	1	19	5	–	–	–
Symplocaceae	2	11	11	–	–	3	2	–	–	–
Ternstroemiaceae	1	9	9	4	1	–	–	1	1	1
Tetrameristaceae	1	1	1	–	–	–	–	1	1	–
Theaceae	4	10	10	1	1	11	4	–	–	–
Thymelaeaceae	4	16	14	1	3	–	–	1	1	1
Tiliaceae	4	19	19	5	4	18	2	–	–	–
Trigoniaceae	1	1	1	–	–	1	1	–	–	–
Urticaceae	2	1	1	–	–	4	1	–	–	–
Verbenaceae	3	7	6	1	1	5	1	–	–	–
Violaceae	1	2	–	–	–	1	1	–	–	–
Vitidaceae (Ampelidaceae)	2	2	1	–	–	–	–	–	–	–
Winteraceae	1	1	–	–	–	–	–	–	–	–
Xanthophyllaceae	1	1	1	–	–	–	–	–	–	–
MONOCOTYLEDONS										
Apostasiaceae	1	1	–	–	–	–	–	–	–	–
Araceae	8	10	–	–	–	–	–	–	–	–
Arecaceae	15	36	5	1	1	–	–	–	–	–
Burmanniaceae	–	–	–	–	–	–	–	1	2	–
Cyperaceae	8	11	–	–	–	–	–	9	12	–
Eriocaulaceae	–	–	–	–	–	–	–	1	1	–
Flagellariaceae	2	2	–	–	–	–	–	1	1	1
Hanguanaceae	1	1	–	–	–	–	–	–	–	–
Liliaceae	3	6	–	–	–	–	–	–	–	–
Marantaceae	1	1	–	–	–	–	–	–	–	–
Orchidaceae	7 (?)	40	–	–	–	–	–	2	2	–
Pandanaceae	2	9	–	–	–	–	–	–	–	–
Poaceae	2	2	–	–	–	–	–	1	1	–
Restionaceae	1	1	–	–	–	–	–	–	–	–
Xyridaceae	–	–	–	–	–	–	–	1	1	–
Zingiberaceae	1	11	–	–	–	–	–	–	–	–
TOTAL	411	1217	832	232	146	712	221	67	83	35

TABLE 12.2
Bako National Park, Sarawak (25 km northeast of Kuching) — padang heathland

Latitude (Kuching): 1°29′N   **Longitude** (Kuching): 110°20′E
Diagrammatic cross-section (after Brünig, 1961a):

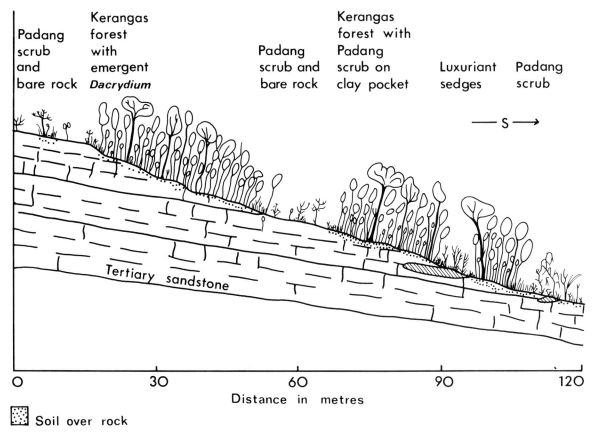

Fig. 12.3. Diagrammatic cross-section showing the distribution of padang scrub and kerangas forest in relation to rock formation in Bako National Park, Sarawak (after Brünig, 1961a and b).

Climate: Afi (Köppen, 1923)

Climatic data (Kuching)	Dec–Feb	March–May	June–Aug	Sept–Nov	Year
Air temperature (°C)					
Mean daily maximum	30.0	31.8	32.6	31.5	31.5
Mean daily minimum	22.2	22.8	22.4	22.4	22.4
Precipitation (mm)	1582	869	610	843	3904
Precipitation (days)	70	63	52	70	255
Relative humidity (%)	75	71	67	72	71

Soils:
1. Grey hydromorphic soil, supporting *Dacrydium–Baeckea–Ploiarium* scrub, 2–3 m tall (primary padang)

A_{00}		thin layer of litter of sedge leaves
A	0–5 cm	slightly purplish-brown, humus, medium sand (moderate roots)
A/C	5–50 cm	greyish-white, slightly silty, medium sand
C	50+ cm	white clay
D		sandstone

TABLE 12.2 *(continued)*

Soils *(continued)*

2. Podzolic yellow earth with gley formation, supporting *Dacrydium–Baeckea–Ploiarium* fire secondary padang, 0.5–1 m tall

A_{00}		almost absent, a few sedge leaves
A_0		almost absent, a few mm of mor between fine roots
A_1	0–5 cm	greyish-brown, mottled with fine rust-coloured spots, humus, silty, weakly loamy to loamy sand (many roots)
B_{21}	5–15 cm	(ochre)-yellow, fine sandy loam (few roots)
B_{22}	15–25 cm	ochre-yellow, fine sandy, clayish loam (roots very rare)
B_{23}	25–30 cm	greyish-white, mottled with small rust coloured spots, coarse sandy clay loam (no roots)
C	30+ cm	whitish-grey, mottled clay

Floristic composition of padang vegetation (primary open-scrub padang and secondary fire padang communities in Bako National Park, Sarawak (Brünig, 1961a, b). (The frequency of each species, in both community types, is indicated as follows: 0=absent; r=rare; p=present; c=common; f=frequent; a=abundant)

		Frequency	
		primary	secondary
TREES AND SHRUBS			
Anisophylleaceae	*Anisophyllea* sp.	r–0	c
	Combretocarpus rotundatus	r–0	p
Barringtoniaceae	*Barringtonia* sp.	0	p
Bonnetiaceae	*Ploiarium alternifolium*	f–0	f
Casuarinaceae	*Casuarina sumatrana*	0	r
Chrysobalanaceae	*Parastemon spicatum*	p–0	c
Clusiaceae	*Calophyllum canum*	c–0	r
	C. fragrans	p–0	r
	C. nodosum	p–0	r
	Cratoxylon glaucum	f–0	f
	Garcinia lanceolata	p–0	p
Dilleniaceae	*Dillenia excelsa*	p–0	p
	D. suffruticosa	0	p
Dipterocarpaceae	*Cotylelobium flavum*	p–r	r
	C. malayanum	r–0	0
	Shorea ovata	r–0	r
	Shorea sp.	r–0	0
	Vatica teysmannia	r–0	r
Elaeocarpaceae	*Elaeocarpus* sp.	0	p
Epacridaceae	*Styphelia malayana*	c–0	0
Ericaceae	*Vaccinium* sp.	r–0	0
Euphorbiaceae	*Baccaurea* sp.	0	p
	Cleistanthus gracilis	p–0	p
	Croton sp.	r–0	0
	Glochidion sp.	p–0	c
	Longetia malayana	p–0	p
Fabaceae	*Ormosia microsperma*	p–0	p
Icacinaceae	*Stemonurus umbellatus*	r–0	p
Ixonanthaceae	*Ixonanthes reticulata*	c–0	c
Melastomataceae	*Melastoma malabathricum*	0	p
	Pternandra coerulescens	p–0	p
Memecylaceae	*Memecylon* sp.	p–0	r
Myricaceae	*Myrica esculenta*	p–0	c
Myrtaceae	*Baeckea frutescens*	f–r	p
	Eugenia tetraptera	c–r	p
	Rhodamnia cinerea	c–0	p
	Rhodomyrtus tomentosus	p–0	p
	Tristania cf. *maingayi*	p–r	c

TABLE 12.2 *(continued)*

		Frequency	
		primary	secondary
TREES AND SHRUBS *(continued)*			
	T. obovata	p–0	c
	Whiteodendron moultonianum	r–0	r
Ochnaceae	*Euthemis engleri*	r–0	r
	Schuurmansiella angustifolia	p–0	0
Podocarpaceae	*Dacrydium beccarii* var. *subelatum*	f–p	r
Potaliaceae	*Fagraea racemosa*	0	p
Rubiaceae	*Canthium didymum*	r–0	r
	Gaertnera vaginans	p–0	p
	Timonius flavescens	p–0	p
	Urophyllum hirsutum	p–0	p
Rutaceae	*Tetractomia beccarii*	p–0	c
Ternstroemiaceae	*Ternstroemia bancana*	p–0	c
Tetrameristaceae	*Tetramerista glabra*	0	p
Thymelaeaceae	*Wikstroemia* sp.	r–0	r
GROUND STRATUM (sedges, grasses, etc.)			
Cyperaceae	*Cyperus javanicus*	r	r
	C. kyllingia	p	p
	Fimbristylis cymosa	p	p
	F. dichotoma	p	p
	F. dura	p	p
	F. pauciflora	c–r	c
	Gahnia tristis	p–r	r
	Hypolytrum nemorum	f–c	a
	Rhynchospora rubra	p–r	r
	Schoenus calostachyus	p	p
	Tetraria borneensis	f–c	a
	Tricostularia undulata	p–0	p
Eriocaulaceae	*Eriocaulon truncatum*	p–0	p
Poaceae	*Isachne confusa*	p–r	p
Xyridaceae	*Xyris borneensis*	p	?
GROUND STRATUM (herbs)			
Burmanniaceae	*Burmannia coelestis*	p	p
	B. disticha	p	p
Droseraceae	*Drosera spathulata*	p	?
Flagellariaceae	*Flagellaria indica*	p	p
Lentibulariaceae	*Utricularia caerulea*	?	p
	U. racemosa	p	?
Nepenthaceae	*Nepenthes albomarginata*	p	p
	N. ampullaria	p	?
	N. gracilis	p	p
	N. mirabilis	p	?
	N. rafflesiana	p	?
Orchidaceae	*Bromheadia findlaysoniana*	f	c
	Spathoglottis chrysantha (?)	p	r
GROUND STRATUM (ferns and lycopods)			
Gleicheniaceae	*Gleichenia linearis*	p	c
Lindsaeaceae	*Schizoloma ensifolium*	c–0	0
Lycopodiaceae	*Lycopodium cernuum*	p	c

REFERENCES

Brünig, E.F., 1961a. *An Introduction to the Vegetation of the Bako National Park*. Sarawak, Rep. Trustees Natl. Parks, 1959–1960. Government Printer, Kuching, Sarawak, 35 pp.

Brünig, E.F., 1961b. *A Guide and Introduction to the Vegetation of the Kerangas Forests and the Padangs of the Bako National Park*. Mimeo. Report, Kuching, Sarawak, 48 pp.

Brünig, E.F., 1968. Some observations on the status of heath forests in Sarawak and Brunei. *Proc. Symp. Rec. Adv. Trop. Ecol.*, 1968: 451–457.

Brünig, E.F., 1970. Stand structure, physiognomy and environmental factors in some lowland forests in Sarawak. *Trop. Ecol.*, 11: 26–43.

Davis, T.A.W. and Richards, P.W., 1933–34. The vegetation of Moraballi Creek, British Guiana: an ecological study of a limited area of tropical rain forest. *J. Ecol.*, 21: 350–384; 22: 106–155.

Haantjens, H.A., 1968. Lands of the Wewak–Lower Sepik Area, Territory of Papua and New Guinea. *C.S.I.R.O. Aust. Land Res. Ser.*, No. 22: 150 pp.

Haile, N.S., McElhinny, M.W. and McDougall, I., 1977. Palaeomagnetic data and radiometric ages from the Cretaceous of West Kalimantan (Borneo), and their significance in interpreting regional structure. *J. Geol. Soc. Lond.*, 133: 133–144.

Johns, R.J. and Stevens, P.F., 1974 (revised). Mount Wilhelm flora, a check list of the species. *P.N.G. Div. Bot. Bull.*, No. 6: 57 pp.

Köppen, W., 1923. *Die Klimate der Erde*. Bornträger, Berlin, 369 pp.

McClure, H.E., 1973. Some aspects of bird migration in Asia. In: A.B. Costin and R.H. Groves (Editors), *Nature Conservation in the Pacific*. Aust. Natl. Univ. Press, Canberra, A.C.T., pp. 149–164.

Paijmans, K., 1975. Explanatory notes to the vegetation map of Papua New Guinea. *C.S.I.R.O. Aust. Land Res. Ser.*, No. 35: 45 pp.

Richards, P.W., 1936. Ecological observations on the rainforest of Mount Dulit, Sarawak. Parts I and II. *J. Ecol.*, 24: 1–37; 340–360.

Richards, P.W., 1957. *The Tropical Rain Forest*. Cambridge University Press, Cambridge, 450 pp.

Sleumer, H., 1964. Epacridaceae. *Flora Malesiana*, Ser. I, 6: 422–444.

Sleumer, H., 1966–67. Ericaceae. *Flora Malesiana*, Ser. I, 6: 469–914.

Smith, A.G. and Briden, J.C., 1977. *Mesozoic and Cenozoic Paleocontinental Maps*. Cambridge University Press, Cambridge, 63 pp.

Van Steenis, C.G.G.J., 1948. Introduction. *Flora Malesiana*, Ser. I, 4: V–XII.

Van Steenis, C.G.G.J., 1949. General considerations. *Flora Malesiana*, Ser. I, 4: XIII–LXIX.

Van Steenis, C.G.G.J., 1957. Outline of vegetation types in Indonesia and some adjacent regions. *Proc. Eighth Pac. Sci. Congr.*, (1953), 4: 61–97.

Van Steenis, C.G.G.J., 1958. Condition and cause in ecological interpretation. *Blumea*, Suppl. 4: 93–95.

Wellman, P. and McDougall, I., 1974. Cainozoic igneous activity in eastern Australia. *Tectonophysics*, 23: 49–65.

Winkler, H., 1914. Die Pflanzendecke Südost-Borneos. *Bot. Jahrb.*, 50 (Suppl.): 188–208.

Chapter 13

NEW ZEALAND HEATHLANDS[1]

C.J. BURROWS, D.R. McQUEEN, A.E. ESLER and P. WARDLE

INTRODUCTION

Soils of very low fertility are widespread in New Zealand (Fig. 13.1). Some of the vegetation is dominated by shrubs, but there are also extensive forest areas, wet peat areas with sedge, fern, restiad or cushion-plant vegetation, and grasslands (above timberline). Some areas of somewhat less infertile soils may carry vegetation hardly distinguishable from heathlands on poor soils; and manuka (*Leptospermum scoparium*) which is one of the most characteristic plants in these and in true heathlands, is also able to inhabit quite fertile, wet soils (Burrows, 1973). Ecotypic variation in this species has not been studied.

Relatively smooth vegetation gradients exist between, for example, forest and scrub or scrub- and sedge-dominated communities on infertile soils, conforming to toposequences, chronosequences, minor variations in soil fertility, or drainage gradients; in consequence, it is often difficult to draw lines between the different structural vegetation types.

Relatively few introduced plants have been successful in invasion of New Zealand heathlands. They commonly include gorse (*Ulex europaeus*), several species of *Hakea*, Spanish heath (*Erica lusitanica*) and several species of *Juncus*.

The family Ericaceae is unimportant in the flora of New Zealand heathlands. The families and genera of common native species found in scrub, sedge, restiad, fern, grass or cushion heathland are listed in Table 13.1.

The heathlands to be considered here include (a) those of the podzol and gley-podzol (kauri gumland) soils of the Northland Peninsula, (b) those on raised peat in the Waikato district; (c) those on volcanic ejecta in the Central Volcanic Plateau area of the North Island; (d) those on peat soils and gleys in the mountains of both main islands; (e) those on podzolised and gleyed soils of lower montane and hill country of the southern North Island and Nelson–Marlborough; (f) those on gley, gley-podzol and peat soils of glacial outwash

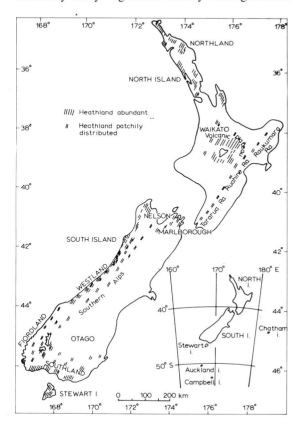

Fig. 13.1. Distribution of heathlands and related shrublands in New Zealand.

---

[1] Manuscript completed February, 1977.

TABLE 13.1

Genera of common native species in New Zealand heathlands

DICOTYLEDONS

Apiaceae	*Hemiphues* (h)
Asteraceae	*Cassinia* (s1), *Celmisia* (h), *Olearia* (s), *Senecio* (s, h)
Donatiaceae	*Donatia* (h3)
Droseraceae	*Drosera* (h)
Epacridaceae	*Archeria* (s1), *Cyathodes* (s, d1, d2), *Dracophyllum* (s, d2), *Epacris* (s1), *Pentachondra* (d1)
Ericaceae	*Gaultheria* (s, d), *Pernettya* (d1)
Gentianaceae	*Gentiana* (h)
Haloragaceae	*Gonocarpus* (h)
Hydrocotylaceae	*Centella* (h)
Lentibulariceae	*Utricularia* (h)
Menyanthaceae	*Liparophyllum* (h)
Myrsinaceae	*Myrsine* (s, d)
Myrtaceae	*Leptospermum* (s, t1), *Metrosideros* (t), *Neomyrtus* (s1)
Rhamnaceae	*Pomaderris* (s), northern North Island only
Rubiaceae	*Coprosma* (s, d1), *Nertera* (h)
Scrophulariaceae	*Euphrasia* (h), *Hebe* (s, some 1)
Spigeliaceae	*Mitrasacme* (h)
Stylidiaceae	*Forstera* (h), *Oreostylidium* (h), *Phyllachne* (h3)
Thymelaeaceae	*Pimelea* (s, d)

MONOCOTYLEDONS

Agavaceae	*Phormium* (h)
Centrolepidaceae	*Centrolepis* (h3), *Gaimardia* (h3)
Cyperaceae	*Baumea* (h2), *Carex* (h), *Carpha* (h), *Eleocharis* (h), *Gahnia* (h4), *Lepidosperma* (h2), *Oreobolus* (h3), *Schoenus* (h2), *Scirpus* (h), *Tetraria* (h2), *Uncinia* (h)
Liliaceae	*Astelia* (h), *Dianella* (h), *Herpolirion* (h)
Orchidaceae	*Calochilus* (h), *Orthoceras* (h), *Thelymitra* (h)

Poaceae	*Chionochloa* (h4), *Microlaena* (h), *Notodanthonia* (h2)
Restionaceae	*Empodisma* (h2), *Sporadanthus* (h2)

GYMNOSPERMS

Podocarpaceae	*Dacrydium* (s, d, t1), *Phyllocladus* (s2), *Podocarpus* (s, t2)

FERNS, LYCOPODS

Blechnaceae	*Blechnum*
Gleicheniaceae	*Gleichenia*
Lindsaeaceae	*Lindsaea*
Lycopodiaceae	*Lycopodium*
Schizaeaceae	*Schizaea*

MOSSES

Dicnemonaceae	*Eucamptodon*
Dicranaceae	*Campylopus*, *Dicranoloma*
Sphagnaceae	*Sphagnum*

LICHENS

Cladoniaceae	*Cladia*, *Cladonia*

Additional genera of tall scrub or forest on poor soils include:

DIOCOTYLEDONS

Araliaceae	*Pseudopanax* (s, t)
Cunoniaceae	*Weinmannia* (t)
Elaeocarpaceae	*Elaeocarpus* (t)
Escalloniaceae	*Quintinia* (t)
Fagaceae	*Nothofagus* (t)

GYMNOSPERMS

Araucariaceae	*Agathis* (t2)
Cupressaceae	*Libocedrus* (t1)

Explanations of symbols: t, trees; s, shrubs; d, dwarf shrubs; h, herbs. 1, microphyllous; 2, more or less sclerophyllous; 3, cushion-forming; 4, tussock-forming.

surfaces, moraines, old marine deposits and hills in northwest Nelson and Westland; (g) those on peat and gley soils in Southland — South Otago, Fiordland and Stewart Island; and (h) those on peat on the more distant islands including the Chatham, Auckland and Campbell groups (Fig. 13.1). The soil types, climatic régimes, landforms, history and vegetation, thus, are diverse. The history of the sites often extends back at least to the end of the last glaciation and, in some places, much longer. During this time there has been continuous soil leaching and gleying or peat

accumulation in a cool to warm temperate, humid environment (>1000 to about 7500 mm of annual average precipitation). According to the Köppen (1923) climatic type system, all of New Zealand falls into the **Cfb** category, but the Thornthwaite system gives a better model of the real diversity of the climate (Garnier, 1950), demonstrating that the more westerly and higher regions are the most humid (with precipitation often 2500 mm per annum and rising to 7500 mm in Central Westland and Fiordland). The broken terrain and considerable altitude of the mountain ranges of the main

New Zealand islands often cause climate, land-form types and corresponding vegetation types to be separated into many small units.

Fire has often been associated with the development of heathland, but not invariably so. Although some heathlands were induced by natural fires long before the Polynesian settlers arrived here, about 1000 years ago, and extensive areas have originated since, some heathlands originated as a result of natural soil processes before any human disturbance occurred. Others have been induced after European settlement in the 1840s and 1850s.

The study of heathlands is not far advanced in New Zealand, especially with respect to the ecophysiology of the plants. In the following account we attempt to draw together and briefly summarize existing information on the nature of our heathlands so that comparisons may be made with heathland vegetation elsewhere.

General information on New Zealand and localities may be obtained from McLintock (1960). Soil descriptions and analytical data are given in Soil Bureau bulletins (Soil Bureau, 1968; Soil Bureau Staff, 1954, 1968). Climate data are given by Garnier (1958) and the New Zealand Meteorological Service (1966). Though considerably out of date, Cockayne (1928) is still a general reference to New Zealand vegetation.

HEATHLANDS OF THE GUMLANDS IN NORTHLAND (A.E. Esler)

In the 1840s, *Leptospermum* stands covered 300 000 ha on flat to rolling land where mor-forming kauri (*Agathis australis*; Araucariaceae) formerly grew. The kauri stands left deposits of gum in the soil, and hence these areas are known as *gumlands*. Some of the heathlands almost certainly originated after Polynesian settlement and were maintained by repeated fires, some may be older. Bog areas with sedge vegetation were also present. In the nineteenth and early twentieth centuries the gumlands were dug over for the valuable kauri gum, and fire was used deliberately to maintain low vegetation. In the absence of fire, many gumlands areas began to revert to secondary forest. By the 1930s improved farming techniques enabled the podzolic soils to be used for agriculture and little gumland heath now remains.

The soils are derived from deeply weathered sandstones and claystones on gentle topography, up to 330 m altitude. They are very acid, low in phosphorus, and, beneath a thin siliceous topsoil, have a cemented horizon which forms a barrier to roots, air and moisture.

Leptospermum scoparium predominates in the vegetation (Appendix, Site 1) with epacrids (*Cyathodes*, *Dracophyllum*) and *Pomaderris kumeraho*, sedges (species of *Schoenus*, *Baumea*, *Tetraria* and *Lepidosperma*) and the fern *Gleichenia circinata* also important. Few introduced species invade gumlands. They include *Hakea* (Proteaceae), *Pultenaea*, *Oxylobium* and *Psoralea* (Fabaceae) and *Acacia* (Mimosaceae) (Esler and Rumball, 1975).

THE WAIKATO BASIN RAISED BOGS (D.R. McQueen)

Lowland raised peat bogs of the Waikato Basin were formerly extensive — up to 23 000 ha — but have largely been cleared for farming. Some remaining areas were described by Campbell (1964). The mature vegetation is dominated by the tall restiad *Sporadanthus traversii* and *Leptospermum scoparium*. *Empodisma minus*, *Schoenus brevifolius* and *Baumea teretifolia* form a close subcanopy cover (Appendix, Site 2).

Some of the bogs originated in shallow lakes, others on impermeable, weathered tephra alluvium. The Moanatuatua bog has a maximum peat depth of some 14 m.

HEATHLANDS OF THE VOLCANIC PLATEAU, CENTRAL NORTH ISLAND (D.R. McQueen)

Central volcanoes

Mount Tongariro (2005 m), Mount Ruapehu (2823 m) and Mount Ngauruhoe (2312 m) have erupted tephra (ash) and lava at intervals through the Pleistocene (Stevens, 1974), and the latter two are still active at times. The ejecta are andesitic–basaltic. The volcanoes rise above regional treeline (1250 m), and forest on and around the volcanoes has frequently been disturbed, both by vulcanism and Polynesian fires. On such areas, as well as above timberline, epacrid and manuka heathland occurs on immature soils from andesitic

Fig. 13.2. Andesite ash heathland, east side Mount Tongariro, North Island, New Zealand, *c*. 1500 m altitude. Principal species: *Dracophyllum recurvum*, *Celmisia spectabilis*, *Chionochloa rubra*, *Epacris alpina*. (Photo I.A.E. Atkinson.)

debris (Fig. 13.2). The dominant shrubs are *Dracophyllum* spp. and *Leptospermum scoparium*. On older, deep, tephra-based soils woody species decrease in importance and the tall tussock grass *Chionochloa rubra* dominates as soils become increasingly gleyed and podzolised. There are complex mosaics of dwarf heathland (with low shrubs and *Racomitrium lanuginosum*), herbaceous communities, tall shrub and grass heathland (Appendix, Site 3). The adventive *Calluna vulgaris* (Ericaceae) occurs in places.

Rhyolitic pumice from Taupo eruptions

During the Pleistocene several very large, violent eruptions of rhyolitic tephra (commonly called pumice) occurred (Kohn, 1973). The most recent (paroxysmal) eruption of this type was the Taupo event of 130 AD (Stevens, 1974). Large areas of the tephra occur *in situ* near the source, Lake Taupo, and there is much alluvium derived from it in the surrounding district. Vegetation was almost all destroyed over an area of 1.8 million ha, and pumice to at least 15 cm depth is the surface cover over 2.2 million ha (Vucetich and Pullar, 1963). Forest reoccupied most well-drained areas below timberline, but much of it was destroyed by Polynesian fires from the tenth century (Nicholls, 1963). More burning after European settlement left a mosaic of forest remnants, *Leptospermum* seral scrub on slopes and *Dracophyllum subulatum–Poa laevis* heathland on flat country (Appendix, Site 4 and Fig. 13.3). Since 1920 the extent of this has been greatly reduced by plantation forestry (with *Pinus radiata*) and agriculture.

Taupo pumice soils are not inherently chemically infertile, but the acid humus of *Dracophyllum subulatum* apparently makes nutrients unavailable. The soils are drought-prone and the area is subject to summer frosts, so that heathland is maintained, and forest reinvasion is very slow (Druce, 1952; Poole, 1953; Van't Woudt, 1953).

Fig. 13.3. Pumice heathland, lower Taupo area, North Island, New Zealand. Foreground: deep pumice frost flat with *Dracophyllum subulatum Poa laevis, Pimelea prostrata, Notodanthonia* sp. and *Polytrichum juniperinum*. Background: slope dominated by *Leptospermum scoparium* and some *L. ericoides*. (Photo D.R. McQueen.)

HEATHLANDS OF OLD FOREST SOILS IN SOUTHERN NORTH ISLAND AND NORTHERN SOUTH ISLAND (D.R. McQueen and C.J. Burrows)

Forest areas with poor podzolic soils were burnt before and after European settlement, often repeatedly. Succession to forest would occur if burning is discontinued (Druce, 1957) but in some sites would be very slow, especially above 370 m (southern North Island), on truncated soils, where *Leptospermum scoparium* and *Dracophyllum filifolium* are prominent (Appendix, Site 5). In northern South Island early Pleistocene gravels and strongly weathered schist from sea level to about 800 m, often quite steep, bear heathlands (Fig. 13.4) in which *L. scoparium, Erica lusitanica*, *Pteridium esculentum* and *Ulex europaeus* are important (Appendix, Site 6). If undisturbed these slowly revert to forest.

SUBALPINE AND ALPINE HEATHLANDS OF THE MOUNTAINS (C.J. Burrows and D.R. McQueen)

Above timberline (1220 m in North Island and northern South Island, 900 m in southern South Island), especially on the wetter western side of both main islands, occurs scrub vegetation, some of which may be regarded as heathland (Fig 13.5). The least fertile and wetter sites have gley soils, but peats may predominate in mountains where hard rocks slow down physical rock weathering and erosion. In the North Island also, the slopes are mantled to a greater or lesser degree with tephra.

These soils are inhabited by dense scrub, in which *Dracophyllum* species and other genera are prominent, or open scrub–tall tussock and forb mosaics where *Chionochloa* spp., sedges and broad-leaved herbs are important (P. Wardle, 1960, 1962; Elder, 1965; J.A. Wardle, 1972). The composition of the vegetation varies, depending on the occurrence of species of local distribution, but there is a basic similarity imparted by the presence of some widespread species (Appendix, Sites 7 and 8). The heathland scrub merges into scrub of different composition on more fertile sites. The band of subalpine scrub is widest where *Nothofagus* forest is absent.

Below timberline low forests occupy similar soils. They are dominated by *Nothofagus menziesii* and *N. solandri* or small podocarps, *Dacrydium biforme, Podocarpus hallii, Phyllocladus alpinus* and *Libocedrus bidwillii* (Cupressaceae), and accompanied by many species from the subalpine shrub heathlands.

In the subalpine zone, on sites which have been stabilized for long periods (radiocarbon dates range from 5000 to 10 000 years B.P.), bogs occur, usually on shallow, infertile peat. Only in Fiordland and Southland is there well-marked development of blanket peat near and above timberline.

Chionochloa rubra is a common dominant in lower subalpine bogs (except in Fiordland) along with species of *Dacrydium, Dracophyllum, Empodisma, Gleichenia* and *Sphagnum*. Otherwise they are mainly cushion or sedge bogs. Species of

344

Fig. 13.4. Induced heathland dominated by *Leptospermum scoparium*, *Erica lusitanica* and *Cassinia leptophylla* (to 2 m), Port Underwood, Marlborough, New Zealand, 150 m. The vegetation, last burnt about 30 years ago, also contains scattered individuals of the forest margin species *Pteridium esculentum*, *Hebe salicifolia*, *Coriaria arborea*, *Coprosma robusta*, *Pseudopanax crassifolium* and in open areas *Gahnia setifolia*, *Lepidosperma australe* and *Drosera auriculata*. (Photo C.J. Burrows.)

Donatia, Oreobolus, Centrolepis, Carpha, Celmisia and dwarf epacrids (*Cyathodes, Pentachondra*) are prominent (cf. Dobson, 1975).

Near the main dividing range in both islands, where precipitation is high (2500 mm per annum or more), subalpine bogs grade into alpine grassland on shallow, strongly leached and gleyed mineral soils, or, in Fiordland, on peat. This vegetation ranges from scrubline to about 1800 m (northern South Island) or 1500 m (southern South Island).

The composition of the vegetation is very variable but common dominants are, in the North Island, *Chionochloa rubra* and *C. flavescens*, in the northern third of the South Island, *C. australis* or *C. rubra*, in the middle South Island, *C. crassiuscula* and stunted *C. pallens*, and, in Fiordland, *C. crassiuscula, C. acicularis* and *C. teretifolia*, the two latter on peat of varying depths.

Plateaux near and above timberline, on hard,

acid rocks in west Nelson, carry complex mosaics of forest, bog, and shrub and grass heathlands (Mason and Moar, 1955).

THE NORTHWEST NELSON AND WESTLAND PAKIHI
(C.J. Burrows and P. Wardle)

Pakihi is a Maori word meaning open grass country or barren land. On surfaces up to about 1000 m, in many parts of the northwest and west of the South Island, which otherwise were heavily forested in 1850, extensive tracts (totalling over 300 000 ha) of such country occur (Holloway, 1954; Rigg, 1962; Mark and Smith, 1975; P. Wardle, 1977). Some appear to be natural and longstanding. Others originated after forest was burnt by Polynesian settlers or after logging or burning by European settlers. Pakihi is now used as

Fig. 13.5. Undisturbed subalpine heathland scrub, Tararua Range, southern North Island, New Zealand, 940 m altitude. The vegetation, 2 to 2.5 m tall, contains *Dacrydium biforme*, *Phyllocladus alpinus*, *Olearia colensoi*, *Dracophyllum filifolium*, *Coprosma pseudocuneata*, *Pimelea gnidia*, *Phormium cookianum*, *Chionochloa flavescens* and *Astelia nervosa*. (Photo P. Wardle.)

a term for vegetation of such areas, dominated by *Leptospermum scoparium* and/or by shorter sedge (*Baumea* spp.)–fern (*Gleichenia circinata*)–restiad (*Empodisma minus*) communities (Appendix, Sites 9 and 10; Figs. 13.6 and 13.7). The soils are always strongly gleyed and usually very infertile, and there may be no peat or a thin to deep peat cover over mineral soil. Pakihi is best developed on level or gently rolling terrain, but it can occur on slopes up to 30°. The landforms range from terrace surfaces, which originally were glacial outwash, to moraines, raised marine sediments including beach gravels, dunes and estuarine deposits, dissected hill country, and plateaux formed on hard bedrock.

In the absence of man, or disturbance by natural fires, it may be postulated that pakihi originated in at least two ways. Firstly, on fresh surfaces such as glacial outwash or moraines a sequence of vegetation development, similar to that described by

Chavasse (1962) or Stevens (1968), probably took place. At the close of glaciation, succession proceeded from scrub→mixed broadleaved angiosperm forest→gymnosperm (*Dacrydium cupressinum*) forest. This may have taken from about 200 to 500 years, perhaps more if the climate was cool. A slow process of soil maturation took place, with continued leaching of nutrients, impedance of drainage by weathering and compaction, or by formation of an iron pan, and ultimate strong gleying. In places, the presence of loess on the surface enhanced impedance of drainage (Ross and Mew, 1975). Mature forest dominated by *Dacrydium cupressinum* often occupies very infertile soils and, in places, is known to have been on the same sites for some 10000 years (Moar, 1971). In some sites, probably because of local conditions of excessive waterlogging or excessive nutrient impoverishment, forest degeneration gave rise to a

Fig. 13.6. Air view of the margin of a large natural, undisturbed pakihi on a slightly raised peat dome, Saltwater Forest, Westland, New Zealand, 16 m altitude. The open pakihi is dominated by *Empodisma minus–Gleichenia circinata–Baumea teretifolia*, with scattered short *Leptospermum scoparium* and *Dracophyllum palustre* (to 1 m) over *Dacrydium laxifolium, Drosera spathulata, Cyathodes empetrifolia, Lycopodium ramulosum, Dicranoloma billardieri, Sphagnum cristatum* and *S. falcatulum*. At the margin is a band of tall *Leptospermum* (3 m) over *Gahnia rigida* and *Empodisma*, grading into *Dacrydium colensoi, Phyllocladus alpinus, Leptospermum, Quintinia acutifolia* and *Weinmannia racemosa* (to 12 m) which, in turn, grades into moderately dense *Dacrydium cupressinum* forest (to 25 m) with *Weinmannia* and *Quintinia* over *Phyllocladus* and *Neomyrtus pedunculata*. Behind is a similar smaller pakihi. (Photo J. Johns, N.Z. Forest Service.)

semi-pakihi or bog forest community (Fig. 13.8) with low *Dacrydium cupressinum*, the small trees *D. intermedium* and *D. colensoi*, and the shrub *Phyllocladus alpinus*, all indicative of very low soil fertility (Appendix Site 11). *Leptospermum scoparium* is also often present. Peat, if present, is usually shallow. Further degeneration on the wettest sites gave rise to *Leptospermum*–sedge–fern–restiad communities, usually underlain by shallow peat.

The second way in which pakihi is believed to have originated naturally, without the aid of fire, is

that wet areas (probably mesotrophic or eutrophic shallow lakes or tarns originally), which were colonized by marsh plants soon after their formation at the close of glaciation, gradually accumulated peat. Eventually, oligotrophic conditions prevailed but the peat (ultimately up to 10 m deep) was too wet for colonization by forest which surrounded the site. The vegetation which could be supported was a short *Leptospermum*–sedge–fern–restiad community with marginal stands of tall *Leptospermum, D. colensoi, D. intermedium* and *Phyllocladus*. Climatic variations in the Holocene appear to have caused temporary drying and wetting periods, during which advance and retreat of marginal forest occurred. Such processes are believed to have occurred on glacial deposits which originated during the last glaciation, ending some 13 000 years ago (Suggate, 1965). Some surfaces in Westland, with soils of very low fertility, originated during earlier glaciations or interglacials. Although they often carry pakihi vegetation now, their history is not clear.

Pakihi within 30 km of the coast in the north-western part of the South Island is developed on a wide variety of substrates, mainly on raised marine deposits. Presumably, after glaciation, forest and bog development on these was similar to that described above. The composition of the vegetation here is very like that of the Northland gumlands. The presence of beech (*Nothofagus* spp.) north of the Taramakau River and from Paringa southward, causes some variations in the present-day vegetation patterns of bog forest, or forest marginal to open pakihi. *Nothofagus solandri* and *N. menziesii* are commonly present and the surrounding taller forests are dominated by these two, with or without *Dacrydium cupressinum*, and *N. fusca* and *N. truncata* (northern Westland only). It may be that the beeches are among the species which have led to podzolisation and gleying, giving rise to pakihi conditions (Mew, 1975).

It is notable that there is evidence from Westland that pakihi vegetation formed a kind of treeless (and possibly shrubless) moorland during the last glaciation. A pollen diagram, believed to demonstrate the transition from interglacial to glacial conditions, shows a clear sequence in which forest species were replaced by pakihi species (but not *Leptospermum*) (Dickson, 1973). There is pollen-analytical evidence that pakihi vegetation persisted

Fig. 13.7. An old induced pakihi with a hummock and pool system, Virgin Flat, Westland, New Zealand, 125 m altitude. There is a thin, inorganic, strongly gleyed and leached soil. The vegetation, last burnt about 10 years ago, consists of *Baumea teretifolia*, *Tetraria capillaris*, *Gleichenia circinata* and *Empodisma minus* with scattered short *Leptospermum scoparium* and *Epacris pauciflora* (to 50 cm) over *Centrolepis ciliata*, *Drosera spathulata*, *Liparophyllum gunnii*, *Lycopodium ramulosum* and *Campylopus* sp. (Photo C.J. Burrows.)

during later phases of the glaciation (Dr. N.T. Moar, pers. comm.).

Present-day disturbance of the predominant forest (dominated by *Nothofagus* spp. or *Dacrydium cupressinum*) on gley soils, or of the bog forest dominated by smaller *Dacrydium* species and *Phyllocladus alpinus*, gives rise to open pakihi (Appendix, Site 10). On the slightly more fertile and better-drained sites, dense *Leptospermum* stands develop (Fig. 13.9) and in places there are plentiful indications of return to forest. In others, either this process is very slow or the *Leptospermum* is in equilibrium with environmental conditions. On wetter and the most infertile sites the predominant plants are species of *Baumea*, *Empodisma minus* and *Gleichenia circinata*. The common causes of development of induced pakihi are burning or logging and burning. The sedge–fern–restiad vegetation or immature

Leptospermum stands may be maintained by repeated fires. It is evident that the Polynesian people burned tracts of forest on which extensive areas of pakihi developed, but a radiocarbon date for a *Dacrydium colensoi* stump (5950 ± 70 years B.P. — sample N.Z. 3939) from Westport, northern Westland, shows that pakihi there had originated long ago, almost certainly as a result of natural fires. Whatever the cause and however long ago induced pakihi originated, the resultant vegetation is generally similar, but there are temporal differences depending on the stage of vegetation maturity which each site has reached, local spatial differences related mainly to drainage variations, and regional differences allied to the overall distribution patterns of species important in the vegetation.

Transitions from semi-pakihi forest (Hughes, 1975) into mature beech (*Nothofagus*) forest or tall podocarp forest dominated by *rimu* (*Dacrydium cupressinum*) occur on many terrace surfaces in Westland where drainage (though impeded) is sufficient to allow regeneration of the *Dacrydium* or *Nothofagus* species. Intergrades also occur between *Podocarpus dacrydioides* forest and semi-pakihi forest (Mr. K. Smith, pers. comm.).

Heavy fertilizer applications have been used to convert some pakihi areas to farms or plantation forest, but the basic difficulties of structureless soil, lack of clay colloids, and poor drainage make it uncertain whether these will succeed in the long term, although there has been initial success in some instances.

SOUTHERN SOUTH ISLAND AND STEWART ISLAND
(C.J. Burrows)

Natural heathland is extensive above timberline, on peat, in valley bogs (often raised bog complexes)

and in some bogs on peat-infilled coastal lagoons (Fig. 13.10). Very extensive heathlands developed as a result of Polynesian burning. Some were seral to forest, with *Leptospermum scoparium* dominant, and others (with this species or *Chionochloa rubra* dominant) were relatively stable (Appendix, Sites 12–14).

The mineral soils, in these instances, are gleyed and podzolised.

The climate is cool, cloudy and windy. Dense, natural shrub and open shrub-bog heathlands on plateaux-like mountain tops of Southland and South Otago (about 900–1500 m) usually contain species of *Dacrydium*, *L. scoparium*, *Dracophyllum* and *Libocedrus*. They grade into bog with *Empodisma*, *Chionochloa rubra*, *Donatia* and *Sphagnum* (Johnson et al., 1977). The inland valley and basin bogs have *L. scoparium* and *Dracophyllum oliveri* over *Empodisma* and *Baumea* spp. (Burrows and Dobson, 1972).

Fig. 13.8. Undisturbed vegetation complex on strongly gleyed and leached soils, Waikukupa Forest, Westland, New Zealand, 425 m altitude. The wet basin in the foreground carries *Chionochloa rubra* and *Dacrydium bidwillii* (to 1 m) over *Empodisma minus*, *Gleichenia circinata*, *Dacrydium laxifolium*, *Carpha alpina* and *Cyathodes empetrifolia*. Behind is a band of scrub (to 3 m) with *Leptospermum scoparium*, *Dacrydium colensoi* and *D. biforme* over *Empodisma* and *D. intermedium* × *D. laxifolium*, grading, on the hill, into taller (5 to 8 m) bog heath forest containing *D. intermedium*, over *L. scoparium* and *Gahnia procera*. (Photo P. Wardle.)

Fig. 13.9. Induced *Leptospermum scoparium* heathland, 90 m altitude. Twelve Mile Bluff, Westland, New Zealand, logged and burnt about 50 to 80 years ago. The main cover consists of *Leptospermum* 2 to 4 m high with some *Cyathodes fasciculata*, *Dracophyllum longifolium*, *Epacris pauciflora* and scattered small *Weinmannia racemosa* and *Metrosideros umbellata*. In open places *Baumea teretifolia*, *B. tenax*, *Dianella nigra*, *Gahnia rigida*, *Phormium cookianum*, *Gleichenia circinata* and *Lycopodium laterale* are common. In the left background undisturbed forest contains *Nothofagus truncata*, *N. solandri*, *Dacrydium cupressinum*, *D. intermedium*, *Weinmannia*, *Metrosideros* and *Quintinia acutifolia*. (Photo C.J. Burrows.)

The coastal bog heathlands in Southland (Kelly, 1968) have, as well as extensive *Leptospermum* stands, a great deal of cushion vegetation in which *Oreobolus pectinatus* and *Donatia* are dominant, and other communities which resemble the inland valley bogs (Appendix, Site 14).

Above timberline (550 m) on Stewart Island (Appendix, Site 15) is a wide belt of scrub heathland with similar composition to those on the Southland or other South Island mountains, but this grades upward through 300 m into a vegetation resembling transitions between cushion bog and wet alpine grass heathland of the South Island mountains (Cockayne, 1909; Wells and Mark, 1966). This forms mosaics with low scrub heathland (on better drained sites). Species of *Donatia*, *Oreobolus*, *Phyllachne*, *Carpha*, *Astelia*, *Dracophyllum* and *Chionochloa* are abundant. *Dracophyllum* heathland is well-developed on Big South Cape Island (Fineran, 1973).

On fluvio-glacial terraces, along the south coast of west Southland (Holloway, 1954) and at low levels in Stewart Island (Cockayne, 1909), is developed a complex of bog forest on gley soils or peat, very similar to the semi-pakihi of Westland.

At West Cape, on raised shore platforms, with peat lying on bedrock and gravel, is a distinctive heathland complex in which occur bog forest, dense to open shrub heathland, tall *Chionochloa* grass heathland and cushion bog (Wardle et al., 1973).

Fig. 13.10. Undisturbed vegetation, Richters Rock bog, western Southland, New Zealand, 245 m altitude. Scattered *Leptospermum scoparium* (to 3 m) and some *Dacrydium bidwillii* (middle, to 1 m) overlies *Empodisma minus, Baumea tenax, B. rubiginosa, Oreobolus strictus, Dicranoloma billardieri, Sphagnum cristatum, S. falcatulum* and *Cladonia leptoclada*. (Photo C.J. Burrows.)

CHATHAM, AUCKLAND AND CAMPBELL ISLANDS
(C.J. Burrows)

On the island groups south and east of the South Island of New Zealand, cool, cloudy, wind-swept oceanic climates have led to the development of deep peat soils. Some of these appear to be fertile, influenced by influx of nutrients in sea spray, and often by faecal enrichment by seals, penguins, petrels and other sea birds; and they are aerated by burrowing by petrels. Only the larger islands have vegetation which may be regarded as heathland. A high degree of endemism in the flora indicates long isolation, but the woody vegetation of all but Chatham Island was probably mainly derived from the New Zealand mainland after the end of the last glaciation. On the Chathams, 800 km east of the South Island, radiocarbon dates from basal peat show that, in some areas, the history of heathland on peat goes back more than 40 000 years (Dr. J. Dodson, pers. comm.).

Chatham Islands

Peat, 1–10 m deep or more, is developed over most of the undulating, plateau-like surface of the main island (rising to 270 m), some 40 000 ha or more of which can be regarded as heathland (Cockayne, 1928; Wright, 1959). The vegetation was disturbed by Polynesian burning and, since about 1850, very profoundly by European fires and farm practices so that no undisturbed vegetation remains and very little resembling the original forest-like *Dracophyllum arboreum* heathland. After fire there is development of *Pteridum esculentum* and *Cyathodes* vegetation, followed by *D. paludosum* and, in the absence of further disturbance, *D. arboreum*.

At the lower altitudes are raised bogs and basin bogs of low fertility. Some carry the restiad *Sporadanthus traversii*, and others support species of *Juncus, Carex, Gleichenia* and *Sphagnum*.

Auckland Islands and Campbell Island

The Auckland and Campbell Islands, 480 km and 640 km south of the South Island and 540 and 600 m high, respectively, have an extensive development of blanket peat, 6 m or more deep, in many areas clearly influenced by seals, penguins, albatrosses and petrels and also by salt spray. (Appendix, Site 16). The peat covers about 40000 ha and 8000 ha, respectively, but heathland proper is less extensive.

On Auckland Island, above a discontinuous belt of *Metrosideros* forest, to an altitude of 150 m, a heathland in which *Dracophyllum longifolium* is prominent extends to 300m. This forms mosaics with tall grass–herb communities (*Chionochloa antarctica*, *Pleurophyllum* spp. and others). *Oreobolus* cushion bog communities are extensive, with composition similar to those of the South Island (Moar, 1958; Leamy and Blakemore, 1960; Godley, 1965).

Campbell Island has a heathland scrub from sea level to 150 m. The dominant plant species in this is *Dracophyllum scoparium*, and it grades upward into herb communities and *Chionochloa antarctica* grassland. Bog communities are present, similar to those on Auckland Island (Laing, 1909; Sorenson, 1954).

CONCLUSION

Heathland is of long standing in New Zealand, as is witnessed by the large number of plant species well-adjusted to living in these conditions, and the evidence for its presence during glacial periods and earlier interglacials. There is, in fact, a strong likelihood of its presence in the area during the Tertiary, for some of its components are known from the Tertiary fossil record (Fleming 1975) and other aspects of the biology of some of the species indicate this also (Wardle, 1968).

Despite the similarities between, for example, the North Auckland gumlands and the Westland pakihi, or heathland throughout New Zealand dominated by *Leptospermum*, or characterized by the presence of one or another *Dracophyllum* species, there is a very complex pattern of heathland in the country. A great deal of research is needed to determine the history of particular heathland types and to unravel the details of ecology and eco-physiology of the plants to find out how they can inhabit such poor soils. New Zealand is a natural laboratory for such studies because there are still many areas little affected by human interference.

APPENDIX I: SITE DESCRIPTIONS OF NEW ZEALAND HEATHLANDS

Climate and soil data do not always come from exactly the same location as the vegetation records; soil analyses, in general, were performed using the methods listed by Metson (1956).

SITE 1

Northland, New Zealand

Leptospermum (gumland) heathland, Kaikohe

Latitude: 35°26'S   **Longitude:** 172°49'E
Altitude: 300 m   **Landform:** Undulating lowland
Fire: 10 years before 1976

Climate

Climatic data[1]	Summer (Dec–Feb)	Autumn (March–May)	Winter (June–Aug)	Spring (Sept–Nov)	Year (Jan–Dec)
Bright sunshine (h)	–	–	–	–	2013
Max. air temp. (°C)	26.0	21.0	15.8	19.5	20.1
Min. air temp (°C)	12.9	10.9	6.2	8.6	11.9
Precipitation (mm)	307.3	424.2	530.9	355.6	1618.0
Rain (days)	32	42	56	44	174
Frost (days ≤0°C)	–	1.4	21.5	4.2	28.1
Gale (days)	0.2	0.4	1.0	0.1	1.7

[1] Nearest climate station Kerikeri (altitude 73 m).

SITE 1 *(continued)*

Soil

0–10 cm   dark grey silt loam, friable, weak granular structure (pH 4.1, citric sol P zero exch. Ca. 1.5 meq 100 g^{-1})

10–28 cm white to pale grey silt loam, very firm, cemented in places, massive (pH 4.1, citric sol P 1.6 μg g^{-1}, exch. Ca 0.9 meq 100 g^{-1})

28–68 cm grey clay with strong brown mottling, very firm, prismatic to blocky structure (pH 4.2, citric sol P 0.9 μg g^{-1}, exch. Ca 0.7 meq 100 g^{-1})

< 68 cm   grey clay, very firm, weak coarse nut structure (pH 4.3); Parent material — massive siltstone

Vegetation

SHRUBS			*Schoenus brevifolius* (to 1 m)
Epacridaceae	*Dracophyllum lessonianum* (to 100 cm)		*S. tendo*
	Cyathodes fasciculata		*Tetraria capillaris*
	Epacris pauciflora		
Myrtaceae	*Leptospermum scoparium* (to 80 cm)	OTHER HERBS	
Proteaceae	*Hakea sericea* (adventive)	Liliaceae	*Dianella nigra*
Rhamnaceae	*Pomaderris kumeraho*	Haloragaceae	*Gonocarpus sp.*
GRAMINOIDS		FERNS, LYCOPODS	
Cyperaceae	*Baumea teretifolia* (to 70 cm)	Gleicheniaceae	*Gleichenia circinata*
	Lepidosperma australe	Lycopodiaceae	*Lycopodium laterale*
		Schizaeaceae	*Schizaea fistulosa*

SITE 2

Waikato Basin, New Zealand

Restiad heathland, Moanatuatua bog (Campbell, 1964)

Latitude: 37°50′S   **Longitude:** 175°15′E
Altitude: 40 m   **Landform:** Raised bog in undulating lowland
Fire: More than 20 years before 1976   **Soil:** 2 m + brown fibrous peat

Vegetation

SHRUBS		LYCOPODS	
Epacridaceae	*Epacris pauciflora*	Lycopodiaceae	*Lycopodium laterale*
Myrtaceae	*Leptospermum scoparium* (to 1 m)		*L. serpentinum*
GRAMINOIDS		MOSSES	
Cyperaceae	*Baumea teretifolia*	Dicranaceae	*Campylopus kirkii*
	Schoenus brevifolius		
Restionaceae	*Empodisma minus* (to 0.5 m)		
	Sporadanthus traversii (to 1.8 m)		

SITE 3

Central Volcanic Plateau, New Zealand

Shrub–grass heathland, Mounts Ruapehu–Tongariro (McQueen, unpubl.)

Latitude: 39°10′S   **Longitude:** 175°35′E
Altitude: 1100–1800 m **Landform:** Volcano slope interfluves, valley flanks
Fire: No indication of burning

Climate

Climatic data[1]	Summer (Dec–Feb)	Autumn (March–May)	Winter (June–Aug)	Spring (Sept–Nov)	Year (Jan–Dec)
Bright sunshine (h)			no records		
Max. air. temp. (°C)	16.2	12.1	6.3	10.6	11.3
Min. air temp. (°C)	6.4	3.9	−0.7	2.0	2.9
Precipitation (mm)	658	659	812	785	2914
Rain (days)	–	–	–	–	196
Frost (days <0°C)	10.9	30.1	60.6	38.1	140.7
Snow (days)	0.6	3.0	9.6	4.0	17.2
Gales (days)			no records		

[1] Nearest climate station Chateau Tongariro (altitude 1119 m) (Anonymous, 1970).

Soil

 0–10 cm very dark brown sandy silt
10–35 cm medium greyish-brown silty loam
>35 cm   medium yellow brown silt loam
Parent material — andesitic volcanic ash

Vegetation

SHRUBS

Epacridaceae	*Dracophyllum filifolium* (90 cm)
	D. recurvum (20 cm)
	Epacris alpina (20 cm)
Podocarpaceae	*Podocarpus nivalis* (30 cm)
Scrophulariaceae	*Hebe tetragona* (45 cm)

DWARF SHRUBS

Epacridaceae	*Pentachondra pumila*
Myrsinaceae	*Myrsine nummularia*
Podocarpaceae	*Dacrydium laxifolium*
Rubiaceae	*Coprosma cheesemanii*

GRAMINOIDS

| Poaceae | *Chionochloa rubra* (45 cm) |
| | *Notodanthonia setifolia* |

OTHER HERBS

| Apiaceae | *Anisotome aromatica* |
| Asteraceae | *Celmisia spectabilis* |

MOSSES

| Grimmiaceae | *Racomitrium lanuginosum* |

SITE 4

Central Volcanic Plateau, New Zealand

Shrub–grass heathland, Tokoroa (McQueen, 1961)

Latitude: 38°15′S   **Longitude:** 175°52′E
Altitude: 370 m   **Landform:** Plain on pumice alluvium
Fire: 12–14 years before 1976

Soil (Vucetich, 1978)

Profile	Description	pH	citric sol P (μg g^{-1})	Organic carbon (%)	N%	C/N	Cation exchange capacity (meq 100 g^{-1})	Exchangeable Ca+Mg+K (meq 100 g^{-1})
0 – 8 cm	very dark brown sand, pumice cobbles, pebbles in lower part	5.4	80	7.5	0.44	17	18.1	3.0
8 – 16 cm	dark brown sand	5.7	30	2.7	0.17	16	7.3	1.0
16 – 66 cm	pale grey massive sand on pale grey loose pumice gravel	6.2	10	0	0.01		3.5	1.8

Parent material — rhyolitic pumice of Taupo eruption.

Vegetation

SHRUBS		GRAMINOIDS	
Epacridaceae	*Dracophyllum subulatum* (100–200 cm)	Cyperaceae	*Lepidosperma australe*
	Cyathodes juniperina	Poaceae	*Holcus lanatus* (adventive)
Myrtaceae	*Leptospermum scoparium* (to 1 m)		*Notodanthonia gracilis*
Scrophulariaceae	*Hebe stricta*		*Poa laevis*

DWARF SHRUBS		OTHER HERBS	
Asteraceae	*Raoulia albosericea*	Asteraceae	*Celmisia gracilenta*
Epacridaceae	*Cyathodes fraseri*		*Hypochoeris radicata* (adventive)
Thymelaeaceae	*Pimelea prostrata*	Geraniaceae	*Geranium sessiliflorum*

SITE 5

Southern North Island, New Zealand

Closed *Leptospermum* heathland, seral to forest, Rimutaka summit (McQueen, unpubl.)

Latitude: 41°10′S   **Longitude:** 175°15′E
Altitude: 540 m   **Landform:** Broad ridge
Fire: Possibly 90 years before 1976

Soil

1–0 cm	litter
0–1 cm	dark brown fibrous humus
1–3 or 5 cm	dark grey-brown silt loam (pH 4.5, organic C 5.56%, total N 0.21%, C/N ratio
	26.5, cation exchange capacity 17.3 meq 100 g^{-1}, exch. Ca + Mg + K 1.59 meq 100 g^{-1})
5–25 cm	light grey-brown silt loam
>25 cm	iron-stained yellowish-brown silt loam

Parent material — weathered, washed loess or greywacke bedrock

Vegetation

SHRUBS		FERNS, LYCOPODS	
Cunoniaceae	*Weinmannia racemosa*	Blechnaceae	*Blechnum procerum*
Epacridaceae	*Cyathodes fasciculata*	Lycopodiaceae	*Lycopodium scariosum*
Myrtaceae	*Leptospermum scoparium* (to 200 cm)		

SITE 6

Northeastern Marlborough, New Zealand

Leptospermum–Erica heathland, seral to forest, Ocean Bay

Latitude: 41°20′S   **Longitude:** 174°05′E
Altitude: 152 m   **Landform:** Convex slope on ridge above steep hillside
Fire: Burnt about 30 years before 1976

Climate

Climatic data[1]	Summer (Dec–Feb)	Autumn (March–May)	Winter (June–Aug)	Spring (Sept–Nov)	Year (Jan–Dec)
Bright sunshine (h)	–	–	–	–	2012
Max. air temp. (°C)	19.4	16.4	11.5	15.0	15.6
Min. air temp. (°C)	12.5	10.2	5.9	8.5	9.2
Precipitation (mm)	271.8	292.1	386.0	299.7	1249.7
Rain (days)	–	–	–	–	159
Frost (days <0°C)	–	1.4	13.4	3.1	17.9
Snow (days)	–	0.1	0.9	0.3	1.3
Gales (days)	7.6	6.8	5.4	10.5	30.3

[1] Nearest climate station Wellington (altitude 126 m).

Soil

0–10 cm	dark grey-brown silt loam, medium coarse crumb structure, charcoal (pH 4.8, citric sol. P 20 μg^{-1}, exch. Ca 3 meq 100 g^{-1}, exch. Mg 41 meq 100 g^{-1}, exch. K 5 meq 100 g^{-1}).
10–70 cm	light yellow-brown stony silt loam, slightly gleyed, with well-weathered stones; weak coarse nut to blocky structure
>70 cm	weathered yellowish schist

Vegetation

SHRUBS		GRAMINOIDS	
Asteraceae	*Cassinia leptophylla* (to 200 cm)	Cyperaceae	*Gahnia setifolia*
Ericaceae	*Erica lusitanica* (adventive, to 100 cm)		*Lepidosperma australe*
Myrtaceae	*Leptospermum scoparium* (to 200 cm)	Poaceae	*Notodanthonia gracilis*

SITE 6 *(continued)*

OTHER HERBS

Droseraceae   *Drosera auriculata*

Scattered through the dense stand are patches of *Pteridium esculentum*, *Ulex europaeus* (adventive) and individual shrubs and trees of *Hebe salicifolia*, *Coriaria arborea*, *Coprosma robusta*, *Pseudopanax arboreus* (maximum height 2 m) and *Weinmannia racemosa* (maximum height 50 cm).

SITE 7

Westland, New Zealand

Mixed subalpine scrub heathland, Mount Te Kinga

Latitude: 42°40′S   **Longitude:** 171°30′E
Altitude: 1050 m   **Precipitation:** about 4500 m per annum
Landform: Broad ridge leading to summit of mountain
Fire: No evidence of burning

Soil

0–3 cm   litter
3–8 cm   greasy, dark brown, fibrous humus
8–18 cm   dark purplish-grey, organic silt loam, compact, firm, weak crumb structure
18–33 cm   grey silt loam, compact, firm, almost structureless
33–43 cm   dull yellow-brown silt loam, firm, weak nutty structure
>43 cm   fine granite gravel
Parent material — weathered granite

Vegetation

SHRUBS		GRAMINOIDS	
Araliaceae	*Pseudopanax colensoi*	Cyperaceae	*Schoenus pauciflorus*
Asteraceae	*Cassinia vauvilliersii*	Poaceae	*Chionochloa flavescens*
	Olearia colensoi		*C. rubra*
	Senecio elaeagnifolius		
Epacridaceae	*Archeria traversii*	**OTHER HERBS**	
	Dracophyllum uniflorum (100 cm)	Agavaceae	*Phormium cookianum*
	D. longifolium	Asteraceae	*Celmisia* spp.
Ericaceae	*Gaultheria crassa*	Liliaceae	*Astelia nervosa*
Myrsinaceae	*Myrsine divaricata*		
Pittosporaceae	*Pittosporum crassicaule*	**FERNS**	
Podocarpaceae	*Dacrydium biforme* (150 cm)		
	Podocarpus nivalis (50 cm)	Blechnaceae	*Blechnum minus*
	Phyllocladus alpinus (150 cm)		
Rubiaceae	*Coprosma pseudocuneata*		
	C. ciliata		

SITE 8

Fiordland, New Zealand

Mixed subalpine scrub heathland, Lake Mike

Latitude: 45°49′S   **Longitude:** 166°55′E
Altitude: 850 m   **Precipitation:** about 7500 mm per annum
Landform: Ice-mamillated minor cirque
Fire: No evidence of burning
Soil: 30 cm medium brown fibrous peat on bedrock (gneiss)

Vegetation

SHRUBS		GRAMINOIDS	
Araliaceae	*Pseudopanax colensoi*	Poaceae	*Chionochloa acicularis*
	P. lineare		*C. crassiuscula*
Asteraceae	*Olearia colensoi*		
	Senecio elaeagnifolius	OTHER HERBS	
Epacridaceae	*Archeria traversii*	Liliaceae	*Astelia petriei*
	Dracophyllum uniflorum (150 cm)		
	D. menziesii	FERNS	
Fagaceae	*Nothofagus menziesii* (stunted, to 200 cm)	Blechnaceae	*Blechnum minus*
Podocarpaceae	*Dacrydium biforme* (150–200 cm)		
Rubiaceae	*Coprosma serrulata*		
	C. crenulata		

SITE 9

Westland, New Zealand

Restiad–fern–sedge *pakihi*, Virgin Flat

Latitude: 41°52′S   **Longitude:**171°35′E
Altitude: 125 m   **Landform:** Alluvial terrace surface
Fire: Last burnt about 10 years before 1976

Soil

0–25 cm pale grey, compact, structureless sandy silt, becoming sandier with depth, wet, soft at top, firmer near bottom
25–30 cm brown, gritty sandy silt, firm
30–35 cm pale yellowish sand on granite cobbles (thin iron pans, at intervals, can be seen in road cuts in the alluvium)
Parent materials — granite alluvium

Vegetation

SHRUBS		Droseraceae	*Drosera spathulata*
Epacridaceae	*Epacris pauciflora* (scattered, height to 50 cm)	Menyanthaceae	*Liparophyllum gunnii*
Myrtaceae	*Leptospermum scoparium* (scattered, height to 50 cm)	FERNS, LYCOPODS	
		Gleicheniaceae	*Gleichenia circinata*
GRAMINOIDS		Lycopodiaceae	*Lycopodium ramulosum*
Cyperaceae	*Baumea teretifolia* (height to 30 cm)		
	Tetraria capillaris	MOSSES	
Poaceae	*Notodanthonia nigricans*	Dicranaceae	*Campylopus* sp.
Restionaceae	*Empodisma minus*		*Dicranoloma billardieri*
		Sphagnaceae	*Sphagnum cristatum*
OTHER HERBS			
Asteraceae	*Celmisia dubia*		
Centrolepidaceae	*Centrolepis ciliata*		

SITE 10

Westland, New Zealand

Leptospermum–sedge–fern heathland, 12 Mile Bluff

Latitude: 42°18′S   **Longitude:** 171°18′E
Altitude: 90 m   **Landform:** Ancient uplifted shoreline, interfluve between deep gorges
Fires: Induced from poor *Dacrydium–Nothofagus–Weinmannia–Quintinia* forest 50–80 years previously; possibly not burnt since

Climate

Climatic data[1]	Summer (Dec–Feb)	Autumn (March–May)	Winter (June–Aug)	Spring (Sept–Nov)	Year (Jan–Dec)
Bright sunshine (h)	–	–	–	–	1745
Max. air temp. (°C)	19.0	16.5	12.0	15.3	15.8
Min. air temp. (°C)	12.4	9.6	5.1	8.6	8.9
Precipitation (mm)	632.5	581.7	607.1	642.6	2560.8
Rain (days)	–	–	–	–	193
Frost (days ⩽0°C)	0.2	4.0	22.6	4.1	30.9
Snow (days)	–	–	0.1	–	0.1
Gales (days)	–	0.8	0.2	0.2	1.2

[1] Nearest climate station Greymouth (altitude 3 m).

Soil

 0–10 cm   dark brown sandy silt, friable, structureless
10–35 cm   pale grey-brown friable, structureless sandy silt
>35 cm   sandstone pebbles, over bedrock (conglomerate)

Vegetation

SHRUBS

Cunoniaceae	*Weinmannia racemosa* (scattered)
Epacridaceae	*Cyathodes fasciculata*
	Dracophyllum longifolium
	Epacris pauciflora
Myrtaceae	*Leptospermum scoparium* (height to 200 cm)
	Metrosideros umbellata
Thymelaeaceae	*Pimelea gnidia*

GRAMINOIDS

Cyperaceae	*Baumea teretifolia*
	B. tenax
	Gahnia rigida (100 cm)
Restionaceae	*Empodisma minus* (sparse)

OTHER HERBS

Agavaceae	*Phormium cookianum* (100 cm)
Droseraceae	*Drosera spathulata*
Liliaceae	*Dianella nigra*
Orchidaceae	*Thelymitra* sp.

FERNS, LYCOPODS

Blechnaceae	*Blechnum procerum*
Gleicheniaceae	*Gleichenia circinata* (50 cm)
Lycopodiaceae	*Lycopodium laterale*
Schizaeaceae	*Schizaea fistulosa*

MOSSES

Dicnemonaceae	*Eucamptodon inflatus*
Dicranaceae	*Campylopus introflexus*
	Dicranoloma billardieri

SITE 11

Westland, New Zealand

Bog forest (semi-*pakihi*), Westland National Park

Latitude: 43°20′S   **Longitude:** 170°07′E
Altitude: 120 m   **Landform:** Depressions, flat sites, undulating moraines
Fire: No indication of burning

Soil

0–12 cm dark brown, organic silt loam becoming less organic with depth, soft, structureless (pH 4.3, citric sol P 4 $\mu g \cdot g^{-1}$, exch. Ca 2.5 meq 100 g^{-1})

12–20 cm pale brown silt loam, firm, structureless (pH 4.4, citric sol P 0.6 $\mu g\ g^{-1}$, exch. Ca 0.3 meq 100 g^{-1})

20–50 cm light bluish-grey silt, very firm, massive (pH 4.5, citric sol P 0.8 $\mu g\ g^{-1}$, exch. Ca 1.4 meq 100 g^{-1})

>50 cm   dark red-brown, stony sandy silt, firm, massive, on schist gravel

Parent material — grey schist alluvium and/or loess

Vegetation

TREES		GRAMINOIDS	
Myrtaceae	*Leptospermum scoparium*	Cyperaceae	*Gahnia procera*
Podocarpaceae	*Dacrydium biforme* (height to 10 m)		*Uncinia rupestris*
	D. colensoi (height to 10 m)	Restionaceae	*Empodisma minus* (bog margin)
	D. intermedium (scattered)		
	Phyllocladus alpinus		
		HERBS	
STUNTED TREES — SAPLINGS		Iridaceae	*Libertia pulchella*
Cunoniaceae	*Weinmannia racemosa*	Liliaceae	*Astelia nervosa*
Elaeocarpaceae	*Elaeocarpus hookeranus*	Philesiaceae	*Luzuriaga parviflora*
Escalloniaceae	*Quintinia acutifolia*		
Myrtaceae	*Metrosideros umbellata*		
Podocarpaceae	*Podocarpus hallii*	**FERNS, LYCOPODS**	
		Blechnaceae	*Blechnum procerum*
SHRUBS		Gleicheniaceae	*Gleichenia circinata*
Araliaceae	*Pseudopanax colensoi*		*G. cunninghamii*
	P. simplex	Hymenophyllaceae	*Hymenophyllum multifidum*
Epacridaceae	*Cyathodes juniperina*	Lycopodiaceae	*Lycopodium ramulosum*
	Dracophyllum palustre		
Myrtaceae	*Neomyrtus pedunculata*		
Rubiaceae	*Coprosma colensoi*	**MOSSES**	
	C. foetidissima	Sphagnaceae	*Sphagnum* spp.

SITE 12

Southland, New Zealand

Shrub–restiad heathland, Ajax Hill

Latitude: 46°25′S   **Longitude:** 169°18′E
Altitude: 680 m   **Landform:** Undulating plateau surface
Fire: No indication of burning
Soil: 100+ cm brown fibrous peat (pH 3.9)

Vegetation

SHRUBS		GRAMINOIDS	
Epacridaceae	*Dracophyllum longifolium* (200 cm)	Cyperaceae	*Oreobolus strictus*
Myrtaceae	*Leptospermum scoparium* (200 cm)	Restionaceae	*Empodisma minus*
Podocarpaceae	*Dacrydium biforme* (occasional)		
		MOSSES	
DWARF SHRUBS		Dicranaceae	*Dicranoloma billardieri*
Epacridaceae	*Cyathodes empetrifolia*		
		LICHENS	
CUSHION PLANTS		Cladoniaceae	*Cladia retipora*
Donatiaceae	*Donatia novae-zelandiae*		*C. sullivanii*

SITE 13

Southland, New Zealand

Open *Leptospermum*–sedge–restiad heathland, Richters Rock bog, Manapouri

Latitude: 45°45′S   **Longitude:** 167°32′E
Altitude: 245 m   **Landform:** Peat-filled basin
Fire: No indication of burning

Soil
 0–10 cm   dark yellowish-brown fibrous peat
10–60+cm dark reddish-brown fibrous peat

Vegetation

SHRUBS		FERNS	
Myrtaceae	*Leptospermum scoparium* (height to 300 cm)	Gleicheniaceae	*Gleichenia circinata*
		MOSSES	
		Dicnemonaceae	*Eucamptodon inflatus*
		Dicranaceae	*Dicranoloma billardieri*
GRAMINOIDS		Sphagnaceae	*Sphagnum cristatum*
Cyperaceae	*Baumea rubiginosa* (height to 40 cm)		*S. falcatulum*
	B. tenax (height to 30 cm)		
	Oreobolus strictus	LICHEN	
Restionaceae	*Empodisma minus*	Cladoniaceae	*Cladonia leptoclada*
			Cladia retipora

SITE 14

Southland, New Zealand

Cushion bog heathland, Waituna Lagoon

Latitude: 46°30′S   **Longitude:** 168°34′E
Altitude: 3 m   **Landform:** Slight depression in almost level peat bog.
Fire: Probably burnt some years ago but little noticeable effect
Soil: 100+ cm brown fibrous peat, many plant roots in top 30 cm

Climate

Climatic data[1]	Summer (Dec–Feb)	Autumn (March–May)	Winter (June–Aug)	Spring (Sept–Nov)	Year (Jan–Dec)
Bright sunshine (h)	–	–	–	–	1661
Max. air temp. (°C)	18.0	12.1	10.1	14.8	14.4
Min. air temp. (°C)	8.4	5.0	1.0	4.4	4.6
Precipitation (mm)	279.4	299.7	246.4	261.6	1087
Rain (days)	–	–	–	–	206
Frost (days ≤0°C)	7.3	32.9	55.4	25.9	114.2
Snow (days)	–	0.7	1.8	1.3	3.2
Gales (days)	2.7	2.9	1.7	2.3	9.6

[1] Nearest climate station Invercargill (altitude 1 m).

Vegetation

SHRUBS

Myrtaceae	*Leptospermum scoparium* (scattered) (height to 30 cm)

DWARF SHRUBS

Epacridaceae	*Cyathodes pumila*
	Pentachondra pumila
Ericaceae	*Pernettya macrostigma*

CUSHION PLANTS

Centrolepidaceae	*Centrolepis ciliata*
	Gaimardia setacea
Cyperaceae	*Oreobolus pectinatus*
Donatiaceae	*Donatia novae-zelandiae*

GRAMINOIDS

Cyperaceae	*Carpha alpina*
Poaceae	*Microlaena thomsonii*
Restionaceae	*Empodisma minus*

OTHER HERBS

Apiaceae	*Hemiphues suffocata*
Droseraceae	*Drosera binata*
	D. spathulata
Haloragaceae	*Gonocarpus micranthus*
Lentibulariaceae	*Utricularia monanthos*
Liliaceae	*Herpolirion novae-zelandiae*
Orchidaceae	*Thelymitra* sp.
Rubiaceae	*Nertera scapanioides*
	N. balfouriana
Stylidiaceae	*Oreostylidium subulatum*

FERNS, LYCOPODS

Gleicheniaceae	*Gleichenia circinata*
Lindsaeaceae	*Lindsaea linearis*
Lycopodiaceae	*Lycopodium ramulosum*
Schizaeaceae	*Schizaea fistulosa*

MOSSES

Sphagnaceae	*Sphagnum cristatum*
	S. falcatulum

SITE 15

Stewart Island, New Zealand

Cushion bog heathland, Table Hill

Latitude: 47°03′S   **Longitude:** 167°52′E
Altitude: 860 m   **Landform:** Gentle slope of mountainside
Fire: No indication of burning
Soil: Peat, details not recorded

Vegetation

DWARF SHRUBS

Epacridaceae	*Dracophyllum politum*
	Pentachondra pumila
Myrsinaceae	*Myrsine nummularia*
Rubiaceae	*Coprosma pumila*
Scrophulariaceae	*Hebe laingii*

CUSHION PLANTS

Asteraceae	*Celmisia argentea*
	Raoulia goyenii
Centrolepidaceae	*Centrolepis ciliata*
Cyperaceae	*Oreobolus impar*
	O. pectinatus
Donatiaceae	*Donatia novae-zelandiae*
Stylidiaceae	*Phyllachne colensoi*

GRAMINOIDS

Cyperaceae	*Carpha alpina*
	Oreobolus strictus
Poaceae	*Chionochloa pungens*
	Microlaena thomsonii
	Notodanthonia nigricans

OTHER HERBS

Apiaceae	*Aciphylla traillii*

Asteraceae	*Celmisia gracilenta*
	Senecio bellidioides
	S. scorzoneroides
Droseraceae	*Drosera arcturi*
	D. spathulata
	D. stenopetala
Gentianaceae	*Gentiana linearis*
Liliaceae	*Astelia linearis*
	Bulbinella gibbsii
Ranunculaceae	*Caltha novae-zelandiae*
Scrophulariaceae	*Euphrasia dyeri*

LYCOPODS

Lycopodiaceae	*Lycopodium fastigiatum*

MOSSES

Dicranaceae	*Campylopus* sp.
	Dicranoloma billardieri
Grimmiaceae	*Racomitrium lanuginosum*
Hedwigiaceae	*Rhacocarpus purpurascens*

LICHENS

Cladoniaceae	*Cladia aggregata*
	C. retipora
	C. sullivanii
	Cladonia leptoclada

SITE 16

Auckland Islands

Scrub–grass heathland, main island

Latitude: 50°50′S   **Longitude:** 166°10′E
Altitude: 250 m   **Landform:** Gentle slope

Climate

Climatic data	Summer (Dec–Feb)	Autumn (March–May)	Winter (June–Aug)	Spring (Sept–Nov)	Year (Jan–Dec)
Bright sunshine (h)		No record			
Max. air temp. (°C)	12.6	10.6	8.3	10.0	10.6
Min. air temp. (°C)	6.5	5.4	3.1	3.9	5.0
Precipitation (mm)	871.7	629.2	508.5	415.8	2098.3
Rain (days)	–	–	–	–	331
Frost (days ≤ 0°C)	–	3	6	1	10
Snow (days)	1	6	10	9	26
Gales (days)	17	19	14	23	73

SITE 16 *(continued)*

Soil

Profile	Description	pH	citric sol P (μg g^{-1})	Organic C (%)	N (%)	Cation exchange capacity (meq 100 g^{-1})	BS (%)	Exchangeable Ca+Mg+K (meq 100 g^{-1})
0–10 cm	dark reddish brown peat	4.0	0.12	56	1.17	121	23	28
10–90 cm	dusky red peat	3.9	0.10	56	0.83	190	24	46

Vegetation

SHRUBS

Araliaceae	*Pseudopanax simplex*
Asteraceae	*Cassinia vauvilliersii*
Epacridaceae	*Dracophyllum longifolium* (height to 150 cm)
Myrsinaceae	*Myrsine divaricata*
Myrtaceae	*Metrosideros umbellata* (stunted)
Rubiaceae	*Coprosma ciliata*
Scrophulariaceae	*Hebe odora*

DWARF SHRUBS

Epacridaceae	*Cyathodes empetrifolia*

Rubiaceae	*Coprosma pumila*

GRAMINOIDS

Poaceae	*Chionochloa antarctica*

CUSHION PLANTS

Centrolepidaceae	*Centrolepis ciliata*
Cyperaceae	*Oreobolus pectinatus*

OTHER HERBS

Droseraceae	*Drosera stenopetala*
Liliaceae	*Astelia subulata*

REFERENCES

Anonymous, 1970. Summaries of climatological observations to 1970. *N.Z. Meteorol. Serv. Misc. Publ.*, No. 143: 77 pp.

Burrows, C.J., 1973. The ecological niches of *Leptospermum scoparium* and *L. ericoides* (Angiospermae: Myrtaceae). *Mauri Ora* 1: 5–12.

Burrows, C.J. and Dobson, A.T., 1972. Mires of the Manapouri–Te Anau Lowlands. *Proc. N.Z. Ecol. Soc.*, 19: 75–99.

Campbell, E.O., 1964. The restiad peat bogs of Motumaoho and Moanatuatua. *Trans. R. Soc. N.Z. (Bot.)*, 2: 219–27.

Chavasse, C.G., 1962. Forest, soils and land forms of Westland. *N.Z. For. Serv. Inf. Ser.*, No. 43: 14 pp.

Cockayne, L., 1909. *Report on a Botanical Survey of Stewart Island.* Department of Lands, Wellington, 68 pp.

Cockayne, L., 1928. *The Vegetation of New Zealand.* Engelmann, Leipzig, 456 pp.

Dickson, M., 1973. Palynology of a Late Oturi interglacial and early Otira glacial sequence from Sunday Creek, Westland, New Zealand. *N.Z. J. Geol. Geophys.*, 15: 590–598.

Dobson, A.T., 1975. Vegetation of a Canterbury subalpine mire complex. *Proc. N.Z. Ecol. Soc.*, 22: 67–75.

Druce, A.P., 1952. The vegetation of western Taupo. *N.Z. Sci. Rev.*, 10: 89–91.

Druce, A.P., 1957. Botanical survey of an experimental catchment, Taita, New Zealand. *N.Z. Dep. Sci. Ind. Res., Bull.*, No. 124: 81 pp.

Elder, N.L., 1965. Vegetation of the Ruahine Range. *Trans. R. Soc. N.Z. (Bot.)*, 3: 13–166.

Esler, A.E., and Rumball, P.J., 1975. Gumland vegetation at Kaikohe, Northland, New Zealand. *N.Z. J. Bot.*, 13: 425–36.

Fineran, B.A., 1973. A botanical survey of seven Muttonbird islands, Southwest Stewart Island. *J. R. Soc. N.Z.*, 3: 475–526.

Fleming, C.A., 1975. The geological history of New Zealand and its biota. In: G. Kuschel (Editor), *Biogeography and Ecology in New Zealand.* W. Junk, The Hague, pp. 1–86.

Garnier, B.J., 1950. The climates of New Zealand, according to Thornthwaite's classification. In: *New Zealand Weather and Climate. N.Z. Geogr. Soc. Misc. Ser.*, No. 1: 84–104.

Garnier, B.J., 1958. *The Climate of New Zealand.* Edward Arnold, London, 191 pp.

Godley, E.J., 1965. Notes on the vegetation of the Auckland Islands. *Proc. N.Z. Ecol. Soc.*, 12: 57–63.

Holloway, J.T., 1954. Forests and climate in the South Island of New Zealand. *Trans. R. Soc. N.Z.*, 82: 329–410.

Hughes, H.R., 1975. Regeneration of semi-pakihi forest, Granville State Forest, Westland. *Beech Res. News*, No. 3: 9–15.

Johnson, P.N., Mark, A.F. and Baylis, G.T.S., 1977. Vegetation at Ajax Hill, south-east Otago, New Zealand. *N.Z. J. Bot.*, 15: 209–220.

Kelly, G.C., 1968. Waituna Lagoon, Foveaux Strait. *Bull. Wellington Bot. Soc.*, 35: 8–19.

Kohn, B.T., 1973. *Some Studies of New Zealand Quaternary Pyroclastic Rocks*. Thesis, Victoria University, Wellington, 340 pp.

Köppen, W., 1923. *Die Klimate der Erde*. Bornträger, Berlin, 369 pp.

Laing, R.M., 1909. The chief plant formations and associations of Campbell Island. In: *The Subantarctic Islands of New Zealand*. Government Printer, Wellington, pp. 482–92.

Leamy, M.L. and Blakemore, L.C., 1960. The peat soils of the Auckland Islands. *N.Z. J. Agric. Res.*, 3: 526–546.

McLintock, A.H., 1960. *A Descriptive Atlas of New Zealand*. Government Printer, Wellington, 109 pp.

McQueen, D.R., 1952. Succession after forest fires in the southern Tararua Mountains. *Bull. Wellington Bot. Soc.*, 24: 10–19

McQueen, D.R., 1961. Indigenous-induced vegetation and *Pinus radiata* on volcanic ash soils. *Proc. N.Z. Ecol. Soc.*, 8: 1–14.

Mark, A.F. and Smith, P.M.F., 1975. A lowland vegetation sequence in South Westland: Pakihi bog to mixed beech–podocarp forest. Part 1: The principal strata. *Proc. N.Z. Ecol. Soc.*, 22: 76–92.

Mason, R. and Moar, N.T., 1955. Notes on the vegetation and the flora of Mount Augustus, Buller County. *N.Z. J. Sci. Technol.*, 37A: 175–186.

Metson, A.J., 1956. Methods of chemical analysis for soil survey samples. *N.Z. Dep. Sci. Ind. Res. Soil Bur. Bull.*, 12: 1–208.

Mew, G., 1975. Soil in relation to forest type in beech forests in the Inangahua Depression, West Coast, South Island. *Proc. N.Z. Ecol. Soc.*, 22: 42–50.

Moar, N.T., 1958. Contributions to the Quaternary history of the New Zealand flora. I. Auckland Island peat studies. *N.Z. J. Sci.*, 1: 449–465.

Moar, N.T., 1971. Contributions to the Quaternary history of the New Zealand flora. 6. Aranuian pollen diagrams from Canterbury, Nelson and North Westland, South Island. *N.Z. J. Bot.*, 9: 80–145.

New Zealand Meteorological Service, 1966. *Summaries of Climatological Observations at New Zealand Stations to 1960*. Government Printer, Wellington, 59 pp.

Nicholls, J.L., 1963. Vulcanicity and indigenous vegetation in the Rotorua district. *Proc. N.Z. Ecol. Soc.*, 10: 58–65.

Poole, A.L., 1953. The vegetation (Western Taupo). *Proc. N.Z. Ecol. Soc.*, 2: 13.

Rigg, H.H., 1962. The pakihi bogs of Westport, New Zealand.

Trans. R. Soc. N.Z. (Bot.), 1(7): 91–108.

Ross, C.W. and Mew, G., 1975. Current thoughts on some Westland gley podzols — should they be re-classified? *N.Z. Soil News*, 23: 78–86.

Soil Bureau, 1968. Soils of New Zealand, Parts 1, 2, 3. *N.Z. Dep. Sci. Ind. Res. Soil Bur. Bull.*, No. 26: 127 pp.

Soil Bureau Staff, 1954. General survey of the soils of North Island, New Zealand. *N.Z. Dep. Sci. Ind. Res. Soil Bur. Bull.*, No. 5: 286 pp.

Soil Bureau Staff, 1968. General survey of the soils of South Island, New Zealand. *N.Z. Dep. Sci. Ind. Res. Soil Bur. Bull.*, No. 27: 404 pp.

Sorenson, J.H., 1954. Ecology of the Subantarctic Islands: Botanical factors. *Proc. N.Z. Ecol. Soc.*, 2: 17–18.

Stevens, G.R., 1974. *Rugged Landscape*. A.H. and A.W. Reed, Wellington, 286 pp.

Stevens, P.R., 1968. *A Chronosequence of Soils Near the Franz Josef Glacier*. Thesis, Lincoln College, Canterbury, 389 pp.

Suggate, R.P., 1965. Late Pleistocene geology of the northern part of the South Island, New Zealand. *N.Z. Geol. Surv. Bull.*, No. 77: 91 pp.

Van't Woudt, B.D., 1953. Factors affecting the growth of exotic conifers in the Taupo pumice country. *N.Z. Sci. Rev.*, 11: 45–48.

Vucetich, C.G., 1978. Soils, agriculture and forestry of Waiotapu Region, Central North Island, New Zealand. *N.Z. Soil Bur. Bull.*, No. 31, 99 pp.

Vucetich, C.G. and Pullar, W.A., 1963. Ash beds and soils in the Rotorua district. *Proc. N.Z. Ecol. Soc.*, 10: 65–72.

Wardle, J.A., 1972. The composition and structure of the forests and shrublands of the Grey catchment headwaters. *N.Z. For. Serv. Prot. For. Branch Rep.*, No. 112: 47 pp.

Wardle, P., 1960. The subalpine scrub of the Hokitika catchment, Westland. *Trans. R. Soc. N.Z.*, 88: 47–61.

Wardle, P., 1962. Subalpine forest and scrub in the Tararua Range. *Trans. R. Soc. N.Z. (Bot.)*, 1: 77–89.

Wardle, P., 1968. Evidence for an indigenous pre-Quaternary element in the mountain flora of New Zealand. *N.Z. J. Bot.*, 6: 120–125.

Wardle, P., 1977. Plant communities of Westland National Park, New Zealand. *N.Z. J. Bot.*, 15: 323–398.

Wardle, P., Mark, A.F. and Baylis, G.T.S., 1973. Vegetation and landscape of the West Cape District, Fiordland, New Zealand. *N.Z. J. Bot.*, 11: 599–626.

Wells, J.A. and Mark, A.F., 1966. The altitudinal sequence of climax vegetation on Mt Anglem, Stewart Island. Part 1: the principal strata. *N.Z. J. Bot.*, 4: 267–282.

Wright, A.C.S., 1959. Soils of Chatham Island (Rekohu). *N.Z. Dep. Sci. Ind. Res. Soil Bur. Bull.*, No. 19: 59 pp.

Chapter 14

EUROPEAN HEATHLANDS[1]

C.H. GIMINGHAM, S.B. CHAPMAN and N.R. WEBB

INTRODUCTION

Heathlands have long played a prominent part in the West European scene. From early times the languages of countries on the Atlantic fringe of the Continent contained words to describe them, notably "bruyère" (Romance origin), "lande" (Celtic origin) and "Heide" or "heath" (Gothic origin). Elsewhere in Europe related vegetation types were to be found on mountains, but in this western region from Scandinavia southwards to northern Spain (Fig. 14.1) heathlands were extensive also in the lowlands and played an important

part in the lives of people. As a result, the heathland landscapes of Europe (Fig. 14.2, 14.3) have caught the attention of painters, poets and writers as well as scientists, and have led to the evolution of unique patterns of land use, with their special traditions and tools. It was to this open, largely tree-less country on acid soils of low fertility, so often dominated by heather (*Calluna vulgaris*), that the word "heath" was first applied; though now as the name of a formation, "heathland" refers to vegetation of similar physiognomy throughout the world.

For many years, however, there was controversy about the origins and status of European' heathlands. There has also been continuing argument about the ecological consequences of a long history of use for grazing, associated in some parts with regular burning as a means of vegetation management. This system of land use has resulted in ecosystems which are of particular interest because they are composed of naturally occurring, not cultivated, species and are systematically managed for herbivore production. This, together with the intrinsic interest of heathlands from the viewpoints of plant sociology, physiological ecology and community dynamics, has placed them amongst the most intensively studied ecosystems in Europe. It is the intention of this chapter, while providing a general account of European heathlands, to reflect the results of some aspects of these studies.

In recent years there has been a further development, namely a widespread and rapid decline, throughout much of Western Europe, in the area occupied by heathland vegetation. Increasingly, heathlands have gone out of production as grazing

Fig. 14.1. Map showing the lowland heathland region of Western Europe.

[1] Manuscript completed February, 1977

Fig. 14.2. Heathland landscape in Western Europe. Heath vegetation in northwestern Scotland, largely dominated by *Calluna vulgaris*. (Photo C.H. Gimingham.)

Fig. 14.3. Heathland with scattered tall shrubs (*Calluna vulgaris* with *Juniperus communis*), Lüneburger Heath, northern Germany. (Photo C.H. Gimingham.)

lands, and the traditional combination of low productivity with minimal inputs is being replaced by more intensive use for agriculture or forestry. While in this way soil improvement and enhanced production may be achieved, a distinctive type of landscape of great scientific, recreational and aesthetic value is in danger of extinction. This can be prevented only by an effective policy for the conservation of examples of the entire range of heathland communities throughout Western Europe. The results of research on heathland ecosystems must now be applied to the development of systems of management for nature conservation.

COMMUNITY STRUCTURE

Most European heathlands may be described as low (canopy at 1 m or less above ground) or dwarf (25 cm or less) heathland, with ericoid shrubs generally forming dense cover. As in other regions, they may conveniently be divided into three categories according to broad differences in habitat:

(1) dry-heathlands
(2) wet-heathlands
(3) mountain and arctic heathlands

The first two belong to the lowland, temperate parts.

European heathlands

(1) Dry-heathlands

The community structure most commonly equated with heathland is best expressed in the dry-heathlands, developed on freely drained, nutrient-poor substrata where, for any reason, trees are very sparse or lacking (Fig. 14.2). Where tall shrubs such as *Juniperus communis*, *Ulex europaeus* or *Sarothamnus scoparius* are present, though sparse, a tall open shrubland with heath is frequent, as in parts of southwestern Sweden, Denmark, The Netherlands, northern Germany (Fig. 14.3) and France. However, in the majority of heathlands neither trees nor tall shrubs contribute significant cover, leaving ericaceous species to form a dense stratum, less than 1 m in height, usually to the exclusion of all other life forms at this level (Fig. 14.4A). *Calluna vulgaris*, being a vigorous and

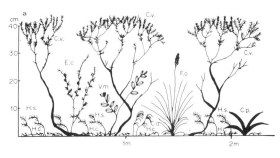

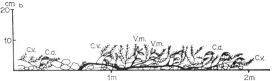

Fig. 14.4. Profile diagrams illustrating the structure of: A. *Calluna vulgaris*-dominated dry-heathland with *Erica cinerea* and *Vaccinium myrtillus*. B. Mountain dwarf heathland (*Calluna vulgaris* and *Vaccinium myrtillus*).
Legend: C.a.=*Cladonia arbuscula*; C.p.=*Carex pilulifera*; C.v.=*Calluna vulgaris*; E.c.=*Erica cinerea*; F.o.=*Festuca ovina*; H.c.=*Hypnum cupressiforme*; H.s.=*Hylocomium splendens*; V.m.=*Vaccinium myrtillus*.

relatively long-lived species (see p. 387) which includes the whole of the West European heathland region within the limits of its distribution (Fig. 14.5), is a widespread dominant. While this plant is sometimes associated with other Ericaceae of similar stature, (e.g. *Erica cinerea*, *Vaccinium myrtillus*), it is replaced as dominant only in habitats at the margins or beyond the limits of its ecological range.

Below this low-shrub canopy there may be a discontinuous stratum of partially shade-tolerant, often creeping dwarf shrubs [e.g. *Vaccinium vitis-idaea* (Fig. 14.6), *Empetrum nigrum*, *Arctostaphylos uva-ursi*] with some herbs and graminoid plants, reaching to about 20 cm above the surface. Robust bryophytes (e.g. *Pleurozium schreberi*) and a few rosette or creeping vascular plants may form another layer at 5 to 10 cm, while smaller bryophytes or lichens constitute a ground stratum. The extent to which these subordinate strata are developed and their density depend both on the habitat and on the age structure of the stand of the dominant species (Gimingham, 1972).

(2) Wet-heathlands

In acid oligotrophic habitats, wet-heathland occurs on soils with impeded drainage or on

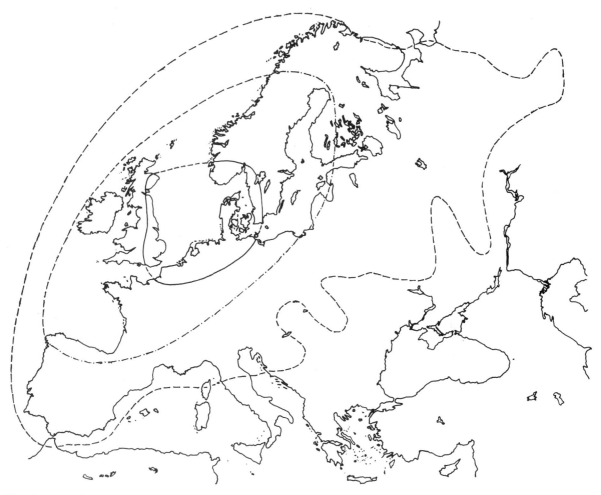

Fig. 14.5. Map of the main area of distribution of *Calluna vulgaris*; outlying stations omitted.
Legend: - - - - - - , approximate limits of distribution; - · - · - · - · - · - , area within which *Calluna* is commonly a community dominant; ————— , approximate area of ecologically optimal habitats. (After Beijerinck, 1940, modified.)

continually moist organic substrata. These communities occupy an intermediate position on the vegetational gradient from dry-heathland to mires on saturated peat; the wet end of this sequence is covered in Volume 4 of this series. Trees are seldom present on wet-heathlands in the more oceanic parts of the region, though elsewhere they may occur sparsely. Tall shrubs, represented commonly by *Salix* species, are sometimes present as isolated individuals or in small clumps, and the low shrubs are normally less dense than in dry-heathlands.

In habitats of moderate moisture status, where the surface at least is aerated for part of the summer, a type sometimes described as "humid heath" occurs, in which *Calluna vulgaris* is mixed with *Erica tetralix*, and while grass and sedge species are abundant there is little *Sphagnum* moss. In the wet-heathlands on permanently moist soils, the contribution of *Calluna* is much reduced and often confined to hummocks (Rutter, 1955). Its place is increasingly taken by other ericaceous shrubs such as *Erica tetralix* or *Ledum palustre*, and by *Myrica gale* on very wet peat. These shrubs generally produce an open-heathland, with considerable admixture of tall grass such as *Molinia caerulea* or tufted sedges (e.g. *Eriophorum* sp., *Trichophorum caespitosum*). There is normally a dense ground stratum composed chiefly of *Sphagnum* species, associated with various other bryophytes and small herbs. Lichens are largely absent.

Fig. 14.6. *Vaccinium vitis-idaea*, occupying the centre of a degenerate *Calluna vulgaris* bush (northeastern Scotland). (Photo C.H. Gimingham.)

(3) Mountain and arctic heathlands

Under the more extreme climatic conditions beyond the normal altitudinal and latitudinal limits of forest, dwarf heathland is widespread. Here dwarf shrubs (Ericaceae, *Salix* spp. etc.) form a low mat, often no more than 10 cm in height (Fig. 14.4B), of entwined stems and branches, sometimes dense and continuous but often restricted to stripes or patches. Graminoid species and a variety of bryophytes and lichens contribute to the community. Vegetation of similar structure occurs in exposed coastal localities. The arctic heathlands are covered more fully by Bliss in Chapter 15.

Life forms

The range of life forms represented in a typical dry-heathland is indicated by the examples given in Table 14.1. The predominant category is that of the low shrub with very small (leptophyll), evergreen, sclerophyllous, more or less cylindrical leaves, such as *Calluna vulgaris* (Fig. 14.7), *Erica* spp., and *Empetrum* spp., all of which can behave as dwarf shrubs (<25 cm) as well. In addition, there are low or dwarf shrubs with flat, sclerophyllous leaves in the nanophyll (or bordering on microphyll) category, such as *Vaccinium vitis-idaea* (Fig. 14.6) and *Arctostaphytos uva-ursi*. These examples are evergreen, but some are deciduous, or largely deciduous (retaining only some of the youngest leaves in winter) such as *Vaccinium myrtillus*, *Arctous alpinus*, and non-ericaceous plants such as *Salix repens* (Salicaceae), *Myrica gale* (Myricaceae) and *Genista anglica* (Fabaceae). The latter, like the species of *Ulex* which belong to heathland communities [*U. europaeus*, *U. gallii* and *U. minor* (Fig. 14.13)], possesses spines, but otherwise spines are not characteristic of European heathland species.

Among the hemicryptophytes, graminoid herbs (both caespitose and rhizomatous) are generally represented, many being evergreen but a few deciduous. Some heathland communities occupy-

TABLE 14.1

Analysis of life-forms of vascular plant species in three examples of West European heathland communities (numbers of species in each class)

	Dinnet Muir, Aberdeenshire, NE Scotland	Mästocka, Halland. SW Sweden	Wilsede Berg, Lüneburger Heath, N Germany
	Herb-rich *Calluna–Arctostaphylos* heathland	*Calluna–Vaccinium–Arctostaphylos* heathland	*Calluna–Genista* heathland
PHANEROPHYTES			
Small trees (2–10 m)	1	2	1
Low shrubs (25–100 cm)	2	1	1
CHAMAEPHYTES			
Creeping dwarf shrubs, stems often intertwined to form mats	1	4	1
Dwarf shrubs with erect or ascending stems (<25 cm)	2	3	3
HEMICRYPTOPHYTES			
(1) Graminoid herbs			
(a) Caespitose. tillering herbs, evergreen	5	3	3
(b) Sparingly tufted, more or less rhizomatous, evergreen	1	1	0
(2) Forbs			
(a) with erect or scrambling stems, dying back in winter	5	2	0
(b) seasonal rosettes	1	3	0
(c) creeping	1	1	0
(d) evergreen	3	1	0
HETEROTROPHIC			
Twining parasites	0	0	1
Total number of species	22	21	10

ing soils of at least moderate nutrient status are described as "herb-rich", indicating the presence of a fair number of forbs as well as graminoids, ranging from creeping and scrambling types to rosettes, both evergreen and seasonal. Geophytes, mostly dying back in winter to underground tubers or rhizomes, are also represented. Annuals are extremely few.

A few species of heathland communities, are partial root parasites (e.g. *Melampyrum pratense*, *Pedicularis sylvatica*), and a twining vascular parasite (*Cuscuta epithymum*) is widespread, except in northern heathlands. Many European heathlands are noteworthy for their rich flora of either bryophytes or lichens. About 75% of the total number of species of green plants recorded in dry-

heathlands are bryophytes or lichens; an even greater percentage has been recorded in wet-heathlands (Table 14.2). Agaric fungi are also numerous.

Relationships with other types

The communities of dry-heathlands are floristically closely related to those of tall scrub and woodlands on similar soils, including woods with *Picea abies*, *Pinus sylvestris*, *Quercus* species, and *Fagus sylvatica*. To the south of the heathland region they merge into Mediterranean scrub types, where, apart from tall species of *Erica*, the communities are generally characterised by taller shrubs with larger leaves. There is also a close

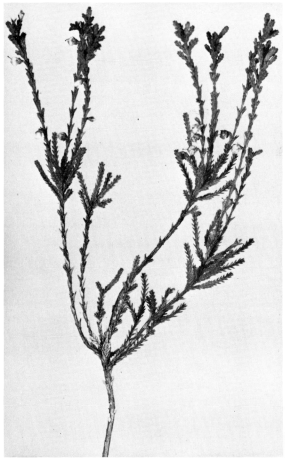

Fig. 14.7. Twig of *Calluna vulgaris* (natural size), showing two complete annual increments. (Photo E. Middleton.)

relationship to grassland, which in the absence of trees may take the place of heathland under heavy grazing, or on soils of higher nutrient status, or with increasing continentality of climate.

Wet-heathlands link closely with the vegetation of acid mires (see Volume 4 of this series), for the shrubs give place to Poaceae (*Molinia caerulea, Nardus stricta*) or Cyperaceae (*Eriophorum* spp., *Trichophorum caespitosum*) under conditions which encourage active peat formation. This transition may also be promoted by grazing and burning.

Dwarf heathland is characteristic of the low-alpine zone on European mountains (or low- and middle-alpine in the more oceanic parts; McVean and Ratcliffe, 1962). Under still more severe conditions it is replaced by bryophyte- or lichen-dominated vegetation, or by discontinuous "fell-

field" communities. Similarly, in the north, tundra heathlands give place to grass–sedge tundra, lichen tundra or moss tundra.

Root systems

On podzols, as well as on poorly aerated organic soils, heathland vegetation is shallow-rooting. The bulk of the finely branched root system of *Calluna vulgaris* (much of which is derived adventitiously from buried stems) is usually confined to the upper 20 cm of the soil profile (Chapman, 1970). A limited number of roots may penetrate more deeply, especially on freely drained, sandy soils, but where either a hard pan or a waterlogged layer is encountered they may form an intertwined mat at this level (Gimingham, 1972).

Many of the other species also root pre-dominantly in the upper 20 cm (e.g. *Erica cinerea, E. tetralix, Ulex minor, Agrostis* spp., *Festuca* spp.), but some extend to lower levels especially in sandy or gravelly soils (e.g. *Deschampsia flexuosa, Carex arenaria*) and in the deeper peats (e.g. *Molinia caerulea, Eriophorum* spp., *Trichophorum caespitosum*, all of which may extend to 60 cm or more).

Many heathland species, though not *Calluna vulgaris*, are rhizomatous [e.g. *Erica tetralix, Vaccinium myrtillus*, (Fig. 14.8)]. Others have tuberous perennating organs (*Potentilla erecta, Listera cordata*). The majority of the shrubs are mycotrophic, while *Myrica gale* and species such as *Genista* spp., *Ulex* spp. and *Lathyrus montanus* (Fabaceae) have nitrogen-fixing root nodules.

ECOLOGICAL DISTRIBUTION

Climate

The area of Western Europe indicated in Fig. 14.1 may be termed the "heathland region" for it is here that heathland vegetation occupies extensive tracts of lowland, whereas elsewhere it is confined to upland habitats. The climate of this region belongs to Köppen's (1923) **Cfb** category, which indicates moist, temperate conditions with mild winters and relatively long spring and autumn periods (Table 14.3). The mean temperature of the warmest month is normally less than 22°C, while at least four months have means above 10°C.

TABLE 14.2

Species-area data (mean number of species per quadrat) for five British Heathlands (after Hopkins, 1955 and pers. comm., 1978)

Locality	Species-area data	Quadrat size (m²)							
		0.01	0.25	1	4	16	25	100	400
Matley heathland	all species	8.6	16.7	18.2	22.1	25.0	25.7	28.0	34.0
	angiosperms	3.9	6.4	6.7	7.4	8.2	8.4	9.5	12.0
Shropshire heathland	all species	6.4	11.9	13.2	16.2	20.0	20.8	25.0	32.0
	angiosperms	1.8	2.6	3.2	3.7	4.6	5.0	6.3	8.0
Yorkshire heathland	all species exl. lichens	1.6	2.7	4.7	8.0	12.9	14.1	20.3	24.0
	angiosperms	1.1	1.4	1.9	2.4	3.6	3.6	5.3	7.0
Pennine bog-heathland	all species	6.7	14.3	17.0	26.1	33.8	35.3	41.3	46.0
	angiosperms	2.1	2.8	3.3	4.3	5.4	5.6	6.3	7.0
Perthshire bog-heathland	all species	9.6	17.5	19.5	23.7	27.7	29.5	35.8	48.0
	angiosperms	3.5	5.5	6.3	6.8	7.4	7.5	8.5	10.0

Matley heathland, 3.2 km east of Lyndhurst in the New Forest — dominated by *Calluna vulgaris*, *Erica tetralix* and *Molinia caerulea*. Total number of species present: 12 angiosperms, 11 bryophytes, and 11 lichens.

Shropshire heathland, on Long Mynd about 3.2 km southwest of Church Stretton — dominated by *Calluna vulgaris*, burnt about five years before the site was sampled. Total number of species present: 8 angiosperms, 11 bryophytes, and 13 lichens.

Yorkshire heathland, about 13 km west of Richmond — dominated by *Calluna vulgaris*, not burnt for seven to twelve years before sampling. Total number of species present: 7 angiosperms, 17 bryophytes and more than 9 lichens.

Pennine bog-heathland, Moor House National Nature Reserve, about 14.5 km south of Alston — part of a blanket bog at an altitude of 550 m with *Calluna vulgaris*, *Eriophorum vaginatum* and several species of *Sphagnum* frequent. Total number of species present: 7 angiosperms, 7 *Sphagnum* spp., 20 other bryophytes, and 12 lichens.

Perthshire bog-heathland, near Rannoch station, on an area of bog at an altitude of 290 m, burnt a few years before it was sampled. The most frequent species were *Calluna vulgaris*, *Erica tetralix*, *Narthecium ossifragum*, *Trichophorum caespitosum*, *Eriophorum angustifolium*, several species of *Sphagnum*, several liverworts and the lichens *Cladonia impexa* and *C. uncialis*. Total number of species present: 10 angiosperms, 7 *Sphagnum* spp., 17 other bryophytes, and 14 lichens.

However, heathlands occur only in the more strongly oceanic parts of the **Cfb** zone, where the annual rainfall is relatively high and well spread throughout the year (*c.* 600 to 1100 mm, falling on more than 115 rain days per year). Heathlands are, however, best developed under conditions intermediate between those of subcontinental régimes and those described as "hyper-oceanic", which are experienced in the extreme west of southern Norway, in parts of northwest Scotland and in west Ireland, where except on the steeper slopes peatlands are more characteristic than heathlands.

Many of the ericaceous shrubs and other species typical of European heathlands show more or less pronounced oceanic affinities in their geographical distribution, notably *Erica cinerea*, *E. tetralix*, and to a lesser extent *Calluna vulgaris* and *Vaccinium myrtillus*. Bannister (1970) interprets this as a requirement for mild winters, based (at least in the first two species mentioned above) on susceptibility to winter drought. Certainly mild winters, as well as the absence of extended periods of high evaporation stress in summer, characterise the region in which these plants are among the dominants of heathland vegetation.

The climatic conditions under which dwarf heathland develops on European mountains are, typically, those of the low-alpine zone (*c.* 1200 to 1500 m). In the more oceanic parts, however, as in Britain, persistent cloud and mist, low summer temperatures and a short growing season all combine to confine the dwarf heathlands to lower altitudes, for instance from 360 to 850 m in northern Scotland (Poore and McVean, 1957). Both here and in the arctic, heaths are tolerant of high levels of exposure to wind, but tend to be eliminated where wind prevents the formation of a protective snow cover in winter. On the other hand, heath communities do not survive late snow lie and are not found under snow patches.

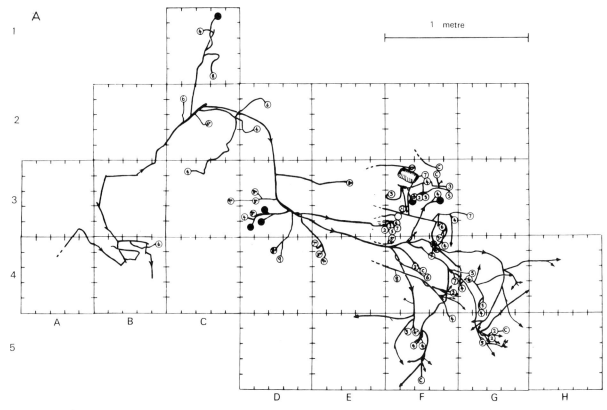

1 metre

Fig. 14.8. *Vaccinium myrtillus.* A. Excavated rhizome system *in situ*. Open circles with figures give the age in years of living branches, closed circles are dead branches of unknown age. B. Rhizome and shoot system (winter condition). (Figure from Flower-Ellis, 1971.)

5 cm

Soils

As in other parts of the world, European heathlands generally belong to acid substrata of low nutrient status. These include stabilised siliceous sand, various types of podzolic soils derived from freely drained parent materials including fluvio-glacial sands and gravels, glacial tills and weathered rock debris, oligotrophic brown earth soils, humic gleys, ranker soils and peat (Table 14.4). With some exceptions, heathlands are absent from soils rich in exchangeable nutrients, especially calcium.

The freely drained substrata give rise to dry-heathlands, which are generally associated with podzolic soil profiles. *Calluna vulgaris* and its associated ericaceous shrub species produce abundant litter, rich in polyphenols, which is relatively

TABLE 14.3

Climatic means from selected localities in the lowland heathland region of Western Europe

	Locality	Mean temperature (°C)		Rainfall mean for year (cm)	Rain days mean for year	Mean relative humidity at mid-day in the dryest month (%)
		January	July			
Norway:	Stavanger	1.4	15.3	108.5	153[1]	64 (May)
Sweden:	Simlångsdalen	−1.4	17.0	103.6	–	
Denmark:	Studsgaard	−0.6	16.1	79.0	133[1]	62 (May)
	Aarhus	−0.6	16.7	67.6	117[1]	59 (May)
Great Britain:	Aberdeen	3.0	14.2	83.8	199[2]	70 (June)
	York	3.6	16.7	62.7	179[2]	76 (June)
	Plymouth	6.4	15.8	96.0	183[2]	72 (April, May)
Netherlands:	Groningen	1.4	17.5	74.7	203[3]	59 (May)
Germany:	Lüneburg	0.0	17.2	61.2	123[1]	55 (May)
France:	Caen	4.7	17.2	68.1	155[3]	–
	Bordeaux	5.3	20.6	82.1	164[3]	54 (August)
Spain:	Santander	9.2	19.2	112.0	177[3]	71 (March)

The definition of "rain days" varies from country to country. Suffixes indicate bases of calculation as follows: [1] days with 0.1 cm rain, or more; [2] days with 0.025 cm rain, or more; [3] days with 0.01 cm rain, or more.
Table extracted from Gimingham (1972) where references are given to primary sources of data.

slow to decompose and forms an accumulation of acid raw humus (*mor*) on the soil surface. This is avoided by deep-burrowing earthworms and consequently is not readily mixed with the mineral material. However, humic acids are taken up as water drains downwards following the frequent falls of rain, and iron, aluminium and other ions are mobilised and transported to lower levels. According to the proportions in which these are deposited in the B horizon, the soils may be described as iron podzols, humus podzols or iron-humus podzols. In heath podzols considerable amounts of humic material may be seen in the A horizon, sometimes forming a distinct A_1, at others staining the whole eluvial horizon. The formation of a hard pan in the B horizon is also a common feature of heathland podzols: this may be a humus pan as in the deep podzols of the northern German or Danish plains, or a thin iron pan as in many British moorland soils.

Heathlands may also develop on oligotrophic brown earths (Coombe and Frost, 1956), but the dominance of *Calluna vulgaris* seems always to lead to acidification (Grubb et al., 1969) and eventual podzolisation. Dimbleby (1952, 1962) has presented evidence from Yorkshire, northern England, of the conversion of brown-earth soils to podzols as a result of replacement of deciduous forest by heathland in Bronze Age times (c. 1000 B.C.). Even certain eutrophic brown soils or calcium-rich substrata such as limestone or chalk soils may sometimes bear heathland, presumably when surface leaching has been sufficiently pronounced to permit the establishment of heathland species. "Chalk heath" is a well-known community including both calcicolous and heathland species (Grubb et al., 1969).

Acid soils with impeded drainage lead often to the development of wet-heathland, which is associated with the formation of thick layers of peaty humus. This type of heathland is also found at the margins of peat bogs and on wet hillsides in high-rainfall areas. Actively growing, deep saturated peat is beyond the ecological range of heathland vegetation, but where peat deposits have begun to dry out or have been subject to artificial drainage, so that the surface is aerated at least in summer, the original bog community may be replaced either by wet-heathland, or where surface drying is advanced, by luxuriant *Calluna vulgaris* heathland.

The soils of mountain or arctic dwarf heathland are usually rankers[1] and tundra rankers.

---
[1] A ranker is defined by Kübiena (1953) as a "soil formation low in lime, whose humus horizon lies immediately on the parent material, which consists usually of lime-deficient siliceous or silicate rocks".

In the majority of heathland soils pH lies in the range between 3.4 and 6.5 and the C/N ratio is high. The soils generally show relatively low levels of exchangeable cations, and are almost invariably described as deficient in phosphorus (Table 14.4). Appliction of phosphatic fertiliser leads to increase in the concentration of phosphorus in the vegetation, and in most ericaceous species to a marked improvement in flowering. The soils also frequently appear to be nitrogen-deficient, since addition of nitrogen fertiliser leads to increased shoot growth in many species, but this is presumably an expression of low rates of mineralisation of nitrogen rather than any deficiency in total nitrogen content.

ORIGINS AND STATUS

The climatic and edaphic environment of the lowland heathlands of Western Europe, described above, is clearly one which could, in most instances, equally well support forest. Indeed, remnants of "natural" forest are to be found side by side with heathland: boreal coniferous forest in the northern part of the heathland region (Norway, parts of Sweden and Scotland), and oakwood (*Quercus* spp.) or beechwood (*Fagus sylvatica*) to the south. Furthermore, examples of heathlands being invaded by trees are common (Fig. 14.9). The explanation of the origins of heathlands therefore becomes the explanation of the lack of trees.

In the case of mountain dwarf heathlands, tundra heathlands and certain heathland communities of exposed coastal sites this may be attributed to severe climatic conditions. Other heathlands are seral, for example dune heathlands and those on drying peat surfaces (pp. 386–387), and owe their existence to successional processes which, in the absence of any deflecting factor, would be expected to continue towards the development of woodland. Tansley (1939) even went so far as to suggest that some of the inland heathlands of southern England represent a stage in the post-glacial succession of vegetation, the further development of which was arrested by human intervention (burning and grazing).

In some instances, heathlands may be edaphically determined. This might apply to certain wet-

Fig. 14.9. Heathland invaded by *Betula pubescens*. Dinnet Muir, Scotland. (Photo C.H. Gimingham.)

TABLE 14.4

Soil analyses from examples of heathland on various types of substratum (Scotland)

Horizons		Sample depth (cm)	Soil separates			Organic fraction				Exchangeable cations (meq 100 g^{-1})						% Base saturation	Phosphorus mg P$_2$O$_5$ 100 g^{-1}		pH
			% sand	% silt	% clay	% C	% N	C/N	% OM	Ca	Mg	Na	K	H	Total		total	acetic soluble	
I. Coastal dune heathland: Sands of Forvie, Aberdeenshire (altitude 15 m)																			
A$_1$	dark brown humic material with sand grains	2–6	80.3	1.4	4.8	8.89	0.25	35.6	15.3	1.35	0.69	0.39	0.54	20.4	23.37	12.7	66	3.0	4.20
A$_2$	bleached layer: whitish grey sand	11–12	91.3	0.6	5.5	1.21	0.07	13.7	2.1	0.15	0.08	0.09	0.04	nil	0.36	100	30	2.0	4.80
B$_2$	yellow orange stained sand	23–27	93.9	0.5	5.2	0.24	0.02	12.0	0.4	0.15	0.09	0.04	<0.005	nil	0.28	100	30	3.0	5.61
	gleyed boulder clay underlying the blown sand	40–44	64.9	10.8	21.1	2.98	0.15	19.9	5.1	3.60	0.97	0.30	0.25	10.0	15.12	33.4	71	1.6	6.05
II. Podzol with thin iron pan: Cairn o' Mount, Kincardineshire (granite boulder clay) (altitude 380 m)																			
H	black decomposed raw humus	18–23	n.d.	n.d.	n.d.	54.0	1.27	41.0	93.0	1.375	3.64	n.d.	0.232	44.3	49.55	10.6	150	1.39	4.17
A$_2$	bleached loamy coarse sand	25–36	82.4	11.28	6.32	2.94	0.124	28.0	5.06	<0.15	<0.08	n.d.	0.019	5.5	5.52	0.35	80	0.98	4.39
A$_{2g}$	gleyed bleached layer	43–53	77.9	10.9	11.2	3.94	0.166	28.0	6.78	0.281	<0.08	n.d.	0.052	11.2	11.53	2.86	60	0.896	4.63
	Iron Pan (0.2 cm)																		
B$_3$	indurated layer	53–63	74.1	13.6	12.3	n.d.	n.d.	n.d.	n.d.	0.313	<0.08	n.d.	0.066	3.75	4.13	9.21	80	0.523	4.86
B$_3$/C	weakly indurated layer	79–91	86.6	7.1	6.3	n.d.	n.d.	n.d.	n.d.	0.132	<0.08	n.d.	0.051	1.56	1.74	10.4	80	0.612	4.83
C	loose parent material	102–117	84.5	7.95	7.55	n.d.	n.d.	n.d.	n.d.	1.91	<0.08	n.d.	0.064	1.71	3.68	53.5	70	0.613	4.86

III. Drained deep peat: Sourhope Farm, Roxburghshire (altitude c. 400 m)

	Depth																
	0–10	—	—	59.50	1.520	39.2	86.9	4.95	9.19	0.52	1.25	109.74	125.65	12.7	167	14.2	3.39
	10–20	—	—	57.40	1.540	37.2	98.8	2.48	8.20	1.08	0.75	114.27	126.78	9.8	127	10.6	3.35
	20–30	—	—	58.70	1.378	42.6	101.0	2.86	10.89	1.19	0.42	130.00	145.36	10.6	98	5.5	3.40
	41–51	—	—	63.00	1.520	41.4	109.0	1.79	11.24	1.10	0.19	133.40	147.72	9.6	69	1.7	3.50
	61–71	—	—	61.40	1.169	52.5	106.0	1.07	9.65	1.09	0.14	128.90	140.85	8.1	57	0.9	3.56
	91–102	—	—	50.40	1.551	32.4	86.9	<0.15	7.79	0.95	0.14	115.00	123.88	7.2	50	0.7	3.65
	122–130	—	—	48.10	1.398	34.4	84.0	<0.15	1.32	0.42	0.09	117.80	119.63	1.5	67	—	4.00

IV. Mountain heathland on relatively freely drained sub-alpine soil: Morven, Aberdeenshire (meta-basic igneous till) (altitude 618 m)

	Depth																
black peaty humus, few bleached quartz grains	5–13	n.d.	n.d.	n.d.	17.47	1.061	16.5	30.01	1.31	0.91	0.22	0.30	64.44	67.18	4.1	379	1.9	4.30
dark brown mealy amorphous humus, many angular stones	18–25	n.d.	n.d.	n.d.	17.39	0.925	18.8	29.39	nil	0.20	0.14	0.20	48.18	48.72	1.1	552	1.0	4.82
dark reddish brown gritty loam	33–38	77.8	11.1	4.8	4.07	0.257	15.8	6.99	nil	0.04	0.07	0.05	14.90	15.06	1.1	477	4.4	4.89
brown stony and gritty sand loam	46–53	73.4	14.3	9.9	1.82	0.114	16.0	3.13	nil	0.29	0.10	0.08	3.33	3.80	12.4	430	36.7	5.00
brown stony firm sandy loam	69–76	64.7	23.2	9.1	2.90	0.138	21.0	4.98	nil	0.06	0.17	0.24	3.81	4.28	11.0	622	52.9	5.31

Analyses by Soil Survey of Scotland. Macaulay Institute for Soil Research.

I, II and III extracted from Gimingham (1960).

n.d. = not determined.

heathlands on very poorly drained organic sub-strata, but it is difficult to demonstrate that, given a suitable climatic régime, any soil capable of supporting wet-heathland could not support some type of woodland. It was claimed by Müller (1924) that the widespread dry-heathlands on the glacial outwash plains of Jutland, Denmark, owe their existence in large measure to failure of forest development due to the presence of indurated soil horizons dating from the late glacial tundra phase. However, this view has had to be revised in the light of more recent investigations.

Even as early as 1892 it was held by Krause that heathlands in northern Germany occupied areas formerly covered by forest, and this view was further developed by Graebner (1901). Shortly afterwards, Smith (1902, 1911) showed that Graebner's interpretation could be extended to Scotland, where the frequent presence of tree stumps, roots, branches and bark preserved in humus or peat beneath present-day heathland vegetation offered good evidence of its derivation from former woodland. Since then this hypothesis has been amply supported by evidence from pollen analysis in heathland areas in Norway (Faegri, 1940), Sweden (Malmström, 1937, 1939), Denmark (Iversen, 1941, 1949; Jonassen, 1950), The Netherlands (Waterbolk, 1957), France (Duchaufour, 1948, 1956) and Britain (Godwin, 1944a, 1948, 1956; Mitchell, 1951, 1956; Morrison, 1959; Dimbleby, 1962, 1965; Smith and Willis, 1962). In all these areas the former existence of forest is confirmed, and clear indication given of its decline and replacement by vegetation dominated by Ericaceae.

This still leaves the cause of the change unexplained. Graebner (1901) thought that the replacement of forest by heathland could be explained largely as a natural consequence of continued soil podzolisation in an Atlantic type of climate. Although recognising that the process had been greatly accelerated by human intervention, this amounts to a view of heathland as at least potentially a "climax" vegetation type. Faegri (1940) maintained that, at least in southwest Norway, the prime cause was climatic change towards increased oceanicity, occurring at the onset of the Subatlantic period. This, in his view, constituted a climatic predisposition towards change, which may have been actualised or acceler-

ated by human interference. This explanation, as applied to the extreme north of the heathland region, has received support from Bøcher (1943) with reference to the Faeröerne Islands (Faeroes), and Durno (1958), whose diagrams for northern Scotland suggest that changes of this kind began before man is likely to have had a significant influence.

Under the milder climatic conditions of much of the region, however, it is difficult to account in this way for the widespread decline of forest and its replacement by heathland. Further, the evidence from pollen analysis establishes beyond doubt that this change was by no means synchronous throughout the region, and in numerous localities bore no particular relation to the onset of the Subatlantic period. Often, the vegetational change is associated with signs of human activity — charcoal layers, artefacts, and pollen of cereals, and grasses or agricultural weeds such as Plantago lanceolata. Instances of conversion from forest to heathland have now been found from as early as the late Neolithic period (c. 2500 B.C. onwards) in Denmark (Iversen, 1941, 1949), The Netherlands (Waterbolk, 1957) and southeast England (Godwin, 1944a, b). They become more frequent in Bronze Age times, and are very commonly associated with the start of the Iron Age following 500 B.C., which broadly coincided with the onset of the Subatlantic period. There are also many well-documented cases of later origin of heathlands, extending throughout the historical period up to late in the nineteenth century.

Although the coincidence of a period of increased human impact with a significant climatic shift makes it difficult to disentangle these potential causes, the evidence for ascribing the origins of heathlands very largely to human influence, throughout the greater part of the region, is now overwhelming. Pollen analysis has revealed the details of the first temporary inroads into the forest, known by the Danish term landnam, which widely preceded more permanent clearance. Landnam is heralded by a decline in the values for total tree pollen, often associated with charcoal deposits indicating the use of fire. This is immediately followed by evidence of a phase of agriculture (pollen of cereals and weeds of cultivation); then by an increase in the pollen of Ericaceae and Poaceae signifying a temporary stage of heathland or grassland prior to return of woodland as the area was abandoned. This sequence of events may be repeated several times, representing successive periods of occupation, before a more lasting retreat of the forest is shown. In heathland areas this is accompanied by a massive and persistent rise in ericaceous pollen.

Landnam sequences clearly demonstrate the ability of trees to recolonise once an area was abandoned. Any lasting replacement of forest by heathland therefore implies continued use of the land by man and the maintenance of some form of management. It was, in fact, because the open areas, whether heathland or grassland, were found to be valuable for the increasing stocks of domestic herbivores (mainly cattle and sheep) that they were retained in this form by grazing management supplemented, as necessary, by burning.

At first, only a small proportion of the extensive forest cover of Europe was destroyed to make way for heath or grassland, mainly in coastal and lowland districts in the neighbourhood of settlements. However, during historical times the process gathered monumentum as settlements and populations expanded. For example, in Denmark, Iversen (1964, 1969) has shown that certain heathlands originated in Viking times (e.g. 740 and 830 A.D.), and there is evidence of considerable expansion of heathland both in England and Europe between 1100 and 1200 A.D. Land was required for cultivation and grazing, but the forests also suffered from demands for timber for both constructional purposes and for fuel. In particular, the use of charcoal for iron smelting resulted in forest destruction being carried into the more remote and upland areas. The resulting open country provided opportunities for increasing the scale of pastoral agriculture. With the advent of hardy breeds of cattle and, more especially, sheep, there was a progressive extension northwards in Europe of the use of the steadily expanding heathlands for grazing. Nowhere was this more apparent than in Britain, where the practice of grazing large flocks of sheep (mainly the Cheviot and Black-faced breeds) on heathlands, to the exclusion of all other domestic animals, extended from northern England into Scotland in the late eighteenth and nineteenth centuries.

Throughout Western Europe the heathlands proved useful for pasturing grazing animals, and were especially valuable as a source of winter forage (Romell, 1951, 1952). For this reason, they were maintained and managed up to the latter part of the nineteenth century, but from that time onwards increasing intensification of agriculture in the more fertile districts reduced the value of heathlands. Throughout most of Western Europe and Scandinavia they have been progressively "reclaimed" for cultivation or afforestation. Only in northern England and Scotland are they still used extensively for sheep farming.

A further factor contributing to the survival of heathlands in Britain is the presence of the red grouse (*Lagopus lagopus scoticus*). Open heathland constitutes the natural habitat of this game bird, which increased considerably in numbers as a consequence of the expansion of heathlands. The interest of sportsmen in grouse shooting has more than compensated for some decline in the profitability of sheep farming. This has ensured, for the time being at least, the retention of large tracts of heathland in the upland districts of England and Scotland, which are managed as "grouse moors".

FIRE

The effects of regular management of European heathlands, particularly those in Britain, by fire are more fully discussed in Chapter 49. However, brief reference must be made here to fire as a factor of great importance in regard to both the origins and the continuance of many European heathlands. It is reasonable to suppose that, before human intervention was significant, localised fires due to natural causes were amongst the factors responsible for the occurrence of open patches in the forests of Western Europe, which on suitable soils were occupied by vegetation akin to that of heathland. After forest clearance, the value of open heathland for grazing purposes could only be sustained if trees were prevented from re-invading and *Calluna vulgaris* (the chief grazing plant) was kept in a productive condition. For reasons to be discussed in Ch. 26, Volume B, these objectives have generally been achieved by burning, and to a greater or lesser extent most West European heathlands (other than those at high altitudes or on the coast) have been subjected, from time to time, to fire (Fig. 14.10).

Fig. 14.10. Management of *Calluna*-heathland in Scotland by burning in March. (Dead fronds of *Pteridium aquilinum* in foreground.) (Photo C.H. Gimingham.)

Occasional burning probably has only a minor effect on the floristic composition of heathland communities. However, in so far as this factor contributes to the maintenance of heathlands, the vegetation may be said to be adapted to fire and to show some of the characteristics of a fire climax. The widespread dominance of *Calluna vulgaris* may, at least in part, result from the fact that it regenerates quite readily from stem bases which, especially when buried in litter or humus, may survive the passage of fire, and there is some evidence that seed germination is improved by short periods of heat pre-treatment (Whittaker and Gimingham, 1962).

When burning is relatively frequent, there is a considerable effect on floristic composition, which may be simplified and impoverished. This is particularly evident in Britain, where *Calluna vulgaris*, sometimes with an understorey of *Erica cinerea*,

may be virtually the only low shrub species. The remaining flora consists largely of relatively fire-resistant species such as *Potentilla erecta* (which has an underground tuberous rootstock), *Vaccinium myrtillus* or *Erica tetralix* (with rhizomes). Many typical heathland species are fire-sensitive (or sensitive to the combination of burning with grazing) and are consequently reduced or lacking: examples are *Juniperus communis*, *Genista anglica* and *Polypodium vulgare*. The diversity of bryophytes is also reduced, though a few remain as constant species of fire-managed heathland, e.g. *Pohlia nutans* (Shimwell, 1975). On the other hand, regular burning often encourages the development of lichen-rich heathland communities (Ward, 1970, 1971a, b), although again certain species are fire-sensitive.

Heathland communities in which vascular species other than *Calluna vulgaris* are very poorly represented are therefore generally indicative of repeated burning. They have frequently been named *Callunetum vulgaris* (e.g. McVean and Ratcliffe, 1962), but it is usually possible to find fragments of stands which have escaped such severe treatment and are less impoverished. These generally indicate that the managed vegetation has been derived from one or other of the heathland community types to be described in the next section (Shimwell, 1975).

COMPOSITION — REGIONAL VARIATION

In general, European heathland communities are not floristically rich. None the less, their geographical area embraces an extensive range of climatic and edaphic régimes, resulting in considerable variation in community composition. This has been the subject of a large number of investigations, not least because the Zürich–Montpellier and Scandinavian phytosociological systems evolved in Europe with the result that, in common with other vegetation types, much discussion has centred on the problems of classifying heathland communities. While a substantial measure of success has been achieved in the search for an acceptable treatment of heathlands according to the Braun-Blanquet system, it is also evident that heathlands provide an excellent example of continuous variation in floristic composition and repay study on this basis. Both approaches have contributed conspicuously to an understanding of regional variation in the composition of heathlands (Gimingham, 1961, 1969, 1972).

An outline of the main trends of variation follows, using the primary division into dry-heathlands, wet-heathlands and mountain heathlands introduced on p. 367. In view of the fact that agreement is still lacking on the details of a phytosociological hierarchy, it is not possible to present any one system as generally accepted. However, correspondences will be indicated, as appropriate, to those categories which have received widest currency.

Dry-heathlands

Since 1949 these have very generally been included in a class named Nardo-Callunetea (Preising, 1949), though a number of authors have adhered to an earlier category, Calluno-Ulicetea, established by Braun-Blanquet and Tüxen in 1943. Each of these titles gives some indication of the major floristic affinities of the class. Within this, the bulk of European dry-heathlands are commonly assigned to the order Calluno-Ulicetalia (Tüxen, 1937), but certain communities in the extreme south of the heathland region are separated under Erico-Ulicetalia (Braun-Blanquet et al., 1964).

Throughout the dry and mesophilous heathlands of low and middle altitudes in Western Europe, *Calluna vulgaris* is an almost universal component, often dominant. However, this is inclined to give a false suggestion of floristic uniformity because, as indicated in a preliminary way by the division into the two orders referred to above, there are substantial differences between northern and southern heathlands, also between western (Atlantic) heathlands and those in more easterly situations (De Smidt, 1967). These trends follow climatic gradients, but superimposed on them are variations in composition related to the moisture and nutrient status of the soils and other factors such as management by burning. In addition, maritime influences are reflected in floristic composition, and dune heathlands, although related to the type characteristic of the geographical location, contain distinctive species [such as *Carex arenaria* and *Ammophila arenaria* (Fig. 14.11)].

Leaving aside for separate mention heathlands of the more highly oceanic, Atlantic seaboard of Europe, and starting in the north, a widely distributed type has been described as "boreal heather moor" (Birse, 1976), in which, in addition to *Calluna vulgaris*, species having northern patterns of distribution are strongly represented, notably *Vaccinium myrtillus*, *V. vitis-idaea* (Fig. 14.7), *V. uliginosum*, *Empetrum nigrum* and mosses such as *Hylocomium splendens* and *Pleurozium schreberi*. This type of community occurs in the Faeröerne islands, southern Norway, southern Sweden, Denmark, Scotland and northern England, becoming increasingly confined to north-facing slopes in districts further south, as for example in northern Germany and The Netherlands.

While *Calluna vulgaris* is very frequently the structural dominant, this is not always the case and one of the other ericaceous species may assume this

Fig. 14.11. *Empetrum nigrum* (right foreground) and *Calluna vulgaris* on stabilised sand-dune, accompanied by *Ammophila arenaria*, *Festuca rubra* and lichens. (Photo C.H. Gimingham.)

rôle. Their contribution to the community varies considerably: for example *Vaccinium myrtillus* is particularly associated with the more northerly and oceanic heathlands from Norway to northern Germany, becoming dominant even to the exclusion of *Calluna vulgaris* on exposed, windswept ridges in northern England. These communities may contain a number of species with strongly oceanic affinities, for example *Blechnum spicant* and, particularly in southwestern Norway and Britain, *Erica cinerea*.

Vaccinium vitis-idaea contributes to many of these heathlands in southern Norway, Sweden (Malmer, 1965) and Scotland (Gimingham, 1964), and is especially characteristic of the inland heathlands of Jutland (Denmark), where the more strongly oceanic species tend to be reduced. Very similar heathland-types occur in upland districts of Belgium (see p. 386). Throughout the areas mentioned, *Empetrum nigrum* is a frequent component, sometimes dominant. It appears to be best represented in the heathlands of rather more oceanic regions than those of which *Vaccinium vitis-idaea* is characteristic, notably along the western fringes of southern Sweden, Jutland (Denmark), and The Netherlands. In these regions it also contributes strongly to coastal dune heaths,

and corresponding communities occur on the coasts of Scotland.

A further species of importance in some examples of boreal heathland is *Arctostaphylos uva-ursi*, which is described by Bøcher (1943) as a "boreal dry soil plant" and tends to occupy soils of rather higher nutrient status, with less raw humus accumulation, that those of the communities already mentioned. It occurs under strongly oceanic conditions in southwestern Norway and is a constant of a well-defined community from the eastern central uplands of Scotland (investigated by Ward, 1970, 1971a, b), but it is also well represented in certain heathlands of southwestern Sweden (province of Halland) and Denmark where subcontinental conditions are approached.

Bøcher (1943) divided these communities into two alliances — Myrtillion boreale and Empetrion boreale. Owing however to the complex variation in composition, they have commonly been united in one alliance, sometimes termed Myrtillion (Bridgewater, 1970) but perhaps more commonly Empetrion boreale, as in Birse (1976), or Empetrion nigri (Schubert, 1960).

In the region of southern Scandinavia (Damman, 1957), northern Germany, and The Netherlands there is overlap, or intergradation, between this boreal heathland type and a group of communities generally united in the alliance Calluno-Genistion (Duvigneaud, 1944). In the latter, species of northerly distribution are reduced in quantity, or lacking, and others with more southerly affinities are prominent. Among these are *Genista anglica* (a plant which also belongs to some of the boreal heathlands, for example in Scandinavia and Scotland, where it is specially abundant in the *Arctostaphylos uva-ursi* communities), particularly in the more oceanic districts, *Genista pilosa* in suboceanic areas, and *G. germanica* and *G. tinctoria* in subcontinental parts. Another widespread characteristic species is *Cuscuta epithymum*. In general the bryophyte flora of these heathlands is poorer than that of boreal heathlands, whereas the lichen component may be diverse. Northern outliers of this heathland type occur in southern Sweden, Denmark and southeastern England, but it becomes increasingly extensive in northern Germany, The Netherlands and Belgium, and intergrades with other types in northern France.

All these communities may include varying amounts of a tall shrub component, notably *Juniperus communis* (Fig. 14.3). This species, however, is poorly represented in most British heathlands (perhaps due to the long history of grazing and burning), and tends to become less prominent southwards from The Netherlands. *Ulex europaeus* is also widely present in heathlands, and (especially in parts of northern France and Britain) may be very abundant, with a tendency to become dominant to the extent of completely excluding heathland vegetation. Its status in heathland vegetation is, however, uncertain, and has probably been enhanced by soil disturbance. *Sarothamnus scoparius* is a further tall shrub species which contributes quite widely to heath communities, particularly those in a sector extending from the south of The Netherlands to northern France, and has been used in naming an alliance Sarothamnion scopariae (Preising, 1949, quoting Tüxen, unpublished).

Heathland communities of the more highly oceanic western margins of the region have certain distinctive features. While they have generally been included in the same order as those surveyed above, a recent treatment (Géhu, 1975) separates them under Ulicetalia minoris, with the com-

munities belonging mainly to the north, northwest and more easterly areas (described above) constituting the order Vaccinio-Genistetalia. In the most westerly parts of southern Norway *Erica cinerea* is very strongly represented in the heathland communities. Although they share a number of plants with heathlands of the northerly Empetrion boreale, the presence of a group of species having markedly western distribution patterns, such as *Carex binervis*, is distinctive. Closely similar communities occur in northern and northwestern Scotland (Fig. 14.12) and in Ireland. Continuing southwards, *Erica cinerea* remains characteristic of strongly oceanic heathland types, especially in northern France, but also in southwestern France and northern Spain. However, from Ireland and Wales southwards, low-growing species of *Ulex* become an integral part of these communities (Fig. 14.13). In the most oceanic parts (Ireland, Wales, southwestern England, northern and northwestern France) *Ulex gallii* is important. *Ulex minor* replaces *U. gallii* eastwards from the county of Dorset in southern England, and is also very prominent in the oceanic heathlands of northern France, extending west and south where it is often mixed with *U. gallii*. All these oceanic heathlands are commonly linked in the one alliance Ulicion minoris (=Ulicion nanae; Duvigneaud, 1944), but some authors recognise a separate alliance, Ulicion gallii (Des Abbayes and Corillion, 1949).

An additional ericaceous species which has an outlying station in the extreme southwest of Britain, *Erica vagans*, becomes significant in many of the oceanic heathland communities of northern France and extends southwards into Spain. In the southwest of France and in the Basque and Cantabrian regions of Spain (Dupont, 1975) this species, together with *Daboecia cantabrica* (which also has outlying stations in the British Isles, in western Ireland), forms part of a community of great floristic richness which has been variously linked with the Ulicion minoris or with an alliance named Ericion umbellatae (Braun-Blanquet et al., 1952) (=Cistion hirsutae; Braun-Blanquet et al., 1964). This lack of uniformity in practice emphasises the difficulties in achieving a satisfactory hierachy of categories, especially in view of the fact that the latter alliance is commonly placed in a different order, the Erico-Ulicetalia, from that to

Fig. 4.12. Oceanic, maritime heathland-type containing *Calluna vulgaris*, *Erica cinerea* (in flower) and low-growing *Juniperus communis*. North coast of Scotland (Photo C.H. Gimingham.)

Fig. 14.13. *Calluna vulgaris–Ulex minor* heathland, Dorset, south England. (Photo C.H. Gimingham.)

which the former is assigned. The order Erico-Ulicetalia comprises southerly heathland and scrub types which contain a variety of ericaceous species with predominantly southern, often Mediterranean patterns of distribution, such as *Erica umbellata*, *E. mackaiana*, *E. australis* ssp. *aragonensis*, *E. arborea*, *E. mediterranea*. These communities occur mainly in southwestern France, and in the coastal lowlands of northern and northwestern Spain, extending into Portugal. In the latter areas, the heath species are increasingly associated with taller, larger leaved shrubs, for example species of *Cistus*. When these predominate in the vegetation, the term scrub (e.g. maquis-scrub) applies instead of heathland.

Many of the heathland communities referred to in this section show variations in composition related to the nutrient status of the soil. This bears in particular on the prominence of grass species which, in addition to a variety of forbs, increase in importance as nutrient status increases. Communities in which this trend is most evident are often described as "herb-rich" or "species-rich", and plants such as *Viola riviniana*, *Thymus drucei*, *Lathyrus montanus*, *Hypericum pulchrum*, *Lotus corniculatus* and *Campanula rotundifolia* may be cited as indicators.

Wet-heathlands

In all parts of the heathland region there are gradients from dry, freely drained soils through moist soils with varying degrees of impedence of drainage, to those which are waterlogged for varying periods of the year, leading often to peat formation. The corresponding gradients in floristic composition have been divided into categories such as "dry", "mesophilous", "humid" and "wet". Most of the dry-heathland community types mentioned in the foregoing section have "mesophilous" (or "moist"), and "humid" variants. For example, within the Ulicion nanae in northern France and southern England there is a distinctive "mesophilous" community (Géhu, 1975) containing *Erica ciliaris* along with *Calluna vulgaris* and *Erica tetralix*. In similar habitats in western France, *Erica scoparia* may be associated with this group of species. Where the description "humid" has been applied, *Erica tetralix* is abundant, in association with species derived from the "dry" end

of the gradient such as *Calluna vulgaris*, *Ulex minor* or *U. gallii* etc., and others from the "wet" end such as *Molinia caerulea*.

Communities which are properly described as wet-heathland, however, are commonly separated into a completely different class, Oxycocco-Sphagnetea. As the title indicates, the class is distinguished by the important role of *Sphagnum* spp. Within it various names have been given to groups of wet-heathland communities, generally stressing the significance of *Erica tetralix* [e.g. order Ericetalia tetralicis (Moore, 1968); alliance Ericion tetralicis (Schwickerath, 1933)]. In addition to the plants already mentioned, there are varying contributions from graminoid species such as *Molinia caerulea*, *Eriophorum vaginatum*, *Juncus squarrosus* and *Trichophorum caespitosum*, while *Polytrichum commune* is often present, forming large clumps. Geographical variation in composition is, in general, rather less in wet- than in dry-heathlands. However, in the northern part of the region *Empetrum nigrum* is a conspicuous component of some wet-heathlands, while *Erica ciliaris* and *E. scoparia* distinguish southern wet-heathlands.

Mountain heathlands

Heathlands at the higher altitudes are distinctive in structure (see pp. 367, 369), and also to a considerable extent in composition. There may be an arctic–alpine element in the flora, in addition to a number of species such as *Vaccinium myrtillus*, *V. vitis-idaea*, and *V. uliginosum*, which as well as being typical of the more northerly lowland heathlands are tolerant of subalpine and low-alpine conditions. These occur widely in heathlands on European mountains.

It is possible here to give only a very general indication of the extensive range of dwarf heathland communities. The transition between subalpine scrub (fringing the upper limit of forest) and low-alpine dwarf-shrub heathland is often the habitat of a dwarf juniper (*Juniperus communis* ssp. *nana*) community, with *Calluna vulgaris* and other ericaceous species such as *Arctostaphylos uva-ursi* or *Arctous alpinus*. Communities of this type occur from Scandinavia and Britain to the mountains of southern Europe, and have been grouped in an alliance Juniperion nanae (Braun-Blanquet et al.,

1939). Dwarf *Calluna vulgaris* is widely dominant in heathlands above the natural timberline, although it disappears at the higher altitudes. A community in which *Calluna vulgaris* and the moss *Rhacomitrium lanuginosum* together form a low, dense carpet is found on mountains under the oceanic conditions of northern Britain (giving place to *Rhacomitrium–Empetrum* heathland at higher altitudes), while dwarf *Calluna–Vaccinium* (*V. vitis-idaea* and *V. myrtillus*) heathlands are more widespread, in both oceanic and suboceanic areas, for example in Scandinavia (Nordhagen, 1928, 1936), Britain, Germany (Harz, Sauerland), Belgium and France (including *Calluna–Vaccinium myrtillus* communities in the Pyrenees). Avoiding the more oceanic regions are heathlands with *Phyllodoce coerulea* (suboceanic–subcontinental, Bøcher, 1943), and *Cassiope tetragona* (subcontinental–continental). On soils of relatively high nutrient status herb-rich subalpine heathlands occur, for example the *Calluna–Antennaria dioica* association.

A further factor which profoundly affects the composition of mountain heathlands is the duration of snow cover (Dahl, 1956). This is particularly evident on examination of the range of communities lying above the limits of *Calluna vulgaris* as a dominant. In exposed areas where snow cover is thin and of relatively short duration, *Loiseleuria procumbens* is often the chief species, giving its name to an alliance, Loiseleurieto-Vaccinion, which is represented in almost all the major mountain systems of Europe. In similar situations on calcareous soils *Dryas octopetala* (Rosaceae) heathlands are found, in which this species is often mixed with *Empetrum hermaphroditum*. The latter, along with *Vaccinium myrtillus* and *V. uliginosum* forms communities, frequently rich in lichens, where snow cover lasts longer. A very similar community, generally rich in bryophytes, is associated with snow beds. In the Alps and other mountains of southern Europe the series is completed by a taller and particularly handsome community containing *Rhododendron* species (notably *R. ferrugineum*) where the protective snow cover lasts longest. The *Vaccinium* and *Rhododendron* heathlands are commonly grouped in Rhodoreto-Vaccinion, belonging, with the Loiseleurieto-Vaccinion, to the order Vaccinio-Piceetalia (class Vaccinio-Piceetea). This arrangement expresses the floristic relationships between montane forest vegetation and the dwarf heathlands above its limits.

SUCCESSION

Only those heathlands which belong to altitudes above the timberline, and perhaps certain types of wet-heathland on peat and exposed maritime heathlands, can be regarded as stable or semi-stable. Most of the rest of the European heathlands are either seral or are perpetuated as heathlands by continued management.

Successions involving heathland communities are of two main types. The first comprises vegetation and habitat changes in peat bogs. These may result from long-term shifts of climate, leading to progressive reduction in the ombrogenous supply of water to bog surfaces. Analysis of peat deposits indicates successional trends involving increases in the quantities of *Erica tetralix* and *Calluna vulgaris*, with the establishment of wet-heathland in place of bog. Continuance of this trend has led, particularly in more continental regions, to replacement of heathland by scrub or woodland on the drained and aerated peat, but reversal of the climatic trend (such as occurred at the Subboreal–Subatlantic transition) brought about in the more oceanic areas a return of the bog vegetation, usually with dominance of *Sphagnum* species.

The second main type of succession which includes heathland is that on non-calcareous coastal sand dunes. In many parts of the Atlantic coasts of Europe, heathland develops on stabilised dunes, when sand accretion has largely stopped and the pH has fallen to about 6.5 or less. At this stage, *Calluna vulgaris* and in some areas *Erica cinerea* (western Britain, northern France) or *Empetrum nigrum* (Scotland, western Scandinavia, The Netherlands), can colonise the patches of bare sand which remain among the shoots of species such as *Ammophila arenaria*, *Carex arenaria* and *Festuca rubra* (Fig. 14.11). In view of the rarity of seedlings of the heathland plants, it seems probable that suitable conditions for forward colonisation by heathland species in a developing dune system occur infrequently, and that the succession proceeds intermittently rather than continuously. It may also be considerably affected by rabbit-grazing in some areas (Gimingham, 1972). In time, where seed-parents of shrubs and trees are present, the succession may proceed to scrub and woodland, but this is often prevented by human interference.

In the same way, most lowland heathlands are potentially dynamic and, if management ceases, may be invaded by tall shrubs or trees (Fig. 14.9). The most rapid invaders are often *Betula* spp. (*B. pubescens*, *B. pendula*), *Juniperus communis* and *Sorbus aucuparia*. In appropriate parts of the heathland region other shrubs which colonise heathlands include *Sarothamnus scoparius*, *Frangula alnus*, *Prunus spinosa* and *Crataegus monogyna*, and tree invaders may be *Quercus* spp. (*Q. robur*, *Q. petraea*) or *Fagus sylvatica*. However, the time required for the entry of tall shrub or tree species depends not only on the rate of arrival of their seeds, but also on the structure of the heathland community concerned. In the case of *Calluna* communities, this varies considerably in relation to the age structure of the stand and the growth phase of individual plants of *Calluna vulgaris*.

CALLUNA VULGARIS

Because *Calluna vulgaris* is for various reasons so often the dominant species in European heathlands (Gimingham, 1960), a regional survey is incomplete without special mention of its influence on the structure and dynamics of these communities. In all but the mountain and wet-heathlands, the life span of an individual plant normally lasts between 30 and 40 years. During this time the processes of growth and maturation lead to various changes in the general morphology of the plant (Fig. 14.14), which were first described by Watt (1955) under the following headings:

The pioneer phase. In the early stages of development the shape of the young plant is more or less pyramidal, with rather regular branching from the axis of a single 'leader'. However, the apex of the leading shoot seldom continues growth for more than two seasons, when it is replaced by two or more branches arising just below the tip (Fig. 14.7), while other laterals become equivalent to the leader, producing a radiating, bushy type of

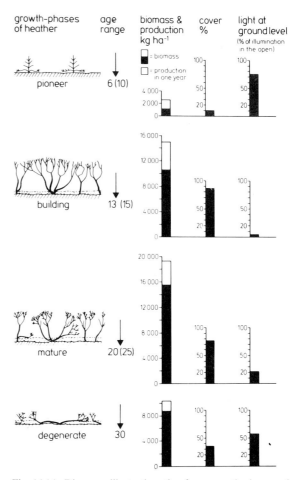

Fig. 14.14. Diagram illustrating the four growth-phases of *Calluna vulgaris*, and associated changes in biomass and production, cover and illumination at ground level.

growth. In normal habitats, the plant has lost the appearance of a pioneer before it is 6 years old. By this time, the scattered or clustered pioneer plants have expanded, and cover has increased, often approaching 100%.

The building phase follows, normally until the plant is rather over 15 years old. Productivity of peripheral green shoots is high, as is the ratio between the weight of these and that of the woody framebranches, though this decreases with age. Flowering is profuse. Isolated plants are hemispherical in shape, but in a dense stand a canopy forms at about 30 to 40 cm above the ground, borne on branched woody stems. In either case very little light penetrates to ground level and vigorous heath in this phase excludes most other species.

The mature phase, lasting often until the plant is well over 20 years old, is characterised by some reduction in extension growth at the periphery, where the leaf-bearing shoots become more condensed and usually darker in colour. The middle branches begin to separate, in time forming a central gap which allows increased illumination at ground level, and air circulation within the bush.

The degenerate phase is marked by further collapse of the central branches (often under the weight of winter snow) and their death, progressively from the middle of the bush outwards, extending the gap (Fig. 14.15). For a time, the peripheral branches may remain alive, because frequently they have become partially buried in moss, litter or humus and have produced adventitious roots. Dead branches remain visible in the centre of the patch,

surrounded by a ring of living shoots which may, temporarily, remain quite dense. Eventually the whole dies back.

In a *Calluna*-dominated heathland, where the population of this species is uneven-aged, plants of each of these growth phases exist side by side producing an uneven, patchy structure (Watt, 1947). The effects of *Calluna vulgaris* on the microhabitat change very substantially as it passes through the sequence of growth phases, altering the possibilities for other species to co-exist with the dominant in a given patch (Barclay-Estrup, 1970, 1971). As a result, on any one patch, as *Calluna vulgaris* passes through its sequence of phases there are accompanying changes in the diversity and cover contribution of other species (Fig. 14.16).

The number of associated species and their cover usually reach their maximum while *Calluna vulgaris* is in the pioneer phase, for often there are patches of bare ground available for colonization,

Fig. 14.15. Degenerate bush of *Calluna vulgaris*. The central branches are all dead, creating a gap which is colonised by bryophytes and lichens (*Parmelia physodes* on old branches of *Calluna*). (Photo C.H. Gimingham.)

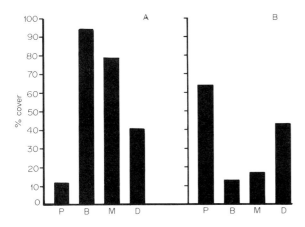

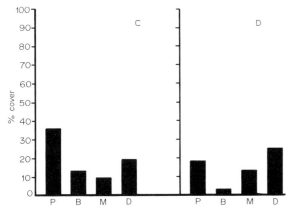

Fig. 14.16. Histograms showing the cover contribution of various categories of the vegetation in areas occupied by *Calluna vulgaris* in each of its growth-phases (data from nine 1-m² quadrats). From Barclay-Estrup and Gimingham (1969). A. *Calluna vulgaris*. B. Other dwarf-shrubs. C. Graminoid herbs. D. Bare ground and lichens.
Legend: *P*=pioneer phase; *B*=building phase; *C*=mature phase; *D*=degenerate phase.

TABLE 14.5

Number of species of vascular plants per 4 m² in *Calluna vulgaris* stands of various ages: Perthshire, Central Scotland

Phase	Age of *Calluna* stand since last fire (years)	Number of species of vascular plants
Pioneer	3	15
Late pioneer	5–6	22
Building	8	16
Late building	13–14	11
Late mature	*c.* 25	14

while spreading plants may grow into the spaces between the young individuals of *Calluna* (Table 14.5). As the latter coalesce and enter the building phase, they begin to exert their greatest effect and the contribution of other species declines, almost to the point of extinction. When, however, the canopy begins to thin out in the centre of the bush with the onset of the mature phase, the first plants to recolonise the ground below are shade-tolerant bryophytes such as *Hypnum cupressiforme* ssp. *ericetorum*. Later, as the gap develops, the patch of this moss may be invaded by others, such as *Pleurozium schreberi* and *Hylocomium splendens*. When the degenerate phase is reached, not only is

there normally a period in which both bryophytes and lichens increase in abundance and cover, but also various vascular plants spread in or colonise, such as *Arctostaphylos uva-ursi*, *Vaccinium* spp., *Empetrum nigrum*, *Deschampsia flexuosa* and *Dryopteris dilatata*.

Evidence has been presented (Watt, 1947; Barclay-Estrup and Gimingham, 1969) for regarding this process as, in many instances, cyclical (Fig. 14.17). On occasion, if bare ground is exposed in the centre of a degenerate bush, young plants of *Calluna* may be able to establish before or along with other species, fairly rapidly re-occupying the space. More frequently, however, the developing community of other species prevents this but after they have progressed through their own 'developmental sequences their cover declines and this is the opportunity for *Calluna* to re-invade and commence the cycle anew. Allowing for up to 30 years of occupancy of a site by an individual of *Calluna*, and perhaps nearly as much again by other species, the whole cycle may sometimes require over 50 years for completion. Such cycles are therefore only in evidence in heathlands where management by burning or grazing is absent or minimal, but under these conditions the number of species participating in a heathland community, in association with dominant *Calluna vulgaris*, may be quite large.

It is especially in the gap phase (i.e. the period extending from the onset of degeneracy in an old bush to the arrival of new pioneer plants in the gap), that associated species have their chief opportunity. By the same token, it is in this part of the cycle that there is opportunity for foliage of a competitor such as *Pteridium aquilinum* to appear (Watt, 1955), or for the establishment of seedlings of shrubs or trees. When this occurs further

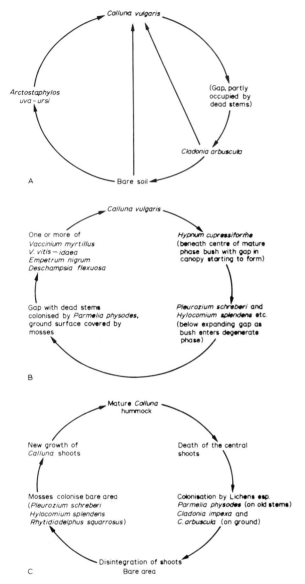

Fig. 14.17. Examples of cyclical change in *Calluna*-dominated heathland communities. A. In a *Calluna vulgaris–Arctostaphylos uva-ursi* community, east central Scotland (after Watt, 1947). B. In a *Calluna vulgaris–Vaccinium* community. C. In a dune heathland.

changes are likely to be successional in nature rather than cyclical.

Managed stands, unlike those just discussed, are even-aged, because regeneration following a fire is normally uniform and synchronous. Where vegetative regeneration predominates the pioneer phase is abbreviated and the stand may pass in a very few years into the building phase. It is the objective of

burning to maintain as much as possible of a heathland area in the building phase, and to prevent any stand progressing into the mature or degenerate condition. Hence, much of the area is occupied by what is virtually a monoculture of *Calluna vulgaris* in its most vigorous and exclusive condition. It is, therefore, not necessarily the direct effects of fire which alone explain the floristic impoverishment of managed heathlands (p. 380), but also the indirect (competitive) effects of *Calluna* monocultures in the building phase.

A similar comment may be applied to the prevention of invasion by shrubs and trees in heathlands. Fire, in itself, often creates ideal conditions for the germination of their seeds and establishment of seedlings (e.g. *Betula* spp., Table 14.6; *Pinus sylvestris*). However, if re-establishment of a *Calluna* canopy is rapid and thorough, there is high mortality of tree seedlings (Table 14.6). When this effect is reinforced by grazing, elimination may be complete; but otherwise some may survive (especially when a good seed year has produced seedlings in large numbers on a burnt patch).

TABLE 14.6

Numbers of individuals (seedlings and young trees) of birch (*Betula pendula*) in areas of 20 m^2 in *Calluna vulgaris* stands of various age: Dinnet Muir, Aberdeenshire, northeast Scotland (the stands in each site were close together, on level ground and similar soil; they were within 15 m of mature birch trees producing abundant seed)

Calluna vulgaris stands		Numbers of *Betula pendula* (means of two replicates ± standard error)
phase	age (years)	
Site 1		
(Recently burnt)	1–2	55.0 ± 21.0
Late pioneer	8+	6.5 ± 0.5
Building	10–12	3.0 ± 0
Early mature	15	0.5 ± 0.5
Degenerate	>18	1.0 ± 1.0
Site 2		
(Recently burnt)	1–9	268.0 ± 162.0
Late pioneer	7–9	6.5 ± 0.5
Building	11–13	2.5 ± 0.5
Early mature	14	1.0 ± 0
Degenerate	>20	0.5 ± 0.5

Data from Dr. W.K. Gong.

PRODUCTION AND ORGANIC MATTER ACCUMULATION

When used in its strictest sense the term heathland applies to areas where some member of the Ericaceae dominates the vegetation. In a wider sense however, the term is often used to describe a particular kind of countryside or scenery that includes a number of associated types of vegetation such as bracken (*Pteridium aquilinum*), gorse (*Ulex* spp.), birch scrub (*Betula* spp.) and a range of peatland and mire communities. Although it would be interesting to discuss patterns of primary production and accumulation of organic matter in relation to succession, soil, climate and land use for such a range of vegetation types, the most that can be attempted for European heathlands at the present is to make comparisons between areas dominated by the common heather (*Calluna vulgaris*). The following account will concentrate upon areas of mineral soils rather than soils which are primarily organic in content.

Estimates of primary production may have been undertaken by different workers for a number of reasons, and to a great extent these reasons determine the data that are available for discussion and comparison. Production data may have been obtained for comparisons of apparent efficiency or function between types of vegetation, or between similar types occurring at different sites. Reliable estimates of standing crop and primary production are an essential prerequisite for the construction of nutrient budgets and analysis of energy flow, both of which are important in attempting to understand the functional relationships within any biological community. In a system such as heathland where nutrients are in short supply the measurement of inputs, losses and storage are especially important, and an understanding of such factors is of value when considering the possible effects of management processes that may well have long-term consequences for the overall dynamics and survival of the system.

There is an obvious set of relationships between production, decomposition, grazing and the resultant accumulation of organic matter and nutrients in the ecosystem. Such relationships between litter production, loss and accumulation in forests (Jenny et al., 1949; Nye, 1961; Olson, 1963) and blanket bog (Gore and Olson, 1967) have been examined using mathematical models that assume the attainment or existence of steady-state conditions. Where such conditions can be assumed it simplifies matters, but in the case of heathland ecosystems it soon becomes apparent that the majority of sites are by no means in such a state. The application of Watt's development phases (Watt, 1947) to heathland studies (Barclay-Estrup, 1970; Chapman et al., 1975a, b; Chapman and Webb, 1978; see also pp. 387 390) demonstrates this point. Production and associated processes within the heathland ecosystem are clearly related to this developmental sequence and any results obtained must therefore be considered and discussed against the age of the vegetation. As the standard definitions of the heathland development phases (see Gimingham, 1972, pp. 125–127) are subjective, and levels of production are not constant throughout any single phase, it would be more satisfactory to sample a series of sites of different ages and make use of growth or accumulation curves (Chapman et al., 1975a, b). This approach however can only be used where a sufficiently complete age series exists, and many prevailing management practices and accidental fires make older stands of heather rare or nonexistent in many areas.

The mode of growth and pattern of shoot development in *Calluna vulgaris*, as described by Mohamed and Gimingham (1970), has been important in determining the methods used and the particular parameters measured by workers investigating the primary production of heath and moorland areas. The current year's growth of shoot material is in the form of new lateral green material, leading long shoots that may bear flowers, and recognisable growth increments to previous years' short shoots. Short shoots may persist for up to four years, depending upon the locality and the particular site. This mode of growth and pattern of shoot survival results in a litter-producing system that is dependent upon up to four years of green shoot production. The flow and interrelationships of organic matter through *Calluna* in the heathland system are shown in Fig. 14.18, where the compartments on the upper line represent the components of net primary production.

Whilst estimates of primary production and organic matter accumulation have been made on a number of European heathlands, only relatively

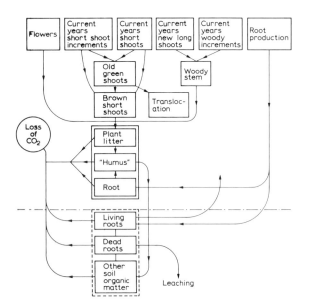

Fig. 14.18. Box model representing the transfer of organic matter in the growth of *Calluna vulgaris*. (From Chapman and Webb, 1978.)

little published work allows direct comparisons to be made at the level of total above-ground net production. In a number of cases the work was carried out with the principal object of assessing production of potentially edible material for animals such as sheep or grouse (*Lagopus lagopus scoticus*).

Above-ground standing crop

The relationship between the age of the vegetation and the weight of the above-ground standing crop for one particular area is shown in Fig.14.19 where an age sequence obtained from lowland heathland in southern England has been fitted with a Gompertz growth curve (Fig. 14.19) from an age of three years onwards. Chapman et al. (1975a) have shown that the deviation from such a fitted curve during the first three years is due to the pattern of green shoot production and survival. It can be seen from the data shown in Fig. 14.19 that, if rates of increment of standing crop are to be obtained from consecutive estimates of the standing crop, then variable and even negative results may well be obtained, unless considerable effort is taken in selection of size and numbers of samples. To some extent this problem is reduced where an age series can be sampled and an averaged estimate

of the rate of increase in the standing crop obtained. This latter approach however is often restricted by the availability of older aged stands of heather, demonstrated in Fig. 14.20 where the majority of determinations of above-ground standing crop from a range of heathlands in Europe are from sites less than twenty years old. It is difficult to make many comparisons from data such as those shown in Fig 14.20, especially when confidence limits of the individual means are taken into account. When only younger stands of heathland are considered (up to *c.* 15 years) an impression is gained of surprising uniformity in rates of increase of standing crop. Differences between the fast and slow rates of recovery following different intensities of burning at Kerloch Moor (Miller and Watson, 1974b) are almost as great as those shown by the data from all the other sites included in Fig. 14.20.

When trying to compare heathland sites on the basis of standing crop, or indeed any other

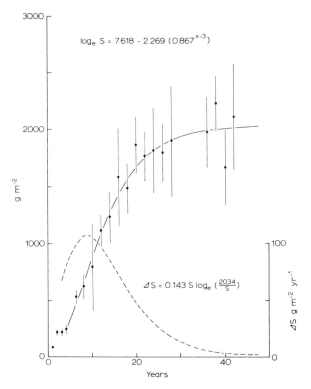

Fig. 14.19. Standing crop (continuous line) and rate of increment (pecked line) of the above-ground components of *Calluna* from Dorset heathlands in relation to their age with 95% confidence limits (from Chapman et al., 1975a).

TABLE 14.7

Data sources and symbols used for derivation of graphs

	Scotland	
✹	Glen Muick, Aberdeen	Moss (1969)
◉	Lochnagar, Aberdeen	Moss (1969)
✳	Corndavon, Aberdeen	Moss (1969)
▼	Kerloch Moor, Kincardineshire	Moss (1969), Miller and Watson (1978), G.R. Miller (pers. comm.)
▽	Glensaugh, Kincardineshire	Grant (1971)
☆	Elsick, Kincardineshire	Barclay-Estrup (1970)
△	Cairngorms	Summers (1972)
◇	North Cairn o'Mount Kincardineshire	Robertson and Davies (1965)
◆	Polworth Moss, Berwickshire	Robertson and Davies (1965)
▲	Listonshiels, Midlothian	Robertson and Davies (1965)
	Northern England	
■	Blanchland Moor, Northumberland	Robertson and Davies (1965)
□	Teesdale, Durham	Bellamy and Holland (1966)
☆	Moor House, Westmorland	Forrest (1971), Forrest and Smith (1975)
	Eastern England	
✸	Cavenham Heath, West Suffolk	Chapman (unpubl.)
✩	Lakenheath Warren, West Suffolk	Chapman (unpubl.)
✡	Westleton Heath, East Suffolk	Chapman (unpubl.)
	Southern England	
●	Dorset Heathlands	Chapman (1967), Chapman et al. (1975a, b)
○	New Forest, Hampshire	Chapman (unpubl.)
★	Woodbury Common, Devon	Chapman (unpubl.)
❋	Dartmoor, Devon	Chapman (1967, unpubl.)
✳	Exmoor, Somerset	Chapman (unpubl.)
☆	Penhallow Moor, Cornwall	Chapman (unpubl.)
★	Chapel Porth, Cornwall	Chapman (unpubl.)
	Sweden	
☆	Skanör	Tyler et al. (1973)

parameter of performance or production, it is difficult to separate the effects of climate from those of soil. In the British Isles, heathlands in the south are mostly lowland and upon infertile soils where the larger part of the nutrient capital is contained within the organic matter of the litter and root zone (Chapman, 1970). Heathlands in the upland areas of Britain occur mostly in Scotland and the north of England, where climatic conditions are generally less favourable for plant growth but the soils are often more fertile. The apparent similarity of the heathland growth curves from different areas may therefore partly be the result of interaction of climate and soil nutrient factors. The information shown in Fig. 14.20 includes results from two different granite areas, from Dartmoor in the southwest of England (altitude *c*. 420 m), and from Kerloch Moor in eastern Scotland (altitude *c*. 150 m). The mean summer temperatures on Dartmoor are in the order of 2°C higher than at Kerloch Moor and the standing crops at comparable ages from Dartmoor are also consistently higher (Fig. 14.21).

Litter production

Estimates of litter production by heathland vegetation have been published by only a few authors and are shown in Fig. 14.23. Litter production has been measured by the use of litter traps (Cormack and Gimingham, 1964; Chapman, 1967; Forrest, 1971; Tyler et al., 1973; Chapman et al., 1975a; Forrest and Smith, 1975) and estimates of potential litter production (Fig. 14.22) based upon shoot and flower production have been described by Chapman et al. (1975a). In this latter work it was thought that such estimates of potential litter production were more reliable than trapping results from younger and more open heathland, where loss due to wind blow was shown to be significant.

Calluna litter is composed of four main components: woody material, long shoots, short shoots and floral parts that include seed capsules. Litter fall is almost negligible from young plants until the second or third growing season, hence the characteristic "bump" at the start of the growth curve shown in Fig. 14.19. Litter production increases until rates of about 200 to 250 g m^{-2} yr^{-1} are reached after about twenty years. Whilst some

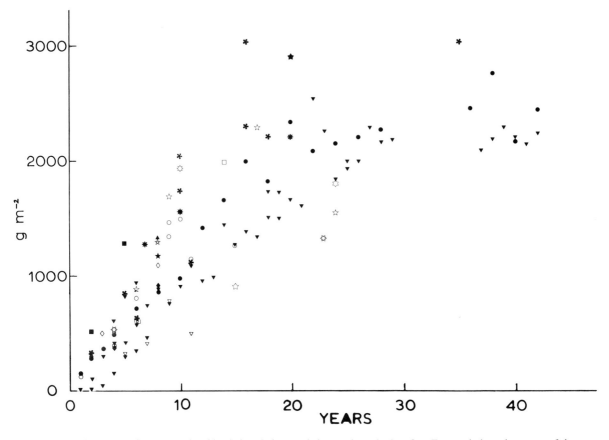

Fig. 14.20. Standing crop of European heathlands in relation to their age since the last fire. For symbols and sources of data see Table 14.7.

litter may be shed in all months of the year, there are two periods of higher fall. The first of these is in October and November due to the increase in short shoot material and the second in February (Cormack and Gimingham, 1964; Chapman et al., 1975a) when most seed capsules are shed. The work of Forrest (1971) on primary production by blanket bog vegetation at Moor House in the Pennines of northern England presents rather a different pattern where maximum shoot loss was between June and October. At that site snow cover effectively prevents litter fall during the winter months. Litter production by *Calluna* shows a good relationship with the above-ground standing crop (Fig. 14.23). The individual data obtained by different workers are shown in Fig. 14.23 with the relationship obtained from the smoothed data shown in Fig. 14.22; the latter suggest a more curvilinear relationship, due to the increase in the

relative amount of wood to shoot material in older heather.

Net above-ground primary production

A number of problems involved in trying to make comparisons between primary production of different heathland areas have been mentioned. These include variation with age, the differences in parameters measured by particular workers, and differences in their methods. Despite these problems there are four ways in which comparisons can be made.

(a) Age. It is unlikely that sufficient age sequences such as that published by Chapman et al. (1975a) (Fig. 14.24) will become available in the very near future to enable sites and areas to be compared in this way. Apart from the work involved, the

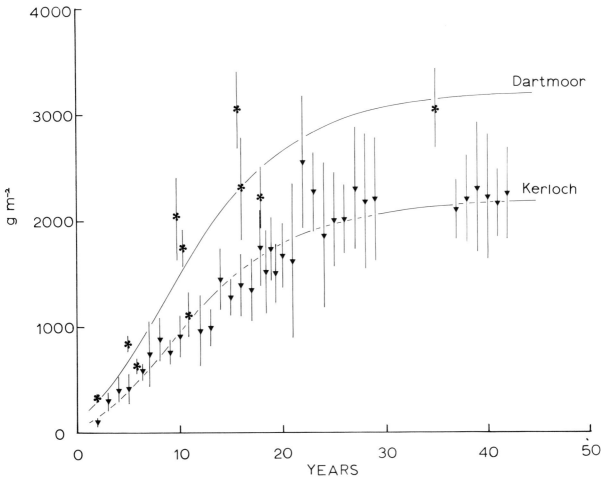

Fig. 14.21. Standing crop of above-ground vegetation from Dartmoor (Chapman, 1967, and unpubl.) and Kerloch Moor (G.R. Miller, pers comm.) in relation to age since the last fire.

scarcity of older stands makes this approach difficult. A more possible approach may be to make comparisons based upon some relative parameter that shows less variation with age than does total above-ground production. Relative production per unit weight of green material is such a parameter that might be worth considering. When calculated from the Dorset data relative production varies less with age than production expressed upon a simple ground-area basis. If relative estimates of production were based upon photosynthetic area, not easily measured in the case of *Calluna*, and allowance made for shoot age and interception of light by the heather canopy, an even less age-dependent parameter might be obtained.

(b) Developmental phases. The results obtained by different workers that can be designated to a particular developmental phase are summarised in Table 14.8. Despite the fact that some of the data relate only to shoot and flower production, a picture emerges where net above-ground production increases in the pioneer (or post-burn phase of Chapman and Webb, 1978), and the building phase, reaching a maximum in mature heather, and finally showing reduced values in the degenerate stage. The values reported by Barclay-Estrup (1970) are somewhat higher than those of other workers and are probably due to the fact that he was investigating shoot and flower production in the different developmental phases of a mixed-age site, and therefore sampled quadrats having

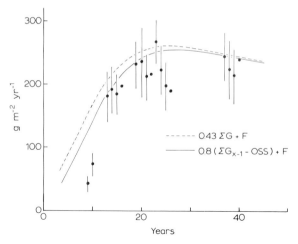

Fig. 14.22. Litter production by *Calluna* from Dorset heathlands in relation to their age. The point data ($\pm 95\%$ confidence limits) are from litter traps, the curves are indirect estimates of litter production (from Chapman et al., 1975a).

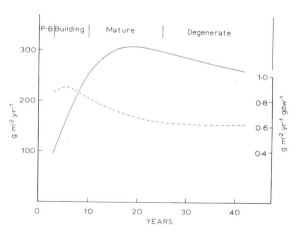

Fig. 14.24. Net above-ground production (solid line), and relative production (pecked line) of Dorset heathlands in relation to their age.

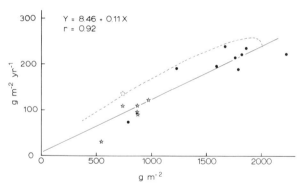

Fig. 14.23. Litter production by *Calluna* in relation to the above-ground standing crop. For symbols and sources of data see Table 14.7. The dotted line is derived from indirect estimates shown in Fig. 14.22.

particular physiognomic characteristics. In comparison with other workers who have sampled randomly within stands of a particular age his results might be expected to emphasise the features of a particular developmental phase, whilst those of the other workers provide more average values.

(c) Corrected cover basis. Miller and Watson (1978) have compared the estimates of flower and shoot production by *Calluna* that have been obtained by a number of workers in relation to altitude of the sites, after correcting the data to equivalent weights that would have been obtained

from 100% cover of *Calluna*. The information shown in Fig. 14.25 is based upon Miller and Watson's work, but with some additions and a consequently revised regression line. Comparisons based solely upon flower and shoot production must be treated with caution, as they may only reflect variations in the relative amounts of wood and shoot production. However, the production of green matter at Kerloch Moor (G.R. Miller, pers. comm.) and in Dorset (Chapman et al., 1975a) are very similar, as are the standing crop curves, so that it seems unlikely that such differences exist between these two sites.

(d) Growing season. A further way in which data can be standardised for comparison is by making allowance for the differences in growing season at different sites. This approach has been used by Summers (1972) and by Heal and Perkins (1976), but problems arise regarding the measurement and definition of the growing season, and more information is required before such an approach can be used with any degree of confidence.

Root production

In all discussions of primary production the question of root production arises, and very few data are available upon which any really meaningful statements or comparisons can be made. It must be remembered however that comparisons based solely upon above-ground estimates of net

TABLE 14.8

Above-ground production and standing crop data from some European heathlands according to the developmental phase of the vegetation

Site	Mean age (years)	Σ vegetation production ($g\ m^{-2}\ yr^{-1}$)	*Calluna* production ($g\ m^{-2}$)	Σ standing crop ($g\ m^{-2}$)	*Calluna* standing crop ($g\ m^{-2}$)	Reference
Elsick Heath						Barclay-Estrup (1970)
pioneer or post-burn	5.7	276*	149*	889	287	
building	9.0	471*	442*	1702	1508	
mature	17.1	393*	364*	2305	1924	
degenerate	24.0	195*	141*	1561	1043	
Dorset Heathlands						derived from Chapman
pioneer or post-burn	1.5	160	*c.* 90	291	–	et al. (1975a)
building	6.75	230	193	811	573	
mature	17.5	323	298	1862	1502	
degenerate	32.5	298	278	2425	1966	
Skanör Heath, Sweden						Tyler et al. (1973)
pioneer or post-burn	–	–	–	–	–	
building	–	–	–	–	–	
mature	*c.* 15	307	232	920	741	
degenerate	–	–	–	–	–	
Moor House						Forrest (1971)
pioneer or post-burn	–	–	–	–	–	
building	8	407	168	1510	740	
mature	–	–	–	–	–	
degenerate	–	–	–	–	–	
Kerloch Moor						Miller and Watson (1978)
pioneer or post-burn	4.5	–	168	–	420	
building	13.5	–	210	–	1180	
mature	23.5	–	270	–	2000	
degenerate	39.5	–	239	–	2200	

*Shoot and flower production only.

production assume constant ratios of above- to below-ground production, and any preliminary conclusions made may well need modification when reliable estimates of root production become available.

Litter accumulation

Litter accumulation is the result of interaction between litter production and litter loss. The weights of plant litter accumulated at a number of different dry heathland sites are shown in Fig. 14.26. Whilst the scatter is considerable there is a clear correlation with the age of the site since burning, although there are insufficient data to make any comparisons between sites, or between upland and lowland areas. While the combined data in Fig. 14.26 do not show any evidence of accumulation reaching a steady state, the results from Chapman et al. (1975b) on the Dorset heathlands that are shown in Fig. 14.27 indicate maximum values after about thirty years and do suggest an approach towards some form of steady state. On these heathlands it was shown that rates of decomposition derived from litter bags could be combined with estimates of litter production to predict levels of litter accumulation. When this was done the predictive curves (*A* and *B* in Fig. 14.27) are seen to differ significantly from the observed values. This discrepancy was explained by the

398

C.H. GIMINGHAM ET AL.

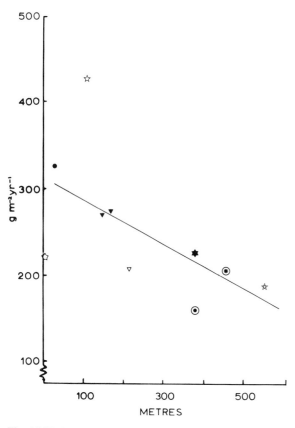

Fig. 14.25. Production of shoots and flowers by *Calluna* in relation to altitude (from Miller and Watson, 1978, with additions). For symbols and sources of data see Table 14.7.

invasion of the litter layer by roots once the vegetation had reached an age of about 10 years. From then onwards the proportion of recognisable root material increases until it reaches about 0.4 after forty years. The root invasion of the litter appears to be related to canopy closure and the consequent maintenance of higher and more constant humidity at the soil surface. This root development is accompanied by changes in the overall nutrient dynamics of the litter layer that take place from about twenty years onwards. Up to this time the accumulation of nutrients in the litter is about the same as the rate of nutrient supply by litter production, but from then the rate of nutrient accumulation decreases rapidly and nutrients are lost from the litter layer (Fig. 14.28). In the absence of reliable root production data it is difficult to interpret the results fully, but there are clearly important changes when nutrients held in the litter layer become available, with important

consequences for the overall nutrient régime of the vegetation.

Nutrient budgets

The cycling, sources, and losses of mineral nutrients in an ecosystem such as heathland, where they are in short supply, are of obvious importance to the management and maintenance of the habitat (these are discussed at length by Groves in Vol. B, Ch. 16). A number of studies have investigated the nutrient capital contained within various components of the heathland system (Thomas et al., 1945; Robertson and Davies, 1965; Chapman, 1967; Chapman, 1970; Tyler et al., 1973). Inputs to heathland and moorland systems in the form of precipitation have been described by various authors (Gorham, 1958; Allen et al., 1968; Gore, 1968).

Losses of nutrients that take place during heathland fires have been assessed by Allen (1964), Chapman (1967), and Evans and Allen (1971). The distribution and changes in the nutrient content of heathland soils after fires have been investigated by Allen et al. (1969) and by Hansen (1969). Losses from a Pennine moorland ecosystem were estimated from stream-flow analyses and other data by Crisp (1966). An overall nutrient budget for lowland heathland in southern England was constructed by Chapman (1967) who showed that, except for nitrogen and phosphorus, losses incurred during burning could be replenished by nutrient inputs from rainfall over a twelve-year period (see also Vol. B, Ch. 26, pp. 251–252 and Fig. 26.2). It is also possible that phosphorus may be replenished by rainfall, but when nutrient contents in precipitation are low they may merely represent local redistribution rather than true inputs to the system. For example dust, pollen grains etc. may be derived from closely adjacent vegetation so that unfiltered rain samples give inflated estimates of nutrient input, whereas filtered samples may well provide underestimates. The maintenance of nitrogen levels by fixation plants such as *Ulex* spp., and nutrient losses by leaching, need more study before complete nutrient balances can be assessed.

Breakdown and decomposition of litter

Litter is that material which has been shed by the aerial parts of the *Calluna* plants and which lies

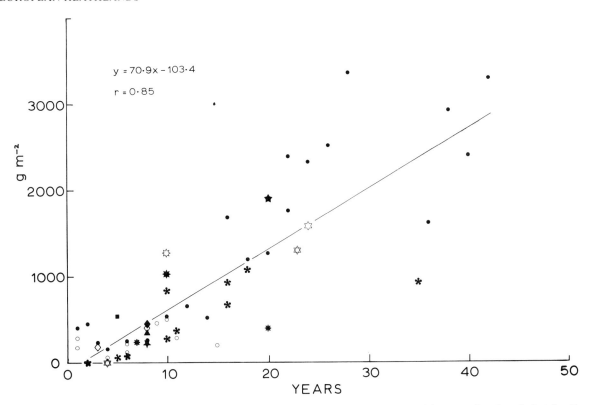

Fig. 14.26. Standing crop of accumulated litter on European heathlands in relation to the age of the vegetation since the last fire. For symbols and sources of data see Table 14.7.

upon the soil surface. Strictly speaking the plant also contributes to the litter with dead roots but the rate at which this occurs is difficult to measure, particularly on heathlands where the litter layer consists of an intimate mixture of fine living and dead roots as well as the dead aerial parts of the plant.

The phrase *litter breakdown* will be used to describe the physical changes which take place in the litter by the action of such factors as wind, water, frost and removal or comminution by animals, all of which cause the fragmentation of the material. *Decomposition* is the chemical degradation of the litter arising from the metabolic activity of soil organisms and causing its chemical simplification and the release of nutrients and energy. In many instances these two processes cannot be separated and are associated with physical movements of the litter; where this happens the term *litter loss* is more appropriate.

Table 14.9 gives the percentages per unit dry weight of the main nutrient elements in freshly fallen and accumulated *Calluna* litter, green

Calluna shoots and oak litter (Carlisle et al., 1966). Comparisons between the figures of Moss (1967, 1969) and of Chapman (1967) for green *Calluna* shoots and those for freshly beaten *Calluna* litter from plants in southern England (N.R. Webb, unpubl.) show that, on death, there is a drop in those nutrients that are readily leached, and a rise in the nitrogen content. Chapman et al. (1975b) give figures of 53.32% and 1.16% for the carbon and nitrogen contents, respectively, of brown short shoots (the main constituents of fresh litter), figures that suggest a C/N ratio of 46:1 for the newly fallen litter, a value indicating that it should decay readily.

The organic constituents of fresh *Calluna* litter together with comparative data for green heather shoots are presented in Table 14.10. The soluble carbohydrate content of fresh litter is 20% of that for the green shoots. This reduction in concentration may occur by leaching, utilisation of simple carbohydrates by leaf-surface fungi, or the withdrawal by the plant of carbohydrate for storage. Satchell and Lowe (1967) give values of 0.7

400

C.H. GIMINGHAM ET AL.

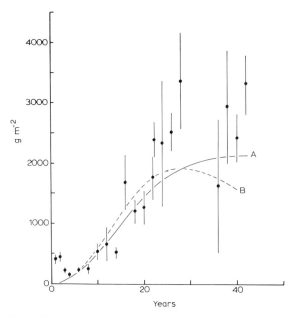

Fig. 14.27. The accumulation of organic matter in the litter layer of Dorset heathlands in relation to their age. Mean values (±95% confidence limits) are from field data. Curves represent predicted accumulation from a constant decay rate (A) and from a linearly increasing rate (B). (From Chapman et al., 1975b).

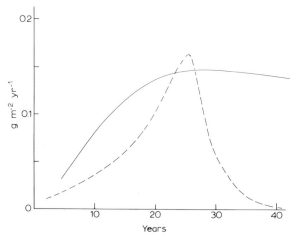

Fig. 14.28. The rates of accumulation (pecked line) and input of phosphorus (solid line) to the litter in relation to the age of the heathland since being burnt (based on Chapman et al., 1975b).

to 4.5% for the soluble carbohydrate content of ten species of temperate trees, and 0.7 to 2.0% for that of three coniferous species. The carbohydrate content of *Calluna* is higher than that of the conifers, and only exceeded by that of six of the trees. From this point of view *Calluna* litter, with a

fairly low soluble carbohydrate content, would not be the best substrate for organisms causing decay. Data are also available for the soluble tannin content of a similar range of litters. It has been shown that the content of polyphenols of plant litter affects its palatability to soil fauna (Heath and King, 1964; Satchell and Lowe, 1967). *Calluna* has a fairly high polyphenol content, which may explain its rather slow rate of breakdown and decomposition (Bell, 1974).

The commonest approach to studying the loss of *Calluna* litter has employed the litter-bag technique. A variety of *Calluna* litter types have been used in these experiments; Cormack and Gimingham (1964) used fresh litter, Chapman (1967) used the accumulated litter from the soil surface, and N.R. Webb (unpubl.) in Dorset has compared both fresh and accumulated litter. Only results from Moor House (Heal and Perkins, 1976) and Dorset have extended over several seasons; all other studies have been for one year. The results from Moor House and Dorset are similar, showing a weight loss of from 15 to 20% in the first year. At the Sands of Forvie, Cormack and Gimingham (1964) recorded rates between 6 and 15%. These weight losses are less than those of the litter in temperate deciduous woodland where Anderson (1973) found that up to 70% of the litter may have been lost after two seasons.

Litter-bag data are notoriously difficult to interpret. The presence of the bag considerably modifies the micro-environment of the decomposing litter (Anderson, 1973), but on heathland the main problem arises from an invasion of the bags by roots similar to that previously described for litter on the soil surface in the old phases. In Dorset up to 15% roots have been recorded in the litter in the bags after 2½ years. Root invasion may be stimulated by humidity changes and by the release of nutrients from the decaying litter. On Dorset heathland the bags are soon invaded by the litter fauna and within one season contain species similar to those in the adjacent litter.

Nutrient loss from litter bags on Dorset heathlands indicate losses of about 60% of the potassium and magnesium after four years, and 40% of the sodium, but hardly any loss of calcium and phosphorus. Indeed the concentration of phosphorus like that of nitrogen may even increase with time. There is almost no decrease over one year in

TABLE 14.9

Percentage (in dry matter) of the principal plant nutrients in *Calluna* and its litter

Type	Nutrient:	Na	K	Ca	Mg	P	N	C	Ash	C/N	Reference
Green *Calluna* shoots		0.047	0.43	0.40	0.16	0.081	1.18	–	3.51	–	Moss (1969)
Green *Calluna* shoots		0.056	0.44	0.68	0.24	0.083	1.28	53.3	–	41.6	Chapman et al. (1975b)
Fresh *Calluna* litter		0.031	0.104	0.52	0.081	0.071	1.09	–	4.05	–	Webb (unpubl.)
Accumulated *Calluna* litter		0.023	0.078	0.27	0.042	0.072	1.14	–	–	–	Webb (unpubl.)
Oak litter		0.029	0.095	0.69	0.11	0.055	1.04	51.3	3.30	49.8	Carlisle et al. (1966)

TABLE 14.10

Percentages (in dry matter) of the main organic components of *Calluna* litter and shoots

Organic constituent	Type:	Fresh *Calluna* litter	Green *Calluna* shoots
Soluble carbohydrate		2.7	13.3
Holocellulose		39.5	–
α-Cellulose		14.5	–
Soluble tannin		1.26	–
Crude fat		10.2	8.6
Lignin		35.6	–
Reference		Webb (unpubl.)	Moss (1969)

Mangenot (1966) has shown from *in vitro* studies that *Calluna* and *Vaccinium* have leaves of a similar quality as substrates for micro-organisms. The succession of the microflora on decomposing *Calluna* leaves was followed but the flora was poorer than that associated with typical herbaceous types. Peptone- and pectin-decomposing yeasts were shown to be important in the decomposition of *Calluna* litter. However, a full study of the processes of decay and the involvement of micro-organisms of *Calluna* on heathland is required.

THE INVERTEBRATES

The invertebrate fauna of heathland has not received attention comparable with that of the floral components of the association. Until recent studies few attempts have been made to correlate changes in the fauna with the marked seral succession of the plants. Such changes can easily be seen from the pioneer or post-burn phases, in which the very open plant community has a poorer fauna, to the older, mature and degenerate phases which often have a rich and diverse fauna. As the growth cycle of the heather proceeds the spatial structure and the microclimate of the vegetation change, increasing the range of habitats which are available for exploitation by animals. At this stage the microclimatic changes should be considered since they have an over-riding influence on the development of the fauna.

the soluble carbohydrate content, a decrease of 10 to 15% in both the holocellulose and α-cellulose contents, and less than 10% decrease in the soluble tannin content. From Table 14.9 it can be seen that much of the soluble carbohydrate content of short shoots of *Calluna* is lost before they fall as litter. The persistent tannin content, which is not reduced by weathering or decomposition probably renders the litter unpalatable to many organisms.

Densities of animal populations in soil and litter are smaller on heathland than in woodland. Their contribution to litter decay is likely to be small particularly since the larger species (earthworms and millipedes), which in woodland contribute significantly to breakdown of the litter (Raw, 1967; Satchell, 1967), are mostly absent. Chapman and Webb (1978) have estimated that cryptostigmatid mite populations of a square metre consume from 0.1 to 1.0 g dry weight of material each year, which represents about 1% of the annual litter fall; clearly the contribution of the litter fauna is small.

The microflora associations in peat beneath *Calluna* have been studied extensively (Heal and Perkins, 1976), but there have been few studies of microflora associated with *Calluna* on heathland.

On heathland, microclimatic studies have been carried out by Delany (1953), Stoutjesdijk (1959), Gimingham (1964), and Barclay-Estrup (1971), and the important aspect of temperatures, especially those occurring during heathland fires, have been investigated by Whittaker (1961), Kenworthy (1963) and Webb (unpubl.). The results of Barclay-Estrup (1971) for a Scottish heathland show that the pioneer phase is characterised by extremes, especially of temperature. In this phase the humidity over the soil surface and beneath the plant is low, the plants are small and scattered and there is no litter layer. In some instances in a post-burn phase there may be a residual litter layer that has not been burnt by the fire, but this is soon dispersed by wind. In the pioneer and post-burn phases the habitats available for animals are few. Soil- and litter-dwelling forms are able to survive fires (Merrett 1976; Webb, unpubl.) since the litter layer acts as an effective insulator to the high temperatures reached in the vegetation (Whittaker, 1961; Webb, unpubl.). In the building phase conditions begin to stabilise, temperature fluctuations are less; as the canopy closes, the humidity beneath the plants begins to rise, and by this stage a litter layer has accumulated. Similar conditions prevail during the mature phase, but in the degenerate phase microclimate is more variable as the canopy opens. However, by this time the plants have reached considerable size, providing an important structural feature and protecting the litter layer.

Heather is a particularly difficult habitat to sample for invertebrates. Sweep-netting is difficult because of the woody nature of the vegetation and one of the most satisfactory methods is the vacuum sweep-net; however reliable density estimates are difficult to obtain from these methods. Likewise pitfall trapping is effective for recording the presence of surface active animals but it is completely unreliable in assessing relative numbers of species. Heat extraction of the soil and litter faunas usually provides reliable density estimates. It is however difficult to prescribe methods that are of value in assessing the faunas of heathlands and effective in measuring seral changes, and it is important to recognise that the efficiency of many sampling methods may vary from one growth phase to another because of changes in the structure of the vegetation. This may result in misleading impressions of changes, especially those of relative density, taking place in the fauna. The possibilities of using a wide range of methods have yet to be explored, and a manual such as Southwood (1966) should be consulted.

The earliest study of the insect fauna of heathland was that of Richards (1926) who, in an extensive survey of Oxshott Common, Surrey, recorded the insects found in a number of wet- and dry-heathland habitats, which represented the succession from burnt heathland to pine woodland. He came to the conclusion that there was a characteristic fauna associated with *Calluna* wherever it grew, and that the same fauna was associated with *Erica cinerea* and *E. tetralix*. It was Richards' view that the insect fauna was controlled by the plant rather than by any special edaphic or physiological conditions. The most important factors affecting the distribution of animals associated with *Calluna* were their powers of dispersal and their edaphic needs. A number of the species associated with *Calluna* were flightless, and as a consequence were absent from the early stages of colonisation. Richards noticed also that there were many species that occurred only in the older stands of *Calluna*, and that most of the insects found in the heather lived either beneath the plant or on the ground. It is this stratum of the habitat that is mostly highly developed in the older stands, accounting for the richer insect fauna.

The Studland Peninsula in Dorset was the site of an extensive ecological survey by Diver (Diver and Good, 1934; Merrett, 1971) from 1931 to 1939. This area contains salt marsh, dunes, dune slack, as well as dry-heathland, wet-heathland and bog. Diver was interested in the interactions between the plants and animals of this area, and in papers on Orthoptera (Diver and Diver, 1933), Syrphidae, Mollusca, Crambidae and Pyralidae (reviewed by Merrett, [1971]) he described the species present, their range tolerances with regard to factors such as wetness, vegetation structure, type of plant community and geology, and observations on the behaviour, especially that associated with feeding.

The next major study was that of Delany (1956) who examined the animal communities associated with pioneer heathland on three sites in the southwest of England. His study extended over fourteen months and was developed mainly as an adjunct to a study on the ecology of the thysanuran *Dilta littoralis*. He studied localities in the New Forest in Hampshire, the Pebble Bed Commons in south Devon, and on the island of Lundy in the Bristol Channel. He distinguished a mesofauna composed of animals which were over

5 mm in length, which he sampled by hand collecting from quadrats of 0.5 m², and a microfauna of animals under 5 cm in length which he sampled with a Tullgren funnel. He concluded that the mesofauna was made up mostly of species that occurred widely although there were a few that were associated only with heather. The Hemiptera had the greatest number of species associated with heather while the Coleoptera and Araneae had few species. Much the same picture emerged for the microfauna which he showed to consist of species which had a wide occurrence, especially the mites. Delany discussed his results in relation to those of Richards (1926), with which they were very similar. The mesofauna of pioneer heathland consisted mainly of species adapted to a wide range of habitats and it might be expected that, as the heathland matured, the number of species characteristic of this community type would increase. It is hardly surprising that the early colonisers of pioneer heathland were species of wide habitat tolerances.

Moore (1962), surveying the conservation problems of Dorset heathlands, used ten indicator species. One of each pair was restricted to heathland and the other was of widespread occurrence. He used the pair of dragonflies (Odonata) *Ceriagrion tenellum*, restricted to heathland and bog pools, and *Pyrrhosoma nymphula* which occurs widely; and the butterflies (Lepidoptera) *Plebeius argus*, restricted to heathland, and *Eumenis semele* which is widespread. The ability of these indicator species to withstand changes in land use and to survive in isolated areas of heathland was assessed.

The heathlands of southeast Dorset have been the subject of intensive invertebrate studies emanating from Furzebrook Research Station. Studies have concentrated mainly on the patterns of distribution, changes in seasonal abundance, and recovery after fires, rather than on overall surveys. Brian (1964) and Brian et al. (1976) have made detailed investigations into the populations of ants (Hymenoptera: Formicidae), Merrett (1967, 1968, 1969 and 1976) has studied the spiders (Araneae), and Webb (in prep.) the soil fauna. Most of this work has been done in Dorset on Hartland Moor National Nature Reserve, almost the whole of which was burnt in 1959. It was two years later in 1961 that Brian commenced his study of the ants. A survey site of 8 ha, covering both

dry- and wet-heathland communities, was surveyed from 157 sampling points, at which records were made of thirteen species of ants present at sugar baits, altitude, soil moisture, soil organic matter and integrated soil temperature, the plant species present and the percentage of bare ground (Brian, 1964). All of the variables were highly correlated, and the data were subjected to principal component analysis. The four commonest species of ant were *Lasius alienus*, *Lasius niger*, *Tetramorium caespitum* and *Formica fusca*. The results indicated that *L. niger* inhabited the cool low wet areas, with plant cover of 83% composed mainly of grasses (*Molinia caerulea* and *Agrostis setacea*). The commonest heathland plants of this area were *Erica tetralix* and *Erica ciliaris*, and *Calluna* was present. Also living in this association were the ants *Myrmica scabrinodis* and *M. ruginodis*.

Formica fusca preferred the low but drier areas, although with a substantial amount of organic matter, and a vegetation cover of 85%, but in these habitats, which are similar to those of *Lasius niger*, *Calluna* was more abundant. The habitat of *Tetramorium caespitum* differed less than that of *Formica fusca* from *Lasius niger*. *Tetramorium caespitum* preferred the higher, warmer areas with a plant cover of about 62%. Grasses were less frequent and the vegetation consisted mainly of *Calluna* with some *Erica cinerea*. *Lasius alienus* lived on high, dry, warm and sparsely vegetated heathland with the lowest soil organic matter content. The interactions of the ants with the changes in vegetation are summarised in Fig. 14.29. This pattern of distribution on heathland has been tested experimentally by Elmes (1971), who transplanted whole colonies of *Lasius niger* into the habitats of *Tetramorium caespitum* and *Lasius alienus*. Elmes showed that colonies, transplanted as controls in other areas of wet-heathland and cool dry-heathland, survived well, but *L. niger* was unable to survive in regions typically occupied by *L. alienus*, and he considered these two sympatric species to be isolated by ecological as well as behavioural differences. This mechanism was reinforced in territorial selection by fertilised colony-founding queens.

Ten years after the first survey, Brian et al. (1976) surveyed the same area again in order to assess the changes in the ant populations. During this time the vegetation had developed through the post-

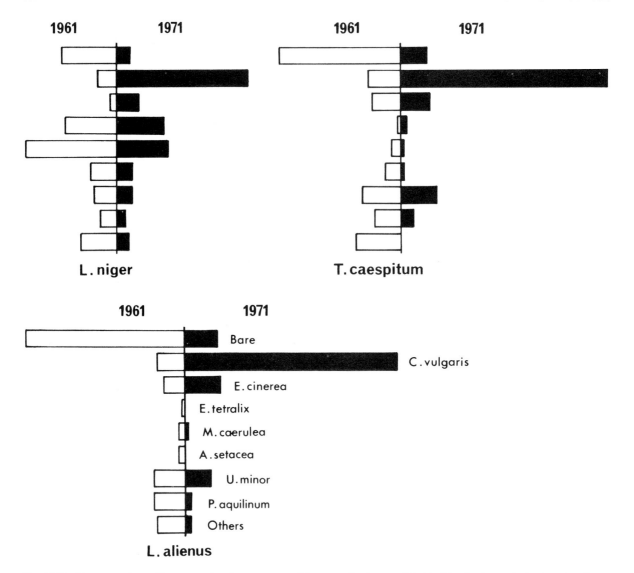

Fig. 14.29. The proportions of bare ground and components of the vegetation in areas inhabited by the three principal species of ants on a southern English heathland two years after burning and twelve years after burning (after Brian, 1964, and Brian et al., 1976).

burn phase to the late building phase. The most striking change was in the amount of bare ground which had decreased from 36% to 6%. *Calluna* cover had now increased from 10% to 54%, that of *Ulex minor* had hardly changed (10 to 11%) and species of *Erica* had become more common. Once again principal components were extracted in order to examine the distribution of ants in relation to features of soil and vegetation. The relative position that the ants occupied in the component space was similar to that ten years earlier. *Tetramorium caespitum* had increased from 17 to 24%

and *Formica fusca* from 6 to 15%, while *Lasius niger* and *L. alienus* had declined from 28% to 19% and from 42% to 24% respectively. *Formica fusca* had gained most at the expense of *Lasius alienus*. Brian et al. (1976) considered that the spread of the vegetation probably altered the competitive balance in favour of *Tetramorium caespitum* since this species is a seed feeder and can build nest mounds to avoid shading by the growing plants. Species such as *Lasius alienus* which require a high level of insolation decline as the vegetation spreads. Fire thus re-establishes a sparse plant cover which is more suitable for *L. alienus*, and Brian et al. (1976) point out that these two species co-exist though being adapted to opposite phases of the burning cycle.

In a series of papers, Merrett (1967, 1968, 1969) has investigated, by continuous pitfall trapping for three years, the distribution and phenology of 195 species of spider on the same heathland. The changes which took place in populations of ground-living spiders following burning were described by Merrett (1976) from two adjacent areas, which were monitored continuously by pitfall traps for five years. Trapping was commenced in one area a week after it had been burnt, and at the same time in an adjacent area not subject to the burn. Both areas had been burnt five years previously, so that after five years' trapping results were available for the first ten years' growth of the heather. Vegetation and bare ground were recorded annually and the changes in the fauna correlated. The activity of the ground-living species was not markedly affected until the canopy closed and a litter layer started to build up at an age of about ten years. A total of 109 species were recorded, but only 51 of these occurred in sufficient numbers for analysis. Merrett was able to supplement these results with observations from other nearby sites. He recognised six categories:

(1) Pioneer species which declined in numbers quickly and which were absent when the plant cover had increased to 90%.

(2) Pioneer species which persisted ten years or more and showed a gradual decline in number over the ten years of sampling.

(3) Species which reached maximum numbers between five and ten years and whose numbers built up gradually over the ten years. Studies from other areas suggested that these species would decline in the 10-year to 15-year age classes.

(4) Species which reached peak numbers after ten years.

(5) Species typical of mature heathland and which were divided into two classes; web spinners, living mainly in the canopies of *Calluna* and *Ulex*, and ground-living species.

(6) Species whose period of peak abundance was uncertain.

The soil fauna of these heathlands has been examined by Webb (1972), Chapman and Webb (1978) and Webb (in prep.). Three sites were examined by Webb (1972): degenerate heather which was at least thirty years old; mature heather fifteen years old; and building heather eight years old. A total of 35 species of oribatid mites (Cryptostigmata) were recorded but the numbers were similar from each of the sites — 26, 23 and 24 respectively. Most species occurred at all sites. Surprisingly, species of Phthiracaridae, which are usually associated with decaying wood and thus with the older stands, occurred in the younger stands, probably because they possess a thick exoskeleton protecting them from desiccation. There seemed to be little distinction between the number of detritus-feeding species and the number of fungal-feeding species. Differences, if any, are probably reflected in differences in the relative densities of the species. A later study (Webb, unpubl.) examined six sites which were respectively freshly burnt and 1, 4, 8, 15 and 27 years old since burning. In addition, the temperatures occurring in the litter and upper layers of the soil were recorded during experimental burning of heathland. The results were similar to those of Whittaker (1961) for a Scottish moor. The temperature recorded just beneath the litter surface was 65°C but at depths greater than 1 cm no temperature change could be detected, the fire passing over the surface too quickly. It seems likely that few soil and litter animals, except those at the surface, were killed during a heathland fire. The litter layers, as Whittaker (1961) and Whittaker and Gimingham (1962) have pointed out, possess extremely good insulating properties. The development of fauna closely follows that of the vegetation, and particularly the build-up of the litter layer. The litter remaining after a fire is gradually lost by the action of wind over the first three to four years after the fire. During this time the soil and litter fauna decline, with species of Phthiracaridae persisting the longest. The subsequent development of the fauna is slow and is impeded not only by the slow development of the litter layer but by its dryness. It is only when the mature phase is reached and the canopy closes that humidity rises and becomes more constant, and the fauna is able to develop.

Despite the extensive studies which have been described for the heathlands of southern Britain, no comprehensive survey of invertebrates has been undertaken which relates the distribution of the insects to the growth phases of *Calluna*. By contrast the fauna of *Calluna*-dominated areas in the uplands of northern Britain have been well studied. But comparisons are difficult since not only the seral changes but the geographical differences in the distribution of the fauna have to be considered. Most of the southern heathlands lie within an area of particularly mild climate, and within the British distributions of many continental species at the northern limits of their ranges. Thus by comparison with the more northerly upland areas, the fauna of southern heathlands are rich and diverse.

At Moor House in the northern Pennines, extensive studies of the invertebrate fauna have been carried out over a number of years, latterly much of the work being under the auspices of the International Biological Programme. An early review of some of this work was made by Cragg (1961) and an extensive review has been compiled by Coulson and Whittaker (1978). It is not proposed to review this work further; it is strictly outside the scope of this chapter, since at this locality *Calluna* is mostly rooting in an organic and not a mineral soil.

At Scottish sites on a mineral soil, investigations have been carried out by Barclay-Estrup (1974) and Miller (1974). The study of Barclay-Estrup (1974) was essentially a preliminary one. Pitfall samples were analysed for a period of a little over a year. The greatest abundance of arthropods was in the degenerate and pioneer phases. In the latter case, the most abundant group was the ants, which prefer the higher levels of insolation afforded by the pioneer phase. Spiders declined initially after the pioneer phase but then numbers built up gradually towards a maximum in the degenerate phase. The myriapods by contrast are most abundant in the building and mature phases. Barclay-Estrup (1974) considered that the main microclimatic influence on the fauna was the humidity change associated with the closing of the canopy in the early building phase.

Miller (1974) made an extensive study of the invertebrates associated with the growth phases of *Calluna* on a Scottish moor. He investigated a mixed-age community representing all the growth phases of the heather and another which consisted of even-aged stands produced by burning management. At each of these sites he examined four habitats: the soil surface and within the litter; the litter surface; the canopy of the vegetation; and the

atmosphere above the canopy. The sampling methods involved pitfall trapping, sticky traps, heat extraction of soil and litter cores and vacuum sweep netting of the canopy. In addition features of the vegetation were measured to enable correlations with the animal populations to be made.

Miller (1974) concluded that in unmanaged heather of mixed age there was little change in either size or diversity of the invertebrate populations during the phases of the growth cycle. By contrast heather managed by burning and of an even age had population densities in the pioneer phase which were highest, and in which diversity was second highest. The greatest diversity was found in different phases depending on habitat. Burning resulted in lower diversity of faunas in the litter and the atmosphere above the plants, an increase in that on the litter surface and almost no change in the canopy. Thus, it would appear that, in a mixed-aged unmanaged *Calluna* community there is little difference between the faunas of the different phases, but in managed even-aged stands, although the composition of the fauna may be similar from phase to phase, densities are highest in the older phases.

These results are comparable with the results from the south of England, where the majority of heather communities are even-aged stands which have originated from a past history of management by burning. The pattern of faunal change in these stands is similar to that in Scotland. In general, the richest fauna occurs in the older, especially the late mature and degenerate, phases of the heather, which offer the greatest habitat diversity to the animals. It is only here that any characteristic heathland fauna can be recognised. The pioneer and post-burn phases, which last for about ten years, have an interesting fauna, composed mostly of colonising species and species which are adapted to conditions of bare ground or high levels of insolation, which disappears when the canopy closes. The subsequent fauna is more dependent on the equitable microclimatic conditions which become established in the older phases of the *Calluna*.

Herbivores

Vertebrate herbivores are not very numerous in European heathlands. This possibly reflects several of the special features of these heaths: their relatively recent origin, the low productivity of non-woody material, and the fact they have in many instances been systematically managed for the production of domestic herbivores. Mention has already been made of the pasturing of sheep and cattle (see p. 379): this section will deal briefly with naturally occurring herbivores.

Among mammals, perhaps the only one which is particularly characteristic of heathland is the mountain hare (alpine hare, blue hare), *Lepus timidus*. This species occupies upland and northern heathlands, often in considerable numbers, and is very largely dependent on *Calluna vulgaris* which in winter constitutes nearly 100% of its diet (Hewson, 1962). The density of mountain hares on upland heathlands varies considerably, ranging for example from few or none to as many as 38 per km^2 in spring. Densities above about 8 per km^2, however, are rare and seem to be associated with heathlands on soils of relatively high nutrient status with correspondingly enhanced concentrations of nutrients in the young, edible shoots of *Calluna*. It is noteworthy in this context that hares have been found to feed preferentially on *Calluna* which has been enriched by the addition of fertiliser (Miller, 1968).

The brown hare (*Lepus europaeus*) and the rabbit (*Oryctolagus cuniculus*) are to be found in lowland heaths, though both belong primarily to other types of vegetation and generally prefer grass and other herbage to *Calluna*. Rabbits, however, will graze *Calluna* intensively, particularly where it is close to their burrows (Farrow, 1917, 1925) and not too tall. Repeated grazing of young heather plants by rabbits causes very dense growth of new shoots and the development of compact, ball-shaped plants or a low spreading mat.

Red deer (*Cervus elaphus*) also make considerable use of *Calluna vulgaris*, although grazing and browsing on many other kinds of vegetation as well. Originally a forest animal, with the expansion of heathland and moorland in upland Britain red deer have become adapted to living in a largely tree-less environment, normally seeking the shelter of woodland only in winter. They belong mainly to the high ground and to the wetter, western districts, rather than to *Calluna*-dominated dry-heathlands at low altitudes on the eastern side of the country, though they may spread into these in winter. Because of their importance for sport dating from about the middle of the nineteenth century, very large stocks were built up, often to the detriment of their grazing ranges and to neighbouring farms or plantations. Densities are still very high, compared to other parts of Western Europe, and deer stalking and shooting remains a valuable source of revenue to Highland estates. Promising experiments are also being conducted on deer "farming" in heathland areas of eastern Scotland.

Game birds constitute another group of heathland herbivores. In Scandinavia the willow grouse (*Lagopus lagopus*) is one of these, although not particularly closely associated with heathland. However, the British subspecies *Lagopus lagopus scoticus*, the red grouse, is pre-eminently a bird of heathlands (p. 379), feeding almost exclusively on *Calluna vulgaris* in winter. Even in summer *Calluna* usually accounts for over 50% of the diet, the remainder of which is composed largely of other heath species such as leaves and fruit of *Vaccinium myrtillus*, *Empetrum nigrum*, *Oxycoccus* spp., *Erica* spp., etc. Red grouse are territorial birds, the size of a breeding territory normally ranging from 2 to 5 ha (extremes 0.2 to 13.2 ha; Miller and Watson, 1974a). As with hares, the breeding stock of grouse varies in relation to the nutrient content of the edible shoots of their food plants, being greater in areas where the nutrient status of the soil is relatively high. On the poorer soils it can be shown that application of nitrogenous fertilisers increases the breeding stock of an area, and the same has been reported for phosphatic fertiliser on bog peat in western Ireland. There is also evidence that the density of grouse is related to the proportion of young stands of *Calluna* (2 to 3 years old, for instance when the concentration of nutrients in the shoots is high) in an area, so long as these are in relatively small patches, allowing the inclusion of both young and old *Calluna* in a territory (see Vol. B, Ch. 26). This relationship between population density and the area of young stands of *Calluna* is the prime reason for the management of "grouse moors" by burning (see Vol. B, Ch. 26).

Mountain and arctic heathlands support a related species, the ptarmigan (*Lagopus mutus*). *Calluna vulgaris*, where present, again provides a substantial proportion of the food of this species, but other plants are generally included according to availability, for example *Vaccinium myrtillus* and *Empetrum* spp. in Britain, or *Salix herbacea* in Iceland (Watson, 1964). Heathland bordering on forest provides a habitat for black grouse (*Lyrurus tetrix*), which ranges northwards from Belgium and The Netherlands to Scandinavia and Britain, as well as occurring in upland areas further south.

Estimates have been made by Miller and Watson (1974b) of the biomass, production and cropping of the most important herbivores on heathland in northeastern Scotland. Their figures are quoted in Table 14.11 as an indication of the low annual yield.

Insectivores

The rather numerous invertebrate fauna of heathland (pp. 401–406) attracts a variety of insectivores, few of which however occupy heathland exclusively or even preferentially. These include lizards (e.g. the viviparous lizard (*Lacerta vivipera*), common shrews (*Sorex araneus*) and in grassy areas wood mice and common voles (*Apodemus sylvaticus*, *Microtus agrestis*), and a variety of birds, notably the meadow pipit (*Anthus pratensis*), stonechat (*Saxicola torquata*) and wood-lark (*Lullula arborea*) among others. Frogs (*Rana temporaria*) and newts (*Triturus vulgaris*) are encountered on wet-heathlands.

Some species, however, are particularly characteristic of lowland sandy heathlands. These include the sand lizard (*Lacerta agilis*), the smooth snake (*Coronella austriaca*) and the Dartford warbler (*Sylvia undata*), all becoming increasingly scarce as the habitats available to them disappear (Gimingham, Vol. B, Ch. 26). The Dartford warbler favours localities in which *Ulex europaeus* occurs abundantly amongst old *Calluna vulgaris*, a type of community generally resulting from a mosaic of not too frequent, small fires. It is a local resident on certain heathlands in southern England where its population has dwindled during the past century; allied races extend southwards from the Channel Islands and France.

Several ground-nesting species of birds move into open heathland for breeding, such as the curlew (*Numenius arquata*) and nightjar (*Caprimulgus europaeus*).

Predators

The larger herbivores of heathland have few predators. Foxes (*Vulpes vulpes*) and stoats (*Mustela erminea*) both feed on rabbits and hares as well as many of the smaller animals. Some of the birds of prey which hunt over heathlands and moors may sometimes take grouse or other game birds and their young: these include the hen harrier (*Circus cyaneus*), buzzard (*Buteo buteo*), and in the

TABLE 14.11

Estimates of biomass, production and cropping of four vertebrate herbivores on heathland in northwest Scotland

Crop	Average density in spring (number per ha)	Biomass (kg ha^{-1})	Production of young (number ha^{-1} yr^{-1})	Annual crop taken by man (number ha^{-1})	(kg ha^{-1})
Sheep	0.30	13.6	0.24	0.24	5.7
Red deer	0.089	6.1	0.015	0.014	1.1
Mountain hare	0.16	0.43	0.50	see note 1	
Red grouse	0.65	0.41	0.89	0.40	0.25

Table from Miller and Watson (1974a), where references are given to primary sources of data.

These data are averages from a small number of study areas in northeast Scotland and so should not be interpreted too widely. For example, there are few sheep or hares in most of the Cairngorm Mountains.

[1] Mountain hares are seldom exploited and so it is impossible to calculate a reliable figure for the annual crop. However, data from a few moors where hares are particularly abundant and where bag records are kept suggest that an average annual yield of about 0.7 kg ha^{-1} is obtained there.

more remote coastal or mountain districts the peregrine falcon (*Falcon peregrinus*) and golden eagle (*Aquila chrysaetos*).

The hooded crow (*Corvus cornix*) in the north of Scotland and northwards from The Netherlands and Belgium, and the carrion crow (*Corvus corone*) in the rest of the heathland region, will eat the eggs of ground-nesting birds including the grouse, as well as various small animals and carrion. Of these predators, the hen harrier is perhaps the most distinctly associated with heathlands, where it nests among *Calluna vulgaris*. A related species, Montagu's harrier (*Circus pygargus*), sometimes favours wet-heathland areas.

There are, in addition, various predators which take a variety of small mammals and birds. Adders (*Vipera berus*) may be very plentiful in heathland, both in dry sandy soils and also, to some extent, in wet heath. Wild cat (*Felis sylvestris*), in the more remote northern districts, hunt rabbits as well as smaller prey. Several hawks and owls include small mammals and birds, as well as insects, etc., in their diet: most typical of heathland are the merlin (*Falco columbarius*), the kestrel (*Falco tinnunculus*) and the short-eared owl (*Asio flammeus*), which hunts by day.

ACKNOWLEDGEMENTS

Grateful acknowledgement is made to Dr. G.R. Miller for making available his unpublished data upon the performance of *Calluna* at Kerloch Moor, and to Dr. G. Tyler for additional data regarding a site at Skanör in Sweden.

The agreement of the respective publishers to the reproduction of the following figures is acknowledged: Figs. 14.1, 14.4A, 14.9, 14.13 and 14.14 from *An Introduction to Heathland Ecology* (Gimingham, 1975), Oliver and Boyd, Edinburgh. Figs. 14.5, 14.7, 14.16 and 14.17 from *Ecology of Heathlands* (Gimingham, 1972), Chapman and Hall, London. Fig. 14.18 from *Production Ecology of British Moors and Montane Grasslands* (O.W. Heal and D.F. Perkins, Editors), Springer-Verlag, Berlin.

The permission of J.G.K. Flower-Ellis, and the Department of Forest Ecology and Forest Soils, Royal College of Forestry, Stockholm, to reproduce Fig. 14.8 is gratefully acknowledged.

The permission of the Editors of the *Journal of Ecology* is acknowledged in connection with the reproduction of Figs. 14.19, 14.22, 14.27, 14.28, and of the Editors of the *Journal of Animal Ecology* in connection with Fig. 14.29.

REFERENCES

Allen, S.E., 1964. Chemical aspects of heather burning. *J. Appl. Ecol.*, 1: 347–367.

Allen, S.E., Carlisle, A., White, E.J. and Evans, C.C., 1968. The plant nutrient content of rainwater, *J. Ecol.*, 56: 497–504.

Allen, S.E., Evans, C.C. and Grimshaw, H.M., 1969. The distribution of mineral nutrients in soil after heather burning. *Oikos*, 20: 16–25.

Anderson, J.M., 1973. The breakdown and decomposition of sweet chestnut (*Castanea sativa* Mill.) and beech (*Fagus sylvatica* L.) in two deciduous woodland soils. *Oecologia*, 12: 251–274.

Bannister, P., 1970. The annual course of drought and heat resistance in heath plants from an oceanic environment. *Flora*, 159: 105–123.

Barclay-Estrup, P., 1970. The description and interpretation of cyclical processes in a heath community. II. Changes in biomass and shoot production during the *Calluna* cycle. *J. Ecol.*, 58: 243–249.

Barclay-Estrup, P., 1971. The description and interpretation of cyclical processes in a heath community. III. Microclimate in relation to the *Calluna* cycle. *J. Ecol.*, 59: 143–166.

Barclay-Estrup, P., 1974. Arthropod population in a heathland as related to cyclical changes in the vegetation. *Entomol. Mon. Mag.*, (1973) 109: 79–84.

Barclay-Estrup, P. and Gimingham, C.H., 1969. The description and interpretation of cyclical processes in a heath community. I. Vegetational change in relation to the *Calluna* cycle. *J. Ecol.*, 57: 737–758.

Beijerinck, W., 1940. *Calluna*: a monograph on the Scotch heather. *Verh. K.Ned Akad. Wet. Tweede Sectie.*, 38: 1–180.

Bell, M.K., 1974. Decomposition of herbaceous litter. In: C.H. Dickinson and G.J.F. Pugh (Editors), *Biology of Plant Litter Decomposition*, Academic Press, London, pp. 37–67.

Bellamy, D.J. and Holland, P.J., 1966. Determination of the net annual aerial production of *Calluna vulgaris* (L.) Hull, in northern England. *Oikos*, 17: 272–275.

Birse, E.L., 1976. *Plant Communities and Soils of the Lowland and Southern Upland Regions of Scotland*. Macaulay Institute for Soil Research, Aberdeen, 226 pp.

Bøcher, T.W., 1943. Studies on the plant geography of the North-Atlantic heath formation. II. Danish dwarf shrub communities in relation to those of Northern Europe. *Biol. Skr.*, 2: 1–129.

Braun-Blanquet, J. and Tüxen, R., 1943. Übersicht der höheren Vegetationseinheiten Mitteleuropas. *Comm. Station Int. Géobot. Médit. Alp.*, 84 (Montpellier).

Braun-Blanquet, J., Sissingh, G. and Vlieger, J., 1939. *Prodromus der Pflanzengesellschaften. 6 Klasse der Vaccinio-Piceetea.* Montpellier.

Braun-Blanquet, J., Roussine, N. and Nègre, R., 1952. *Les groupements végétaux de la France méditerranéenne.* Centre Nationale de la Recherche Scientifique, Paris, 297 pp.

Braun-Blanquet, J., Pinto da Silva, A.R. and Rozeira, A., 1964. Résultats de trois excursions géobotaniques à travers le Portugal septentrional et moyen. III. Landes à cistes et éricacées. *Agron. Lusit.,* 23: 229–313.

Brian, M.V., 1964. Ant distribution in a southern English heath. *J. Anim. Ecol.,* 33: 451–461.

Brian, M.V., Mountford, M.D., Abbott, A. and Vincent, S., 1976. The changes in ant species distribution during ten years post-fire regeneration of a heath. *J. Anim. Ecol.,* 45: 115–133.

Bridgewater, P., 1970. *Phytosociology and Community Boundaries of the British Heath Formation.* Thesis, University of Durham, Durham, 199 pp.

Carlisle, A., Brown, A.H.F. and White, E.J., 1966. Litter fall, leaf production and the effects of defoliation by *Tortrix viridana* in a sessile oak (*Quercus petraea*) woodland. *J. Ecol.,* 54: 65–85.

Chapman, S.B., 1967. Nutrient budgets for a dry heath ecosystem in the south of England. *J. Ecol.,* 55: 677–689.

Chapman, S.B., 1970. The nutrient content of the soil and root system of a dry heath ecosystem. *J. Ecol.,* 58: 445–452.

Chapman, S.B. and Webb, N.R., 1978. The productivity of *Calluna* heathland in southern England. In: O.W. Heal and D.F. Perkins (Editors), *Production Ecology of Some British Moors and Montane Grasslands.* Springer-Verlag, Berlin, pp. 247–262.

Chapman, S.B., Hibble, J. and Rafarel, C.R., 1975a. Net aerial production by *Calluna vulgaris* on lowland heath in Britain. *J. Ecol.,* 63: 233–258.

Chapman, S.B., Hibble, J. and Rafarel, C.R., 1975b. Litter accumulation under *Calluna vulgaris* on a lowland heathland in Britain. *J. Ecol.,* 63: 259–271.

Coombe, D.E. and Frost, L.C., 1956. The nature and origin of the soils over the Cornish serpentine. *J. Ecol.,* 44: 605–615.

Cormack, E. and Gimingham, C.H., 1964, Litter production by *Calluna vulgaris* (L.) Hull. *J. Ecol.,* 52: 285–297.

Coulson, J.C. and Whittaker, J.B., 1978. Ecology of moorland animals. In: O.W. Heal and D.F. Perkins (Editors), *Production Ecology of Some British Moors and Montane Grasslands.* Springer-Verlag, Berlin, pp. 52–93.

Cragg, J.B', 1961. Some aspects of the ecology of moorland animals. *J. Anim. Ecol.,* 30: 205–234.

Crisp, D.T., 1966. Input and output of minerals for an area of Pennine moorland: The importance of precipitation, drainage, peat erosion and animals. *J. Appl. Ecol.,* 3: 327–348.

Dahl, E., 1956. *Rondane: Mountain Vegetation in South Norway and Its Relation to the Environment.* Norske Videnskap Akademie, Oslo, 374 pp.

Damman, A.W.H., 1957. The south Swedish *Calluna* heath and its relation to the Calluneto-Genistetum. *Bot. Not.,* 110: 363–398.

Delany, M., 1953. Studies on the microclimate of *Calluna* heathland. *J. Anim. Ecol.,* 22: 227–239.

Delany, M., 1956. The animal communities of three areas of pioneer heath in south-west England. *J. Anim. Ecol.,* 25: 112–126.

Des Abbayes, H. and Corillion, R., 1949. Sur la répartition d'*Ulex gallii* Planch. et d'*Ulex nanus* Sm. dans le massif armoricain. *C.R. Seánces Soc. Biogéogr.,* 228–229: 86–89.

De Smidt, J.T., 1967. Phytogeographical relations in the North West European heath. *Acta Bot. Neerl.,* 15: 630–647.

Dimbleby, G.W., 1952. Soil regeneration on the north-east Yorkshire moors. *J. Ecol.,* 40: 331–341.

Dimbleby, G.W., 1962. The development of British heathlands and their soils. *Oxford For. Nem.,* 23: 1–121.

Dimbleby, G.W., 1965. Post-glacial changes in soil profiles. *Proc. R. Soc. B.,* 161: 355–362.

Diver, C. and Diver, P., 1933. Contributions towards a survey of the plants and animals of South Haven Peninsula, Studland heath, Dorset. III. Orthoptera. *J. Anim. Ecol.,* 2: 36–69.

Diver, C. and Good, R., 1934. The South Haven Peninsula Survey (Studland Heath, Dorset): general scheme of survey. *J. Anim. Ecol.,* 3: 129–132.

Duchaufour, P., 1948. Recherches écologiques sur la chênaie atlantique française. *Ann. Ecol. Nat. Eaux For., Nancy,* 11: 1–332.

Duchaufour, P., 1956. Note sur les phases de la podzolisation sur grès vosgiens. *Trans. 6th Int. Congr. Soil Sci,,* E: 367–370.

Dupont, P., 1975. Les limites altitudinales des landes atlantiques dans les montagnes Cantabriques (nord de l'Espagne). In: J.M. Géhu (Editor), *Colloques Phytosociologiques. II. La végétation des landes d'Europe Occidentale. Lille — 1973.* Cramer, Vaduz, pp. 47–58.

Durno, S.E., 1958. Pollen analysis of peat deposits in Eastern Sutherland and Caithness. *Scot. Geogr. Mag.,* 74: 127–135.

Duvigneaud, P., 1944. Les genres *Cetraria, Umbilicaria* et *Stereocaulon* en Belgique. *Bull. Soc. R. Bot. Belg.,* 81: 58–129.

Elmes, G.W., 1971. An experimental study on the distribution of heathland ants. *J. Anim. Ecol.,* 40: 495–499.

Evans, C.C. and Allen, S.E., 1971. Nutrient losses in smoke produced during heather burning. *Oikos,* 22: 149–154.

Faegri, K., 1940. Quartärgeologische Untersuchungen im westlichen Norwegen. II. Zur spätquartären Geschichte Jaerens. *Berg. Mus. Årb., 1939–40, Naturvid. Rekke.,* 7: 1–201.

Farrow, E.P., 1917. On the ecology and vegetation of Breckland. III. General effects of rabbits on the vegetation. *J. Ecol.,* 5: 1–18.

Farrow, E.P., 1925. On the ecology of the vegetation of Breckland. VIII. Views relating to the probable former distribution of *Calluna* heath in England. *J. Ecol.,* 13: 121–125.

Flower-Ellis, J.G.K., 1971. Age structure and dynamics in stands of bilberry (*Vaccinium myrtillus* L.). *Avd. Skogsekol., Stockholm,* 9: 1–108.

Forrest, G.I., 1971. Structure and production of north Pennine blanket bog vegetation. *J. Ecol.,* 59: 453–479.

Forrest, G.I. and Smith, R.A.H., 1975. The productivity of a range of blanket bog vegetation types in the northern Pennines. *J. Ecol.,* 63: 173–202.

Géhu, J.-M., 1975. Essai pour un système de classification phytosociologique des landes Atlantiques Françaises. In: J.-M. Géhu (Editor), *Colloques Phytosociologiques. II. La végétation des landes d'Europe Occidentale. Lille* — *1973.* Cramer, Vaduz, pp. 361–377.

Gimingham, C.H., 1960. Biological flora of the British Isles: *Calluna vulgaris* (L.) Hull. *J. Ecol.*, 48: 455–483.

Gimingham, C.H., 1961. North European heath communities: a "network of variation". *J. Ecol.*, 49: 655–694.

Gimingham, C.H., 1964. Dwarf-shrub heaths. In: J.H. Burnett (Editor), *The Vegetation of Scotland.* Oliver and Boyd, Edinburgh, pp. 232–287.

Gimingham, C.H., 1969. The interpretation of variation in north European dwarf-shrub communities. *Vegetatio*, 17: 89–108.

Gimingham, C.H., 1972. *Ecology of Heathlands.* Chapman and Hall, London, 266 pp.

Gimingham, C.H., 1975. *An Introduction to Heathland Ecology.* Oliver and Boyd, Edinburgh, 124 pp.

Godwin, H., 1944a. Neolithic forest clearances. *Nature, Lond.,* 153: 511.

Godwin, H., 1944b. Age and origin of the 'Breckland' heaths of East Anglia. *Nature, Lond.,* 154: 6.

Godwin, H., 1948. Studies of the post-glacial history of British vegetation. X. Correlation between climate, forest composition, prehistoric agriculture and peat stratigraphy in sub-boreal and sub-atlantic peats of the Somerset levels. *Philos. Trans. R. Soc., B,* 233: 275–286.

Godwin, H., 1956. *History of the British flora.* Cambridge University Press, Cambridge, 384 pp.

Gore, A.J.P., 1968. The supply of six elements by rain to an upland peat area. *J. Ecol.,* 56: 483–496.

Gore, A.J.P. and Olson, J.S., 1967. Preliminary models for accumulation of organic matter in an *Eriophorum/Calluna* ecosystem. *Aquilo Ser. Bot.,* 6: 297–313.

Gorham, E., 1958. The influence and importance of daily weather conditions in the supply of chloride, sulphate and other ions to freshwaters from atmospheric precipitation. *Philos. Trans. R. Soc., B,* 241: 147–178.

Graebner, P., 1901, 1925. *Die Vegetation der Erde. Vol. 5. Die Heide Norddeutschlands.* Engelmann, Leipzig, 277 pp. (1901: 1st ed.; 1925: 2nd ed.).

Grant, S.A., 1971. Interactions of grazing and burning on heather moors, 2. Effects on primary production and level of utilization. *J. Br. Grassland Soc.,* 26: 173–181.

Grubb, P.J., Green, H.E. and Merrifield, R.C.J., 1969. The ecology of chalk heath: its relevance to the calcicole-calcifuge and soil acidification problems. *J. Ecol.,* 57: 175–212.

Hansen, K., 1969. Edaphic conditions of Danish heath vegetation and the response to burning-off. *Bot. Tidssk.,* 64: 121–140.

Heal, O.W. and Perkins, D.F., 1976. I.B.P. studies on montane grassland and moorlands. *Philos. Trans. R. Soc., B,* 274: 295–314.

Heath, G.W. and King, H.G.C., 1964. Litter breakdown in deciduous forest soils. *8th Int. Congr. Soil Sci.,* 3: 979–987.

Hewson, R., 1962. Food and feeding habits of the mountain hare, *Lepus timidus scotius,* Hilzheimer. *Proc. Zool. Soc. Lond.,* 139: 515–526.

Hopkins, B., 1955. The species-area relations of plant communities. *J. Ecol.* 43: 409–426.

Iversen, J., 1941. Landnam i Danmarks Stenalder. *Dan. Geol.*

Unders., Raekke 2, 66: 1–67.

Iversen, J., 1949. The influence of prehistoric man on soil fertility. *Dan. Geol. Unders., Raekke 4,* 6: 1–25.

Iversen, J., 1964. Retrogressive vegetational succession in the postglacial. *J. Ecol.,* 52 (Suppl.): 59–70.

Iversen, J., 1969. Retrogressive development of a forest ecosystem demonstrated by pollen diagrams from fossil mor. *Oikos, Suppl.,* 12: 35–49.

Jenny, H., Gessel, S.P. and Bingham, F.T., 1949. Comparative study of decomposition rates of organic matter in temperate and tropical regions. *Soil Sci.,* 68: 419–432.

Jonassen, H., 1950. Recent pollen sedimentation and Jutland heath diagrams. *Dan. Bot. Ark.,* 13: 1–168.

Kenworthy, J.B., 1963. Temperatures in heather burning. *Nature, Lond.,* 200: 1226.

Köppen, W., 1923. *Die Klimate der Erde.* Bornträger, Berlin, 369 pp.

Krause, E.H.L. 1892. Die Heide. *Bot. Jahrb.,* 14.

Kübiena, W.L., 1953. *The soils of Europe.* Thomas Minby, London, 318 pp.

McVean, D.N. and Ratcliffe, D.A. 1962. *Plant Communities of the Scottish Highlands.* H.M.S.O. London. 445 pp.

Malmer, N. 1965. The South-western dwarf-shrub heaths. In: The plant cover of Sweden. *Acta Phytogeogr. Suecica,* 50: 123–130.

Malmström, C. 1937. Tönnersjöhedens försökspark i Halland. *Medd. Statens. Skogsförskningsinst.,* 30: 323–528.

Malmström, C. 1939. Hallands skogar under de senaste 300 åren. *Medd. Statens Skogsförskningsinst.,* 31: 171–300.

Mangenot, F. 1966. Étude microbiologique des litières. *Bull. Écol. Nat. Sup. Agron. Nancy,* 8: 113–125.

Merrett, P. 1967. The phenology of spiders on heathland in Dorset I. Families Atypidae, Dyseridae, Gnaphosidae, Clubionidae, Thomisidae, Salticidae. *J. Anim. Ecol.,* 36: 363–374.

Merrett, P. 1968. The phenology of spiders on heathland in Dorset. Families Lycosidae, Pisauridae, Agelenidae, Mimetidae, Theridiidae, Tetragnathidae, Argiopidae. *J. Zool., Lond.,* 156: 239–256.

Merrett, P. 1969. The phenology of linyphiid spiders on heathland in Dorset. *J. Zool., Lond.,* 157: 289–307.

Merrett, P. 1971. *Captain Cyril Diver (1892–1969): a Memoir.* Furzebrook Research Station, Wareham, 58 pp.

Merrett, P. 1976. Changes in the ground-living spider fauna after heathland fires in Dorset. *Bull. Br. Arachnol. Soc.,* 3: 214–221.

Miller, B.J.F. 1974. *Studies of Changes in the Populations of Invertebrates Associated with Cyclical Process in Heathland.* Thesis, University of Aberdeen, Aberdeen, 177 pp.

Miller, G.R., 1968. Evidence for selective feeding on fertilized plots by red grouse, hares and rabbits. *J. Wildl. Manage.,* 32: 849–853.

Miller, G.R. and Watson, A., 1974a. Heather moorland: a man-made ecosystem. In: A. Warren and F.B. Goldsmith (Editors), *Conservation in Practice.* Wiley, London and New York, pp. 145–166.

Miller, G.R. and Watson, A., 1974b. Some effects of fire on vertebrate herbivores in the Scottish highlands. *Proc. Annu. Tall Timbers Fire Ecol. Conf.,* 13: 39–64.

Miller, G.R. and Watson, A., 1978. Heather productivity and its relevance to the regulation of red grouse populations. In: O.W. Heal and D.F. Perkins (Editors), *Production Ecology of Some British Moors and Montane Grasslands*. Springer-Verlag, Berlin, pp. 277–285.

Mitchell, G.F., 1951. Studies in Irish Quaternary deposits, No. 7. *Proc. R. Irish. Acad.*, 53B: 111–206.

Mitchell, G.F., 1956. Post-Boreal pollen diagrams from Irish raised bogs. *Proc. R. Irish Acad.*, 57B: 185–251.

Mohamed, B.F. and Gimingham, C.H., 1970. The morphology of vegetative regeneration in *Calluna vulgaris*. *New Phytol.*, 69: 743–750.

Moore, J.J., 1968. A classification of the bogs and wet heaths of northern Europe. (Oxycocco-Sphagnetea Br.–Bl. et Tx. 1943.) In: R. Tüxen (Editor), *Pflanzensoziologische Systematik*. W. Junk, Den Haag, pp. 306–320.

Moore, N.W., 1962. The heaths of Dorset and their conservation. *J. Ecol.*, 50: 369–391.

Morrison, M.E.S., 1959. Evidence and interpretation of "landnam" in the North-East of Ireland. *Bot. Not.*, 112: 185–204.

Moss, R., 1967. Probable limiting nutrients in the main food of Red Grouse *Lagopus scoticus*. In: K. Petrusewicz (Editor), *Secondary Production of Terrestrial Ecosystems*. Polish Academy of Sciences, Warsaw, pp. 369–379.

Moss, R., 1969. A comparison of red grouse (*Lagopus l. scoticus*) stocks with the production and nutritive value of heather (*Calluna vulgaris*). *J. Anim. Ecol.*, 38: 103–122.

Müller, P.E., 1924. Bidrag til de jydske hedesletters naturhistorie. *K. Dan. Vid. Selsk. Biol. Medd.*, 4.

Nordhagen, R., 1928. *Die Vegetation und Flora des Sylenegebietes*. J. Dybwad, Oslo, 612 pp.

Nordhagen, R., 1936. Versuch einer Einteilung der subalpinen–alpinen Vegetation Norwegens. *Berg. Mus. Årb. Naturvid. Rekke*, 7.

Nye, P.H., 1961. Organic matter and nutrient cycles under moist tropical forest. *Plant Soil*, 13: 333–346.

Olson, J.S., 1963. Energy storage and the balance of producers and decomposers in ecological systems. *Ecology*, 44: 322–331.

Poore, M.E.D. and McVean, D.N., 1957. A new approach to Scottish mountain vegetation. *J. Ecol.*, 45: 401–439.

Preising, E., 1949. Nardo-Callunetea. Zur Systematik der Zwergstrauch-Heiden und Magertriften Europas mit Ausnahme des Mediterrangebietes, der Arktis und der Hochgebirge. *Mitt. Flor.–Soz. Arbeitsgem.*, N.F., 1: 12–25.

Raw, F., 1967. Arthropoda (except Acari and Collembola). In: A. Burges and F. Raw (Editors), *Soil Biology*. Academic Press, London, pp. 323–362.

Richards, O.W., 1926. Studies on the ecology of English heaths. III. Animal communities of the felling and burn successions at Oxshott Heath, Surrey. *J. Ecol.*, 14: 244–281.

Robertson, R.A. and Davies, G.E., 1965. Quantities of plant nutrients in heather ecosystems. *J. Appl. Ecol.*, 2: 211–219.

Romell, L.–G., 1951. Liens landskap och mulens. *Sver. Nat.*, 42: 9–18.

Romell, L.–G., 1952. Heden. In: *Natur i Halland*. Stockholm, pp. 331–347.

Rutter, A.J., 1955. The composition of wet-heath vegetation in relation to the water-table. *J. Ecol.*, 43: 507–543.

Satchell, J.E., 1967. Lumbricidae. In: A. Burges and F. Raw (Editors), *Soil Biology*. Academic Press, London, pp. 259–322.

Satchell, J.E. and Lowe, D.G., 1967. Selection of leaf litter by *Lumbricus terrestris*. In: O. Graff and J.E. Satchell (Editors), *Progress in Soil Biology: Proceedings of the Colloquium on Dynamics of Soil Communities, Braunschweig–Völkenrode 1966*. North-Holland, Amsterdam, pp. 102–119.

Schubert, R., 1960. Die zwergstrauchreichen azidiphilen Pflanzengesellschaften Mitteldeutschlands. *Pflanzensoziologie*, 11: 1–235.

Schwickerath, M., 1933. Die Vegetation der Landkreise Aachen und ihre Stellung in nördlichen Westdeutschland. *Aachener Beitr. Heimatkd.*, 13.

Shimwell, D.W., 1975. Man-induced changes in the heathland vegetation of central England. In: J.–M. Géhu (Editor), *Colloques Phytosociologiques. II. La végétation des landes d'Europe Occidentale. Lille–1973*. Cramer, Vaduz, pp. 59–71.

Smith, A.G. and Willis, E.H., 1962. Radiocarbon dating of the Fallahogy landnam phase. *Ulster J. Archaeol.*, 24/5: 16–24.

Smith, W.G., 1902. The origin and development of heather moorland. *Scott. Geogr. mag.*, 18: 587–597.

Smith, W.G., 1911. Scottish Heaths. In: A.G. Tansley (Editor), *Types of British Vegetation*. Cambridge University Press, Cambridge, pp. 113–116.

Southwood, T.R.E., 1966. *Ecological Methods*. Chapman and Hall, London, 391 pp.

Stoutjesdijk, Ph., 1959. Heaths and inland dunes of the Veluwe. *Wentia*, 2: 1–96.

Summers, C.F., 1972. *Aspects of production in Montane Dwarf Shrub Heaths*. Thesis, University of Aberdeen, Aberdeen, 163 pp.

Summers, C.F. 1978. Production in montane dwarf shrub heaths. In: O.W. Heal and D.F. Perkins (Editors), *Production Ecology of Some British Moors and Montane Grasslands*. Springer-Verlag, Berlin, pp. 263–276.

Tansley, A.G., 1939. *The British Islands and Their Vegetation*. Cambridge University Press, Cambridge, 930 pp.

Thomas, B., Eskritt, J.R. and Trinder, N., 1945. Minor elements of common heather (*Calluna vulgaris*). *Emp. J. Exp. Agric.*, 13: 93–99.

Tüxen, R., 1937. Die Pflanzengesellschaften Nordwest Deutschlands. *Mitt. Flor.–Soz. Arbeitsgem.*, N.F., 5: 155–174.

Tyler, G., Gullstrand, C., Holmquist, K. and Kjellstrand, A., 1973. Primary production and distribution of organic matter and metal elements in two heath ecosystems. *J. Ecol.*, 61: 1, 251–268.

Ward, S.D., 1970. The phytosociology of *Calluna–Arctostaphylos* heaths in Scotland and Scandinavia. I. Dinnet Moor, Aberdeenshire. *J. Ecol.*, 58: 847–863.

Ward, S.D., 1971a. The phytosociology of *Calluna–Arctostaphylos* heaths in Scotland and Scandinavia. II. The north-east Scottish heaths. *J. Ecol.*, 59: 679–696.

Ward, S.D., 1971b. The phytosociology of *Calluna–Arctostaphylos* heaths in Scotland and Scandinavia. III. A critical examination of the Arctostaphyleto-Callunetum. *J. Ecol.*, 59: 697–712.

Waterbolk, H.T., 1957. Pollenanalytisch onderzoek van twee noordbrabantse tumuli. In: G. Beex, *Twee grafheuvels in Noord-Brabant*. Bijdr. Stud. Brabantse Heem., 9: 34–39.

Watson, A., 1964. The food of ptarmigan (*Lagopus mutus*) in Scotland. *Scot. Nat.*, 71: 60–66.

Watt., A.S., 1947. Pattern and process in the plant community. *J. Ecol.*, 35: 1–22.

Watt, A.S., 1955. Bracken versus heather, a study in plant sociology, *J. Ecol.*, 43: 490–506.

Webb, N.R., 1972. Cryptostigmatid mites recorded from heathland in Dorset. *Entomol. Mon. Mag.*, 107: 228–229.

Whittaker, E., 1961. Temperature in heath fires. *J. Ecol.*, 49: 709–715.

Whittaker, E. and Gimingham, C.H., 1962. The effects of fire on regeneration of *Calluna vulgaris* (L.) Hull from seed. *J. Ecol.*, 50: 815–822.

Chapter 15

ARCTIC HEATHLANDS[1]

L.C. BLISS

INTRODUCTION

Shrub vegetation, with species as an important component, is very common in the Low Arctic. As the closed taiga (boreal) forest thins out to form an open woodland (lichen or shrub woodland), the number of individuals of deciduous and evergreen shrub species and the cover they provide increases. Beyond these broad open woodlands, often 100 to 300 km wide, is a narrowed zone (10 to 50 km wide) of scattered trees and tree islands within a low-shrub/dwarf-shrub heathland or dwarf-shrub heathland/cottongrass tussock tundra: the forest-tundra. At higher latitudes or altitudes, forest-tundra gives way to arctic and alpine tundra respectively. These lands are often, but not always, dominated by low shrub vegetation with heath species comprising a significant part. Tundra eco-systems are dealt with in detail in Volume 3 of this series; but the arctic heathlands, which merge into the forest-tundra and northern boreal forests, have a great deal in common with the heathlands of the north temperate zone.

The evergreen and deciduous shrubs common to these lands belong to the hypoarctic flora *sensu* Yurtsev (1972) characteristic of the subalpine and Low Arctic across northeastern Asia and north-western North America. Most of the heathland species have their centre of distribution in eastern Asia and Alaska, the region called "Beringia" by Hultén (1937). As used here, Arctic refers to those lands in the Northern Hemisphere beyond the climatic limit of trees in average upland sites. The limits of the High and Low Arctic are presented in Fig. 15.1. "Tundra" here refers to those cold treeless lands where plants including cryptogams cover 80 to 100% of the soil surface and where

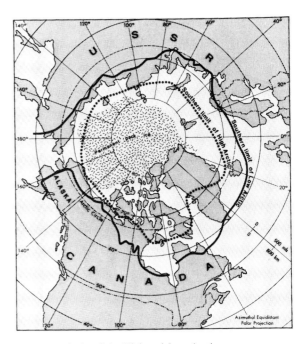

Fig. 15.1. Limits of the High and Low Arctic.

shrub species of varying height are common except in wetlands on peats (fibrisols, spodosols). The objectives of this chapter are: (1) to describe the arctic plant communities in which heathland species predominate; and (2) to describe some of the climatic, soil, autecological and primary-production characteristics of these communities. As used here "heath" includes species within the Ericaceae, Empetraceae, and Diapensiaceae.

---

[1] Manuscript completed December, 1976.

415

CLIMATE AND SOILS

Arctic lands are characterized by long, cold, dark winters and short summers with continuous light. Tundras with heathland vegetation as a prominent feature are confined to the Low Arctic and are more prevalent near the forest-tundra. The southern boundary of the Arctic was long thought to be described by the 10°C July isotherm, but more recent studies show that the summer position of the Arctic Front delimits the transition from boreal forest to arctic tundra (Bryson, 1966) and that annual net radiation at this contact is roughly 18 000 to 19 000 cal^{-2} (7.5 to 8.0×10^8 J m^{-2}) across Canada and 16 000 cal cm^{-2} (6.7×10^8 J m^{-2}) across Alaska (Hare and Ritchie, 1972). These approaches stress the integration of several environmental parameters in setting the limits of the boreal forest and arctic biomes.

Mean temperature for July, the warmest month, is generally 8° to 11°C in the Low Arctic, decreasing to 4° to 6°C in the High Arctic where heath vegetation is minor. Precipitation is low, generally 150 to 400 mm per annum with 40 to 60% of this coming as rain (Table 15.1). With nearly constant and often strong winds, snow is reworked so that exposed slopes and ridges, with cushion plants of *Dryas*, *Saxifraga* and dwarf prostrate shrubs of *Salix* and *Vaccinium*, have a thin snow cover (2 to 10 cm) that is nearly blown free at times and that melts in May or June. Snow-protected habitats with 20 to 50 cm of snow, habitats that melt by early June, often contain heathland communities. The deep snowbeds (1 to 3 m deep) that melt in July typically contain *Cassiope*-dominated heathland communities.

The growing season (from snow melt and start of leaf growth to 50% fall coloration of leaves) is generally 3 to 3.5 months at most Low Arctic stations, decreasing to 2 to 2.5 months in the High Arctic where heathland vegetation occurs. About 50% of the annual radiation is received before snow melts and the number of months with mean temperature above 0°C is generally 3.5 to 5.0 (Table 15.1).

Soil temperatures are generally 3° to 5°C from late June through mid-August at a depth of 10 cm, where most roots of the heath species occur (Bliss, 1956; Haag, 1974). The unfrozen layer (active layer) above the permafrost table is generally 30 to 50 cm at maximum thaw in many sites with heathland vegetation (Bliss, 1956; Hernandez, 1973; Wein and Bliss, 1974).

In many Low Arctic habitats there are only general correlations between topography, soils and plant communities (Tedrow and Cantlon, 1959). The relationships are more precise with well-drained soils than with poorly drained soils. Where

TABLE 15.1

Climatic data for selected arctic stations where dwarf-shrub heathland vegetation occurs (data are from various sources)

Station	Latitude (°N)	Elevation (m)	Temperature (°C)			Precipitation (mm)		Number of months with mean temp. above 0°C
			mean monthly		mean annual	June–Aug.	annual	
			Feb.	July				
Katzebue, Alaska	66°52′	3	−20.0	11.5	− 6.7	109	208	5
Umiat, Alaska	69°22′	105	−32.5	12.0	−11.8	78	148	4.5
Tuktoyaktuk, N.W.T.	69°27′	18	−29.1	10.3	−10.7	64	130	4.5
Baker Lake, N.W.T.	64°18′	12	−32.8	10.8	−12.2	86	213	4.5
Frobisher Bay, N.W.T.	63°45′	21	−25.3	8.0	−8.9	149	415	5
Holman, Victoria Island	70°44′	9	−30.9	7.5	−12.2	68	163	4.5
Truelove Lowland, Devon Island	75°33′	10	−35.8	3.9	−15.4	50	176	3
Ivigtut, West Greenland	61°12′	30	−7.1	9.9	0.8	255	1133	6
Godhaven, West Greenland	69°16′	11	−13.0	8.1	−0.9	–	395	5
Scoresby Sund, East Greenland	70°30′	17	−15.0	6.0	−6.3	–	380	3.5
Isfjord Radio, Spitsbergen	78°10′	–	−11.4	4.8	−3.7	–	395	4
Haugastol, Norway	60°31′	988	−9.5	10.7	−0.1	207	669	5.5
Kresty Village, U.S.S.R.	71°25′	30	−27.9	11.4	−12.3	120	344	4

heathland communities occur on well-drained soils of upper slopes and ridge tops, the soils are weakly developed brunisols ("Arctic Brown" *sensu* Tedrow and Hill, 1955) with a weakly developed reddish-brown to dark yellow-brown B-horizon. In many areas of Alaska, Yukon, and the western North-west Territories, soils of low-shrub/dwarf-shrub heathland and dwarf-shrub heathland/cottongrass tussock communities are gleysols ("Arctic Soils") or regosols with a thin organic layer (2 to 5 cm) over an imperfectly drained mineral soil (Hanson, 1953; Tedrow et al., 1958; Hettinger et al., 1973; Janz, 1974. In more poorly drained sites, fibrisols ("Half Bog") with a 10 to 20 cm organic layer over a gleysol may support a heathland community. Soils under heath-dominated communities have a low pH (4.0 to 6.0) and are low in percent base saturation (Table 15.2). Janz (1974) reported that, in the Mackenzie Delta region, heathland vegetation including *Dryas* was better developed on soils low in phosphorus and potassium and that a negative correlation exists between heathland vegetation (sclerophylly) and tissue phosphorus content.

PLANT COMMUNITIES

Low-shrub/dwarf-shrub heathland tundra

Within the Low Arctic, heath species are generally a component of plant communities rather than the sole dominants. Although the number of arctic heath species is not large (*c.* 25) the species in a given habitat (2 to 6) are often very common in numbers, cover, and biomass.

Low-shrub tundra [hypoarctic tundra *sensu* Yurtsev (1972), southern arctic tundra *sensu* Aleksandrova (1970), northern subarctic tundra *sensu* Andreev (1966)] covers extensive areas across Eurasia, and limited areas in southern Greenland, eastern Canada, east of the Mackenzie River Delta, northern Yukon Territory, and northern and western Alaska.

North-America
Low Arctic. The deciduous low shrubs (40 to 60 cm) *Betula nana* ssp. *exilis*, *Salix glauca*, and *S. pulchra* form an open canopy (20 to 30% cover) with an understorey of *Carex bigelowii* and

TABLE 15.2

Soil characteristics for various arctic heathland communities

Horizon	Depth (cm)	Soil color	pH	Organic matter (%)	Ca	Mg	K	P	N (%)	Reference
					(meq 100 g^{-1})			μg g^{-1}		
Low-shrub–dwarf-shrub heathland, Nigerdlikasik, (Greenland)										
A$_h$	0–4	greyish brown	4.2	10	0.38	0.44	0.23	33	–	Hansen (1969)
A$_1$	4–30	greyish orange	5.0	5	0.12	0.07	0.07	64	–	
B$_2$	30–35	orange grey	5.2	2	0.23	0.04	0.04	260	–	
Low-shrub–dwarf-shrub heathland, Tasiussakasik, (Greenland)										
A$_h$	0–5	dark brown	4.2	41	2.67	2.60	0.54	63	–	Hansen (1969)
A$_e$	5–20	orange grey	4.7	2	0.24	0.34	0.08	79	–	
B$_1$	20–30	reddish brown	5.3	1	0.10	0.07	0.06	250	–	
B$_2$	30–50	light brown	5.5	1	0.12	0.07	0.04	275	–	
Dwarf-shrub heathland, Truelove Lowland, Devon Island (N.W.T.)										
FH	5–0	–	8.2	24	70	21	217*	5	1.6	Walker and Peters (1977)
Ah$_1$	0–5	–	6.9	2	7.1	2.9	24*	2	0.1	
Ah$_2$	5–31	–	7.4	4	21.3	7.4	23*	1	0.3	
Low-shrub–dwarf-shrub heathland, Tuktoyaktuk (N.W.T.)										
H	0–10	–	5.7	–	–	–	157*	12	0.5	Haag (1974)
B/C	10–20	–	4.8	–	–	–	113*	1	0.5	

*ppm:

Eriophorum vaginatum ssp. *spissum*. Various combinations of *Ledum palustre* ssp. *decumbens*, *Vaccinium uliginosum*, *V. vitis-idaea* ssp. *minus*, *Empetrum hermaphroditum*, *Arctous alpina*, *A. rubra*, and *Cassiope tetragona* comprise the dwarf-shrub heathland layer (10 to 20 cm). A nearly complete ground cover includes the mosses *Hylocomium splendens*, *Aulacomnium turgidum*, *Polytrichum juniperinum*, *Sphagnum* spp. and the lichens *Cetraria nivalis*, *C. cucullata*, *C. islandica*, *Cladonia gracilis*, and *Cladina mitis* (Hanson, 1953; Hettinger et al., 1973; Corns, 1974). At Cape Thompson, Johnson et al. (1966) reported that in snow-accumulation sites where snow melts by early to mid-July, heaths of *Ledum palustre* ssp. *decumbens*, *Vaccinium uliginosum*, *V. vitis-idaea* ssp. *minus*, *Arctous alpina*, and *Cassiope tetragona* dominate with scattered plants of *Luzula nivalis*, *Hierochloe alpina*, *Polygonum bistorta*, *Anemone narcissiflora*, and small amounts of lichens. *Cassiope tetragona* occurs in nearly pure stands where snow remains until late July.

In areas of polygons, the result of ice wedges in these permafrost regions, the flat-topped polygons (3 to 6 m diameter) have cushion plants of *Dryas integrifolia* and *Salix phlebophylla*, with tufts of *Kobresia myosuroides* and *Hierochloe alpina*, in the dry soils of the tops where there is little winter snow cover. In the troughs (15 to 20 cm deep) between polygons where snow accumulates, the typical heath plants *Ledum*, *Vaccinium*, *Arctous*, and *Cassiope* occur (Johnson et al., 1966).

In the Mackenzie Delta region shrub tundra is the dominant feature, occupying 55 to 90% of the land (Corns, 1974). Within these shrub communities and the upland sedge communities, heathland species are very common. The low shrub-heath with varying combinations of *Betula nana* ssp. *exilis*, *Salix glauca* ssp. *acutifolia* and *S. pulchra* averaged 12% cover. The dwarf-heath shrubs (*Empetrum*, *hermaphroditum*, *Ledum palustre* ssp. *decumbens*, *Vaccinium uliginosum*, *V. vitis-idaea* ssp. *minus*, and *Arctous alpina*) averaged 23% cover. In shrub communities with scattered clumps of *Alnus crispa* (1 to 2 m high), the percentage cover of the heath shrubs remained about the same. As in many other Low Arctic tundras, these same heath species provide much of the ground cover on the tops of raised center polygons, the main microsites of true heathland communities without the partial

canopy cover of the low- to medium-height species of *Salix*, *Betula* and in some areas *Alnus crispa*.

In the District of northern Keewatin, vegetation is generally sparse due to the lack of soil development as a result of severe Pleistocene glaciation and the cold continental climate with strong winds and little winter snow cover. At Peely Lake (60°03′N) gravel summits and fell fields have patches of heathland in which *Ledum palustre* ssp. *decumbens*, *Vaccinium vitis-idaea* and *Empetrum hermaphroditum* are most important, with lesser amounts of *Cassiope tetragona*, *Vaccinium uliginosum* and *Arctous alpina*. *Betula glandulosa* and *Salix brachycarpa* ssp. *niphoclada* are very minor (Larsen, 1972). This shows the great reduction in low-shrub and even dwarf-shrub species as one approaches the High Arctic on the Precambrian Shield. In the Chesterfield Inlet area, west of Hudson Bay, northern Quebec, and Lake Harbour area, Baffin Island, there are limited areas of *Betula glandulosa* and *Salix cordifolia*. Heathlands occur in warmer, protected sites, dominated by *Empetrum hermaphroditum*, *Vaccinium vitis-idaea*, *V. uliginosum*, *Cassiope tetragona*, *Ledum palustre* ssp. *decumbens*, *Arctous alpina*, *Carex bigelowii*, *Salix herbacea*, *S. uva-ursi*, and *Luzula confusa* (Polunin, 1948).

High Arctic. In the Canadian High Arctic (arctic islands and District of Keewatin, N.W.T.) (Fig. 15.1) heathland vegetation is a very minor component. On steep granite south-facing slopes near Holman, Victoria Island (71°N, 118°W) mixed heathlands of *Cassiope tetragona*, *Vaccinium uliginosum*, ssp. *microphyllum*, *Arctous alpina*, *Rhododendron lapponicum*, *Salix arctica* and *S. reticulata* occur with a few scattered clumps of *Salix alaxensis* and *S. pulchra* (Bliss and Svoboda, 1979). These slopes are snow covered in winter, but are not snow-bed habitats.

Elsewhere in the southern islands and northward to Devon, Axel Heiberg and Ellesmere Islands, *Cassiope tetragona* dominates in snow-bed habitats along with lesser amounts of *Salix arctica*, *Dryas integrifolia*, *Carex misandra* and several forb and moss species (Beschel, 1970; Brassard and Longton, 1970; Bliss et al., 1977; Bliss and Svoboda, 1979). Snow melts in these habitats in early July as opposed to late-melt snow-bed communities (mid- to late July) where there are few forbs and mosses. *Vaccinium uliginosum* ssp. *microphyllum* is the only other heath species that extends north of 75° and it occurs only in small patches in isolated, warm, moist, and snow-protected habitats. In the Truelove Lowland, Devon Island (75° 33′N, 84° 40′W) snow-bed communities dominated by *Cassiope tetragona* occur in granite rock outcrops that occupy about 12% of the lowland (43 km²) (Bliss et al., 1977). Elsewhere at these latitudes such communities seldom account for more than 0.5 to 2% of the more completely vegetated landscape and are totally lacking in most areas. Soils are regosols with a dark brown A-horizon grading to the C-horizon (Walker and Peters, 1977).

Greenland

In southwestern Greenland (62° to 66°N), low-shrub vegetation with varying combinations of *Salix glauca* ssp. *callicarpea*, *Betula glandulosa* and *B. nana* occur. More locally *Alnus crispa*, *Sorbus groenlandica* and *Juniperus communis* ssp. *nana* are found on warmer slopes, snow covered in winter (Böcher, 1954, 1959; Hansen, 1969). Rich heathlands occur on steep, moist slopes that are warm microsites in summer and that contain soils with high amounts of organic matter (Table 15.2). Weakly developed podzols (brunisols?) are found under heathlands near the coast dominated by *Empetrum hermaphroditum*, where precipitation is higher (Hansen, 1969). Within the heathland there are various combinations of *Cassiope tetragona*, *Ledum palustre* ssp. *decumbens*, *Vaccinium uliginosum* ssp. *microphyllum*, *V. vitis-idea* ssp. *minus*, *Phyllodoce caerulea*, *Empetrum hermaphroditum* and *Rhodendron lapponicum*. In these heaths, *Betula nana*, *Carex rupestris*, *C. bigelowii* and various species of forbs and lichens, occur in varying amounts. Where late snow melt occurs, *Harrimanella hypnoides* and *Salix herbacea* predominate (Böcher, 1954, 1959).

In central western Greenland (Disko Island, 69°N and to 74°N), dwarf-shrub heath vegetation becomes a minor component of the landscape and is poor in species. Within this region all heath species listed for southwestern Greenland reach their northern limit with the exception of *Cassiope tetragona* and *Vaccinium uliginosum* (Sørensen, 1943).

Eastern Greenland is generally colder, winter snow depth is greater and there are fewer land masses free from ice than in western Greenland.

The result is less well-developed vegetation in terms of species richness and plant-community diversity than along the west coast. The only large land-areas free of ice are in the Fiord Region (70° to 77°N) compared with continuous ice-free land in western Greenland from 60° to 72°N. The result is only limited areas of heath-shrub tundra (Seidenfaden and Sørensen, 1937; Oosting, 1948; Raup, 1971).

In the Mesters Vig District (73°N), Raup (1971) reported that "heath-tundra" occupied about 10% of the land and an additional 19% was occupied by scattered heath species and lichens. As elsewhere, varying combinations of heath species occur, depending upon their physiological tolerance limits, of which there is all too little information. The most commonly found species in these heathlands include *Cassiope tetragona*, *Vaccinium uliginosum* ssp. *microphyllum* and *Salix arctica*. Less widely distributed are *Arctous alpina*, *Betula nana*, *Rhododendron lapponicum* and *Dryas octopetala* (Raup, 1971). Most heath tundra is found on well-drained, relatively stable slopes, underlain by "Arctic Brown" (brunisolic) soils (Ugolini, 1966). Snow-bed communities are dominated by *Cassiope tetragona* and often include *Salix herbacea* and *Dryas octopetala*. *Betula nana* is more important in the warmer and drier (more continental) inland fiord sites than near the cool maritime climate of coastal areas (Oosting, 1948). Similar patterning of shrub and herbaceous species occurs in western Greenland.

Fenno-Scandinavia

In the mountains of Fenno-Scandinavia, *Betula pubescens* ssp. *tortuosa* generally forms timberline. Above this, the alpine zone is divided into three sections or belts; dwarf-heath shrubs are important in the lower two (Dahl, 1956; Kilander, 1965). The dwarf-shrub heathland communities of the low alpine belt (1050 to 1350 m) typically melt out by late June to early July and include *Betula nana*, *Empetrum hermaphroditum*, *Phyllodoce caerulea*, *Salix herbacea*, *Vaccinium myrtillus*, *V. uliginosum*, *V. vitis-idaea*, *Arctous alpina*, *Loiseleuria procumbens*, *Juncus trifidus*, and *Carex bigelowii*. *Vaccinium myrtillus* is one of the most abundant species and forms the upper limit of the low alpine belt (Rønning, 1960; Kilander, 1965). The middle alpine belt (1350 to 1600 m) is also characterised by heathland. Important species included *Loiseleuria procumbens*, *Phyllodoce caerulea*, *Empetrum herma-phroditum* and *Vaccinium vitis-idae*. *Harrimanella hypnoides* and *Salix herbacea* are common in snow-bed sites. In the northern mountains *Cassiope tetragona* is the most important dwarf-heath species in the middle alpine belt (Rønning, 1960; Gjaerevoll and Bringer, 1965).

In northern Finland the higher mountains have an open birch forest of *Betula pubescens* ssp. *tortuosa* with an understorey of *Empetrum herma-phroditum*, *Vaccinium vitis-idaea*, *V. uliginosum*, *V. myrtillus*, and *Ledum palustre*. Above 330 m there is a sparse birch scrub with a dwarf-shrub heath-land dominated by *Vaccinium vitis-idaea* and *Empetrum hermaphroditum* with lesser amounts of *V. uliginosum* and *Arctous alpina* (Kallio, 1975).

In general, iron podzol or humus podzol soils are found under the heathland communities and are low in available nutrients (Dahl, 1956; Kallio, 1975).

Although Svalbard (Spitsbergen) is far north (78°N), its climate is less severe than latitude would indicate because of marine climate influence. In snow-bed sites, a dwarf-shrub heathland domi-nated by *Cassiope tetragona*, *Dryas octopetala* and *Salix polaris* occurs with lesser amounts of Carex rupestris and *Saxifraga oppositifolia*.

U.S.S.R.

Large areas of the highlands and plains of the Kola Peninsula are covered with dwarf shrub tundra. The spotted (soil boil[1]) alpine tundra above 1000 m is dominated by *Loiseleuria pro-cumbens*, the cushion plants or subshrubs *Silene acaulis* and *Dryas octopetala*, and the shrubs *Cassiope tetragona*, *Betula nana*, *Arctous alpina*, *Vaccinium vitis-idaea*, V. uliginosum, *Salix reti-culata* and *S. polaris* (Chepurko, 1972). Where soil moisture is more available, *Betula nana*, the heath species listed above and *Empetrum hermaphroditum* predominate.

Shrub tundra extends in a nearly unbroken belt across the European and Asiatic Arctic (Aleks-androva, 1970). In the southern Subarctic shrub tundra (low-shrub tundra *sensu* Bliss, 1979), *Betula nana*, *Salix glauca* and *S. phylicifolia* form a layer at 70 to 80 cm. A lower stratum is formed by *Vaccinium uliginosum*, *V. vitis-idaea*, *V. myrtillus*, *Empetrum hermaphroditum*, *Arctous alpina*, *Ledum palustre*, and *Carex globularis*. Here, as elsewhere,

there is a well-developed layer of the mosses *Polytrichum strictum*, *P. commune*, *Pleurozium schreberi* and the lichens *Cladonia sylvatica*, *C. rangiferina*, *C. uncialis* and *Peltigera apthosa* (Aleksandrova, 1970).

Areas of dwarf-shrub heath tundra with soil boils (spot medallions) consist of *Carex lugens*, *Vaccinium uliginosum*, *Arctous alpina*, *Empetrum hermaphroditum*, *Ledum palustre*, and *Loiseleuria procumbens* (Vikhireva-Vasilkova et al., 1964).

At Agapa in the West Taimyr (71°N, 89°E) in a wetland shrubby tundra, *Betula nana*, *Salix glauca*, *S. pulchra*, *S. lanata*, *S. reptans* and *Vaccinium uliginosum* predominate (Vassilijevskaja et al., 1975).

The dwarf-shrub communities within the forest-tundra near Sverdlovsk (67°N, 66°E) are domi-nated by *Vaccinium vitis-idae*, *V. uliginosum*, *Empetrum hermaphroditum*, *Arctous alpina*, *Loise-leuria procumbens*, *Diapensia lapponica* and *Carex hyperborea* (=*C. bigelowii*) (Gorchakovsky and Andreyashkina, 1972). In the northern Ural Moun-tains and Polar Urals, lower elevation tundras are often dominated by low-shrub tundra of *Betula nana*, *Salix glauca*, *S. pulchra*, *S. lanata* and lesser amounts of *S. phylicifolia*. Important heathland species include *Cassiope tetragona*, *Harrimanella hypnoides*, *Arctous alpina*, *Vaccinium vitis-idaea*, *V. uliginosum* ssp. *microphyllum*, and *Ledum palustre* ssp. *decumbens*. Minor heaths are *Vaccinium myrtillus*, *Loiseleuria procumbens*, and *Phyllodoce caerulea*. The latter two species are absent for the Siberian Arctic (Igoshina, 1969).

In the far east of Siberia, in the mountains of the Chukchi Peninsula, heathland vegetation plays a prominent role. On the upper slopes where closed vegetation is found (200 to 250 m), dwarf-shrub heathlands of *Diapensia obovata*, *Dryas punctata*, *Loiseleuria procumbens*, *Rhododendron kamtsch-aticum*, *Arctous alpina*, and *Vaccinium vitis-idaea* are found with an abundance of lichens and mosses. Heathlands at lower elevations (100 to 200 m) where there is considerable snow cover in winter have continuous carpets of *Betula nana* ssp. *exilis*, *Empetrum hermaphroditum*, *Ledum decum-bens*, and *Vaccinium uliginosum*. In snow beds the

[1] "Soil boil", "frost boil" and "spot medallion" refer to soils where there is sufficient churning each year to prevent plant growth except round the edges. Soil boils are typically 50 to 200 cm across.

arctic–alpine species *Cassiope tetragona* and *Phyllodoce caerulea* occur. Within the heathlands several species of dwarf willow occur — *Salix chamissonis*, *S. cuneata*, *S. phlebophylla*, and *S. reticulata*. There are no willow scrub thickets along rivers and streams as occur elsewhere in the tundra zone of the U.S.S.R. (Tikhomirov and Gavrilyuk, 1969).

Dwarf-shrub heathland/cottongrass tussock tundra

Communities dominated by tussocks of *Eriophorum vaginatum* and dwarf heaths cover large areas of western and northern Alaska (Hanson, 1953; Churchill, 1955; Bliss, 1956; Britton, 1957; Johnson et al., 1966) and the Yukon Territory (Hettinger et al., 1973; Wein and Bliss, 1974), but are quite limited in the Northwest Territories (Larsen, 1965; Corns, 1974). Cottongrass is the characteristic species although not always the dominant. Various combinations of *Ledum palustre* ssp. *decumbens*, *Empetrum hermaphroditum*, *Vaccinium uliginosum* ssp. *microphyllum* and *V. vitis-idaea* comprise the major heath species along with *Betula nana*, *Carex bigelowii* and several species of forbs. The common mosses include *Hylocomium splendens*, *Dicranum elongatum*, *Aulacomnium turgidum*, *A. palustre*, and *Tomenthypnum nitens*. The lichens *Cetraria cucullata*, *C. nivalis*, *Cladonia rangiferina*, *C. amaurocraea*, *Dactylina arctica* and *Thamnolia vermicularis* are common (Bliss, 1956; Johnson et al., 1966; Hettinger et al., 1973).

These communities are typically found on slopes and plots with imperfectly drained soils (gleysols — "Arctic Soils"). These soils have an organic layer of 2 to 15 cm over a gleyed mineral soil. Soil nutrient level and percent base saturation are generally low (Tedrow, 1963).

Similar plant communities dominated by dwarf-shrub heath species and *Eriophorum vaginatum* extended across the Asiatic tundra. These tundras occur on the Taimyr Peninsula (Pavlova, 1969; Polozova, 1970) and form a wide belt across eastern Siberia (Kriuchkov, 1968; Yurtsev, 1972).

LIFE CYCLE STRATEGY OF HEATH SHRUBS

Arctic heath species are a part of the Arcto-Tertiary flora that must have evolved in a cool to cold temperate climate in relation to open grown coniferous forests. These were probably upland areas as the continents drifted north in mid- to late Tertiary time (Löve and Löve, 1975).

The heaths are slow-growing species that have a variety of conservative features. Most species develop rhizomes that parallel the soil surface at a depth of 3 to 8 cm, from which arise short roots that extend to a depth of 5 to 10 cm (Bliss, 1956), mostly within the shallow peat layer. Most of these roots function for two to three years, but the rhizomes are long-lived and reach lengths of 30 to 60 cm. Gleysols with little or no surface peat predominate in the cottongrass tussock/dwarf-shrub heathland communities, thus most roots are in the cold, wet mineral soil. The rooting environment is within relatively warm soil horizons (3 to 8°C) (Bliss, 1956; Haag, 1974) compared with 1 to 2°C at 20 to 30 cm depths, yet nutrients have limited availability because of general slow decomposition of organic matter, low nitrogen-fixation rates and cold soils limiting phosphorus metabolism (Haag, 1974). Haag found that cold soils did not limit nitrogen uptake in the native species as has been reported for agronomic species in the U.S.S.R. Nutrient cycling is low and these evergreen-leaved species appear to have evolved mechanisms to recycle nutrients within this organic mat so that the limited nutrient supply is conserved (Haag, 1974; Svoboda, 1977). An indication of this is the coriaceous, evergreen leaves that function two to three years in several species (e.g. *Vaccinium vitis-idaea*, *Ledum palustre*, *Cassiope tetragona*, etc.) in the Low Arctic, probable translocation of nitrogen and phosphorus to stems and roots before leaf fall (Reiners and Reiners, 1970), and limited shoot growth per year (1 to 3 cm).

The nutrient status of heath shrubs before and after tundra fires showed that the content of nitrogen, phosphorus and potassium was significantly higher in *Ledum palustre* ssp. *decumbens* following regrowth, but *Vaccinium uliginosum* shoots were significantly higher only in phosphorus after regrowth (Wein and Bliss, 1973). Of the heath shrubs sampled at four locations in Alaska and the Northwest Territories, annual shoot production of *Ledum* spp., *Vaccinium vitis-idaea*, *V. uliginosum* and *Arctous* spp. was generally significantly reduced following fire. However, the individual shoots of *Ledum palustre* ssp. *decumbens* averaged 4 to 7 cm growth per year for four years after the burn at Inuvik, N.W.T. compared with 1 to 2 cm per year in control shoots (Wein and Bliss, 1973).

These species generally flower and fruit abundantly although mature, viable seed may be limited. Germination rates for seed of some heath species are notoriously low. Bliss (1962) reported no germination in light or dark for *Empetrum hermaphroditum*, *Vaccinium vitis-idaea* ssp. *minus*, and *V. uliginosum*, yet germination was 50% or more in the light and nil in the dark for *Cassiope tetragona* and *Ledum palustre* ssp. *decumbens*.

A major characteristic of arctic and alpine heathland species is their low rate of photosynthesis (Hadley and Bliss, 1964; Johnson and Tieszen, 1976). These species with evergreen leaves, including *Dryas*, have high stomatal resistances, low rates of water uptake, and low rates of carbon dioxide uptake (Mayo et al., 1973; Courtin and Mayo, 1975). One can assume that these conservative physiological characteristics include relatively low rates of protein synthesis and storage

(Hadley and Bliss, 1964) and that this favours production of structural tissue requiring little nutrient uptake and therefore making the relatively low photosynthetic rates [2 to 10 mg CO_2 (g dry wt.)$^{-1}$ hr^{-1})] more efficient (Haag, 1974). Johnson and Tieszen (1976) reported higher rates of photosynthesis for *Ledum palustre* ssp. *decumbens*, *Cassiope tetragona* and *Dryas integrifolia* than for *Vaccinium vitis-idaea* spp. *minus*. Maximum rates of photosynthesis at 15°C leaf temperature and saturating light were 17 and 18 mg CO_2 (g dry wt)$^{-1}$ hr^{-1} for one- and two-year-old leaves of *Ledum* and only 10 mg CO_2 (g dry wt)$^{-1}$ hr^{-1} in leaves of the current year.

One of the best integrative measures of the adaptive physiology of species is their net annual dry matter production. A few data are available in both Low Arctic and High Arctic heath-dominated communities. The data presented in Table 15.3 show that production rates are generally low compared with temperate region communities, yet are some of the highest rates when compared with other arctic plant communities.

TABLE 15.3

Phytomass (alive) and net annual primary production (g m^{-2}) of low-shrub–dwarf-shrub heathland, dwarf-shrub heathland, and dwarf-shrub heathland/cottongrass tussock communities in the Arctic

Plant community	Location	Phytomass			Net production			Authority
		vascular plants		cryptogams	vascular plants		cryptogams	
		above-ground	below-ground total		above-ground	below-ground total		
		heaths others			heaths others			
Low-shrub heathland	Tuktoyaktuk, N.W.T.	– –	–	–	62	–	–	Haag (1974)
Lichen heathland	Hardangervidda, Norway	54 8	191	377	75 13	100	88	Østbye et al. (1975)
Dwarf-shrub heathland	Kola Peninsula, U.S.S.R.	555	1228	–	101	134	–	Chepurko (1972)
Dwarf-shrub heathland	Kola Peninsula, U.S.S.R.	170	476	–	20	30	24	Chepurko (1972)
Dwarf-shrub heathland	Devon Island, N.W.T.	102 43	671	–	12 5	90	20	Bliss et al. (1977)
Dwarf-shrub heathland intermixed with cottongrass tussock	Dempster Highway, Yukon Terr.	36 75	1524*	302	16 23	–	132	Wein and Bliss (1974)
	Eagle Creek, Alaska	104 145	2008*	89	46 26	–	34	Wein and Bliss (1974)

* Including dead and alive material.

SUMMARY

In summary, dwarf-heath shrubs play a very important role in the plant communities of the Low Arctic and a minor one in the High Arctic. The species generally grow on shallow peats, gleysols, or brunisols, all with low nutrient availability. The species are conservative in their physiology (water loss, photosynthetic rates, nutrient status) and dry matter production. As a result their ecological role in the Arctic is similar to that of temperate region heathland communities characteristic of nutrient- and water-deficient habitats.

REFERENCES

Aleksandrova, V.D., 1970. The vegetation of the tundra zones in the USSR and data about its productivity. In: W.A. Fuller and P.G. Kevan (Editors), *Productivity and Conservation in Northern Circumpolar Lands. IUCN New Ser.*, No. 16: 93–114.

Andreev, V.N., 1966. Peculiarities of zonal distribution of the aerial and underground phytomass on the East European Far North. *Bot. Zh. Kyyiv*, 51: 1401–1411 (in Russian).

Beschel, R.E., 1970. The diversity of tundra vegetation. In: W.A. Fuller and P.G. Kevan (Editors), *Productivity and Conservation in Northern Circumpolar Lands, IUCN New Ser.*, No. 16: 83–92.

Bliss, L.C., 1956. A comparison of plant development in microenvironments of arctic and alpine tundras. *Ecol. Monogr.*, 26: 303–337.

Bliss, L.C., 1962. Seed germination in arctic and alpine species. *Arctic*, 11: 180–188.

Bliss, L.C., 1979. North American and Scandinavian tundras. In: L.C. Bliss, O.W. Heal and J.J. Moore (Editors), *Tundra Ecosystems, A Comparative Analysis*. Cambridge University Press, in press.

Bliss, L.C. and Svoboda, J., 1979. Plant communities and plant production on Banks and Victorian Islands, N.W.T. (in preparation).

Bliss, L.C., Kerik, J. and Peterson, W., 1977. Primary production of dwarf shrub heath communities, Truelove Lowland. In: L.C. Bliss (Editor), *Truelove Lowland, Devon Island Canada: A High Arctic Ecosystem*. University of Alberta Press, Edmonton, Alta., pp. 217–224.

Böcher, T.W., 1954. Oceanic and continental vegetational complexes in Southwest Greenland. *Medd. Grønl.*, 148(1): 336 pp.

Böcher, T.W., 1959. Floristic and ecological studies in Middle West Greenland. *Medd. Grønl.*, 156(5): 68 pp.

Brassard, G.R. and Longton, R.E., 1970. The flora and vegetation of Van Hauen Pass, Northwestern Ellesmere Island. *Can. Fld. Nat.*, 84: 357–364.

Britton, M.E., 1957. Vegetation of the arctic tundra. In: H.P. Hansen (Editor), *Arctic Biology*. Oregon State University Press Corvallis, Ore., pp. 22–61.

Bryson, R.A., 1966. Airmass streamlines and the boreal forest. *Geogr. Bull.*, 8: 228–269.

Chepurko, N.L., 1972. The biological productivity and the cycle of nitrogen and ash elements in the dwarf shrub tundra ecosystems of the Khibina Mountains (Kola Peninsula). In: F.E. Wielgolaski and T. Rosswall (Editors), *Proceedings IV International Meetings on the Biological Productivity of Tundra*. Tundra Biome Steering Committee, Stockholm, pp. 236–247.

Churchill, E.D., 1955. Phytosociological and environmental characteristics of some plant communities in the Umiat region of Alaska. *Ecology*, 36: 606–627.

Corns, I.G.W., 1974. Arctic plant communities east of the Mackenzie Delta. *Can. J. Bot.*, 52: 1730–1745.

Courtin, G.M. and Mayo, J.M., 1975. Arctic and alpine plant water relations. In: F.J. Vernberg (Editor), *Physiological Adaptations to the Environment*. Intext Ed. Publ., New York, N.Y., pp. 201–224.

Dahl, E., 1956. Rondanr Mountain vegetation in south Norway and its relation to environment. *Skr. Det. Nor. Vidensk,–Akad., Mat-Nat. Kl.*, No. 3: 374 pp.

Gjaerevoll, O. and Bringer, K.G., 1965. Plant cover of the alpine regions. In: *The Plant Cover of Sweden. Acta Phytogeogr. Suecica*, 50: 257–268.

Gorchakovsky, P.L. and Andreyashkina, N.I., 1972. Productivity of some shrub, dwarf shrub and herbaceous communities of forest-tundra. In F.E. Wielgolaski and T. Rosgwall (Editors), *Proceedings IV. International Meetings on the Biological Productivity of Tundra*. Tundra Biome Steering Committee, Stockholm, pp. 113–116.

Haag, R.W., 1974. Nutrient limitations to plant production in two tundra communities. *Can. J. Bot.*, 52: 103–116.

Hadley, E.B. and Bliss, L.C., 1964. Energy relationships of alpine plants on Mt. Washington, New Hampshire. *Ecol. Monogr.*, 34: 331–357.

Hansen, K., 1969. Analyses of soil profiles in dwarf-shrub vegetation in South Greenland. *Medd. Gronl.*, 178(5): 33 pp.

Hanson, H.C., 1953. Vegetation types in northwestern Alaska and comparisons with communities in other arctic regions. *Ecology*, 34: 111–140.

Hare, F.K. and Ritchie, J.C., 1972. The boreal bioclimates. *Geogr. Rev.*, 62: 333–365.

Hernandez, H., 1973. Natural plant recolonization of surficial disturbances, Tuktoyaktuk Peninsula Region, Northwest Territories. *Can. J. Bot.*, 51: 2177–2196.

Hettinger, L., Janz, A. and Wein, R.W., 1973. *Vegetation of the Northern Yukon Territory, 1*. Biol. Rep. Ser. Can. Arctic Gas Study Ltd., Calgary, Alta., 171 pp.

Hultén, E., 1937. *Outline of the History of Arctic and Boreal Biota During the Quaternary Period*. Bokforlags Aktiebolaget Thule, Stockholm, 168 pp.

Igoshina, K.N., 1969. Flora of the mountain and plain tundras and open forests of the Urals. In: B.A. Tikhomirov (Editor), *Vascular Plants of the Siberian North and the Northern Far East*. Izdatel'stvo Nauka, Leningrad (Israel Program Sci. Transl., Jerusalem), pp. 102–344.

Janz, A., 1974. *Topographic and Site Influences on Vegetation, Soil and Their Nutrients East of the Mackenzie Delta*. Thesis, Department of Botany, University of Alberta, Edmonton, Alta., 68 pp.

Johnson, A.W., Viereck, L.A., Johnson, R.E. and Melchior, H., 1966. Vegetation and flora. In: N.J. Wilimovsky and J.N. Wolfe (Editors), *Environment of The Cape Thompson Region, Alaska*. U.S. Atomic Energy Comm., Div. Tech. Info., Washington, D.C., pp. 277–254.

Johnson, D.A. and Tieszen, L.L., 1976. Aboveground biomass allocation, leaf growth, and photosynthesis patterns in tundra plant forms in arctic Alaska. *Oecologia*, 24: 159–173.

Kallio, P., 1975. Finland. In: T. Rosswall and O.W. Heal (Editors) *Structure and Function of Tundra Ecosystems*. Ecol. Bull. N.F.R., No. 20 Swed. Natl. Sci. Res. Council, Stockholm, pp. 193–223.

Kilander, S., 1965. Alpine zonation in the southern part of the Swedish Scandes. In: *The Plant Cover of Sweden. Acta Phytogeogr. Suecica*, 50: 78–84.

Kriuchkov, V.V., 1968. Tussock tundras. *Bot. Zh., Kyyiv*, 53: 1716–1730 (in Russian).

Larsen, J.A., 1965. The vegetation of the Ennadai Lake Area, N.W.T. Studies in subarctic and arctic bioclimatology. *Ecol. Monogr.*, 35: 37–59.

Larsen, J.A., 1972. The vegetation of Northern Keewatin. *Can. Fld. Nat.*, 86: 45–72.

Löve A. and Löve, D., 1975. Cryophytes, polyploidy and continental drift. *Phytocoenologia*, 2: 54–65.

Mayo, J.M., Despain, D.G. and Van Zinderen Bakker Jr., E., 1973. CO_2 assimilation by *Dryas integrifolia* on Devon Island, Northwest Territories. *Can. J. Bot.*, 51: 581–588.

Oosting, H.J., 1948. Ecological notes on the flora of East Greenland and Jan Mayen. In: L.A. Boyd (Editor), *The Coast of Northeast Greenland. Am. Geogr. Soc., Spec. Publ.*, No. 30.

Østbye, E., Berg, A., Blehr, O., Espeland, M., Gaare, E., Hagen, A., Hesjedal, O., Hagvar, S., Kjelvik, S., Lien, L., Mysterud, I., Sandhaug, A., Skar, H.J., Skartveit, A. Skre, O. Skogland, T., Solhøy, T., Stenseth, N.C. and Wielgolaski, F.E., 1975. Hardangervidda, Norway. In: T. Rosswall and O.W. Heal (Editors), *Structure and Function of Tundra Ecosystems*. Ecol. Bull, N.F.R. No. 20. Swed. Natl. Sci. Res. Council, Stockholm, pp. 225–264.

Pavlova, E.B., 1969. Vegetal mass of the tundra of Western Taimyr. *Vestn. Mosk. Univ.*, Ser. 6, 5: 62–67 (in Russian).

Polozova, T.G., 1970. Biological features of *Eriophorum vaginatum* L. as a tussock-former (based on observations in tundras of western Taimyr). *Bot. Zh., Kyyiv*, 55: 431–442.

Polunin, N., 1948. Botany of the Canadian Eastern Arctic. *Natl. Mus. Can. Bull.*, No. 104: 304 pp.

Raup, H., 1971. Vegetation of the Mesters Vig District, Northeast Greenland. *Medd. Grønl.*, 194(3): 48 pp.

Reiners, W.A. and Reiners, N.M., 1970. Energy and nutrient dynamics of forest floors in three Minnesota forests. *J. Ecol.*, 58: 497–519.

Rønning, O.I., 1960. The vegetation and flora north of the Arctic Circle. In: *Norway North of 65*. Oslo. University Press, Oslo, pp. 50–72.

Seidenfaden, G. and Sørensen, T., 1937. The vascular plants of Northeast Greenland from 74°30′ to 79°00′ N lat. and a summary of all species found in East Greenland. *Medd. Grønl.*, 101(4): 215 pp.

Sørensen, T., 1943. The flora of Melville Bugt. *Medd. Grønl.*, 124(5): 305 pp.

Svoboda, J., 1977. Energy and production of raised beach communities. In: L.C. Bliss (Editor), *Truelove Lowland, Devon Island Canada: A High Arctic Ecosystem*. University of Alberta Press, Edmonton, Alta., pp. 185–216.

Tedrow, J.C.F., 1963. Arctic soils. In: *Permafrost International Conference Proceedings*. NAS-NRC Publ. 1287: 50–55.

Tedrow, J.C.F. and Cantlon, J.E., 1959. Concepts of soil formation and classification in arctic regions. *Arctic*, 11: 166–179.

Tedrow, J.C.F. and Hill, D.E., 1955. Arctic brown soil. *Soil Sci.*, 80: 265–275.

Tedrow, J.C.F., Drew, J.V., Hill, D.E. and Douglas, L.A., 1958. Major genetic soils of the Arctic Slope of Alaska. *J. Soil Sci.*, 9: 33–45.

Tikhomirov, B.A. and Gavrilyuk, V.A., 1969. The flora of the Bering Coast of Chukchi Peninsula. In: B.A. Tikhomirov (Editor), *Vascular Plants of the Siberian North and the northern Far East*. Izdatel'stvo, Leningrad (Israel Program Sci. Transl., Jerusalem), pp. 74–105.

Ugolini, F.C., 1966. Soils of the Mesters Vig District, Northeast Greenland. 1. The arctic brown and related soils. *Medd. Grønl.*, 176(1): 25 pp.

Vassilyevskaya, V.D., Ivanov, V.V., Bogatyrev, L.G., Pospelova, E.B., Schalaeva, N.M. and Greishina, L.A., 1975. Agapa, USSR. In: T. Rosswall and O.W. Heal (Editors), *Structure and Function of Tundra Ecosystems*, Ecol. Bull. N.F.R., No. 20. Swed. Natl. Sci. Res. Council, Stockholm, pp. 141–158.

Vikhireva-Vasilkova, V.V., Gavrilyuk, V.A. and Shamurin, V.F., 1964. Aerial and underground mass of some dwarf shrub communities of Koryak Land. *Prob. Sev.*, No. 8 (in Russian).

Walker, D.B. and Peters, T., 1977. Soils of the Truelove Lowland and plateau. In: L.C. Bliss (Editor), *Truelove Lowland, Devon Island Canada: A High Arctic Ecosystem*. University of Alberta Press, Edmonton, Alta., pp. 31–61.

Wein, R.W. and Bliss, L.C., 1973. Changes in arctic *Eriophorum* tussock communities following fire. *Ecology*, 54: 845–852.

Wein, R.W. and Bliss, L.C., 1974. Primary production in arctic cottongrass tussock tundra communities. *Arct. Alp. Res.*, 6: 261–274.

Yurtsev, B.A., 1972. Phytogeography of Northeastern Asia and the problem of Transberingian floristic interrelations. In: A. Graham (Editor), *Floristics and Paleofloristics of Asia and Eastern North America*. Elsevier, Amsterdam, pp. 19–54.

Chapter 16

ALPINE SCRUB AND HEATHLAND COMMUNITIES IN JAPAN[1]

R.L. SPECHT (Compiler)

GENERAL

The alpine regions of Japan are characterised by carpet-like stands of *Pinus pumila*. This species is widely distributed in the alpine zone from central Honshu northwards. Its southern limit lies on Mount Tekari-dake, towards the southern end of the Akaishi Mountain Range, while its western limit is represented by Mount Haku san. In the north, it often drops down almost to sea level, such as on Rebun Island, northern Hokkaido, as well as in the central and northern Kuril Islands and the northernmost part of Sakhalin. *Pinus pumila* scrub is widely distributed outside Japan in continental northeast Asia, from northern Korea, through Manchuria, the Amur district of eastern Siberia and Kamchatka and westwards to Lake Baykal (Ishizuka, 1974).

The hilltops, ridges and windblown slopes of the alpine region are snow-free relatively early, and are exposed to the strong effects of wind and desiccation. Alpine desert and windward grassland develop under the most exposed conditions, while alpine dwarf-shrub heathland vegetation, having the appearance of a dark-greenish mat, flourishes in more protected and relatively stable habitats. Windward grassland develops in the most severely windswept areas around mountain summits and ridges, replacing the alpine dwarf-shrub heathland. Strong erosion and desiccation by wind are the prime factors in its development (Ishizuka, 1974). Alpine heathlands of Japan are essentially similar to the arctic heathlands found in the mountains of the Chukchi Peninsula of Far Eastern Siberia (see Bliss in Chapter 15 above).

Information on the topographic and climatic relationships of both the alpine scrub (*Pinus pumila–Vaccinium* spp. alliance) and alpine dwarf-shrub heathland (*Arcterica nana–Loiseleuria procumbens* association) is given below. The floristic composition of these plant communities is then tabulated.

---

[1] Manuscript completed July, 1977.

Diagrammatic cross section (after Ishizuka, 1974):

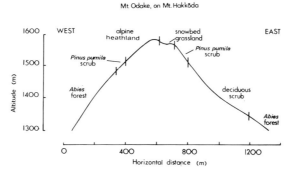

Fig. 16.1. Diagrammatic cross-section, Mount Hakkōda (Japan).

Climate: The Montane Zone of central Japan, dominated by cool-temperate broad-leaved deciduous forests

Climatic data[1]	Winter (Dec–Feb)	Spring (March–May)	Summer (June–Aug)	Autumn (Sept–Nov)	Year (Jan–Dec)
Max. air temp. (°C)	3	12	23	17	14
Min. air temp. (°C)	−9	−2	14	6	3
Precipitation (mm)	107	246	503	363	1219
Precipitation (days)	17	26	40	30	113
Relative humidity (%)					
at 0600 hours	90	89	96	95	93
at 1400 hours	53	53	70	67	61

[1] Climate station: Oiwake (36°20′N, 138°33′E, 1006 m).

The mean annual temperature on the summit of Mount Hakkōda (40°40′N, 140°50′E, 1585 m) would be about 1.5°C, assuming an adiabatic lapse rate of *c*.5°C per 1000 m altitude.

Species composition of alpine scrub: *Pinus pumila–Vaccinium* spp. alliance.

(1) *Pinus pumila–Ledum palustre* association in Hokkaido and northeastern Honshu.

(2) *Pinus pumila–Cetraria crispa* association on the upper slopes and ridges of the Hida, Kiso and Akaisha Mountain Ranges, central Honshu, and in the Hidaka Mountain Range, Hokkaido.

(3) *Pinus pumila–Rhododendron brachycarpum* association mainly in the mountains on the Sea of Japan side of central and northern Honshu.

(4) *Pinus pumila–Rubus pedatus* association (including *Vaccinium ovalifolium, Tripetaleia bracteata, Gaultheria miqueliana*) in central Honshu and Hokkaido.

GYMNOSPERMS

Cupressaceae	*Juniperus communis* var. *nipponica*
Pinaceae	*Pinus pumila*
Taxaceae	*Taxus cuspidata* var. *nana*

ANGIOSPERMS

Aceraceae	*Acer tschonoskii*
Aquifoliaceae	*Ilex rugosa*
Betulaceae	*Alnus maximowiczii*
Cornaceae	*Cornus canadensis*
Ericaceae	*Gaultheria miqueliana, Ledum palustre* var. *diversipilosum, Rhododendron brachycarpum, Tripetaleia bracteata*
Rosaceae	*Rubus pedatus, Sorbus commixta, S. mat-*

	sumurana, S. sambucifolia
Vacciniaceae	*Vaccinium ovalifolium, V. vitis-idaea*

MOSSES

Dicranaceae	*Dicranum fuscescens, D. majus*
Hylocomiaceae	*Hylocomium splendens*
Hypnaceae	*Hypnum plicatulum, Ptilium crista-castrensis*
Entodontaceae	*Pleurozium schreberi*

LICHENS

Cladoniaceae	*Cladonia alpestris, C. rangiferina*
Parmeliaceae	*Cetraria crispa*

Species composition of alpine dwarf-shrub heathland: *Arcterica nana–Loiseleuria procumbens* association of the Hida, and Akaishi Mountains, central Honshu

ANGIOSPERMS

Asteraceae	*Leontopodium* spp.
Diapensiaceae	*Diapensia lapponica* var. *obovata*
Empetraceae	*Empetrum nigrum* var. *japonicum*
Ericaceae	*Arcterica nana, Arctous alpina* var. *dilatatum, Loiseleuria procumbens*
Vacciniaceae	*Vaccinium uliginosum*

LICHENS

Parmeliaceae	*Cetraria* spp.

REFERENCE

Ishizuka, K., 1974. Mountain vegetation. In: M. Numata (Editor), *The Flora and Vegetation of Japan*. Kodansha, Tokyo–Elsevier, Amsterdam, pp. 173–210.

Chapter 17

APPALACHIAN BALDS AND OTHER NORTH AMERICAN HEATHLANDS[1]

R.H. WHITTAKER

INTRODUCTION

To a North American, the word "heath" has a simple and clearly bounded meaning: a plant that is a member of the family Ericaceae (or the order Ericales). The term heath or heathland is applied much less to plant communities, partly because of the limited area of communities dominated by Ericaceae in North America. There are, however, some distinctive and interesting ericaceous communities. I shall first describe what are probably the best known of these, the "heath balds" and then try to summarize the broader geographic occurrence of North American heathlands.

APPALACHIAN HEATH BALDS

Occurrence and status

The heath balds are a most attractive and justly celebrated feature of the vegetation of the southern Appalachian Mountains, particularly of Great Smoky Mountains National Park (Cain, 1930; Whittaker, 1956). The balds occur on many of the ridges and rocky points of the mountains from near their major summits (1800–2000 m) down to about 1400 m (with a few bald areas down to 1200 m or below) (Fig. 17.1). Although there are hundreds of balds, their total area is limited; they are scattered islands of shrubland in the vast and largely continuous forest mantle of the Great Smoky Mountains. Their most characteristic sites are steep ridges, but they occur also on the few areas of outcropping rock at high elevations (particularly on Mount Le Conte), and on a few more rounded summits (such as Brushy Mountain). It is not easy

to state why they occur on certain ridges and not on others, but correlations can be noted: the balds are most common on ridges on the west slope of the Great Smoky Mountains and have their greatest extents on the south- and west-facing slopes below those ridges. Occurrence of balds is favored by combinations of high elevation, steepness of slope, and strong exposure to the prevailing westerly winds.

In all cases they are surrounded by forest that itself has a heath undergrowth. At high elevations the balds are in contact with spruce–fir forests (*Picea rubens* and *Abies fraseri*) that cover most of the mountain surface above about 1370 m in the northeast half of the Great Smoky Mountains. Many of the high-elevation balds are dominated by *Rhododendron catawbiense*, and this species also occurs as a fairly dense tall (2–4 m) understorey in a "spruce heath" forest type on some steep, south-facing slopes. The lower-elevation balds are dominated by varying mixtures of *Rhododendron maximum*, *Kalmia latifolia*, and other shrub species that occur in two other forest or woodland types that contact many of the balds — pine heathland and oak–chestnut heath. South- and southwest-facing slopes (on the west side of the range) mostly support pine heathlands above 1200 m, with small trees of *Pinus pungens* in open stand above a fairly dense low heath stratum (0.5–1.5 m) dominated by varying combinations of *R. maximum* and *K. latifolia* with vacciniaceous shrubs (*Vaccinium pallidum*, *Gaylussacia baccata*, and others). As this pine heathland or woodland is followed toward

[1] Manuscript completed June, 1977.
Contribution from research supported by the National Science Foundation.

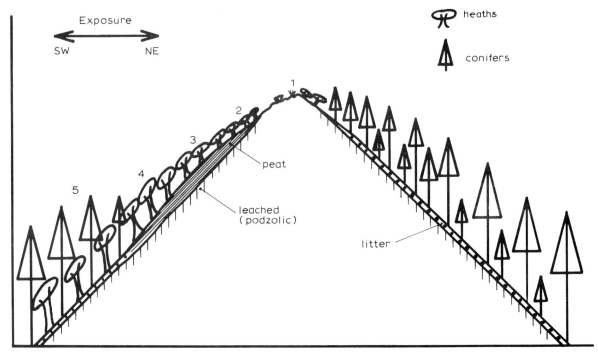

Fig. 17.1. Heath Bald Pattern, Great Smoky Mountains, Tennessee (35°40′N, 83°25′W).

Commu- nities	Type	Soil	Dominant species	
			higher elevation	lower elevation
1	exposed rock and open-heathland	none to lithosol	*Leiophyllum lyonii* *Deschampsia flexuosa*	*Leiophyllum lyoni*[1]
2	low-heathland (<1 m)	peat over podzol	*Rhododendron carolinianum*	mixed heathland: *Rhododendron catawbiense,* *Rhododendron maximum,*
3	mid-heathland (1–1.5 m)	same	*Rhododendron catawbiense*	*Kalmia latifolia,* etc.
4	high-heathland (2–4 m)	same	same shrub species as preceding, but trees often entering the stand	
5	heath-forest	mor over podzol or podzolic	same shrubs become undergrowth of a forest dominated by: *Picea rubens*	*Quercus prinus* or other species, depending on exposure

[1] Occurs only on some sharp and exposed ridges.

Climate (can only be inferred from limited data on elevation trends, see Shanks, R.E., 1954. Climates of the Great Smoky Mountains. *Ecology*, 35: 354–361; all values approximate.)

	Higher elevation (2000 m)	Lower elevation (1600 m)
Mean annual precipitation (cm)	230	200
Mean monthly temperatures (°C)		
January	0.5	1.0
July	15.5	17.0
annual	6.7	8.1

	Higher elevation (2000 m)	Lower elevation (1600 m)
Thornthwaite (1948) climatic indices		
moisture index	285	242
thermal efficiency	20.9	23.4
climatic province	ArC′$_2$b′$_4$	ArB′$_1$a′
	(perhumid microthermal)	(perhumid cool-mesothermal)
Köppen climatic types	**Cfc**	**Cfb**

lower elevations, it changes continuously with replacement of pine species (*P. pungens* by *P. rigida* and then *P. virginiana*), with increasing representation of oaks (*Quercus coccinea*, *Q. prinus*, and others), and decreasing coverage of heaths into the pine and pine–oak forests of low elevations (500–600 m). On many west- and east-facing slopes the oak–chestnut heathland type occurs with *Q. prinus* and the dead remains of *Castanea dentata* rising, with some other tree species, above a dense, tall heath understorey (2–4 m) dominated by *R. maximum* and *K. latifolia* with *Clethra acuminata*, *Lyonia ligustrina*, and a lower shrub layer of *Gaylussacia ursina*. The shrub floras of the heath balds, and of the forests with which they are in contact, are similar; but the shrub strata of the forests are usually richer in species and (except in the pine heathlands) taller. One can thus think of a heath flora for the mountains that is concentrated in the balds and that radiates outward from the balds to dominate, with additional species, the understoreys of a wide range of forest types.

One can also think of the balds as derived from such forests by exclusion of the tree stratum. Cain (1930), in his study of the balds, judged that they were produced from forests when the trees were destroyed by fire, windfall, or landslide, permitting the heathland species already present in these sites to close in and largely exclude tree reproduction by density of cover and soil effects. Others, noting that the balds are subject to fire (as are the pine heathlands and oak–chestnut heathlands), have thought them stages of fire succession. As Cain (1930) notes, however, there is almost no evidence of successful tree invasion of some of the balds — those in more typical or extreme bald sites. It is likely that the balds are not simply a successional type but a range of communities of different status:

(1) Balds of rock outcrops at highest elevations

seem clearly to be communities of primary succession from open rock through scattered tufts of *Deschampsia flexuosa* and a *Leiophyllum lyoni* mat to a closed shrub canopy of *Rhododendron carolinianum* or *R. catawbiense*. Some, but not all, of the latter heath balds are successional toward forest.

(2) Trees are invading the lower edges of some of the balds. As in many contacts of shrubland with forest, fire extends the range of shrub dominance at the expense of trees which can, as a secondary succession, re-invade to form heath forest again.

(3) Apart from these, the central portions of many of the balds appear stable and to consist of old shrub root systems capable of repeated regrowth following fires. Whittaker (1956) judged that these nuclear balds were self-maintaining, fire-adapted, topographic climaxes on their steep and exposed sites. Likely reasons for their occurrence include combinations of wind exposure, soil leaching and infertility, and summer drought (in sites in which soil water drains downward without replacement from above) that favor shrubs over trees.

Physiognomy and floristics

Two major gradients affect the characteristics of the heath balds. There is, first, a physiognomic (and, partly, developmental) gradient from the open *Deschampsia* and *Leiophyllum* stages, through closed heathlands of low (0.5 m) and medium (1–1.5 m) canopy heights to tall heathlands (2–4 m) that are successional to forest or form the understorey of oak–chestnut heathland and some other forest types (Table 17.1). Soils of the balds (Cain, 1930, 1931) vary from rock with soil pockets in the open stage, to a shallow dry peat, to a well-developed podzol overlain by peat up to 0.5 m depth in closed balds; the latter soils gradate into

TABLE 17.1

Measurements for North American heath and heathland forest communities

	Rock surface, *Deschampsia–Leiophyllum*	*Leiophyllum lyoni* mat, Mount Le Conte	*Rhododendron carolinianum* bald	*Rhododendron catawbiense* bald	Open mixed heathland, Brushy Mountain	Mixed heath bald, Peregrine Peak	Mixed heath bald, Brushy Mountain	Mixed heath bald, Rocky Spur
Sample number[1]	1	2	3	4	5	6	7	8
Site								
elevation (m)	2010	2010	2010	2010	1500	1430	1500	1560
exposure	SW	SW	SW	NE	SE	SW	NE	SW
inclination (°)	30	20	32	20	15	35	15	20
Canopy height[2] (m)								
trees	–	–	–	–	–	–	–	–
shrubs	–	0.5	1.3	2.5	0.7	1.5	1.8	1.7
Coverage[3] (%)								
trees	–	–	–	–	–	–	–	–
arborescent shrubs	–	–	–	–	–	–	–	–
lower shrubs	20	100+	190	230	136	168	234	246
herbs	30	0	0	2	40	2	4	0
lichens and moss	30	2	5	45	34	40	2	20
Clipping weights[4] (g m^{-2})								
shrubs	8	307	212	264	120	171	267	279
herbs	11	0	0	0	26	0	0.4	0
thallophytes above ground	24	2	10	52	9	98	2	42
Net production[5] (g m^{-2} yr^{-1})								
trees	–	–	–	–	–	–	–	–
shrubs above ground	12	460	409	491	234	380	592	629
Biomass[6] (kg m^{-2})								
trees	–	–	–	–	–	–	–	–
shrubs	0.025	1.2	1.8	2.5	.85	2.0	2.8	3.7
Vascular plant species								
per 100 m^2	7	2	2	5	13	7	10	5
per 1000 m^2	–	–	–	–	–	–	–	–
Simpson dominance index[7]	0.76	1.00	0.99	0.98	0.30	0.45	0.46	0.54

[1] Samples 1–17 are from the Great Smoky Mountains, Tennessee (Whittaker, 1963, 1966), 60+ and 120+ from Long Island, New Y (Whittaker and Woodwell, 1969, and unpublished), 75+ and 80+ from Mendocino County California (Westman, 1975; Westman Whittaker, 1975). The last four columns are means of 5 samples each.

[2] Volume-weighted mean height for trees, mean canopy-top level for shrubs.

[3] Individual-point foliage cover.

[4] Current twigs with leaves for shrubs, current above-ground growth for herbs, biomass for thallophytes; all weights oven-dry.

[5] Above-ground net primary productivity, determined by clipping samples, forest growth measurements, and dimension analysis techiques.

[6] Above-ground plant biomass, dry weights, from dimension analysis techniques.

[7] $C = \Sigma p_i^2$, in which p_i = above-ground net production by species as decimal fractions of total above-ground net production for shrub strat

* Sample area 0.25 ha.

High mixed heath bald, Brushy Mountain	Pinus virginiana forest, Pittman Center	Pinus pungens heathland, Greenbrier Pinnacle	Quercus prinus heathland, Bullhead Trail	Tsuga–Rhododendron heath-forest	Picea–Rhododendron heath-forest	Oak–pine forest, Brookhaven	Pine plains, West Hampton	Cupressus pygmaea pygmy forest	Pinus muricata–Rhododendron heath-forest
9	11	13	14	16	17	60+	120+	75+	80+
1490	550	1340	970	1280	1740	25	15	140	100
SE	SW	SW	W	NNE	ESE	–	–	–	–
14	26	22	32	30	40	level	level	level	level
–	17	12	10	34	23	7.6	1.5	1.2	21
4.7	3.0	0.7	3.5	5.2	4.5	0.6	0.5	–	–
–	182	96	38	166	126	126	78	36	76
276	–	–	212	66	118	–	–	–	55
12	43	196	60	50	24	106	22	49	59
0	6	68	16	0	0	3	19	1	16
12	0	0	30	22	58	0	–	42	2
397	18	83	145	58	72	39	–	31	36
0	2	36	1	0	0	2	–	0.5	22
12	2	4	24	12	75	tr.	–	–	–
–	950	210	220	850	610	796	–	234	906
983	40	173	318	172	202	61	–	72	161
–	13.0	5.2	4.0	49	30	6.4	–	2.4	40
10.8	0.12	0.57	2.4	2.1	2.1	0.16	–	0.25	1.5
9	–	–	–	–	–	–	5.5	8.2	12*
–	32	20	23	5	7	18	–	–	–
0.29	0.40	0.39	0.24	0.74	0.66	0.40	–	0.46	0.60

the podzols and podzolic soils of the forest heathlands. Samples 3 to 8 in Table 17.1 are typical heathlands of the middle height range; and the individual-point coverages[1] of the shrub stratum of these are in the range from 200 to 250%, the same range as for well-developed, closed forests. These typical heathlands are both closed and dense, and one forces one's way through the canopy of interlocking branches only with effort. Travel through the heath balds is by trail; cross-country movement through them ranges from difficult to impossible. Quadrat measurements are often made by crawling beneath the canopy. Since the leaves

[1] Individual-point coverage expresses the number of contacts on foliage of plant individuals, recorded by vertical point quadrats at 100 sample sites. At any sample site, the point quadrat may contact foliage of more than one species, thus the total contacts for all species may total more than 100%.

are predominantly broad-sclerophyll, the balds resemble in physiognomy dense types of maquis or chaparral.

A floristic gradient relates to elevation. Many of the high-elevation balds have a single evergreen shrub dominant, *Rhododendron carolinianum* or *R. catawbiense*, and almost no other vascular plants. Toward lower elevations the shrub canopy is increasingly mixed, with the evergreen *R. catawbiense*, *R. maximum*, and *Kalmia latifolia* dominant, but with deciduous heaths (*Vaccinium corymbosum*, *Gaylussacia baccata*) and other shrubs (*Pyrus melanocarpa*, *Viburnum cassinoides*) also present. Compositions of a number of heath bald samples are given by Cain (1930) and Whittaker (1963). The herb stratum, though it may have 20 to 40% coverage in some of the open-heathlands, has near zero coverage in the typical, closed heathlands. The major species is *Gaultheria procumbens*; other herbs are *Galax aphylla*, *Epigaea repens*, *Medeola virginiana*, *Trillium undulatum*, and a small hemiparasitic annual, *Melampyrum lineare*. The herb stratum, such as it is, is thus dominated by ericaceous chamaephytes — the prostrate, evergreen "ground heaths" *Gaultheria procumbens*, *Galax* (Diapensiaceae), and the less common *Epigaea*. Some of the high-elevation balds have fairly high coverage of lichen and moss (Table 17.1).

Other floristic aspects of the balds may be noted. Although deciduous species are present, the balds are predominantly evergreen-sclerophyll in both the shrub canopy and the sparse herb layer. The shrub leaves are entire, mainly oblong or elliptic, and in most species in the microphyll size range (2.2–20 cm²) (Cain, 1930). Leaf sizes of the shrub dominants decrease with increasing elevation, however, as is shown in Table 17.2; the leaves are also generally smaller in the balds than in forest undergrowth. Mean leaf persistence decreases with elevation, and from forest to bald in a given species. Cain (1930) records the Raunkiaer life forms for 54 species (some of them peripheral to the balds): microphanerophytes 48%, nanophanerophytes 22%, chamaephytes 4%, hemicryptophytes 15%, geophytes 9%, and a therophyte 2%. This phanerophyte predominance occurs in a temperate forest climate, in which hemicryptophytes normally predominate in life-form spectra. Among the characteristic bald species mentioned here there are no hemicryptophytes (if *Deschampsia* is set aside as successional and *Galax* is interpreted as a chamaephyte). Species diversities for the balds range from low to very low (Table 17.1; and Whittaker, 1965); in this they are in polar contrast

TABLE 17.2

Size and mean persistence of leaves of shrub dominants in heathland balds and as understorey plants of forest at various altitudes in the Great Smoky Mountains (Whittaker, 1962)

	Elevation (m)	Leaf size (cm²)	Leaf blade dry weight/area (mg cm⁻²)	Leaf persistence (yr)
Rhododendron maximum				
(a) in bald	1500	38	16.9	2.9
(b) in undergrowth (of cove forest)	430	62	11.2	4.6
Rhododendron catawbiense				
(a) in balds	1430	24	13.3	2.2
	2010	25	12.1	1.3
(b) in undergrowth (of *Picea rubens* forest)	1770	50	9.3	2.6
Rhododendron carolinianum				
(a) in bald (mid-elevation)	1430	9	10.4	1.1
(b) in bald (summit)	2110	6	9.8	1.1
Kalmia latifolia				
(a) in bald	1430	9	16.6	1.6
(b) in undergrowth (of *Quercus prinus–Castanea dentata* heath)	975	12	10.0	1.9
Leiophyllum lyoni				
(a) in bald	1500	0.2	4.0	2.2

to Southern Hemisphere heathlands. Fidelity of the bald species is low; all the plant species of the balds can be found in forests and most of them occur extensively in forests. Only one species of the open stage (*Leiophyllum lyoni = Dendrium prostratum*) is largely limited to the balds, and this species is also the only southern Appalachian endemic in the bald flora.

Productivity and biomass

Some data on productivity of heath balds and of shrub strata in heath forests are given in Table 17.1. Net primary productivity of the balds ranges from 400 to 600 g m^{-2} yr^{-1} above ground and probably from 600 to 900 g m^{-2} yr^{-1} including below-ground production, a range typical of closed, temperate shrublands (Whittaker, 1963). Sample 9, with a productivity of 980 g m^{-2} yr^{-1} above ground, hence more than 1200 g m^{-2} yr^{-1} total, is in the range characteristic of temperate forests (Whittaker, 1966); this sample is a high heathland thought to be successional to forest. The distribution of net production, and of biomass, in dominants is shown in Table 17.3. Biomass accumulation ratios (above-ground biomass/annual net above-ground production) range from 4 to 6 for the more typical heathlands and are about 11 for the high heathland (Whittaker, 1963); above-ground shrub-stratum biomasses are from 1.2 to 3.7 kg m^{-2} for the typical heathlands, 10.8 kg m^{-2} for the high heathland (Table 17.1). These closed heathlands of humid forest climates have much higher productivity and biomass than low, open Australian heathlands of drier climates (Specht et al., 1958).

TABLE 17.3

Mean percentage distribution of net above-ground production and biomass in dominants of typical middle-elevation heathlands (Whittaker, 1962)

	Net production	Biomass
Stem wood	12	41
Stem bark	2	7
Branch wood and bark	16	30
Older leaves	15*	13
Current twigs and leaves	49	8
Fruits	6	1

* Growth after first summer.

Geographic relations

The most extensive heath balds are in the northeast half of the Great Smoky Mountains. The southwest half of the range lacks both spruce–fir forests and heath balds; the latter are in some sense replaced there by grassy balds (dominated by *Danthonia compressa*), which are much more limited in area and number and of more uncertain successional status. The grassy balds may represent oak openings of forest cleared of trees by early settlers; the controversy regarding their origin is discussed by Camp (1931), Wells (1937), Whittaker (1956), Billings and Mark (1957), Mark (1958) and others. Both heath balds and grassy balds occur at elevations well below those at which a true alpine timberline would be expected. Northward from the Great Smoky Mountains heath balds are less extensive but occur locally in the Appalachian Mountains; Roan Mountain (Brown, 1941) has both grassy and *Rhododendron catawbiense* bald areas. Pine forests with heaths prominent in the understorey, dominated by *Pinus rigida* or other species, are more widespread in mountains and on the Coastal Plain; *Rhododendron maximum*, *Kalmia latifolia*, and other heath bald species occur in the understorey of oak forests northward to southern New England. Pine barrens with heaths occur also inland to Wisconsin (*P. banksiana*, *P. resinosa*; *V. angustifolium*, *G. baccata*; *Myrica asplenifolia*, *Ceanothus ovatus*, *Andropogon gerardii*, etc.) (Curtis, 1959; Vogl 1970) and the bluffs of the Mississippi River, south to Florida and west to eastern Texas (Braun, 1950).

Heath balds far north of the Great Smoky Mountains occur in the Mahoosuc Mountains of Maine, the easternmost (and therefore most maritime) ridge of the White Mountains (Fahey, 1976). *Kalmia angustifolia* is the major species in a pattern gradating from dwarf heathland (0.25 m) on ridge crests through low heathland (0.25 to 0.75 m) on slopes to heathland 0.5 to 1.0 m tall at the edge of the forest (of *Picea mariana* and *Abies balsamea*). Ericaceous species (*K. angustifolia*, *Ledum groenlandicum*, *Chamaedaphne calyculata* var. *angustifolia*, *Vaccinium uliginosum* var. *alpinum*, *V. angustifolium*, *K. polifolia*, and *V. vitis-idaea* var. *minus*) make up 75% of the shrub canopy; *Rhododendron canadense* occurs on the bald edge and *Empetrum nigrum* on ridge crests, and *Viburnum cassinoides* is present. The sparse herb stratum includes the ground heath *Gaultheria hispidula*. Most of these species are of wide occurrence in other communities — forest undergrowth, bog shrub borders, and shrub tundra. Only a few, non-ericaceous species (*V. cassinoides*, *Trillium undulatum*) and no major species appear in the lists for the balds of both the Great Smoky and the Mahoosuc Mountains. They are floristically quite different, but between them they relate the Appalachian balds to many eastern forests and (the Mahoosuc balds) to the transcontinental bog and tundra heathlands.

OTHER NORTH AMERICAN HEATHLANDS

North America has no heathland formation characterizing a geographic area, as the southwest

Australian heathland and South African *fynbos* do, no extensive low heathlands resembling those of northwest Germany, and limited alpine shrubland to which the term heathland is appropriate. Dwarf-shrub communities occur in the timberline belts of some of the mountains, but are mostly deciduous, dominated by chamaephyte species of *Salix* and *Betula* in some areas, and of *Vaccinium* in others. A number of American heathlands deserve mention, however (Fig. 17.2).

Pine plains

The coastal plain of New Jersey is occupied by a complex of vegetation referred to as "pine barrens" and including oak forests, extensive pine forests, and cypress (*Chamaecyparis thyoides*) bogs (Harshberger, 1916; Robichaud and Buell, 1973;

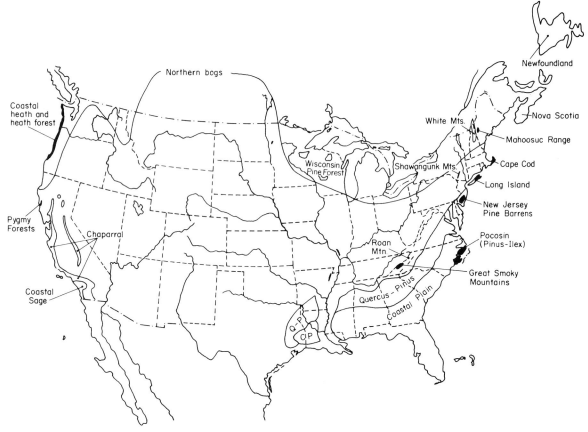

Fig. 17.2. The United States and part of Canada, showing occurrence of heathlands and some related communities in North America. Pocosin communities are discussed in this book by Christensen (Chapter 18), and the pygmy forest by Westman (Chapter 20). Arctic heathlands are discussed by Bliss (Chapter 15); these occur widely in the north and in the major mountain ranges and are not indicated on the map. The line marked "Northern bogs" is an approximate southern limit for the occurrence of these, but individual bogs occur south of this line.

Forman, 1979; and Little, Chapter 19). Soils are predominantly sandy and the vegetation is subject to fires. Related vegetation on Long Island, New York, includes a community-gradient from deciduous oak forest through oak–pine forest (*Quercus alba*, *Q. coccinea*, *P. rigida*) with an undergrowth in which heaths are prominent (*Vaccinium vacillans*, *V. angustifolium*, *Gaylussacia baccata*, *Kalmia angustifolia*, *Myrica pensylvanica* and *Gaultheria procumbens*), to pine–oak scrub (*P. rigida*, a shrubby oak, *Q. ilicifolia*, and the heaths) to the "pine plains" (Harshberger, 1916; McCormick and Buell, 1968; Olsvig et al., 1979). The last are distinctive miniature pine forests in which *P. rigida* grows only 1 to 2 m tall with contorted stems, low-spreading branches, short internodes, and serotinous cones. The tree oaks of the oak–pine forest (Whittaker and Woodwell, 1969) are absent, and the major species other than pine in the Long Island pine plains are *Q. ilicifolia*, *Gaylussaccia baccata*, and the prostrate heaths *Arctostaphylos uva-ursi* and *Gaultheria procumbens*. A few herbs, including *Melampyrum lineare*, occur with very low coverage. The whole pattern of vegetation has been swept by fires; but, despite the arguments of Andresen (1959) there seems little doubt that the *P. rigida* of the true pine plains is genetically different from the erect *P. rigida* of the oak–pine forests. Fires, which affect all these types, are not the sufficient cause of the contrast between the miniature forests and the oak–pine forests, some of which are of the same age since fire (Good and Good, 1975). The community-gradient described probably involves increasing soil nutrient poverty and acidity, with the pine plains occurring on some of the most deficient, coarse-sandy soils (Olsvig et al., 1979). Although the scrub canopy (about 60–80% coverage, Table 17.1) is formed by a pine and an oak, it may be as reasonable to apply the concept of heathland community to the pine plains as to an Australian mountain heathland dominated by a small species of *Casuarina* (Whittaker, 1979). Like some other heathlands the pine plains occur on sandy soil in a moderately humid, coastal climate. Similar dwarf *P. rigida* heaths (without *Q. ilicifolia*) occur in the Shawangunk Mountains of southeastern New York on a quartzite conglomerate capping the mountains.

Maritime Province heathlands

Northward from the pine barrens, lowland heathlands occur in areas where coniferous forests (*Picea mariana*, *Abies balsamea*, *Pinus strobus*) are climax in the Maritime Provinces of Canada. Strang (1970) describes for southern Nova Scotia a pattern of four community types: (1) *Corema conradii* (Empetraceae) dominant on tops of ridges and knolls with shallow soils, with *Arctostaphylos uva-ursi*, *Gaultheria procumbens*, and *Cladonia* spp.; (2) *Gaylussacia baccata* dominant on slopes and most extensive, with *Vaccinium angustifolium*, *Kalmia angustifolia*, *Aralia nudicaulis*; (3) *Rhododendron canadense* dominant in depressions with peat over rocks, with *Chamaedaphne calyculata*, *K. angustifolia*, *Ledum groenlandicum*, *Sphagnum* spp.; and (4) mixed deciduous shrubs along water courses. The heathland soil is a bouldery, ortstein podzol derived from coarse granitic tills. A strong correlation was observed between *Gaylussacia* dominance and hardpan formation, especially on coarse soils. The pollen record and other evidence suggested a shift from open *Pinus* woodland to *Betula* and *Alnus* with a shrub understorey some 2000 years ago, in response to climatic deterioration. Fire, since the occupation of the area by man about 900 years ago, has fostered the dominance of *Gaylussacia* and formation of a cemented pan. The heathland seems now a stable community on its shallow, infertile soil and a poor prospect for afforestation. Damman (1971) describes the heathland of *Kalmia angustifolia* that is extensive in eastern and southern Newfoundland. The area once supported forest, but this heathland also appears now to maintain itself indefinitely.

Arctic–alpine dwarf-shrub heathland

Alpine tundra heathland communities are apparently best developed in the Presidential Range of New Hampshire, the principal range of the White Mountains (Bliss, 1963, and see also Chapter 15). The dwarf-shrub heathland is dominated by *Vaccinium uliginosum*, *V. angustifolium*, *V. vitis-idaea* var. *minus*, and *Ledum groenlandicum;* *Empetrum eamesii* ssp. *hermaphroditum*, *Kalmia polifolia*, and *Arctostaphylos alpina* are sometimes present with other non-heath species (*Cornus canadensis*, *Maianthemum canadense*, and

Trientalis borealis). The shrubs form dense mats 0.1 to 0.3 m tall with 96% plant cover at timberline just above and in the lee of continuous krummholz. Above the dwarf-shrub heathland more extensive slopes of the mountains are occupied by heath–rush communities, dominated by clumps of *Juncus trifidus* with low, scattered plants of *V. vitis-idaea* var. *minus*, *V. uliginosum*, *Potentilla tridentata*, and smaller amounts of *Carex bigelowii* and *Diapensia lapponica*. Similar dwarf-shrub heathlands occur in windswept sites in the southern part of the arctic zone in Canada; among the most important of their species are *Empetrum nigrum*, *Arctostaphylos alpina*, *Vaccinium vitis-idaea* var. *minus*, *Ledum decumbens*, *Rhododendron lapponicum*, *V. uliginosum*, *Betula nana* ssp. *exilis*, and *Salix arctica*. In areas of longer snow cover some of these species occur with *Cassiope tetragona* and *Phyllodoce caerulea* in dwarf heathland 10 to 15 cm tall (Knapp, 1965). Across the continent from New Hampshire, in coastal Alaskan mountains, heath mats mostly less than 0.3 m thick occur on cold, dry, glaciated surfaces, and from timberline up to 1500 m; *Cassiope mertensiana*, *C. stelleriana*, and *Empetrum nigrum* dominate many of these mats, *Phyllodoce glanduliflora*, *Vaccinium caespitosum*, and *V. uliginosum* also occur (Cooper, 1942; Heusser, 1960). Similar heathlands extend south through British Columbia into Washington and Oregon (Jones, 1936; Brink, 1959; Heusser, 1960), with *Phyllodoce empetriformis*, *Cassiope mertensiana*, and at lower elevations *Vaccinium deliciosum* as dominants in the Cascade Mountains (Franklin and Dyrness, 1969). Some alpine and subalpine heath species occur further south in the Sierra Nevada of California, but in this drier range these heaths do not dominate extensive communities.

Bog shrub communities

South of the New Jersey pine barrens, pocosins or bog shrub communities dominated by *Cyrilla racemiflora* and the heaths *Zenobia pulverulenta* and *Lyonia lucida* occur on the Coastal Plain from Virginia to Florida with an area of most concentrated occurrence indicated in Fig. 17.2; these communities are further described in this volume by Christensen (Chapter 18).

Bog successions in more northern forest climates lead through shrub stages, often arranged as belts of shrubs of increasing height surrounding the open bog of *Carex*, *Eriophorum* and *Sphagnum*. Low-shrub stages in different areas of eastern bogs include some of the following species: *Chamaedaphne calyculata* var. *angustifolia*, *Ledum groenlandicum*, *Andromeda polifolia* and *A. glaucophylla*, *Kalmia polifolia* and *K. angustifolia*, *Vaccinium oxycoccos*, *V. macrocarpon*, and *V. uliginosum*, *V. angustifolium*, *Myrica gale*, and *Rubus chamaemorus*. Beyond this predominantly ericaceous low-shrub belt (or belts) in deciduous forest climates a tall-shrub belt occurs, often including *Vaccinium corymbosum* along with *Rhododendron canadense* and non-ericaceous shrubs (*Rhus*, *Cornus*, *Ilex*, *Salix*, *Betula*, *Alnus*, *Pyrus*) (Osvald, 1928, 1954). Some of the heaths of the low-shrub belt occur across the north of the North American continent in borders of bogs in the taiga and tundra. In Pacific Coast bogs from Alaska to Washington *Ledum groenlandicum*, *Andromeda polifolia*, *Empetrum nigrum*, *Kalmia polifolia*, *Vaccinium uliginosum* and *V. oxycoccos* var. *ovatifolium* occur with the western heath, *Gaultheria shallon* (Osvald, 1933; Heusser, 1960). In these Pacific Coast bogs the deciduous tall-shrub belt characteristic of eastern bogs is often lacking, and replaced by trees (that may be stunted and slow-growing) of *Pinus contorta*, *Betula papyrifera* var. *occidentalis*, *Tsuga heterophylla*, and *Thuja plicata* (of which the last two dominate regional climax forests).

Coastal heathlands

A strip of vegetation affected by wind and salt spray extends along the Pacific Coast of Washington and northern Oregon. Inland of the dune and strand communities, dense thickets, 1 to 3 m tall, occur with *Gaultheria shallon* as the dominant species (Heusser, 1960; Franklin and Dyrness, 1969). Like the heath balds, some of these thickets are fairly impenetrable, and they are composed largely of broad-sclerophylls; the *Gaultheria* leaves are evergreen, broadly ovate, and fairly large (mesophyll, around 30 to 50 cm²). Species typically present are heaths (*G. shallon*, *Vaccinium ovatum*, *Rhododendron macrophyllum*, *Arctostaphylos columbiana*, and *A. uva-ursi*), *Myrica californica*, and *Rubus spectabilis*. *V. ovatum* becomes important in wind-swept sites, while stands on wetter sites include *Ledum glandulosum* var. *columbianam* and the non-heaths *Salix hookeriana*, *Alnus rubra*, *Spiraea douglasii*, and *Lonicera involucrata*. Seedlings of *Pinus contorta* and *Picea sitchensis* may be present; and on the inland edge of the heath thicket these trees form forests, often stunted and wind-trimmed, in which the heaths dominate the understorey. *G. shallon* is one of the most important undergrowth species of the northwestern forests; *V. ovatum* and other species mentioned are of wide occurrence in forest undergrowth. Southward along the Oregon coast, in drier climates, the *Gaultheria* heathland and heath forest are replaced by other coastal vegetation — herb and low-shrub communities grading into a taller shrubland including heath species (*Rhododendron occidentale*, *Arctostaphylos columbiana*, *G. shallon* and *V. ovatum*) among other shrubs (*Ceanothus thyrsiflorus*, *Garrya elliptica*, *Alnus sinuata*, and *Ribes menziesii*) (Franklin and Dyrness, 1969). The Pacific Coastal heathland is a more extensive and more impressive community than the heath-like (but predominantly non-ericaceous) dwarf-shrub communities on sand dunes in eastern North America, with dominants (in different stands and areas): *Hudsonia tomentosa*, *H. ericoides*, *Corema conradii*, *Lechea maritima*, *L. tenuifolia*, and *Arctostaphylos uva-ursi* (Knapp, 1965).

Pygmy forests

Still farther south on coastal terraces of Mendocino County, California, a community-gradient extends from *Sequoia sempervirens* giant forests, through mixed evergreen forest (*Pseudotsuga menziesii* with broad-sclerophyll trees), pine forest or forest–heathland (*Pinus muricata* with a tall understorey, 3 to 5 m high, of *Rhododendron macrophyllum* and *Myrica californica*), and a more open pine heathland, to a pygmy forest of *Pinus contorta* ssp. *bolanderi* and *Cupressus pygmaea* (McMillan, 1956; Jenny et al., 1969; Westman, 1975; Westman and Whittaker, 1975). In extreme stands the *Cupressus* is dominant, its "trees" 1 to 2 m tall; the undergrowth is predominantly of low heath — *Vaccinium ovatum* (an evergreen species, here only about 0.5 m tall), *Arctostaphylos columbiana* and *Ledum glandulosum* (0.5 to 1.0 m), *Gaultheria shallon* (0.1 to 0.2 m in

these stands), and *Arctostaphylos nummularia* (0.1 m). The heath species that also occur in the adjoining forests are dwarfed in the pygmy forest; the endemic *P. contorta* ssp. *bolanderi* is a small pine species, and the endemic *A. nummularia* is a dwarf heath. The whole vegetation pattern from *Sequioa* forest to pygmy forest is fire-affected. The striking gradation of communities relates, however, not simply to fire but to a gradient of soils (Gardner and Bradshaw, 1954, Jenny ct al., 1969; Westman, 1975) from forest brown earth in the *Sequoia* forest to an extreme podzol — intensely leached and acid and underlain by an impermeable hardpan — in the pygmy forest. Above the hardpan the siliceous soil is waterlogged in winter and dust-dry in summer. It seems fair to regard the pygmy forest, like the pine plains, as a heath community of an extreme soil in a maritime climate. Related, if less severely stunted, communities of *Pinus* and *Cupressus* with heaths occur on acid soils and serpentines farther south in California (Mason, 1946; McMillan, 1956). Fire-adapted pine heathlands of *Pinus attenuata* with *Arctostaphylos* species occur on serpentine from the Siskiyou Mountains of Oregon (Whittaker, 1960) to the Santa Ana Mountains of southern California (Vogl, 1973).

California chaparral

The *Pinus* and *Cupressus* communities in California relate also to the chaparral. Proceeding southward, toward drier but still maritime climates, Californian vegetation forms an extended physiognomic gradient: mixed evergreen forest (Whittaker, 1960), broad-sclerophyll forest, mixed broad-sclerophyll shrubland or chaparral, chamise (*Adenostoma fasciculatum*) chaparral, soft chaparral or coastal sage scrub, (*Artemisia californica*, *Salvia* spp.) and semi-desert types extending into Baja California. The chaparral (the part of the gradient from broad-sclerophyll forest to coastal sage) shows some broad relationships in its physiognomy and climate to Mediterranean *maquis* and *garigue* and to Southern Hemisphere formations — the South African *fynbos* and the Australian heathlands and mallee as well as the Chilean *matorral* (Mooney and Dunn, 1970). Although chaparral is not heathland as understood in this volume, some relationships deserve comment. Knapp (1965), with European comparisons in mind, refers to the coastal sage community as low shrub-heathland; but the type might better be termed shrub steppe or even semi-desert. The mixed hard chaparral is dominated by diverse combinatuons of species including shrubs of the genera *Quercus*, *Cercocarpus*, *Ceanothus*, *Rhamnus*, *Heteromeles*, *Prunus*, and *Arctostaphylos* (Cooper, 1922; Sampson, 1944; Wells, 1962; Knapp, 1965; Hanes, 1971). In some situations, particularly middle and higher elevations in mountains on acid soils from the San Jacinto Mountains and Coast Ranges to the Sierra Nevada and Mount Shasta, the heath genus *Arctostaphylos* is dominant (Wilson and Vogl, 1965; Vogl and Schorr, 1972) and, although *Adenostoma fasciculatum* (Rosaceae) is not a heath, it is ericoid in its minute, linear, terete, evergreen leaves. The broad-sclerophyll and linear-sclerophyll leaf types, which often occur in the same community in Southern Hemisphere heathlands, are in California largely segregated into the two major phases of the "hard chaparral": the former in the various types of mixed chaparral, and the latter in the chamise and redshanks (*Adenostoma sparsifolium*) chaparral (Hanes, 1965).

CONCLUSION

A few points on the North American heathlands suggest themselves.

(1) North America has few communities fitting the classical, European concept of heathland: stands of small-leaved (leptophyll), evergreen ericoid dwarf-shrubs forming a canopy usually below 0.5 m, with trees and tall shrubs very sparse or lacking (Graebner, 1901; Gimingham, 1961, 1970, 1972). The *Phyllodoce empetriformis–Cassiope mertensiana* low-alpine shrublands of the Cascade Mountains (Franklin and Dyrness, 1969) may best correspond to the North European concept of heathland. Although leptophyll shrublands occur, they seem marginal among the American heathlands, and of their dominants (*Leiophyllum*, *Phyllodoce*, *Cassiope*, *Empetrum*, *Corema*, *Hudsonia*, *Adenostoma*) only the first three are ericaceous, *Empetrum* and *Corema* are members of the related family Empetraceae. North American heathlands are primarily communities with evergreen broad-sclerophyll shrubs, secon-

darily with deciduous, nanophyll and often vacciniaceous shrubs.

(2) Epacridaceae, Restionaceae, and other Southern Hemisphere heathland components are lacking, as are the major European heath genera *Erica* and *Calluna*. Among characteristic heath genera in North America are the holarctic *Rhododendron* and *Arctostaphylos*, the nearctic *Kalmia* and *Gaylussacia* (the last also in South America), and the widespread *Vaccinium* and *Gaultheria*. Species of *Empetrum* and *Myrica* often occur with these and might be considered honorary heaths.

(3) Heathlands bracket the North American continent. A thin and interrupted line of heathlands extends parallel to the Atlantic Coast, across the continent on the North, and down the Pacific Coast. On each side of the continent and in the north heath species spread more widely in forest undergrowth — primarily (but not only) dry forests of acid soils in the East, and many coniferous forests in the west from northwestern maritime forests with dense, tall, evergreen heaths to subalpine forests in the Rocky Mountains with low, deciduous *Vaccinium*.

(4) True heathlands form a small part of North American vegetation; they occur primarily as localized communities of extreme situations. As on other continents, heathlands seem most characteristic of acid soils in humid or subhumid, maritime climates (the Great Smoky Mountains are continental but have a distinctive climate with two rainy seasons).

(5) These are marginal to forests or are forest-derivatives — shrublands in which trees are excluded or of drastically reduced stature. The characteristic heathlands of acid soils in forest climates are floristically linked with other communities marginal to forest: shrublands of timberline and tundra, bog edges, some windswept coasts, some serpentine soils, and the chaparral complex of drier, Mediterranean climates.

(6) The North American heathland communities are floristically poor. They are of low species diversity compared with southern hemisphere heathlands and North American forests, and they are poor in endemic species. The few narrowly endemic species occur in the most extreme heathland types. It seems likely that the North American heathlands are young as types of communities,

derived from forest heathlands in adaptation to extreme sites, in contrast to the old, rich, regional heathlands of Australia and South Africa.

REFERENCES

Andresen, J.W., 1959. A study of pseudo-nanism in *Pinus rigida* Mill. *Ecol. Monogr.*, 29: 309–332.

Bliss, L.C., 1963. Alpine plant communities of the Presidential Range, New Hampshire, *Ecology*, 44: 678–697.

Billings, W.D. and Mark, A.F., 1957. Factors involved in the persistence of montane treeless balds. *Ecology*, 38: 140–142.

Braun, E.L., 1950. *Deciduous Forests of Eastern North America*. Blakiston, Philadelphia, Pa., 596 pp.

Brink, V.C., 1959. A directional change in the subalpine forest-heath ecotone in Garibaldi Park, British Columbia. *Ecology*, 40: 10–16.

Brown, D.M., 1941. The vegetation of Roan Mountain: a phytosociological and successional study. *Ecol. Monogr.*, 11: 61–97.

Cain, S.A., 1930. An ecological study of the heath balds of the Great Smoky Mountains. *Butler Univ. Bot. Stud.*, 1: 176–208.

Cain, S.A., 1931. Ecological studies of vegetation of the Great Smoky Mountains of North Carolina and Tennessee. I. Soil reaction and plant distribution. *Bot. Gaz.*, 91: 22–41.

Camp, W.H., 1931. The grass balds of the Great Smoky Mountains of Tennessee and North Carolina. *Ohio J. Sci.*, 31: 157–165.

Curtis, J.T., 1959. *The Vegetation of Wisconsin: An Ordination of Plant Communities*. University of Wisconsin, Madison, Wis., 657 pp.

Cooper, W.S., 1922. The broad-sclerophyll vegetation of California. *Carnegie Inst, Wash, Publ.*, 319: 1–124.

Cooper, W.S., 1942. Vegetation of the Prince William Sound region, Alaska, with a brief excursion into post-Pleistocene climatic history. *Ecol. Monogr.*, 12: 1–22.

Damman, A.W.H., 1971. Effect of vegetation changes on the fertility of a Newfoundland forest site. *Ecol. Monogr.*, 41: 253–270.

Fahey, T.J., 1976. The vegetation of a heath bald in Maine. *Bull. Torrey Bot. Club*, 103: 23–29.

Forman, R.T.T. (Editor), 1979. *Pine Barrens: Ecosystem and Landscape*. Academic Press, New York, N.Y., in press.

Franklin, J.F. and Dyrness, C.T., 1969. Vegetation of Oregon and Washington. *U.S. Dep. Agric., Forest Serv. Res. Pap.*, PNW-80: 1–216 (Pacific Northwest Forest and Range Experiment Station, Portland, Ore.).

Gardner, R.A. and Bradshaw, K.E., 1954. Characteristics and vegetation relationships of some podzolic soils near the coast of northern California. *Proc. Soil Sci. Soc. Am.*, 18: 320–325.

Gimingham, C.H., 1961. North European heath communities: a "network of variation." *J. Ecol.*, 49: 655–694.

Gimingham, C.H., 1970. British heathland ecosystems: the outcome of many years of management by fire. *Proc. Tall Timbers Fire Ecol. Conf.*, 10: 293–321.

Gimingham, C.H., 1972. *Ecology of Heathlands*. Chapman and Hall, London 266 pp.

Good, R.E. and Good, N.F., 1975. Growth characteristics of two populations of *Pinus rigida* Mill. from the Pine Barrens of New Jersey. *Ecology*, 56: 1215–1220.

Graebner, P., 1901. *Die Heide Norddeutschlands. Vegetation der Erde*, 5. Engelmann, Leipzig, 277 pp.

Hanes, T.L., 1965. Ecological studies on two closely related chaparral shrubs in southern California. *Ecol. Monogr.*, 35: 213–235.

Hanes, T.L., 1971. Succession after fire in the chaparral of southern California. *Ecol. Monogr.*, 41: 27–52.

Harshberger, J.W., 1916. *The Vegetation of the New Jersey Pine Barrens*. Sower, Philadelpia, Pa. (reprinted by Dover, New York, 1970).

Heusser, C.J., 1960. Late-Pleistocene environments of North Pacific North America. *Am. Geogr. Soc. Spec. Publ.*, 35: 1–308.

Jenny, H., Arkley, R.J. and Schultz, A.M., 1969. The pygmy forest-podsol ecosystem and its dune associates of the Mendocino coast. *Madroño*, 20: 60–74.

Jones, G.N., 1936. A botanical survey of the Olympic Peninsula, Washington. *Univ. Wash. Publ. Biol.*, 5: 1–286.

Knapp, R., 1965. *Die Vegetation von Nord- und Mittelamerika und der Hawaii-Inseln*. Fischer, Jena, 373 pp.

McCormick, J. and Buell, M.F., 1968. The plains; pigmy forests of the New Jersey Pine Barrens, a review and annotated bibliography. *Bull. N.J. Acad. Sci.*, 13: 20–34.

McMillan, C., 1956. The edaphic restriction of *Cupressus* and *Pinus* in the Coast Ranges of central California. *Ecol. Monogr.*, 26: 177–212.

Mark, A.F., 1958. The ecology of the southern Appalachian grass balds. *Ecol. Monogr.*, 28: 293–336.

Mason, H.L., 1946. The edaphic factor in narrow endemism. *Madroño*, 8: 209–226; 241–257.

Mooney, H.A., and Dunn, E.L., 1970. Convergent evolution of mediterranean-climate evergreen sclerophyll shrubs. *Evolution*, 24: 292–303.

Olsvig, L.S., Cryan, J.F. and Whittaker, R.H., 1979. Gradients from pine plains to oak-pine in the Pine Barrens. In: R.T.T. Forman (Editor), *Pine Barrens: Ecosystem and Landscape*. Academic Press, New York, N.Y., in press.

Osvald, H., 1928. Nordamerikanska mosstyper. *Sven. Bot. Tidskr.*, 22: 377–391.

Osvald, H., 1933. Vegetation of the Pacific Coast bogs of North America. *Acta Phytogeogr. Suecica*, 5: 1–33.

Osvald, H., 1954. The vegetation of Argyle Heath in southern Nova Scotia. *Nytt Mag. Bot.*, 3: 171–182.

Robichaud, B. and Buell, M.F., 1973. *Vegetation of New Jersey: A Study of Landscape Diversity*. Rutgers University, New Brunswick, N.J., 340 pp.

Sampson, A.W., 1944. Plant succession on burned chaparral lands in northern California. *Bull. Calif. Agric. Exp. Station*, 685: 1–144.

Specht, R.L., Rayson, P. and Jackman, M.E., 1958. Dark Island Heath (Ninety-Mile Plain, South Australia). VI. Pyric succession: Changes in community composition, coverage, dry weight, and mineral nutrient status. *Aust. J. Bot.*, 6: 59–88.

Thornthwaite, C.W., 1948. An approach toward a rational

classification of climate. *Geogr. Rev.*, 38: 55–94.

Strang, R.M., 1970. The ecology of the rocky heathlands of western Nova Scotia. *Proc. Tall Timbers Fire Ecology Conf.*, 10: 287–292.

Vogl, R.L., 1970. Fire and the northern Wisconsin pine barrens. *Proc. Tall Timbers Fire Ecol. Conf.*, 10: 175–209.

Vogl, R.J., 1973. Ecology of knobcone pine in the Santa Ana Mountains, California. *Ecol. Monogr.*, 43: 125–143.

Vogl, R.J. and Schorr, P.K., 1972. Fire and manzanita chaparral in the San Jacinto Mountains, California. *Ecology*, 53: 1179–1188.

Wells, B.W., 1937. Southern Appalachian grass balds. *J. Elisha Mitchell Sci. Soc.*, 53: 1–26.

Wells, P.V., 1962. Vegetation in relation to geological substratum and fire in the San Luis Obispo quadrangle, California. *Ecol. Monogr.*, 32: 79–103.

Westman, W.E., 1975. Edaphic climax pattern of the pygmy forest region of California. *Ecol. Monogr.*, 45: 109–135.

Westman, W.E. and Whittaker, R.H., 1975. The pygmy forest region of northern California: studies on biomass and primary productivity. *J. Ecol.*, 63: 493–520.

Whittaker, R.H., 1956. Vegetation of the Great Smoky Mountains. *Ecol. Monogr.*, 26: 1–80.

Whittaker, R.H., 1960. The vegetation of the Siskiyou Mountains, Oregon and California. *Ecol. Monogr.*, 30: 279–338.

Whittaker, R.H., 1962. Net production relations of shrubs in the Great Smoky Mountains. *Ecology*, 43: 357–377.

Whittaker, R.H., 1963. Net production of heath balds and forest heaths in the Great Smoky Mountains. *Ecology*, 44: 176–182.

Whittaker, R.H., 1965. Dominance and diversity in land plant communities. *Science*, 147: 250–260.

Whittaker, R.H., 1966. Forest dimensions and production in the Great Smoky Mountains. *Ecology*, 47: 103–121.

Whittaker, R.H., 1979. Vegetational relationships of the Pine Barrens. In: R.T.T. Forman (Editor), *Pine Barrens: Ecosystem and Landscape*. Academic Press, New York, N.Y., in press.

Whittaker, R.H. and Woodwell, G.M., 1969. Structure, production and diversity of the oak–pine forest at Brookhaven, New York. *J. Ecol.*, 57: 155–174.

Wilson, R.C. and Vogl, R.J., 1965. Manzanita chaparral in the Santa Ana Mountains, California. *Madroño*, 18: 47–62.

Chapter 18

SHRUBLANDS OF THE SOUTHEASTERN UNITED STATES[1]

N.L. CHRISTENSEN

INTRODUCTION

The southeastern United States is usually thought of as an area dominated by forests; warm-temperate to subtropical forests to the south, pine forests in the Coastal Plain, and broad-leaf deciduous forests in the Piedmont and mountains. However, ecosystems dominated by shrubs are common in the mountains as heath balds or "slicks", and in the coastal plain as evergreen shrub bogs or "pocosins." These communities are quite distinct from each other in both environment and biota. The heath balds are discussed by Whittaker in Chapter 17.

The vegetation in many of the poorly drained interstream areas on the lower terraces of the Atlantic Coastal Plain is characterized by a dense cover of evergreen shrubs with widely spaced, emergent pond pines (*Pinus serotina*)[2]. Shrub stature may exceed 4 m and pines may reach to 20 m (Fig. 18.1), however, in many areas the pine is quite stunted and essentially part of the shrub layer (Fig. 18.2). Wells (1942) commented that shrub-bogs communities were among the most neglected with respect to ecological research, and our knowledge of vegetational variation within the environmental and geographic range of this eco-system remains scant. Furthermore, virtually no information is available on functional properties of these ecosystems such as nutrient cycling and productivity.

PHYSIOGRAPHY

Coastal Plain areas supporting evergreen shrub-bog vegetation extend as far north as Virginia and south into south-central Florida. Although segments of the Coastal Plain reach elevations over 200 m, most of the Plain, particularly areas where shrubs dominate, lies between 100 m and sea level. The primary surface feature which distinguishes this area from other provinces is its relatively poorly weathered mantle of Cenozoic sediments. As many as seven terraces and scarps are recognized on the Coastal Plain corresponding to shorelines of Pleistocene interglacial seas. The older terraces are characterized by rolling hills while the younger, more seaward terraces have relatively little topographic relief.

Shrub bogs occur in three somewhat distinct topographic situations (Woodwell, 1956):

(1) The so-called Carolina bays are elliptical depressions scattered from Virginia to northern Florida being most common in southeastern North Carolina and northern South Carolina. They vary in size from 100 m to several kilometers along their long axis. Their elliptical shape and the uniform southwest–northeast orientation of their axes have led some to suggest that they are scars of an intense meteor shower (Wells and Boyce, 1953; Murray, 1961). Others (Johnson, 1944) feel they are the result of karst-type activities.

(2) Evergreen shrubs are commonly encountered on poorly drained upland flats, sometimes abruptly bordering xeric sandhill vegetation (Wells and Shunk, 1931). Woodwell (1956) found that such areas are frequently underlain by semiplastic clay which supports a perched water table for much of the year.

(3) In the sandhills area of North and South

---

[1] Manuscript completed September, 1976.
[2] Nomenclature follows Radford et al. (1968).

Fig. 18.1. Evergreen shrub bog in the Green Swamp, North Carolina, approximately fifteen years since burning. The dominant shrubs here are *Cyrilla racemiflora* and *Lyonia lucida*. The emergent tree is pond pine, *Pinus serotina*.

Carolina, the troughs between sand ridges may support evergreen shrub bogs. This alteration of sandhills and pocosin is frequently referred to as "ridge and bay" vegetation.

CLIMATE

Seasonal variations in climate for three locations in the Coastal Plain are listed in Table 18.1. In general, this area is typified by a warm-temperate to subtropical climate. The length of the growing season varies considerably from north to south. North Carolina has approximately 200 consecutive frost-free days, whereas northern Florida averages 300 during the growing season.

Precipitation, almost entirely as rainfall, occurs year-round with a peak during the summer as a result of convectional and monsoonal storms.

Summer precipitation increases to the south, whereas winter precipitation from cyclonic storms is more abundant to the north. Three- to four-week drought periods are most common in spring and autumn, and it is during these periods that fires are most likely.

HYDROLOGY

Wells (1946) suggested that the single most important factor controlling shrub-bog vegetation is the length of time during which the substrate is saturated with water or the hydroperiod. No quantitative data are available for this factor; however, the hydroperiod in shrub-bog communities is agreed to be relatively long (Wells and Shunk, 1928; Wells, 1942, 1946; Woodwell, 1956). In better drained areas, frequent fires favor

Fig. 18.2. Shrub bog in the Green Swamp approximately three years after burning. The dominant shrub is *Zenobia pulverulenta*.

grass–sedge or savanna vegetation. In wet areas with a well-defined drainage, standing water for much of the year, or a more mineral substrate, swamp forests are favored. Wells (1946) found that growth of shrub species was retarded in areas extensively flooded for long periods. Woodwell (1958), however, found that *Pinus serotina* was quite tolerant of extended flooding.

SOILS

The single unifying feature of soils supporting evergreen shrub-bog vegetation is peat accumulation. The depth to mineral soil may be several meters, particularly in the elliptical bays. However, the peat layer rarely exceeds 30 cm in the sandhill pocosins (Woodwell, 1956). The pH of these soils is quite low (3.0–4.0) which Wells (1942) suggests is

indicative of low calcium. Caughey (1945) suggests that calcium deficiency may be more responsible for the sclerophyllous characteristics of these evergreen shrubs than drought stress. Using nutrient enrichment techniques, Woodwell (1958) found that growth of oats on pocosin soils was limited by nitrogen, phosphorus and calcium deficiencies; however, growth of pond pine was limited only by nitrogen and phosphorus. The low pH and poor aeration create a reducing environment which may render both nitrogen and phosphorus less available. Woodwell (1958) found that *Pinus serotina* could utilize, and perhaps required, ammonium nitrogen. Barnes and Naylor (1959) showed that this species could utilize many forms of organic nitrogen. The low availability of nitrogen may also be the result of the high carbon to nitrogen (C/N) ratios in these soils. Under such conditions soluble nitrogen compounds are

TABLE 18.1

Seasonal variations in climate at three locations in the southeastern United States Coastal Plain.

Climatic data	Location	Winter (Dec–Feb)	Spring (Mar–May)	Summer (June–Aug)	Autumn (Sept–Nov)	Annual average
Solar Radiation[1] (J cm^{-2} day^{-1})	SC	1097	2053	2184	1465	1700
Percentage of possible sunshine	NC	59.3	68.0	66.0	64.3	65.0
	SC	58.0	67.0	62.6	59.0	63.0
	FL	57.3	66.3	58.0	52.0	59.0
Mean daily maximum air temperature (°C)	NC	15.0	22.8	30.6	25.3	23.3
	SC	16.4	24.4	31.7	24.8	24.4
	FL	19.6	26.5	32.8	26.6	26.4
Mean daily minimum air temperature (°C)	NC	3.0	10.8	21.0	12.8	11.9
	SC	3.9	11.8	21.4	12.8	12.5
	FL	7.5	14.4	22.5	16.3	15.2
Precipitation (mm)	NC	232.56	249.84	451.2	297.36	1230.76
	SC	208.32	250.08	463.2	258.24	1179.84
	FL	181.92	252.24	500.64	345.84	1280.64
Frost days (days <0°C)	NC	40	5	0	3	48
	SC	29	3	0	4	37
	FL	11	1	0	1	11

[1] Solar radiation data available only for Charleston, S.C.
NC=Wilmington. N.C., latitude 34°16′N, longitude 77°55′W, elevation 9 m; SC=Charleston, S.C., latitude 32°54′N, longitude 80°02′W, elevation 12 m; FL=Jacksonville, Fla. latitude 30°25′N, longitude 81°39′W, elevation 8 m. Data are from U.S. Department of Commerce, Weather Bureau national Climatic Summaries.

thought to be immobilized by heterotrophic soil microbes. Woodwell (1958) found C/N ratios ranging from 57 to 110.

The most productive soils associated with shrub bogs are clays and loams overlain by a shallow peat layer. Increased thickness of the peat layer, or coarser textured mineral soils are usually associated with less productive sites (Woodwell, 1956).

VEGETATION

General description

Woodwell (1956) compared vegetational composition and structure in 54 shrub bogs in North and South Carolina, and thus provided the most thorough phytosociological study of this ecosystem to date. He distinguished three associations named for the dominant shrub species: (1) *Cyrilla racemiflora*; (2) *Lyonia lucida;* and (3) *Zenobia pulverulenta*. The constituent species, their average cover values, and percent constancy for each association are listed in Table 18.2. Data for many of the less common shrubs and herbs were not actually in Dr.

Woodwell's (1956) thesis, but were obtained directly from his field notes[1]. These associations are differentiated primarily on the basis of relative dominance of constituent species rather than abrupt changes in the presence or absence of particular species. Wells (1946) intensively sampled a large continuous shrub-bog area (c. 20 000 ha) in Pender County, North Carolina. He divided the area into vegetational zones roughly corresponding to Woodwell's associations. Quantitative data for each zone were not published, however, total coverage values for each species over the entire area are shown in Table 18.2. Harper (1906, 1914) and Wright and Wright (1932) listed species occurring in shrub-bog areas of Georgia and Northern Florida, however, quantitative data are missing. Nevertheless, most of the species and all of the genera listed in Table 18.2 are common in these areas.

Woodwell (1956) found that the associations dominated by *Cyrilla racemiflora* and *Lyonia lucida* were segregated geographically. From southeastern North Carolina northward, *C. racemiflora* is

[1] I thank Dr. George Woodwell and Dr. C.W. Ralston for graciously loaning me those notes.

TABLE 18.2

Variations in species composition of shrub bogs from Woodwell (1956) and Wells (1946)

Woodwell's data are divided into 3 associations: (1) the *Cyrilla racemiflora* association represented by 17 stands; (2) the *Lyonia lucida* association represented by 18 stands; (3) the *Zenobia pulverulenta* association represented by 19 stands. The values presented are the average cover percentages for that association. Values in parentheses are the percentage of stands in which the species was observed. The Wells (1946) data are from the Holly Shelter Bay, Pender County, N.C., and represent actual percent cover based on line intercepts. Values in parentheses are frequencies or percentage of transects in which the species was recorded.

Data from:	Woodwell 1956)			Wells (1946
Association:	*Cyrilla*	*Lyonia*	*Zenobia*	
SMALL TREES (<10 m)				
Aceraceae				
Acer rubrum	0.9 (29.4)	2.0 (11.7)	2.9 (31.6)	*
Magnoliaceae				
Magnolia virginiana	2.0 (41.2)	7.0 (55.6)	5.4 (68.4)	1.10 (20)
Nyssaceae				
Nyssa sylvatica var. *biflora*	3.4 (41.2)	1.8 (27.7)	2.0 (10.5)	0.09 (4)
Theacae				
Gordonia lasianthus	* (17.6)	* (16.6)	* (47.3)	4.78 (44)
SHRUBS				
Aquifoliaceae				
Ilex coriacea	19.0 (76.5)	12.8 (66.7)	10.3 (78.9)	2.21 (28)
I. glabra	8.3 (64.7)	0.5 (33.3)	6.7 (31.6)	3.13 (27)
Caprifoliaceae				
Viburnum nudum	–	–	–	0.08
Clethraceae				
Clethra alnifolia	1.1 (47.1)	12.3 (44.4)	3.3 (57.9)	2.48 (24)
Cyrillaceae				
Cyrilla racemiflora	48.7 (100)	11.0 (50.0)	9.4 (47.3)	32.17 (83)
Ericaceae				
Cassandra calyculata	–	–	4.6 (26.3)	0.34 (15)
Gaylussacia dumosa	–	–		0.05 (4)
G. frondosa	–	–	* (15.8)	2.50 (34)
Kalmia angustifolia var. *caroliniana*	–	–	1.8 (52.6)	0.48 (18)
Lyonia ligustrina	* (5.8)	–	* (10.5)	0.40 (5)
L. lucida	11.8 (88.2)	44.1 (100)	7.4 (100)	6.845 (74)
L. mariana	* (11.7)	* (11.1)	* (5.2)	*
Rhododendron viscosum var. *serrulatum*	–	–	–	0.100 (3)
Ericaceae				
Vaccinium atrococcum	–	–	* (5.2)	–
V. corymbosum	* (17.6)	* (11.1)	* (31.5)	0.315 (11)
Zenobia pulverulenta	0.7 (41.2)	*	38.7 (100)	28.29 (94)
Hammamelidaceae				
Fothergilla gardenii	* (11.7)	–	* (26.3)	*
Lauraceae				
Persea borbonia	1.1 (47.1)	4.9 (66.6)	4.1 (73.7)	0.880 (29)
Myricaceae				
Myrica cerifera	* (17.6)	* (22.2)	* (15.8)	*
M. heterophylla	* (11.7)	–	* (21.1)	–
Rosaceae				
Rubus sp.	–	* (11.1)	–	–
Sorbus arbutifolia	–	–	* (10.5)	0.09 (7)
Saxifragaceae				
Itea virginica	* (11.7)	–	* (10.5)	0.425 (13)

TABLE 18.2 *(continued)*

Data from:	Woodwell 1956)			Wells (1946
Association:	*Cyrilla*	*Lyonia*	*Zenobia*	
HERBS				
Bryophyta				
Sphagnaceae				
Sphagnum spp.	0.7 (47.1)	1.3 (66.6)	1.4 (42.1)	*
Pteridophyta				
Blechnaceae				
Woodwardia virginica	3.2 (76.5)	3.1 (83.3)	1.6 (78.9)	4.91 (83)
Osmundaceae				
O. regalis var. *spectabilis*	–	* (16.6)	–	*
Pteridaceae				
Pteridium aquilinum	* (11.7)	* (16.6)	* (5.2)	–
Antophyta				
Asteraceae				
Pluchea foetida	–	–	* (15.8)	–
Cyperaceae				
Carex spp.	–	* (16.6)	* (31.6)	–
Liliaceae				
Smilax laurifolia	1.5 (82.3)	1.5 (77.8)	1.7 (84.2)	5.97 (93)
Melastomataceae				
Rhexia spp.	* (5.8)	* (11.2)	* (5.2)	*
Poaceae				
Andropogon virginicus	–	* (11.2)	* (10.5)	–
Arundinaria gigantea	–	–	* (15.8)	1.66 (17)
Panicum sp.	* (11.7)	* (5.5)	–	–
Sarraceniaceae				
Sarracenia flava	* (11.7)	* (16.6)	* (10.5)	*
S. minor	–	* (11.2)	–	–
S. purpurea	* (11.7)	–	–	*

Asterisk indicates species was present but not sufficiently abundant for cover estimate.
Dash indicates species was not recorded.

the dominant shrub along with *Ilex coriacea*. Well's (1946) study was done in this area. In the southeastern counties of North Carolina, *Cyrilla* and *Lyonia lucida* are codominants with *L. lucida* becoming more prevalent farther south. Although *Cyrilla* is common into Florida, it does not usually dominate shrub-bogs south of North Carolina. *Ilex coriacea* is also a codominant in the southern shrub bogs. In some areas, *Ilex* may be more common than either *Cyrilla* or *Lyonia*.

Both Woodwell and Wells agree that *Zenobia pulverulenta* is indicative of recently disturbed areas, particularly burns. Associated with *Zenobia* are *Clethra alnifolia*, *Kalmia angustifolia*, *Ilex glabra* and *Cassandra calyculata*. Woodwell attributed the dominance of *Zenobia* following fire to its ability to sprout rapidly. On sites favorable to

rapid growth, *Zenobia* is overtopped by *Cyrilla* and *Lyonia* within five years; however, *Zenobia* may survive over fifteen years on poor sites. *Woodwardia* and *Arundinaria* are particularly common in these successional areas. Species composition of this association is quite similar to that of bog margins, particularly where bogs abut onto flatwoods or savannas. This is undoubtedly due to the more frequent burning of these marginal areas. The *Zenobia* stage is not a necessary seral stage in the development of *Cyrilla* or *Lyonia* communities, and these species may regenerate directly following fire under certain conditions (Woodwell, 1956).

Three species, *Pinus serotina*, *Smilax laurifolia*, and *Woodwardia virginica* were found at nearly every study site regardless of association. *Pinus serotina* averaged 346 trees ha^{-1} in the *Cyrilla* association, 320 trees ha^{-1} in the *Lyonia* association and 167.5 trees ha^{-1} in the *Zenobia* association. Although its cover value is never particularly high, *S. laurifolia* is perhaps the most noticed plant in the shrub bog. Because of its long sharp spines and vine-like growth form, it renders many shrub bogs impassable without the aid of a machete. *Woodwardia* is most common along the margins of shrub bogs and following disturbance. Small trees, particularly *Acer rubrum*, *Gordonia lasianthus*, and *Magnolia virginiana*, are common in shrub bogs. These plants are usually associated with locally elevated areas.

High species richness is in most ecosystems associated with the most favorable sites. Woodwell (1956) suggests this may not be the case in shrub bogs. The most productive sites favor rapid growth of *Cyrilla*, *Ilex*, and *Lyonia*, which outcompete other shrubs, particularly *Zenobia*. Poorer sites seem to favor co-existence of several species. This observation is borne out by the fact that 33 species are recorded for the *Zenobia* association, while only 25 and 26 are recorded for the *Cyrilla* and *Lyonia* associations respectively.

Successional relations

No thorough study of successional relations in shrub bogs has been done to date. The most important form of disturbance in these areas is fire which occurs only during periods of protracted drought. However, during the past century logging has had a considerable impact on many shrub areas. Following removal of the shrub cover, herb diversity increases dramatically. As mentioned previously, *Woodwardia* becomes much more prevalent along with broomsedge, *Andropogon virginicus*, and switch cane, *Arundinaria gigantea*. In moister areas *Andropogon* becomes less prevalent, and *Carex* spp. and *Sphagnum* spp. become dominant (Wells, 1946). Insectivorous plants, particularly the pitcher plants, *Sarracennia* spp., may be quite abundant. Herbs such as *Lachnanthes caroliniana* (Haemodoraceae) and *Rhexia* spp. (Melastomataceae) may also be quite common in the first two years following disturbance.

As previously discussed, *Zenobia* becomes dominant during the first year following disturbance, and is subsequently replaced by *Cyrilla* or *Lyonia*.

The rate of this replacement is largely dependent on site quality. On poor sites *Zenobia* may retain dominance indefinitely. *Pinus serotina* seedlings are abundant following fires, however, much of the re-establishment is from epicormic sprouts arising from the bole of mature trees.

Wells and Shunk (1928) and Penfound (1952) suggested that shrub bogs were themselves successional and would eventually be replaced by swamp forests dominated by Atlantic white cedar (*Chamaecyparis thyoides*). This seems particularly true of many of the shrub bogs in large swampy areas where the white cedar has been intensively logged. Slowly decaying boles of cedar are sometimes found in the peat in these areas. However, there is no indication that the smaller pocosins of the Carolina bays or the sandhills were ever anything or will ever be anything other than shrub bogs.

Floristic relationships

The floristics of evergreen shrub bogs have not been thoroughly studied; however, a few generalizations are apparent. The genera and species of these areas have both tropical and temperate affinities. The Cyrillaceae and Clethraceae are, for example, primarily tropical families whereas many of the genera of Ericaceae found in shrub bogs appear to be of temperate origin.

A number of genera and a few species have disjunct distributions, being found in coastal-plain shrub bogs, skipping the Piedmont and recurring in the southern Appalachians. These include *Leucothoe axillaris*, *Vaccinium corymbosum* and the genus *Fothergilla*. This disjunction may simply reflect the lack of suitable habitats in the Piedmont.

Relationship to other Coastal Plain ecosystems

Swamp forests. In areas typified by standing water for large portions of the year, wet areas with a definite drainage pattern, or areas with less organic accumulation, shrub bogs are replaced by swamp forest. These forest areas are dominated by variety of hardwood species, including *Nyss Acer*, as well as Atlantic whit' (*Chamaecyparis thyoides*) and cypress *distichum* and *T. ascendens*). Man' dominated by shrub bogs wer' swamp forest. Intense logging and the lowering of water tables for silvicultural purposes may have been responsible for this conversion.

Savannahs and flatwoods. In areas with lower water tables or shorter hydroperiods than shrub bogs, the vegetation changes to a pine savannah and, with even dryer conditions, to pine flatwood. Pine savannas are typically dominated by longleaf pine (*P. palustris*), with a diverse understorey of herbs and forbs. The ecotone between the savannas and bogs may be quite abrupt, sue to the effects of fire. Savannah fires occur with a frequency of approximately four to eight years (Christensen, 1976). Such fires kill shrub seedlings encroaching on the savanna but fail to burn into the shrub bog, on account of the higher foliage moisture content. The burn cycle in shrub bogs appears to be closer to twenty to forty years. The importance of fire in the

maintenance of this ecotone is further evidenced by the abundance of species such as *Zenobia pulverulenta*, *Kalmia angustifolia* and *Arundinaria gigantea* on the pocosin margin. These species are also dominant throughout shrub bogs following fire.

Sandhill and sand-scrub vegetation. In areas typified by particularly coarse sands, such as the ridge and bay areas of the Carolina sandhills or the sand ridges of Florida, shrub bogs may abut onto quite xeric sandhill or sand-scrub vegetation. The sandhill vegetation, described in detail by Wells and Shunk (1931), is dominated by long-leaf pine, turkey oak (*Quercus laevis*) and wiregrass (*Aristida stricta*). This vegetation type is common throughout the southeastern United States on coarse sands. Sand-scrub communities occur on similar substrates, but are floristically distinct and unique to the remnants of Pleistocene sand dunes in central

Fig. 18.3. Saw-palmetto flatwood near the Okefinokee swamp, Georgia. The dominant tree is *Pinus palustris*. The most conspicuous shrubs include *Serenoa repens* (saw palmetto), *Ilex glabra*, *Quercus chapmanii* and *Kalmia hirsuta*.

Florida. This vegetation is somewhat shrubbier that that of sandhills, and is dominated by sand pine (*Pinus clausa*) and several scrub oaks. The presence of *Ilex glabra* and palmetto (*Serenoa repens*) suggest a relationship to the palmetto flatwoods of Georgia and Florida. The relationships between sand scrub and other vegetation types are discussed by Laessele (1958).

Palmetto flatwoods. In large areas of southern Georgia and Florida, drier conditions and nutrient poor soils result in a transition from shrub bog to palmetto flatwoods. These areas could qualify as a tree-heathland. They are dominated by a nearly closed pine canopy (usually *Pinus palustris*) with a dense shrub understorey (Fig. 18.3). The most conspicuous, though not necessarily most abundant, "shrub" is the saw palmetto. Other species include several held in common with shrub bogs, such as *Ilex glabra* and *Kalmia hirsuta*. Scrub oaks, *Quercus chapmanii* and *Q. virginiana*, are also abundant. The transition from palmetto flatwood to shrub bog may again be quite abrupt, apparently resulting from variations in the fire cycle in these communities.

REFERENCES

Barnes, R.L. and Naylor, A.W., 1959. Effect of various nitrogen sources on growth of isolated rates of *Pinus serotina*. *Physiol. Planta*, 12: 82–89.

Caughey, M.G., 1945. Water relations of pocosin or bog shrubs. *Plant Physiol.*, 20: 671–689.

Christensen, N.L., 1976. The role of fire in the stability of southeastern coastal plain ecosystems. In: *Proc. Symp. Prescribed Fire*. U.S. Dep. Agric. For. Serv., Atlanta, Ga., pp. 17–24.

Harper, R.M., 1906. A phytogeographical sketch of the Altamaba Grit Region of the coastal plain of Georgia. *Ann. N.Y. Acad. Sci.*, 17: 1–415.

Harper, R.M., 1914. Geography and vegetation of Northern Florida. *Fla. Geol. Surv. Ann. Rep.*, 6: 163–437.

Johnson, D., 1944. Mysterious craters of the Carolina Coast. *Am. J. Sci.*, 5: 247–255.

Laessele, A.M., 1958. The origin and successional relationships of sandhill vegetation and sand-pine scrub. *Ecol. Monogr.*, 28: 361–387.

Murray, G.E., 1961. *Geology of the Atlantic and Gulf Coastal Province of North America*. Harper, New York, N.Y., 692 pp.

Penfound, W.T., 1952. Southern swamps and marshes. *Bot. Rev.*, 18: 413–446.

Radford, A.E., Ahles, H.E. and Bell, C.R., 1968. *Manual of the Vascular Flora of the Carolinas*. University of North Carolina Press, Chapel Hill, N.C., 1183 pp.

Wells, B.W., 1942. Ecological problems of the southeastern United States coastal plain. *Bot. Rev.*, 8: 533–561.

Wells, B.W., 1946. Vegetation of Holly Shelter Wildlife Management Area. *N.C. Dep. Conserv. Dev, Bull.*, 2: 40 pp.

Wells, B.W. and Boyce, S.G., 1953. Carolina Bays: Additional data on their origin, age and history. *J. Elisha Mitchell Sci. Soc.*, 69: 119–141.

Wells, B.W. and Shunk, I.V., 1928. A southern upland grass-sedge bog. An ecological study. *N.C. Agric. Exp. Station. Tech. Bull.*, No. 32: 73 pp.

Wells, B.W. and Shunk, I.V., 1931. The vegetation and habitat factors of the coarser sands of the North Carolina coastal plain: An ecological study. *Ecol. Monogr.*, 1: 465–520.

Woodwell, G.M., 1956. *Phytosociology of Coastal Plain Weylands of the Carolinas*. Thesis, Duke University, Durham, N.C., 50 p. (unpublished).

Woodwell, G.M., 1958. Factors controlling growth of pond pine seedlings in organic soils of the Carolinas. *Ecol. Monogr.*, 28: 219–236.

Wright, A.H. and Wright, A.A., 1932. The habitats and composition of the vegetation of Okefinokee Swamp, Georgia. *Ecol. Monogr.*, 2: 109–232.

Chapter 19

THE PINE BARRENS OF NEW JERSEY[1]

S. LITTLE

INTRODUCTION

On the outer Coastal Plain of southern New Jersey is an area of about 520 000 ha that is known as the Pine Barrens (Fig. 19.1). The Pine Barrens have long attracted the attention of botanists, foresters, soil scientists, and other people interested in the biological sciences, because certain members of the flora and fauna are unique, and the composition and appearance of the vegetation here contrast markedly with that of the predominantly deciduous forests of the inner Coastal Plain and the Piedmont.

The term "Barrens" implies unsuitability for economic production of most agricultural crops — not an area barren of vegetation. In fact, on very little area of the Pine Barrens are shrubs the dominant vegetation. The floristic and morphological attributes of all these plant communities, however, fall within the definition of heathlands and related vegetation given in Chapter 1 of this volume. Gimingham (1972) defines heathland as areas in which dominant plants are dwarf shrubs, often evergreen and in the Ericaceae. Trees or tall shrubs occur rarely if at all, and the dwarf shrubs are often 10 to 20 or sometimes 30 cm or more tall. In this narrow European definition, relatively rare spots of heathland occur in the Barrens.

The Pine Barrens do include several members of the Ericaceae in shrubby understoreys of forest stands, and also include areas where trees less than 3.4 m tall are the dominant vegetation. On about 5000 ha where trees tend to remain below that height, evidence indicates that killing fires are the primary cause of the dwarf stature of the vegetation.

Stone (1911) estimated that 565 species of plants

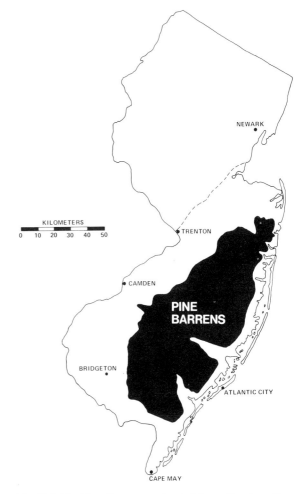

Fig. 19.1 The Pine Barrens are located in southeastern New Jersey, on the Coastal Plain (Stone, 1911). The northern boundary of the Coastal Plain extends northeastward from Trenton (Kümmel, 1940).

---

[1] Manuscript completed October, 1976.

occurred in the Barrens — including 179 obvious introductions from other districts. McCormick (1970) estimated the presently known vascular flora at 800 species, varieties, and forms. Two of these are believed to grow only in the New Jersey Pine Barrens: *Leiophyllum buxifolium* and *Breweria pickeringii* var. *caesariensis*. However, a variety of each plant does grow elsewhere (Fernald, 1950). Fourteen kinds of northern plants reach their southern limits or southernmost Coastal Plain limits in the Pine Barrens, and 109 kinds of southern plants reach their northern limits in the Barrens (McCormick, 1970). McCormick listed one salamander, three frogs, one lizard, and six snakes as also reaching their northern limit of distribution in the Pine Barrens.

Present-day vegetation of the Barrens reflects intensive use and misuse of the land during the 300 years since settlements by Europeans. Heavy cutting of wood products provided much of the material for early exports to England and the West Indies, as well as to nearby cities. Cook stated in 1857 that many forested swamps of Cape May County had been cut over twice, and some three times, since settlement. Until 1865 wood was the primary fuel for industries and homes, both in New Jersey and in nearby cities such as New York and Philadelphia. Forests of the Barrens were the source of fuel for local use; and because the area is near both Philadelphia and New York, and because much of the Barrens were near water suitable for transportation of wood products, they were also a major supplier of those cities.

Since 1865 the cutting has gradually decreased, but even in recent years the forests of the Barrens have supplied pine pulpwood, some fuel wood and lumber, and such white-cedar products as boat boards, paneling, rustic furniture, fence posts, boxes, and crates. By now, many sites have been subjected to at least five clear cuttings. In addition, large wildfires have occurred in the Barrens since the seventeenth century (Lutz, 1934; Little, 1950); and during one weekend of 1963 six wildfires burned about 65 400 ha (Banks and Little, 1964). Wildfires and heavy cuttings have combined to keep many upland forests on a 25- to 50-year rotation since 1800 (Little and Moore, 1945).

The area of Pine Barrens forests has shrunk through the years, initially through agricultural use of certain soils for vegetables and fruits, later through the use of some sites for cultivated cranberries and blueberries, and in recent years through residential use. The last is particularly noticeable on the edges of the Barrens, but is beginning to spread throughout, particularly along highways and waterways.

GEOLOGY, TOPOGRAPHY, AND SOILS

The Pine Barrens occur in the outer Coastal Plain, much on sands, gravels, and clays (to a minor extent) of the Cohansey Formation, deposited by the sea during the Miocene or Pliocene (Kümmel, 1940; Widmer, 1964). In places Beacon Hill gravel of the Pliocene caps the Cohansey Formation, especially on hills. During later periods, probably during interglacial stages of the Pleistocene, non-glacial deposits of fluvial, estuarine, or marine origin developed in the Pensauken, Bridgeton, and Cape May Formations (Kümmel, 1940).

Much of the Pine Barrens not only has low elevation, but lacks steep slopes. High points are generally less than 70 m.

Several soil series occur here: mostly Lakewood and Evesboro on excessively drained uplands, Downer and Woodmansie on well-drained uplands, Lakehurst and Klej on moderately well-drained sites, and Atsion and Berryland on poorly and very poorly drained sites. Stream valleys have shallow deposits of organic soils. Mineral soils were formerly classified as podzols or podzolic, and now as Entisols, Ultisols, and Spodosols (Markley, 1971).

Forest floors are mors of varying depths: on long unburned sites 6 to 8 cm in oak–pine uplands, 15 to 20 cm on poorly and very poorly drained sites (Burns, 1952; Little and Moore, 1953).

Physical composition of the soils varies. Sands often form 90 to 95% of upland A horizons (Burns, 1952). Gravel may form 9 to 36% of B and C horizons (Lutz, 1934). Clay contents may be 2 to 3% of A horizons, 6 to 12% of B and C horizons; silt may form 6 to 10% of such horizons, or in some soils silt and clay combined form only 2% of the same horizons (Lutz, 1934; Andresen, 1959).

Pine Barrens soils are acid and low in nutrients. The range in pH may be 3.5 to 5.5 in upland soils (Lutz, 1934; Burns, 1952; Andresen, 1959), while the average pH of organic soils is about 4.0 (Waksman et al., 1943). Nitrogen content may be 0.02 to 0.06% in A horizons and less in the other horizons, exchangeable calcium in A horizons 19 to 146 ppm, potassium 3 to 15 ppm (Lutz, 1934; Burns, 1952), and phosphorus 0.5 ppm (Andresen, 1959).

VEGETATION

Differences in soil-moisture conditions, in land use, and in history of disturbance have created great differences in the composition and productivity of vegetation in the Pine Barrens.

Upland sites

Successional stages

Stages of secondary succession unaffected by fire are relatively rare in the Pine Barrens, both because abandonment of upland fields has not been extensive in recent years and because fires have been very common. However, as in other areas, herbaceous vegetation dominates recently abandoned fields. In a 0.4 ha field, formerly an orchard and then plowed without further cultivation, McCormick and Buell (1957) found 25 species of grasses and sedges, 30 species of forbs, and 5 woody species five months after plowing. *Panicum capillare*, *P. virgatum*, and *Cyperus filiculmis* were the most important plants, but none formed more than 9% of the cover. Other authors, including Stone (1911), stress the dominance of other grasses, *Andropogon scoparius* and *A. virginicus*, in old fields before woody plants dominate the site.

Near most abandoned fields there is an adequate supply of pine seed, so pine seedlings become established, usually *Pinus rigida* or *P. echinata*. Hardwoods soon invade these stands; and, when pines are 38 years old, there may be more than 1900 hardwood individuals 1.4 cm or more in diameter at breast height (1.4 m) per hectare (Little, 1973). In the absence of fire and other disturbances that favor pines, they are replaced by a mixture of hardwoods in which *Quercus* (mostly *Q. velutina*, *Q. alba*, *Q. prinus*, and *Q. coccinea*) and *Carya* spp. are dominant (Little and Moore, 1949; Buell and Cantlon, 1950; Little, 1973).

Productivity of pine stands of seedling origin is markedly more than that of hardwoods in yield of merchantable wood products. Basal areas of 40-year-old stands of *P. echinata* or *P. rigida* may be twice that of hardwoods of similar age (37 versus 18 m^2 ha^{-1}). Merchantable volumes of stems to a minimum diameter of 10 cm may be about 92 m^3 ha^{-1} in stands predominantly of oaks, about 200 m^3 ha^{-1} in stands of *P. echinata* or *P. rigida* of that age (Moore, 1939; Moore and Waldron, 1940).

Fire-induced types

As a result of past cuttings — and especially of past fires and the absence of previous cultivation — most upland forests fall into two broad types; (1) oak–pine, in which stems of *Pinus rigida* and *P. echinata* grow in mixture with oaks, chiefly *Q. velutina*, *Q. alba*, *Q. prinus*, and *Q. coccinea*, the pines often forming an open overstorey over younger hardwood sprouts; and (2) pine–scrub oaks, in which the pines are typically *P. rigida* and the oaks are *Q. ilicifolia*, *Q. marilandica*, and *Q. prinoides* (Little, 1946).

Composition of the two types appears to be due to differences in fire history. *P. echinata* and the tree oaks (*Q. velutina*, *Q. alba*, etc.) do not bear viable seed until stems are twenty years old or more, while *P. rigida*, *Q. ilicifolia*, *Q. marilandica*, and *Q. prinoides* are precocious — the first two often producing viable seed on 3-year-old open-grown sprouts. Consequently, pine–scrub oak stands occupy areas that have been subjected to frequent fires of killing intensity over a long period (Little and Somes, 1964). These areas tend to be the drier soils, but there is closer correlation of this vegetation with frequent killing fires than with soil series. Oak–pine stands occur on Lakewood as well as other upland soils.

Oak–pine. Proportions of pine to oak and among the various species of oak vary. In some stands *P. echinata* is the predominant pine, in others *P. rigida*. Though *Q. velutina* is the most common oak, in some stands *Q. prinus*, *Q. coccinea*, or some other species may predominate (Stephenson, 1965; and personal observations). Basal area also varies appreciably, in older stands at least from 10 to 19 m^2 ha^{-1}; and pines can form as little as 15% of the basal area, or as much as nearly 100% in areas recently burned by wildfires — where for several

454   S. LITTLE

Fig. 19.2 Oak–pine stands unburned for many years. The typical cover of *Vaccinium vacillans*, *Gaylussacia baccata*, and *G. frondosa* is more conspicuous in summer (A) than in winter (B). Spots covered by mosses and lichens, or only by litter, frequently occur amidst the shrub cover. Pines in these stands are, of course, more noticeable in winter than in summer.

years resistant pines may make up nearly all the phanerophytes taller than 10 m.

Open space in the tree canopy varies at least from 13 to 30% (Stephenson, 1965) — much more just after a severe fire or cutting. Light intensity measured at 2 m in a stand having 16% open space in the tree canopy was 29% of full sunlight (Buell and Cantlon, 1950).

Open space in the shrub layer may vary from 31 to 86% (Stephenson, 1965). In many stands ericaceous shrubs form 90% of the cover (Fig. 19.2). *Gaylussacia baccata* and *Vaccinium vacillans* are the most common shrubs. *Gaylussacia frondosa*, *Gaultheria procumbens*, *Kalmia latifolia*, and *K. angustifolia* are other members of the Ericaceae occurring in these stands; but usually they form little of the cover, and sometimes they are absent. Other shrubs of variable occurrence include *Quercus ilicifolia*, *Q. prinoides*, *Myrica pensyl-*

vanica, *Comptonia peregrina*, *Smilax glauca*, and *S. rotundifolia*. Only *Q. ilicifolia* is often a tall (> 2 m) shrub. *K. latifolia*, *G. frondosa*, and *Myrica* are often mid-height (1–2 m) shrubs, while *G. baccata*, *K. angustifolia*, *C. peregrina*, *Q. prinoides*, and especially *V. vacillans* are low (< 1 m) shrubs. *Smilax* is a woody vine, *Gaultheria* a chamaephyte (< 25 cm tall).

In typical oak–pine stands, herbaceous species are rare, widely scattered, and often depauperate. In one area the most abundant species were *Andropogon scoparius*, *Panicum commonsianum*, *Pteridium aquilinum*, *Tephrosia virginiana*, *Carex pensylvanica*, *Cypripedium acaule*, *Panicum virgatum*, *Baptisia tinctoria*, and *Stipa avenacea* (Buell and Cantlon, 1950). In uncut stands having a good cover of ericaceous shrubs, herbs contribute very little to the cover — often less than 0.5% (Buell and Cantlon, 1953).

Mosses and lichens may be absent or cover up to 48% of the ground (Moul and Buell, 1955; Bernard, 1963). In an area where most of the overstorey trees were 55 years old or older, the cover of mosses and lichens was 7% in uncut controls, 30% in uncut plots subject to frequently prescribed burns, and 44% in frequently burned plots cut four years previously. Lichens were dominant in recently cut areas, mosses in the uncut plots (Buell and Cantlon, 1953). *Dicranum scoparium, Ceratodon purpureus,* and *Polytrichum juniperinum* formed 98% of the moss cover in one area, while lichens were *Cladonia* spp. and *Parmelia* sp. (Moul and Buell, 1955).

Light winter fires, when properly used in prescribed burning, do not reduce growth nor cause significant mortality of the oak–pine overstorey (Little, 1946; Somes and Moorhead, 1950), but they do change understorey and forest floor conditions — increasing the proportion of bare ground and the cover of herbs, mosses, and lichens while decreasing shrub cover. Cover of *Gaylussacia* is greatly reduced, and that of *Vaccinium vacillans* to a far lesser extent, by several prescribed fires or one severe wildfire (Buell and Cantlon, 1953; Laycock, 1967). The rhizome of *V. vacillans* almost always grows in mineral soil, and its roots penetrate to depths of 64 to 120 cm, whereas rhizome and roots of *G. baccata* are much shallower — so fire naturally reduces the cover of *G. baccata* far more (Laycock, 1967). Herbaceous cover often increases far more than the 5% recorded in uncut stands on Lakewood soil (Buell and Cantlon, 1953). Where light is favorable, as along roadsides or after thinning or heavy cutting of the overstorey, relatively high cover (15 to 50%, almost 100% in spots) develops — especially on Evesboro or heavier soils. *Carex pensylvanica* may be the chief component, but on Evesboro or heavier soils forbs are common: *Solidago* spp; *Asclepias tuberosa; Gerardia* spp.; *Liatris graminifolia;* legumes such as *Tephrosia virginiana, Baptisia tinctoria, Cassia nictitans,* and several species of *Lespedeza;* and others (Little, 1974).

Associated with the changed understorey of herbaceous plants is a change in the amount of pine regeneration. *Pinus rigida* and *P. echinata* need a seedbed of mineral soil or a very thin forest floor for successful establishment of seedlings (Little and Moore, 1949). Most seedlings of both species

develop basal crooks that bring dormant buds against or into mineral soil on upland sites; thus protected, the buds produce sprouts if the stems are killed by fire (Little and Somes, 1956; Little and Mergen, 1966). Improved seedbed and increased light favor the establishment of pine regeneration before harvest cutting, as well as after that and the discontinuance of prescribed burning, so the number of pines in young stands tends to increase with the number of burns before harvest cutting (Little and Moore, 1950; Little, 1964). In areas subject to frequent prescribed burns, conspicuous understoreys of pine regeneration develop near roads.

In oak–pine stands that are beyond the sapling stage and have not been subject to prescribed burns, the life-form spectra are similar to those of the oak stand studied by Archard and Buell (1954): phanerophytes formed 99% of the cover; chamaephytes, hemicryptophytes, geophytes, and therophytes were absent or formed less than 1%.

Biomass values for the uplands of the New Jersey Pine Barrens are not available, but for a comparable oak–pine stand 43 years old, Woodwell et al. (1975) gave the dry-matter biomass above and below ground as 6560 and 3630 g m^{-2}, and the net annual productivity as 859 and 337 g m^{-2}. In comparing their Long Island (New York) stand with certain hardwood forests of New Hampshire, Belgium, and France, they found it had the lowest nitrogen uptake and concentration and was also low in uptake and concentration of calcium, phosphorous, and potassium. In the New Jersey Pine Barrens, oak–pine stands grow only about a third as much merchantable wood per hectare as pine stands of seedling origin on comparable sites (Moore, 1939).

Pine–scrub oak. Oak–pine stands grade into stands dominated by *Pinus rigida* and scrub oaks. Among the scrub oaks, *Q. prinoides* is usually the least common and often absent, while the relative importance of *Q. marilandica* and *Q. ilicifolia* varies. Personal observations and Lutz's (1934) data suggest that one species sometimes occurs alone; in many areas both occur; but in more than half of the areas *Q. ilicifolia* is more abundant than *Q. marilandica.*

Basal area of pine–scrub oak stands varies from little to at least as much as in oak–pine stands. One stand had 13.8 m² ha⁻¹ of basal area, a light intensity at the 2 m level of 68% of full sunlight, and open space in the tree canopy of 58% (Buell and Cantlon, 1950). In one pine–scrub oak stand, the cover of phanerophytes was 93%; hemicryptophytes 7%; chamaephytes, geophytes, and therophytes were almost absent (Archard and Buell, 1954).

In the shrub layer open space may vary from 0 to 50%, but is often 24 to 30% (Buell and Cantlon, 1950; Stephenson, 1965). Tall and mid-height shrubs, such as *Quercus ilicifolia*, *Q. prinoides*, *Kalmia latifolia*, and *Gaylussacia frondosa*, may have covers of 0 to 100%. Low shrubs such as *G. baccata* and *Vaccinium vacillans* and chamaephytes such as *Hudsonia ericoides* and *Arctostaphylos uva-ursi* may have covers of 10 to 90% (Andresen, 1959; Stephenson, 1965).

Herbaceous plants are often rarer in the pine–scrub oak areas than in oak–pine stands, but the same species may occur (Buell and Cantlon, 1950; and personal observations). Mosses and lichens of pine–scrub oak stands are also similar to the species recorded in oak–pine, their cover often ranging from 0 to 14% (Bernard, 1963) but in spots

to 50%. Lutz (1934) found 1 fern, 6 grasses, 3 sedges, 22 forbs, 3 mosses and 14 lichen species in his plots within this type. As in oak–pine stands, prescribed winter fires may reduce the shrub cover and increase the cover of herbs, mosses, and lichens; but, because the shrub oaks sprout vigorously, changes in the shrub and other layers are usually not so marked as in oak–pine stands.

Productivity of merchantable wood is low — partly because of the openness of canopies and partly because of past wildfires that deformed tree crowns or killed trees before they became merchantable, and in many areas created stands of slow-growing sprouts from old stools. Even where trees reach merchantable size, the yield per hectare during a 40-year period is about 10% of that from pine stands of seedling origin (Little, 1964).

Shrublands

The area of the Pine Barrens about which most has been published is called the "Plains"; where vegetation tends to remain below 3.4 m tall (Fig. 19.3). McCormick and Buell (1968) defined Plains vegetation as characterized by:

Fig. 19.3. This Plains area has sprouts about five years old. Dominant species are *Pinus rigida*, *Quercus marilandica*, *Q. ilicifolia*.

(1) Predominance of a closed-cone[1] race of *Pinus rigida* and nearly complete absence of *P. echinata* and all oaks except *Q. marilandica* and *Q. ilicifolia*.

(2) Predominance of multiple sprouts from large, irregularly shaped stools that are much older than the current stems.

(3) Large number of sprout clumps, 3400 to 13 600 ha^{-1}, of which 25 to 65% are *P. rigida*, and few seedlings.

(4) Trees usually less than 3.4m tall, typically 0.6 to 1.5m tall.

(5) *Gaylussacia baccata* and *Vaccinium vacillans* the most abundant and evenly distributed shrubs. *Kalmia angustifolia*, *K. latifolia*, *Comptonia peregrina*, and *Leiophyllum buxifolium* are widely distributed but only locally abundant. *Corema conradii* is prominant in places. Such chamaephytes as *Gaultheria procumbens*, *Arctostaphylos uva-ursi*, and *Epigaea repens* are widely distributed, but contribute little to the bulk of the vegetation. At present the Plains vegetation is considered to be just an extreme form of the pitch pine–scrub oak type (Little and Somes, 1964; McCormick and Buell, 1968).

Estimates of the area in Plains vegetation have varied greatly through the years: 9110 ha (Harshberger, 1916), 4969 ha (Gaskill, 1921), 6070 ha (Moore, 1939), and 4945 ha (McCormick and Buell, 1968). Differences seem due partly to inconsistencies in definition and partly to changes produced by fires and by regrowth in the intervals between estimates (McCormick and Buell, 1968).

The cause of the low stature of Plains vegetation has been the subject of many hypotheses: poverty of the site (Smock, 1892); hard pan (Gifford, 1895); soil organisms (Harshberger, 1916); such climatic factors as strong winds (Harshberger, 1916; Seifriz, 1953); or a high content of soluble aluminum in the soil, causing toxicity (Joffe and Watson, 1933; Tedrow, 1952). However, those hypotheses were disproved by Lutz (1934) and Andresen (1959). For a more complete review of hypotheses and publications, see McCormick and Buell (1968).

The primary cause of the low stature of the vegetation appears to be frequent wildfires, possibly at 8-year intervals (Lutz, 1934; Andresen, 1959) over a long period, but there are secondary factors. One is the severe competition between and within sprout clumps. Density varies: (1) in a 27-

year-old stand of *P. rigida*, *Q. ilicifolia*, and *Q. marilandica*, 39 520 stems ha^{-1} in 13 300 sprout clumps, which included 3160 *P. rigida* stems > 1.4 cm in diameter (at 1.4 m height) (Little and Somes, 1964); (2) in a 9-year-old stand 16 730 stems ha^{-1} in 2200 sprout clumps of *P. rigida* and 1040 clumps of *Q. marilandica* (McCormick and Buell, 1968). Severe competition within sprout clumps is also typical: *P. rigida* may have 3 to 249 sprouts per stool, with an average of 50, a year after a killing fire (Little, 1946). Another factor is the age of stools. Many may be forty to sixty years old when the last set of sprouts started (Little and Somes, 1964), and Andresen (1959) showed that sprouts from younger stools would be more upright and more vigorous.

Recent evidence suggests that genetics may be a secondary factor. In plots converted to local *P. rigida* seedlings, growth of trees appeared to be initially normal, but, once above 2 m in height, stems became noticeably crooked. When stems were about fifteen years old and maximum heights were 4.3 m, many of the trees had flat tops with no well-defined terminal shoot. In the same study area, planted stems of *P. strobus* reached a maximum height of 5.5m in eleven years, a growth rate that would produce trees 18 to 24m tall in fifty years (Little, 1972). Good and Good (1975) collected cones from Barrens and Plains areas (three to six trees only), and found the Plains progeny had a larger percentage of poor, shrubby forms and was more precocious in cone production at the age of 5.5 years than the Barrens progeny. Thus, current evidence is that the fire history has favoured a race of *P. rigida* that has predominantly closed cones and poor form, is a vigorous sprouter, and matures early.

The characteristic of closed cones is not universal in Plains areas, nor is it limited thereto. Rare trees in our Plains plots had open cones. Closed cones are borne by *P. rigida* stems up to at least 20m tall in areas at least 12km outside the Plains areas, but within the Barrens. Ledig and Fryer (1972) found closed cones on occasional trees of *P. rigida* in Virginia, West Virginia, and Steuben County, New York; on 50% of sampled trees in one stand on Long Island, New York; and on all six

---

[1] These closed, also called serotinous, cones require extra heat, as from a fire, to open.

trees sampled in one stand of central Pennsylvania. The closed-cone characteristic seems to vary, both locally and throughout the range of *P. rigida*, with the frequency of past severe fires.

Heathlands

Patches of heathland occur on upland sites in the Barrens, mostly along highways or airport runways (Fig. 19.4), in areas disturbed by bombing, and sometimes in abandoned borrow pits. Some of these spots may be dominated by *Vaccinium vacillans* and *Gaylussacia baccata*, but more common on very dry soils are dense patches of *Arctostaphylos uva-ursi* or open patches dominated mostly by *Hudsonia ericoides*. Any such spots of heathland apparently owe their existence to disturbances that eliminated trees and tall shrubs and favored the lower-growing plants (mowing road edges, for instance).

Lowland sites

Successional stages

Secondary succession on lowland sites, like the uplands, tends to be from coniferous to hardwood stands. On moderately well to very poorly drained soils (the pine lowlands), the trend is from relatively pure overstoreys of *Pinus rigida*, probably to *Quercus* species on some moderately well-drained sites, but on other such sites and on poorly and very poorly drained soils to *Acer rubrum*, *Nyssa sylvatica*, *Magnolia virginiana*, and *Ilex opaca* (Little, 1946, 1950, 1964). In swamps on the organic soils *Chamaecyparis thyoides* is the most common invader of clearings, but *Pinus rigida*, *Betula populifolia*, and *Acer rubrum* may also occur. Again the most shade-tolerant species are *Acer*, *Nyssa*, *Magnolia*, and *Ilex* (Little, 1950).

Deep-burning fires provide suitable conditions

Fig. 19.4. Heathland along an airport runway. Here tree growth was eliminated, and the area is mowed periodically. *Arctostaphylos uva-ursi* occurs in the center foreground. Shrubs immediately back of the shovel are first-year sprouts of *Vaccinium vacillans* and *Gaylussacia baccata*.

for the establishment of *P. rigida* seedlings on poorly drained and very poorly drained sites. Such fires eliminate much of the competing vegetation and favor the establishment of large numbers of *P. rigida* seedlings — about 17 000 per ha in one study (Little and Moore, 1953, Little, 1964). The effectiveness of such fires seems due to killing of basal dormant buds, as well as rhizomes, in the organic mat. Rhizomes of such shrubs as *Amelanchier, Clethra, Gaylussacia, Ilex glabra, Kalmia, Lyonia ligustrina,* and *Pyrus melanocarpa* occur in this mat (Laycock, 1967).

Subclimax stands of *Chamaecyparis* have been favored by several types of disturbances: clear-cuttings, certain fires, extensive windthrow, and flooding. Effectiveness of such disturbances in favoring the establishment of new stands of *Chamaecyparis* has varied greatly. Partial cuttings of *Chamaecyparis* overstoreys favor the development of dense understoreys of shrubs and hardwoods, thus hastening succession (Little, 1950).

Pine lowlands

Most present forests on the moderately well to very poorly drained soils are dominated by *Pinus rigida*. Many of the stems are relatively young slow-growing sprouts, probably from old stools. In addition, many existing stems of *P. rigida* have been deformed by wildfires that killed the foliage and many twigs, forcing the development of new branch and terminal shoots from dormant buds (Little, 1959, 1964). An impression is created that on these sites *P. rigida* reaches heights of only 5 to 10 m. However, in stands unburned by wildfires since the establishment of seedlings, relatively large trees can be found: up to 49 cm in diameter at breast height and 31 m tall. In part because sprouts predominate, the productivity of pine lowlands is unknown. Cover of the tree canopy varies greatly, but is often between 30 and 60%.

Many shrub species — more than 20 shrubs and woody vines (McCormick, 1970) — occur on these sites, but the composition and height of the shrub layer vary with soil-moisture conditions. On the drier spots of moderately well-drained soils, *Quercus ilicifolia, Hudsonia ericoides, Carex* spp., *Andropogon* spp., *Xerophyllum asphodeloides,* and *Pteridium aquilinum* dominate the cover; but these grade rapidly into a dense cover of such shrubs as *Kalmia angustifolia, Lyonia mariana,* and

Chamaedaphne calyculata. Also common are *Gaylussacia frondosa, G. baccata, G. dumosa, Pyrus melanocarpa, Ilex glabra,* and along the edge or in open spots *Leiophyllum buxifolium.* Scattered throughout, but dominating the understorey of very poorly drained sites, are tall shrubs such as *Clethra alnifolia, Vaccinium corymbosum, V. atrococcum, Rhododendron viscosum, Lyonia ligustrina* (Little and Moore, 1953; McCormick, 1970). The shrub layer on poorly drained sites may be about 1 m tall, on very poorly drained sites 2 to 3 m; and, especially on the poorly drained soils and under somewhat open tree canopies of the very poorly drained sites, the cover is almost complete. Above-ground biomass values of the shrub layer may be similar to that of certain North Carolina stands (Wendel et al., 1962): from about 500 g m^{-2} in shrubs 1 m tall to about 1500 where shrubs are 2 to 3 m tall.

In some lowland sites the herbaceous layer is well developed as the result of past disturbances. The most typical species are *Xerophyllum asphodeloides* and *Pteridium aquilinum*; but, where shrubs are scarce or absent, such plants as *Polygala* spp., *Drosera* spp., *Rhexia virginica,* and certain orchids may be common on the poorly and very poorly drained portions. Mosses and lichens found in the drier portions are the same as occur on upland sites, and in wet portions the same as occur in the swamps. *Sphagnum* species are the most conspicuous mosses, forming nearly solid mats in wet spots.

Forested swamps

Forested swamps vary greatly in the composition and density of the various layers of vegetation. Unthinned *Chamaecyparis* stands fifty to sixty years old usually have a relatively complete crown canopy, sparse intermediate layers, but a good floor cover of mosses, with some liverworts and herbaceous plants (Fig. 19.5). In contrast, hardwood stands often have a more open canopy above 8 m, more cover of hardwood trees and shrubs in the intermediate layers, and relatively few herbaceous plants. Extent of the area in hollows (where water usually stands) varies greatly; and where the hollows are large, they may create holes in the various layers.

Fig. 19.5. A. A stand of *Chamaecyparis thyoides* on a swamp site. B. Interior of a well-stocked stand mostly of *Chamaecyparis* about 55 years old. Note the sparse cover in the layers between the canopy and 8 cm above the swamp surface.

Differences between the two types of relatively mature stands are shown by the following data (Little, 1947):

	Chamaecyparis	*Acer–Nyssa–Magnolia*
Cover (%):		
> 7.6 m	80–85	65–70
4.6–7.6 m	5	15–20
1.84–4.5 m	5–20	30–40
8–183 cm	1–14	17–18
< 8 cm	22–44	23–33
Light intensity (% of full sunlight) at:		
1.2 m	7–9	3–6
9 cm	5–6	1

Chamaecyparis stands are composed of many stems per hectare — regenerating stands may have up to 145 400 *Chamaecyparis* and 15 700 hardwood stems greater than 30 cm in height, sapling stands up to 20 000 stems above 1.8 m, and 65- to 80-year-old stands may have 2000 to 4000 overstorey stems (Korstian and Brush, 1931; Little, 1947; Bamford and Little, 1960). Basal areas are high: 58 to 66m^2

ha^{-1} in unthinned stands 60 years old (Little, 1947; Bamford and Little, 1960). Because of the density, volumes are also high: 60-year-old stands with overstorey stems 17m tall contain 497m^3 ha^{-1} (Korstian and Brush, 1931).

In contrast, in hardwood swamps the basal area of trees (26–35 m^2 ha^{-1}) and the overstorey cover are relatively low, and cover in intermediate layers is high. The trees tend to be of all sizes and ages (to over 100 years), and the understorey of tall shrubs is very prominent (Little, 1947, 1950).

In 16 plots sampling different forest conditions on swamp sites, 25 species of shrubs, 11 vines and subshrubs, 32 herbs, 5 ferns, 1 club moss, 5 liverworts, 18 mosses, and 21 lichen species were found. Shrubs occurring in 10 or more plots were *Clethra alnifolia*, *Vaccinium corymbosum*, *Rhododendron viscosum*, *Gaylussacia frondosa*, and *Leucothoe racemosa*. Two ferns, *Woodwardia virginica* and *Osmunda cinnamomea*, were found in 7 or 8 plots, and the most frequent herb was *Carex collinsii* in 5 plots. Certain herbs such as *Pogonia ophioglossoides*, *Sarracenia purpurea*, *Drosera rotundifolia*, *Trientalis borealis*, and *Bartonia paniculata* were found in *Chamaecyparis* stands, but not in the hardwood stands studied. Liverworts occurring in 12 to 15 plots were *Calypogeia trichomanis*, *Microlepidozia sylvatica*, and *Odontoschisma prostratum*. *Sphagnum* species tend to dominate many spots, and two of the four species found, *S. magellanicum* and *S. recurvum*, occurred in 11 or 12 plots. *Dicranum flagellare* was also found in 11 plots, but no lichens occurred in more than 7 plots. The most common were *Cladonia calycantha* and *C. incrassata*, occurring on the lower boles of *Chamaecyparis* (Little, 1951). In 60-year-old *Chamaecyparis* stands, the density of *Chamaecyparis* seedlings 1 to 3 years old ranged from 700000 to 2460000 ha^{-1} (Little, 1950).

Grasslands and shrublands

In the pine lowlands and swamps there are many spots, mostly 0.5 ha or less in size, that lack a tree canopy. These are ponds, grasslands or shrublands; and they owe their existence to past disturbances — mining of bog ore, deep-burning fires, creation and abandonment of cranberry bogs and reservoirs, and removal of "turf" (shrubs and roots) to stabilize dam or road slopes (McCormick, 1970; Little, 1974).

Sedge and grass marshes occur in some openings along streams and occupy some abandoned cranberry bogs. Stone (1911) described such areas as having *Danthonia epilis*, *Panicum ensifolium*, *Rhynchospora* spp., *Scleria minor*, *Sarracenia purpurea*, *Eriocaulon* spp., *Narthecium americanum*, *Tofieldia racemosa*, *Lophiola americana*, *Calopogon pulchellus*, *Pogonia ophioglossoides*, and beds of *Sphagnum*.

Shrublands occur in the upper portions of swamps or along the edges and in depressions within the pine lowlands. In most such areas *Chamaedaphne calyculata* is the principal shrub, often forming dense pure cover (Fig. 19.6). Clumps of *Vaccinium*, mostly *V. corymbosum*, may occur, especially in the drier portions. In places the cover of *Chamaedaphne*, often about 1m tall, encircles an open pond or grassy marsh. Under the *Chamaedaphne* is a dense mat of *Sphagnum* species. Such shrublands seem to be due to fires that burned the organic mat deeply enough so that water generally stands on the surface (Little, 1950; McCormick, 1970), providing conditions suitable for *Chamaedaphne*, which Rigg (1940) and Wright (1941) found to be the first shrub to invade quaking bogs. Observations indicate that *Chamaedaphne* may dominate such sites for fifty years; but when the *Sphagnum* mat is high enough to provide suitable seedbeds for *Chamaecyparis*, *Acer*, and other tree species, the *Chamaedaphne* gradually yields its dominance.

CONCLUSION

Areas of somewhat similar vegetation occur elsewhere in the eastern United States. Stands of *Pinus rigida* and associated *Quercus* species occur on Long Island (New York), in southeastern Massachusetts, and in several other states either on sand plains or on the shallow soils of ridges and upper slopes. Stands of *Chamaecyparis* and associated species occur northward and southward, mostly in sandy portions of coastal sections.

Stands similar to Pine Barrens lowlands dominated by *P. rigida* are not so common, although those of *Pinus serotina* in eastern North Carolina have strong resemblances. However, the Pine Barrens of New Jersey are unique both in their blend of southern and northern species and in their relatively extensive area.

Fig. 19.6. *Chamaedaphne calyculata* predominant in a narrow swamp, probably because a deep-burning fire consumed enough of the organic soil so that water usually stands on the surface. Tall shrubs near the edge are mostly *Vaccinium corymbosum*.

The Pine Barrens occur on soils characterized by high contents of coarse sand and gravel and by low contents of nutrients. Although average amounts of precipitation are adequate, available-water capacity of the soils is low, and vegetation especially of uplands is periodically subjected to severe drought stress. Soils, climate, heavy use and misuse since the seventeenth century, and especially the severe fires have shaped the present composition and appearance of Pine Barrens vegetation. However, even though members of the Ericaceae dominate shrub composition, and even though tree overstoreys often appear scrubby, little of the Barrens qualifies today as heathland and shrubland. Existing areas of heathland and shrubland do show that their extent could be greatly enlarged through expansion of severe treatments similar to those creating the present areas.

REFERENCES

Andresen, J.W., 1959. A study of pseudo-nanism in *Pinus rigida* Mill. *Ecol. Monogr.*, 29: 309–332.

Archard, H.O. and Buell, M.F., 1954. Life-form spectra of four New Jersey pitch pine communities. *Torrey Bot. Club Bull.*, 81: 169–175.

Bamford, G.T. and Little, S., 1960. Effects of low thinning in Atlantic white-cedar stands. *U.S.Dep. Agric. For. Serv. Northeast. For. Exp. Station. Res. Note* 104: 4 pp.

Banks, W.G. and Little, S., 1964. The forest fires of April 1963 in New Jersey can point the way to better protection and management. *Proc. Soc. Am. For.*, 1963: 140–144.

Bernard, J.M., 1963. Forest floor moisture capacity of the New Jersey Pine Barrens. *Ecology*, 44: 574–576.

Buell, M.F. and Cantlon, J.E., 1950. A study of two communities of the New Jersey Pine Barrens and a comparison of methods. *Ecology*, 31: 567–586.

Buell, M.F. and Cantlon, J.E., 1953. Effects of prescribed burning on ground cover in the New Jersey pine region. *Ecology*, 34: 520–528.

Burns, P.Y., 1952. Effect of fire on forest soils in the Pine Barren region of New Jersey. *Yale Univ. School For. Bull.*, 57: 50 pp.

Cook, G.H., 1857. *Geology of the County of Cape May, State of Jersey.* N.J. Geol. Surv., 208 pp.

Fernald, M.L., 1950. *Gray's Manual of Botany.* American Book Co., New York, N.Y., 8th ed., 1632 pp.

Gaskill, A., 1921. Report of the State Forester. *N.J. Dep. Conserv. Dev. Annu. Rep.*, 1921: 53–70.

Gifford, J., 1895, A preliminary report on the forest conditions of South Jersey. *N.J. Geol. Surv. Annu. Rep. State Geol.*, 1894: 245–286.

Gimingham, C.H., 1972. *Ecology of Heathlands.* Chapman and Hall, London, 266 pp.

Good, R.E. and Good, N.F., 1975. Growth characteristics of two populations of *Pinus rigida* Mill. from the Pine Barrens of New Jersey. *Ecology*, 56: 1215–1220.

Harshberger, J.W., 1916. *The Vegetation of the New Jersey Pine-Barrens.* Christopher Sower Co., Philadelphia, Pa. 329 pp.

Joffe, J.S. and Watson, C.W., 1933. Soil profile studies: V. Mature podzols. *Soil Sci.*, 35: 313–329.

Korstian, C.F. and Brush, W.D. 1931. Southern white cedar. *U.S. Dep. Agric. Tech. Bull.*, 251:75 pp.

Kümmel, H.B., 1940. The geology of New Jersey, *N.J. Dep. Conserv. Dev. Geol. Ser. Bull.*, 50: 203 pp.

Laycock, W.A., 1967. Distribution of roots and rhizomes in different soil types in the Pine Barrens of New Jersey. *U.S. Geol. Surv. Prof. Pap.*, 563–C: 29 pp.

Ledig, F.T. and Fryer, J.H., 1972. A pocket of variability in *Pinus rigida. Evolution*, 26: 259–266.

Little, S., 1946. The effects of forest fires on the stand history of New Jersey's pine region. *U.S. Dep. Agric. For. Serv. Northeast. For. Exp. Station For. Manage. Pap.*, 2: 43 pp.

Little, S., 1947. *Ecology and Silverculture of Whitecedar and Associated Hardwoods in Southern New Jersey.* Thesis, Yale Univ, New Haven, Conn., 277 pp.

Little, S., 1950. Ecology and silviculture of whitecedar and associated hardwoods in southern New Jersey. *Yale Univ. School For. Bull.*, 56: 103 pp.

Little, S., 1951. Observations on the minor vegetation of the Pine Barren swamps in southern New Jersey. *Torrey Bot, Club Bull.*, 78: 153–160.

Little, S., 1959. Silvical characteristics of pitch pine. *U.S. Dep. Agric. For. Serv. Northeast. For. Exp. Station, Station Pap.*, 119: 22 pp.

Little, S., 1964. Fire ecology and forest management in the New Jersey pine region. *Proc. Tall Timbers Fire Ecol. Conf.*, 3: 34–59.

Little, S., 1972. Growth of planted white pines and pitch seedlings in a South Jersey plains area. *N.J. Acad. Sci. Bull.*, 17(2): 18–23.

Little, S, 1973. Eighteen-year changes in the composition of a stand of *Pinus echinata* and *P. rigida* in southern New Jersey. *Torrey Bot. Club Bull.*, 100: 94–102.

Little, S., 1974. Wildflowers of the Pine Barrens and their niche requirements. *N.J. Outdoors*, 1(3): 16–18.

Little, S. and Mergen, F., 1966. External and internal changes associated with basal-crook formation in pitch and shortleaf pines. *For. Sci.*, 12: 268–275.

Little, S. and Moore, E.B., 1945. Controlled burning in South Jersey's oak-pine stands. *J. For.*, 43: 499–506.

Little, S. and Moore, E.B., 1949. The ecological role of prescribed burns in the pine–oak forests of southern New Jersey. *Ecology*, 30: 223–233.

Little, S. and Moore, E.B., 1950. Effects of prescribed burns and shelterwood cutting on reproduction of shortleaf and pitch pine. *U.S. Dep. Agric. For. Serv. Northeast. For. Exp. Station, Station Pap.*, 35: 11 pp.

Little, S. and Moore, E.B., 1953. Severe burning treatment tested on lowland pine sites. *U.S. Dep. Agric. For. Serv. Northeast. For. Exp. Station, Station Pap.*, 64: 11 pp.

Little, S. and Somes, H.A., 1956. Buds enable pitch and shortleaf pines to recover from injury. *U.S. Dep. Agric. For. Serv. Northeast. For. Exp. Station, Station Pap.*, 81: 14 pp.

Little, S. and Somes, H.A., 1964. Releasing pitch pine sprouts from old stools ineffective. *J.For.*, 62: 23–26.

Lutz, H.J., 1934. Ecological relations in the pitch pine plains of southern New Jersey. *Yale Univ. School For. Bull.*, 38: 80 pp.

McCormick, J., 1970. The Pine Barrens: a preliminary ecological inventory. *N.J. State Mus. Rep.*, 2: 100 pp.

McCormick, J. and Buell, M.F., 1957. Natural revegetation of a plowed field in the New Jersey Pine Barrens. *Bot. Gaz.*, 118: 261–264.

McCormick, J. and Buell, M.F., 1968. The Plains: pygmy forests of the New Jersey Pine Barrens, a review and annotated bibliography. *N.J. Acad. Sci. Bull.*, 13(1): 20–34.

Markley, M.L., 1971. *Soil Survey of Burlington County, New Jersey.* U.S. Dep. Agric. Soil Conserv. Serv., Washington, D.C., 120 pp.

Moore, E.B., 1939. *Forest Management in New Jersey.* N.J. Dep. Conserv. Dev., 55 pp.

Moore, E.B. and Waldron, A.F., 1940. A comparison of the growth of oak and pine in southern New Jersey. *N.J. Dep. Conserv. Dev. Div. For. Parks Tech. Note.*, 10: 6 pp.

Moul, E.T. and Buell, M.F., 1955. Moss cover and rainfall interception in frequently burned sites in the New Jersey Pine Barrens. *Torrey Bot. Club Bull.*, 82: 155–162.

Rigg, G.B., 1940. Comparisons of the development of some sphagnum bogs of the Atlantic coast, the interior, and the Pacific coast. *Am. J. Bot.*, 27: 1–14.

Seifriz, W., 1953. The oecology of thicket formation. *Vegetatio*, 4: 155–164.

Smock, J.C., 1892. Geological work in the southern part of the State. *N.J. Geol. Surv. Annu. Rep. State Geol.*, 1891: 4–10.

Somes, H.A. and Moorhead, G.R., 1950. Prescribed burning does not reduce yield from oak-pine stands of southern New Jersey. *U.S. Dep. Agric. For. Serv. Northeast. For. Exp. Station, Station Pap.*, 36: 19 pp.

Stephenson, S.N., 1965. Vegetation change in the Pine Barrens of New Jersey. *Torrey Bot. Club Bull.*, 92: 102–114.

Stone, W., 1911. The plants of southern New Jersey with especial reference to the flora of the Pine Barrens and the geographic distribution of the species. *N.J. State Mus. Annu. Rep.*, 1910: 21–828.

Tedrow, J.C.F., 1952. Soil conditions in the Pine Barrens of New Jersey. *Bartonia*, 26: 28–35.

Waksman, S.A., Schulhoff, H., Hickman, C.A., Cordon, T.C.

and Stevens, S.C., 1943. The peats of New Jersey and their utilization. *N.J. Dep. Conserv. Dev. Geol. Ser. Bull.*, 55 (part B): 278 pp.

Wendel, G.W., Storey, T.G. and Byram, G.M., 1962. Forest fuels on organic and associated soils in the Coastal Plain of North Carolina. *U.S. Dep. Agric. For. Serv. Southeast. For. Exp. Station, Station Pap.* 144: 46 pp.

Widmer, K., 1964. *The Geology and Geography of New Jersey.* D Van Nostrand, New York, N.Y., 193 pp.

Woodwell, G.M., Whittaker, R.H. and Houghton, R.A., 1975. Nutrient concentrations in plants in the Brookhaven oak–pine forest. *Ecology*, 56: 318–332.

Wright, K.E., 1941. The great swamp. *Torreya*, 41: 145–150.

Chapter 20

CALIFORNIAN COASTAL FOREST HEATHLANDS[1]

WALTER E. WESTMAN

INTRODUCTION

Sandy, acid, poorly drained soils support a range of heath species along the west coast of North America. Whittaker has provided a brief description of the coastal heathlands of Oregon and Washington, and of the pygmy conifer forests of central California, in Chapter 17. In the present chapter, I discuss the nature of the heathland vegetation along the central Californian coast in greater detail, focusing in particular on the role of soil chemistry in influencing the heathland patterns.

ECOSYSTEM STRUCTURE

A diagrammatic cross-section of the vegetation and edaphic pattern across a belt of coast 4 km wide at Mendocino, California, is illustrated in Fig. 20.1, with accompanying details in Table 20.1. Along 40 km of coastline centred on Mendocino, five wave-cut graywacke terraces uplifted during the early to mid-Pleistocene (120 000–1 000 000 years B.P.; Gardner, 1967) remain in the flat, terraced position in which they were found. Except for small areas near Santa Cruz, Monterey and Santa Barbara (Wahrhaftig and Birman, 1965; Gardner, 1967), the remaining coastal terraces of California have been tilted, eroded and obscured by slip faulting and other crustal movements. The sandstone terrace soils are slowly-draining, a feature aggravated by the formation of a B_{ir} ironpan horizon approximately 30 cm below the surface. The podzolization on the coastal terraces has proceeded to a greater degree than on other landforms in the region, and these terraces (with

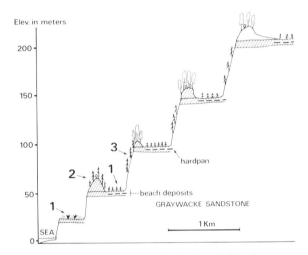

Fig. 20.1. Coastal land system, Mendocino, California.

the exception of the first, which is grass-covered) support the most predominantly heath-composed vegetation. The heath species *Vaccinium ovatum* is the second most dominant plant on the sites in terms of biomass (Westman and Whittaker, 1975), though two endemic dwarf coniferous trees (*Pinus contorta* ssp. *bolanderi* and *Cuppressus pygmaea*) dominate the above-ground structure of these pygmy forests. On the leading edges of upper terraces, eolian sand dunes of greater permeability have accumulated, which support a forest of *Pinus muricata* with a *Rhododendron* heath understorey of greater biomass. The ravines and terrace slopes, being well-drained and rapidly weathering, support a eutrophic and still more massive vegetation of coniferous trees (*Pseudotsuga menziesii, Sequoia*

[1] Manuscript completed November, 1976.

Contribution from research supported by a National Science Foundation grant to R.H. Whittaker. H. Jenny kindly provided data on litter fall.

TABLE 20.1

Coastal land system, Mendocino, California (Lat. 39°27'N, Long. 123°48'W)

Climate **Csb** (Köppen, 1923)

Climatic data (Fort Bragg, Calif.)	Summer (June–Aug)	Autumn (Sept–Nov)	Winter (Dec–Feb)	Spring (March–May)	Annual mean (or total)
Air temperature (°C)					
Mean daily maximum (°C)	18	18	14	15	16.8
Mean daily minimum (°C)	10	8	5	7	7.2
Precipitation (mm)	20	190	530	230	970
Fog at 8 a.m. (% of days)	48	29	9	19	26

Soil data

Land unit	Land form	Soil	Vegetation
1	marine-derived terrace	ground-water podzol (sideraquod), Blacklock, Aborigine soil types	grassland on first terrace; pygmy conifer forest and heath (*Pinus contorta* ssp. *bolanderi*, *Cupressus pygmaea*, *Vaccinium ovatum*) on upper terraces
2	sand dune	Podzol (haplorthod) Noyo soil type	bishop pine forest with heath understorey (*Pinus muricata*; *Rhododendron macrophyllum*, *Vaccinium ovatum*, *Gaultheria shallon*)
3	ravines and wave-cut terrace slopes	forest brown earths, alluvial soils and mild podzols, e.g. Hugo, Ferndale, Caspar soil types	redwood–Douglas fir and mixed evergreen forests with heath understorey (*Sequoia sempervirens*, *Pseudotsuga menziesii*, *Rhododendron macrophyllum*)

References: Gardner (1967), Jenny et al. (1969), Westman (1975).

sempervirens, Abies grandis, Picea sitchensis, Tsuga heterophylla) with scant heath understorey.

Although the distinction in stature between forest types on the three landforms in this region makes the vegetation/soil pattern particularly striking, the same or related species recur on patches of similar landscape elsewhere along the western coast of North America. *Vaccinium ovatum* is found from British Columbia in Canada south to the Mexican border (Abrams, 1951). *Cupressus pygmaea* is limited to the coastal podzols of the Mendocino region and to a sandstone cliff in southern Mendocino County at Anchor Bay, but it is only one of several examples of local endemism and edaphic restriction among the ten species of *Cupressus* in California (Wolf and Wagener, 1948; McMillan, 1952, 1956). Frequently these cypresses are found, as in Mendocino, in association with species of *Pinus*, and a variety of heath species. In

the Mendocino region, the perennial, evergreen shrub species of Ericaceae extending across the set of podzolic soils are *Arctostaphylos columbiana* var. *columbiana, A. glandulosum, A. nummularia, Gaultheria shallon, Ledum glandulosum* ssp. *columbianum* and var. *olivaceum, Rhododendron macrophyllum, Vaccinium ovatum* var. *saporosum* and var. *ovatum*, and *Vaccinium parvifolium*. On the terrace soils, the heath species are mostly low shrub phanerophytes (*G. shallon* is a chamaephyte); as podzolization decreases, the species increasingly occur as mid-height shrub phanerophytes. The soils and vegetation/soil pattern in this region have been discussed by Gardner (1958, 1967), Gardner and Bradshaw (1954), McMillan (1956), Jenny et al (1969) and Westman (1971, 1975). An account of the heathland vegetation in the coast redwood forest communities may be found in Waring and Major (1964) and Westman (1975).

ENVIRONMENTAL STRESSES

As with the pine–oak forest heathlands of eastern North America (Whittaker and Woodwell, 1968; see also Whittaker's Chapter 17) and the *Eucalyptus* forest heathlands of southeast Queensland (Westman and Rogers, 1977; see also Specht in Chapter 6), the central Californian coastal heathland vegetation is subject to frequent fires. Like these other heathlands, the understorey vegetation on Mendocino terraces is characterized by an abundance of root crowns and other underground storage structures from which re-sprouting can occur after fire. Unlike *Pinus rigida* and oak species of eastern North America, and *Tristania conferta* and certain eucalypt saplings in Queensland, however, the tree species on Mendocino terraces do not resprout from basal crowns after fire.

In addition to fire, the Mendocino terrace vegetation is subject to severe edaphic stresses which contribute to the dwarfing of vegetation on the site. Pines of 2 to 3 m height would normally be 15 to 25 years old on the Pine Barrens of New Jersey, but 70 to 100 years old in Mendocino (Westman, 1971). Whether the major edaphic factors contributing to the stunting on Mendocino terraces are nutritional alone, or involve drainage or micro-organism interactions as well, remains unresolved. Nevertheless there is clear experimental and historical evidence to indicate that the dwarf stature of at least some species on Mendocino terraces is not genetically determined. McMillan

showed in a series of fertilizer (McMillan, 1956) and transplant studies (McMillan, 1959, 1964) that *Cupressus pygmaea* responds vigorously in height growth to increased nutrient status of the soils. Mathews (1929) reported individuals of *C. pygmaea* as tall as 50 m, and with girths at breast height of 3 m and more, growing in dune swales in the region. *Vaccinium ovatum* similarly increases in stature as podzolization decreases along the gradient in the Mendocino region from a shrub height of 0.3 m to 1 m or taller, and other heath species respond similarly. There is some evidence for ecotypic adaptation to low nutrient status in the endemic pine, *Pinus contorta* ssp. *bolanderi*, however, which does not occur beyond the terrace soils, and fails to produce better growth on nutrient-enriched soils (McMillan, 1956).

VEGETATION/SOIL RELATIONS

In Fig. 20.2, the foliar cover (measured as overlapping line transect interceptions by each plant of the species in each stratum, and summed, on 61 sites ranging from 0.02 to 0.10 ha) of ericaceous species in the Mendocino region is plotted against pH of the surface soil horizon. The figure emphasizes the changing dominance by heath species as soils become increasingly acid. Westman (1975) found, in a study of twelve soil nutrient criteria, together with soil texture, soil organic matter, depth to hardpan, depth of rooted soil profile, distance from coast, elevation, and

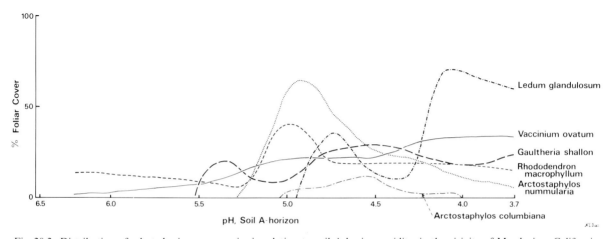

Fig. 20.2. Distribution of selected ericaceous species in relation to soil A-horizon acidity, in the vicinity of Mendocino, California. Importance values of species as sum of individual overlap % foliar cover in herb, shrub and tree strata in 61 stands.

available water holding capacity, that soil pH was the best single predictor of vegetation change in the region. This finding suggests the possibility that soil pH might be directly affecting vegetative growth, either by affecting the availability of essential elements, increasing the toxicity of some (e.g. aluminium at lower pH), or by affecting micro-organisms indirectly involved in mineralization and nutrient uptake.

At the present time resolution of the role of pH in nutrient response of the heath is limited by inadequate information. Levels of soil aluminum would be sufficiently high (60–75 $\mu g\, g^{-1}$) on dune and terrace soils to inhibit plant growth in low pH soils, if the heath species were similar to certain crop species (see, for instance, Ligon and Pierre, 1932) in their tolerances to aluminum, but tolerances of the Californian heath species to aluminum levels are not known. Early laboratory studies on

the effect of pH on nitrifying bacteria suggested that nitrate formation was completely inhibited below pH 4.0 (Waksman, 1936; Lutz and Chandler, 1946), but more recent field studies have found that nitrate release may occur during warmer seasons in soils with pH as low as 3 to 3.7 (Davy and Taylor, 1974; Runge, 1974). Nitrogen occurs increasingly in the ammonium form as soil pH decreases (Davy and Taylor, 1974); but the relative preferences of the Mendocino heath species for ammonium vs. nitrate nitrogen are not known. *Vaccinium corymbosum*, a species not present in the region, has been shown to grow better with nitrogen in the ammonium form (Cain, 1952; Oertli, 1963; Ballinger, 1966; Dijkshoorn, 1969). If similar preferences were exhibited by other species of *Vaccinium*, pH effects on nitrogen mineralization would seem unlikely to play a predominant role in explaining vegetative stunting.

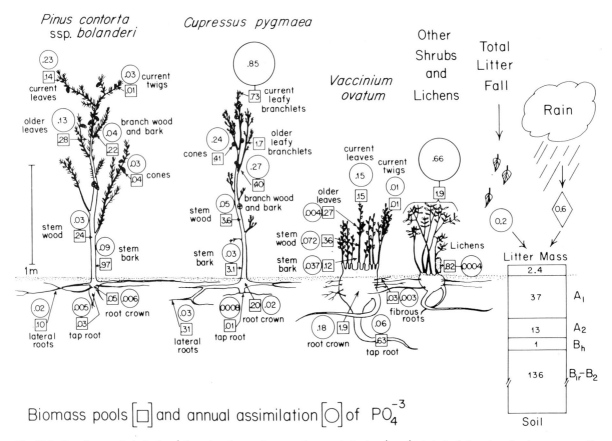

Fig. 20.3. Standing stocks in kg ha^{-1} (boxes) and annual net uptake rates in kg ha^{-1} yr^{-1} (circles) of phosphate ion in pygmy conifer heaths in Mendocino, California. Litterfall nutrient values are for the entire ecosystem, and are minimum estimates using average nutrient concentrations in litter mass. Rainfall nutrient values from Allen et al. (1968) for coastal U.K. were used.

All ericaceous species in the Mendocino region exhibit a marked tendency to concentrate a number of essential plant nutrients (calcium, magnesium, sodium, potassium, copper, zinc and manganese) in their foliage, relative to the surface soil, as the concentration of the nutrient in the soil decreases (data of Westman, 1971). The full explanation for this observation is unknown, but the phenomenon is consistent with the suggestion that some other edaphically related factor, such as poor soil aeration, may be a major factor inhibiting growth.

Fig. 20.3 illustrates the pattern of storage and assimilation of phosphate-phosphorus in the Mendocino pygmy forest (data of Westman, 1971, 1978). The heath species *Vaccinium ovatum* accounts for a larger proportion of total phosphate storage on an areal basis than *Pinus*, primarily through its massive root crown biomass. It is also noteworthy that rainfall can contribute an input to the ecosystem which is substantial relative to existing nutrient stocks. As nutrient levels in the soil increase, Mendocino heath forests show a decreased rate of nutrient cycling, larger standing stocks, and a greater tendency to store nutrients in the standing vegetation (Westman, 1971, 1978). The relative dominance of ericaceous species also decreases. There is thus suggestive evidence that the heath species of the western United States are favored on soils of low nutrient status, a generalization which appears increasingly apt for heathlands of very different floristic origin in certain other parts of the world (e.g. Australia; Specht et al., 1977).

REFERENCES

Abrams, L., 1951. *Illustrated Flora of the Pacific States, 3.* Stanford University Press, Stanford, Calif., 866 pp.

Allen, S.E., Carlisle, A., White, E.J. and Evans, C.C., 1968. The plant nutrient content of rainwater. *J. Ecol.*, 56: 497–504.

Ballinger, W.E., 1966. Soil management, nutrition, and fertilizer practices. In: P. Eck and N. Childers (Editors), *Blueberry Culture*. Rutgers University Press, New Brunswick, N.J., pp. 132–178.

Cain, J.C., 1952. A comparison of ammonium and nitrate nitrogen for blueberries. *Proc. Am. Soc. Hort. Sci.*, 59: 161–166.

Davy, A.J. and Taylor, K., 1974. Seasonal patterns of nitrogen availability in contrasting soils in the Chiltern Hills. *J. Ecol.*, 62: 793–807.

Dijkshoorn, W., 1969. The relation of growth to the chief ionic constituents of the plant. In: I.H. Rorison (Editor),

Ecological Aspects of the Mineral Nutrition of Plants. Blackwell, Oxford, pp. 201–213.

Gardner, R.A., 1958. Soil–vegetation associations in the redwood–Douglas–fir zone of California. In: *First North American Forest Soils Conf.* Agric. Exp. Station, Mich. State Univ., East Lansing, Mich., pp. 86–101.

Gardner, R.A., 1967. *Sequence of Podzolic Soils Along the Coast of Northern California.* Thesis. University of California, Berkeley, Calif., 236 pp.

Gardner, R.A. and Bradshaw, K.E., 1954. Characteristics and vegetation relationships of some podzolic soils near the coast of northern California. *Proc. Soil Sci. Soc. Am.*, 18: 320–325.

Jenny, H., Arkley, R.J. and Schultz, A.M., 1969. The pygmy forest–podsol ecosystem and its dune associates of the Mendocino coast. *Madroño*, 20: 60–74.

Köppen, W., 1923. Die Klimate der Erde. Bornträger, Berlin, 369 pp.

Ligon, W.S. and Pierre, W.H., 1932. Soluble aluminum studies: II. Minimum concentrations of aluminum found to be toxic to corn, sorghum, and barley in culture solutions. *Soil Sci.*, 34: 307–321.

Lutz, H.J. and Chandler, R.F., 1946. *Forest Soils.* Wiley and Sons, New York, N.Y., 514 pp.

Mathews, W.C., 1929. Measurements of *Cupressus pygmaea* Sarg. on the Mendocino "Pine Barrens" or "White Plains." *Madroño*, 1: 216–218.

McMillan, C., 1952. The third locality for *Cupressus abramsiana*, Wolf. *Madroño*, 11: 189–194.

McMillan, C., 1956. The edaphic restriction of *Cupressus* and *Pinus* in the Coast Ranges of central California. *Ecol. Monogr.*, 26: 177–212.

McMillan, C., 1959. Survival of transplanted *Cupressus* in the pygmy forests of Mendocino County, California. *Madroño*, 15: 1–4.

McMillan, C., 1964. Survival of transplanted *Cupressus* and *Pinus* after thirteen years in Mendocino County, California. *Madroño*, 17: 250–254.

Oertli, J.J., 1963. Effect of form of nitrogen and pH on growth of blueberry plants. *Agron. J.*, 55: 305–307.

Runge, M., 1974. Die Stickstoff-Mineralisation im Boden eines Sauerhumus-Buchenwaldes. Teil II: Die Nitratproduktion. *Oecol. Plant.*, 9: 219–230.

Specht, R.L., Connor, D.J. and Clifford, H.T., 1977. The heath–savannah problem: the effect of fertilizer on sand-heath vegetation of North Stradbroke Island, Queensland. *Aust. J. Ecol.*, 2: 179–186.

Wahrhaftig, C. and Birman, J.H., 1965. The Quaternary of the Pacific Mountain System in California. In: H.E. Wright, Jr. and D.G. Frey (Editors), *The Quaternary of the United States*. Princeton University Press, Princeton, N.J., pp. 299–340.

Waksman, S.A., 1936. *Humus. Origin, Chemical Composition and Importance in Nature.* William and Wilkins, Baltimore, Md., 494 pp.

Waring, R.H. and Major, J., 1964. Some vegetation of the California coastal redwood region in relation to gradients of moisture, nutrients, light, and temperature. *Ecol. Monogr.*, 34: 167–215.

Westman, W.E., 1971. *Production, Nutrient Circulation, and*

Vegetation–Soil Relations of the Pygmy Forest Region of Northern California. Thesis. Cornell University, Ithaca, N.Y., 411 pp.

Westman, W.E., 1975. Edaphic climax pattern of the pygmy forest region of California. *Ecol. Monogr.*, 45: 109–135.

Westman, W.E., 1978. Patterns of nutrient flow in the pygmy forest region of northern California. *Vegetatio*, 36: 1–15.

Westman, W.E. and Rogers, R.W., 1977. Biomass and structure of a subtropical eucalypt forest, North Stradbroke Island. *Aust. J. Bot.*, 25: 171–191.

Westman, W.E. and Whittaker, R.H., 1975. The pygmy forest region of northern California: studies on biomass and primary productivity. *J. Ecol.*, 63: 493–520.

Whittaker, R.H. and Woodwell, G.M., 1968. Dimension and production relations of trees and shrubs in the Brookhaven forest, New York. *J. Ecol.*, 56: 1–25.

Wolf, C.B. and Wagener, W.W., 1948. The New World cypresses. *El Aliso*, 1: 1–444.

Chapter 21

MURI AND WHITE SAND SAVANNAH IN GUYANA, SURINAM AND FRENCH GUIANA[1]

A. COOPER

INTRODUCTION

The geographical area of Guiana (Fig. 21.1) is found along the northern coast of tropical South America, between the Amazon and Orinoco river deltas (Beard, 1953). Within its boundaries are the three countries Guyana, Surinam and French Guiana, all with broadly similar geology, geography and natural history. Widely distributed throughout the region are bleached, sandy deposits (Schols and Cohen, 1953; Bleackley and Khan, 1963) mostly of Pliocene or Pleistocene age (Bleackley, 1957; Heyligers, 1963).

Although these deposits are derived from material of a variable lithology, they often consist of almost pure quartz sand (Dujardin, 1959) and the soils which they form are usually extremely infertile (Hamilton, 1945; Stark et al., 1959; F.A.O., 1966). The sands support a range of vegetation types from evergreen tropical forest (Davis and Richards, 1933, 1934; Schulz, 1960) to xeromorphic woodland and scrub (Benoist, 1925; Fanshawe, 1952; Heyligers, 1963) and edaphic savannah (Heyligers, 1963). Similar plant communities, found in equatorial Amazonia and Sarawak, are described by Klinge and Medina in Chapter 22 and by Specht and Womersley in Chapter 12.

The structure and composition of these communities is determined mainly by soil moisture conditions and man's activities (Fanshawe, 1952; Heyligers, 1963; Van Donselaar-ten Bokkel Huinink, 1966). The natural scrub vegetation is sclerophyllous and consists of two well-defined types, characteristic of soils with good and bad drainage respectively. Scrub from the drier soils is known as *muri* (tropical lowland sandy heathland) but that from the wetter soils, being generally lower and much more open, is usually referred to as wet white-sand savannah (tropical lowland wet-heathland or edaphic savannah).

The vegetation of bleached sandy soils throughout Guiana is usually very sensitive to damage by fire, especially during periods with rainfall lower than usual. In extreme cases, recurrent fires in the multi-layered communities are responsible for the origin and maintenance of anthropogenic savannah (Lindeman and Moolenaar, 1959; Heyligers, 1963). The tree and shrub component of the vegetation is progressively eliminated until the forb and herb component, to some extent adventive (Cooper, 1976), becomes dominant.

GEOMORPHOLOGY

Precambrian gneisses and schists, together with intrusives of various ages form the continental basement rocks of the Guiana Shield (Williams et al., 1967). These are overlain by the Roraima sandstone Formation (Bracewell, 1927) whose peneplained and much dissected surface gives rise to the Pakaraima Mountains. In front of the Shield (Fig. 21.1), lowland sedimentary material belonging to the Berbice, Coropina and Demerara Formations (Schols and Cohen, 1953) is spread out like a fan, with its axis between the Berbice and Courantyne Rivers (Bleackley, 1956).

The sandy deposits of the Berbice Formation (Fig. 21.1) were derived from the Roraima Formation, and consist mainly of littoral-deltaic sediments (Bleackley, 1957) and washed-out detritic sands (Boyé, 1959). The proto-Berbice was a

---

[1] Manuscript completed October, 1976.

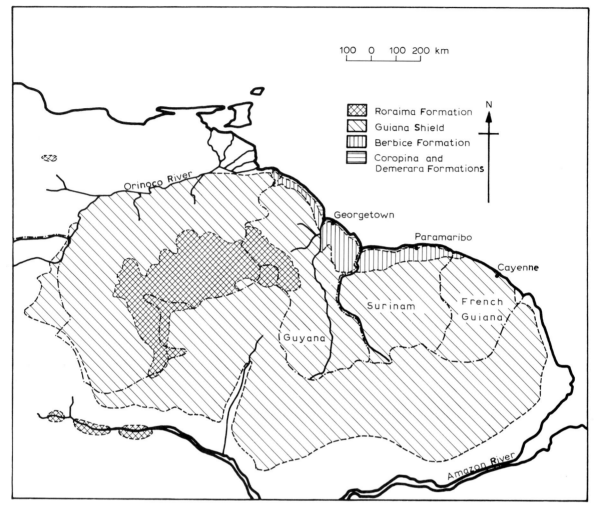

Fig. 21.1. The geology and geography of northeastern South America and Guiana.

major river at the time of greatest deposition. It drained large areas of the Guiana Shield and Roraima Formation and formed a huge deltaic basin which corresponded to the extent of the Berbice Formation. After its deposition, continental tilting, related to Andean orogenic movements (Van der Hammen, 1961), elevated the Berbice Formation to about 130 m above sea level inland, but only about 2 m above sea level near the present-day coast (Bleackley, 1956). Tilting was responsible for the origin of the existing river system, which is determined mainly by fault patterns of the Guiana Shield and the flat nature of the original deltaic deposition surface. During the Eemian interglacial, podzolization of the surface deposits of the Berbice Formation took place (Heyligers, 1963) but pedogenesis was interrupted by an erosional phase in the pre-Flandrian period of the Würm glacial. This resulted in the disruption of illuvial hard pan and the surface deposition from 1 to 1.5 m of sandy material, thus further modifying the topography and determining its present form.

The Coropina and Demerara coastal clay formations (Fig. 21.1) were deposited during successive marine transgressions and regressions during the Riss, Eemian, Würm and Holocene periods (Van der Hammen, 1974). Stranded beach ridges, sand spits and sand bars, laid down as sandy riverain deposits, are distinctive features of both formations (Bakker, 1951; Bleackley, 1957).

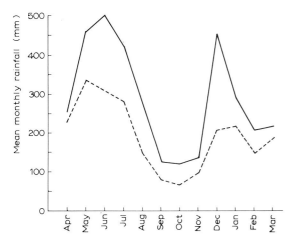

Fig. 21.2. Mean monthly rainfall (mm) in Guyana (———) and Surinam (-----).

CLIMATE

Guyana has a tropical climate with two wet and two dry seasons. In coastal regions, major and minor rainfall peaks occur in June and December respectively (Fig. 21.2). They give rise to a long rainy season from May to mid-August, a long dry season from mid-August to November, a short rainy season from December to January and a very short dry season from February to April. There is great variation in the duration and intensity of rainfall, with annual rainfall variation reaching 50% of the mean, and monthly rainfall variation reaching over 100% of the mean at times (Fanshawe, 1952). Prolonged drought occurs approximately once every fourteen years, due to diminution of rainfall in the dry seasons and failure of the rains during the major wet season.

Mean daily shade temperature on the coast varies only slightly from 26.1°C and has a daily range of about 9.5°C. Relative humidity is usually high and rarely falls below 60% during the day, even during dry seasons. The air remains almost saturated for long periods during the rainy season. The northeast Trade Winds blow for eight to nine months every year, from September to April; during the coolest months, from January to March, they also blow at nights. Violent winds of long duration — that is, hurricanes — are unknown. Daily average sunshine throughout the year is just over 6 h but reaches 10 h during the dry seasons.

In Surinam, which is nearer the equator, mean daily shade temperature is 27.1°C, but the climate is drier. Humidity and total hours of sunshine are related to rainfall (Schulz, 1960). In northern Surinam, annual precipitation is between 2000 and 2400 mm (Fig. 21.2). There is a long rainy season from April to August, a long dry season from August to December and a short dry season from February to April. Failure of the short dry season occurs occasionally, when the short rainy season passes directly into the long rainy season.

SOILS

Soils supporting plant communities of stunted sclerophyllous scrub in Guiana fall mainly into one of the types: bleached sand; podzol; or groundwater podzol (Appendices I and II). On the Berbice Formation, bleached sands are associated with the better drained deposits, whilst groundwater podzols are generally found where drainage is poor (Bakker, 1951; Van der Eyk, 1957; Bleackley and Khan, 1963; Heyligers, 1963; Klinge, 1968). Most groundwater podzols are found near the coast on sediments which, principally because of their low elevation, are not deeply incised with creeks and rivers (Bleackley, 1956). As a result, they easily become waterlogged during rainy seasons (Heyligers, 1963) despite their coarse sandy nature. Periodic drying of surface soil horizons caused by climatic seasonality leads to the deposition of an illuvial humus horizon (Andriesse, 1969) at depths of between 50 and 150 cm, in the zone of groundwater fluctuation. Bleached sands and, more rarely, groundwater podzols on the Berbice Formation often have illuvial horizons which are relics of former pedogenic eras (Bleackley and Khan, 1963; Heyligers, 1963). The illuvial horizon is usually cemented into hard pan by various proportions of organic matter and of aluminium and iron oxides, but on bleached sands the hard pan does not affect soil drainage since it is not extensive and, where it occurs, is generally at great depth.

In the coastal plain region of Guiana, areas of podzolic soils are associated with the higher, more extensive sand ridges of the Demerara clay Formation (Bakker, 1951; Lindeman, 1953; Bakker, 1954; Bleackley, 1957). Bleached sand overlies an illuvial humus horizon which is cemented into an iron pan and which lies at a depth of from 70 to 90 cm. Sandy riverain deposits of bleached sand overlying illuvial silt and subject to seasonal flooding have also been described by Welch (1972).

Little is known of the predominantly sandy soils of the plateau mountains of the Roraima Formation. Podzols, probably formed under the influence of topographic groundwater, have been described from low-elevation outliers in Surinam (Kramer and Van Donselaar, 1970). They are relatively shallow soils (c. 90 cm), and the sandstone from which they are derived is badly drained. For the most part, soils of the Roraima Formation at higher elevations are thin, and scattered in patches and pockets between exposed rock (Fanshawe, 1952). The sand is white or red in colour, and in places the organic matter content is great (Van Donselaar, 1965), giving rise to predominantly peaty soils in any depressions (Beard, 1953).

VEGETATION

Xeromorphic scrub (muri)

Fanshawe (1952) defined xeromorphic scrub as "scrubby vegetation with few species, often twisted, bent or semi-prostrate". The shrubby trees which are present "are 6–8 feet [i.e. 1.8–2.5 m] tall with an occasional emergent up to 15 feet [i.e. 4.5 m] high". The vegetation "occurs widely scattered or concentrated in small patches with bare soil or rock in between". Fanshawe specified "muri" as the form of xeromorphic scrub found on "white sands" in Guyana and described it as "roughly circular patches of scrubby vegetation separated by bare or sparsely-grassed sand". The most common emergent in Guyana was given as *Clusia nemorosa* which occurred in the "centre of each circle" surrounded by a ring of shrubs, the outer belt of which consisted of semi-prostrate shrubs (Table 21.1).

In Surinam, Van Donselaar-ten Bokkel Huinik (1966) described a community consisting of "dome-shaped bushes" (bush islands) "surrounded by nearly bare, white sandy soil". The basic structure of the bush islands was the same as that

of the muri given by Fanshawe (1952) but more emergent and shrub species were listed. Since the vegetation of Surinam is closely related to that of French Guiana (Lanjouw, 1936; Beard, 1953) and judging by summaries given by Benoist (1925) and Boyé (1959), it seems likely that a plant community similar to the xeromorphic scrub of Guyana and Surinam also occurs in French Guiana.

Xeromorphic scrub is characteristic of the deepest and most well-drained, coarse sandy soils (bleached sands) of watersheds (usually forested) in Guyana and Surinam (Hughes, 1946; Fanshawe, 1952; Lindeman and Moolenaar, 1959). The transition from forest to scrub is gradual, giving rise to the now accepted view that xeromorphic scrub is a natural community whose structure and composition is determined primarily by edaphic factors, and in particular by soil moisture status (Fanshawe, 1952; Beard, 1953; Heyligers, 1963; Van Donselaar-ten Bokkel Huinink, 1966). Pulle (1906) and Lanjouw (1936) have emphasized the role of leaching processes in the origin of xeromorphic scrub, and Lanjouw (1936), Gonggrijp and Burger (1948) and Welch (1972) have suggested that recurrent fires are, to a greater or lesser extent, responsible. Both factors, however, appear to be only of secondary importance in comparison with soil moisture conditions in naturally occurring xeromorphic scrub (Heyligers, 1963; Van Donselaar-ten Bokkel Huinink, 1966).

Heyligers (1963), in a survey of the plant communities of a limited area of "white-sand" soil in Surinam, was strongly in favour of the edaphically determined continuum hypothesis to explain the transition from xeromorphic forest to xeromorphic woodland then to xeromorphic scrub. At the same time he showed that these plant communities were very much influenced by fire and he demonstrated that, in some cases, fire was entirely instrumental in the origin of xeromorphic scrub from xeromorphic woodland. Van Donselaar-ten Bokkel Huinink (1966) regarded xeromorphic scrub as a naturally occurring climax vegetation. She suggested that, if the influence of seasonal drought did not show such massive cyclic fluctuation, then a type of closed scrub or xeromorphic woodland would develop. The occurrence of extreme drought, probably a result of short-term cylic climatic change, would tend to prevent this by precipitating the death of many shrub and tree

TABLE 21.1

Common species of areas of xeromorphic scrub in Guyana and Surinam

---

Fanshawe (1952): Xeromorphic scrub from the lowland region of Guyana

Emergent trees
Clusia nemorosa

Surrounding shrubs
Byrsonima crassifolia

Houmiri floribunda
Pagamea capitata
P. guianensis
Retiniphyllum schomburgkii
Ternstroemia punctata

Heyligers (1963): *Houmiri* variant of the *Ternstroemia–Matayba* scrub at Jodansavanne, Surinam

Predominant species of upper and middle strata
Clusia fockeana
Conomorpha magnoliifolia
Houmiri floribunda
Licania incana
Ormosia costulata
Pagamea capitata
Retiniphyllum schomburgkii
Swartzia bannia
Ternstroemia punctata

Associate species principally of upper strata
Bombax flaviflorum
Clusia nemorosa
Ilex jenmani
Mapouria chloranthra
Matayba opaca
Pagamea guianensis
Pisonia sp.
Protium heptaphyllum
Trattinnickia burseriflora

Associate species of lower strata
Calycolpus revolutus
Myrcia sylvatica

Climbers
Bredemeyera densiflora var. *glabra*
Doliocarpus calinea
Tetracera asperula

Principal species of herb and moss stratum
Actinostachys pennula
Axonopus attenuatus
Borreria suaveolens
Catasetum sp.
Epidendrum nocturnum
Lagenocarpus weigeltii
Schizaea incurvata
Trachypogon plumosus

---

species. Surviving individuals or groups of plants could then act as regeneration centres giving the typical muri-type physiognomy. Fire fitted into her cyclic regeneration hypothesis in that it would have similar effects to those of extreme drought.

Fanshawe (1952) reported the widespread occurrence of xeromorphic scrub as a result of fire on sandstone outcrops of the Roraima Formation which, "if given the chance would probably develop into woodland". The dual nature of the cyclic regeneration hypothesis explains earlier confusion in the literature as to the relative roles of soil moisture, fire, and leaching in the origin of xeromorphic scrub, and accounts for the similarity of emergent muri tree species and the tree species found in the surrounding woodland and forest. Heyligers (1963) has further clarified the effects of recurrent fires on bleached sands by demonstrating their effectiveness in progressively destroying

woody plants, and allowing their replacement with fire- and drought-resistant grasses, sedges and forbs to give a very sparse and poor anthropogenic savannah vegetation.

White sand savannah

On bleached sandy soils where groundwater reaches the surface for long periods during the rainy season, and then sinks beyond the reach of plant roots during the following dry season, there is a very sparse savannah vegetation, principally of grasses, sedges and forbs (Heyligers, 1963). Therophytes and rosette species are common (Van Donselaar-ten Bokkel Huinink, 1966) along with the sedge *Lagenocarpus tremulus* and species from the Xyridaceae and Lentibulariaceae (Van Donselaar, 1965). Stunted sclerophyllous shrubs are found occasionally, but their development is controlled by alternating periods of soil waterlogging and desiccation, which Beard (1944) proposed as being inhibitory to tree and shrub growth. Seasonal soil waterlogging prevents penetration of the soil by their vertical roots, and therefore they die when water supplies cannot be reached during periods of drought (Van Donselaar-ten Bokkel Huinink, 1966). Maximum cover of the savannah species varies from about 40 to 60% (Heyligers, 1963; Van Donselaar-ten Bokkel Huinink, 1966), and a layer of blue-green algae or amorphous organic matter (Heyligers, 1963) may cover large areas of the soil surface.

Within the savannah, many small bush islands, often not exceeding 1 to 2 m, but sometimes over 10 m in diameter and 1.5 m in height are found; occasional emergents may reach 3 m in height. The bush islands may be classed as the *Comolia vernicosa* variant of the *Clusia fockeana–Scleria pyramidalis* scrub described by Heyligers (1963). The soils of the bush islands show an abrupt transition from, and contrast strongly with, the almost structureless soils of the open savannah. They lie above the level of the highest water table, capping the slight topographic rises on which they are situated (Heyligers, 1963). Van Donselaar-ten Bokkel Huinink (1966), however, did not observe any significant differences in relative elevation of the soil surface between bush islands and savannah in Surinam. The species composition of savannah and bush islands is shown in Table 21.2.

In Guyana, the water table occurs at the same relative elevation in the bush islands and in the surrounding savannah, indicating no major differences in the drainage pattern (pers. obs.). However, the mineral soil within the bush islands is generally covered with a thick layer of leaf litter, contains large quantities of organic matter (Heyligers, 1963), and has a greater percentage sand, silt and clay content than the surrounding savannah soil. Where the savannah topography is relatively flatter, bush islands are few and widely spaced, but together give the impression of continuous scrub from a distance (Heyligers, 1963). This is reflected in recorded values for percentage cover by shrub species ranging from 15 (Heyligers, 1963) to 100 (Van Donselaar-ten Bokkel Huinink, 1966).

The size, number and species composition of bush islands in any area is related to the extent of microrelief features, the survival of their constituent shrubs and trees being related to the volume of soil which does not become waterlogged in rainy seasons (pers. obs.). The ameliorating effects of bush island shrubs on microclimate allows a more diverse flora (Heyligers, 1963) and fauna (Hoevers, 1967) than in the savannah. Of great significance to island stability is the contribution of litter to increasing the level of the soil surface and to protecting the soil from leaching and erosion. The characteristic nest building and feeding activities of the termite fauna (*Nasutitermes*) also helps to maintain the greater percentage silt and clay content of the bush island soils.

In Guyana, regeneration of bush island species takes place within the bush islands, but to only a limited extent in the savannah. The success of bush island shrubs outside the bush islands is very poor indeed. Here, regeneration occurs only on small microtopographic rises formed around tussocky savannah species by a rain-splash mechanism and termite activity. Seedling success outside bush islands may also be related to the incidence of a combination of atypically favourable rainy and dry seasons related to long-term climatic oscillation. Favourable seasonal combinations would lead to a reduction in mean water-table height and duration of soil waterlogging during rainy seasons, and a reduction of the incidence of severe drought during dry seasons.

Savannah and bush islands together make up an ecosystem which Heyligers (1963) described as "wet white-sand savannah" and which he recognised as being edaphically determined. Similar savannahs are found throughout Guiana — in other parts of Surinam (Lanjouw, 1936; Lindeman, 1953; Van Donselaar, 1965; Van Donselaar-ten Bokkel Huinink, 1966), in French Guiana (Benoist, 1925) and in Guyana (Fanshawe, 1952; Bleackley and Khan, 1963). They are found principally on groundwater humus podzols derived from sediments of the Berbice Formation in the coastal region of Guiana. As outlined earlier, the landscape is very flat and is drained by numerous, shallow black-water creeks, which themselves carry a type of sclerophyllous marsh woodland (Fanshawe, 1952). An edaphically determined continuum links this woodland with the savannah bush island complex of the watersheds.

Wet, white sand savannah is not confined to soils derived from the coastal sediments of the Berbice

TABLE 21.2

Common species in areas of wet white sand savannah in Surinam

Van Donselaar (1965): The savannah type *Xyridio-Paspaletum pulchelli* from the coastal region of Surinam (those species which were found only in one relevé have been omitted)

Species of open savannah	Species of bush islands
Bulbostylis conifera	*Bactris campestris*
Comolia lythrarioides	*Cassytha filiformis*
Curtia tenuifolia	*Clusia fockeana*
Leptocoryphium lanatum	*Comolia vernicosa*
Lisianthus coerulescens	*Conomorpha magnoliifolia*
Mesosetum loliiforme	*Doliocarpus calinea*
Paepalantus polytrichoides	*Houmiri balsamifera* var. *guianensis*
Panicum micranthum	*Lagenocarpus tremulus*
Paspalum polychaetum	*Licania incana*
P. pulchellum	*Marlieria montana*
Polygala adenophora	*Philodendron latifolium*
Rhynchospora barbata var. *glabra*	*Retiniphyllum schomburgkii*
R. graminea	*Scleria stipularis*
R. tenuis	*Tetracera asperula*
Sauvagesia sprengelii	*Tibouchina aspera*
Syngonanthus simplex	*Xyris guianensis*

Heyligers (1963): *Comolia* variant of *Clusia–Scleria* bush islands, Jodansavanne, Surinam

Predominant species of upper and middle strata	*Lisianthus uliginosus*
Bactris campestris	*Marlieria montana*
Clusia fockeana	*Micania* sp.
Conomorpha magnoliifolia	*Miconia ciliata*
Houmiri floribunda	*Panicum nervosum*
Licania incana	
Pagamea capitata	Climbers
Retiniphyllum schomburgkii	*Cassytha filiformis*
Scleria pyramidalis	*Doliocarpus calinea*
Ternstroemia puncatata	*Tetracera asperula*
Associate species, principally of upper strata	Principal species of herb and moss stratum
Bombax flaviflorum	*Actinostachys pennula*
	Cladonia sandstedei
Associate species of lower strata	*Lagenocarpus amazonicus*
Comolia vernicosa	*Lindsaea stricta* var. *parvula*
Croton hostmannii	*Radiella nana*
Lagenocarpus tremulus	*Sphagnum antillarum*

Formation. Welch (1972) described a vegetation type resembling wet, white sand savannah from the upper Demerara River in Guyana, on soils consisting of a thin layer of bleached sand overlying alluvial silt and subject to flooding after heavy rain. Lindeman (1953) and Van Donselaar (1965) refer to savannah on the largest of the sand ridges of the Coropina and Demerara Formations in the coastal region of Surinam. The iron-pan podzols of these ridges are badly drained, and are occupied by a flora very similar to that on wet, white sand savannah.

The flora and soil of the plateau mountains of the Roraima sandstone Formation are similar to those of the edaphic savannahs of the coastal region of Guiana (Fanshawe, 1952; Van Donselaar, 1965). The once more.extensive, and less elevated, Roraima Formation probably carried a pre-Pleistocene savannah vegetation, which acted as a distribution centre for the migration of plant species to the redistributed sandstone sediments of the Berbice Formation. Van Donselaar (1965) proposed a bridge hypothesis to explain the disjunct distribution of the Roraima and Berbice

Formation savannah floras. My own view, how-ever, is that the existing, relatively small areas of wet, white sand savannahs in lowland Guyana are relict, and once formed part of a much more extensive edaphic savannah vegetation covering very much larger areas of the Berbice Formation, and possibly dating to the Riss glacial period.

Van Donselaar (1965) summarised the existing floristic and ecological literature dealing with the plateau vegetation of the Roraima Formation. Plant communities have been variously described as "a type of wet paramo vegetation growing in peat bog" (Beard, 1953), "heath-like" (Fanshawe, 1952) and "savannah" (Jenman, 1882), whilst their soils are generally thin and sandy with greater or lesser amounts of peaty organic matter.

CLASSIFICATION

A scheme for the classification of the climax vegetation of tropical America was proposed by Beard (1944, 1955) who followed the classical Anglo-American system of the Clements–Tansley school. His basic unit of classification was the plant association, a floristic grouping with consistent dominants. Associations were grouped into for-mations according to their structure and physiog-nomy and these were in turn arranged into formation series on the basis of environmental factors. Fanshawe (1952) expanded and con-solidated this system which he used to classify the vegetation of Guyana (then British Guiana).

Both classificatory systems were extensive, but suffered from ambiguity of terminology and lack of floristic detail, especially in relation to the stunted sclerophyllous vegetation of Guyana. A major deficiency was the separation of predominantly one-layered vegetation types (savannah) from multi-layered vegetation types which were related both floristically and ecologically.

Heyligers (1963) and Van Donselaar (1965) classified the vegetation of bleached sandy soils in Surinam using the survey techniques and metho-dology of the Braun-Blanquet school. Heyligers (1963) produced a classificatory system, on a local basis only, which he did not test in the field in other parts of Surinam or Guiana as a whole, and as a result he preferred not to rank his phytosociologi-cal units. His classificatory system recognised the dynamic nature of, and the ecological relationships

between, the various plant communities which he delimited, and it included woodland, scrub, bush island and savannah vegetation types. The applica-bility of his scheme throughout Surinam and other parts of Guiana has since been partially validated by Van Donselaar (1965, 1969), Van Donselaar-ten Bokkel Huinink (1966) and Kramer and Van Donselaar (1970), although it is unlikely that it will remain unchanged in any revision to ensure its wider application.

Van Donselaar (1965) produced a formal classi-ficatory system for the savannah vegetation of Surinam which can be applied throughout Guiana (Van Donselaar, 1970) but he did not formalise a system to include savannah bush islands and surrounding scrub and woodland communities.

APPENDIX I: PROFILE DESCRIPTIONS OF BLEACHED SANDS AND GROUNDWATER PODZOLS IN GUYANA AND SURINAM

Heyligers (*1963*): Deeply drained bleached sand, Jodansavanne, Surinam

0–8.5 cm	"micropodzol", viz. 0–1 cm: bleached, light reddish sand; 1–2.5 cm: darker red-dish, somewhat finer sand; 2.5–8.5 cm: bleached sand
8.5–40 cm	light to medium, reddish brown, sub-angular sand; between 15 and 18 cm a layer of roots up to 1 cm thick
40–75 cm	lighter coloured sand, but with spots coloured as the layer above, these spots mainly around roots
75–210 cm	white, somewhat finer textured sand; about eight brownish spots around roots per m²; these roots are partly living, partly dead
210–deeper than 730 cm	sharp-edged, white sand; sand becoming coarse and more densely packed at about 300 cm; between 450 and 550 cm big quartz pebbles up to 3 cm, and some dark quartz fragments up to 0.8 cm; below 550 cm sand somewhat less coarse, down to 270 cm with some dead roots; below the sand a kaolinitic clay

Dujardin (*1959*): Deeply drained bleached sand, Sandhills, Guyana

0–5 cm	humus and sand
5–25 cm	grey, humus-stained sand
25–30 cm	transition zone
30–914 cm	creamy-white sand
below 914 cm	dark brown sandy clay passing into yel-lowish brown and bluish-grey clays

APPENDIX I *(continued)*

Bleackley and Khan (1963): Deeply drained bleached sand with iron pan, Maria Elizabeth Mine, Mackenzie, Guyana

0–40 cm	spoil from bauxite workings
40–55 cm	sandy clay, very dark grey; sample 1 ⎫ strong contamination
55–70 cm	sand, very dark grey; sample 2 ⎬ by overlying spoil
70–105 cm	bleached white sand sample 3
105–125 cm	bleached white sand; sample 4
125–140 cm	transitional into humic layer; sample 5
140–165 cm	black humic loamy sand; sample 6
165–180 cm	transitional loamy sand, very hard but moist; sample 7
180–200 cm	gleyed sandy loam resting on iron pam 1 cm thick; sample 8
201–231 cm	light red sandy loam; sample 9

Van der Eyk (1957): Deeply drained bleached sand with deep hard pan, lowland Surinam

0–2 cm	white, loose, moderately coarse sand
2–30 cm	greyish brown, loose, moderately coarse sand
30–55 cm	light brownish grey, loose moderately coarse sand gradually merging into next horizon
50–285 cm	white, loose, moderately coarse sand
285–305 cm	very dark brown, moderately coarse sand cemented into a hard pan by organic matter
305–310 cm	very dark brown, moderately coarse sand with rounded gravel, constituting the lower part of the hard pan
310 cm and below	greyish brown, non-sticky, plastic clay, rich very fine mica plates; below the clay is weathered schist

Heyligers (1963): Representative of poorly drained groundwater podzol, Jodansavanne, Surinam

0–3 cm	reddish sand on top of a thin grey layer of charcoal
3–25 cm	light reddish brown sand, grey-spotted with many roots and rootlets; some of the roots thicker and running horizontally; sample 2 taken at 10 cm

25–75 cm	light reddish brown sand with some greyish spots; many rootlets; sample 3 taken at 40 cm
75–170 cm	nearly white sand with some darker spots round rootlet concentrations; thin rootlets down to the hard pan; sand becoming coarser with increasing depth; sample 4 taken at 90 cm; sample 5 at 170 cm; sample 5a at 220 cm; on the hard pan a very thin layer of beige-grey silty sand
170–180 cm	blackish brown, not coarse sand, moderately strongly cemented (sample 6), fading into dark brown
180–190 cm	dark and coffee-brown sand, here more, there less strongly cemented (sample 7); in both layers rootlets and roots to 8 mm diameter; Sometimes (reformed?) fractures in which rootlets have penetrated; also dead roots causing black spots
190–220 cm	light rusty brown, strongly cemented, coarser sand (sample 8), darker in places with dead rootlets, fading into light brown
220–240 cm	light brown, cemented sand (sample 9)
240–255 cm	still lighter brown (sample 10) with some darker spots
255–263 cm	fairly abrupt transition into dark brown, coarse, less strongly cemented sand (sample 11)
263–270 cm	blackish-brown, but slightly cemented, coarse sand with dead roots and other organic detritus (sample 12)
Subsoil: 270–280 cm	loose, grey, coarse sand (sample 13)
280–deeper than 460 cm	loose, glistening white, very coarse sand; sample 14 taken at 285 cm; 15 taken at 325 cm

Van der Eyk (1957): Poorly drained groundwater podzols, lowland, Surinam

0–30 cm	greyish brown, very friable, light sandy loam
30–45 cm	white, very friable, fine sandy loam
45–65 cm	white, fine sandy loam of very close packing
65–85 cm	dark reddish brown hard pan, consisting of sericite containing 'silt loam', cemented by iron oxides; many brown and grey mottles
85–265 cm and below	mottled white and brown, kaolin-like "silt loam", rich in sericite; some very coarse sand in the upper part

APPENDIX II: SOIL CHEMICAL AND PHYSICAL PROPERTIES

Bleackley and Khan (*1963*): Deeply drained bleached sand with iron pan, Maria Elizabeth Mine, Mackenzie, Guyana (see Appendix I for profile description)

Sample number	Depth (cm)	% Sand	% Silt	% Clay	pH	%C	Fe $\mu g\ g^{-1}$	Al $\mu g\ g^{-1}$	P $\mu g\ g^{-1}$
1	40–55	51	2	41	5.2	3.8	9500	480	3
2	55–70	82	3	8	5.4	3.4	3500	260	3
3	70–105	92	1	7	5.6	0.2	0	0	4
4	105–125	87	2	10	5.8	0.1	0	0	4
5	125–140	88	1	8	5.3	2.0	350	50	6
6	140–165	83	2	9	5.1	3.2	1000	260	7
7	165–180	80	1	12	4.9	3.8	200	520	7
8	180–200	78	4	16	5.1	1.3	400	520	9
9	201–321	79	2	18	5.1	0.3	2000	320	6

Heyligers (*1963*): Representative of poorly drained groundwater podzol, Jodansavanne, Surinam (See Appendix I for profile description)

Sample number	Median grain size (μm)	Specific surface	Percentage by weight of particles $<50\ \mu$m	Organic matter content[1] (% C in dry soil)	Composition of the extract obtained with HCl (0.1 eq l^{-1}) ($\mu g\ g^{-1}$ dry soil) Al	Fe	Ca	Mg	K	Na	P	NH_4	NO_3
2	300	44	1	0.3	180	20	30	70	240	30	10	20	10
3	270	45	4	0.1	50	10	10	40	90	10	tr.	20	10
4	290	49	2	0.1	20	5	10	90	70	10	tr.	10	10
5	280	56	4	0.1	20	10	10	100	120	10	tr.	10	10
5	340	53	4	0.0	20	10	10	90	200	20	tr.	10	10
6	290	68	9	1.2	6400	30	10	20	390	100	10	20	20
7	–	–	–	1.1	9200	90	60	30	390	40	20	20	10
8	310	53	4	0.5	8100	20	130	5	40	20	tr.	40	10
9	–	–	–	0.2	7200	20	120	5	40	20	20	40	10
10	330	67	9	0.1	9100	80	240	5	80	30	tr.	30	10
11	–	–	–	0.7	8300	90	80	5	50	30	50	40	10
12	420	44	4	1.0	4900	20	40	5	110	30	20	40	10
13	440	28	tr.	0.2	1600	20	30	5	120	30	10	40	20
14	–	–	–	0.1	50	20	20	60	40	10	10	60	10
15	1230	12	tr.	0.1	70	10	40	30	20	10	10	30	10

[1] By weight.

REFERENCES

Andriesse, J.P., 1969. A study of the environment and characteristics of tropical podzols in Sarawak (East Malaysia). *Geoderma*, 2: 201–227.

Bakker, J.P., 1951. Bodem en bodemprofielen van Suriname, in het bijzonder van de noordelijke savannestrook. *Landbouwk. Tijdschr.*, 63: 397–391.

Bakker, J.P., 1954. Über den Einfluss von Klima, jüngerer Sedimentation und Bodenprofilentwicklung auf die Savannen Nord-Surinams (Mittelguyana). *Erdkunde*, 8: 89–112.

Beard, J.S., 1944. Climax vegetation in tropical America. *Ecology*, 25: 127–158.

Beard, J.S., 1953. The savanna vegetation of northern tropical America. *Ecol. Monogr.*, 23: 149–215.

Beard, J.S., 1955. The classification of tropical American vegetation types. *Ecology*, 36: 89–100.

Benoist, R., 1925. La végétation de la Guyane française Il — Les savanes. *Bull. Soc. Bot. Fr.*, 72: 1066–1076.

Bleackley, D., 1956. The geology of the superficial deposits and coastal sediments of British Guiana. *Br. Guiana Geol. Surv. Bull.*, 30: 46 pp.

Bleackley, D., 1957. Observations on the germorphology and geological history of the coastal plain of British Guiana. *Br. Guiana Geol. Surv.*, Summ., 1: 21 pp.

Bleackley, D. and Khan, E.J.A., 1963. Observations on the white sand areas of the Berbice Formation, British Guiana. *J. Soil Sci.*, 14: 44–51.

Boyé, M., 1959. New data on the coastal sedimentary formations in French Guiana 1. The Quaternary and the problem of detritic white sand. *Proc. 5th Inter-Guiana Geol. Conf.*, pp. 145–159.

Bracewell, S., 1927. *Report on Preliminary Geological Survey of the Potaro-Ireng District*. Geological Survey of British Guiana, 17 pp.

Cooper, A., 1976. An ecological and phytogeographic study on the embankment flora of the Linden Highway, Guyana. *Carib. J. Sci.*, 15, in press.

Davis, T.A.W. and Richards, P.W., 1933. The vegetation of Moraballi Creek, British Guiana: an ecological study of a limited area of tropical rain forest. Part I. *J. Ecol.*, 21: 350–384.

Davis, T.A.W. and Richards, P.W., 1934. The vegetation of Moraballi Creek, British Guiana: an ecological study of a limited area of tropical rain forest. Part II. *J. Ecol.*, 22: 106–133.

Dujardin, R.A., 1959. Report on British Guiana white sand as a possible source of glass sand. *Br. Guiana Geol. Surv. Min. Resour. Pamphlet*, 8: 14 pp.

Fanshawe, D.B., 1952. The vegetation of British Guiana. A preliminary review. *Imp. For. Inst. Oxford, Inst. Pap.*, 29: 96 pp.

Food and Agriculture Organisation of the United Nations, 1966. *Report on the Soil Survey Project of British Guiana*. Rome, FAO/FS, 19BRG.

Gonggrijp, J.W. and Burger, D., 1948. *Bosbouwkundige studiën over Surinam*. PUDOC, Wageningen, 262 pp.

Hamilton, R., 1945. *Bijdrage tot de bodemkundige kennis van (Nederlands) West-Indië (Tropengronden I)*. Thesis, University of Utrecht, Utrecht, 58 pp.

Heyligers, P.C., 1963. Vegetation and soil of a white-sand savanna in Suriname. *Verh. K. Ned. Akad. Wet.*, 54: 148 pp.

Hoevers, L., 1967. Herpetological collections from the Atkinson–Maduni–Laluni area. East Demerara. Guyana. *Timehri*, 43: 34–50.

Hughes, J.H., 1946. Forest resources. In: V. Roth (Editor), *Handbook of Natural Resources of British Guiana*. The "Daily Chronicle" Ltd., Georgetown, 243 pp.

Jenman, G.S., 1882. Remarks on the aspects and flora of the Kaiteur Savanna. *Timehri*, 1: 229–250.

Klinge, H., 1968. *Report on Tropical Podzols*. F.A.O. Rome, MR/69832.

Kramer, K.U. and Van Donselaar, J., 1970. A sketch of the vegetation and flora on the Koppel Savanna near Tafelberg, Suriname. *Verh. K. ned. Akad. Wet.*, 59: 496–524.

Lanjouw, J., 1936. Studies on the vegetation of the Suriname savannahs and swamps. *Ned. Kruidk. Arch.*, 46: 823–851.

Lindeman, J.C., 1953. The vegetation of the coastal region of Suriname. *Meded. Bot. Lab. Herb. Rijks-Univ. Utrecht*, 113: 135 pp.

Lindeman, J.C. and Moolenaar, S.P., 1959. Preliminary survey of the vegetation types of northern Suriname. In: *The Vegetation of Suriname*, 1(2). Van Eedenfonds, Amsterdam, 145 pp.

Pulle, A.A., 1906. *An Enumeration of the Vascular plants known from Surinam*. University of Leyden, Leyden, 555 pp.

Schols, H. and Cohen, H., 1953. Progress in the geological map of Surinam. *Geol. Mijnb.*, Ser. 2, 15: 142–151.

Schulz, J.P., 1960. Ecological studies on rain forest in Northern Suriname. *Verh. K. Ned. Akad. Wet.*, Ser. 2, 53: 267 pp.

Stark, J., Rutherford, G.K., Spector, J. and Jones, T.A., 1959. *Soil and Land-Use Surveys, No. 6, British Guiana*. Imp. Coll. Trop. Agric., Trinidad, 24 pp.

Van der Eyk, J.J., 1957. *Reconnaissance Soil Survey in Northern Surinam*. Thesis, Landbouwhogeschool, Wageningen, 98 pp.

Van der Hammen, T., 1961. The Quaternary climatic changes of northern South America. *Ann. N.Y. Acad. Sci.*, 95: 676–683.

Van der Hammen, T., 1974. The Pleistocene changes of vegetation and climate in tropical South America. *J. Biogeogr.*, 1: 1–36.

Van Donselaar, J., 1965. An ecological and phytogeographic study of northern Surinam savannas. *Wentia*, 14: 163 pp.

Van Donselaar, J., 1970. Observations on savanna vegetation-types in the Guianas. *Vegetatio*, 17: 271–312.

Van Donselaar-ten Bokkel Huinink, W.A.E., 1966. Structure, root systems and periodicity of savanna plants and vegetation in northern Surinam. *Wentia*, 17: 162 pp.

Welch, I.A., 1972. Vegetation types of Guyana. *For. Bull., Min. Mines For., Guyana*, N.S., No. 4: 42 pp.

Williams, E., Cannon, R.T. and McConnell, R.B., 1967. The folded Precambrian of northern Guyana related to the geology of the Guiana Shield. *Geol. Surv. Guyana Rec.*, 5: 60 pp.

Chapter 22

RIO NEGRO CAATINGAS AND CAMPINAS, AMAZONAS STATES OF VENEZUELA AND BRAZIL[1]

H. KLINGE and E. MEDINA

INTRODUCTION

In southern Venezuela and the adjacent part of Brazil a dense sclerophyllous forest (20 to 30 m high), known as *caatinga*, is developed on the white sandy soils characteristic of the upper reaches of the Rio Negro. Further south, the caatinga grades into a dense sclerophyllous scrub (15 to 17 m high) known as *campina*. In some habitats the campina is even more depauperate, being as low as 1.5 m in height; these stunted communities are termed *low campina* or *bana* (in southern Venezuela). Caatinga and campina have been described by Spruce (1908), Ducke and Black (1954), Ferri (1960), Rodriques (1961), Takeuchi (1961, 1962), Vieira and Filho (1962), Hueck (1966), and Klinge et al. (1977). Aubréville (1961) proposed for these vegetation types the term "fourrés et forêts basses amazoniens sur sables blancs".

While typical Amazon caatinga occurs in the upper Rio Negro Basin with monthly rainfall of at least 100 mm (Keses, 1956; Walter and Medina, 1971; Heuveldop, 1976), the campina is found farther south where the climate is characterised by a short dry season (Walter and Lieth, 1960–67). Thus, the Rio Negro caatinga is restricted to near-equatorial regions, while campinas are characteristic of the lower limit of the equatorial region.

Caatingas and campinas grow on soils which show the characteristic profile of temperate spodosols (Anonymous, 1967; Soil Survey Staff, 1975); such soils cover a rather large area in equatorial Amazonia and Guiana (Klinge, 1966, 1968) and are associated with black-water rivers (Richards, 1941; Sioli, 1954; Klinge, 1967).

The mechanism of spodosol formation under tropical vegetation is not fully understood. The existence of egg-cup podzols under *Agathis australis* in New Zealand points to the importance of the chemical composition of litter in the formation of raw humus (Gibb, 1964). Exudate from the litter causes the destruction of the original soil minerals (except quartz and some other stable minerals); the resulting products migrate to a greater depth, thus developing the spodosol profile.

GENERAL REMARKS

Rio Negro caatinga and campina are plant associations dominated by trees. Herbs and low shrubs are generally rare in these communities, except for epiphytes which are abundant in the campina. The few vegetation analyses carried out in campina and Amazon caatinga show clearly that a few tree species are dominant or tend to dominance, whilst in the mixed forest on different soils, often similarly low in nutrients (Klinge, 1976a), no single tree species becomes dominant (Richards, 1957).

However, shrubland may be found here and there within the campina or Amazon caatinga. It is locally known as low campina or low Amazon caatinga (*caatinga baixa, campina baixa* of Portuguese-speaking authors); in southern Venezuela it bears the name of "bana". It is so far not clear whether the occurrence of bana, as well as caatinga and campina, amidst a relative luxuriant vegetation is due exclusively to a particular water economy of the soils (which do not necessarily occupy the lowest ground in the area) or due to an interaction between water régime (flooding followed by complete desiccation of upper soil layers) and low availability of nutrients, mainly phos-

---

[1] Manuscript completed October, 1976.

phorus. The dominance of broad-sclerophyllous, evergreen species, is supposed to be the consequence of selection of species with extremely low phosphorus requirements (Loveless, 1961), and it has been shown that broad-sclerophylls have a lower ratio of nitrogen to phosphorus than mesophytic, deciduous leaves (Montes and Medina, 1977).

Ferri (1960 and 1961), in a study on transpiration of a number of plants from a low caatinga in northwestern Brazil, concluded that the limiting factor in this vegetation is not water stress, but the severe oligotrophy of the soil environment. Nevertheless, water stress seems to be an important selective factor, since most species of bana show a high degree of leaf tilting, which contributes to reduce the radiation load during rainless days.

STRUCTURE AND FLORA

Amazon caatinga and campina

Ducke and Black (1954) described the species richness of Amazon caatinga as highest when compared to any other Hylean vegetation and mentioned particularly the high degree of endemism in the caatinga flora. In their opinion genuine Amazon caatinga has no direct affinity with any other Amazon forest nor savannah land, except for the campina.

The heath-like character of campina and low caatinga is locally given by species of *Drosera*, *Lycopodium*, *Sphagnum*, bromeliads and *Cladonia*.

The height of Amazon caatinga is usually between 20 and 30 m, there being often no definite "*a*" stratum, i.e. the main canopy is not overtopped by emergents. Campinas are generally somewhat lower in height, and bana and related shrubland may be as low as 1.5 m in height. Even small spots of barren white sand may occur.

Tree diameters are generally below 30 cm but may, in exceptional individuals, reach about 50 cm. The low height, the relatively high stocking density and the medium diameters, give the Amazon caatinga its peculiar aspect which can also be observed in the campinas. In the latter, tortuous trees are found, a feature absent in the Amazon caatinga where slender and straight boles predominate.

A conspicuous floristic element of the Amazon caatinga is the tree genus *Eperua* (Caesalpiniaceae). *E. leucantha* may be dominant or co-dominant as in the upper Rio Negro. *E. purpurea* may be dominant in other forests on grey-white loamy sand showing almost no signs of podzolisation. In the wallaba forests of Guiana the dominant species is *E. falcata*.

Manaus[1] campina

The uppermost stratum is somewhat above 15 m in height and consists of *Aldina latifolia* varying in height from 15 to 17 m, and in diameter at breast height from 19 to 49 cm (Takeuchi, 1961). Its density is low (0.02 ind. m^{-2}). The branches of this species are heavily covered by epiphytes. The main canopy is dominated by *Clusia insignis* (frequency 0.13 ind. m^{-2}), followed by *Annona* sp. (frequency 0.075 ind. m^{-2}). *C. insignis* is also found in neighbouring mixed forest, where it grows much taller and thicker. Other species in the main canopy of the campina are *Eperua purpurea*, *Talisia esculenta*, *Macrolobium arenarium*, *Aniba hostmanniana*, *Heisteria guianensis*, and *Erythrina corallodendron*.

The lower stratum consists of bushes (*Pagamea duckei*, *Cinchona* sp., *Miconia tomentosa*, *Miconia* sp., *Croton* sp., *Neea oppositifolia*, and the palm *Amylocarpus inermis*). The latter has a frequency of 0.11 ind. m^{-2}, whilst that of *Pagamea duckei* is 0.64 ind. m^{-2}. Below the bush layer, the undergrowth is composed of a large number of seedlings of species occurring in the higher strata. Altogether, a sample of 200 m^2 contained 190 trees, 107 shrubs, 147 herbs, 31 lianas and 22 palms.

High caatinga

In high caatinga of northwestern Brazil (Takeuchi, 1962) the forest 20 to 25 m tall is dominated by *Eperua leucantha* (0.115 ind m^{-2}) and other leguminous trees like *E. purpurea*, *E. rubiginosa*, *Aldina discolor*, *Tachigalia* sp., and *Lucuma* sp. (Sapotaceae), the latter species having much lower frequencies than that of *E. leucantha*. The lower stratum has a density of 2.18 trees m^{-2},

---

[1] Manus (3°06′S, 60°00′W) is a town on the confluence of the Amazon River and the Rio Negro in Brazil.

about 50% of which are *Eperua* spp. In the undergrowth, *Anthurium* is conspicuous.

In the same area, Rodrigues (1961) studied a high caatinga dominated by *Micrandra crassipes*, *Eperua leucantha*, *E. purpurea* and *Peltogyne catingae*. The tallest tree measured 29 m in height. The density of trees in the upper layer was 28 with 679 in the lower stratum, in an area of 375 m². *Anthurium* was again conspicuous in the ground flora. It was observed that the plant species of the immediately adjacent mixed forest did not mix with those of the caatinga, or vice versa.

A characteristic feature of the Amazon caatinga is the abundance of mosses. Bryophytes are also comspicuous in the *wallaba* forest of Guiana (Richards, 1954).

A plot of Amazon high caatinga measuring 10 ha (including a piece of low caatinga/bana) is being studied near San Carlos de Rio Negro (1°54′N, 67°06′W), south of the Casiquiare River (Brünig, 1976; Heuveldop, 1976; Brünig and Heuveldop, 1976; Klinge, 1976b; Klinge et al., 1977). Whilst in the high caatinga (Tables 22.1 and 22.2) the maximum diameter of the dominant *Micrandra spruceana* is 41.7 cm and maximum height is 25.4 m, this species contributes only 1% of total fresh above-ground phytomass of the low caatinga/bana. Here, where the maximum height is 11 m and maximum diameter is 11.8 cm, ten additional species each constitute between 5 and 10% of total above-ground phytomass, but *Eperua* is lacking completely. These additional species also occur in the high caatinga as well as where the high caatinga grades into low caatinga, but locally may form a higher percentage of phytomass.

CONCLUSIONS

Considerable areas of infertile bleached· sand, sandy podzolic soils and groundwater podzols are found in the adjacent Amazonas States of southern Venezuela and northwestern Brazil. Similar in-

TABLE 22.1.

Composition of total above-ground phytomass in high and low Amazon caatinga, at San Carlos de Rio Negro

| | Percentage | | | |
	Micrandra spruceana	*Manilkara* sp.	*Eperua leucantha*	other tree species
High Amazon caatinga (max. height 25.4 m)	45	8	14	33
High Amazon caatinga grading into low Amazon caatinga	14	<1	1	85
Low Amazon caatinga (max. height 11 m)	1	<1	0	99

TABLE 22.2

Stocking density, total above-ground phytomass of tree and bryophyte vegetation in high and low Amazon caatinga, at San Carlos de Rio Negro

| | Stocking density (individuals m^{-2}) | | | Total above-ground phytomass (kg m^{-2}) | | |
| | 1 cm dbh[1] | 1–10 cm dbh | 10 cm dbh | trees | | mosses |
				total	leaves	
High Amazon caatinga (max. height 25.4 m)	21.14	1.00	0.11	42.67	1.89	0.48
High Amazon caatinga grading into low caatinga	23.23	0.98	0.09	67.44	2.58	0.09
Low Amazon caatinga (max. height 11 m)	39.66	1.75	0.06	17.00	1.31	0.34

[1] dbh: diameter at breast height.

fertile sandy soils are found in the Berbice Formation of littoral-deltaic sediments and detritic sand of Plio-Pleistocene age which are found in coastal Guyana and Surinam from near the coast (circa 2 m altitude) inland to about 130 m above sea level (see Cooper in Chapter 21). These tropical spodosols are virtually absent on continental Africa, apart from isolated localities on the east coast (Giesecke, 1930) and on some islands like Pemba and Madagascar (Chaminade, 1949; Calton, 1955).

Identical, infertile sandy soils are found in lowland Malesia, where the soils and associated plant communities have been investigated in detail in Sarawak and Brunei on the island of Borneo (Brünig, 1974). Small areas of groundwater podzols occur in New Guinea and northeastern Australia (see Specht and Womersley in Chapter 12, and also Lavarack and Stanton, 1977).

In all the above-mentioned tropical lowland localities where infertile sandy soils are found, the plant communities are characterised by dense, sclerophyllous canopies. On well-drained sites in regions of high rainfall, the communities are usually dominated by medium-height (20 to 40 m), slender (below 30 cm), and straight trees. As shown in Table 22.3, this oligotrophic closed-forest is termed caatinga in the Amazonas States, wallaba forest in Guiana, and *Heidewald* or kerangas forest in Borneo.

In drier localities, or on seasonally waterlogged groundwater podzols, the oligotrophic closed-forests become progressively stunted to form closed-scrubs (termed campina in the Amazonas States; muri scrub in Guyana; padang scrub in Borneo) of twisted trees and shrubs, 2 to 8 m tall; or heathlands (low campina or bana in the Amazonas States, bush island vegetation in Guiana; padang heathland in Borneo) of low shrubs less than 2 m tall.

In all these tropical localities, the sequence of sclerophyllous closed-forest to closed-scrub to heathland appears to be governed by increasing water stress induced either by drier climatic conditions or by seasonal waterlogging.

Sclerophyllous, evergreen species dominate all sections of the sequence from closed-forest to closed-scrub to heathland regardless of the degree (or lack) of water stress. It appears that sclerophylly is the consequence of selection of species which can survive on the extremely low phosphate levels in these oligotrophic soils.

As found in many other parts of the world, both the total number of species and the number of

TABLE 22.3

Structural relationships of sclerophyllous plant communities of the Rio Negro/Amazon River area in South America, compared with similar tropical areas in the Guianas and in Borneo

Formation	Rio Negro/Amazon River	Guianas	Borneo
Mesophyllous closed-forest	mixed forest	mixed forest (24 m, emergents 42 m)	mixed dipterocarp forest
Sclerophyllous closed-forest	(1) high caatinga (20–30 m) (2) low caatinga, campina (10–20 m)	wallaba forest (30 m)	*Heidewald* Heath-forest Kerangas forest (40 m)
Sclerophyllous closed-scrub	low campina, bana (3–10 m)	muri scrub (2–3 m)	padang scrub (2–7 m)
heathland	low campina, bana (1.5–3 m)	bush islands (<1.5 m)	padang heathland (<2 m)
References	Klinge and Medina, this Chapter	Davis and Richards (1933, 1934) Richards (1957) Schulz (1960) Cooper, Chapter 21, this Volume	Winkler (1914) Richards (1957) Brünig (1974) Specht and Womersley Chapter 12, this Volume

endemic species recorded in the Amazonian sclerophyllous vegetation are greater than noted in adjacent vegetation on more fertile soils (Ducke and Black, 1954). The oligotrophic bana, campina and caatinga communities of Amazonia are structurally homologous with heathlands, sclerophyllous closed-scrub, and sclerophyllous closed-forest respectively in other areas of the world. But no representative of the order Ericales or of related heathland families (see Specht in Chapter 1) is present as a sclerophyllous tree or shrub in the Amazonian vegetation (Table 22.4). Species of *Drosera*, *Lycopodium* and *Sphagnum* in the understorey appear to be the only tenuous links with the flora of sclerophyllous communities of the rest of the world.

The absence of representative heathland elements in the tropical, lowland, sclerophyllous vegetation of the Amazonas States needs comment. Typical heathland genera[1] (such as *Gaultheria* 84 spp.; *Gaylussacia* 59 spp.; *Vaccinium* 50 spp.) have been recorded in subalpine/alpine localities in the Cordillera de los Andes of Bolivia, Brazil, Colombia, Ecuador and Peru and in the Guiana Massif of Venezuela and Guyana. A similar lack of typical heathland families and genera is characteristic of lowland areas of tropical Africa — but this is not surprising as few extensive areas of oligotrophic soils are found in the humid, tropical part of this continent (see Killick in Chapter 4). On the other side of the world, heathland genera of temperate Australia penetrate into the lowland as well as the alpine habitats of humid, tropical Malesia (see Specht and Womersley in Chapter 12).

It seems probable that many of the characteristic heathland families and genera evolved on oligotrophic soils under humid tropical conditions (see Specht in Chapter 1). In lowland tropical areas, the temperature optimum for growth would have been high; a gradient in temperature optima would be expected in species evolving in tropical lowland to alpine altitudes. In southern Africa and Australia where a Mediterranean-type climate now exists, a mixture of heathland species is found with temperature optima ranging from tropical to cool temperate (see Kruger in Vol. B, Ch. 1, and Specht et al. in Vol. B, Ch. 2). It may be argued that, during the Mesozoic, no extensive areas of oligotrophic soils existed in the lowland warm-temperate to tropical climate in the central part of the African land mass. It seems probable that oligotrophic soils were found only in cooler highland areas thus favouring the migration of heathland species with low temperature optima across Africa into the land masses which are now tropical South America, Europe and North America (see Specht in Chapter 1). Consequently, as no immigration of suitable biotypes was possible from outside, a unique suite of sclerophyllous species evolved to occupy the lowland oligotrophic soils of the tropical Guianas and Amazonas.

TABLE 22.4

Typical genera of shrubs and small trees common in the caatinga and campina vegetation of Amazonia

Annonaceae	*Annona*
Arecaceae	*Amylocarpus*
Caesalpiniaceae	*Aldina, Eperua, Macrolobium, Peltogyne, Tachigalia*
Clusiaceae	*Clusia*
Euphorbiaceae	*Croton, Micranda*
Fabaceae	*Erythrina*
Lauraceae	*Aniba*
Melastomataceae	*Miconia*
Nyctaginaceae	*Neea*
Olacaceae	*Heisteria*
Rubiaceae	*Cinchona, Pagamea*
Sapindaceae	*Talisia*
Sapotaceae	*Lucuma, Manilkara*

REFERENCES

Anonymous, 1967.*Soil Classification. A Comprehensive System. 7th Approximation.* U.S. Government Printing Office, Washington, D.C., 265 pp.

Aubréville, A., 1961.*Étude écologique des principales formations végétales du Brésil et contribution à la connaissance des forêts de l'Amazonie brésilienne.* Centre Technique Forestier, Nogent-sur-Marne, 268 pp.

Brünig, E., 1974. *Ecological Studies in the Kerangas Forest of Sarawak and Brunci.* Borneo Literature Bureau, Kuching, Sarawak, 237 pp.

Brünig, E., 1976. Variation der Struktur im Regenwald von San Carlos de Rio Negro. *Amazoniana*, 6: 275–277.

Brünig, E.F. and Heuveldop, J., 1976. Structure and functions in natural and man-made forests in the humid tropics. In: Norwegian IUFRO Congress Committee (Editor), *Proc. XVI IUFRO World Congress, Division I*: 500–511.

---

[1] Data collected in *Index Kewensis* (Hooker and Jackson, 1895) and *Supplementa* I–XV (1886–1970).

Calton, W.E., 1955. Some east African soil complexes I. In: *Trans 5th Int. Congr. Soil Sci., Leopoldville 1954,* 4: 62–65.

Chaminade, R., 1949. La pédogénese et les types de sols à Madagascar. *Bull. Agric. Congo Belge,* 40: 303–308.

Davis. T.A.W. and Richards, P.W., 1933. The vegetation of Moraballi creek, British Guiana: An ecological study of a limited area of tropical rain forest. I. *J. Ecol.,* 21: 350–384.

Davis, T.A.W. and Richards, P.W., 1934. The vegetation of Moraballi creek, British Guiana: An ecological study of a limited area of tropical rain forest. II. *J. Ecol.,* 22: 106–155.

Ducke, A. and Black, G.A., 1954. Notas sobre a fitogeografia da Amazonia brasileira. *Bol. Téc. Inst. Agron. Norte,* 29: 1–62.

Ferri, M.G., 1960. Contribution to the knowledge of the ecology of the "Rio Negro caatinga" (Amazon). *Bull. Res. Coun. Israel,* 8: 195–208.

Ferri, M.G., 1961. Problems of water relations of some Brazilian vegetation types, with special consideration of the concepts of xeromorphy and xerophytism. *UNESCO Arid Zone Res.,* 16: 191–197.

Gibbs. H.S., 1964. Soils of Northland. In: *Natural Resources Survey Part III.* Owen, Wellington, pp. 25–28.

Giesecke, F., 1930. Tropische und subtropische Humus und Bleicherdebildungen. In: E. Blanck (Editor), *Handbuch der Bodenlehre, 4.* Springer, Berlin, pp. 184–224.

Heuveldop, J., 1976. Erste Ergebnisse bestandesmeteorologischer Untersuchungen im Regenwald von San Carlos de Rio Negro. *Amazoniana,* 6: 299–300.

Hooker, J.D. and Jackson, R.B., 1895. *Index Kewensis, I and II.* Clarendon Press, Oxford, 1268 pp.; 1299 pp. (also *Index Kewensis, Supplementum* I–XV; 1886–1970).

Hueck, K., 1966. *Die Wälder Südamerikas. Ökologie, Zusammensetzung und wirtschaftliche Bedeutung.* Fischer, Stuttgart, 422 pp.

Keses, P.A., 1956. El clima de la región de Rio Negro venezolano (Territorio Federal Amazonas). *Mem. Soc. Nal. La Salle,* 16: 268–312.

Klinge, H., 1966. Verbreitung tropischer Tieflandspodole. *Naturwissenschaften,* 53: 442–448.

Klinge, H., 1967. Podzol soils: A source of blackwater rivers in Amazonia. In: H. Lent (Editor), *Atas Simposio Biota Amazonica, Belém 1966,* 3: 117–125.

Klinge, H., 1968. *Report on Tropical Podzols.* F.A.O., Rome, 88 pp. (1st draft).

Klinge, H., 1976a. Nährstoffe, Wasser und Durchwurzelung von Podsolen und Latosolen unter tropischen Regenwald bei Manaus/Amazonien. *Biogeographica,* 7: 45–58.

Klinge, H., 1976b. Die Phytomasse dominanter Baumarten einer amazonischen Caatinga. *Amazoniana,* 6: 327–328.

Klinge, H., Medina, E. and Herrera, R., 1977. Studies in the ecology of Amazon caatinga forest in southern Venezuela. I. Introduction. *Acta Cient. Venez.,* 28: 270–276.

Lavarack, P.S. and Stanton, J.P., 1977. Vegetation of the Jardine River Catchment and adjacent coastal areas. *Proc. R. Soc. Qld.,* 88: 39–48.

Loveless, A.R., 1961. A nutritional interpretation of sclerophylly based on differences in the chemical composition of sclerophyllous and mesophytic leaves. *Ann. Bot.,* 25: 168–184.

Montes, R. and Medina, E., 1977. Seasonal changes in nutrient content with different ecological behavior. *Proc. III Symp. Trop. Ecol., Zaire. Geo-Eco-Trop.,* 1: 295–307.

Richards, P.W., 1941. Lowland tropical podsols and their vegetation. *Nature, Lond.,* 148: 129–131.

Richards, P.W., 1954. Notes on the bryophyte communities of lowland tropical rain forest, with special reference to Moraballi creek, British Guiana. *Vegetatio,* 5/6: 319–328.

Richards, P.W., 1957. *The Tropical Rain Forest. An Ecological Study.* University Press, Cambridge, 450 pp.

Rodrigues, W.A., 1961. Aspectos fitossociologicos das catingas do Rio Negro. *Bol. Mus. para. "E. Goeldi",* 15: 1–41.

Schulz, J.P., 1960. Ecological studies on rain forest in northern Suriname. *Verh. K. Ned. Akad. Wet., Afd. Nat.,* Ser. 2, 53: 1–267.

Sioli, H., 1954. Gewässerchemie und Vergänge in den Böden im Amazonas-gebiet. *Naturwissenschaften,* 41: 456–457.

Soil Survey Staff, 1975. *Soil Taxonomy: A Basic System of Soil Classification for Making and Interpreting Soil Surveys.* U.S. Dep. Agric. Handbook No. 436. U.S. Government Printing Office, Washington, D.C., 754 pp.

Spruce, R., 1908. *Notes of a Botanist on the Amazon and the Andes.* MacMillan, London, Vols. 1 and 2, 518 pp. and 542 pp.

Takeuchi, M., 1961. The structure of the Amazonian vegetation. III. Campina forest in the Rio Negro region. *J. Fac. Sci. Univ. Tokyo, Section III, Bot.,* 8: 27–35.

Takeuchi, M., 1962. The structure of the Amazonian vegetation. IV. High campina forest in the upper Rio Negro. *J. Fac. Sci., Univ. Tokyo, Section III, Bot.,* 8: 279–288.

Vieira, L.S. and Filho, J.P.S.O., 1962. As caatingas do Rio Negro. *Bol. téc. Inst. Agron. Norte,* 42: 1–32.

Walter, H. and Lieth, H., 1960–67. Klimadiagramm-Weltatlas. Fischer, Jena.

Walter, H. and Medina, E., 1971. Caracterización climática de Venezuela sobre la base de climadiagramas de estaciones particulares. *Bol. Sco. Venez. Cienc. Nat.,* 29: 211–240.

Winkler, H., 1914. Die Pflanzendecke Südost-Borneos. *Bot. Jahrb.,* 50: 188–208.

Chapter 23

SOUTHERN OCEANIC WET-HEATHLANDS (INCLUDING MAGELLANIC MOORLAND)[1]

D.M. MOORE

INTRODUCTION

In the cool-temperate zone of South America and the southern Atlantic Ocean, wet-heathland as strictly defined — communities dominated by heath-like shrubs up to about 1 m in height and growing on peat — occurs in Tierra del Fuego, southern Patagonia, the Falkland Islands and the Tristan da Cunha/Gough Island Archipelago. Like the north temperate wet-heathlands, such southern heathlands develop under exposed, cool, oceanic conditions and, furthermore, are dominated by *Empetrum rubrum*, which is closely related to the *Empetrum* species important in the Northern Hemisphere. The climatic conditions also support a number of associations in which heath-like (i.e. with ericoid leaves) plants are not important but which are structurally, and often, floristically, similar to the *Empetrum* communities so that, as in this account, they are usefully and customarily included in any consideration of southern heathland. In many of these communities cushion-forming dwarf shrubs are prominent and, because of their circum-Antarctic affinities, give a distinctively Southern Hemisphere stamp to the physiognomy of the vegetation. The southern oceanic wet-heathlands, then, encompass associations very reminiscent of those in, for example, Western Europe (see Gimingham et al., in Chapter 14) and others whose affinities lie across Antarctica to the New Zealand region (see Burrows et al., in Chapter 13).

The wet-heathlands described here grade into the bogs and mires covered in Volume 4 of this series; and, as higher altitudes or latitudes are reached, they give place to the tundra communities described in Volume 3.

CLIMATE AND SOILS

Southern oceanic heathlands and Magellanic moorland occur in regions with relatively small seasonal differences in temperature and precipitation (Table 23.1), both tending to be higher in summer and autumn, while extremes of temperature ($-7°$C Islas Evangelistas, 24.9°C Tristan da Cunha) are infrequent. Although rainfall is generally high, it varies considerably in the areas supporting heathland or Magellanic moorland vegetation (Wace, 1960), ranging from yearly totals of almost 5000 mm on Isla Wellington, Chile (Holdgate, 1961) to about 600 mm in the Falkland Islands (Table 23.1). Significantly, parts of north western Tierra del Fuego and Patagonia east of the Andes also support communities dominated by *Empetrum rubrum* and with other species, such as *Blechnum pennamarina*, *Oxalis enneaphylla* and *Marsippospermum grandiflorum*, characteristic of the southern oceanic heathlands. These communities, however, are structurally and floristically obviously part of the Patagonian steppe vegetation and the low rainfall of the areas they occupy, 326 mm per annum (Pisano Valdes, 1971) or less, suggests that the Falkland Islands approach the lower rainfall limit for oceanic heathlands. It seems clear that the prevailing cloud cover and exposure to the frequent strong westerly winds are important in providing conditions for the development of the heaths. This is well demonstrated in the western Chilean archipelago where valleys can provide sheltered conditions in which *Nothofagus* woodland can ascend almost to the highest elevations capable of supporting plant life, while elsewhere

---

[1] Manuscript completed December, 1976.

TABLE 23.1

Summary of climatic data for representative localities supporting southern oceanic wet-heathland and/or Magellanic moorland

		Gough Island (40°19′S, (9°57′W)	Falkland Is., Port Stanley (51°42′S, 57°52′S)		Chile, Cabo San Isidro (53°47′S, 70°58′S)		Chile, Islas Evangelistas (52°25′S, 75°06′S)	
Mean monthly	Spring	247.2	37.8		64.2		241.7	
precipitation	Summer	247.2	66.8		75.6		254.1	
(mm)	Autumn	371.1	51.5		86.0		215.2	
	Winter	278.6	47.3		66.4		219.1	
Annual precipitation	(mm)	3225.0	609.7		876.5		2791.5	
Mean monthly	Spring	10.3*	7.8	1.7	9.0	2.8	7.3	3.3
maximum and	Summer	14.0*	12.3	5.3	11.8	5.3	9.8	6.0
minimum	Autumn	12.3*	8.1	2.8	9.1	3.0	9.1	4.9
temperatures (°C)	Winter	9.8*	4.1	0.5	5.2	1.7	6.3	2.3
Annual mean hours sun per day		2.8	4.2		–		–	
Mean annual days snow		0	55		33		28	
Source		Wace (1961)	Pepper (1954)		Pisano (1972), Knoch (1930)		Pisano (1970), Knoch (1930)	
Köppen climate		**Cfbi**	**ETf**		**ETf**		**ETfi**	

* Only mean monthly temperatures available.

Magellanic moorland or feldmark vegetation is dominant.

As can be seen from Appendix I, developed mineral soils do not generally underlie southern oceanic heathlands or Magellanic moorland. Peats of varying thickness and lithosols are almost universal; the particular communities depend on the degree of soil wetness, which results, at least in part, from the drainage — in turn dependent upon the topography and underlying rocks. The upper levels of the soil, at least, are generally rich in incompletely decomposed plant remains, with organic content exceeding 60%, and rarely less acid than pH 5.0.

VEGETATION AND FLORA

Characteristically, the southern oceanic wet-heathlands are dominated by *Empetrum rubrum*, with which are associated a number of widespread south temperate species such as *Blechnum pennamarina*, *Cardamine glacialis*, *Apium australe*, *Tetroncium magellanicum* and *Nertera depressa*. Perhaps because of the ecological amplitude of the constituent species, or the species-poverty of the communities in these regions, or because they result from the interaction of several environmental variables, wet-heathland communities appear to occupy a central role in the areas where they occur, merging with all the other formations present (Wace and Dickson, 1965; Moore, 1960), which may, in a sense, be considered as more specialized types developed in response to particular environmental factors, such as shelter, exposure and drainage (Fig. 23.1). Furthermore, there are clear links with communities dominated by dwarf shrubs in other climatic zones, such as those with *Empetrum rubrum* in the more continental steppe of east-Patagonia and north western Tierra del Fuego (see p. 489) and the *Acaena magellanica* association of subantarctic conditions in South Georgia (Walton, 1976) and Kerguelen (Cour, 1959). In addition, the non-availability of pulvinate dwarf shrubs in truly oceanic islands, such as those of the Tristan da Cunha/Gough Archipelago, obviously places strictures on the life-form composition of their heathland communities.

Although they intergrade floristically and structurally, it is convenient to consider the communities under three headings: dwarf-shrub heathland, grass and grass–sedge heathland, and Magellanic moorland. Nomenclature follows Moore (1968, 1974) and Wace and Dickson (1965).

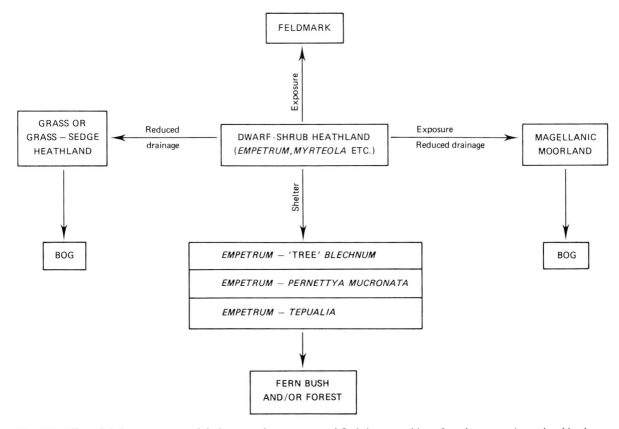

Fig. 23.1. Effect of shelter, exposure and drainage on the structure and floristic composition of southern oceanic wet-heathlands.

Dwarf-shrub heathland

As can be seen from Table 23.2, more than 75% of the biomass of these communities is provided by low or dwarf shrubs, of which *Empetrum rubrum* is the most important, although others, such as *Myrteola nummularia* and *Pernettya pumila*, may be co-dominant or locally dominant. The heathland is developed over peat of varying depths (usually from 10 to 50 cm), which may be dry or wet as long as water run-off is possible, and the growth form of the *Empetrum* often differs in response to the amount of run-off and hence root aeration (Wace, 1961). In the drier areas, except the Tristan da Cunha Archipelago, cushion plants are frequently prominent and some, such as *Bolax gummifera*, may be locally dominant (Moore, 1968). With increasing altitude and exposure this drier heathland grades into feldmark, in which the same species are present but the communities are more open exposing large areas of mineral soil

(Moore, 1975). Under wetter conditions *Sphagnum* becomes important and often co-dominant with the *Empetrum*, although not in the Falkland Islands, which lack these mosses. Once drainage is impeded to the extent that standing water becomes visible this heathland grades into bog. Caespitose graminoid herbs are usually present, often prominently so, as in Tristan da Cunha/Gough Island Archipelago; where their occurrence rises towards 50% of the biomass the communities can be taken as comprising an ecotone between dwarf shrub and grass or grass–sedge heathland and usually reflect increased shelter and, sometimes, groundwater.

The importance of the small tree ferns *Blechnum magellanicum* and *B. palmiforme*, which are often locally dominant in the heathland (Wace, 1961; Moore, 1968), probably reflects a stage towards bush or forest (Wace and Holdgate, 1958; Wace, 1961), although this is not reached in the treeless Falkland Islands. This certainly seems to be the case with the *Empetrum–Pernettya mucronata*

TABLE 23.2

Oceanic dwarf-shrub wet-heathland

	Leaf type and size[1]	Percentage total canopy volume				
		Site I[2]	Site II[2]	Site III[2]	Site IV[2]	Site V[2]
Evergreen mid-height shrubs (100–200 cm)						
Blechnaceae						
Blechnum magellanicum	BS (Mac)	–	–	+	+	8.3
Desfontainiaceae						
Desfontainia spinosa	BS (Mic)	–	–	–	–	20.8
						29.2
Evergreen low shrubs (50–100 cm)						
Blechnaceae						
Blechnum palmiforme	BS (Mac)	54.4	19.7	–	–	–
Empetraceae						
Empetrum rubrum	NS (Eri–Lep)	28.2	54.7	59.7	88.9	52.1
Ericaceae						
Pernettya mucronata	BS (Nan)	–	–	–	+	18.7
		82.6	**74.4**	**59.7**	**88.9**	**70.8**
Evergreen creeping dwarf shrubs						
Asteraceae						
Baccharis magellanica	BS (Nan)	–	–	11.9	–	–
Ericaceae						
Pernettya pumila	BS (Lep–Nan)	–	–	4.1	4.4	–
Myrtaceae						
Myrteola nummularia	BS (Lep–Nan)	–	–	0.4	–	–
				16.4	**4.4**	
Deciduous creeping dwarf-shrubs						
Blechnaceae						
Blechnum penna-marina	BS (Nan–Mic)	**2.0**	+	**6.8**	–	+
Evergreen pulvinate dwarf shrubs						
Asteraceae						
Abrotanella emarginata	BS (Lep)	–	–	0.2	–	–
Hydrocotylaceae						
Azorella filamentosa	BS (Lep)	–	–	0.3	–	–
A.lycopodioides	BS (Nan)	–	–	0.5	–	–
Thymelaeaceae						
Drapetes muscosus	BS (Nan)	–	–	1.1	–	–
				2.1		
Evergreen caespitose graminoid herbs						
Juncaceae						
Luzula alopecurus	Gra	–	–	+	6.7	–

TABLE 23.2 *(continued)*

	Leaf type and size[1]	Percentage total canopy volume				
		Site I[2]	Site II[2]	Site III[2]	Site IV[2]	Site V[2]
Poaceae						
Deschampsia aff. *kingii*	Gra	5.6	9.2	–	–	–
D. flexuosa	Gra	–	–	1.9	–	–
Spartina arundinacea	Gra	–	13.1	–	–	–
		5.6	22.3	1.9	6.7	
Deciduous caespitose graminoid herbs						
Poaceae						
Agrostis aff. *magellanica*	Gra	4.0	–	–	–	–
A. magellanica	Gra	–	–	1.7	–	–
		4.0		1.7		
Rhizomatous graminoid herbs						
Juncaceae						
Marsippospermum grandiflorum	Gra	–	–	2.0	+	–
Deciduous forbs						
Apiaceae						
Apium australe	H (Mes)	+	3.3	–	–	–
Caryophyllaceae						
Cerastium arvense	H (Lep)	–	–	2.0	–	–
Rubiaceae						
Galium antarcticum	H (Lep)	–	–	1.6	–	–
			3.3	3.6		
Creeping forbs						
Haloragaceae						
Gunnera magellanica	H (Mes)	–	–	2.4	–	–
Lycopodiaceae						
Lycopodium magellanicum	NS (Lep)	–	–	1.0	–	–
				3.4		
Seasonally green geophytes						
Iridaceae						
Sisyrinchium filifolium	Gra	–	–	1.0	–	–
Seasonal annuals						
Poaceae						
Aira praecox	Gra	–	–	1.0	–	–
Ground mosses						
Sphagnaceae						
Sphagnum sp.	H (Lep)	5.6	–	–	–	–

TABLE 23.2 *(continued)*

[1] Abbreviations for leaf type and size:

BS	=	broad-sclerophyll	Lep	= Leptophyll	Mic	= Microphyll
Eri	=	Ericoid	Mac	= Macrophyll	Nan	= Nanophyll
Gra	=	Graminoid	Mes	= Mesophyll	H	= Herbaceous
NS	=	narrow-sclerophyll				

Site I: Gough Island, Hag's Tooth, 40°19′S, 9°57′W; low closed mat heath over wet peat (Wace, 1961).
Site II: Gough Island, Rockhopper Point; low cushion closed-heathland (Wace, 1961).
Site III: Falkland Islands, East Falkland, Port Stanley; 51°42′S, 57°52′W; low cushion closed-heathland (Skottsberg, 1913; Moore, 1968, unpubl.).
Site IV: Tierra del Fuego, Islas Wollaston, Isla Otter, 55°36′S, 67°30′W; low closed-heathland (Skottsberg, 1916).
Site V: Puerto Cutter Cove, Canal Jerónimo, Magallanes, Chile, 52°23′S, 72°35′W; mid-height shrub heathland (Pisano Valdes, 1970).

Species not included in the Table (<1% canopy volume):

Site I: Evergreen creeping dwarf shrubs — (Rosaceae) *Acaena sarmentosa*. Evergreen caespitose graminoid herbs — (Cyperaceae) *Carex thouarsii*, *Uncinia compacta*. Deciduous forbs — (Polypodiaceae) *Elaphoglossum succisifolium*. Seasonal rosette forbs — (Asteraceae) *Lagenophora nudicaulis*. Ground mosses — *Rhacomitrium lanuginosum*, *R. loriforme*, *Dicranoloma hariotii*, *D. imponens*, *Lepyrodon alaris*, *Ptychomnium densifolium*. Ground hepatics — *Lepicolea ochroleuca*.

Site II: Evergreen caespitose graminoid herbs — (Cyperaceae) *Scirpus sulcatus* and *S. thouarsii*. Deciduous forbs — (Polypodiaceae) *Elaphoglossum succisifolium*.

Site III: Evergreen creeping dwarf shrubs — (Rosaceae) *Acaena magellanica*, (Ericaceae) *Gaultheria antarctica*. Evergreen pulvinate dwarf shrubs — (Hydrocotylaceae) *Bolax gummifera*, (Santalaceae) *Nanodea muscosa*. Evergreen caespitose graminoid herbs — (Cyperaceae) *Carex fuscula*, *Oreobolus obtusangulus*, (Poaceae) *Cortaderia pilosa*, *Festuca contracta*. Rhizomatous graminoid herbs — (Juncaceae) *Juncus scheuchzerioides*. Deciduous forbs — (Brassicaceae) *Cardamine glacialis*, (Gentianaceae) *Gentianella magellanica*. Seasonal rosette forbs — (Asteraceae) *Aster vahlii*, *Lagenophora nudicaulis*, *Leuceria suaveolens*, *Taraxacum gilliesii*. Creeping forbs — (Lobeliaceae) *Pratia repens*, (Rubiaceae) *Nertera depressa*.

Site IV: Evergreen small trees (2–10 m) — *(Winteraceae) Drimys winteri*. Evergreen low shrubs (25–50 cm) — (Saxifragaceae) *Escallonia serrata*, (Berberidaceae) *Berberis ilicifolia*, (Asteraceae) *Chiliotrichium diffusum*. Evergreen creeping dwarf shrubs — (Philesiaceae) *Luzuriaga marginata*. Deciduous forbs — (Aspleniaceae) *Asplenium dareoides*. Ground mosses — *Brachythecium rutabulum*, *Dicranum aciphyllum*, *Ptychomnium densifolium*. Ground hepatics — *Aneura tenax*, *Jamesoniella oenops*, *Plagiochila remotidens*. Ground lichens — *Cladonia aggregata*, *Sticta endochrysea*.

Site V: Evergreen small trees (2–10 m) — (Cupressaceae) *Pilgerodendron uvifera*. Evergreen low shrubs (50–100 cm) — (Berberidaceae) *Berberis ilicifolia*, (Asteraceae) *Chiliotrichium diffusum*. Deciduous low shrubs (50–100 cm) — (Onagraceae) *Fuchsia magellanica*. Scrambling plants — (Philesiaceae) *Philesia magellanica*, (Cornaceae) *Griselinia ruscifolia*. Creeping forbs — (Rosaceae) *Rubus geoides*.

communities so common in Tierra del Fuego and western Patagonia (Pisano Valdes, 1970, 1972, 1973), and with the *Empetrum–Tepualia stipularis* association, which occurs in western Chile from the Estrecho de Magallanes (Straits of Magellan) north to Isla Wellington (49°10′S) (Holdgate, 1961), where heathland gives way to *Nothofagus* woodland.

Grass and grass–sedge heathland

The caespitose graminoid herbs present, with varying degrees of importance, in the dwarf-shrub heathland comprise 70% or more of the biomass in these communities (Table 23.3). Graminoid-heathland seems to be a response to sheltered conditions, as on Gough Island (Wace, 1961), where it probably also reflects a decrease in drainage, since it frequently occurs around bog margins; the large areas of *Cortaderia* heathland in the Falkland Islands certainly result from increased waterlogging rather than exposure (Moore, 1968). The peat underlying this heathland is coarse and fibrous, and usually much less compact than that under dwarf-shrub heathland. On Gough Island it rarely exceeds a depth of 20 cm, but in the Falkland Islands often reaches 1 m or more. Wetter facies of the graminoid-heathland have Cyperaceae, such

TABLE 23.3

Oceanic grass and grass–sedge heathland

	Leaf type and size[1]	Percentage total canopy volume	
		Site I[2]	Site II[2]
Evergreen creeping dwarf shrubs			
Asteraceae			
Baccharis magellanica	BS (Nan)	4.0	–
Deciduous creeping dwarf shrubs			
Blechnaceae			
Blechnum penna-marina	BS (Nan–Mic)	2.0	3.8
Evergreen (±) caespitose graminoid herbs			
Cyperaceae			
Carex caduca	Gra	2.8	–
Scirpus sp.	Gra	–	7.6
Poaceae			
Cortaderia pilosa	Gra	84.0	–
Deschampsia flexuosa	Gra	2.4	–
D. aff. *kingii*	Gra	–	28.4
Glyceria sp.	Gra	–	9.5
		89.2	**45.5**
Deciduous caespitose graminoid herbs			
Poaceae			
Agrostis carmichaelii	Gra	–	**22.7**
Deciduous forbs			
Apiaceae			
Apium australe	H (Mes)	–	**11.4**
Creeping forbs			
Ranunculaceae			
Ranunculus carolii	H (Mic)	–	**6.0**
Seasonal annuals			
Poaceae			
Aira caryophyllea	Gra	2.4	–
A. praecox	Gra	2.0	–
		4.4	
Ground bryophytes			
		–	**10.6**

[1] Abbreviations for leaf type and size:

BS	= broad-sclerophyll	Mes	= Mesophyll
Gra	= Graminoid	Mic	= Microphyll
NS	= narrow-sclerophyll	Nan	= Nanophyll
H	= Herbaceous		

[2] Site I: Falkland Islands, East Falkland, Port Darwin, 51°51′S, 58°56′W (Skottsberg, 1913; Moore, 1968, unpubl).

Site II: Gough Island, Edinburgh Peak, 40°19′S, 9°57′W (Wace, 1961).

Miscellaneous species:

Site I: Evergreen caespitose graminoid herbs — (Cyperaceae) *Carex fuscula*. Deciduous caespitose graminoid herbs — (Poaceae). *Festuca magellanica, Trisetum spicatum*. Deciduous forbs — (Oxalidaceae) *Oxalis enneaphylla*, (Apiaceae) *Oreomyrrhis hookeri*. Creeping forbs — (Lycopodiaceae) *Lycopodium magellanicum*. Seasonal rosette forbs — (Asteraceae) *Chevreulia lycopodiodes*

Site II: Evergreen caespitose graminoid herbs — (Cyperaceae) *Carex thouarsii, Uncinia compacta*. Deciduous forns — (Brassicaceae) *Cardamine glacialis*, (Aspidiaceae) *Dryopteris aquilina*. Creeping forbs — (Rubiaceae) *Nertera depressa*.

as *Carex fuscula* and *Oreobolus obtusangulus* (Falkland Islands) or *Carex insularis* and *Uncinia compacta* (Gough Island), as local dominants or co-dominants. This reinforces the impression that this heath is partly transitional between dwarf-shrub heathland and peat bog, with which it often interdigitates in a complex mosaic. Graminoid-heathland is not very widespread in southern South America but the montane herbaceous association, dominated by *Poa* and *Juncus*, described by Pisano Valdes (1970) from Canal Jerónimo in southern Chile, is very similar to the facies of grass heathland found near its altitudinal limit of *c.* 200 m. in the Falkland Islands.

Magellanic moorland

In the high-rainfall (2000 to 5000 mm) areas of westernmost Chile from the Golfo de Peñas (48°S) southwards to Cabo de Hornos (Cape Horn) is found a series of dwarf-shrub communities which, collectively, have been termed Magellanic moorland (Godley, 1960). In sheltered areas an evergreen woodland, dominated by *Nothofagus betuloides, Drimys winteri, Maytenus magellanica* and *Pilgerodendron uvifera*, is developed but exposure to the fierce westerly gales results in Magellanic moorland, dominated by cushion-forming dwarf shrubs, which give rise to a blanket peat the prevalence of which is one of the most characteristic features of the Magellanic moorland zone. The moorland seems to result from a combination of high rainfall, low temperatures, exposure, poor drainage and intractable igneous rock, and is consequently largely confined to Andean diorites of the wettest outer island zone of the southern Chilean archipelagos.

Detailed data for Magellanic moorland are not yet available but the most important species are the hard, cushion-forming *Donatia fascicularis* (Donatiaceae), *Caltha dioneifolia* (Ranunculaceae), *Astelia pumila* (Liliaceae), *Gaimardia australis* (Centrolepidaceae), *Bolax caespitosa*, (Apiaceae), *Phyllachne uliginosa* (Stylidiaceae), *Drapetes muscosus* (Thymelaeaceae), the dwarf, prostrate gymnosperm *Dacrydium fonckii* and the coriaceous-leaved graminoids *Tetroncium magellanicum* (Juncaginaceae) and *Uncinia kingii* (Cyperaceae), each of which can be dominant or co-dominant in the communities included in Magellanic moorland (Pisano Valdes, 1972; U. Eskuche, pers. comm., 1976). Amongst the associates of the various dominants listed above the following are probably the most consistently encountered: evergreen creeping dwarf-shrubs — *Acaena pumila* (Rosaceae), *Myrteola nummularia* (Myrtaceae), *Pernettya pumila*, *Gaultheria antarctica* (Ericaceae): evergreen caespitose graminoid herbs — *Carpha alpina* var. *schoenoides*, *Oreobolus obtusangulus*, *Schoenus*

antarcticus (Cyperaceae), *Marsippospermum grandiflorum* (Juncaceae), *Tapeinia pumila* (Iridaceae); rosette forbs — *Perezia magellanica* (Asteraceae), *Drosera uniflora* (Droseraceae); creeping forbs — *Tribeles australis* (Saxifragaceae), *Gunnera lobata* (Haloragaceae); partial root parasites, small forbs — *Nanodea muscosa* (Santalaceae).

As pointed out by Godley (1960), the closest affinities of the Magellanic moorland are with the New Zealand cushion bogs, with which they share many of the dominant genera, but there is also a similarity to the hard *Astelia pumila* bogs of the Falkland Islands and central Tierra del Fuego. These latter are in many ways intermediate between the Magellanic moorland and the dwarf-shrub heathland considered earlier, and reinforce the structural and floristic similarity between them. Such a bog at Fjordo Relander, south-central Tierra del Fuego (54°23'S, 70°0'W), studied by Roivainen (1954), had a biomass largely made up of evergreen creeping dwarf shrubs (27%), principally *Empetrum rubrum*, *Pernettya pumila* and *Myrteola nummularia*, and evergreen pulvinate dwarf shrubs (52%), principally *Astelia pumila* and *Donatia fascicularis*; caespitose graminoid herbs (4%), creeping forbs (8%) and ground mosses (8%) made up most of the rest of the vegetation.

APPENDIX I: CHARACTERISTICS OF REPRESENTATIVE SOILS

MAGELLANIC MOORLAND

(1)   *Astelia pumila* community, Puerto Edén, Isla Wellington, Chile, 49°10'S, 74°27'W (Holdgate, 1961)

0–140 cm	reddish peat	water content 82% total weight / organic content 96% dry weight / pH 3.8
<140 cm	coarse sand	water content 21% / organic content 9% / pH 4.5

(2)   *Astelia–Tetroncium* community, Peninsula Brunswick, Magallanes, Chile, 53°30'S, 71°30'W (Pisano Valdes, 1973)

O$_1$ 0–25 cm   partly or totally decomposed vegetation; pale brown; pH 4–4.5
O$_2$ 25–100 cm   fibrous peat, red; pH <4.0
D < 100 cm   impermeable, compacted glacial sand or rock

DWARF-SHRUB HEATHLAND

(1)   *Empetrum/Tepualia* mosaic, Puerto Edén, Isla Wellington, Chile, 49°10'S, 74°27'W (Holdgate, 1961)

0–10 cm	dryish loose humus	water content 78% total weight / organic content 89% dry weight / pH 3.7
10–55 cm	black peat	
55–160 cm	compact lighter peat	
< 160 cm	grit	water content 22% / organic content 6% / pH 4.5

APPENDIX I *(continued)*

DWARF-SHRUB HEATHLAND *(continued)*

(2) *Empetrum–Pernettya mucronata*, Bahia Morris, Isla Capitan Aracena, Tierra del Fuego, Chile, 54°15'S, 71°00'W (Pisano Valdes, 1972).

A_0	0–15 cm	black or dark grey, much organic material; pH ±4.5
A_1	15–20 cm	greyish
A_2	20–35 cm	dark reddish grey, accumulation of clay
B_3	< 35 cm	rubble

ACKNOWLEDGEMENTS

I thank the Royal Society of London and the Natural Environment Research Council for support, and Dr. U. Eskuche, Corrientes, Argentina, for access to preliminary phytosociological data of the International Botanical Transect of southern Patagonia.

REFERENCES

Cour, P., 1959. Flore et végétation de l'Archipel de Kerguelen. *Terres Aust. Antarct. Fr.*, Nos. 8–9: 3–40.

Godley, E.J., 1960. The botany of southern Chile in relation to New Zealand and the Subantarctic. *Proc. R. Soc. Lond., Ser. B*, 152: 457–475.

Holdgate, M.W., 1961. Vegetation and soils in the south Chilean islands. *J. Ecol.*, 49: 559–580.

Knoch, K., 1930. Sudamerika. In: W. Köppen and R. Geiger (Editors), *Handbuch der Klimatologie, 2. Amerika.* Borntraeger, Berlin, 349 pp.

Moore, D.M., 1968. The vascular flora of the Falkland Islands. *Br. Antarct. Surv. Sci. Rep.*, 60: 1–202.

Moore, D.M., 1974. Catálogo de las plantas vasculares nativas de Tierra del Fuego. *Anal. Inst. Patagonia, Punta Arenas (Chile)* 5(1–2): 105–121.

Moore, D.M., 1975. The alpine flora of Tierra del Fuego. *Anal. Inst. Bot. A.J. Cavanilles*, 32(2): 419–440.

Pepper, J., 1954. *The Meteorology of the Falkland Islands and Dependencies, 1944–1950.* Falkland Islands and Dependencies Meteorological Service, London, 249 pp.

Pisano Valdes, E., 1970. Vegetación del área de los fiordos Toro y Cóndor y Puerto Cutter Cove. *Anal. Inst. Patagonia, Punta Arenas (Chile)*, 1(1): 41–57.

Pisano Valdes, E., 1971. Estudio ecológico preliminar del parque nacional "Los Pingüinos (Estrecho de Magallanes). *Anal. Inst. Patagonia, Punta Arenas (Chile)*, 2(1–2): 76–92.

Pisano Valdes, E., 1972. Comunidades vegetales del área de Bahía Morris, Isla Capitán Aracena, Tierra del Fuego (Parque Nacional "Hernando de Magallanes"). *Anal. Inst. Patagonia, Punta Arenas (Chile)*, 3(1–2): 103–130.

Pisano Valdes, E., 1973. Fitogeografia de la Peninsula Brunswick, Magallanes. I. Comunidades Meso-higromorficas e higromorficas. *Anal. Inst. Patagonia, Punta Arenas (Chile)*, 4(1–3): 141–206.

Roivainen, H., 1954. Studien über die Moore Feuerlands. *Ann. Bot. Soc. "Vanamo"*, 28(2): 1–205.

Skottsberg, C.J.F., 1913. A botanical survey of the Falkland Islands. *K. Sven. Vet. Akad. Handl.*, 50(3): 1–129.

Skottsberg, C.J.F., 1916. Die Vegetationsverhältnisse längs der Cordillera de los Andes S. von 41°S. *K. Sven. Vet. Akad. Handl.*, 56(5): 1–366.

Wace, N.M., 1960. The botany of the southern oceanic islands. *Proc. R. Soc. Lond., Ser. B.*, 152: 475–490.

Wace, N.M., 1961. The vegetation of Gough Island. *Ecol. Monogr.*, 31: 337–367.

Wace, N.M. and Dickson, J.H., 1965. The terrestrial botany of the Tristan da Cunha Islands *Philos. Trans. R. Soc. Lond., Ser. B.*, 249: 273–360.

Wace, N.M. and Holdgate, M.W., 1958. The vegetation of Tristan da Cunha, *J. Ecol.*, 46: 593–620.

Walton, D.W.H., 1976. Dry matter production in *Acaena* (Rosaceae) on a Subantarctic island. *J. Ecol.*, 64: 399–415.